Exercise Evaluation and Prescription—2nd Edition

Exercise Evaluation and Prescription—2nd Edition

Editors

Cristina Cortis
Andrea Fusco
Carl Foster

MDPI

Basel • Beijing • Wuhan • Barcelona • Belgrade • Novi Sad • Cluj • Manchester

Editors

Cristina Cortis
University of Cassino and
Lazio Meridionale
Cassino
Italy

Andrea Fusco
University "G. D'Annunzio"
of Chieti-Pescara
Chieti
Italy

Carl Foster
University of Wisconsin
La Crosse, WI
USA

Editorial Office
MDPI AG
Grosspeteranlage 5
4052 Basel, Switzerland

This is a reprint of articles from the Special Issue published online in the open access journal *Journal of Functional Morphology and Kinesiology* (ISSN 2411-5142) (available at: https://www.mdpi.com/journal/jfmk/special_issues/Exercise_Evaluation2).

For citation purposes, cite each article independently as indicated on the article page online and as indicated below:

Lastname, A.A.; Lastname, B.B. Article Title. *Journal Name* **Year**, *Volume Number*, Page Range.

ISBN 978-3-7258-1919-5 (Hbk)
ISBN 978-3-7258-1920-1 (PDF)
doi.org/10.3390/books978-3-7258-1920-1

Contents

About the Editors

Cristina Cortis

Cristina Cortis is currently a full professor in sports science at the Department of Human Sciences, Society, and Health of the University of Cassino and Lazio Meridionale, Italy. She previously obtained a Ph.D. degree in 2009 in sport and health sciences at the University of Rome "Foro Italico", Italy. She was visiting researcher at the Human Performance Laboratory of the University of Wisconsin-La Crosse, USA, and University of Connecticut, USA. Prof. Dr. Cortis was recognized as a fellow of the American College of Sports Medicine in 2020 and a fellow of the European College of Sport Science in 2024.

Andrea Fusco

Andrea Fusco is currently an assistant professor in movement science at the Department of Medicine and Aging Sciences of the University "G. d'Annunzio" of Chieti-Pescara, Italy. He obtained a Ph.D. degree in the field of sport and health sciences at the University of Cassino and Lazio Meridionale, Italy, receiving the title of *Doctor Europaeus*. He was also visiting researcher at the Human Performance Laboratory of the University of Wisconsin-La Crosse (USA) and the Biomechanics Laboratory in the Department of Sport Science and Kinesiology of the University of Salzburg, Austria.

Carl Foster

Carl Foster is professor emeritus in exercise and sports science at the University of Wisconsin-La Crosse. He received his Ph.D. (1976) from the University of Texas at Austin. From 1977 to 1998, he was professor of medicine in the University of Wisconsin-Medical School at Sinai Samaritan Medical Center in Milwaukee, after which he moved to University of Wisconsin-La Crosse. He is a fellow of the American College of Sports Medicine and the American Association of Cardiovascular and Pulmonary Rehabilitation and has been President (2005-2006) and Treasurer (2015-2021) of the American College of Sports Medicine. He previously received the Citation Award from the American College of Sports Medicine (2009), the ML Pollock Established Investigator Award from the American Association of Cardiovascular and Pulmonary Rehabilitation (2010), and the American College of Sports Medicine Honor Award (2024). From 1989 to 2003, he was an Associate Editor of *Medicine and Science in Sports and Exercise*, and from 2009 to 2013, he was Editor-in-Chief of the *International Journal of Sports Physiology and Performance*. From 1983 to 2002, he was the Chair of the Sports Medicine and Sports Science Committee of U.S. Speed Skating.

Journal of
Functional Morphology and Kinesiology

MDPI

Editorial

Exercise Evaluation and Prescription—Second Edition

Carl Foster [1], Cristina Cortis [2,*] and Andrea Fusco [2]

1 Department of Exercise and Sport Science, University of Wisconsin-La Crosse, La Crosse, WI 54601, USA
2 Department of Human Sciences, Society and Health, University of Cassino and Lazio Meridionale, 03043 Cassino, Italy
* Correspondence: c.cortis@unicas.it

Introduction

In the first volume of "Exercise Evaluation and Prescription" in the Journal of Functional Morphology and Kinesiology [1], we presented the case that prescribing exercise was integrally linked to a prior evaluation of exercise capacity and psychophysiological markers of exercise performance (including maximal oxygen uptake, maximal strength, blood lactate, ventilation, and the Rating of Perceived Exertion (RPE)). The link between evaluation and prescription provided a rational basis for prescribed exercise. In volume two, this link is extended to markers of the response to, and recovery from, systematic exercise training.

This extension is grounded in general adaptation syndrome (GAS), described by Hans Seyle 75 years ago [2]. In his study, there was a predictable response to any imposed stress. If the stress was too large, or too frequently applied, there was a predictably negative outcome (exhaustion). On the other hand, in the presence of a stressor that was neither too large nor too frequent, there was adaptation such that the organism could subsequently tolerate greater levels of stress. Although Seyle conceptualized GAS as a generic response to any stressor, it has become particularly associated with the adaptive response to exercise training.

One way of understanding the adaptive response to training is using the 'training impulse' (TRIMP) model of Banister et al. [3], who discovered that exercise training produces changes in both fitness and fatigue [4,5]. The TRIMP model was originally designed to work using training heart rate (HR) and time, but more recent studies have shown that it could work using RPE and time, resulting in the session RPE (sRPE) [6–9]. More recent studies have suggested that it could be understood using responses during the warm-up for subsequent exercise bouts [10], questionnaires [11], HR variability (HRV) [12–14], or training intensity distribution [15–17]. In any case, the process of monitoring training using any tool appears to be a useful method of evaluating the exerciser (athlete fitness rehabilitation) and optimizing exercise prescription [18].

This volume presents several papers, written from the perspective of optimizing training programs by better understanding the purpose and process of evaluating exercise capacity, either to better prescribe exercise training or to better understand the outcome of exercise training programs. A total of 30 papers are published, including twenty-five original articles, four reviews, and one case report; the papers focus on team sports (soccer, volleyball, handball, rugby, and basketball), individual sports (cycling, running, weightlifting, and Paralympic powerlifting), diseases (in overweight, obese, and postmenopausal women with type 2 diabetes mellitus (T2DM); individuals with pulmonary embolism; individuals with cardiovascular disease or its risk factors; individuals receiving β-blocker treatment; and breast cancer survivors), healthy individuals, and murine models.

Several studies were carried out in collegiate soccer players [12–14]. Aiming to compare accumulated workloads between starters and reserves in collegiate soccer, Jagim et al. [12] reported a greater distance covered by starters throughout the season, resulting

Citation: Foster, C.; Cortis, C.; Fusco, A. Exercise Evaluation and Prescription—Second Edition. *J. Funct. Morphol. Kinesiol.* **2023**, *8*, 5. https://doi.org/10.3390/jfmk8010005

Received: 5 December 2022
Accepted: 20 December 2022
Published: 23 December 2022

in almost double the training load compared to reserves. The authors suggested the management of player workloads in collegiate soccer to address potential imbalances between starters and reserves throughout a season. In National Collegiate Athletics Association Division I women's soccer, Askow and colleagues [13] showed that the sRPE-derived training load was strongly associated with Global Positioning System-derived measures of external load; these relationships were stronger during match play, with acceleration load and total distance exhibiting the strongest relationship with sRPE-derived training load. Finally, Redd et al. [14] found improvements in certain lower leg muscle tensiomyography measures in male soccer players in response to different warm-up protocols, including small-sided games and dynamic and plyometric exercises, regardless of the specific warm-up activities used.

Regarding the most used tests in soccer players and referees, using a meta-analysis, Grgic et al. [15] reviewed the effects of mental fatigue based on performance in the Yo-Yo test and Loughborough soccer passing and shooting tests. According to their results, mental fatigue negatively impacts endurance-based running performance, as well as soccer passing and shooting skills. Moreover, Romano et al. [16] evaluated the correlation between the official 6 × 40 m sprint and Yo-Yo Intermittent Recovery Level 1 tests and other field-based tests in male referees, and found that Illinois agility and hand-grip tests could represent simple and low-physical-impact tools for the repeated assessment and monitoring of referee fitness throughout the season. Examining the reliability, validity, sensitivity, and usefulness of the most commonly used field aerobic fitness tests, Bok and Foster [17] reported that the University of Montreal track test and the 30–15 intermittent fitness test were the best solutions for the prescription of long and short high-intensity interval training sessions, respectively.

Investigating the association between three different strength and plyometric exercises and force- and velocity-oriented change-of-direction performance in female handball and soccer players, Falch and colleagues [18] observed correlations between strength and change-of-direction performance, highlighting that the differences in step kinematics between fast and slow performers were mainly found in stronger athletes who were able to carry out a greater workload in a shorter time. Comparing the biomechanical parameters of drop jumps executed on rigid and sand surfaces, Giatsis et al. [19] showed that the compliance of sand decreases the efficiency of the mechanisms involved in the optimization of drop jump performance in male beach volleyball; this suggests that sand can offer injury prevention under the demand for large energy expenditure as it experiences less loading during the eccentric phase of the drop jump. Gómez-Carmona et al. [20] asserted that in semi-professional basketball players, training design must be individualized according to different variables, including physical fitness profiles. From a practical point of view, their results show that the advantages and weaknesses of each athlete can be determined in order to adapt training tasks and game systems to the skills and capabilities of the players. Daly and colleagues [21] documented ex-professional rugby players' understanding of concussions and the analogies they use to describe concussions in order to determine the explicit and implicit pressures of playing professional rugby. The interviews highlighted that players, particularly coaches, did not fully understand the ramifications of concussive injury and other injury risks; this reveals the need to assist coaches in perceiving a concussion as a significant injury and not downplaying its seriousness in contact sports. Describing the conservative treatment of indirect structural muscle injuries, which are the more routinely found and more challenging type, Palermi et al. [22] reviewed therapeutic algorithms for muscle injuries and provided specific exercise rehabilitation for the four main muscle groups (i.e., the hamstrings, quadriceps, adductors, and soleus/gastrocnemius).

In the studies focusing on individual sports [23–27], Leo et al. [23] investigated the effects of COVID-19 restrictions on training and performance physiology measures in U23 elite cyclists, concluding that COVID-19 restrictions did not negatively affect training characteristics and physiological performance measures for a period of <30 days. Boullosa and colleagues [24] monitored the associations between HRV, training load, and performance in a middle-aged recreational female runner during a competitive 20-week macrocycle divided into a first and second mesocycle, in which her best performances over 10 km and 21 km were recorded. Their findings highlighted the practicality of concurrent HRV and sRPE monitoring in recreational runners, with the root mean square of the successive differences in R-R intervals:R-R intervals ratio indicative of specific adaptations.

Sandau et al. [25] examined the predictive validity of a new approach to the estimate one-repetition maximum (1RM) snatch computed from the two-point snatch pull force–velocity relationship, to determine the actual 1RM snatch performance in elite weightlifters. Aidar and colleagues [26] found that the maximum isometric force, rate of force development, time, velocity, and dynamic time had lower values, especially in the initial and intermediate phases in the sticking region, in Paralympic powerlifting athletes. In a subsequent study by the same research group [27] hemodynamic responses in Paralympic bench press powerlifting athletes were analyzed with respect to conventional powerlifting before and up to 60 minutes after training. According to the results of the heart pressure product, there is no risk of hemodynamic overload in conventional or Paralympic powerlifters. Moreover, training promoted a moderate hypotensive effect, with blood pressure adaptation immediately and 60 min after exercise.

Evaluating the effects of an intensive exercise program on health-related outcomes and cardio-metabolic health measures in a group of overweight and obese adults with and without T2DM, Pippi and colleagues [28] found that physical activity level and sitting time did not seem to influence the beneficial effects of exercise intervention. Bentes et al. [29] examined the influence of 12 months of vitamin D supplementation on the components of physical fitness in postmenopausal women with T2DM. Their findings showed that vitamin D supplementation alone was effective in increasing serum vitamin D levels, altering muscle strength, improving muscle function, and preventing and controlling fragility caused by T2DM and aging.

Stavrou et al. [30] assessed the effect of 8 weeks of pulmonary rehabilitation in patients with pulmonary embolism, during unsupervised and supervised pulmonary rehabilitation, on cardiopulmonary exercise testing parameters, sleep quality, quality of life, and cardiac biomarkers. Their findings highlighted that pulmonary rehabilitation may present a safe method of intervention regardless of supervision. Schultz et al. [31] aimed to document the training load accomplished by patients with known cardiovascular disease, or with risk factors likely to cause cardiovascular disease, in a community-based exercise program using both steps/day and the sRPE approach. They concluded that patients in a community clinical exercise program achieve the American College of Sports Medicine guidelines, but accomplish fewer steps than recommended. The same research group [32] developed equations for predicting the peak oxygen uptake (VO_2peak) and ventilatory threshold (VT) during a 6 min walk test, on the basis of walking performance and terminal RPE, in clinically stable patients who took part in a cardiac rehabilitation program. The addition of terminal RPE to the 6 min walk test distance improved the prediction of maximal metabolic equivalents (METs) and METs at VT, which may have practical applications for exercise prescription.

Birnbaumer and colleagues [33] showed that the HR performance curve pattern during incremental cycle ergometer exercise was different in individuals receiving β-blocker treatment compared to that in healthy individuals, suggesting the percentage of maximum power output as a better indicator for exercise intensity prescription in this population. Di Blasio et al. [34] compared the effects of weekly personal feedback, based on objectively measured physical activity, on the trends of both daily sedentary time and physical activity in breast cancer survivors with those of an intervention also including online supervised physical exercise sessions during Italy's COVID-19 lockdown. Their findings showed that the use of an activity tracker and its accompanying app, with the reception of weekly tailored advice and supervised online physical exercise sessions, can elicit proper physical activity in breast cancer survivors in the COVID-19 era.

Rogers et al. [35] investigated whether the anaerobic threshold derived from gas exchange is associated with the transition from a correlated to an uncorrelated random HRV pattern. HR associated with a Detrended Fluctuation Analysis alpha 1 value of 0.5 was closely related to HR at the second VT derived from gas exchange analysis, and has the potential to be a noninvasive marker for training intensity distribution and performance status. Moreover, De Blauw et al. [36] reported similar improvements in cardiovascular function, body composition, and fitness in HRV-guided high-intensity functional training when compared to predetermined high-intensity functional training, despite fewer days at high intensity.

Tyrrell et al. [37] tested a recently developed generalized model [38] to downregulate absolute training intensity in order to account for cardiovascular drift and achieve the desired internal training load using METs during exercise testing and training. Comparing acute responses to three time-matched exercise regimens consisting of sprint interval training, high-intensity interval training, and vigorous continuous training, Benítez-Flores and colleagues [39] suggested that high-intensity interval training and vigorous continuous training accumulate the longest duration at near maximal intensities, which is considered a key factor in enhancing maximal oxygen uptake (VO_2max).

Mayer et al. [40] assessed the surface electromyographic activity of the lumbar extensor muscles during full-range-of-motion, dynamic exercise on a home back extension exercise device at three exercise loads. They concluded that dynamic exercise on a home back extension exercise device is safe and provides a mechanism to progressively activate the lumbar extensor muscles. O' Brien et al. [41] reviewed the literature on the efficacy of flywheel inertia training to increase hamstring strength, and provided general recommendations on valuable flywheel inertia training determinants when integrating this kind of training into a hamstring strengthening program. Finally, Cariati et al. [42] showed that a well-designed aerobic training protocol in terms of speed and speed variation significantly contributes to improving synaptic plasticity and hippocampal ultrastructure, optimizing its benefits in the brains of young murine models.

Given the great success of the first and second editions of this Special Issue, we have launched a third edition, for which we hope to receive contributions focusing on the use of laboratory or field evaluations to generate training advice in patients, healthy people, and athletes.

Conflicts of Interest: The authors declare no conflict of interest.

References

1. Foster, C.; Cortis, C.; Fusco, A. Exercise Evaluation and Prescription. *J. Funct. Morphol. Kinesiol.* **2021**, *6*, 31. [CrossRef] [PubMed]
2. Selye, H. THhe General Adaptation Syndrome and the Diseases of Adaptation. *J. Clin. Endocrinol. Metab.* **1946**, *6*, 117–230. [CrossRef] [PubMed]
3. Banister, E.W.; Calvert, T.; Savage, M.B.T. A Systems Model of Training for Athletic Performance. *Aust. J. Sport. Med.* **1975**, *3*, 57–61.
4. Morton, R.H.; Fitz-Clarke, J.R.; Banister, E.W. Modeling Human Performance in Running. *J. Appl. Physiol.* **1990**, *69*, 1171–1177. [CrossRef] [PubMed]
5. Fitz-Clarke, J.R.; Morton, R.H.; Banister, E.W. Optimizing Athletic Performance by Influence Curves. *J. Appl. Physiol.* **1991**, *71*, 1151–1158. [CrossRef]
6. Foster, C.; Daines, E.; Hector, L.; Snyder, A.C.W.R. Athletic Performance in Relation to Training Load. *Wis. Med. J.* **1996**, *95*, 370–376.
7. Foster, C.; Boullosa, D.; McGuigan, M.; Fusco, A.; Cortis, C.; Arney, B.E.; Orton, B.; Dodge, C.; Jaime, S.; Radtke, K.; et al. 25 Years of Session Rating of Perceived Exertion: Historical Perspective and Development. *Int. J. Sports Physiol. Perform.* **2021**, *16*, 612–621. [CrossRef]
8. Fusco, A.; Knutson, C.; King, C.; Mikat, R.P.; Porcari, J.P.; Cortis, C.; Foster, C. Session RPE During Prolonged Exercise Training. *Int. J. Sports Physiol. Perform.* **2020**, *15*, 292–294. [CrossRef]
9. Fusco, A.; Sustercich, W.; Edgerton, K.; Cortis, C.; Jaime, S.J.; Mikat, R.P.; Porcari, J.P.; Foster, C. Effect of Progressive Fatigue on Session RPE. *J. Funct. Morphol. Kinesiol.* **2020**, *5*, 15. [CrossRef]
10. Rodríguez-Marroyo, J.A.; Villa, J.G.; Pernía, R.; Foster, C. Decrement in Professional Cyclists' Performance After a Grand Tour. *Int. J. Sports Physiol. Perform.* **2017**, *12*, 1348–1355. [CrossRef]

11. Kellmann, M.; Bertollo, M.; Bosquet, L.; Brink, M.; Coutts, A.J.; Duffield, R.; Erlacher, D.; Halson, S.L.; Hecksteden, A.; Heidari, J.; et al. Recovery and Performance in Sport: Consensus Statement. *Int. J. Sports Physiol. Perform.* **2018**, *13*, 240–245. [CrossRef] [PubMed]
12. Jagim, A.R.; Askow, A.T.; Carvalho, V.; Murphy, J.; Luedke, J.A.; Erickson, J.L. Seasonal Accumulated Workloads in Collegiate Women's Soccer: A Comparison of Starters and Reserves. *J. Funct. Morphol. Kinesiol.* **2022**, *7*, 11. [CrossRef]
13. Askow, A.T.; Lobato, A.L.; Arndts, D.J.; Jennings, W.; Kreutzer, A.; Erickson, J.L.; Esposito, P.E.; Oliver, J.M.; Foster, C.; Jagim, A.R. Session Rating of Perceived Exertion (Srpe) Load and Training Impulse Are Strongly Correlated to Gps-Derived Measures of External Load in Ncaa Division i Women's Soccer Athletes. *J. Funct. Morphol. Kinesiol.* **2021**, *6*, 90. [CrossRef] [PubMed]
14. Redd, M.J.; Starling-Smith, T.M.; Herring, C.H.; Stock, M.S.; Wells, A.J.; Stout, J.R.; Fukuda, D.H. Tensiomyographic Responses to Warm-up Protocols in Collegiate Male Soccer Athletes. *J. Funct. Morphol. Kinesiol.* **2021**, *6*, 80. [CrossRef]
15. Grgic, J.; Mikulic, I.; Mikulic, P. Negative Effects of Mental Fatigue on Performance in the Yo-Yo Test, Loughborough Soccer Passing and Shooting Tests: A Meta-Analysis. *J. Funct. Morphol. Kinesiol.* **2022**, *7*, 10. [CrossRef]
16. Romano, V.; Tuzi, M.; Di Gregorio, A.; Sacco, A.M.; Belviso, I.; Sirico, F.; Palermi, S.; Nurzynska, D.; Di Meglio, F.; Castaldo, C.; et al. Correlation between Official and Common Field-Based Fitness Tests in Elite Soccer Referees. *J. Funct. Morphol. Kinesiol.* **2021**, *6*, 59. [CrossRef]
17. Bok, D.; Foster, C. Applicability of Field Aerobic Fitness Tests in Soccer: Which One to Choose? *J. Funct. Morphol. Kinesiol.* **2021**, *6*, 69. [CrossRef]
18. Falch, H.N.; Kristiansen, E.L.; Haugen, M.E.; van den Tillaar, R. Association of Performance in Strength and Plyometric Tests with Change of Direction Performance in Young Female Team-Sport Athletes. *J. Funct. Morphol. Kinesiol.* **2021**, *6*, 83. [CrossRef]
19. Giatsis, G.; Panoutsakopoulos, V.; Kollias, I.A. Drop Jumping on Sand Is Characterized by Lower Power, Higher Rate of Force Development and Larger Knee Joint Range of Motion. *J. Funct. Morphol. Kinesiol.* **2022**, *7*, 17. [CrossRef]
20. Gómez-Carmona, C.D.; Mancha-Triguero, D.; Pino-Ortega, J.; Ibáñez, S.J. Exploring Physical Fitness Profile of Male and Female Semiprofessional Basketball Players through Principal Component Analysis—A Case Study. *J. Funct. Morphol. Kinesiol.* **2021**, *6*, 67. [CrossRef]
21. Daly, E.; White, A.; Blackett, A.D.; Ryan, L. Pressure. A Qualitative Analysis of the Perception of Concussion and Injury Risk in Retired Professional Rugby Players. *J. Funct. Morphol. Kinesiol.* **2021**, *6*, 78. [CrossRef]
22. Palermi, S.; Massa, B.; Vecchiato, M.; Mazza, F.; De Blasiis, P.; Romano, A.M.; Di Salvatore, M.G.; Della Valle, E.; Tarantino, D.; Ruosi, C.; et al. Indirect Structural Muscle Injuries of Lower Limb: Rehabilitation and Therapeutic Exercise. *J. Funct. Morphol. Kinesiol.* **2021**, *6*, 75. [CrossRef] [PubMed]
23. Leo, P.; Mujika, I.; Lawley, J. Influence of COVID-19 Restrictions on Training and Physiological Characteristics in U23 Elite Cyclists. *J. Funct. Morphol. Kinesiol.* **2022**, *7*, 1. [CrossRef] [PubMed]
24. Boullosa, D.; Medeiros, A.R.; Flatt, A.A.; Esco, M.R.; Nakamura, F.Y.; Foster, C. Relationships between Workload, Heart Rate Variability, and Performance in a Recreational Endurance Runner. *J. Funct. Morphol. Kinesiol.* **2021**, *6*, 30. [CrossRef] [PubMed]
25. Sandau, I.; Chaabene, H.; Granacher, U. Predictive Validity of the Snatch Pull Force-Velocity Profile to Determine the Snatch One Repetition-Maximum in Male and Female Elite Weightlifters. *J. Funct. Morphol. Kinesiol.* **2021**, *6*, 35. [CrossRef]
26. Aidar, F.J.; Clemente, F.M.; de Matos, D.G.; Marçal, A.C.; de Souza, R.F.; Moreira, O.C.; de Almeida-Neto, P.F.; Vilaça-Alves, J.; Garrido, N.D.; Dos Santos, J.L.; et al. Evaluation of Strength and Muscle Activation Indicators in Sticking Point Region of National-Level Paralympic Powerlifting Athletes. *J. Funct. Morphol. Kinesiol.* **2021**, *6*, 43. [CrossRef]
27. Aidar, F.J.; Paz, Â.D.A.; Gama, D.D.M.; de Souza, R.F.; Vieira Souza, L.M.; Santos, J.L.D.; Almeida-Neto, P.F.; Marçal, A.C.; Neves, E.B.; Moreira, O.C.; et al. Evaluation of the Post-Training Hypotensor Effect in Paralympic and Conventional Powerlifting. *J. Funct. Morphol. Kinesiol.* **2021**, *6*, 92. [CrossRef]
28. Pippi, R.; Cugusi, L.; Bergamin, M.; Bini, V.; Fanelli, C.G.; Bullo, V.; Gobbo, S.; Di Blasio, A. Impact of BMI, Physical Activity, and Sitting Time Levels on Health-Related Outcomes in a Group of Overweight and Obese Adults with and without Type 2 Diabetes. *J. Funct. Morphol. Kinesiol.* **2022**, *7*, 12. [CrossRef]
29. Bentes, C.M.; Costa, P.B.; Resende, M.; Netto, C.; Dias, I.; da Silveira, A.L.B.; Di Masi, F.; Miranda, H.; de Carvalho, L.M.; Marinheiro, L. Effects of 12 Months of Vitamin d Supplementation on Physical Fitness Levels in Postmenopausal Women with Type 2 Diabetes. *J. Funct. Morphol. Kinesiol.* **2021**, *6*, 87. [CrossRef]
30. Stavrou, V.T.; Griziotis, M.; Vavougios, G.D.; Raptis, D.G.; Bardaka, F.; Karetsi, E.; Kyritsis, A.; Daniil, Z.; Tsarouhas, K.; Triposkiadis, F.; et al. Supervised versus Unsupervised Pulmonary Rehabilitation in Patients with Pulmonary Embolism: A Valuable Alternative in Covid Era. *J. Funct. Morphol. Kinesiol.* **2021**, *6*, 98. [CrossRef]
31. Schultz, K.L.; Foster, C.; Radtke, K.; Bramwell, S.; Cortis, C.; Fusco, A.; Porcari, J.P. Workload Accomplished in Phase III Cardiac Rehabilitation. *J. Funct. Morphol. Kinesiol.* **2021**, *6*, 47. [CrossRef] [PubMed]
32. Porcari, J.P.; Foster, C.; Cress, M.L.; Larson, R.; Lewis, H.; Cortis, C.; Doberstein, S.; Donahue, M.; Fusco, A.; Radtke, K. Prediction of Exercise Capacity and Training Prescription from the 6-Minute Walk Test and Rating of Perceived Exertion. *J. Funct. Morphol. Kinesiol.* **2021**, *6*, 52. [CrossRef] [PubMed]
33. Birnbaumer, P.; Traninger, H.; Sattler, M.C.; Borenich, A.; Hofmann, P. Pattern of the Heart Rate Performance Curve in Subjects with Beta-Blocker Treatment and Healthy Controls. *J. Funct. Morphol. Kinesiol.* **2021**, *6*, 61. [CrossRef]

34. Di Blasio, A.; Morano, T.; Lancia, F.; Viscioni, G.; Di Iorio, A.; Grossi, S.; Cianchetti, E.; Cugusi, L.; Gobbo, S.; Bergamin, M.; et al. Effects of Activity Tracker-Based Counselling and Live-Web Exercise on Breast Cancer Survivors during Italy Covid-19 Lockdown. *J. Funct. Morphol. Kinesiol.* **2021**, *6*, 50. [CrossRef]
35. Rogers, B.; Giles, D.; Draper, N.; Mourot, L.; Gronwald, T. Detection of the Anaerobic Threshold in Endurance Sports: Validation of a New Method Using Correlation Properties of Heart Rate Variability. *J. Funct. Morphol. Kinesiol.* **2021**, *6*, 38. [CrossRef]
36. Deblauw, J.A.; Drake, N.B.; Kurtz, B.K.; Crawford, D.A.; Carper, M.J.; Wakeman, A.; Heinrich, K.M. High-Intensity Functional Training Guided by Individualized Heart Rate Variability Results in Similar Health and Fitness Improvements as Predetermined Training with Less Effort. *J. Funct. Morphol. Kinesiol.* **2021**, *6*, 102. [CrossRef]
37. Tyrrell, T.; Pavlock, J.; Bramwell, S., Cortis, C.; Doberstein, S.T.; Fusco, A.; Porcari, J.P.; Foster, C. Functional Translation of Exercise Responses from Exercise Testing to Exercise Training: The Test of a Model. *J. Funct. Morphol. Kinesiol.* **2021**, *6*, 66. [CrossRef]
38. Foster, C.; Anholm, J.D.; Bok, D.; Boullosa, D.; Condello, G.; Cortis, C.; Fusco, A.; Jaime, S.J.; de Koning, J.J.; Lucia, A.; et al. Generalized Approach to Translating Exercise Tests and Prescribing Exercise. *J. Funct. Morphol. Kinesiol.* **2020**, *5*, 63. [CrossRef]
39. Benítez-Flores, S.; Magallanes, C.A.; Alberton, C.L.; Astorino, T.A. Physiological and Psychological Responses to Three Distinct Exercise Training Regimens Performed in an Outdoor Setting: Acute and Delayed Response. *J. Funct. Morphol. Kinesiol.* **2021**, *6*, 44. [CrossRef]
40. Mayer, J.M.; Udermann, B.E.; Verna, J.L. Electromyographic Analysis of the Lumbar Extensor Muscles during Dynamic Exercise on a Home Exercise Device. *J. Funct. Morphol. Kinesiol.* **2022**, *7*, 26. [CrossRef]
41. O' Brien, J.; Browne, D.; Earls, D.; Lodge, C. The Efficacy of Flywheel Inertia Training to Enhance Hamstring Strength. *J. Funct. Morphol. Kinesiol.* **2022**, *7*, 14. [CrossRef]
42. Cariati, I.; Bonanni, R.; Pallone, G.; Scimeca, M.; Frank, C.; Tancredi, V.; D'arcangelo, G. Hippocampal Adaptations to Continuous Aerobic Training: A Functional and Ultrastructural Evaluation in a Young Murine Model. *J. Funct. Morphol. Kinesiol.* **2021**, *6*, 101. [CrossRef]

Journal of
Functional Morphology and Kinesiology

MDPI

Article

Seasonal Accumulated Workloads in Collegiate Women's Soccer: A Comparison of Starters and Reserves

Andrew R. Jagim [1,2,*], Andrew T. Askow [3], Victoria Carvalho [1], Jason Murphy [2], Joel A. Luedke [1] and Jacob L. Erickson [1]

1 Sports Medicine, Mayo Clinic Health System, Onalaska, WI 54650, USA; victoria.rvalho@gmail.com (V.C.); luedke.joel@mayo.edu (J.A.L.); erickson.jacob@mayo.edu (J.L.E.)
2 Department of Exercise and Sport Science, University of Wisconsin-La Crosse, La Crosse, WI 54601, USA; jmurphy@uwlax.edu
3 Department of Kinesiology and Community Health, University of Illinois at Urbana-Champaign, Urbana, IL 61801, USA; askow2@illinois.edu
* Correspondence: jagim.andrew@mayo.edu

Abstract: Research quantifying the unique workload demands of starters and reserves in training and match settings throughout a season in collegiate soccer is limited. **Purpose:** The purpose of the current study is to compare accumulated workloads between starters and reserves in collegiate soccer. **Methods:** Twenty-two NCAA Division III female soccer athletes (height: 1.67 ± 0.05 m; body mass: 65.42 ± 6.33 kg; fat-free mass: 48.99 ± 3.81 kg; body fat %: $25.22 \pm 4.78\%$) were equipped with wearable global positioning systems with on-board inertial sensors, which assessed a proprietary training load metric and distance covered for each practice and 22 matches throughout an entire season. Nine players were classified as starters (S), defined as those playing >50% of playing time throughout the entire season. The remaining 17 were reserves (R). Goalkeepers were excluded. A one-way ANOVA was used to determine the extent of differences in accumulated training load throughout the season by player status. **Results:** Accumulated training load and total distance covered for starters were greater than reserves ((S: 9431 ± 1471 vs. R: 6310 ± 2263 AU; $p < 0.001$) and (S: 401.7 ± 31.9 vs. R: 272.9 ± 51.4 km; $p < 0.001$), respectively) throughout the season. **Conclusions:** Starters covered a much greater distance throughout the season, resulting in almost double the training load compared to reserves. It is unknown if the high workloads experienced by starters or the low workloads of the reserves is more problematic. Managing player workloads in soccer may require attention to address potential imbalances that emerge between starters and reserves throughout a season.

Keywords: athlete monitoring; workload; GPS; soccer

Citation: Jagim, A.R.; Askow, A.T.; Carvalho, V.; Murphy, J.; Luedke, J.A.; Erickson, J.L. Seasonal Accumulated Workloads in Collegiate Women's Soccer: A Comparison of Starters and Reserves. *J. Funct. Morphol. Kinesiol.* **2022**, *7*, 11. https://doi.org/10.3390/jfmk7010011

Academic Editors: Cristina Cortis, Andrea Fusco and Carl Foster

Received: 22 December 2021
Accepted: 14 January 2022
Published: 16 January 2022

Publisher's Note: MDPI stays neutral with regard to jurisdictional claims in published maps and institutional affiliations.

1. Introduction

The seasonal monitoring of athletes is becoming a popular strategy across athletic departments and within specific teams. The underlying theory of implementing team-based monitoring is to appropriately manage training-related stress and recovery time throughout a season to optimize performance while reducing the risk of injury and mitigating fatigue as much as possible. The specifics of these relationships are likely heavily influenced by the nuances of the methodology used to quantify workload and what performance-based metrics or indices of recovery are being considered. A consensus statement on athlete monitoring was previously published, which provided guidelines on how to assess workload, and associated parameters to consider when drawing comparisons across different sports, monitoring systems, and software programs [1]. Beyond these specifics, the practicality of each sport setting, the specific physiological demands of the sport and competition schedule likely dictate how a monitoring program could be integrated into regular team activities. A recent systematic review found that athletes were at an increased risk for injury during periods of intensified training, changes in acute training load and following a period of higher accumulated workload [2]. The challenge in applied sport science settings is how to identify the specific thresholds that may constitute as excessive workloads, as they are likely specific to the sport, level of competition (or training experience) and the individual athlete.

Depending on the technology available and specific demands of the sport, various metrics may be evaluated to quantify workloads of athletes. Internal workload is often quantified using objective measures of physiological responses (i.e., heart rate, hormonal fluctuations, training impulses, etc.) [1] or subjective measures such as session ratings of perceived exertion (sRPE), perceived recovery status, perceived soreness, etc. [3]. External workload can be quantified using various measures of movement kinematics, derived from accelerometry, global positioning systems (GPS) [4], or local positioning systems (LPS) [1,5]. Each system produces various metrics and parameters used to characterize movement demands such as total distance, high speed running, movement velocity, acceleration, inertial movement units, and number of sprints. Previous work has reported strong relationships between internal and external measures of workload in various sports [6–9]. Certain systems even calculate proprietary metrics to further characterize internal and external workload demands incurred by athletes, some of which are summated throughout a session and reflect both volume and intensity of work (i.e., training load by Polar and player load by Catapult).

Individual sports have unique match demands that are dictated by the rules of play, tactical strategies and fitness level of the athletes. Previous studies have examined the specific match demands across different levels of play in soccer and across each sex [10–13]. In collegiate women's soccer, the typical match results in an average distance of ~9800 m, with approximately 1019 m (~10% of total distance) classified as high speed distance. The overall mean velocity of match play was 63 $m \cdot m^{-1}$ with an average of 15 sprints per match. This distance and speed of play elicited a heart rate response of 142 bpm or 74% of HR max with peak HR values of 197 bpm equating to ~100% HR max and the mean HR was 74.2 ± 6% HR max [13]. As starters typically play >50% of match time throughout the season, it would be safe to assume they are more representative of the mean HR values and other calculations of match demands that are a reflection of the team's performance throughout a match. As a result, starters and non-starters are likely to accrue varying workloads over the course of the season. Therefore, it is necessary to analyze HR and workload separately for starters and reserves.

Currently, limited information is available regarding the accumulation of workloads throughout a season in collegiate soccer and how they differ between starters and reserves. Recently, Curtis et al. [14] were one of the first groups to report on the differences in accumulated workloads between starters and non-starters throughout an NCAA Division I men's soccer season. While starters accumulated substantially more distance and number of accelerations throughout the season, non-starters accumulated more distance and higher TRIMP volumes during training, indicating that non-starters may complete extra work or conditioning during practice compared to starters. Such discrepancies throughout a season may pose challenges to coaches regarding the management of workloads in starters while also providing an adequate and consistent training stimulus for reserves in order to maintain the physiological adaptations required to elicit improvements in performance throughout the season.

It has been previously demonstrated that the accumulative stress of an entire season in NCAA Division I men's [15] and women's [16] soccer leads to significant hormonal perturbations, hematological changes and decrements in aerobic fitness and power, despite lower training loads as the season progressed, in relation to the higher workloads during the pre-season period. Further, post-season declines in soccer-specific fitness parameters appear to be related to the amount of match playing time completed by each player throughout the season [17]. When grouped by starter status, starters appear to experience greater reductions in strength, speed and power compared to reserves following a season in men's collegiate soccer [15]. These trends indicate that players completing higher workloads, which is likely the case in starters, may be at a greater risk for declines in fitness entering the post-season period; however, this has yet to be examined in women's collegiate soccer.

Athlete monitoring is particularly useful when comparing one season to another and examining seasonal outcomes in team success, performance levels, fatigue, and injury rates. By monitoring accumulative workloads throughout a season, practitioners can help guide coaching decisions regarding workload management in starters and reserves. Furthermore, this monitoring could better direct conditioning activities during training sessions for reserves if needed. Appropriate workload management may optimize playing performance leading in to post-season play and reduce risks of injuries throughout the season. Therefore, the purpose of this study was to examine differences in accumulated workloads throughout the season between starters and reserves in collegiate women's soccer.

2. Methods

2.1. Subjects

Twenty-two NCAA Division III collegiate women soccer athletes (height: 1.67 ± 0.05 m; body mass: 65.42 ± 6.33 kg; fat-free mass: 48.99 ± 3.81 kg; body fat %: 25.22 ± 4.78%) participated in this observational study. Athletes who were medically cleared to participate in practice and match play were eligible to participate in this study. Athletes who were not an active member of the women's soccer team, or were currently injured at the start of pre-season, were excluded from participation. For the purposes of player status determination, a threshold of >50% of total match duration for the season was used to designate players as starters (n = 8) or reserves (n = 14) based on previously used methods [13]. Goalkeepers were excluded from this study. All participants provided written consent in accordance to the Institutional Review Board of the University of Wisconsin—La Crosse and Human Subjects Guidelines for Research.

2.2. Study Design

Players were initially invited to an informational meeting prior to the start of the 2019 season during which time details of their participation were explained to them. Demographic information was collected at this time and used to create personalized player profiles in the monitoring system's software platform. Players were equipped with wearable global positioning systems with on-board inertial sensors which assessed heart rate and movement kinematics when worn. Starting during the pre-season training period, all players wore the GPS-based monitoring units throughout the duration of each practice. Players were instructed to position the monitors in place once they were on the field and about to start the warm-up. The monitors were removed at the end of active practice for a total of 47 practices. Players followed the same protocol during all matches at the start of the competitive season for a total of 22 matches. 977 practice files (367 and 610 from starters and reserves, respectively) and 467 match files (172 and 295 from starters and reserves, respectively) were included in the analysis for a total of 1444 unique player sessions. Workload values were then summed at the end of the season for each of the internal and external load variables recorded by the system.

2.3. Athlete Monitoring System

All players were equipped with a GPS-based monitoring system with built-in heart rate monitoring capabilities (Polar TeamPro^{TM} Polar Electro, Oy, Finland). Player demographic information, including age, height and weight, was entered into the proprietary software program associated with the monitoring system which was used to predict aerobic capacity and max heart rate based on age and manufacturer algorithms. The max heart rates (HR) were continually adjusted throughout the pre-season to provide the most accurate and up to date measure of maximal HR. Heart rate zones were used to quantify intensity and defined as: zone 1 = 50–60%, zone 2 = 60–70%, zone 3 = 70–80%, zone 4 = 80–90%, and zone 5 = 90–100%. The software provided a proprietary metric referred to as Training

Load which was calculated from heart rate intensity and duration of activity, presented in arbitrary units. At the end of each training session or match, each sensor was removed from the players, loaded to a docking station, and synced to a cloud-based software program operated by the manufacturer. Data were then exported from this program and later used for analysis.

2.4. Movement Kinematics

The monitoring system provided a count of the frequency of accelerations and decelerations using the following thresholds for categorization: low = $\pm$0.5–1.99 $m \cdot s^{-2}$, moderate = $\pm$2.00–2.99 $m \cdot s^{-2}$, and high = $\pm$3.00–50.0 $m \cdot s^{-2}$ based upon previously used methods [13]. The following thresholds were used for determination of speed walk/stand $\leq$6.99 $km \cdot h^{-1}$, jog = 7.0–14.99 $km \cdot h^{-1}$, run = 15.0–18.99 $km \cdot h^{-1}$, and sprint $\geq$19.00 $km \cdot h^{-1}$. High speed distance (HSD) was a combination of run and sprint speed zones. Sprints were also counted in an accumulating fashion and were defined as any movement resulting in an acceleration >2.8 $m \cdot s^{-2}$. For reference, a detailed summary of the match demands by position has been previously published [13].

2.5. Statistical Analyses

Differences in accumulated training load and movement characteristics between starters and reserves were examined using a repeated measures analysis of variance. When significant main effects or interactions were identified, Bonferroni post hoc analysis were calculated to determine where differences existed. Normal distribution was confirmed via visual inspection of normal Q-Q plots and via assessment of skewness/kurtosis values. Alpha was set at $p < 0.05$ for determination of statistical significance. Pairwise differences were used to calculate Cohen's d (d) effect sizes along with 95% confidence intervals (LB, UB) to determine the magnitude of differences in accumulated workload values. The effect sizes were interpreted using the following criteria: 0.2 = trivial, 0.2–0.6 = small, 0.7–1.2 = moderate, 1.3–2.0 = large, and >2.0 = very large [18]. All data were analyzed using IBM SPSS Statistic for Windows (Version 25.0; IBM Corp., Armonk, NY, USA).

3. Results

Accumulated total distance was significantly greater for starters compared to reserves for the total season (starters: 401.7 $\pm$ 31.9 vs. reserves: 272.9 $\pm$ 51.4 km; $p < 0.001$; d = 2.83 [1.62, 4.03]) and matches (starters: 222.0 $\pm$ 21.3 vs. reserves: 100.9 $\pm$ 38.6 km; $p < 0.001$; d = 3.61 [2.23, 4.98]) as presented in Figure 1A. Accumulated HSD was significantly greater for starters compared to reserves for all sessions (starters: 38.5 $\pm$ 11.9 vs. reserves: 24.8 $\pm$ 8.8 km; $p = 0.006$; d = 1.37 [0.41, 2.33]) and matches (starters: 24.3 $\pm$ 8.0 vs. reserves: 10.2 $\pm$ 6.5 km; $p < 0.001$; d = 2.00 [0.95, 3.05]), as presented in Figure 1B. Accumulated training load was significantly higher in starters compared to reserves for the total season (starters: 9431 $\pm$ 1471 vs. reserves: 6310 $\pm$ 2263; $p = 0.002$; d = 1.54 [0.56, 2.53]) and matches (starters: 5515 $\pm$ 753 vs. reserves: 2392 $\pm$ 1217; $p < 0.001$; d = 2.90 [1.68, 4.12]), respectively as presented in Figure 2A. Accumulated number of sprints during matches (starters: 364.3 $\pm$ 102.8 vs. reserves: 180.9 $\pm$ 78.5; $p < 0.001$; d = 2.09 [1.02, 3.15]) and the season (starters: 700.6 $\pm$ 186.5 vs. reserves: 484.9 $\pm$ 169.4; $p = 0.012$; d = 1.23 [0.29, 2.17]) was significantly higher for starters compared to reserves, respectively, as presented in Figure 2B. Total distance (starters: 179.7 $\pm$ 12.4 vs. reserves: 172.0 $\pm$ 25.5; $p = 0.438$; d = 0.35 [−0.52, 1.23]), HSD (starters: 14.2 $\pm$ 4.2 vs. reserves: 14.6 $\pm$ 3.8; $p = 0.820$; d = 0.10 [−0.77, 0.97]), sprints (starters: 336.4 $\pm$ 90.2 vs. reserves: 304.0 $\pm$ 103.8; $p = 0.470$; d = 0.33 [−0.55, 1.20]), and training load (starters: 3916 $\pm$ 885 vs. reserves: 3918 $\pm$ 1358; $p = 0.998$; d = 0.00 [−0.87, 0.87]) did not differ between starters and reserves during practice sessions throughout the season. Significant and meaningful differences in accumulated time spent in the different heart rate zones and acceleration totals for matches throughout the entire season were observed as presented in Tables 1–3.

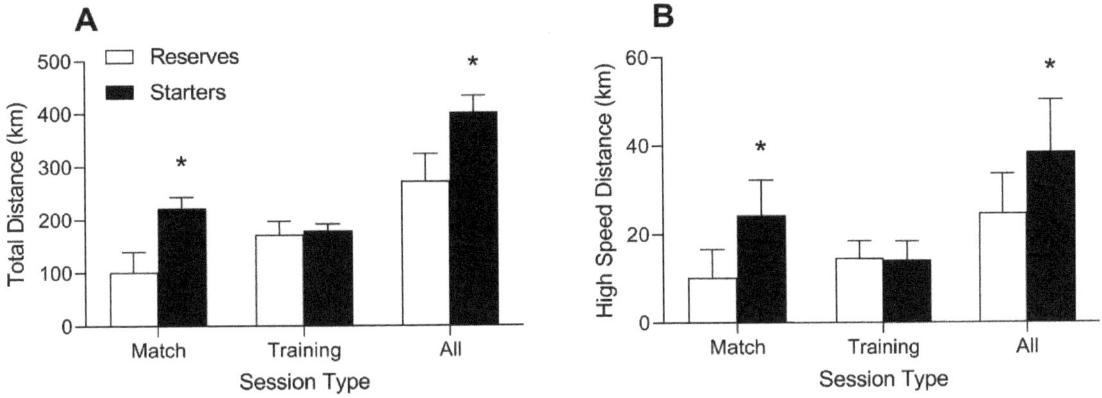

Figure 1. Differences in total distance covered (**A**) and high speed distance (**B**) between starters and reserves in match settings, practice settings, and the cumulative total season. * $p < 0.001$.

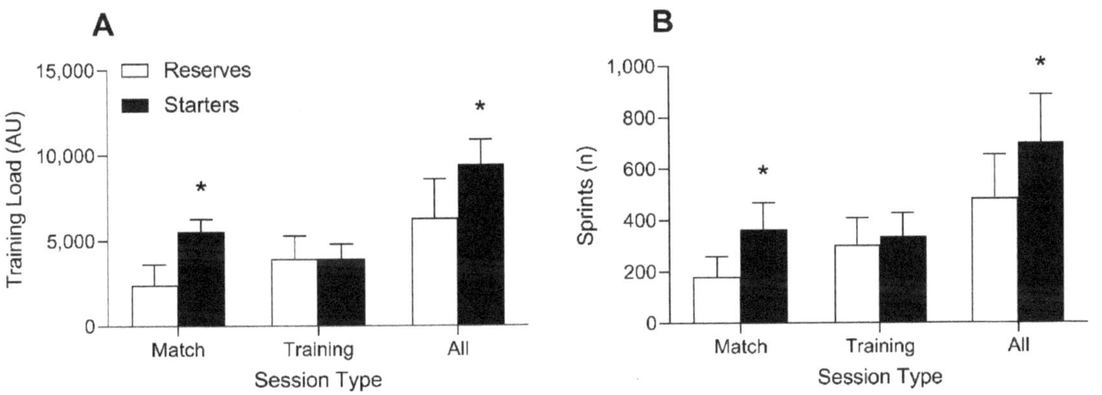

Figure 2. Differences in accumulated training load (**A**) and number of sprints (**B**) between starters and reserves in match settings, practice settings, and the cumulative total season. * $p < 0.001$.

Table 1. Total accumulated workloads for all sessions in NCAA Division III women's soccer by starting status (mean ± SD).

Variable	Starters	Reserves	p	Cohen's d (LB, UB)
VZ1 (km)	188.8 ± 16.4	139.5 ± 23.7	<0.001	2.30 (1.20, 3.40)
VZ2 (km)	174.4 ± 29.6	108.6 ± 24.0	<0.001	2.52 (1.38, 3.67)
VZ3 (km)	27.4 ± 7.3	17.4 ± 5.8	0.002	1.58 (0.59, 2.56)
VZ4 (km)	11.0 ± 5.4	7.4 ± 3.2	0.059	0.89 (−0.02, 1.80)
HRZ1 (min)	1460 ± 278	1473 ± 217	0.906	0.05 (−0.82, 0.92)
HRZ2 (min)	1362 ± 253	1258 ± 386	0.508	0.30 (−0.57, 1.17)
HRZ3 (min)	1161 ± 263	920 ± 255	0.048	0.93 (0.02, 1.85)
HRZ4 (min)	1346 ± 369	870 ± 275	0.003	1.53 (0.55, 2.51)
HRZ5 (min)	923 ± 576	511 ± 251	0.029	1.04 (0.12, 1.96)

Table 1. *Cont.*

Variable	Starters	Reserves	p	Cohen's d (LB, UB)
AZ1 (n)	88,234 ± 8807	65,991 ± 10,932	<0.001	2.17 (1.09, 3.25)
AZ2 (n)	5674 ± 858	3743 ± 985	<0.001	2.05 (0.99, 3.11)
AZ3 (n)	1011 ± 295	728 ± 241	0.023	1.09 (0.16, 2.01)

Velocity Zone Abbreviations: VZ1 (Walk/Stand) ≤6.99 km·h^{-1}; VZ2 (Jog) = 7.0–14.99 km·h^{-1}; VZ3 (Run) = 15.0–18.99 km·h^{-1}; VZ4 (Sprint) ≥19.00 km·h^{-1}. Heart rate zone abbreviations: HRZ1 = 50–60%; HRZ2 = 60–70%; HRZ3 = 70–80%; HRZ4 = 80–90%; HRZ5 = 90–100%. Acceleration Zone Abbreviations: AZ1 = ±0.5–1.99 m·s^{-2}; AZ2 = ±2.00–2.99 m·s^{-2}; AZ3 = ±3.00–50.0 m·s^{-2}.

Table 2. Total accumulated match-day workloads in NCAA Division III women's soccer by starting status (mean ± SD).

Variable	Starters	Reserves	p	Cohen's d (LB, UB)
VZ1 (km)	92.9 ± 12.3	48.8 ± 13.7	<0.001	3.34 (2.02, 4.65)
VZ2 (km)	104.8 ± 19.3	41.9 ± 19.5	<0.001	3.24 (1.95, 4.53)
VZ3 (km)	17.7 ± 5.2	7.4 ± 4.5	<0.001	2.14 (1.06, 3.21)
VZ4 (km)	6.6 ± 3.5	2.7 ± 2.0	0.004	1.46 (0.49, 2.43)
HRZ1 (min)	505 ± 178	677 ± 178	0.042	0.96 (0.05, 1.88)
HRZ2 (min)	506 ± 91	418 ± 183	0.221	0.56 (−0.32, 1.44)
HRZ3 (min)	490 ± 146	281 ± 96	0.001	1.81 (0.79, 2.83)
HRZ4 (min)	842 ± 368	286 ± 161	<0.001	2.19 (1.11, 3.28)
HRZ5 (min)	700 ± 403	269 ± 157	0.002	1.60 (0.61, 2.58)
AZ1 (n)	42,659 ± 5305	23,771 ± 6356	<0.001	3.14 (1.87, 4.41)
AZ2 (n)	3069 ± 545	1358 ± 660	<0.001	2.75 (1.56, 3.94)
AZ3 (n)	529 ± 162	268 ± 125	<0.001	1.87 (0.84, 2.90)

Velocity Zone Abbreviations: VZ1 (Walk/Stand) ≤6.99 km·h^{-1}; VZ2 (Jog) = 7.0–14.99 km·h^{-1}; VZ3 (Run) = 15.0–18.99 km·h^{-1}; VZ4 (Sprint) ≥19.00 km·h^{-1}. Heart rate zone abbreviations: HRZ1 = 50–60%; HRZ2 = 60–70%; HRZ3 = 70–80%; HRZ4 = 80–90%; HRZ5 = 90–100%. Acceleration Zone Abbreviations: AZ1 = ±0.5–1.99 m·s^{-2}; AZ2 = ±2.00–2.99 m·s^{-2}; AZ3 = ±3.00–50.0 m·s^{-2}.

Table 3. Total accumulated practice session workloads in NCAA Division III women's soccer by starting status (mean ± SD).

Variable	Starters	Reserves	p	Cohen's d (LB, UB)
VZ1 (km)	95.9 ± 4.9	90.8 ± 14.8	0.352	0.42 (−0.46, 1.30)
VZ2 (km)	69.5 ± 11.6	66.6 ± 11.0	0.567	0.26 (−0.61, 1.13)
VZ3 (km)	9.7 ± 2.4	10.0 ± 2.5	0.850	0.08 (−0.78, 0.95)
VZ4 (km)	4.5 ± 2.0	4.7 ± 1.6	0.797	0.12 (−0.75, 0.98)
HRZ1 (min)	955 ± 151	796 ± 134	0.019	1.13 (0.20, 2.06)
HRZ2 (min)	856 ± 177	841 ± 226	0.872	0.07 (−0.80, 0.94)
HRZ3 (min)	671 ± 154	639 ± 182	0.686	0.18 (−0.69, 1.05)
HRZ4 (min)	504 ± 134	583 ± 187	0.306	0.47 (−0.41, 1.35)
HRZ5 (min)	223 ± 201	241 ± 154	0.810	0.11 (−0.76, 0.98)
AZ1 (n)	45,576 ± 4104	42,220 ± 6427	0.201	0.59 (−0.30, 1.47)
AZ2 (n)	2606 ± 414	2385 ± 524	0.320	0.45 (−0.43, 1.33)
AZ3 (n)	482 ± 148	459 ± 142	0.723	0.16 (−0.71, 1.03)

Velocity Zone Abbreviations: VZ1 (Walk/Stand) ≤6.99 km·h^{-1}; VZ2 (Jog) = 7.0–14.99 km·h^{-1}; VZ3 (Run) = 15.0–18.99 km·h^{-1}; VZ4 (Sprint) ≥19.00 km·h^{-1}. Heart rate zone abbreviations: HRZ1 = 50–60%; HRZ2 = 60–70%; HRZ3 = 70–80%; HRZ4 = 80–90%; HRZ5 = 90–100%. Acceleration Zone Abbreviations: AZ1 = ±0.5–1.99 m·s^{-2}; AZ2 = ±2.00–2.99 m·s^{-2}; AZ3 = ±3.00–50.0 m·s^{-2}.

4. Discussion

The aim of the current study was to examine differences in accumulated workloads between starters and reserves throughout a collegiate soccer season. Several practically meaningful differences in accumulated measures of external and internal workloads were observed between starters and reserves. Throughout the season, starters covered a greater total distance (401.7 km) and greater distances covered at high speeds (38.5 km) compared to reserves (272.9 and 24.8 km, respectively). Differences in accumulated workload appear to be primarily driven by differences in match playing time, as accumulated match workloads followed a similar pattern as the season totals, whereas few differences in accumulated loads were present when training sessions only were assessed (see Figures 1 and 2). Throughout the season, starters covered a greater distance across all velocity zones, with the exception of sprinting, which was defined as >19.00 km·h^{-1} (Table 1). Similarly, starters recorded a greater number of accelerations in all three acceleration categories (Table 1). These differences in external workload elicited similar discrepancies in internal workloads as starters also spent more time in HR zones 3–5 throughout the season.

When comparing differences in the heart rate-based metric of internal workload (training load), starters again accumulated greater training loads throughout the season compared to reserves. Similar to external workloads, these discrepancies in accumulated measures of internal workload appear to be driven by the differences in match playing time, in which reserves, by definition, did not play as much during match play and therefore slowly accumulated less loads throughout the season. The results of the current study are the first to profile accumulated workloads throughout a season in collegiate women's soccer. Previously Curtis et al. [14] examined accumulated workloads throughout a season in NCAA Division I men's collegiate soccer and reported total distances of ~400 km for starters and ~325 km for reserves. Interestingly, the accumulated total distance for the starters is almost identical to that from the current study indicating the men's and women's collegiate soccer seasons result in comparable total distances covered, despite differences in the level of play (DI vs. DIII). Additionally, and in alignment with findings from the current study, in Division I men, starters also accumulated a higher number of accelerations, covered greater distances in all velocity zones, spent more time at a heart rate intensity of 70–90% of max heart rate and display greater TRIMP values (measure of internal workload) throughout the season compared to reserves [14].

These findings are not surprising considering previous work has indicated comparable match demands between collegiate men and women, regarding distances covered and heart rate response [12,13]. However, the reserves in the study by Curtis et al. [14] accumulated substantially more total distance and TRIMP values throughout the seasons (5 teams monitored over 2 seasons) during training sessions, relative to the starters within the same study. Additionally, the reserves in the current study also did not cover as much distance as those from the study by Curtis et al. [14]. Together, these data suggest that players at the Division I level, may train with greater workloads during practices, or complete a greater number of practices throughout a season compared to Division III. Based on these observations, it appears as though a collegiate men's and women's soccer season may otherwise lead to similar total external workloads throughout a season for starters. One notable difference for internal workload, is that women may spend a greater amount of time at >90% HR max (890 min; HR zone 5) compared to men (586 min), which was actually classified as HR zone 6 in the study by Curtis et al. [14].

It is difficult to discern whether the observed discrepancies in accumulated workloads throughout the season in the current study are an indication that starters are exposed to excessive workloads or reserves are potentially undertraining throughout the season. A limitation of the current study is that neither indices of performance nor biological parameters of training stress or readiness were assessed throughout the season. Therefore, it is difficult to examine any specific relationships between the workloads in the current study and how they may have impacted player performance, fitness levels, physiological adaptations, or hematological changes throughout the season. However, previous research has indicated that greater accumulated workloads throughout a soccer season lead to greater disturbances in hematological, hormonal and autonomic function, and larger reductions in measures of performance [15,19–21]. For example, Walker et al. [21] examined changes in biomarkers throughout a season in collegiate women's soccer and found that athletes with higher training loads and exercise energy expenditure experienced significant physiological changes throughout the course of the season. Mainly, free and total cortisol, prolactin, triiodothyronine (T3), Interleukin-6 (IL-6), creatine kinase, and total iron-binding capacity levels significantly increased throughout the season. Whereas a significant decrease in Omega-3, iron, hematocrit, ferritin and percent transferrin saturation also occurred. Another important finding of this study is that the magnitude of the physiological changes coincided with higher training loads and exercise energy expenditure during the preseason period, which were further intensified by the cumulative effects of the workloads throughout the rest of the season [21]. In a similar study, Huggins et al. [20] also observed clinically meaningful changes in several hematological, hormonal, and inflammatory biomarkers throughout a collegiate soccer season in men. However, to what extent these changes were influenced by individual player loads were not examined, therefore it is difficult to identify how the magnitude of accumulated workload may have exerted any causal influence on such markers. Saidi et al. [22] analyzed 16 elite Tunisian soccer players were over a 12 week period, assessing them at 3 different times, each following various accumulative workloads. The greatest reductions in sprint performance occurred following the most intensive workload period, which also happened to occur at the end of the season, when the accumulative effects of the season may have been a contributing factor. Such physiological changes should be taken into consideration for starters as they are exposed to greater workloads throughout the season and could potentially experience more perturbations in these biomarkers.

There is also the concern that a higher accumulation of workloads throughout a season, may predispose starters to a greater risk of injury. While not assessed in the current study, previous work as suggested that excessive workloads may be associated with greater risks for injuries. The challenge lies within the identification of a threshold by which additional workloads may be deleterious to recovery and performance; and how to identify such a threshold before the line is crossed. Following a 2 year monitoring period in elite youth soccer, authors noted greater relative risks of overall and non-contact injuries following periods of higher accumulative workloads, particularly over 3 and 4 week periods of higher total distances and higher number of accelerations [23]. Similarly, in a study examining weekly training and match loads of 46 elite Australian footballers, the authors reported that as the workload increased, so did the risk of injury. The authors, therefore, recommended that the practice of monitoring and adjusting workloads weekly, may help reduce the risk of injury [24]. These findings are supported by an extensive review conducted by Eckard et al. [25], in which the authors concluded that based on the current evidence there appears to be a strong relationship between training load and injury. Because starters are faced with greater exposures and playing times, they are inherently at a great risk for injury. This elevated risk may be greater at the collegiate level as there is a higher frequency of competition compared to the professional level; however, more evidence is needed to support this hypothesis.

On the contrary, the reduced accumulative workload experienced by the reserves, may also be an area of concern as it may lead to detraining effects; however, previous work in this area has yielded mixed results. At the collegiate [26] and professional level [27], it has been demonstrated that starters were able to maintain and even increase various measures of performance throughout a basketball season, with performance increases tending to coincide with increased playing time. Thereby indicating that regular and sufficient stimulus throughout the season may help elicit performance adaptations throughout the season. Contradictory findings were noted by McLean et al. [28], who reported significant decline in maximal power output during inertial load cycling in starters throughout the season in soccer players, but not in non-starters. In, an earlier study by Kraemer et al. [15], the authors noted significant reductions in vertical jump and sprint performance at the end of a soccer season in starters, which were not observed in non-starters. The authors [15] concluded that because of higher accumulative workloads throughout the season, the starters may have been overtrained, which may have subsequently contributed to the observed reductions in performance. Therefore, it may be a mixture of the two in that the higher accumulative workloads experienced by starters may provide a continued stimulus for further improvements throughout the season but excessive amounts may result in an overtraining effect, in which, athletes may experience decrements in performance and other disturbances in biomarkers. Whereas non-starters may benefit from additional training throughout the season, in order to continue providing regular stimuli likely required to elicit an adaptive response to compensate for the lack of match play. The lack of in-season and post-season performance assessments is a limitation of the current study design. Without these performance assessments, it is difficult to determine if the starters were exposed to excessively high workloads or conversely, if the reserves were not exposed to sufficient workloads required to elicit physiological adaptations or maintain fitness status throughout the season. Future research should employ regular in-season performance testing to evaluate fitness levels, neuromuscular function and overall player readiness to help guide decision making on load management strategies and player readiness.

5. Conclusions

The total accumulative distance covered by starters was almost double that covered by reserves throughout the season. Further, starters accumulated greater HSD, higher training loads, and a greater number of sprints throughout the season for all sessions compared to reserves. Significant discrepancies were also observed for the amount of time starters and reserves spent in different HR zones. Specifically during match play, starters spent most of the match in HR zone 4, while reserve players were found to spend most of their time in HR zone 1, which also contributed to greater accumulative internal workloads (training load) throughout the season for starters. Higher workloads have been linked to more accentuated perturbations of several biomarkers when compared to lower workloads, which may also carry a higher risk of injury. Discrepancies in accumulated workloads throughout the season between starters and reserves have been previously established in men's, and now women's collegiate soccer. However, there have been mixed findings regarding how such workload discrepancies influence the ability to maintain fitness levels throughout a season. Future studies should seek to examine which end of the spectrum is more problematic: the higher workload experienced by starters or the lower workload experienced by reserves. Both extremes may be detrimental to players' health and performance. On one hand, the higher load experienced by starters, may lead to overtraining, accumulative fatigue, inflammation, higher risk of injury and psychological disturbances. On the other hand, the lower load experienced by reserve players may not be sufficient to maintain fitness status throughout a season and would therefore would warrant additional training for reserves throughout the season.

Author Contributions: Conceptualization, A.R.J., J.M., A.T.A., J.A.L. and J.L.E.; Methodology: A.R.J. and A.T.A., Software: A.R.J. and A.T.A.; Investigation: A.R.J., J.M. and A.T.A.; Resources: A.R.J., J.A.L. and J.L.E., Data Curation: A.R.J., J.M. and A.T.A.; Writing—original draft preparation: A.R.J., V.C. and A.T.A. Writing—review and editing: A.R.J., J.M., A.T.A., V.C., J.A.L. and J.L.E. All authors have read and agreed to the published version of the manuscript.

Funding: This research received no external funding.

Institutional Review Board Statement: This study was conducted in accordance with the guidelines set forth in the Declaration of Helsinki and approved by the Institutional Review Board at the University of Wisconsin—La Crosse. Approval number (IRB #19-AJ-707) and the date (16 May 2019).

Informed Consent Statement: Informed consent was obtained from all subjects involved in the study.

Data Availability Statement: The data presented in this study are available on request from the corresponding author.

Conflicts of Interest: The authors declare no conflict of interest.

References

1. Bourdon, P.C.; Cardinale, M.; Murray, A.; Gastin, P.; Kellmann, M.; Varley, M.C.; Gabbett, T.J.; Coutts, A.J.; Burgess, D.J.; Gregson, W.; et al. Monitoring Athlete Training Loads: Consensus Statement. *Int. J. Sports Physiol. Perform.* **2017**, *12*, S2-161–S2-170. [CrossRef]
2. Jones, C.M.; Griffiths, P.C.; Mellalieu, S.D. Training Load and Fatigue Marker Associations with Injury and Illness: A Systematic Review of Longitudinal Studies. *Sport. Med.* **2017**, *47*, 943–974. [CrossRef]
3. Foster, C.; Boullosa, D.; McGuigan, M.; Fusco, A.; Cortis, C.; Arney, B.E.; Orton, B.; Dodge, C.; Jaime, S.; Radtke, K.; et al. 25 Years of Session Rating of Perceived Exertion: Historical Perspective and Development. *Int. J. Sports Physiol. Perform.* **2021**, *16*, 612–621. [CrossRef]
4. Coutts, A.J.; Duffield, R. Validity and reliability of GPS devices for measuring movement demands of team sports. *J. Sci. Med. Sport* **2010**, *13*, 133–135. [CrossRef]
5. Buchheit, M.; Simpson, B.M. Player-Tracking Technology: Half-Full or Half-Empty Glass? *Int. J. Sports Physiol. Perform.* **2017**, *12*, S2-35–S2-41. [CrossRef]
6. Bartlett, J.D.; O'Connor, F.; Pitchford, N.; Torres-Ronda, L.; Robertson, S.J. Relationships between Internal and External Training Load in Team-Sport Athletes: Evidence for an Individualized Approach. *Int. J. Sports Physiol. Perform.* **2017**, *12*, 230–234. [CrossRef]
7. Askow, A.T.; Lobato, A.L.; Arndts, D.J.; Jennings, W.; Kreutzer, A.; Erickson, J.L.; Esposito, P.E.; Oliver, J.M.; Foster, C.; Jagim, A.R. Session Rating of Perceived Exertion (sRPE) Load and Training Impulse Are Strongly Correlated to GPS-Derived Measures of External Load in NCAA Division I Women's Soccer Athletes. *J. Funct. Morphol. Kinesiol.* **2021**, *6*, 90. [CrossRef]
8. Casamichana, D.; Castellano, J.; Calleja-Gonzalez, J.; RomaN, J.S.; Castagna, C. Relationship between indicators of training load in soccer players. *J. Strength Cond. Res.* **2013**, *27*, 369–374. [CrossRef]
9. Iannaccone, A.; Fusco, A.; Skarbalius, A.; Kniubaite, A.; Cortis, C.; Conte, D. Relationship between External and Internal Load Measures in Youth Beach Handball. *Int. J. Sports Physiol. Perform.* **2021**, 1–7. [CrossRef]
10. Mohr, M.; Krustrup, P.; Andersson, H.; Kirkendal, D.; Bangsbo, J. Match Activities of Elite Women Soccer Players at Different Performance Levels. *J. Strength Cond. Res.* **2008**, *22*, 341–349. [CrossRef]
11. Bradley, P.S.; Dellal, A.; Mohr, M.; Castellano, J.; Wilkie, A. Gender differences in match performance characteristics of soccer players competing in the UEFA Champions League. *Hum. Mov. Sci.* **2014**, *33*, 159–171. [CrossRef]
12. Curtis, R.M.; Huggins, R.A.; Looney, D.P.; West, C.A.; Fortunati, A.; Fontaine, G.J.; Casa, D.J. Match Demands of National Collegiate Athletic Association Division I Men's Soccer. *J. Strength Cond. Res.* **2018**, *32*, 2907–2917. [CrossRef]
13. Jagim, A.R.; Murphy, J.; Schaefer, A.Q.; Askow, A.T.; Luedke, J.A.; Erickson, J.L.; Jones, M.T. Match Demands of Women's Collegiate Soccer. *Sports* **2020**, *8*, 87. [CrossRef]
14. Curtis, R.M.; Huggins, R.A.; Benjamin, C.L.; Sekiguchi, Y.; Arent, S.M.; C. Armwald, B.; Pullara, J.M.; West, C.A.; Casa, D.J. Seasonal Accumulated Workloads in Collegiate Men's Soccer: A Comparison of Starters and Reserves. *J. Strength Cond. Res.* **2021**, *35*, 3184–3189. [CrossRef]
15. Kraemer, W.J.; French, D.N.; Paxton, N.J.; Häkkinen, K.; Volek, J.S.; Sebastianelli, W.J.; Putukian, M.; Newton, R.U.; Rubin, M.R.; Gómez, A.L.; et al. Changes in Exercise Performance and Hormonal Concentrations Over a Big Ten Soccer Season in Starters and Nonstarters. *J. Strength Cond. Res.* **2004**, *18*, 121. [CrossRef]
16. McFadden, B.A.; Walker, A.J.; Bozzini, B.N.; Hofacker, M.; Russell, M.; Arent, S.M. Psychological and Physiological Changes in Response to the Cumulative Demands of a Women's Division I Collegiate Soccer Season. *J. Strength Cond. Res.* **2021**, *Publish Ahead*. [CrossRef]
17. Silva, J.R.; Magalhães, J.F.; Ascensão, A.A.; Oliveira, E.M.; Seabra, A.F.; Rebelo, A.N. Individual Match Playing Time During the Season Affects Fitness-Related Parameters of Male Professional Soccer Players. *J. Strength Cond. Res.* **2011**, *25*, 2729–2739. [CrossRef]
18. Hopkins, W.G.; Marshall, S.W.; Batterham, A.M.; Hanin, J. Progressive statistics for studies in sports medicine and exercise science. *Med. Sci. Sport. Exerc.* **2009**, *41*, 3–12. [CrossRef]
19. Sekiguchi, Y.; Huggins, R.A.; Curtis, R.M.; Benjamin, C.L.; Adams, W.M.; Looney, D.P.; West, C.A.; Casa, D.J. Relationship Between Heart Rate Variability and Acute:Chronic Load Ratio Throughout a Season in NCAA D1 Men's Soccer Players. *J. Strength Cond. Res.* **2021**, *35*, 1103–1109. [CrossRef]
20. Huggins, R.A.; Fortunati, A.R.; Curtis, R.M.; Looney, D.P.; West, C.A.; Lee, E.C.; Fragala, M.S.; Hall, M.L.; Casa, D.J. Monitoring Blood Biomarkers and Training Load Throughout a Collegiate Soccer Season. *J. Strength Cond. Res.* **2019**, *33*, 3065–3077. [CrossRef]
21. Walker, A.J.; McFadden, B.A.; Sanders, D.J.; Rabideau, M.M.; Hofacker, M.L.; Arent, S.M. Biomarker Response to a Competitive Season in Division I Female Soccer Players. *J. Strength Cond. Res.* **2019**, *33*, 2622–2628. [CrossRef]
22. Saidi, K.; Ben Abderrahman, A.; Boullosa, D.; Dupont, G.; Hackney, A.C.; Bideau, B.; Pavillon, T.; Granacher, U.; Zouhal, H. The Interplay between Plasma Hormonal Concentrations, Physical Fitness, Workload and Mood State Changes to Periods of Congested Match Play in Professional Soccer Players. *Front. Physiol.* **2020**, *11*, 835. [CrossRef]
23. Bowen, L.; Gross, A.S.; Gimpel, M.; Li, F.-X. Accumulated workloads and the acute:chronic workload ratio relate to injury risk in elite youth football players. *Br. J. Sports Med.* **2017**, *51*, 452–459. [CrossRef]
24. Rogalski, B.; Dawson, B.; Heasman, J.; Gabbett, T.J. Training and game loads and injury risk in elite Australian footballers. *J. Sci. Med. Sport* **2013**, *16*, 499–503. [CrossRef]

25. Eckard, T.G.; Padua, D.A.; Hearn, D.W.; Pexa, B.S.; Frank, B.S. The Relationship between Training Load and Injury in Athletes: A Systematic Review. *Sport. Med.* **2018**, *48*, 1929–1961. [CrossRef]
26. Gonzalez, A.M.; Hoffman, J.R.; Scallin-Perez, J.R.; Stout, J.R.; Fragala, M.S. Performance Changes in National Collegiate Athletic Association Division I Women Basketball Players during a Competitive Season. *J. Strength Cond. Res.* **2012**, *26*, 3197–3203. [CrossRef]
27. Gonzalez, A.M.; Hoffman, J.R.; Rogowski, J.P.; Burgos, W.; Manalo, E.; Weise, K.; Fragala, M.S.; Stout, J.R. Performance Changes in NBA Basketball Players Vary in Starters vs. Nonstarters Over a Competitive Season. *J. Strength Cond. Res.* **2013**, *27*, 611–615. [CrossRef]
28. McLean, B.D.; Petrucelli, C.; Coyle, E.F. Maximal Power Output and Perceptual Fatigue Responses during a Division I Female Collegiate Soccer Season. *J. Strength Cond. Res.* **2012**, *26*, 3189–3196. [CrossRef]

Journal of
Functional Morphology and Kinesiology

MDPI

Article

Session Rating of Perceived Exertion (sRPE) Load and Training Impulse Are Strongly Correlated to GPS-Derived Measures of External Load in NCAA Division I Women's Soccer Athletes

Andrew T. Askow [1,†], Alexa L. Lobato [2,†], Daniel J. Arndts [2,†], Will Jennings [2], Andreas Kreutzer [2], Jacob L. Erickson [3], Phil E. Esposito [2], Jonathan M. Oliver [4], Carl Foster [5] and Andrew R. Jagim [3,5,*]

1 Department of Kinesiology and Community Health, University of Illinois at Urbana-Champaign, Urbana, IL 61801, USA; askow2@illinois.edu
2 Department of Kinesiology, Texas Christian University, Fort Worth, TX 76109, USA; a.lobato@tcu.edu (A.L.L.); d.arndts@tcu.edu (D.J.A.); will.jennings28@tcu.edu (W.J.); a.kreutzer@tcu.edu (A.K.); p.esposito@tcu.edu (P.E.E.)
3 Sports Medicine, Mayo Clinic Health System, Onalaska, WI 54601, USA; erickson.jacob@mayo.edu
4 Department of Biobehavioral Sciences, Teachers College, Columbia University, New York, NY 10027, USA; jo2667@tc.columbia.edu
5 Department of Exercise and Sport Science, University of Wisconsin-La Crosse, La Crosse, WI 54601, USA; cfoster@uwlax.edu
* Correspondence: jagim.andrew@mayo.edu; Tel.: +1-608-392-5280
† A. T. Askow, A. L. Lobato and D. J. Arndts contributed equally to this work.

Citation: Askow, A.T.; Lobato, A.L.; Arndts, D.J.; Jennings, W.; Kreutzer, A.; Erickson, J.L.; Esposito, P.E.; Oliver, J.M.; Foster, C.; Jagim, A.R. Session Rating of Perceived Exertion (sRPE) Load and Training Impulse Are Strongly Correlated to GPS-Derived Measures of External Load in NCAA Division I Women's Soccer Athletes. *J. Funct. Morphol. Kinesiol.* **2021**, *6*, 90. https://doi.org/10.3390/jfmk6040090

Academic Editor: Eleftherios Kellis

Received: 30 September 2021
Accepted: 25 October 2021
Published: 29 October 2021

Abstract: Purpose: The purpose of this study was to determine whether session rating of perceived exertion-derived training load (sRPE-TL) correlates with GPS-derived measures of external load in National Collegiate Athletics Association (NCAA) Division I female soccer athletes. Methods: Twenty-one NCAA Division 1 collegiate women's soccer athletes (11 starters, 10 non-starters; 65.1 ± 7.2 kg, 168.4 ± 7.9 cm, 20.3 ± 1.5 yrs) volunteered to take part in this study. Data for this study were collected over the course of 16 weeks during the 2018 NCAA women's soccer season. External load and heart rate (HR) data were collected during each training session and match during the season. At least 30 min after the end of an activity (e.g., match or practice), athletes were prompted to complete a questionnaire reporting their perceived exertion for the session. sRPE-TL was calculated at the end of the season by multiplying perceived exertion by the respective session duration. Results: sRPE-TL was very strongly correlated with total distance, distance covered in velocity zones 1–3, the number of accelerations in zones 4 and 5, total PlayerLoad™, and PlayerLoad™. For internal load, sRPE-TL correlated very strongly ($0.70 \leq |r| < 0.90$) with Edward's and Bannister's TRIMP and strongly ($0.50 \leq |r| < 0.70$) with duration spent in in heart rate zones 5 and 6 (80–90% and 90–100% max HR, respectively) while correlations with maximum HR (bpm), mean HR (bpm), and mean HR (%) and sRPE-TL were moderate ($0.30 \leq |r| < 0.50$). Conclusions: In NCAA Division I women soccer, sRPE-TL is strongly associated with external measures of workload. These relationships were stronger during match play, with acceleration load and total distance exhibiting the strongest relationship with sRPE-TL.

Keywords: athlete monitoring; internal load; ratings of perceived exertion; soccer

1. Introduction

Progressive overload is critical for maximizing positive adaptations to training. Inadequate overload can cause failure to adapt, while excessive load may increase the risk of injury, illness, non-functional overreaching, or even overtraining syndrome [1]. Thus, load monitoring throughout a season is a crucial component of athlete management. Quantifying training load during sports practice can be challenging, due to the large number of athletes and their unique movement characteristics at any given point throughout the training session or competition. This is particularly true in the setting of team sports, where a given external workload may represent considerable variation in the internal training load, which is likely of more importance as a driver of the adaptive response [2]. A variety of wearable technologies have been developed, of which each offers a unique set of movement kinematics and measures of external workload that can be captured. In particular, wearable global positioning systems (GPS) in combination with inertial sensors have been of particular interest for sport applications due to their ability to quantify kinematic and inertial load metrics in a variety of settings. These systems have a growing body of evidence supporting their use with valid and reliable measures of various parameters of movement kinematics such as velocity, acceleration, and distances covered [3–6] for a variety of different sports [7]. However, these technologies can be cost-prohibitive and require training and knowledge to efficiently interpret data and reduce error. Further, these units only measure external training load, whereas internal load (i.e., the physiological stress imparted by external load) more likely mediates the adaptive response to training and the strain of training and competition experienced by the athlete [2]. Additionally, each GPS-system may have accompanying software programs that utilize various thresholds or proprietary metrics to quantify speed zones, high-speed distances, and various derivations of training load or player load. As a result, comparisons across systems may prove challenging as the resulting parameters may have been calculated differently. Therefore, other methods of assessing training load may offer additional benefits, particularly if they are calculated using consistent methods, and would therefore be able to be compared universally across different settings.

Historically, heart rate (HR)-based methods have been primarily used for quantifying internal training load. However, these methods require the use of wearable technology in order to record HR. Although HR sensors are much more accessible in terms of cost compared to GPS units, they are still susceptible to technical errors resulting in lost data. Further, HR measures of internal load seem to be fairly inadequate in quantifying load during very high-intensity training [8] and may not adequately account for the range of intensities experienced. Furthermore, they may not account for the internal stressors incurred in collision sports, which consist of high-impact collisions and blocking, tackling, or rushing activities. Thus, solely relying on these measures to estimate cumulative load may result in an inaccurate approximation of load, particularly in a highly stochastic sport such as soccer. In response, Foster and colleagues [9] developed the concept of integrating the Rating of Perceived Exertion (RPE) over the duration of a training session or competition to calculate a session RPE (sRPE), for use in retrospective estimation of global perceived exertion for the entire training bout. When multiplied by the duration of an exercise bout, this duration–intensity product yields an sRPE-derived measure of training load (sRPE-TL) as a subjective measure of internal load. This approach has been shown to provide a low-cost and low-tech method of quantifying internal training load in a variety of sport settings [9,10].

Since its introduction, sRPE-TL has been used in a large number of studies and has been shown to correlate well with other measures of internal load [10–14]. However, most training programs continue to prescribe work in terms of external load. Therefore, it is important to understand the relationship between sRPE-TL and measures of external load by examining its validity and utility as a tool for monitoring training load within a particular team sport setting. Several studies have done so and have provided preliminary evidence to indicate that sRPE-TL is highly correlated with total distance [15], accelerometer load [16], and high-speed efforts [16]. Further, a recent meta-analysis indicates that sRPE-TL is more strongly associated with external load compared to traditional HR-based measures of internal load [17]. However, few studies have assessed these relationships in female athletes competing at the collegiate level. Given the physiological differences between males and females and possible sex differences in the reliability of perceived exertion [18], it is important to further examine the relationship between sRPE-TL and measures of external load in females. Furthermore, it is important to examine these relationships across specific sport types as certain competition demands, movement characteristics, and physical abilities of the athletes may influence the magnitude of the relationship observed. This is particularly true considering the relatively liberal substitution rules within collegiate soccer, which differ meaningfully from the "starters continue to play" FIFA rules seen at the professional level. Therefore, the purpose of this study was to determine whether sRPE-TL correlates with GPS-derived measures of external load in National Collegiate Athletics Association (NCAA) Division I female soccer athletes. As a secondary aim, we sought to determine whether sRPE-TL or HR-based internal load was more strongly correlated to the external load. We hypothesized that sRPE-TL would correlate well with measures of external load.

2. Methods

2.1. Subjects

Twenty-one NCAA Division 1 collegiate women's soccer athletes (11 starters, 10 non-starters; 65.1 ± 7.2 kg, 168.4 ± 7.9 cm, 20.3 ± 1.5 yrs.) volunteered to take part in this study. All athletes were screened for health contraindications by the sports medicine staff as part of the team's normal standard of care and, thus, the sole inclusion criterion was to be a member of the team. Goalkeepers were excluded due to the relatively low total distance traveled compared to other positions. Interested athletes were informed of the risks associated and provided written informed consent prior to participation. All procedures involving human subjects were conducted in accordance with the requirements of the Declaration of Helsinki and approved by the Institutional Review Board at Texas Christian University (approval no. 1807-114-1807; approval date 30 July 2018).

2.2. Study Design

The data for this study were collected over the course of 16 weeks during the 2018 NCAA women's soccer season. External load and HR data were collected during each training session and match during the season. At least 30 min after the end of an activity (e.g., match or practice), athletes were prompted to complete a questionnaire reporting their perceived exertion for the session. sRPE-TL was calculated at the end of the season by multiplying perceived exertion by the respective session duration. On match days, the session durations were reflective of total warm-up duration plus the total (actual) on-field time. Subsequently, the relationships between internal load (sRPE-TL and HR) and external load measures were determined to assess the validity of sRPE-TL. A total of 1767 data points from the GPS units and 1105 perceived exertion questionnaires were collected over the course of the season. However, after excluding goal keepers (n = 211), players who left the team early in the season (n = 43), and incomplete datasets (i.e., where either sRPE-TL or external load data were not available; n = 441), a total of 1072 instances of concurrent sRPE-TL and external load data were identified and included in the analysis. The final analysis included both starters and non-starters.

2.3. Preliminary Testing

Two days prior to the beginning of preseason practice, athletes reported to the athletics complex for preliminary testing to determine maximal velocity and HR using methods recommended by Turner and colleagues [19]. For maximal velocity, athletes completed 3 trials of a maximal sprint task with a flying start. Electronic timing gates (Dashr, LLC; Lincoln, NE, USA) were set up 20 m apart and used to calculate maximal velocity. Following the sprint task, athletes completed a Yo-Yo Intermittent Recovery Test Level 1 for the determination of maximal HR, which was later used for the calculation of relative HR variables and HR intensity zones. The test was continued until athletes could no longer complete a 40 m stage in two consecutive attempts. During the test, HR was measured using chest-worn HR sensors (H7; Polar Electro, Kempele, Finland).

2.4. Data Collection Procedures

Over the course of the season, the external load was assessed using commercially available GPS units (OptimEye X4; Catapult Innovations, Melbourne, Australia) sampling at 10 Hz. Prior to the first training session, research personnel fitted athletes with a GPS unit in accordance with manufacturer guidelines. The unit was secured between the athlete's scapulae using a purpose-built cloth garment from the manufacturer. The same GPS unit and garment were used every session for each athlete to minimize the risk of inter-unit variability. Athlete profiles were created within the software program OpenField (Catapult Innovations, Melbourne, Australia), with pertinent demographic data used for each individual and data collected during preliminary testing. Previous research indicated the current GPS monitoring system provides valid and reliable measures of movement kinematic variables [3,5,6]. Specifically, strong relationships ($r = 0.99–1.00$) with small typical error of the estimate (TEE; 0.08–0.38) values have been reported for total distance, and distances at low speeds (0–3 $m{\cdot}s^{-1}$), moderate speeds (3–5 $m{\cdot}s^{-1}$), high speeds (5–7 $m{\cdot}s^{-1}$), and very high speeds (>7 $m{\cdot}s^{-1}$), when compared to criterion measures [5]. There is also evidence indicating a high degree of interunit reliability for peak velocity (ICC = 0.97; Percentage typical error of measurement [%TEM] = 1.6%), and distance covered for low speeds (ICC = 0.97; %TEM = 1.7%) and high speeds (ICC = 0.88; %TEM = 4.8%) [6].

At least 30 min prior to every practice or match, the same research personnel arrived at the athletic complex to prepare the units for data collection and allow for the acquisition of the satellite signal. When athletes arrived and GPS units were connected, external load data were collected using the manufacturer's data acquisition software. In order to ensure that duration-sensitive data were captured accurately, practice and match activities were coded within the software at the time of collection to minimize the inclusion of rest time surrounding practice or between periods. For matches, only time spent during warmups and time on the field were included in the analysis. Half time and time spent on the sidelines were excluded. Following the conclusion of each session, data were uploaded to a cloud-based version of the software for future analysis. Using the software, PlayerLoad™ was automatically calculated as a proprietary metric used to quantify training load. Accumulated PlayerLoad™ can be calculated using the following formula:

$$\text{PlayerLoad}\ (\text{accul})_{t=n} = \sum_{t=0}^{t=n} \sqrt{\left[(\text{fwd}_{t=i+1} - \text{fwd}_{t=i})^2 + (\text{side}_{t=i+1} - \text{side}_{t=i})^2 + (\text{up}_{t=i+1} - \text{up}_{t=i})^2 \right]}$$

where t is time, fwd is forward acceleration, side is sideways acceleration, and up is upwards acceleration. The metric is a volume-based measure of summated external load using arbitrary units. Similarly, total distance traveled and distance traveling at given velocities (band 1 = 0–6 $km{\cdot}h^{-1}$; band 2 = 6–12 $km{\cdot}h^{-1}$; band 3 = 12–18 $km{\cdot}h^{-1}$; band 4 = 18–25 $km{\cdot}h^{-1}$; band 5 = >25 $km{\cdot}h^{-1}$) were calculated for the correlational analysis. Finally, acceleration load, the sum of absolute values (i.e., both positive and negative accelerations) for acceleration (a volume measurement of total speed change activity), was calculated using the software as well.

In order to minimize the effect of fatigue immediately following a session, sRPE data were collected at least 30 min following the end of each session using a smartphone app (TeamBuildr; TeamBuildr, LLC, Landover, MD, USA). The sRPE has been shown to be very robust for time post exercise, around a nominal average of 30 min [10,20]. At this point, athletes were visually presented with the scale previously introduced by Foster et al. [9] and reported their RPE for the session. Following the submission of their responses, data were automatically uploaded to cloud-based software. Any responses submitted on the following day were excluded from further analysis. Session duration was then used to calculate sRPE-TL. To further investigate the relationship between sRPE and other measures of load, all session types were stratified by sRPE into high (sRPE $\geq$ 6) and low (sRPE $\leq$ 5) difficulty sessions. Specifically, this approach was used to determine if differences in various measures of external workload existed when a threshold of sRPE-TL was used to categorically group sessions into "low" and "high" sRPE-TL.

Using previously established methods [14], additional HR-derived measures of internal training load were calculated. Bannister [21] training impulse (TRIMP) scores were calculated using the following formula:

$$TD \times HR_R \times 0.64 \times e^{(1.92 \times HR_R)}$$

where TD is the effective training session duration expressed in min and HR_R is determined with the following equation:

$$\frac{(HR_{TS} - HR_B)}{(HR_{max} - HR_B)}$$

where HR_{TS} is the average training session HR and HR_B is the HR measured at rest. Additionally, Edwards TRIMP scores were also calculated by assessing the product of the accumulated training duration (expressed in minutes) across the 5 HR zones by a coefficient relative to each zone (Zone 1 = 50–60% of HR_{max}; Zone 2 = 60–70% of HR_{max}; Zone 3 = 70–80% of HR_{max}; Zone 4 = 80–90% HR_{max}; Zone 5 = 90–100% of HR_{max}) and then summating the final results [22].

2.5. Statistical Analyses

Following the conclusion of the season, data from the GPS units and smartphone app were combined for analysis. The relationship between internal and external load measures was assessed using linear regression with 95% confidence intervals. Subsequently, differences in measures of internal and external load, stratified by high and low sRPE sessions, were assessed using a multivariate analysis of variance (MANOVA). Normality was assessed via visual inspection of normal Q-Q plots and skewness/kurtosis values. Homoscedasticity was assessed using Levene's Test of Equality of Error Variances. Bonferonni post hoc comparisons were calculated when a significant main effect or interaction was identified. All statistical analyses were conducted using the IBM SPSS Statistics for Windows (version 25.0; IBM Corp., Armonk, NY, USA). The strength of correlation coefficients was classified as trivial ($|r| < 0.10$), weak ($0.10 \leq |r| < 0.30$), moderate ($0.30 \leq |r| < 0.50$), strong ($0.50 \leq |r| < 0.70$), very strong ($0.70 \leq |r| < 0.90$), and nearly perfect ($r \geq 0.90$) [23].

3. Results

Correlation coefficients, *p* values, and 95% confidence intervals for all relationships can be found in Table 1, Table 2, Table 3 below. sRPE-TL was very strongly correlated with total distance, distance covered in velocity zones 1–3, number of accelerations in zones 4 and 5, total PlayerLoad™, and PlayerLoad™. Strong correlations between distance in velocity band 4, high-speed running distance, and accelerations in zones 3 and 6 and sRPE-TL were observed. For internal load, sRPE-TL correlated very strongly ($0.70 \leq |r| < 0.90$) with Edward's and Bannister's TRIMP and strongly ($0.50 \leq |r| < 0.70$) with the duration spent in in heart zones 5 and 6 (80–90% and 90–100% max HR, respectively) while correlations with maximum HR (bpm), mean HR (bpm), and mean HR (%) and sRPE-TL were moderate ($0.30 \leq |r| < 0.50$). When considering event type (i.e., matches vs. practices), relationships were generally stronger in matches. sRPE-TL from match data was nearly perfectly correlated with total distance, Edward's TRIMP, accelerations in velocity zones 4 and 5, and acceleration load, while these relationships were strong for practice.

Table 1. Correlations between sRPE-TL and various training load measures for all samples, games, and practices.

	All (n = 1072)			Games (n = 306)			Practices (n = 766)		
Variable	**r**	**95% CI**	***p***	**r**	**95% CI**	***p***	**r**	**95% CI**	***p***
Dur	0.842	0.827, 0.856	<0.001	0.930	0.916, 0.942	<0.001	0.692	0.659, 0.721	<0.001
sRPE	0.840	0.824, 0.854	<0.001	0.785	0.745, 0.818	<0.001	0.876	0.862, 0.889	<0.001
sRPE-TL	N/A	N/A	N/A	N/A	N/A	N/A	N/A	N/A	N/A
Velocity B1 Dist	0.834	0.818, 0.849	<0.001	0.891	0.869, 0.909	<0.001	0.552	0.510, 0.593	<0.001
Velocity B2 Dist	0.814	0.796, 0.830	<0.001	0.877	0.854, 0.897	<0.001	0.520	0.476, 0.562	<0.001
Velocity B3 Dist	0.741	0.717, 0.762	<0.001	0.772	0.731, 0.807	<0.001	0.424	0.374, 0.471	<0.001
Velocity B4 Dist	0.547	0.511, 0.581	<0.001	0.495	0.421, 0.563	<0.001	0.339	0.285, 0.390	<0.001
Velocity B5 Dist	0.257	0.209, 0.303	<0.001	0.146	0.053, 0.237	0.010	0.150	0.091, 0.207	<0.001
Total Dist	0.841	0.826, 0.855	<0.001	0.905	0.887, 0.921	<0.001	0.583	0.542, 0.621	<0.001
Mean Velocity	0.437	0.395, 0.477	<0.001	0.168	0.075, 0.258	0.003	0.124	0.065, 0.182	<0.001
Max Velocity	0.283	0.236, 0.328	<0.001	0.094	0.000, 0.186	0.101	0.248	0.192, 0.303	<0.001
HSR Dist	0.538	0.502, 0.573	<0.001	0.462	0.385, 0.533	<0.001	0.341	0.287, 0.392	<0.001
Meterage per Minute	0.437	0.395, 0.477	<0.001	0.168	0.075, 0.258	0.003	0.124	0.065, 0.182	<0.001
PlayerLoad™	0.827	0.811, 0.843	<0.001	0.892	0.871, 0.910	<0.001	0.551	0.508, 0.591	<0.001
PlayerLoad™·min^{-1}	0.352	0.307, 0.395	<0.001	0.130	0.036, 0.222	0.023	0.096	0.037, 0.155	0.008
kcal Expenditure	0.817	0.799, 0.833	<0.001	0.871	0.846, 0.892	<0.001	0.541	0.497, 0.581	<0.001
kcal·kg^{-1}	0.844	0.828, 0.858	<0.001	0.906	0.888, 0.922	<0.001	0.593	0.553, 0.630	<0.001
HR B1 Dur	−0.061	−0.111, −0.010	0.047	0.067	−0.027, 0.161	0.240	−0.054	−0.113, 0.006	0.136
HR B2 Dur	−0.123	−0.172, −0.073	<0.001	0.083	−0.011, 0.176	0.147	0.041	−0.019, 0.100	0.261
HR B3 Dur	0.035	−0.015, 0.085	0.251	0.335	0.249, 0.416	<0.001	0.296	0.240, 0.349	<0.001
HR B4 Dur	0.386	0.343, 0.428	<0.001	0.495	0.420, 0.563	<0.001	0.370	0.317, 0.420	<0.001
HR B5 Dur	0.690	0.662, 0.715	<0.001	0.693	0.640, 0.739	<0.001	0.462	0.414, 0.508	<0.001
HR B6 Dur	0.567	0.532, 0.600	<0.001	0.485	0.410, 0.554	<0.001	0.342	0.288, 0.393	<0.001
HR B7 Dur	0.152	0.102, 0.201	<0.001	0.162	0.069, 0.252	0.005	0.086	0.027, 0.145	0.017
Max HR (bpm)	0.350	0.305, 0.394	<0.001	0.374	0.290, 0.452	<0.001	0.270	0.213, 0.324	<0.001
Mean HR (bpm)	0.462	0.422, 0.501	<0.001	0.367	0.282, 0.445	<0.001	0.296	0.241, 0.350	<0.001
Max HR (%)	0.281	0.234, 0.327	<0.001	0.255	0.165, 0.341	<0.001	0.224	0.167, 0.280	<0.001
Mean HR (%)	0.425	0.383, 0.466	<0.001	0.311	0.223, 0.393	<0.001	0.269	0.213, 0.323	<0.001
HR Exertion	0.858	0.845, 0.871	<0.001	0.897	0.877, 0.914	<0.001	0.676	0.642, 0.707	<0.001
Exertion Index	0.786	0.766, 0.805	<0.001	0.848	0.820, 0.873	<0.001	0.481	0.434, 0.526	<0.001
Edward's TRIMP	0.841	0.826, 0.855	<0.001	0.901	0.882, 0.917	<0.001	0.619	0.581, 0.655	<0.001
Bannister's TRIMP	0.808	0.790, 0.825	<0.001	0.855	0.827, 0.878	<0.001	0.557	0.515, 0.597	<0.001
Accel B1 Count	0.062	0.012, 0.112	0.043	0.049	−0.045, 0.143	0.389	−0.002	−0.061, 0.057	0.956
Accel B2 Count	0.143	0.093, 0.192	<0.001	0.078	−0.016, 0.171	0.174	0.031	−0.029, 0.090	0.395
Accel B3 Count	0.512	0.474, 0.548	<0.001	0.527	0.455, 0.592	<0.001	0.241	0.184, 0.296	<0.001
Accel B4 Count	0.854	0.840, 0.867	<0.001	0.916	0.900, 0.930	<0.001	0.600	0.560, 0.636	<0.001
Accel B5 Count	0.852	0.837, 0.865	<0.001	0.914	0.897, 0.928	<0.001	0.594	0.554, 0.631	<0.001
Accel B6 Count	0.504	0.466, 0.541	<0.001	0.658	0.601, 0.708	<0.001	0.252	0.195, 0.307	<0.001
Accel B7 Count	0.093	0.043, 0.142	0.002	0.094	0.000, 0.186	0.101	0.138	0.079, 0.196	<0.001
Accel Load	0.859	0.845, 0.871	<0.001	0.922	0.906, 0.935	<0.001	0.608	0.569, 0.644	<0.001

Velocity B1 = 0–6 km·h^{-1}, velocity B2 = 6–12 km·h^{-1}, velocity B3 = 12–18 km·h^{-1}, velocity B4 = 18–25 km·h^{-1}, velocity B5 = >25 km·h^{-1}; HR B1 = 0–50%, HR B2 = 50–60%, HR B3 = 60–70%, HR B4 = 70–80%, HR B5 = 80–90%, HR B6 = 90–100%, HR B7 = >100%; Accel B1 = −10–−3 m·s^{-2}, accel B2 = −3–−2 m·s^{-2}, accel B3 = −2–−1 m·s^{-2}, accel B4 = −1–0 m·s^{-2}, accel B5 = 0–1 m·s^{-2}, accel B6 = 1–2 m·s^{-2}, accel B7 = 2–3 m·s^{-2}, accel B8 = 3–10 m·s^{-2}.

When correlations between external load and HR-based measures (i.e., Edward's and Bannister's TRIMP) were calculated, results were similar when both matches and practices were considered. Both Edward's and Bannister's TRIMP scores were very strongly correlated with distance in velocity zones 1–3, total distance, total PlayerLoad™, number of accelerations in zones 4 and 5, and total acceleration load. However, Edward's TRIMP was almost perfectly correlated with distance in velocity zone 1, total distance, total

PlayerLoad™, number of accelerations in zone 4 and 5, and total acceleration load for matches, while Bannister's TRIMP score was strongly correlated with these measures.

Table 2. Correlations between Edward's TRIMP and various training load measures for all samples, games, and practices.

	All (n = 1072)			Games (n = 306)			Practices (n = 766)		
Variable	r	95% CI	p	r	95% CI	p	r	95% CI	p
Dur	0.820	0.803, 0.836	<0.001	0.931	0.917, 0.943	<0.001	0.638	0.601, 0.672	<0.001
sRPE	0.608	0.575, 0.638	<0.001	0.603	0.540, 0.660	<0.001	0.447	0.398, 0.494	<0.001
sRPE-TL	0.841	0.826, 0.855	<0.001	0.901	0.882, 0.917	<0.001	0.619	0.581, 0.655	<0.001
Velocity B1 Dist	0.856	0.842, 0.869	<0.001	0.919	0.902, 0.932	<0.001	0.585	0.544, 0.623	<0.001
Velocity B2 Dist	0.849	0.835, 0.863	<0.001	0.882	0.859, 0.901	<0.001	0.647	0.611, 0.680	<0.001
Velocity B3 Dist	0.785	0.765, 0.804	<0.001	0.792	0.754, 0.825	<0.001	0.551	0.508, 0.591	<0.001
Velocity B4 Dist	0.572	0.538, 0.605	<0.001	0.523	0.451, 0.588	<0.001	0.363	0.311, 0.414	<0.001
Velocity B5 Dist	0.277	0.230, 0.323	<0.001	0.167	0.074, 0.257	0.003	0.166	0.107, 0.223	<0.001
Total Dist	0.876	0.863, 0.887	<0.001	0.924	0.909, 0.937	<0.001	0.688	0.656, 0.718	<0.001
Mean Velocity	0.520	0.482, 0.555	<0.001	0.225	0.134, 0.313	<0.001	0.309	0.254, 0.362	<0.001
Max Velocity	0.314	0.268, 0.358	<0.001	0.137	0.044, 0.229	0.016	0.270	0.214, 0.324	<0.001
HSR Dist	0.564	0.529, 0.598	<0.001	0.490	0.415, 0.558	<0.001	0.366	0.313, 0.416	<0.001
Meterage per Minute	0.520	0.482, 0.555	<0.001	0.225	0.134, 0.313	<0.001	0.309	0.254, 0.362	<0.001
PlayerLoad™	0.869	0.856, 0.880	<0.001	0.904	0.886, 0.920	<0.001	0.681	0.647, 0.711	<0.001
PlayerLoad™·min^{-1}	0.453	0.413, 0.493	<0.001	0.176	0.084, 0.266	0.002	0.302	0.247, 0.355	<0.001
kcal Expenditure	0.865	0.852, 0.877	<0.001	0.893	0.873, 0.911	<0.001	0.692	0.660, 0.722	<0.001
kcal·kg^{-1}	0.880	0.868, 0.891	<0.001	0.926	0.911, 0.938	<0.001	0.703	0.671, 0.731	<0.001
HR B1 Dur	−0.335	−0.379, −0.289	<0.001	−0.159	−0.250, −0.066	0.005	−0.472	−0.517, −0.425	<0.001
HR B2 Dur	−0.372	−0.414, −0.328	<0.001	−0.069	−0.162, 0.025	0.227	−0.393	−0.442, −0.342	<0.001
HR B3 Dur	−0.084	−0.134, −0.034	0.006	0.223	0.132, 0.311	<0.001	0.120	0.061, 0.178	<0.001
HR B4 Dur	0.366	0.321, 0.408	<0.001	0.401	0.319, 0.477	<0.001	0.464	0.416, 0.509	<0.001
HR B5 Dur	0.808	0.790, 0.825	<0.001	0.731	0.684, 0.772	<0.001	0.815	0.794, 0.834	<0.001
HR B6 Dur	0.759	0.737, 0.779	<0.001	0.644	0.585, 0.696	<0.001	0.745	0.717, 0.770	<0.001
HR B7 Dur	0.214	0.166, 0.262	<0.001	0.190	0.098, 0.280	<0.001	0.195	0.137, 0.252	<0.001
Max HR (bpm)	0.451	0.410, 0.490	<0.001	0.406	0.324, 0.482	<0.001	0.457	0.409, 0.503	<0.001
Mean HR (bpm)	0.678	0.650, 0.705	<0.001	0.511	0.438, 0.577	<0.001	0.700	0.668, 0.729	<0.001
Max HR (%)	0.457	0.417, 0.496	<0.001	0.347	0.261, 0.427	<0.001	0.534	0.490, 0.575	<0.001
Mean HR (%)	0.696	0.669, 0.721	<0.001	0.503	0.429, 0.570	<0.001	0.761	0.735, 0.785	<0.001
HR Exertion	0.983	0.981, 0.984	<0.001	0.987	0.984, 0.989	<0.001	0.964	0.960, 0.968	<0.001
Exertion Index	0.821	0.804, 0.837	<0.001	0.864	0.838, 0.886	<0.001	0.582	0.541, 0.620	<0.001
Edward's TRIMP	N/A	N/A	N/A	N/A	N/A	N/A	N/A	N/A	N/A
Bannister's TRIMP	0.959	0.955, 0.963	<0.001	0.964	0.957, 0.970	<0.001	0.920	0.910, 0.928	<0.001
Accel B1 Count	0.090	0.040, 0.140	0.003	0.090	−0.004, 0.183	0.116	0.011	−0.048, 0.071	0.757
Accel B2 Count	0.132	0.082, 0.181	<0.001	0.052	−0.042, 0.145	0.365	0.030	−0.029, 0.090	0.404
Accel B3 Count	0.532	0.495, 0.567	<0.001	0.532	0.461, 0.596	<0.001	0.283	0.228, 0.337	<0.001
Accel B4 Count	0.862	0.849, 0.874	<0.001	0.916	0.900, 0.930	<0.001	0.632	0.595, 0.666	<0.001
Accel B5 Count	0.868	0.855, 0.880	<0.001	0.918	0.901, 0.931	<0.001	0.652	0.616, 0.685	<0.001
Accel B6 Count	0.526	0.489, 0.562	<0.001	0.675	0.620, 0.723	<0.001	0.288	0.233, 0.342	<0.001
Accel B7 Count	0.093	0.043, 0.143	0.002	0.140	0.047, 0.231	0.014	0.108	0.049, 0.167	0.003
Accel Load	0.887	0.876, 0.897	<0.001	0.928	0.914, 0.940	<0.001	0.706	0.675, 0.735	<0.001

Velocity B1 = 0–6 km·h^{-1}, velocity B2 = 6–12 km·h^{-1}, velocity B3 = 12–18 km·h^{-1}, velocity B4 = 18–25 km·h^{-1}, velocity B5 = >25 km·h^{-1}; HR B1 = 0–50%, HR B2 = 50–60%, HR B3 = 60–70%, HR B4 = 70–80%, HR B5 = 80–90%, HR B6 = 90–100%, HR B7 = >100%; Accel B1 = −10–−3 m·s^{-2}, accel B2 = −3–−2 m·s^{-2}, accel B3 = −2–−1 m·s^{-2}, accel B4 = −1–0 m·s^{-2}, accel B5 = 0–1 m·s^{-2}, accel B6 = 1–2 m·s^{-2}, accel B7 = 2–3 m·s^{-2}, accel B8 = 3–10 m·s^{-2}.

Table 3. Correlations between Bannister's TRIMP and various training load measures for all samples, games, and practices.

	All (n = 1072)			Games (n = 306)			Practices (n = 766)		
Variable	r	95% CI	p	r	95% CI	p	r	95% CI	p
Dur	0.754	0.732, 0.775	<0.001	0.866	0.840, 0.888	<0.001	0.558	0.515, 0.598	<0.001
sRPE	0.608	0.575, 0.639	<0.001	0.607	0.544, 0.664	<0.001	0.423	0.373, 0.471	<0.001
sRPE-TL	0.808	0.790, 0.825	<0.001	0.855	0.827, 0.878	<0.001	0.557	0.515, 0.597	<0.001
Velocity B1 Dist	0.841	0.826, 0.856	<0.001	0.881	0.858, 0.901	<0.001	0.519	0.474, 0.561	<0.001
Velocity B2 Dist	0.835	0.819, 0.849	<0.001	0.836	0.805, 0.862	<0.001	0.594	0.555, 0.632	<0.001
Velocity B3 Dist	0.794	0.774, 0.812	<0.001	0.783	0.743, 0.817	<0.001	0.509	0.463, 0.552	<0.001
Velocity B4 Dist	0.602	0.569, 0.633	<0.001	0.564	0.496, 0.625	<0.001	0.362	0.310, 0.413	<0.001
Velocity B5 Dist	0.299	0.252, 0.344	<0.001	0.192	0.099, 0.281	<0.001	0.160	0.102, 0.218	<0.001
Total Dist	0.869	0.856, 0.881	<0.001	0.892	0.871, 0.910	<0.001	0.631	0.593, 0.665	<0.001
Mean Velocity	0.579	0.545, 0.611	<0.001	0.308	0.221, 0.391	<0.001	0.315	0.260, 0.367	<0.001
Max Velocity	0.329	0.283, 0.373	<0.001	0.157	0.063, 0.247	0.006	0.271	0.215, 0.326	<0.001
HSR Dist	0.595	0.561, 0.626	<0.001	0.530	0.459, 0.595	<0.001	0.364	0.312, 0.415	<0.001
Meterage per Minute	0.579	0.545, 0.611	<0.001	0.308	0.221, 0.391	<0.001	0.315	0.260, 0.367	<0.001
PlayerLoad™	0.845	0.830, 0.859	<0.001	0.863	0.837, 0.886	<0.001	0.617	0.578, 0.652	<0.001
PlayerLoad™·min^{-1}	0.494	0.455, 0.531	<0.001	0.237	0.146, 0.324	<0.001	0.298	0.243, 0.352	<0.001
kcal Expenditure	0.845	0.830, 0.858	<0.001	0.849	0.820, 0.873	<0.001	0.610	0.572, 0.646	<0.001
kcal·kg^{-1}	0.873	0.860, 0.884	<0.001	0.895	0.874, 0.912	<0.001	0.643	0.607, 0.677	<0.001
HR B1 Dur	−0.254	−0.300, −0.206	<0.001	−0.102	−0.195, −0.008	0.074	−0.354	−0.405, −0.300	<0.001
HR B2 Dur	−0.424	−0.464, −0.382	<0.001	−0.177	−0.267, −0.084	0.002	−0.446	−0.492, −0.397	<0.001
HR B3 Dur	−0.227	−0.274, −0.178	<0.001	0.087	−0.007, 0.180	0.129	−0.062	−0.121, −0.002	0.088
HR B4 Dur	0.283	0.237, 0.329	<0.001	0.314	0.226, 0.396	<0.001	0.400	0.348, 0.448	<0.001
HR B5 Dur	0.756	0.733, 0.777	<0.001	0.662	0.606, 0.712	<0.001	0.742	0.715, 0.768	<0.001
HR B6 Dur	0.792	0.773, 0.810	<0.001	0.693	0.640, 0.739	<0.001	0.756	0.729, 0.780	<0.001
HR B7 Dur	0.293	0.246, 0.338	<0.001	0.313	0.226, 0.396	<0.001	0.287	0.232, 0.341	<0.001
Max HR (bpm)	0.458	0.417, 0.497	<0.001	0.437	0.357, 0.510	<0.001	0.463	0.415, 0.508	<0.001
Mean HR (bpm)	0.737	0.713, 0.759	<0.001	0.613	0.551, 0.669	<0.001	0.766	0.740, 0.790	<0.001
Max HR (%)	0.420	0.378, 0.461	<0.001	0.357	0.272, 0.436	<0.001	0.462	0.414, 0.508	<0.001
Mean HR (%)	0.721	0.696, 0.745	<0.001	0.587	0.521, 0.645	<0.001	0.764	0.738, 0.788	<0.001
HR Exertion	0.955	0.951, 0.960	<0.001	0.966	0.959, 0.971	<0.001	0.903	0.892, 0.914	<0.001
Exertion Index	0.827	0.810, 0.842	<0.001	0.842	0.813, 0.868	<0.001	0.544	0.501, 0.585	<0.001
Edward's TRIMP	0.959	0.955, 0.963	<0.001	0.964	0.957, 0.970	<0.001	0.920	0.910, 0.928	<0.001
Bannister's TRIMP	N/A	N/A	N/A	N/A	N/A	N/A	N/A	N/A	N/A
Accel B1 Count	0.108	0.058, 0.157	<0.001	0.117	0.023, 0.209	0.04	0.008	−0.051, 0.068	0.816
Accel B2 Count	0.160	0.110, 0.208	<0.001	0.073	−0.022, 0.166	0.206	0.053	−0.006, 0.112	0.143
Accel B3 Count	0.550	0.514, 0.584	<0.001	0.564	0.497, 0.625	<0.001	0.261	0.205, 0.316	<0.001
Accel B4 Count	0.822	0.805, 0.837	<0.001	0.852	0.824, 0.876	<0.001	0.554	0.512, 0.594	<0.001
Accel B5 Count	0.833	0.817, 0.848	<0.001	0.861	0.835, 0.884	<0.001	0.581	0.540, 0.619	<0.001
Accel B6 Count	0.513	0.475, 0.549	<0.001	0.667	0.611, 0.716	<0.001	0.257	0.201, 0.312	<0.001
Accel B7 Count	0.092	0.042, 0.142	0.003	0.167	0.074, 0.257	0.003	0.099	0.040, 0.158	0.006
Accel Load	0.861	0.847, 0.873	<0.001	0.883	0.861, 0.903	<0.001	0.626	0.588, 0.661	<0.001

Velocity B1 = 0–6 km·h^{-1}, velocity B2 = 6–12 km·h^{-1}, velocity B3 = 12–18 km·h^{-1}, velocity B4 = 18–25 km·h^{-1}, velocity B5 = >25 km·h^{-1}; HR B1 = 0–50%, HR B2 = 50–60%, HR B3 = 60–70%, HR B4 = 70–80%, HR B5 = 80–90%, HR B6 = 90–100%, HR B7 = >100%; Accel B1 = −10–−3 m·s^{-2}, accel B2 = −3–−2 m·s^{-2}, accel B3 = −2–−1 m·s^{-2}, accel B4 = −1–0 m·s^{-2}, accel B5 = 0–1 m·s^{-2}, accel B6 = 1–2 m·s^{-2}, accel B7 = 2–3 m·s^{-2}, accel B8 = 3–10 m·s^{-2}.

When sessions were stratified and coded as either high or low perceived exertion, significant differences in duration, total distance, PlayerLoad™, and acceleration load between high and low exertion sessions were observed (see Figure 1A–D). Similarly, sessions with high perceived exertion also resulted in higher absolute and relative mean HR, sRPE-TL, Edward's TRIMP, and Bannister's TRIMP values (see Figure 2A–E).

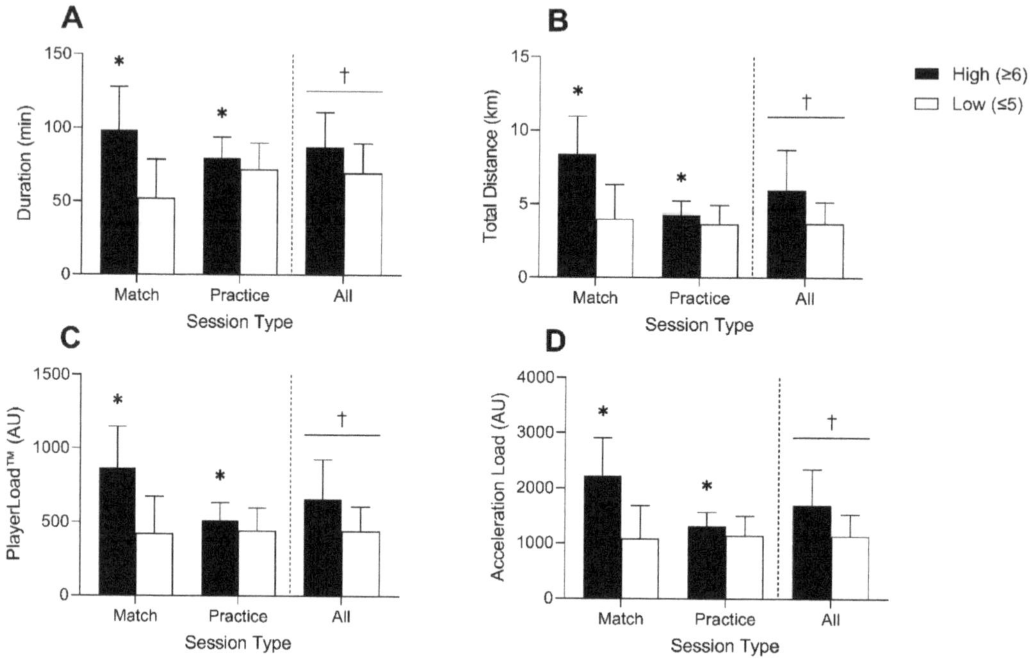

Figure 1. Session duration (**A**), total distance traveled (**B**), PlayerLoad™ (**C**), and acceleration load (**D**) data stratified by high (≥6) or low (≤5) perceived session exertion for matches, practices, and all sessions. * Significant difference between high and low exertion session within session type ($p < 0.001$); † Significant main effect of session exertion ($p < 0.001$).

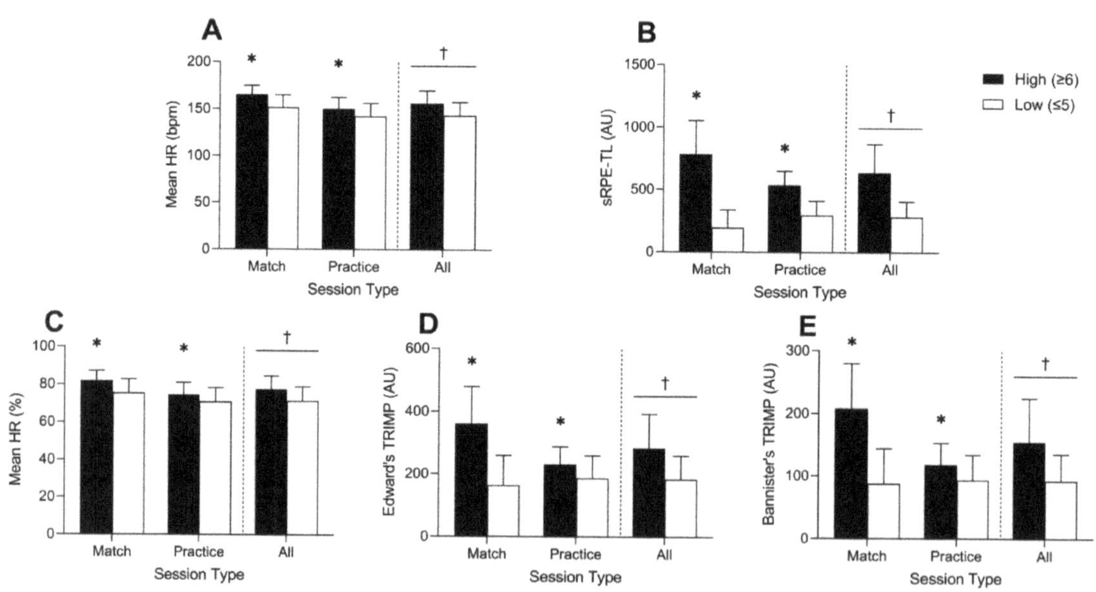

Figure 2. Mean absolute HR (**A**), sRPE-TL (**B**), mean relative HR (**C**), Edward's TRIMP (**D**), and Bannister's TRIMP (**E**) data stratified by high (≥6) or low (≤5) perceived session exertion for matches, practices, and all sessions. * Significant difference between high and low exertion session within session type ($p < 0.001$); † Significant main effect of session exertion ($p < 0.001$).

4. Discussion

The primary aim of the current study was to examine the relationships between sRPE-TL and select measures of external workload in collegiate women's soccer. The primary findings indicate that sRPE-TL was strongly associated with both acceleration load and total distance in both matches and practices, which supports our hypothesis. These relationships were stronger during match play than during practice. Additionally, sRPE-TL was strongly associated with the proprietary PlayerLoad™ metric. As a subjective measure of internal load, sRPE-TL offers a low-cost method to monitor workload throughout a season. This aspect of affordability and minimal equipment need has contributed to its popularity as a monitoring tool within sports. The observed relationship between sRPE-TL and measures of the external workload from the current study aligns with those previously reported in soccer [14,16,24–27] and across a variety of sport types [11,15,28–33]. Previous work in semiprofessional soccer athletes indicated that sRPE-TL was strongly associated with measures of external load such as total distance covered and PlayerLoad™ over 44 training sessions [16]. More recently, Marynowicz et al. [24] observed strong associations between sRPE-TL and total distance, PlayerLoad™, and number of accelerations during an 18-week in-season period in youth soccer athletes. However, small-to-moderate relationships were observed between sRPE-TL and measures of match intensity such as high-speed running distance and the number of impacts. Collectively, these findings suggest that sRPE-TL may better reflect total external workload rather than the intensity of work. However, in the current study, sRPE-TL was found to be strongly associated with acceleration load, which could be classified as an intensity-based measure of external workload volume. It is likely these relationships may be variable based on the fitness level of the athlete and level of competition. An important practical consideration when using sRPE-TL is how the duration is being quantified, specifically during match play. Pustina et al. [34] reported differences in sRPE-TL measures based on how the session duration was defined (i.e., including or excluding warm-ups and half-time, only including on-field playing time, etc.) and found that sRPE-TL calculated using only on-field playing time was the best reflection of external workloads incurred during the match play.

An interesting observation from the current study was that when session types were stratified based on ratings of perceived exertion, differences in internal and external measures of workload were evident. We interpret this as indicating that athletes appear to be able to accurately reflect back on their degree of perceived exertion for a given match or training session, whether the workload metric of interest is external or internal. Additionally, this ability to rate the degree of perceived exertion for external and internal measures remained true whether the metric of interest was pertaining to duration or an accumulative-type metric corresponding to the total amount, duration, or volume of work in addition to metrics that were more reflective of the intensity of work (Figures 1 and 2).

A secondary aim of the current study was to determine which measures of internal load (sRPE-TL or HR-based metrics), were more strongly correlated to the external load. The results of the current study indicate that sRPE-TL was more strongly associated with total PlayerLoad™ compared to simple HR measures (i.e., mean or max HR). However, the accumulative HR zone-based measure, Edward's TRIMP score, appeared to be more strongly associated with select measures of external workload (total distance, PlayerLoad™, kcal, and acceleration load) compared to sRPE-TL (Tables 1 and 2). While exhibiting a stronger relationship to external workload in the current study, TRIMP scores do require the use of HR-monitoring technology as opposed to the subject measure sRPE-TL. Further, when determining the overall utility of sRPE-TL, it is also important to evaluate its relationship with physiological measures of internal load, as it may provide an alternative indicator of internal stress incurred by the athlete. Recently, Costa et al. [26] observed good convergent validity between sRPE-TL and TRIMP scores throughout 6 weeks of in-season competition in female soccer athletes. Interestingly, the authors also noted distinct differences in %HRpeak values observed across pre-identified sRPE ranges, indicating that sRPE is useful in identifying specific thresholds of exertion across a range of HR intensities during soccer competition. Similarly, Impellizzeri et al. [14] reported strong associations between sRPE-TL and three different HR-based methods of quantifying internal load, which is contradictory to findings from the current study. In the current study, sRPE-TL was more strongly associated with measures of external workload (i.e., PlayerLoad™) than with HR measures, indicating that sRPE-TL may be more sensitive to variations in external workload measures rather than a reflection of HR. However, because of the practical nature of sRPE-TL, it may be a more suitable option for team-sport programs that do not have access to team-based HR monitoring systems. Along these same lines, sRPE-TL may also be easier to use when quantifying accumulated workloads across an entire season, when compared to measures of HR. Because of the relationships between sRPE-TL and measures of both external and internal workload, it may also offer a multipurpose metric of overall load as it is a reflection of both the total work completed and the internal physiological stress incurred.

This study is not without limitations. The current study was conducted in collegiate female athletes competing at the NCAA Division I level and, therefore, it is unknown if similar relationships between sRPE-TL and external workload would exist in female and male athletes competing at different levels. Environmental factors (i.e., ambient temperature and humidity) may influence HR-derived measures; however, these external factors were not accounted for in the current analysis. Similarly, lifestyle stressors, dietary intake, and sleep habits may have also influenced the perceived difficulty of training with a subsequent impact on how measures of external workload correlate with sRPE-TL.

5. Conclusions

In NCAA Division I women's soccer, sRPE-TL is strongly associated with both external and internal measures of workload. These relationships were stronger during match play, with acceleration load and total distance exhibiting the strongest relationship with sRPE-TL. sRPE-TL can serve as a valuable, and perhaps superior, tool for monitoring workload in women soccer athletes when GPS-based or measures of HR are not an option, or to be used in conjunction with one another.

Author Contributions: Conceptualization, J.M.O. and A.R.J.; Data curation, A.T.A., A.L.L., D.J.A. and W.J.; Formal analysis, A.T.A., A.L.L., D.J.A., P.E.E., C.F. and A.R.J.; Investigation, A.T.A., A.L.L., D.J.A., W.J. and A.K.; Methodology, J.L.E., P.E.E., J.M.O. and A.R.J.; Project administration, A.T.A., A.L.L., D.J.A., W.J., A.K. and J.M.O.; Supervision, A.K., J.L.E., P.E.E., J.M.O. and A.R.J.; Validation, D.J.A.; Visualization, A.T.A., P.E.E. and C.F.; Writing–original draft, A.T.A., A.L.L. and D.J.A.; Writing–review & editing, W.J., A.K., J.L.E., P.E.E., J.M.O., C.F. and A.R.J. All authors have read and agreed to the published version of the manuscript.

Funding: This research received no external funding.

Institutional Review Board Statement: The study was conducted according to the guidelines of the Declaration of Helsinki, and approved by the Institutional Review at Texas Christian University (approval no. 1807-114-1807; approval date 30 July 2018).

Informed Consent Statement: Informed consent was obtained from all subjects involved in the study.

Data Availability Statement: The data presented in this study are available on request from the corresponding author.

Conflicts of Interest: The authors declare no conflict of interest.

References

1. Meeusen, R.; Duclos, M.; Foster, C.; Fry, A.; Gleeson, M.; Nieman, D.; Raglin, J.; Rietjens, G.; Steinacker, J.M.; Urhausen, A.; et al. Prevention, Diagnosis, and Treatment of the Overtraining Syndrome. *Med. Sci. Sport. Exerc.* **2013**, *45*, 186–205. [CrossRef]
2. Impellizzeri, F.M.; Marcora, S.M.; Coutts, A.J. Internal and External Training Load: 15 Years On. *Int. J. Sports Physiol. Perform.* **2019**, *14*, 270–273. [CrossRef] [PubMed]
3. Roe, G.; Darrall-Jones, J.; Black, C.; Shaw, W.; Till, K.; Jones, B. Validity of 10-HZ GPS and Timing Gates for Assessing Maximum Velocity in Professional Rugby Union Players. *Int. J. Sports Physiol. Perform.* **2017**, *12*, 836–839. [CrossRef] [PubMed]

4. Van Iterson, E.H.; Fitzgerald, J.S.; Dietz, C.C.; Snyder, E.M.; Peterson, B.J. Reliability of Triaxial Accelerometry for Measuring Load in Men's Collegiate Ice Hockey. *J. Strength Cond. Res.* **2017**, *31*, 1305–1312. [CrossRef] [PubMed]
5. Weaving, D.; Whitehead, S.; Till, K.; Jones, B. Validity of Real-Time Data Generated by a Wearable Microtechnology Device. *J. Strength Cond. Res.* **2017**, *31*, 2876–2879. [CrossRef] [PubMed]
6. Johnston, R.J.; Watsford, M.L.; Kelly, S.J.; Pine, M.J.; Spurrs, R.W. Validity and Interunit Reliability of 10 Hz and 15 Hz GPS Units for Assessing Athlete Movement Demands. *J. Strength Cond. Res.* **2014**, *28*, 1649–1655. [CrossRef]
7. Chambers, R.; Gabbett, T.J.; Cole, M.H.; Beard, A. The Use of Wearable Microsensors to Quantify Sport-Specific Movements. *Sports Med.* **2015**, *45*, 1065–1081. [CrossRef] [PubMed]
8. Lisbôa, F.D.; Raimundo, J.A.G.; Salvador, A.F.; Pereira, K.L.; Turnes, T.; Diefenthaeler, F.; Oliveira, M.F.M.D.; Caputo, F. Acute Cardiopulmonary, Metabolic, and Neuromuscular Responses to Severe-Intensity Intermittent Exercises. *J. Strength Cond. Res.* **2019**, *33*, 408–416. [CrossRef] [PubMed]
9. Foster, C.; Florhaug, J.A.; Franklin, J.; Gottschall, L.; Hrovatin, L.A.; Parker, S.; Doleshal, P.; Dodge, C. A New Approach to Monitoring Exercise Training. *J. Strength Cond. Res.* **2001**, *15*, 109–115. [CrossRef] [PubMed]
10. Foster, C.; Boullosa, D.; McGuigan, M.; Fusco, A.; Cortis, C.; Arney, B.E.; Orton, B.; Dodge, C.; Jaime, S.; Radtke, K.; et al. 25 Years of Session Rating of Perceived Exertion: Historical Perspective and Development. *Int. J. Sports Physiol. Perform.* **2021**, *16*, 612–621. [CrossRef]
11. Haddad, M.; Stylianides, G.; Djaoui, L.; Dellal, A.; Chamari, K. Session-RPE Method for Training Load Monitoring: Validity, Ecological Usefulness, and Influencing Factors. *Front. Neurosci.* **2017**, *11*, 612. [CrossRef] [PubMed]
12. Manzi, V.; D'Ottavio, S.; Impellizzeri, F.M.; Chaouachi, A.; Chamari, K.; Castagna, C. Profile of Weekly Training Load in Elite Male Professional Basketball Players. *J. Strength Cond. Res.* **2010**, *24*, 1399–1406. [CrossRef]
13. Alexiou, H.; Coutts, A.J. A Comparison of Methods Used for Quantifying Internal Training Load in Women Soccer Players. *Int. J. Sports Physiol. Perform.* **2008**, *3*, 320–330. [CrossRef] [PubMed]
14. Impellizzeri, F.M.; Rampinini, E.; Coutts, A.J.; Sassi, A.; Marcora, S.M. Use of RPE-Based Training Load in Soccer. *Med. Sci. Sport. Exerc.* **2004**, *36*, 1042–1047. [CrossRef]
15. Wallace, L.K.; Slattery, K.M.; Coutts, A.J. The Ecological Validity and Application of the Session-Rpe Method for Quantifying Training Loads in Swimming. *J. Strength Cond. Res.* **2009**, *23*, 33–38. [CrossRef]
16. Casamichana, D.; Castellano, J.; Calleja-Gonzalez, J.; RomaN, J.S.; Castagna, C. Relationship between Indicators of Training Load in Soccer Players. *J. Strength Cond. Res.* **2013**, *27*, 369–374. [CrossRef]
17. McLaren, S.J.; Macpherson, T.W.; Coutts, A.J.; Hurst, C.; Spears, I.R.; Weston, M. The Relationships Between Internal and External Measures of Training Load and Intensity in Team Sports: A Meta-Analysis. *Sports Med.* **2017**, *48*, 641–658. [CrossRef] [PubMed]
18. Chen, M.J.; Fan, X.; Moe, S.T. Criterion-Related Validity of the Borg Ratings of Perceived Exertion Scale in Healthy Individuals: A Meta-Analysis. *J. Sports Sci.* **2002**, *20*, 873–899. [CrossRef]
19. Turner, A.; Walker, S.; Stembridge, M.; Coneyworth, P.; Reed, G.; Birdsey, L.; Barter, P.; Moody, J. A Testing Battery for the Assessment of Fitness in Soccer Players. *Strength Cond. J.* **2011**, *33*, 29–39. [CrossRef]
20. Christen, J.; Foster, C.; Porcari, J.P.; Mikat, R.P. Temporal Robustness of the Session Rating of Perceived Exertion. *Int. J. Sports Physiol. Perform.* **2016**, *11*, 1088–1093. [CrossRef]
21. Banister, E.W. Modeling Elite Athletic Performance. In *Physiological Testing of Elite Athletes*; Green, H., McDougal, J., Wenger, H., Eds.; Human Kinetics: Champaign, IL, USA, 1991; pp. 403–424.
22. Edwards, S. High Performance Training and Racing. In *The Heart Rate Monitor Book*; Edwards, S., Ed.; Feet Fleet Press: Sacramento, CA, USA, 1993; pp. 113–123.
23. Hopkins, W.G. A scale of magnitudes for effect statistics. *New View Stat.* **2002**, *502*, 411.
24. Marynowicz, J.; Kikut, K.; Lango, M.; Horna, D.; Andrzejewski, M. Relationship Between the Session-RPE and External Measures of Training Load in Youth Soccer Training. *J. Strength Cond. Res.* **2020**, *34*, 2800–2804. [CrossRef]
25. Gaudino, P.; Iaia, F.M.; Strudwick, A.J.; Hawkins, R.D.; Alberti, G.; Atkinson, G.; Gregson, W. Factors Influencing Perception of Effort (Session Rating of Perceived Exertion) during Elite Soccer Training. *Int. J. Sports Physiol. Perform.* **2015**, *10*, 860–864. [CrossRef]
26. Costa, J.A.; Brito, J.; Nakamura, F.Y.; Figueiredo, P.; Rebelo, A. Using the Rating of Perceived Exertion and Heart Rate to Quantify Training Intensity in Female Soccer Players. *J. Strength Cond. Res.* **2019**. [CrossRef] [PubMed]
27. Rodríguez-Marroyo, J.A.; Antoñan, C. Validity of the Session Rating of Perceived Exertion for Monitoring Exercise Demands in Youth Soccer Players. *Int. J. Sports Physiol. Perform.* **2015**, *10*, 404–407. [CrossRef]
28. Lovell, T.W.J.; Sirotic, A.C.; Impellizzeri, F.M.; Coutts, A.J. Factors Affecting Perception of Effort (Session Rating of Perceived Exertion) During Rugby League Training. *Int. J. Sports Physiol. Perform.* **2013**, *8*, 62–69. [CrossRef]
29. Gallo, T.; Cormack, S.; Gabbett, T.; Williams, M.; Lorenzen, C. Characteristics Impacting on Session Rating of Perceived Exertion Training Load in Australian Footballers. *J. Sports Sci.* **2015**, *33*, 467–475. [CrossRef]
30. Scanlan, A.T.; Wen, N.; Tucker, P.S.; Dalbo, V.J. The Relationships Between Internal and External Training Load Models During Basketball Training. *J. Strength Cond. Res.* **2014**, *28*, 2397–2405. [CrossRef]
31. Rago, V.; Vigh-Larsen, J.F.; Deylami, K.; Muschinsky, A.; Mohr, M. Use of Rating of Perceived Exertion–Based Training Load in Elite Ice Hockey Training and Match-Play. *J. Strength Cond. Res.* **2020**. [CrossRef] [PubMed]

32. Scott, T.J.; Black, C.R.; Quinn, J.; Coutts, A.J. Validity and Reliability of the Session-RPE Method for Quantifying Training in Australian Football. *J. Strength Cond. Res.* **2013**, *27*, 270–276. [CrossRef] [PubMed]
33. Lupo, C.; Tessitore, A.; Gasperi, L.; Gomez, M. Session-RPE for Quantifying the Load of Different Youth Basketball Training Sessions. *Biol. Sport* **2017**, *1*, 11–17. [CrossRef] [PubMed]
34. Pustina, A.A.; Sato, K.; Liu, C.; Kavanaugh, A.A.; Sams, M.L.; Liu, J.; Uptmore, K.D.; Stone, M.H. Establishing a Duration Standard for the Calculation of Session Rating of Perceived Exertion in NCAA Division I Men's Soccer. *J. Trainol.* **2017**, *6*, 26–30. [CrossRef]

Journal of
Functional Morphology and Kinesiology

MDPI

Article

Tensiomyographic Responses to Warm-Up Protocols in Collegiate Male Soccer Athletes

Michael J. Redd *, Tristan M. Starling-Smith, Chad H. Herring, Matt S. Stock , Adam J. Wells , Jeffrey R. Stout and David H. Fukuda

Institute of Exercise Physiology and Rehabilitation Science, School of Kinesiology and Physical Therapy, University of Central Florida, Orlando, FL 32816, USA; tstarling-smith@iwp-llc.com (T.M.S.-S.); chadherring@ucf.edu (C.H.H.); matt.stock@ucf.edu (M.S.S.); adam.wells@ucf.edu (A.J.W.); jeffrey.stout@ucf.edu (J.R.S.); David.Fukuda@ucf.edu (D.H.F.)
* Correspondence: redd@ucf.edu

Abstract: The mechanical properties of knee flexors and extensors in 15 collegiate male soccer players following different warm-up protocols [small-sided games (SSG), dynamic (DYN), and plyometric (PLY)] were evaluated. Tensiomyography (TMG) was used to assess contraction time (Tc), delay time (Td) and maximal displacement (Dm) of the rectus femoris (RF) and biceps femoris (BF) of both legs before and after each warm-up, while countermovement jump height variables, 20 m sprint, *t*-test and sit-and-reach were measured following the warm-ups. TMG was analyzed using a three-way [condition × time × leg] ANOVA, while performance variables were analyzed with a repeated measures ANOVA. Main effects of time were observed for BF-Tc ($p = 0.035$), RF-Td ($p < 0.001$), and BF-Td, ($p = 0.008$), and a main effect of condition was seen for RF-Tc ($p = 0.038$). Moreover, participants' 20 m sprint improved following SSG ($p = 0.021$) compared to DYN and PLY. Sit-and-reach was greater following PLY ($p = 0.021$). No significant interactions were noted for the measured TMG variables. Warm-up-specific improvements were demonstrated in sprint speed and flexibility following SSG and PLY, respectively. The present study revealed changes in certain TMG measures following the warm-ups that suggest enhanced response of lower leg muscles regardless of specific activities used.

Keywords: tensiomyography; warm-ups; soccer; knee flexors; knee extensors

Citation: Redd, M.J.; Starling-Smith, T.M.; Herring, C.H.; Stock, M.S.; Wells, A.J.; Stout, J.R.; Fukuda, D.H. Tensiomyographic Responses to Warm-Up Protocols in Collegiate Male Soccer Athletes. *J. Funct. Morphol. Kinesiol.* **2021**, *6*, 80. https://doi.org/10.3390/jfmk6040080

Academic Editors: Cristina Cortis, Andrea Fusco and Carl Foster

Received: 31 August 2021
Accepted: 16 September 2021
Published: 23 September 2021

1. Introduction

Soccer is a sport comprised of intermittent high-intensity bouts of exercise, including maximal sprint efforts, jumping and multiple changes in direction [1,2]. The sport-specific demands of competitive soccer have been identified in a number of studies [3–6]. More specifically, high-level soccer players have been shown to commit 1000–1400 activity changes during a match [2,6–8] with varied high-intensity actions and multiple sprints at near-maximal intensities [8,9]. The preparation of soccer players during training sessions is critical for attaining optimal performance in match play, and the demands of a training session should closely mimic the competitive demands associated with games [10]. Sport coaches and strength coaches alike have sought to optimize player performance through the design and administration of a variety of training sessions. One of the key components to any training program or individual training session is the application of an appropriate warm-up designed to enhance performance [11–13].

Numerous studies have identified the benefits associated with participation in an organized warm-up protocol prior to athletic competition [12–16]. Studies have also shown improvements, such as increased preparedness and improvement in sport performance skills such as sprinting, jumping and change in direction, following training programs that incorporate plyometric exercises and jumps into their warm-up routines [1,12,15,17–22]. The previously identified benefits and improvements include physiological improvements, [1,15,17–19]. While there is some debate on which warm-up protocols produce the best results, there is a consensus that the warmup performed should consist of dynamic movements as well as an active stretching routine for agonist and antagonist muscles groups to reduce risk of injury and improve sport-related performance [12–15,23,24]. Other benefits of a dynamic warm-up include thermogenic effects, including: decreased viscous resistance of joints [12], increased blood flow to working muscles [25,26], increased anaerobic metabolism [14] and improved central nervous system functioning [25]. In support of this, active warm-up routines have been shown to improve performance in activities related to soccer such as sprinting, dribbling and striking a soccer ball [26].

Soccer teams use a variety of warm-up modalities at the start of training sessions such as a series of dynamic movements, reactive work with the ball during small-sided games or the incorporation of explosive plyometric movements into general routines [27]. Generally, these warm-ups have resulted in improved strength and power variables, such as sprint speed and jump height [1,17,18,23,26]. However, while contractile history has been shown to affect the performance of skeletal muscle [28], few studies have examined the contractile properties of muscles following a warm-up session [12,15].

Tensiomyography (TMG) is a non-invasive, indirect measure of superficial skeletal muscles' contractile properties [29]. TMG measures are obtained by stimulating an individual muscle with a small electrical current and then recording a number of variables related to the mechanical properties of muscle contraction, such as: delay time (Td), contraction time (Tc), sustain time (Ts), relaxation time (Tr) and maximal displacement (Dm) [30]. TMG provides a valid and reliable means of immediate feedback [31], with the sensitivity to measure small changes in muscle contractile properties [32]. Subsequently, this method of evaluation has been used to detect neuromuscular status and muscle fatigue [33], which directly impacts muscle activity and performance. In a study by Loturco and colleagues [34], correlations were found between decreased Td, Tc and Dm, as measured by TMG, and greater scores on performance assessments. Pereira and colleagues [35] suggested that velocity of contraction, measured using TMG, may be a sensitive marker for assessing contractile property variations during performance assessments. Furthermore, TMG can be used to assess muscular fatigue by identifying parameters of muscle contraction such as Tc and Dm [36,37]. The aforementioned studies provide support that TMG variables may be affected by varying warm-up protocols.

Therefore, the purpose of this study was to compare the potential differences in the mechanical and physical characteristics of the knee flexors and extensors in collegiate male soccer players following different warm-up protocols. A greater understanding of the effects of various warm-ups on mechanical characteristics of the muscle may help to develop specific warm-up protocols that may lead to improved on-field performance. It is our hypothesis that TMG analysis will reveal that the warm-ups with a more explosive component will have the greatest impact on decreasing Tc and Dm of the muscles, and that the more continuous movement warm-ups may show increased signs of muscular fatigue.

2. Materials and Methods

2.1. Subjects

Fifteen male National Collegiate Athletics Association (NCAA) Division I soccer players ($n = 15$, 20.13 ± 1.26 years, 176.11 ± 7.85 cm, and 78.38 ± 7.50 kg) completed this study (Table 1). A convenience sample of 18 athletes from a single team initially consented; however, 2 athletes did not participate in any of the data collection sessions due to time conflicts. and 1 athlete was removed due to the identification of an ongoing injury that may have become exacerbated with continued participation.

All subjects were familiar with the warm-up protocols performed and the physical assessment tests used for this investigation. The protocols and testing were comprised of warm-ups and physical assessments typically performed by the men's soccer team over the course of the competitive season. All athletes were free from injury and volunteered to participate in this investigation. In an attempt to eliminate the potential for fatigue-related performance decrements, subjects were asked to refrain from any workouts or training sessions during the duration of the protocol. The study was conducted according to the guidelines of the Declaration of Helsinki and approved by the University of Central Florida Institutional Review Board (IRB number SBE-17-13696; approval date: 31 January 2018). All components of the testing procedures were explained to each subject, and informed consent was received before the initial testing session. Neither the subjects nor the assessors were blinded for any of the testing sessions.

Table 1. Means and standard deviations of selected characteristics by player position.

	GK (n = 2)	DF (n = 5)	MF (n = 5)	FW (n = 3)
Age (y)	20.33 ± 0.94	19.60 ± 1.02	20.4 ± 1.62	20.67 ± 0.47
Height (cm)	184.15 ± 3.81	174.24 ± 8.12	175.26 ± 4.25	175.26 ± 10.37
Body Mass (kg)	88.84 ± 5.06	77.82 ± 3.09	77.13 ± 6.12	74.42 ± 9.71
Body Fat (%)	16.15 ± 2.35	11.88 ± 2.01	12.34 ± 4.77	13.13 ± 1.13

GK = goalkeeper; DF = defender; MF = midfielder; FW = forward; n = sample size.

2.1.1. Experimental Design

This study utilized a randomized crossover design to examine the effect of different warm-up protocols on performance and the physical and mechanical characteristics of knee flexors and extensors. The experimental design involved three different warm-up protocols: small-sided games (SSG), dynamic warm-up (DYN) or plyometric warm-up (PLY). The subjects were separated into three groups in order to accommodate weekly scheduling and limit the impact on training sessions. A random-sort procedure in Microsoft Excel was used by the study investigators to assign the order of protocols to each of the three groups. Each group completed the assigned protocol and associated testing at the same time of day for all sessions, while each session was followed by a 24 h off day before completing the next protocol. This 24 h period was intended to minimize fatigue and limit carry-over effects from the previous protocol sessions.

2.1.2. Anthropometrics and Body Composition

Each subject's height was measured with a portable stadiometer (Seca 213; Seca Corp., Chino, CA, USA) and body mass was measured with a calibrated standard medical scale (439 Physician Scale; Detecto, Webb City, MO, USA). Body composition was determined using a bioelectrical impedance analysis (BIA) device (Inbody 770, Biospace Co., Ltd., Seoul, Korea) that determined total fat and fat-free body mass by segmental multifrequency analysis, as recommended by the manufacturer. These anthropometrics and body composition measures were recorded as descriptive values on the subject's first day of testing.

2.1.3. Tensiomyography Measurements

The following procedures were used, as described by Jones and colleagues (2017), for assessment of the musculature of the leg utilizing TMG (TMG Measurement System, TMG-BMC Ltd., Ljubljana, Slovenia). For each muscle examined, the TMG sensor tip was placed at a location determined to measure maximal radial displacement of the selected muscle. All subjects were free from soft tissue or muscular injury at time of examination. Measurements were performed under static and relaxed conditions. Assessment of the RF was conducted with the subject in the supine position and the knee joint fixed at a 120-degree angle (with 180 degrees corresponding to full extension of the knee). The leg measured was positioned on a triangular wedge foam cushion to maintain a fixed knee angle. A digital displacement transducer, which incorporated a spring of 0.17 $N \cdot m^{-1}$, was set perpendicular to the muscle belly to acquire radial displacement of the selected muscle. The site for measurement of the radial displacement of the rectus femoris (RF) was determined by placing a transversal mark with a dermatological pen at 50% of the total length between the greater trochanter and the lateral condoyle of the distal end of the femur. The subject was then instructed to contract the quadriceps muscle of the leg being examined to palpate the rectus femoris and place a line longitudinally across the transversal line, creating an "x" landmark for sensor tip placement. Two square (5 × 5 cm) 2 mm-thick self-adhesive electrodes were placed symmetrically 5 cm (±5 cm) or 3 cm (±3 cm) distal and proximal to the sensor tip. This procedure was completed for both the right and left RF before the subject was repositioned in a prone position to examine posterior muscles of the leg. The site for measurement of the radial displacement of the biceps femoris (BF) was determined by measuring the distance between the ischial tuberosity and the lateral condoyle of the distal end of the femur and placing a transversal mark at 50% of the total length. The subject was then instructed to flex their knee against resistance, placed by the hand of the assessor at the ankle, to palpate the BF; a longitudinal line was drawn across the transversal line, creating an "x" landmark for sensor tip placement. This procedure was then repeated for both the right and left leg. Regarding electrical stimulation procedures, the pulse duration was 1 ms and the initial current amplitude was set at 50 mA. For each test, current amplitude was progressively increased by 10 mA increments until there was no difference in muscle displacement (Dm) or maximal stimulator output 100 mA was reached. For each subject, two consecutive measurements were recorded for each leg and muscle group, and averaged for analysis. Reliability of TMG, including the currently evaluated muscle groups, has been established in a review conducted by Martin-Rodriguez and colleagues [31], who reported high to excellent intraclass correlation coefficient (ICC) values for Dm (0.82–0.99), good to excellent ICC values for Tc (0.70–0.99) and low to excellent ICC values for Td (0.60–0.98). The same review reported standard error of measurement (SEM) values of ±1.0 mm for RF Dm, 1.20 ms for RF Td, 1.90 ms for RF Tc, 0.43 to 1.0 mm for BF Dm, 0.40 to 0.80 ms for BF Td and 1.06 to 5.60 ms for BF Tc.

2.2. *Warm-Up Protocols*

2.2.1. Dynamic (DYN) Warm-Up

The DYN warm-up began with a 5-minute jog around the testing field. The next component was 9 min in duration and completed on the field at a distance of 12 yards, identified by 2 cones (cone A at start line and cone B placed at 12-yard point). The subjects completed a series of DYN patterns from cone A to B, then a jog from B to A with no rest between the DYN movement and jogging. Subjects were allowed a 5-second rest after the jog and before the next dynamic movement pattern began. This alternation between DYN pattern and jogging continued for the entire series of DYN patterns. The DYN warm-up is commonly utilized by the athletes and progressed from simple movements to a series of dynamic stretches and, finally, to more complex movements in multiple planes, including small skips, open the gate, close the gate, tall shuffle, lunge walk, side lunge walk, hamstring walk, knee hug, heel kicks, A-skip, B-skip, C-skip, low shuffle, carioca and lean–fall-sprint. The total activity time of the DYN warm-up was 15 min.

2.2.2. Small-Sided Games (SSG) Warm-Up

The SSG warm-up began with a 5-minute jog around testing field followed by a 5-minute period involving a series of dynamic movement patterns over a distance of 12 yards from cone A to cone B, similar to that of the DYN but with no jogging involved. The final component of the SSG warm-up included players performing "rondo" drills that were common to the athletes, consisting of an individual player defending, or trying to win the ball from, a set of 4 offensive players positioned in a square approximately 7 m × 7 m in dimension. Players rotated positions in this drill, spending 1 min as a defender and 4 min as an offensive player, lasting a total of 5 min. The total activity time of the SSG warm-up was 15 min.

2.2.3. Plyometric (PLY) Warm-UP

The PLY warm-up began with a 5-minute jog around the testing field. The next component was 4 min in duration and completed on the field at a distance of 12 yards, identified by 2 cones (configured similar to the other warm-up protocols.) This component of the PLY warm-up began with the following dynamic movements: small skips, open the gate, close the gate, tall shuffle, lunge walk, side lunge walk and hamstring walk. Unlike the DYN detailed above, these movements were completed from cone A to cone B and back to cone A, with no jogging between movement patterns; however, all subjects were required to observe a 5-second rest at cone A before beginning the next movement. For the final phase of the PLY warm-up, the subject completed a series of plyometric movements utilizing the same 12 yards distance from cane A to cone B and jog back to cone B. Subjects were allowed a 5-second rest after the jog and before the next PLY movement pattern began. The PLY component was commonly utilized by the athletes and included: double leg hops, single leg hops, lateral hops, double leg forward jump for distance, double leg explosive jump for height, single leg bounds, power skips and tuck jumps. The total activity time of the PLY warm-up was 15 min.

2.3. *Performance Testing*

2.3.1. Jump Assessments

Countermovement jump (CMJ) tests were conducted using a validated optical timing system (MyJump V2.1), which recorded video on a tablet computer (iPad, Apple Inc., Cupertino, CA, USA) [38]. During the CMJ test, the subject was instructed to begin by standing with their hands at their waist, and then told to listen for an audible signal, which would alert the subject to bend his knees and maximally jump upward using arms for momentum. This CMJ was performed three times with 1 min of rest between each jump attempt. Variables analyzed during the CMJ included jump height (CMJ-h), flight time (CMJ-ft) and velocity (CMJ-v). Flight time was calculated for each jump after the take-off, and landing was identified in frames of the recorded video. The following equation: $h = t^2 \times 1.22625$ (h = jump height in meters, t = time in seconds) was used, as outlined by Bosco and colleagues [39], to determine jump height. Jump height from the application was found to have excellent reliability (ICC = 0.997, $p < 0.001$) and excellent agreement with countermovement jump height (CMJ-h), as measured using a force platform (ICC = 0.997, $p < 0.001$).

2.3.2. Sprint Speed

Subjects were asked to perform 2 × 20 m maximal sprints on a natural grass surface with 1 min of rest between each sprint trial. Peak sprint speed was assessed by recording video of each sprint using a tablet computer (iPad, Apple Inc., Cupertino, CA, USA) and a mobile application (MySprint Apple Inc., Cupertino, CA, USA). The mobile application was specifically designed to use video analysis to determine the start time and finish time of each sprint using a frame-by-frame method. The tablet computer was mounted to a tripod and set at the 20 m point of the sprint, 18 m from the course, to ensure capturing the start and finish portions of the sprint. The sprint course was marked at 0, 5, 10, 15 and 20 m by vertical poles that were set to account for possible video parallax. That is, the poles were set not exactly at the specific distances (0, 5, 10, 15 and 20 m) but at the point where the subjects were viewed with the tablet computer to have crossed the maker with their hips and were at the target distance. Players were instructed to wear their normal soccer cleats for this test. The better of the 2 sprint trials time was recorded and used for further analyses.

2.3.3. *t*-Test

The *t*-test course was set up with four cones—A, B, C and D. The subject began the *t*-test with both of his feet behind starting point A. The test started with the athlete being given a verbal "Go" command. On this command, the subject sprinted 10 yards forward to point B and touched the cone. Then, he shuffled 5 yards to the left and touched cone C. After that, he shuffled 10 yards to the right and touched cone D and then shuffled 5 yards to the left, back to point B. Then, the subject ran backward passing the finish line at point A. The time started on the player's first movement from point A and stopped when he crossed the finish line. Time was measured using a hand-held stopwatch; this method was shown to have high ICC values (0.988) with electronic timing [40] and all *t*-test trials were timed by the same assessor. Each athlete performed this test two times and the faster time was recoded for further analysis.

2.3.4. Sit-and-Reach

The subject was instructed to sit on the ground with knees straight, legs separated just enough to be comfortable, with the feet placed firmly against the sit-and-reach box. The arms were extended forward with the hands placed palms down on the upper surface of the box, which had a scale for distance printed on its horizontal surface (top side). In this position, the subject reached forward in a slow, consistent motion with no ballistic movement to the position of maximum reach. The test administrator stood close beside the scale and recorded the most distant line held and touched by the fingertips of both hands of the subject. If the hands reached unevenly, the hand reaching the shorter distance was used to determine the score. The score was recorded to the nearest half inch. If the reach appeared to be exactly half-way between two lines, the score was based on the last line actually touched.

2.3.5. Statistical Analysis

All TMG measures data were analyzed using a three-way [condition (DYN vs. SSG vs. PLY) × time (pre-testing vs. post-testing) x leg (left vs. right)] repeated measures analysis of variance (ANOVA) with Holm post hoc analyses performed when appropriate. Due to technical issues during the TMG measurement process, only 14 full sets of data were available for the BF and 11 sets of data for the RF. All performance variable data (n = 15) were analyzed with a one-way repeated measures ANOVA with Holm post hoc analyses performed when appropriate. When violations of sphericity occurred, Greenhouse–Geisser corrections were utilized. Effect sizes were interpreted using Cohen's d, in which d values of 0.8, 0.5 and 0.2 represented large, medium and small effect sizes, respectively [41]. Statistical software (JASP; Version 0.12; JASP Team, Amsterdam, The Netherlands) was used for data analysis. Results were considered significant at an alpha level of $p \leq 0.05$. All data are reported as mean ± standard deviation.

3. Results

3.1. Tensiomyography

The TMG measurements performed pre- (PRE) and post-intervention (POST) are represented in Table 2.

Table 2. Pre- and post-warm-up tensiomyography measurements (collapsed across right and left legs).

Muscle	Measure	Dynamic		Plyometric		Small-Sided Games	
		Pre	Post	Pre	Post	Pre	Post
Rectus Femoris (n = 11)	Tc (ms) #	28.44 ± 4.41	27.01 ± 3.22	28.98 ± 4.60	28.75 ± 4.22	28.15 ± 3.68	27.19 ± 3.49
	Td (ms) *	24.89 ± 1.81	23.50 ± 2.04	25.15 ± 2.43	24.16 ± 1.98	24.46 ± 2.12	23.01 ± 1.39
	Dm (mm)	7.59 ± 1.95	7.21 ± 1.83	7.08 ± 1.97	7.87 ± 2.04	7.56 ± 2.58	7.16 ± 2.32
Biceps Femoris (n = 14)	Tc (ms) *	22.17 ± 3.60	20.71 ± 2.32	22.24 ± 3.65	20.37 ± 2.39	23.51 ± 4.32	21.17 ± 2.73
	Td (ms) *	21.93 ± 1.26	20.82 ± 0.95	21.78 ± 1.26	20.50 ± 1.24	21.90 ± 1.18	20.97 ± 0.99
	Dm (mm)	3.15 ± 0.80	3.06 ± 0.78	3.43 ± 1.11	3.19 ± 0.94	3.49 ± 0.78	3.16 ± 0.80

Tc = contraction time; Td = delay time; Dm = maximal displacement. * main effect for time ($p < 0.05$); # main effect for condition ($p < 0.05$).

No condition × leg × time interaction (F = 0.147, p = 0.755, η_p^2 = 0.014) or main effect of time (F = 3.953, p = 0.075, η_p^2 = 0.283) was noted for RF-Tc measures. However, a significant main effect of condition was observed (F = 3.887, p = 0.038, η_p^2 = 0.280). Follow-up analysis indicated that RF-Tc values were not significantly different between PLY and DYN (p = 0.087, d = 0.710 "medium") or PLY and SSG (p = 0.068, d = 0.745 "medium").

No condition × leg × time interaction ($F = 0.710$, $p = 0.455$, $\eta_p^2 = 0.066$) or main effect for condition ($F = 4.309$, $p = 0.056$, $\eta_p^2 = 0.288$) was noted for RF-Td. However, a significant main effect for time was observed ($F = 29.890$, $p < 0.001$, $\eta_p^2 = 0.749$). Post hoc analysis indicated that RF-Td decreased from PRE to POST (mean difference = 1.275 ms, 95% confidence interval = 0.755 to 1.795 ms).

No significant condition × leg × time interaction ($F = 2.105$, $p = 0.162$, $\eta_p^2 = 0.139$) or main effect for condition ($F = 0.414$, $p = 0.552$, $\eta_p^2 = 0.031$) was noted for BF-Tc. However, a significant main effect of time was observed ($F = 5.537$, $p = 0.035$, $\eta_p^2 = 0.299$). Post hoc analysis indicated that Tc of the BF muscle decreased from PRE to POST (mean difference = 1.890 ms, 95% confidence interval = 0.155 to 3.625 ms).

No condition × leg × time interaction ($F = 0.306$, $p = 0.739$, $\eta_p^2 = 0.023$) or main effect for condition ($F = 0.537$, $p = 0.529$, $\eta_p^2 = 0.040$) was noted for BF-Td. However, a significant main effect of time was observed ($F = 9.749$, $p = 0.008$, $\eta_p^2 = 0.429$). Post hoc analysis indicated that BF-Td decreased from PRE to POST (mean difference = 1.105 ms; 95% confidence interval = 0.341 to 1.870 ms).

No significant condition × leg × time interaction was noted for BF-Dm ($F = 0.708$, $p = 0.502$, $\eta_p^2 = 0.052$) or RF-Dm ($F = 0.445$, $p = 0.541$, $\eta_p^2 = 0.043$). Additionally, no main effects of time ($F = 0.000$, $p = 0.984$, $\eta_p^2 = 0.000$) or condition ($F = 0.072$, $p = 0.931$, $\eta_p^2 = 0.007$) were noted.

3.2. *Performance Testing*

The results for the *t*-test, sprint and sit-and-reach assessments are presented in Table 3. No significant difference was shown between conditions for *t*-test times ($F = 0.943$, $p = 0.402$, $\eta_p^2 = 0.063$). A significant difference was found between conditions for 20 m sprint time ($F = 4.719$, $p = 0.040$, $\eta_p^2 = 0.252$) with a reduction in sprint time following SSG compared to DYN ($p = 0.021$, d = 0.749 "medium"). A nonsignificant change between PLY and DYN for 20 m sprint time was noted ($p = 0.083$) with a "medium" effect size (d = 0.600). A significant difference was found between conditions for sit-and-reach scores ($F = 4.394$, $p = 0.043$, $\eta_p^2 = 0.239$) with PLY, resulting in significantly greater scores than DYN ($p = 0.021$, d = 0.753 "medium").

Table 3. Performance test results ($n = 15$).

Test	Dynamic	Plyometric	Small-Sided Games
20 m Sprint (s)	2.75 ± 0.30	2.64 ± 0.13	2.62 ± 0.15 *
t-Test (s)	10.0 ± 0.27	9.99 ± 0.32	9.94 ± 0.30
Sit-and-Reach (cm)	34.17 ± 6.47	37.60 ± 6.98 †	36.43 ± 7.29
CMJ Height (cm)	59.62 ± 8.10	59.31 ± 8.43	58.92 ± 8.93
CMJ Flight Time (ms)	695.80 ± 46.73	693.80 ± 49.13	691.40 ± 52.44
CMJ Velocity ($m{\cdot}s^{-1}$)	1.71 ± 0.12	1.70 ± 0.12	1.70 ± 0.13

* Significantly lower than plyometric ($p < 0.05$); † Significantly greater than dynamic ($p < 0.05$).

Results for the variables measured during the CMJ are presented in Table 2. Analysis of CMJ performance revealed no differences between conditions for CMJ-h($F = 0.372$, $p = 0.693$, $\eta_p^2 = 0.026$), CMJ-ft ($F = 0.406, p = 0.670$, $\eta_p^2 = 0.028$) or CMJ-v($F = 0.430$, $p = 0.654$, $\eta_p^2 = 0.030$).

4. Discussion

The primary findings of this study revealed that performing the warm-up protocols differentially affected performance assessments and the mechanical characteristics of the knee flexors and extensors in a sample of collegiate male soccer players. In terms of performance assessments, PLY resulted in significant improvements in sit-and-reach when compared to DYN and there was a significant reduction in 20-m sprint time following SSG compared to DYN. Time delay of both the RF and BF, and contraction time of the BF, decreased significantly from pre- to post-warm-up regardless of condition.

This study revealed a significant improvement in delay time, also termed reaction time [32], of the RF and BF from PRE to POST following each warm-up condition. This improvement, or decrease, in delay time following warm-up activities is most likely due to the post-activation potentiation effect generated by the exercises in the warm-up protocols. Previous studies have proposed that initiating neuromuscular potentiation may improve force production and performance [42]. Plyometric and jumping type exercises have been shown to be the most commonly prescribed methods to enhance post-activation potentiation [26]; however, the findings of this study may demonstrate similar acute responses following the exercises used in the three protocols. In support of this observation, improved post-activation potentiation has been postulated to increase the rate of force development in voluntary force production in dynamic muscle contractions [43].

Significant decreases in contraction time of the BF were noted from PRE to POST following each warm-up condition. This may be due in part to the improvements in the central nervous system nerve conduction rate associated with the increased muscle temperature following an active warm-up [25]. While the differences in Tc in the RF were not significant, medium effects were found for PLY compared to DYN ($p = 0.087$, d = 0.710 "medium") and PLY compared to SSG ($p = 0.068$, d = 0.745 "medium"). While it was hypothesized that the plyometric type exercises would uniquely affect contraction time, the reason for this finding is unclear. We are unaware of any studies to date that have examined the effect of such training on contraction time using TMG; therefore, evaluation of plyometric training in this area of muscle mechanical response is warranted. It should also be noted that while the current data exceed previously reported SEM values for BF Tc from Simunic [44], they are below those reported by de Paula Simola [45] et al. and may need to be interpreted with caution.

No significant differences in maximal displacement were noted between warm-up conditions; however, previous studies have reported that this parameter is typically lower in soccer players compared to the general population. Low values of maximal displacement are usually indicative of higher muscle tone [46]. Therefore, based on the movement patterns and training of high-level soccer players, changes in this parameter may be difficult to elicit during a warm-up due to the relatively short duration of exercise.

Assessment of 20 m sprint time revealed a significant improvement following completion of SSG. While these findings are similar to a number of studies showing improvements in sprint time following active warm-ups and dynamic stretching routines [17,21,23,47], they contrast with Gabbett et al. [1], where no significant differences in sprint times were found for players completing an open-skill or closed-skill warm-up. This may be attributed to the specificity of the open-skill tasks examined in basketball players who performed warm-ups of similar distances and intensity regardless of ball possession [1], whereas the present study used a "rondo" drill in which the players intensity and distance were different than the comparison warm-ups. While not significant, a medium effect was noted for improved sprint times in PLY when compared to DYN. Although past studies have examined the effectiveness of dynamic warm-ups for improving sprint time, they have typically examined the difference in dynamic and passive warm-up routines [17,23]. We know of no other study to date that has compared the effect of a dynamic warm-up series vs. a warm-up inclusive of a large plyometric component on agility, sprint and jump performance. Rimmer and Sleivert [48] showed significant improvements in 40 m sprint time (following 8 weeks of sprint-specific plyometric training), while [49] showed improved 20 m sprint time after incorporating plyometric exercise the training programs of elite youth soccer players. The results associated with the latter improvements over short distances were attributed to the reduced ground contact time associated with plyometric exercises, which may have influenced the medium effect size for PLY compared DYN observed in the current study.

We also observed a significant improvement in sit-and-reach performance following PLY compared to DYN. The improved flexibility observed here may be attributed to the increased eccentric component associated with the movement patterns and muscle activation that accompanies repeated jumping type exercises [49]. Recurring bouts of eccentric training have been shown to shown to improve the muscle length–tension curve, which results in improvements in range of motion [50]. The present findings are consistent with previous studies that have shown improvements in flexibility associated with training that incorporates plyometric type exercises [22,49].

We observed no significant differences in measured CMJ variables between PLY, DYN and SSG warm-ups. This may be due to the acute nature of this investigation, in contrast to other studies that have used multiple training sessions incorporating plyometric exercises that have shown improvements in jump performance following warm-ups inclusive of jumping exercises [20,21]. Previous research has shown that skeletal muscle performance is affected by the contractile properties of muscle [28]. However, the single PLY session used in this study may not have increased contractile properties of the muscle associated with repeated activation of the stretch shortening cycle (SSC) induced by jumping-type exercises used in longer duration studies [42,51].

A potential limitation to this study is that the population examined was a homogenous group of male athletes with similar playing experience and competitive level from a single team. Future research should include groups of different playing experience or competitive level, or a control group of non-soccer athletes. Another possible limitation may be the acute nature of the warm-up protocol. Future research should examine whether there may be a difference in response if warm-ups are performed on a regular basis.

5. Conclusions

The warm-up protocols performed in this study represent variations of commonly used exercises for soccer athletes. The DYN warm-up examined herein positively impacted the contractile properties of leg muscles after performing a standardized series of dynamic movements. These exercises were familiar to the athletes and were performed in movement planes similar to those encountered in the competitive environment. Previous research has already established the performance improvements associated with DYN; however, they may have been less evident in the current study due to DYN being compared to SSG and PLY protocols rather than passive stretching or running-only warm-up interventions. The PLY exercises examined were shown to result in improved flexibility, as indicated by the improved sit-and-reach scores, and improved speed in the 20 m sprint, potentially as a result of post-activation potentiation.

The selection of warm-up protocols may be influenced by coaches and/or players attitudes or preferences for specific exercises. Another factor in warm-up selection and administration may be time constraints or the objectives of the coaching staff prior to match play, such as positioning, set piece formation, etc. While extensive research has been conducted on the benefits of warm-ups, there is no consensus on which combination of warm-up activities provide optimal results for sport performance. The disparity in previous findings may be due to the individual responses of athletes to the unique protocols or intensities at which they are performed, as well as differences in neuromuscular properties and fiber type composition. The findings of this study illustrate the benefits of incorporating more plyometric exercises and small-sided game situations to a standard pre-training or pre-competition warm-up. Future research should include studies examining the long-term effects of warm-up administration with larger samples of athletes at different levels of play and/or maturity status.

Author Contributions: Conceptualization, M.J.R., J.R.S. and D.H.F.; methodology, M.J.R., J.R.S., A.J.W., M.S.S. and D.H.F.; formal analysis, M.J.R., C.H.H., T.M.S.-S. and D.H.F.; investigation, M.J.R., T.M.S.-S., C.H.H. and D.H.F.; resources, not applicable; data curation, M.J.R., D.H.F. and J.R.S.; writing—original draft preparation, M.J.R., A.J.W., M.S.S., D.H.F. and J.R.S.; writing—review and editing, M.J.R., A.J.W., M.S.S., D.H.F. and J.R.S.; visualization, M.J.R.; supervision, M.J.R.; project administration, D.H.F. and J.R.S.; funding acquisition, not applicable. All authors have read and agreed to the published version of the manuscript.

Funding: This research received no external funding.

Institutional Review Board Statement: The study was conducted according to the guidelines of the Declaration of Helsinki and approved by the Institutional Review Board (or Ethics Committee) of University of Central Florida.

Informed Consent Statement: Informed consent was obtained from all subjects' parents involved in the study. Child assent was obtained at each testing session.

Data Availability Statement: The data presented in this study are available on request from the corresponding author. The data are not publicly available.

Acknowledgments: Authors would like to acknowledge UCF Athletics Association as well as UCF Men's Soccer coaches and staff, as well as the Sports Science Initiatives program at the University of Central Florida under the Department of Kinesiology in the School of Kinesiology and Physical Therapy.

Conflicts of Interest: The authors declare no conflict of interest.

References

1. Gabbett, T.J.; Sheppard, J.M.; Pritchard-Peschek, K.R.; Leveritt, M.D.; Aldred, M.J. Influence of closed skill and open skill warm-ups on the performance of speed, change of direction speed, vertical jump, and reactive agility in team sport athletes. *J. Strength Cond. Res.* **2008**, *22*, 1413–1415. [CrossRef]
2. Los Arcos, A.; Mendiguchia, J.; Yanci, J. Specificity of jumping, acceleration and quick change of direction motor abilities in soccer players. *Kinesiology* **2017**, *49*, 22–29. [CrossRef]
3. Bangsbo, J. Energy demands in competitive soccer. *J. Sports Sci.* **1994**, *12*, S5–S12. [CrossRef] [PubMed]
4. Reilly, T.; Bangsbo, J.; Franks, A. Anthropometric and physiological predispositions for elite soccer. *J. Sports Sci.* **2000**, *18*, 669–683. [CrossRef] [PubMed]
5. Mohr, M.; Krustrup, P.; Bangsbo, J. Match performance of high-standard soccer players with special reference to development of fatigue. *J. Sports Sci.* **2003**, *21*, 519–528. [CrossRef]
6. Di Salvo, V.; Baron, R.; Tschan, H.; Montero, F.C.; Bachl, N.; Pigozzi, F. Performance characteristics according to playing position in elite soccer. *Int. J. Sports Med.* **2007**, *28*, 222–227. [CrossRef]
7. Bloomfield, J.; Polman, R.; O'Donoghue, P. Physical demands of different positions in FA Premier League soccer. *J. Sports Sci. Med.* **2007**, *6*, 63. [PubMed]
8. Vigne, G.; Gaudino, C.; Rogowski, I.; Alloatti, G.; Hautier, C. Activity profile in elite Italian soccer team. *Int. J. Sports Med.* **2010**, *31*, 304–310. [CrossRef]
9. Stølen, T.; Chamari, K.; Castagna, C.; Wisløff, U. Physiology of soccer. *Sports Med.* **2005**, *35*, 501–536. [CrossRef] [PubMed]
10. Kelly, D.M.; Drust, B. The effect of pitch dimensions on heart rate responses and technical demands of small-sided soccer games in elite players. *J. Sci. Med. Sport* **2009**, *12*, 475–479. [CrossRef]
11. Shellock, F.G.; Prentice, W.E. Warming-up and stretching for improved physical performance and prevention of sports-related injuries. *Sports Med.* **1985**, *2*, 267–278. [CrossRef]
12. Bishop, D. Warm Up I. *Sports Med.* **2003**, *33*, 439–454. [CrossRef]
13. Fradkin, A.J.; Zazryn, T.R.; Smoliga, J.M. Effects of warming-up on physical performance: A systematic review with meta-analysis. *J. Strength Cond. Res.* **2010**, *24*, 140–148. [CrossRef] [PubMed]
14. Febbraio, M.A.; Carey, M.F.; Snow, R.J.; Stathis, C.G.; Hargreaves, M. Influence of elevated muscle temperature on metabolism during intense, dynamic exercise. *Am. J. Physiol.-Regul. Integr. Comp. Physiol.* **1996**, *271*, R1251–R1255. [CrossRef] [PubMed]
15. Bishop, D. Warm Up II. *Sports Med.* **2003**, *33*, 483–498. [CrossRef]
16. Arnason, A.; Andersen, T.E.; Holme, I.; Engebretsen, L.; Bahr, R. Prevention of hamstring strains in elite soccer: An intervention study. *Scand. J. Med. Sci. Sports* **2008**, *18*, 40–48. [CrossRef] [PubMed]
17. Little, T.; Williams, A.G. Effects of differential stretching protocols during warm-ups on high speed motor capacities in professional soccer players. *J. Strength Cond. Res.* **2006**, *20*, 203–207.
18. Holt, B.W.; Lambourne, K. The impact of different warm-up protocols on vertical jump performance in male collegiate athletes. *J. Strength Cond. Res.* **2008**, *22*, 226–229. [CrossRef]
19. Yaicharoen, P.; Wallman, K.; Morton, A.; Bishop, D. The effect of warm-up on intermittent sprint performance and selected thermoregulatory parameters. *J. Sci. Med. Sport* **2012**, *15*, 451–456. [CrossRef] [PubMed]
20. Young, W.B.; Behm, D. Effects of running, static stretching and practice jumps on explosive force production and jumping performance. *J. Sports Med. Phys. Fitness* **2003**, *43*, 21–27. [PubMed]
21. Vetter, R.E. Effects of six warm-up protocols on sprint and jump performance. *J. Strength Cond. Res* **2007**, *21*, 819.
22. da Silva, V.F.N.; da Silva Aguiar, S.; Sousa, C.V.; da Costa Sotero, R.; Souto Filho, J.M.; Oliveira, I.; Mota, M.R.; Simões, H.G.; Sales, M.M. Effects of short-term plyometric training on physical fitness parameters in female futsal athletes. *J. Phys. Ther. Sci.* **2017**, *29*, 783–788. [CrossRef]

23. Fletcher, I.M.; Jones, B. The effect of different warm-up stretch protocols on 20 meter sprint performance in trained rugby union players. *J. Strength Cond. Res.* **2004**, *18*, 885–888. [PubMed]
24. Fletcher, I.M.; Monte-Colombo, M.M. An investigation into the effects of different warm-up modalities on specific motor skills related to soccer performance. *J. Strength Cond. Res.* **2010**, *24*, 2096–2101. [CrossRef] [PubMed]
25. Karvonen, J. Importance of warm-up and cool down on exercise performance. In *Medicine in Sports Training and Coaching*; Karger Publishers: Basel, Switzerland, 1992; pp. 189–214.
26. Gelen, E. Acute effects of different warm-up methods on sprint, slalom dribbling, and penalty kick performance in soccer players. *J. Strength Cond. Res.* **2010**, *24*, 950–956. [CrossRef]
27. Zois, J.; Bishop, D.J.; Ball, K.; Aughey, R.J. High-intensity warm-ups elicit superior performance to a current soccer warm-up routine. *J. Sci. Med. Sport* **2011**, *14*, 522–528. [CrossRef]
28. Sale, D.G. Postactivation potentiation: Role in human performance. *Exerc. Sport Sci. Rev.* **2002**, *30*, 138–143. [CrossRef]
29. Wiewelhove, T.; Raeder, C.; Simola, R.A.d.; Schneider, C.; Döweling, A.; Ferrauti, A. Tensiomyographic markers are not sensitive for monitoring muscle fatigue in elite youth athletes: A pilot study. *Front. Physiol.* **2017**, *8*, 406. [CrossRef] [PubMed]
30. Jones, A.D.; Hind, K.; Wilson, H.; Johnson, M.I.; Francis, P. A standardised protocol for the assessment of lower limb muscle contractile properties in football players using Tensiomyography. *Adv. Skelet. Muscle Funct. Assess.* **2017**, *1*, 13–17.
31. Martín-Rodríguez, S.; Loturco, I.; Hunter, A.M.; Rodríguez-Ruiz, D.; Munguia-Izquierdo, D. Reliability and measurement error of tensiomyography to assess mechanical muscle function: A systematic review. *J. Strength Cond. Res.* **2017**, *31*, 3524–3536. [CrossRef]
32. Rey, E.; Lago-Peñas, C.; Lago-Ballesteros, J. Tensiomyography of selected lower-limb muscles in professional soccer players. *J. Electromyogr. Kinesiol.* **2012**, *22*, 866–872. [CrossRef]
33. Piqueras-Sanchiz, F. In-season analysis of the muscle response speed of knee extensors and flexors in elite futsal players. *Adv. Skelet. Muscle Funct. Assess.* **2017**, *1*, 17–22.
34. Loturco, I.; Gil, S.; de Souza Laurino, C.F.; Roschel, H.; Kobal, R.; Abad, C.C.C.; Nakamura, F.Y. Differences in muscle mechanical properties between elite power and endurance athletes: A comparative study. *J. Strength Cond. Res.* **2015**, *29*, 1723–1728. [CrossRef]
35. Pereira, L.A.; Ramirez-Campillo, R.; Martín-Rodríguez, S.; Kobal, R.; Abad, C.C.C.; Arruda, A.F.S.; Guerriero, A.; Loturco, I. Is Tensiomyography-derived velocity of contraction a sensitive marker to detect acute performance changes in elite team-sport athletes? *Int. J. Sports Physiol. Perform.* **2020**, *15*, 31–37. [CrossRef] [PubMed]
36. García-Manso, J.M.; Rodríguez-Ruiz, D.; Rodríguez-Matoso, D.; de Saa, Y.; Sarmiento, S.; Quiroga, M. Assessment of muscle fatigue after an ultra-endurance triathlon using tensiomyography (TMG). *J. Sports Sci.* **2011**, *29*, 619–625. [CrossRef]
37. Gil, S.; Loturco, I.; Tricoli, V.; Ugrinowitsch, C.; Kobal, R.; Cal Abad, C.C.; Roschel, H. Tensiomyography parameters and jumping and sprinting performance in Brazilian elite soccer players. *Sports Biomech.* **2015**, *14*, 340–350. [CrossRef]
38. Balsalobre-Fernández, C.; Glaister, M.; Lockey, R.A. The validity and reliability of an iPhone app for measuring vertical jump performance. *J. Sports Sci.* **2015**, *33*, 1574–1579. [CrossRef]
39. Bosco, C.; Luhtanen, P.; Komi, P.V. A simple method for measurement of mechanical power in jumping. *Eur. J. Appl. Physiol.* **1983**, *50*, 273–282. [CrossRef] [PubMed]
40. Hetzler, R.K.; Stickley, C.D.; Lundquist, K.M.; Kimura, I.F. Reliability and accuracy of handheld stopwatches compared with electronic timing in measuring sprint performance. *J. Strength Cond. Res.* **2008**, *22*, 1969–1976. [CrossRef]
41. Cohen, J. *Statistical Power Analysis for the Behavioral Sciences*; Academic Press: Cambridge, MA, USA, 2013.
42. Turner, A.P.; Bellhouse, S.; Kilduff, L.P.; Russell, M. Postactivation potentiation of sprint acceleration performance using plyometric exercise. *J. Strength Cond. Res.* **2015**, *29*, 343–350. [CrossRef] [PubMed]
43. Docherty, D.; Hodgson, M.J. The application of postactivation potentiation to elite sport. *Int. J. Sports Physiol. Perform.* **2007**, *2*, 439–444. [CrossRef] [PubMed]
44. Šimunič, B. Between-day reliability of a method for non-invasive estimation of muscle composition. *J. Electromyogr. Kinesiol.* **2012**, *22*, 527–530. [CrossRef]
45. de Paula Simola, R.A.; Raeder, C.; Wiewelhove, T.; Kellmann, M.; Meyer, T.; Pfeiffer, M.; Ferrauti, A. Muscle mechanical properties of strength and endurance athletes and changes after one week of intensive training. *J. Electromyogr. Kinesiol.* **2016**, *30*, 73–80. [CrossRef]
46. Rodríguez-Ruiz, D.; Rodríguez-Matoso, D.; Quiroga, M.E.; Sarmiento, S.; García-Manso, J.M.; da Silva-Grigoletto, M.E. Study of mechanical characteristics of the knee extensor and flexor musculature of volleyball players. *Eur. J. Sport Sci.* **2012**, *12*, 399–407. [CrossRef]
47. Turki, O.; Chaouachi, A.; Behm, D.G.; Chtara, H.; Chtara, M.; Bishop, D.; Chamari, K.; Amri, M. The effect of warm-ups incorporating different volumes of dynamic stretching on 10-and 20-m sprint performance in highly trained male athletes. *J. Strength Cond. Res.* **2012**, *26*, 63–72. [CrossRef] [PubMed]
48. Rimmer, E.; Sleivert, G. Effects of a plyometrics intervention program on sprint performance. *J. Strength Cond. Res.* **2000**, *14*, 295–301.
49. Faigenbaum, A.D.; McFarland, J.E.; Keiper, F.B.; Tevlin, W.; Ratamess, N.A.; Kang, J.; Hoffman, J.R. Effects of a short-term plyometric and resistance training program on fitness performance in boys age 12 to 15 years. *J. Sports Sci. Med.* **2007**, *6*, 519.

50. O'Sullivan, K.; McAuliffe, S.; DeBurca, N. The effects of eccentric training on lower limb flexibility: A systematic review. *Br. J. Sports Med.* **2012**, *46*, 838–845. [CrossRef]
51. Markovic, G.; Mikulic, P. Neuro-musculoskeletal and performance adaptations to lower-extremity plyometric training. *Sports Med.* **2010**, *40*, 859–895. [CrossRef]

Journal of
Functional Morphology and Kinesiology

MDPI

Review

Negative Effects of Mental Fatigue on Performance in the Yo-Yo Test, Loughborough Soccer Passing and Shooting Tests: A Meta-Analysis

Jozo Grgic [1,*], Ivan Mikulic [2] and Pavle Mikulic [2]

1 Institute for Health and Sport, Victoria University, Melbourne, VIC 3011, Australia
2 Faculty of Kinesiology, University of Zagreb, 10000 Zagreb, Croatia; ivan.mikulic@kif.unizg.hr (I.M.); pavle.mikulic@kif.unizg.hr (P.M.)
* Correspondence: jozo.grgic@live.vu.edu.au

Abstract: We aimed to examine the effects of mental fatigue on the Yo-Yo test and Loughborough soccer passing and shooting tests performance using a meta-analysis. The search for studies was performed through eight bibliographic databases (Academic Search Elite, AUSPORT, Cochrane Library, PsycInfo, PubMed/MEDLINE, Scopus, SPORTDiscus, and Web of Science). The methodological quality of the included studies was assessed using the PEDro checklist. A random-effects meta-analysis was performed for data analysis. After reviewing 599 search results, seven studies with a total of ten groups were included in the review. All studies were classified as being of good methodological quality. Mental fatigue reduced the distance covered in the Yo-Yo test (Cohen's *d*: −0.49; 95% confidence interval [CI]: −0.66, −0.32). In the Loughborough soccer passing test, mental fatigue increased the original time needed to complete the test (Cohen's *d*: −0.24; 95% CI: −0.46, −0.03), increased penalty time (Cohen's *d*: −0.39; 95% CI: −0.46, −0.31), and decreased performance time (Cohen's *d*: −0.52; 95% CI: −0.80, −0.24). In the Loughborough soccer shooting test, mental fatigue decreased points per shot (Cohen's *d*: −0.37; 95% CI: −0.70, −0.04) and shot speed (Cohen's *d*: −0.35; 95% CI: −0.64, −0.06). Overall, the findings presented in this review demonstrated that mental fatigue negatively impacts endurance-based running performance as well as soccer passing and shooting skills.

Keywords: data synthesis; cognitive task; mental fatigue; performance; soccer; team-sport

Citation: Grgic, J.; Mikulic, I.; Mikulic, P. Negative Effects of Mental Fatigue on Performance in the Yo-Yo Test, Loughborough Soccer Passing and Shooting Tests: A Meta-Analysis. *J. Funct. Morphol. Kinesiol.* **2022**, *7*, 10. https://doi.org/10.3390/jfmk7010010

Academic Editors: Cristina Cortis, Andrea Fusco and Carl Foster

Received: 14 December 2021
Accepted: 10 January 2022
Published: 13 January 2022

1. Introduction

Mental fatigue is commonly defined as "a psychobiological state caused by prolonged exertion that has the potential to reduce cognitive performance and exercise performance" [1]. Researchers have explored the effects of mental fatigue on performance in various modes of exercise [1]. A seminal study by Marcora et al. [2] reported that mental fatigue induced by a demanding cognitive task negatively impacted performance in cycling to exhaustion at 80% of peak power output. Since this study, research has also examined the effects of mental fatigue on various sport-specific outcomes [3].

For example, in soccer, studies have explored the effects of mental fatigue on performance in the Yo-Yo test, Loughborough soccer passing test, and Loughborough soccer shooting test [4–10]. Briefly, the Yo-Yo intermittent recovery test is a popular test used to evaluate performance in interval running [11–13]. This test determines an individual's capacity to undertake, repeatedly perform, and recover from high-intensity running [11–13]. Due to its structure, the Yo-Yo test is particularly relevant for team sports [11]. While the Yo-Yo test examines endurance performance, the Loughborough soccer passing test and Loughborough soccer shooting test evaluate sport-specific skill levels [14,15] (i.e., passing, controlling, and shooting the ball; Table 1).

Studies evaluated the effects of mental fatigue on performance in these tests, but the findings varied. For example, the mental fatigue-induced reduction in Yo-Yo test performance ranged from small (Cohen's d: -0.21) to very large (Cohen's d: -1.34), making it difficult to establish the true effect in the population [4,5]. In the Loughborough soccer passing test, some studies have reported that mental fatigue negatively impacts performance time, whereas others did not find a significant difference between the control and mental fatigue trials [4,5]. A similar discrepancy has been observed for outcomes such as shot speed in the Loughborough soccer shooting test [4,8].

Table 1. Summary of the main outcomes in the three analyzed tests.

Test	Outcome
Yo-Yo test	• Distance covered
Loughborough soccer passing test	• The original time needed to complete the test • Penalty time (i.e., time accounted for errors made during the test) • Performance time (i.e., original time plus penalty time)
Loughborough soccer shooting test	• Points per shot • Shot speed • Shot sequence time

Several reviews have been published that summarized the effects of mental fatigue on exercise performance [3,16,17]. However, these reviews have not specifically focused on the Yo-Yo test and Loughborough soccer passing or shooting tests. Perhaps more importantly, these reviews did not contain a meta-analysis that pooled the data from all studies on the topic. This would be highly relevant to perform given that some of these studies might have been underpowered to find significant differences. Accordingly, this review aimed to perform a meta-analysis examining the effects of mental fatigue on performance in the Yo-Yo test and Loughborough soccer passing and shooting tests. We hypothesized that mental fatigue would negatively influence performance in these tests.

2. Materials and Methods

2.1. Literature Search Strategy

The search for studies that explored the effects of mental fatigue on performance in the Yo-Yo test, Loughborough soccer passing test, and/or Loughborough soccer shooting test was carried out in two phases. In the first phase, we performed a search through eight different bibliographic databases, including: Academic Search Elite, AUSPORT, Cochrane Library, PsycInfo, PubMed/MEDLINE, Scopus, SPORTDiscus, and Web of Science. In all of these databases, the following search syntax was applied: ("mental fatigue" OR "mentally fatigued") AND ("yo-yo" OR "yoyo" OR "yo yo" OR "intermittent endurance" OR "intermittent recovery" OR "Loughborough soccer passing" OR "Loughborough soccer shooting"). After this phase was completed on 5 September 2021, we then screened the reference list from all studies that were found to satisfy the inclusion criteria. Post reference screening, forward citation tracking (i.e., examining the papers that cited the included studies) using the Google Scholar database was conducted.

2.2. Selection Criteria

Studies were included in the review if they satisfied the following criteria: (1) examined the effects of mental fatigue on performance in the Yo-Yo test, Loughborough soccer passing test, and/or Loughborough soccer shooting test; (2) used a crossover study design that involved a control trial and a mental fatigue trial; and (3) included humans as study participants.

2.3. Data Extraction

We extracted the following data from the included studies: (1) year of study publication and lead author name; (2) participants characteristics (e.g., sex, training status); (3) description of the control and cognitive tasks; (4) performance test and its outcomes; and

(5) mean ± standard deviation for the test outcome(s) from the control and mental fatigue trials. One study did not report these data in the manuscript. For this study [4], on request, the data were received from the corresponding author. Two studies [6,7] presented the data needed for the meta-analysis only in figures. For these two studies, the necessary data were extracted using the Web Plot Digitizer software (https://apps.automeris.io/wpd/) (accessed on 10 September 2021).

2.4. Quality Assessment

The methodological quality of the included studies was assessed using the validated 11-item PEDro checklist [18]. The items on the PEDro checklist evaluate different methodological aspects, including inclusion criteria, randomization, allocation concealment, blinding of participants and assessors, attrition, and data reporting. All items of the PEDro checklist are scored as "1" (criterion is satisfied) or "0" (criterion is not satisfied). The first item is not included in the total score. Therefore, the maximum possible number of points that can be scored is 10. In accordance with previous reviews, the studies were classified studies as poor, fair, good, or excellent quality if they scored ≤3 points, 4–5 points, 6–8 points, and 9–10 points, respectively [19,20].

2.5. Statistical Analysis

The meta-analysis was performed using effect sizes (Cohen's *d*). Cohen's *d* effect sizes were calculated using the mean ± standard deviation data from the control and mental fatigue trials, sample size, and correlation between the trials. Correlation between the trials was not reported in any of the included studies. Therefore, correlation values were estimated using the methodological approach recommended in the Cochrane Handbook [21]. A total of seven meta-analyses were performed, for: distance covered in the Yo-Yo test; time needed to complete the test, penalty time, and performance time in the Loughborough soccer passing test; points per shot, shot speed, and shot sequence time in the Loughborough soccer shooting test. The interpretation of effect sizes was based on the following thresholds: trivial (<0.20), small (0.20–0.49), medium (0.50–0.79), and large (≥0.80) [22]. Negative Cohen's *d* values denote a decrease in performance with mental fatigue. Meta-analyses were performed using the random-effects model [23]. I^2 statistic was used to evaluate heterogeneity. I^2 values were interpreted as low (<50%), moderate (50–75%), and high heterogeneity (>75%). The statistical significance threshold was set at $p < 0.05$. All analyses were performed using the Comprehensive Meta-Analysis software, version 2 (Biostat Inc., Englewood, NJ, USA).

3. Results

3.1. Search Results

There were 40, 243, and 316 search results in the database search, screening of the reference list, and forward citation tracking phases, respectively (Figure 1). Of the search results found in the databases, ten full-text papers were read and six studies were included [4,5,7–10]. One additional study [6] was found in the forward citation tracking. Therefore, this review included a total of seven studies [4–10]. However, one study [4] included three independent groups (i.e., players from under-14, under-16, and under-18). Thus, there was a total of ten groups in the seven included studies.

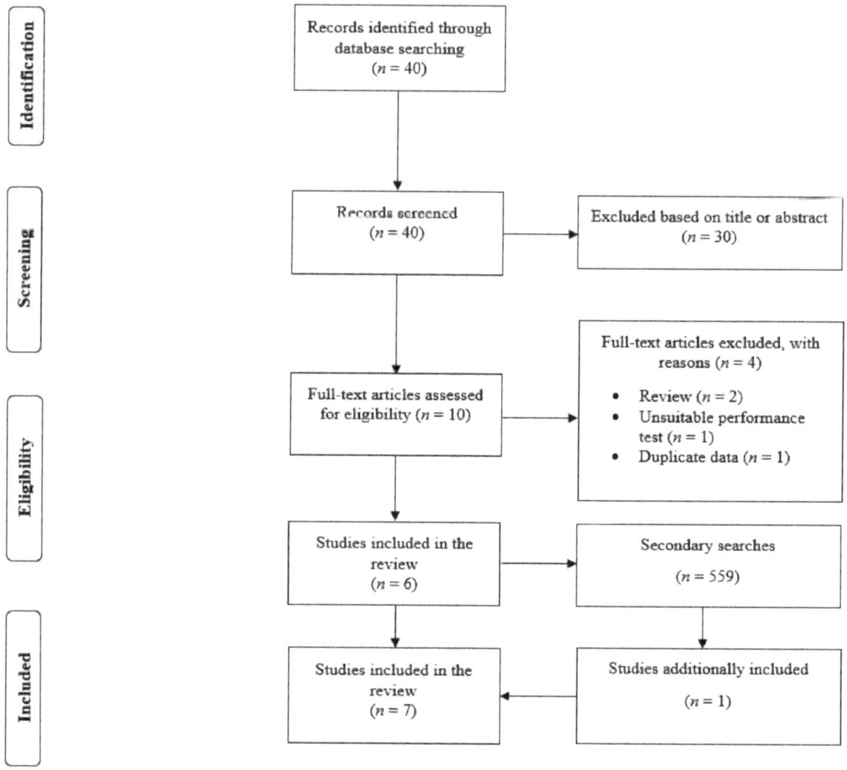

Figure 1. Flow diagram of the search process.

3.2. Summary of Studies

All seven studies explored the effects of mental fatigue on Yo-Yo test performance (10 study groups; Table 2). In all of these studies, the Intermittent Recovery Test Level 1 variant of the Yo-Yo test was used. The pooled number of participants for studies that used the Yo-Yo test was 134. Three studies (5 study groups; $n = 58$) and two studies (4 study groups; $n = 50$) explored the effects of mental fatigue on Loughborough soccer passing test and Loughborough soccer shooting test, respectively. The cognitive task used to induce mental fatigue was the Stroop test in six studies, while one study [5] employed the "Brain It On" application on a smartphone. The duration of the cognitive task was 30 min in all studies. For the control trials, studies used reading of magazines, watching an emotionally neutral video, or no activity. The duration of the control trials for studies that used reading of magazines or watching a video was from 15 to 30 min. Five [4,6–9] studies explored the effects of the cognitive task on mental fatigue. In all these studies, mental fatigue was higher following the cognitive task.

3.3. Methodological Quality

Five studies [5–7,9,10] scored six points on the PEDro checklist, whereas two studies [4,8] scored eight points. Therefore, all included studies were classified as being of good methodological quality (Table 3).

Table 2. Summary of the studies included in the review.

Study	Participants	Control Task	Cognitive Task	Test and Outcomes
Filipas et al. (2021)	36 male soccer players (n = 12 under-14, n = 12 under-16, and n = 12 under-18)	Reading magazines (15 min)	Computerized Stroop test (30 min)	Yo-Yo IR1—distance covered LSPT—original time, distance covered, and performance time LSST—points per shot, shot speed, and shot sequence time
Greco et al. (2017)	16 young male soccer players [a]	"normal activities"	Using the "Brain It On" application on a smartphone (30 min)	Yo-Yo IR1—distance covered LSPT—original time, distance covered, and performance time
Lam et al. (2021)	9 physically active males	No activity	Computerized Stroop test (30 min)	Yo-Yo IR1—distance covered
Penna et al. (2018)	12 handball players	Watching an emotionally neutral video (30 min)	Computerized Stroop test (30 min)	Yo-Yo IR1—distance covered
Smith et al. (2016)	12 moderately trained soccer players and 14 experienced soccer Players [b]	Reading magazines (30 min)	Computerized Stroop test (30 min)	Yo-Yo IR1—distance covered LSPT—original time, distance covered, and performance time LSST—points per shot, shot speed, and shot sequence time
Veness et al. (2017)	10 elite male cricket players	Reading magazines (15 min)	Computerized Stroop test (30 min)	Yo-Yo IR1—distance covered
Weerakkody et al. (2021)	25 male community-level Australian football players	Watching an emotionally neutral video (30 min)	Computerized Stroop test (30 min)	Yo-Yo IR1—distance covered

[a] 8 participants performed the Yo-Yo test and 8 participants performed the LSPT; [b] 12 participants performed the Yo-Yo test and 14 participants performed the LSPT; LSPT: Loughborough soccer passing test; LSST: Loughborough soccer shooting tests; IR1: intermittent recovery level 1.

Table 3. Results from the methodological quality assessment using the PEDro checklist.

Reference	Item 1	Item 2	Item 3	Item 4	Item 5	Item 6	Item 7	Item 8	Item 9	Item 10	Item 11	Total Score
Filipas et al. (2021)	Yes	Yes	Unclear	Yes	No	Yes	Yes	Yes	Yes	Yes	Yes	8
Greco et al. (2017)	Yes	Yes	Unclear	Yes	No	No	No	Yes	Yes	Yes	Yes	6
Lam et al. (2021)	Yes	Yes	Unclear	Yes	No	No	No	Yes	Yes	Yes	Yes	6
Penna et al. (2018)	Yes	Yes	Unclear	Yes	No	No	No	Yes	Yes	Yes	Yes	6
Smith et al. (2016)	Yes	Yes	Unclear	Yes	No	Yes	Yes	Yes	Yes	Yes	Yes	8
Veness et al. (2017)	Yes	Yes	Unclear	Yes	No	No	No	Yes	Yes	Yes	Yes	6
Weerakkody et al. (2021)	Yes	Yes	Unclear	Yes	No	No	No	Yes	Yes	Yes	Yes	6

Yes: criterion is satisfied; No: criterion is not satisfied; Unclear: unable to rate.

3.4. Meta-Analysis Results

Mental fatigue reduced the distance covered in the Yo-Yo test (Cohen's d: −0.49; 95% confidence interval [CI]: −0.66, −0.32; $p < 0.001$; $I^2 = 33\%$; Figure 2).

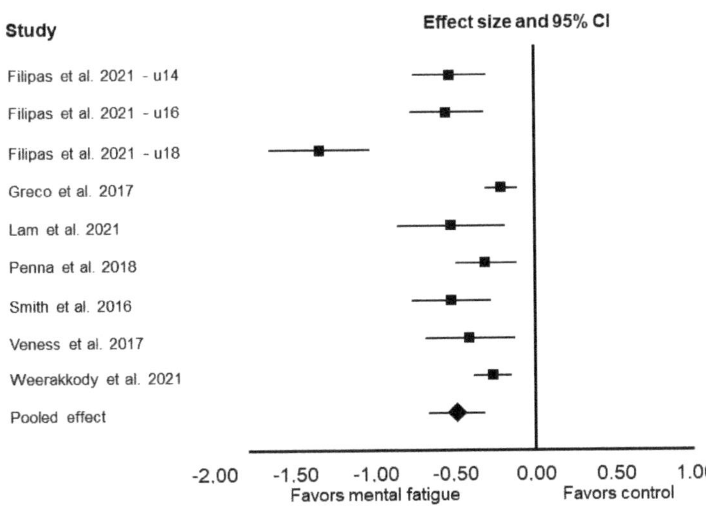

Figure 2. Forest plot presenting the results of the random-effects meta-analysis comparing the effects of control vs. mental fatigue on Yo-Yo test performance. Data are reported as Cohen's d (effect size) and 95% confidence interval (CI). The diamond at the bottom presents the overall effect. The plotted squares denote effect sizes, and the whiskers denote their 95% CIs.

In the Loughborough soccer passing test, mental fatigue increased the original time needed to complete the test (Cohen's d: −0.24; 95% CI: −0.46, −0.03; $p = 0.024$; $I^2 = 0\%$; Figure 3), increased penalty time (Cohen's d: −0.39; 95% CI: −0.46, −0.31; $p < 0.001$; $I^2 = 15\%$), and decreased performance time (Cohen's d: −0.52; 95% CI: −0.80, −0.24; $p < 0.001$; $I^2 = 51\%$).

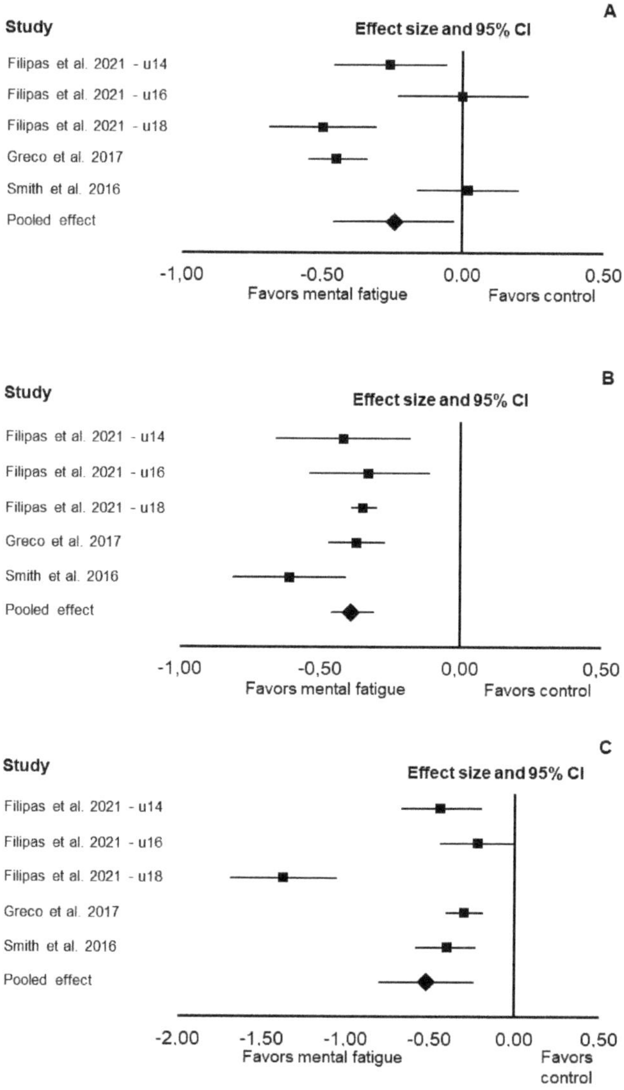

Figure 3. Forest plot presenting the results of the random-effects meta-analysis comparing the effects of control vs. mental fatigue on the original time needed to complete the test (**A**), penalty time (**B**), and performance time (**C**) in the Loughborough soccer passing test. Data are reported as Cohen's d (effect size) and 95% confidence interval (CI). The diamond at the bottom presents the overall effect. The plotted squares denote effect sizes, and the whiskers denote their 95% CIs.

In the Loughborough soccer shooting test, mental fatigue decreased points per shot (Cohen's d: −0.37; 95% CI: −0.70, −0.04; $p = 0.028$; $I^2 = 57\%$; Figure 4) and shot speed (Cohen's d: −0.35; 95% CI: −0.64, −0.06; $p = 0.019$; $I^2 = 0\%$). There was no significant difference between the control and mental fatigue conditions for shot sequence time (Cohen's d: −0.18; 95% CI: −0.52, 0.16; $p = 0.300$; $I^2 = 26\%$).

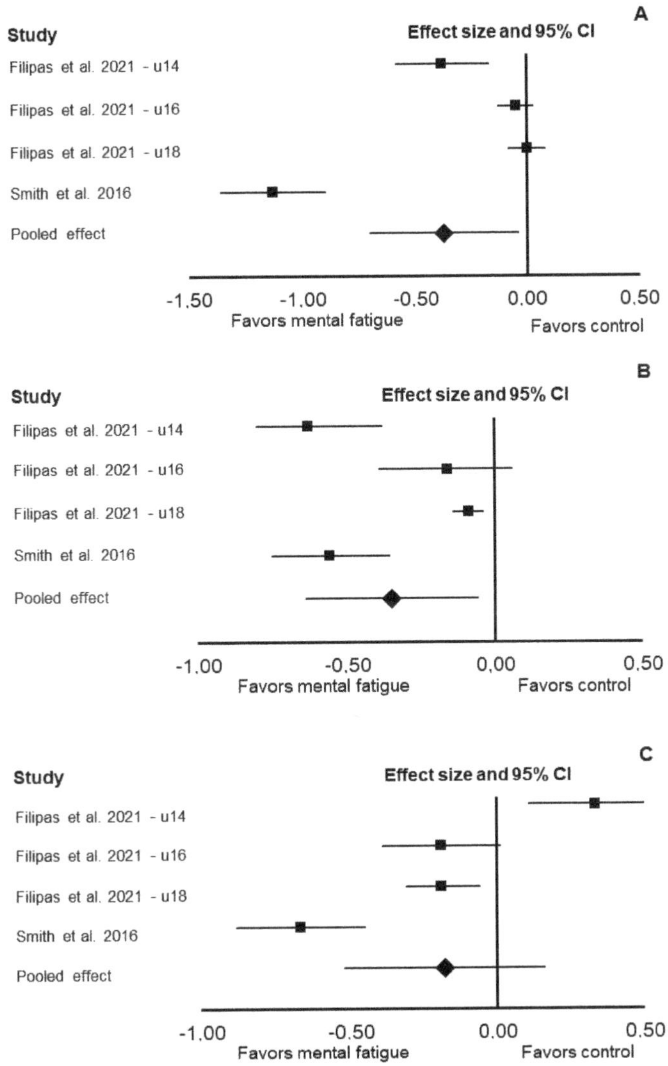

Figure 4. Forest plot presenting the results of the random-effects meta-analysis comparing the effects of control vs. mental fatigue on points per shot (**A**), shot speed (**B**), and shot sequence time (**C**) in the Loughborough soccer shooting test. Data are reported as Cohen's d (effect size) and 95% confidence interval (CI). The diamond at the bottom presents the overall effect. The plotted squares denote effect sizes, and the whiskers denote their 95% CIs.

4. Discussion

In line with our hypothesis, we found that mental fatigue has a negative effect on some aspects of physical and technical exercise performance. Specifically, the distance covered in the Yo-Yo test was lower following the mental fatigue trials. Additionally, it was found that mental fatigue hampered performance in the Loughborough soccer passing test by increasing the original time needed to complete the test and penalty time. As a result, performance time—calculated as original time plus penalty time—was also slower post mental fatigue. Similar to these findings, mental fatigue decreased points per shot and shot speed in the Loughborough soccer shooting test. Overall, these findings demonstrated that mental fatigue negatively impacts endurance-based running performance as well as soccer passing and shooting skills. Most effect sizes were in the range from small to medium and are likely to be practically relevant. These findings are novel, given that previous research on the topic produced conflicting results (i.e., varying effect sizes or differences in findings) [4–10].

Our results add to the body of evidence demonstrating that mental fatigue may hinder endurance-based exercise performance [3,16,17]. Variables such as heart rate, metabolite accumulation, and neuromuscular function are generally not affected by mental fatigue [24]. Therefore, it has been suggested that the negative effect of mental fatigue on exercise performance is due to the increase in the rating of perceived exertion (RPE). However, some studies evaluated RPE values and reported no significant difference between the control and mental fatigue [5,10]. Still, these studies evaluated RPE at test exhaustion, which needs to be considered given the findings that mentally fatigued athletes display higher RPE values during the Yo-Yo test, but not necessarily at the end of the test [4]. Furthermore, there are nuances to mention in the interpretation even if there is no significant difference in RPE values between the trials at test exhaustion. Specifically, despite no significant RPE differences, it should be considered that the total distance covered in the Yo-Yo test in these studies [5,10] was higher in the control condition, showing that participants who were mentally fatigued covered less distance to achieve a similar level of RPE. It has been suggested that the increase in perceived exertion is mediated by cerebral adenosine accumulation during cognitive tasks [24,25]. Indeed, this hypothesis is supported by data indicating that caffeine supplementation—which acts by binding to adenosine receptors—averts the negative effects of mental fatigue on exercise performance [26,27].

The Yo-Yo test has been reported to correlate with several sport-specific outcomes. For example, a positive correlation (r = 0.70–0.81) was found between the distance covered in the Yo-Yo test and the amount of high-intensity running performed during the whole soccer game, during a 5-min period involving peak running intensity, or at the end of each half of a game [11,28,29]. Therefore, it seems likely that the negative effects of mental fatigue on performance in the Yo-Yo test may also translate to sport-specific performance. However, it should be considered that running-based performance in soccer matches is also influenced by a multitude of factors, such as match location, quality of opposition, and match outcome [30]. Therefore, the results obtained during testing may not necessarily be generalized to performance in sports competitions.

Given that the Yo-Yo test is most commonly used for testing (rather than training) purposes, our results highlight that researchers and practitioners should attempt to standardize the cognitive activity before this assessment, particularly when exploring differences between individuals. As demonstrated herein, the negative effect of mental fatigue on performance in the Yo-Yo test may range from small to medium (Cohen's d: −0.49; 95% CI: −0.66, −0.32). This is particularly relevant to mention as these effects are in the range of improvements in Yo-Yo test performance following 3–8 weeks of sprint training or speed endurance training (Cohen's d: 0.30–0.45) [31–33].

In addition to Yo-Yo test performance, this meta-analysis reported that mental fatigue resulted in a lower number of points per shot and shot speed in the Loughborough soccer shooting test. Additionally, mental fatigue negatively affected all of the analyzed outcomes in the Loughborough soccer passing test. Specifically, it seems that mentally fatigued athletes are more prone to making technical errors, as evidenced by the increase in penalty time, which is given for different errors (e.g., missing the target area) [14]. Research has established that mental fatigue increases the attention to irrelevant, compared to relevant stimuli, which might explain poorer technical performance in the Loughborough tests observed herein [34]. Similar to our findings, studies have also reported a lower number of total passes during small-sided games following a mentally fatiguing task [35]. Nevertheless, it remains unclear if these negative effects also translate to performance in sport-specific situations. Some researchers have suggested that the Loughborough soccer passing test might not be a valid test of in-game passing performance, at least among youth players [36]. While this area certainly merits future research, it seems reasonable to suggest that mental fatigue should be avoided before competitions to reduce the likelihood of impairment of technical performance.

There are several limitations of the present review that need to be considered. First, it should be mentioned that all of the included studies involved male participants. Therefore, the results presented herein may not necessarily be generalized to females, and future research in this population is therefore needed. Additionally, the population analyzed among the included studies varied from physically active males to elite athletes (Table 2). The magnitude of the effect observed in most studies was similar, suggesting that the effects of mental fatigue on performance in the three analyzed tests are not population-specific. However, future studies are needed to directly explore the effects of mental fatigue on performance in different populations.

All included studies were classified as being of good methodological quality. Five studies scored 6 points on the PEDro checklist, while two studies scored 8 points. These two studies [4,8] scored more points, given that they also incorporated blinding in their study design. Specifically, these two studies incorporated blinding of assessors measuring the outcomes. Despite the difference in blinding across the included studies, it is interesting to note that generally, all studies reported similar effect sizes with overlapping 95% CIs. This would suggest that the lack of blinding in some studies may not have influenced outcomes.

5. Conclusions

This meta-analysis found that mental fatigue negatively impacts different aspects of physical and technical exercise performance. Specifically, mental fatigue negatively impacted endurance-based running, as evident by a lower distance covered in the Yo-Yo test. Furthermore, outcomes such as the original time needed to complete the test, penalty time, performance time, points per shot, and shot speed in the Loughborough soccer tests were negatively affected by mental fatigue. These findings demonstrate that mental fatigue negatively impacts endurance-based running performance as well as soccer passing and shooting skills. Most effects sizes were in the range from small to medium and are likely to be practically relevant.

Author Contributions: Conceptualization, J.G.; methodology, J.G.; formal analysis, J.G.; investigation, J.G.; writing—original draft preparation, J.G.; writing—review and editing, I.M. and P.M.; visualization, J.G.; funding acquisition, I.M. and P.M. All authors have read and agreed to the published version of the manuscript.

Funding: This research received no external funding.

Institutional Review Board Statement: Not applicable.

Informed Consent Statement: Not applicable.

Data Availability Statement: Data used for the meta-analysis are available on request from the corresponding author.

Conflicts of Interest: The authors declare no conflict of interest.

References

1. Meeusen, R.; Van Cutsem, J.; Roelands, B. Endurance exercise-induced and mental fatigue and the brain. *Exp. Physiol.* **2020**, *106*, 2294–2298. [CrossRef] [PubMed]
2. Marcora, S.M.; Staiano, W.; Manning, V. Mental fatigue impairs physical performance in humans. *J. Appl. Physiol.* **2009**, *106*, 857–864. [CrossRef]
3. Habay, J.; Van Cutsem, J.; Verschueren, J.; De Bock, S.; Proost, M.; De Wachter, J.; Tassignon, B.; Meeusen, R.; Roelands, B. Mental Fatigue and Sport-Specific Psychomotor Performance: A Systematic Review. *Sports Med.* **2021**, *51*, 1527–1548. [CrossRef]
4. Filipas, L.; Borghi, S.; La Torre, A.; Smith, M.R. Effects of mental fatigue on soccer-specific performance in young players. *Sci. Med. Footb.* **2021**, *5*, 150–157. [CrossRef]
5. Greco, G.; Tambolini, R.; Ambruosi, P.; Fischetti, F. Negative effects of smartphone use on physical and technical performance of young footballers. *J. Phys. Educ. Sport* **2017**, *17*, 2495–2501.
6. Lam, H.K.N.; Middleton, H.; Phillips, S.M. The effect of self-selected music on endurance running capacity and performance in a mentally fatigued state. *J. Hum. Sport Exerc.* **2021**, *17*. [CrossRef]
7. Penna, E.M.; Filho, E.; Campos, B.T.; Pires, D.A.; Nakamura, F.Y.; Mendes, T.T.; Lopes, T.; Smith, M.; Prado, L.S. Mental fatigue does not affect heart rate recovery but impairs performance in handball players. *Rev. Bras. Med. Esporte* **2018**, *24*, 347–351. [CrossRef]
8. Smith, M.; Coutts, A.J.; Merlini, M.; Deprez, D.; Lenoir, M.; Marcora, S. Mental Fatigue Impairs Soccer-Specific Physical and Technical Performance. *Med. Sci. Sports Exerc.* **2016**, *48*, 267–276. [CrossRef]
9. Veness, D.; Patterson, S.D.; Jeffries, O.; Waldron, M. The effects of mental fatigue on cricket-relevant performance among elite players. *J. Sports Sci.* **2017**, *35*, 2461–2467. [CrossRef]
10. Weerakkody, N.; Taylor, C.; Bulmer, C.; Hamilton, D.; Gloury, J.; O'brien, N.; Saunders, J.; Harvey, S.; Patterson, T. The effect of mental fatigue on the performance of Australian football specific skills amongst amateur athletes. *J. Sci. Med. Sport* **2021**, *24*, 592–596. [CrossRef]
11. Bangsbo, J.; Iaia, F.M.; Krustrup, P. The Yo-Yo intermittent recovery test: A useful tool for evaluation of physical performance in intermittent sports. *Sports Med.* **2008**, *38*, 37–51. [CrossRef]
12. Krustrup, P.; Mohr, M.; Amstrup, T.; Rysgaard, T.; Johansen, J.; Steensberg, A.; Pedersen, P.K.; Bangsbo, J. The Yo-Yo Intermittent Recovery Test: Physiological Response, Reliability, and Validity. *Med. Sci. Sports Exerc.* **2003**, *35*, 697–705. [CrossRef]
13. Grgic, J.; Oppici, L.; Mikulic, P.; Bangsbo, J.; Krustrup, P.; Pedišić, Ž. Test–Retest Reliability of the Yo-Yo Test: A Systematic Review. *Sports Med.* **2019**, *49*, 1547–1557. [CrossRef] [PubMed]
14. Ali, A.; Williams, C.; Hulse, M.; Strudwick, A.; Reddin, J.; Howarth, L.; Eldred, J.; Hirst, M.; McGregor, S. Reliability and validity of two tests of soccer skill. *J. Sports Sci.* **2007**, *25*, 1461–1470. [CrossRef] [PubMed]
15. Wen, D.; Robertson, S.; Hu, G.; Song, B.; Chen, H. Measurement properties and feasibility of the Loughborough soccer passing test: A systematic review. *J. Sports Sci.* **2018**, *36*, 1682–1694. [CrossRef]
16. Smith, M.R.; Thompson, C.; Marcora, S.M.; Skorski, S.; Meyer, T.; Coutts, A.J. Mental Fatigue and Soccer: Current Knowledge and Future Directions. *Sports Med.* **2018**, *48*, 1525–1532. [CrossRef] [PubMed]
17. Van Cutsem, J.; Marcora, S.; De Pauw, K.; Bailey, S.; Meeusen, R.; Roelands, B. The effects of mental fatigue on physical per-formance: A systematic review. *Sports Med.* **2017**, *47*, 1569–1588. [CrossRef]
18. Maher, C.G.; Sherrington, C.; Herbert, R.D.; Moseley, A.M.; Elkins, M. Reliability of the PEDro scale for rating quality of ran-domized controlled trials. *Phys. Ther.* **2003**, *83*, 713–721. [CrossRef]
19. Grgic, J.; Pickering, C. The effects of caffeine ingestion on isokinetic muscular strength: A meta-analysis. *J. Sci. Med. Sport* **2019**, *22*, 353–360. [CrossRef]
20. Grgic, J. Caffeine ingestion enhances Wingate performance: A meta-analysis. *Eur. J. Sport Sci.* **2018**, *18*, 219–225. [CrossRef]
21. Higgins, J.P.T.; Deeks, J.J.; Altman, D.G. On behalf of the Cochrane Statistical Methods Group. In *Imputing Standard Deviations for Changes from Baseline*; Chapter 16.1.3.2; Higgins, J.P., Green, S., Eds.; Cochrane Collaboration: Chichester, UK, 2011.
22. Cohen, J. A power primer. *Psychol. Bull.* **1992**, *112*, 155. [CrossRef]
23. Borenstein, M.; Hedges, L.V.; Higgins, J.P.T.; Rothstein, H.R. A basic introduction to fixed-effect and random-effects models for meta-analysis. *Res. Synth. Methods* **2010**, *1*, 97–111. [CrossRef]
24. Martin, K.; Meeusen, R.; Thompson, K.G.; Keegan, R.; Rattray, B. Mental Fatigue Impairs Endurance Performance: A Physiological Explanation. *Sports Med.* **2018**, *48*, 2041–2051. [CrossRef]
25. Pageaux, B.; Lepers, R.; Dietz, K.C.; Marcora, S.M. Response inhibition impairs subsequent self-paced endurance performance. *Eur. J. Appl. Physiol.* **2014**, *114*, 1095–1105. [CrossRef]
26. Franco-Alvarenga, P.E.; Brietzke, C.; Canestri, R.; Goethel, M.F.; Hettinga, F.; Santos, T.M.; Pires, F.O. Caffeine improved cycling trial performance in mentally fatigued cyclists, regardless of alterations in prefrontal cortex activation. *Physiol. Behav.* **2019**, *204*, 41–48. [CrossRef] [PubMed]
27. Grgic, J. Effects of Caffeine on Resistance Exercise: A Review of Recent Research. *Sports Med.* **2021**, *51*, 2281–2298. [CrossRef]
28. Krustrup, P.; Bangsbo, J. Physiological demands of top-class soccer refereeing in relation to physical capacity: Effect of intense intermittent exercise training. *J. Sports Sci.* **2001**, *19*, 881–891. [CrossRef] [PubMed]
29. Krustrup, P.; Mohr, M.; Ellingsgaard, H.; Bangsbo, J. Physical Demands during an Elite Female Soccer Game: Importance of Training Status. *Med. Sci. Sports Exerc.* **2005**, *37*, 1242–1248. [CrossRef]

30. Teixeira, J.; Leal, M.; Ferraz, R.; Ribeiro, J.; Cachada, J.; Barbosa, T.; Monteiro, A.; Forte, P. Effects of Match Location, Quality of Opposition and Match Outcome on Match Running Performance in a Portuguese Professional Football Team. *Entropy* **2021**, *23*, 973. [CrossRef]
31. Mohr, M.; Krustrup, P.; Nielsen, J.J.; Nybo, L.; Rasmussen, M.K.; Juel, C.; Bangsbo, J. Effect of two different intense training regimes on skeletal muscle ion transport proteins and fatigue development. *Am. J. Physiol. Regul. Integr. Comp. Physiol.* **2007**, *292*, R1594–R1602. [CrossRef] [PubMed]
32. Vorup, J.; Tybirk, J.; Gunnarsson, T.P.; Ravnholt, T.; Dalsgaard, S.; Bangsbo, J. Effect of speed endurance and strength training on performance, running economy and muscular adaptations in endurance-trained runners. *Graefe's Arch. Clin. Exp. Ophthalmol.* **2016**, *116*, 1331–1341. [CrossRef] [PubMed]
33. Rogan, S. Comparison of Two Kinds of Endurance Training Programs on the Effects of the Ability to Recover in Amateur Soccer Players. *Asian J. Sports Med.* **2015**, *6*, e22585. [CrossRef]
34. Boksem, M.A.; Meijman, T.F.; Lorist, M.M. Effects of mental fatigue on attention: An ERP study. *Cogn. Brain Res.* **2005**, *25*, 107–116. [CrossRef]
35. Trecroci, A.; Boccolini, G.; Duca, M.; Formenti, D.; Alberti, G. Mental fatigue impairs physical activity, technical and deci-sion-making performance during small-sided games. *PLoS ONE.* **2020**, *15*, 0238461. [CrossRef] [PubMed]
36. Serpiello, F.R.; Cox, A.; Oppici, L.; Hopkins, W.G.; Varley, M.C. The Loughborough Soccer Passing Test has impractical criterion validity in elite youth football. *Sci. Med. Footb.* **2017**, *1*, 60–64. [CrossRef]

Journal of
Functional Morphology and Kinesiology

MDPI

Article

Correlation between Official and Common Field-Based Fitness Tests in Elite Soccer Referees

Veronica Romano [1,*], Manuel Tuzi [1], Ada Di Gregorio [1], Anna Maria Sacco [1], Immacolata Belviso [1], Felice Sirico [1], Stefano Palermi [1], Daria Nurzynska [2], Franca Di Meglio [1], Clotilde Castaldo [1], Angelo Pizzi [3] and Stefania Montagnani [1]

1 Department of Public Health, University of Naples "Federico II", 80131 Naples, NA, Italy; tuzimanuel@gmail.com (M.T.); adadigregorio.nutrizione@gmail.com (A.D.G.); annamaria.sacco@unina.it (A.M.S.); immacolata.belviso@unina.it (I.B.); felice.sirico2@unina.it (F.S.); stefano.palermi@unina.it (S.P.); franca.dimeglio@unina.it (F.D.M.); clotilde.castaldo@unina.it (C.C.); montagna@unina.it (S.M.)

2 Department of Medicine, Surgery and Dentistry "Scuola Medica Salernitana", University of Salerno, 84081 Baronissi, SA, Italy; dnurzynska@unisa.it

3 Italian Association of Referees (AIA, Associazione Italiana Arbitri)–FIFA Referees Medical Committee, 00185 Rome, RM, Italy; angelopizzi@alice.it

* Correspondence: veronica.romano@unina.it

Abstract: Official tests are used to assess the fitness status of soccer referees, and their results correlate with match performance. However, FIFA-approved tests expose the referees to high physical demands and are difficult to implement during the sportive year. The aim of our study was to evaluate the correlation between the 6 × 40-m sprint and Yo-Yo Intermittent Recovery Level 1 (IR1) official tests and other field-based tests that require no or little equipment, are not time-consuming, and impose low physical demands. All tests were performed by male referees from the Regional Section of the Italian Referee Association (n = 30). We observed a strong correlation between 6 × 40-m sprint and Illinois agility tests (r = 0.63, p = 0.001) and a moderate correlation between Yo-Yo IR1 and hand-grip strength in the dominant (r = 0.45, p = 0.014) and non-dominant hand (r = 0.41, p = 0.031). Interestingly, only a moderate correlation (r = −0.42, p = 0.025) was observed between the FIFA official tests, 6 × 40-m sprint and Yo-Yo IR1. These results suggest that Illinois agility and hand-grip tests could represent simple and low-physical-impact tools for repeated assessment and monitoring of referee fitness throughout the sportive season.

Keywords: soccer; referee; fitness; Illinois agility test; hand-grip strength

Citation: Romano, V.; Tuzi, M.; Di Gregorio, A.; Sacco, A.M.; Belviso, I.; Sirico, F.; Palermi, S.; Nurzynska, D.; Di Meglio, F.; Castaldo, C.; et al. Correlation between Official and Common Field-Based Fitness Tests in Elite Soccer Referees. *J. Funct. Morphol. Kinesiol.* **2021**, *6*, 59. https://doi.org/10.3390/jfmk6030059

Academic Editors: Cristina Cortis, Andrea Fusco and Carl Foster

Received: 5 May 2021
Accepted: 22 June 2021
Published: 1 July 2021

1. Introduction

The referee is in overall control of the soccer game. In addition to the considerable psychological and cognitive demands placed on referees during games [1], an extensive load is imposed on their cardiovascular and musculoskeletal systems. However, it is the training, performance, and injury prevention of soccer players that have been extensively studied over the last few decades, with only limited scientific studies dedicated to fitness monitoring and injury prevention in referees. Importantly, referees are usually not full-time professionals, they are on average 15–20 years older than players, and they cannot normally be substituted. During the game, they perform a mix of walking and running activity of low, medium, and high intensity [2,3]. Hence, the physical fitness of elite soccer referees is of fundamental importance for effective officiating.

National and international soccer referees' associations routinely assess the performance of elite-level officials. Several field-based tests have been used for years to this aim. Following the reports on the poor validity of the 2 × 50-m sprint, 2 × 200-m sprint, and 12-min Cooper tests [4,5] in measuring the referees' match-related physical capacity, the Fédération Internationale de Football Association (FIFA) introduced a 6 × 40-m sprint test and a high-intensity 150-m interval test as official fitness tests for men and women referees [6]. The intensity of the latter was subsequently reduced in the latest guidelines, which now recommend 40 × 75-m run/25-m walk intervals [7].

While the validity of the 6 × 40-m sprint test and its correlation with referees' match performance were confirmed [8], only weak evidence exists for the high-intensity 150-m interval test [9], with comparison studies suggesting the advantage of the Yo-Yo Intermittent Recovery Level 1 (Yo-Yo IR1) test [10]. Subsequently, the validity of the latter test in determining the maximal activation of the aerobic system through intermittent exercise was observed [8]. As a result, the Yo-Yo IR1 can now be used in addition to the official tests as a method of assessing the aerobic fitness of referees, according to FIFA [6].

The fitness assessment is carried out at the beginning of the sportive year, and it defines the eligibility of referees for participation at the international, national, and regional levels. It is only rarely performed during the season. For one thing, the tight schedule can make it difficult for the active referees to participate in official assessments; indeed, even the training sessions between matches are often unsupervised and scheduled at the discretion of the participant. For another, the recommended tests are strenuous and physically demanding. In particular, the 6 × 40-m sprint test requires a maximal anaerobic muscular activation and the Yo-Yo IR1 test imposes muscular exhaustion. For these reasons, referees' physical fitness for participation cannot be easily assessed during the training periods or intervals between the officiated soccer matches. Hence, simpler, less time-consuming, and less physically demanding tests could be potentially more feasible and useful for fitness screening during the sportive season.

The aim of this study was to evaluate the correlation between the results of the official fitness tests performed by AIA referees, namely 6 × 40-m sprint and Yo-Yo IR1, and other common tests aimed at the evaluation of several domains of physical fitness, such as: the hand-grip strength (HGS) test, which evaluates explosive strength in the upper limbs; the sit-and-reach (SaR) test, which assesses flexibility; the Illinois agility (IA) test, which assesses agility; and the standing long jump (SLJ) and standing quintuple jump (SQJ) tests, which both evaluate explosive strength in the lower limbs. Evidence of a valid correlation between these tests could allow the introduction of a fitness evaluation protocol that is easy to perform, consumes little time, and has low impact on subsequent physical activity. Such a protocol, then, could be used to evaluate referees during the sportive season to monitor their fitness level and guarantee the best possible performance and injury prevention during officiated matches.

2. Materials and Methods

2.1. Participants

The participants were enrolled in the study on a voluntary basis and represented a convenience sample of the Regional Section (Lazio) of the Italian Referee Association (Associazione Italiana di Arbitri, or AIA). All participants were male referees in the regional category of officiating, who participate in two to three training sessions per week and officiate one soccer match per week from September to May. Referees with active painful musculoskeletal complaints or with a history of a painful condition or surgery within the previous 6 months were excluded.

The study was approved by the Department Review Board for ethical concerns. All participants received an exhaustive explanation of the study protocol and objective. Each participant provided written informed consent prior to participation in the study and acknowledged that they cannot be identified via this paper because all data were made anonymous.

2.2. Study Protocol

The study protocol is graphically represented in Figure 1.

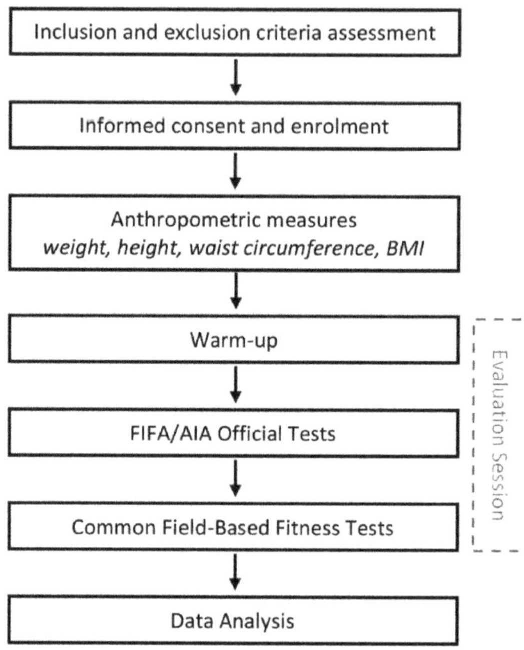

Figure 1. Study protocol flowchart.

The study took place in September 2020. The participants were instructed to avoid vigorous physical activity during the previous 48 h and not to drink any caffeinated beverages during the previous 24 h before the evaluation session. A medical specialist in sport medicine obtained the medical history of each participant. Another doctor collected anthropometric data, including sex, age, dominant side, height, weight, and waist circumference. The body mass index was calculated in the standard way in kilograms per square meter. Each fitness evaluation session was performed on an artificial soccer field in one day and divided into three stages. The participants were familiarized with the testing procedures and verbally encouraged during test performance.

All activities were conducted according to the FIFA guidelines [6]. The participants were not allowed to wear spiked track shoes. The first stage consisted of a 20-min warm-up. A professional fitness coach, together with a sports medicine doctor, supervised these activities. The second stage comprised FIFA/AIA fitness tests. Each participant performed the 6 × 40-m sprint test, followed by the Yo-Yo IR1 test, with an 8-min interval between the two. The time of the fastest sprint during the first test, recorded in seconds, and the maximum distance covered during the second test, in meters, were used in the analysis. After completion of the official tests, the participants were allowed 15 min of rest. During this time, they could drink water freely, but no food or other beverages were allowed. The subsequent third stage of evaluation comprised the following field-based fitness tests: HGS, SLJ, SQJ, SaR, and IA test.

The HGS was measured with a digital dynamometer (Dynex, MD systems, Inc., Westerville, OH, USA) in the dominant and non-dominant arm, while the subject was standing with their shoulders in neutral position and their elbow flexed at 90 degrees [11]. Three measurements for each arm were made, with a recovery period of 30 s between repetitions. The best result obtained for each arm, expressed in kilograms, was considered for analysis.

The SLJ was performed with a two-foot take-off and landing, with swinging of the arms and bending of the knees allowed [12]. The distance was measured from the baseline to the point where the back of the heel nearest to the take-off line landed on the ground. The participants repeated the jump three times, and the longest distance was used for subsequent analysis. The SQJ required five consecutive two-foot jumps from a standing start. At each landing phase, the feet were aligned before the subsequent jump, based on the foot nearest to the baseline. After the last jump, the total distance from the start line was recorded in centimeters.

For the SaR test, the participant sat on the ground with knees fully extended and legs together. The soles of the feet were placed against a box, with the 23-cm point at the level of the contact. With arms extended and one hand placed on top of the other, the participant reached forward as far as possible without flexing their knees or moving their feet [13]. The final position of the hands, reached at the fourth trial, was recorded in centimeters from a measuring scale placed on top of the box.

The IA test was performed using the procedures outlined by Negra et al. [14]. The corners of a 10 m × 5 m field were marked with cones. Another four cones were placed down the center of the rectangle, 3.3 m apart. The participant remained prone on the ground with their hands at the shoulder level and their chin touching the starting line. On a verbal command, the referee got up and ran along the previously indicated course, turning between the cones. The completion time, expressed in seconds, was used for subsequent analysis.

2.3. Statistical Analysis

The results of the FIFA/AIA official tests were correlated with those of other field-based fitness tests using the Pearson correlation coefficient. All statistical analyses were performed with STATA software (StataCorp. v.12, College Station, TX, USA) by a researcher who was not involved in the data collection. The anthropometric measurements and the results of the fitness tests were considered as continuous data. Normality of the data distribution was confirmed using the Shapiro–Wilk test, and the results were reported as mean ± standard deviation. The level of statistical significance was set at $p < 0.05$. The datasets generated and analyzed during the current study are available from the corresponding author on reasonable request.

The null hypothesis was that no correlation exists between FIFA/AIA official tests and common field-based fitness tests, corresponding to a Pearson correlation coefficient = 0. The alternative 2-tailed hypothesis was that a large correlation was present between variables, corresponding to a Pearson correlation coefficient of at least 0.5. Admitting a type I error of 5% and a power of 80%, the minimum sample size, calculated according to Hulley et al. [15], was 29 subjects. The correlation was considered absent at $r < 0.1$. Higher values of the correlation coefficient were graded according to the following scale: weak, when r = 0.1–0.29; moderate, when r = 0.3–0.49; strong, when r = 0.5–0.69; very strong, when r = 0.7–0.9; and nearly perfect, when $r > 0.9$.

3. Results

Thirty-five referees volunteered to participate and were assessed for enrollment criteria. Five were excluded due to recent musculoskeletal injuries (three referees suffered an ankle sprain in the previous 2 months, one referee underwent a meniscal repair 4 months earlier, and one referee complained of unspecific muscular symptoms in his right calf during the previous week). As a result, the study included 30 male referees.

The mean age of the participants was 22.24 years (SD 1.8, range 20–26 years). The anthropometric characteristics are presented in Table 1, while the summary results of the FIFA/AIA official tests and other field-based fitness tests are reported in Table 2.

Table 1. Anthropometric characteristics of the participants (n = 30).

Characteristic	Mean	SD	Range
Height (cm)	180.36	6.33	168–191
Weight (kg)	73.67	7.38	59.2–89
Waist circumference (cm)	78.35	5.97	67–93
BMI (kg/m^2)	22.65	2.07	18.58–27.58

Table 2. Results of FIFA/AIA official tests and other field-based fitness tests (n = 30).

	Mean	SD	Range
	FIFA/AIA official tests:		
6 × 40-m sprint (s)	5.63	0.18	5.34–6.03
YO-YO IR1 (m)	1678.62	292.67	1320–2280
	Other field-based fitness tests:		
HGS, dominant hand (kg)	46.14	5.55	35.65–60
HGS, non-dominant hand (kg)	42.36	5.21	33.85–58.5
SLJ (cm)	225.88	17.52	201.5–265
SQJ (cm)	1093.03	85.59	926–1239
SaR (cm)	19.69	8.93	1.5–38
IA (s)	16.37	0.78	15.35–18.44

Table note: Yo-Yo IR1 = Yo-Yo Intermittent Recovery test, SLJ = standing long jump, SQJ = standing quintuple jump, SaR = sit-and-reach test, IA = Illinois Agility, HGS = hand-grip strength.

The anthropometric measures of our study participants did not show significant correlations with the results of the FIFA/AIA official or other field-based fitness tests. The correlation coefficients between the results of FIFA/AIA official and other field-based fitness tests are reported in Table 3.

Table 3. Correlation coefficients between the results of FIFA/AIA official tests and other field-based fitness tests.

Test	6 × 40-m Sprint	YO-YO IR1
HGS, dominant	−0.30 (0.112)	0.45 (0.014) *
HGS, non-dominant	−0.18 (0.351)	0.41 (0.031) *
SLJ	0.04 (0.850)	0.02 (0.917)
SQJ	0.06 (0.772)	0.16 (0.401)
SaR	−0.01 (0.982)	0.28 (0.137)
IA	0.63 (0.001) *	−0.29 (0.119)

Table note: Data are reported as correlation coefficient, r (significance level, p). * denotes statistically significant correlation ($p < 0.05$).

There was a strong positive correlation between the 6 × 40-m sprint and IA tests (r = 0.63, p = 0.001). A moderate positive correlation was observed between the Yo-Yo IR1 and HGS tests in the dominant (r = 0.45, p = 0.014) as well as non-dominant hand (r = 0.41, p = 0.031). Moreover, a moderate negative correlation was observed (r = −0.42, p = 0.025) between the two FIFA/AIA official tests, 6 × 40-m sprint and Yo-Yo IR1, indicating that the referees able to cover a longer distance in the Yo-Yo IR1 test were able to perform the 6 × 40-m test in a shorter time (Figure 2).

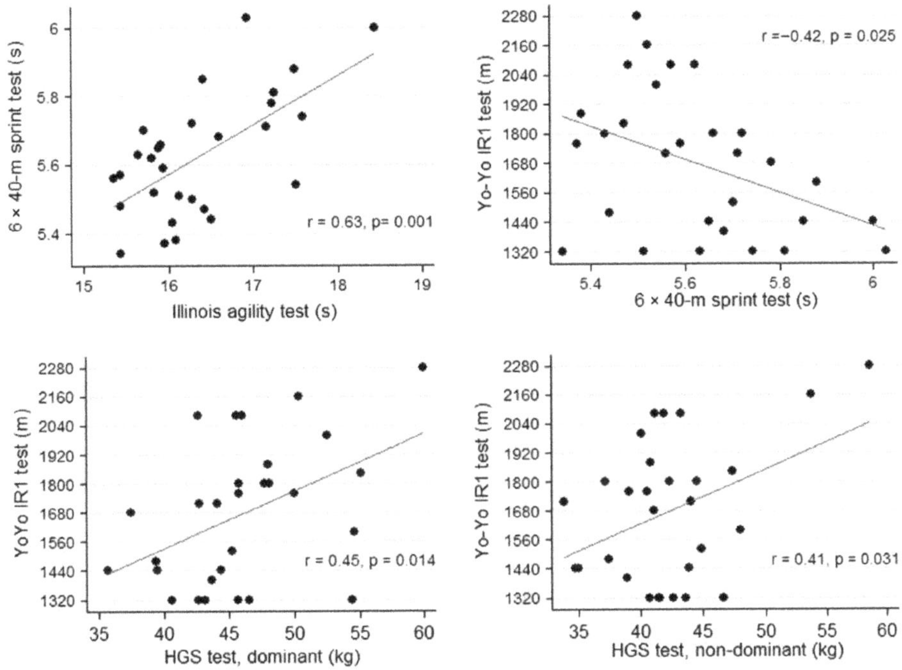

Figure 2. Scatter plots of the significant correlations between the results of FIFA/AIA official tests and other field-based fitness tests.

There was also a very strong correlation between the dominant and non-dominant sides in the HGS test (r = 0.72, $p < 0.001$). Finally, the results of the SLJ, SQJ, and SaR tests did not correlate with any of the FIFA/AIA official tests (r < 0.1).

4. Discussion

This study aimed to examine the correlation between two FIFA/AIA official tests used for the yearly evaluation of the fitness level of international and regional soccer referees and other field-based fitness tests, which could be potentially implemented to monitor referee fitness during the sportive season. We have demonstrated that the results of the 6 × 40-m sprint test correlated strongly with the results of the IA test, while the Yo-Yo IR1 test correlated with HGS in both the dominant and non-dominant hand. Interestingly, there was only a moderate correlation between the two FIFA/AIA-approved tests, namely the Yo-Yo IR1 and 6 × 40-m sprint tests.

FIFA recommends that two tests, 6 × 40-m sprint and high-intensity 150-m interval tests, be performed during official fitness screening of the soccer referees at least once a year. Two other FIFA-approved tests, dynamic Yo-Yo or Yo-Yo IR1, can be used optionally. The 6 × 40-m sprint test is relatively simple and evaluates the ability to perform repetitive sprints. The Yo-Yo IR1 is technically more complex and requires a specific learning phase from both referees and evaluators. Nevertheless, Yo-Yo IR1 is the most commonly used test for monitoring the ability to cope with intermittent exercise in team sports [9]. Performance in this test was related to high-intensity action during the match, maximum oxygen consumption (VO2 max), and heart rate response during high-intensity intermittent activity [16]. The same was not observed for the 150-m interval test [8]. Accordingly, AIA chose the 6 × 40-m sprint and Yo-Yo IR1 tests, whose results correlated highly with the referees' performance during matches, as a method to evaluate the level of fitness of its referees and allow them to officiate in a specific category.

Due to the large number of activities scheduled for the referees during the sportive year, the official tests are usually performed at the beginning of the season. Their implementation during the year is generally infeasible. First, they require a specific place and time allocation; second, these tests impose a high physical demand on the referees. From the above observations, it follows that other field-based fitness tests need to be applied if the assessment of referee fitness was to take place during the sportive season, in between the scheduled activities.

The field-based fitness tests evaluated in the present study are commonly used under different physiological [14,17,18] and pathological conditions [19,20] and their validity has been extensively proved in other studies [10,21,22]. Together, they are able to assess several aspects of fitness status, and they are often performed in sequence, a so-called fitness battery, to cover a wide range of fitness domains. For the scope of our study, however, we chose to analyze the correlation between the FIFA/AIA official tests and the individual alternative tests, rather than all of them as a fitness battery. Mainly, this was to avoid increasing the physical strain of the task, which, in line with the hypothesis of the study, should be performed during the sportive season, in between scheduled officiating activities. The main advantage of these tests is their simplicity, as they require no or little equipment, they are not time-consuming (taking less than 5 min to complete), and they impose low physical demand.

The results of our study indicate that several common field-based fitness tests correlate significantly with the FIFA/AIA official tests. In particular, a strong positive correlation was present between the 6 × 40-m sprint test and the IA test. Although this correlation could seem obvious, as both tests are speed-related and are reported in seconds, they do not assess the same fitness domain. In the 6 × 40-m sprint, the referee has to sprint straight ahead, while the IA test requires a more complex pattern of running, with several changes in direction and combined phases of acceleration and deceleration. For these reasons, the 6 × 40-m sprint test is commonly considered a speed test, while the IA test is universally interpreted as an agility test. Undoubtedly, the characteristics of the IA test are highly pertinent to the actual activity performed by the referees during the match. Therefore, the observed high correlation between the results of these two tests is particularly relevant for testing our hypothesis.

Another correlation observed in the study was that between the Yo-Yo IR1 and HGS tests, for both the dominant and non-dominant side. The "two-sidedness" of this correlation can be in large part explained by a high correlation of HGS between both sides in the same subject, even though the scores for the dominant hand were usually significantly higher than for the non-dominant hand. The correlations were positive, which means that the referees able to cover a longer distance in the Yo-Yo IR1 test are able to reach higher scores in the HGS test, or vice versa. Apparently, this correlation is less intuitive. Indeed, the Yo-Yo IR1 test is an interval test covering resistance and lower limb strength—that is, musculoskeletal fitness and cardiovascular fitness—while the HGS test merely assesses the strength of the distal upper limb. Nevertheless, previous studies highlighted the correlation between the HGS test and exercise capacity, whether in healthy subjects [23], elderly people [24], or patients with pathological conditions [25,26]. Interestingly, the results of the HGS test correlated with the 6-min walking test and incremental shuttle walking test, which was developed to assess the functional capacity of patients with chronic airway obstruction. The latter one is indeed similar to the Yo-Yo IR1 test in structure. Singh et al. [27] observed a significant correlation between the HGS and incremental shuttle walking tests and concluded that the HGS test is the main determinant of patient performance.

Our study is not without limitations. The inclusion criteria were restricted to a specific regional category of referees with similar anthropometric characteristics. All participants were males aged 20–26 years, and they officiated in the same category. Thus, they were engaged in a similar athletic activity, consisting of training sessions and matches, throughout the sportive year. Although these characteristics guarantee the homogeneity of the cohort, they may limit the external validity and the generalization of the results. Although the sample size was limited, it was calculated to be statistically valid. As for the study protocol, all tests were performed during the same day and repeated in the same sequence on different days. Counterbalancing measures were not considered in the study design. The warm-up and the recovery phases between the tests were included to limit the impact of muscular fatigue on physical performance. Thus, in the authors' view, a bias able to influence the results of the tests can be excluded.

5. Conclusions

In conclusion, the results of the present study show that a significant correlation exists between the 6 × 40-m sprint and IA tests and between the Yo-Yo IR1 and HGS tests. It is not our intention, however, to suggest that the IA and HGS tests should be substituted for the official FIFA/AIA tests. Instead, given the difference in the technical, physical, and physiological demands of these tests, they could be applied to the assessment of referee fitness in a complementary manner: 6 × 40-m sprint and Yo-Yo IR1 tests at the beginning of the sportive year for referee qualification and categorization, and IA and HGS tests during the sportive year for the systematic or even random evaluation of the referees' fitness status during the whole season. At the same time, the latter two could be used in studies aimed at monitoring the impact of training sessions and officiated matches on referee fitness during the sportive year.

Author Contributions: Conceptualization, V.R. and A.D.G.; Data curation, S.P.; Formal analysis, A.M.S., F.S.; Investigation, I.B., D.N.; Methodology, M.T.; Project administration, F.D.M.; Resources, V.R.; Supervision, S.M.; Validation, A.P.; Visualization, C.C.; Writing—original draft, V.R., A.D.G., S.P., A.M.S., F.S., I.B., M.T., S.M., A.P.; Writing—review and editing, D.N., F.D.M. and C.C. All authors have read and agreed to the published version of the manuscript.

Funding: This research received no external funding.

Institutional Review Board Statement: The study was conducted according to the guidelines of the Declaration of Helsinki. The project identification code is 154/17 (date 28 July 2017).

Informed Consent Statement: Informed consent was obtained from all subjects involved in the study.

Data Availability Statement: The datasets generated and analyzed during this study are available from the corresponding author upon reasonable request.

Acknowledgments: The authors wish to thank the Italian Association of Referees (AIA, Associazione Italiana Arbitri) and Marcello Nicchi, the Regional Committee of the AIA (CRA, Comitato Regionale Arbitri) in Lazio and Luca Palanca, all members of the section Roma 1 of this Committee, and the athletic referent for the Lazio region, Gianluca Perna, for their general administrative assistance. We also extend our thanks to all the referees who participated in the study. Their support of our study does not constitute an endorsement of its results or the content of this article.

Conflicts of Interest: The authors declare no conflict of interest.

References

1. Castagna, C.; Abt, G.; D'Ottavio, S. Physiological aspects of soccer refereeing performance and training. *Sports Med.* **2007**, *37*, 625–646. [CrossRef] [PubMed]
2. Bizzini, M.; Junge, A.; Bahr, R.; Helsen, W.; Dvorak, J. Injuries and musculoskeletal complaints in referees and assistant referees selected for the 2006 FIFA World Cup: Retrospective and prospective survey. *Br. J. Sports Med.* **2009**, *43*, 490–497. [CrossRef] [PubMed]
3. D'Ottavio, S.; Castagna, C. Physiological load imposed on elite soccer referees during actual match play. *J. Sports Med. Phys. Fit.* **2001**, *41*, 27–32.
4. Castagna, C.; Abt, G.; D'Ottavio, S. Relationship between fitness tests and match performance in elite soccer referees. *J. Strength Cond Res.* **2002**, *16*, 231–235. [PubMed]
5. Mallo, J.; Navarro, E.; Garcia-Aranda, J.M.; Gilis, B.; Helsen, W. Activity profile of top-class association football referees in relation to performance in selected physical tests. *J. Sports Sci.* **2007**, *25*, 805–813. [CrossRef]
6. 2018 Refereeing International Lists Zurich: Fédération Internationale de Football Association. 2018. Available online: https://resources.fifa.com (accessed on 20 February 2019).

7. Fitness Tests for Match Officials. Zurich: Fédération Internationale de Football Association. 2020. Available online: https://resources.fifa.com (accessed on 27 May 2021).
8. Weston, M.; Castagna, C.; Helsen, W.; Impellizzeri, F. Relationships among field-test measures and physical match performance in elite-standard soccer referees. *J. Sports Sci.* **2009**, *27*, 1177–1184. [CrossRef] [PubMed]
9. Krustrup, P.; Bangsbo, J. Physiological demands of top-class soccer refereeing in relation to physical capacity: Effect of intense intermittent exercise training. *J. Sports Sci.* **2001**, *19*, 881–891. [CrossRef]
10. Bangsbo, J.; Iaia, F.M.; Krustrup, P. The Yo-Yo intermittent recovery test: A useful tool for evaluation of physical performance in intermittent sports. *Sports Med.* **2008**, *38*, 37–51. [CrossRef]
11. Leyk, D.; Gorges, W.; Ridder, D.; Wunderlich, M.; Rüther, T.; Sievert, A.; Essfeld, D. Hand-grip strength of young men, women and highly trained. *Eur. J. Appl. Physiol.* **2006**, *99*, 415–421. [CrossRef]
12. Kale, M.; Aşçi, A.; Bayrak, C.; Açikada, C. Relationships among jumping performances and sprint parameters during maximum speed phase in sprinters. *J. Strength Cond Res.* **2009**, *23*, 2272–2279. [CrossRef]
13. Wells, K.F.; Dillon, E.K. The Sit and Reach—A Test of Back and Leg Flexibility. *Res. Q* **1952**, *23*, 115–118. [CrossRef]
14. Negra, Y.; Chaabene, H.; Amara, S.; Jari, S.; Hammami, M.; Hachana, Y. Evaluation of the Illinois Change of Direction Test in Youth Elite Soccer Players of Different Age. *J. Hum. Kinet.* **2017**, *58*, 215–224. [CrossRef] [PubMed]
15. Hulley, S.B.; Cummings, S.R.; Browner, W.S.; Grady, D.; Newman, T.B. *Designing Clinical Research: An Epidemiologic Approach*, 4th ed.; Lippincott Williams & Wilkins: Philadelphia, PA, USA, 2013; p. 79.
16. Ingebrigtsen, J.; Bendiksen, M.; Randers, M.B.; Castagna, C.; Krustrup, P.; Holtermann, A. Yo-Yo IR2 testing of elite and sub-elite soccer players: Performance, heart rate response and correlations to other interval tests. *J. Sports Sci.* **2012**, *30*, 1337–1345. [CrossRef] [PubMed]
17. Schmitz, B.; Klose, A.; Schelleckes, K.; Jekat, C.M.; Krüger, M.; Brand, S.M. Yo-Yo IR1 vs. incremental continuous running test for prediction of 3000-m performance. *J. Sports Med. Phys. Fit.* **2017**, *57*, 1391–1398.
18. Pizzigalli, L.; MichelettiCremasco, M.; La Torre, A.; Rainoldi, A.; Benis, R. Hand grip strength and anthropometric characteristics in Italian female national basketball teams. *J. Sports Med. Phys. Fit.* **2017**, *57*, 521–528.
19. Knudsen, A.W.; Naver, A.; Bisgaard, K.; Nordgaard-Lassen, I.; Becker, U.; Krag, A.; Slinde, F. Nutrition impact symptoms, handgrip strength and nutritional risk in hospitalized patients with gastroenterological and liver diseases. *Scand. J. Gastroenterol.* **2015**, *50*, 1191–1198. [CrossRef] [PubMed]
20. Lee, M.R.; Jung, S.M.; Bang, H.; Kim, H.S.; Kim, Y.B. Association between muscle strength and type 2 diabetes mellitus in adults in Korea: Data from the Korea national health and nutrition examination survey (KNHANES) VI. *Medicine* **2018**, *97*, e10984. [CrossRef]
21. Dodds, R.M.; Syddall, H.E.; Cooper, R.; Benzeval, M.; Deary, I.J.; Dennison, E.M.; Der, G.; Gale, C.R.; Inskip, H.M.; Jagger, C.; et al. Grip strength across the life course: Normative data from twelve British studies. *PLoS ONE* **2014**, *9*, e113637. [CrossRef]
22. Lockie, R.G.; Post, B.K.; Dawes, J.J. Physical Qualities Pertaining to Shorter and Longer Change-of-Direction Speed Test Performance in Men and Women. *Sports* **2019**, *7*, 45. [CrossRef]
23. Dourado, V.Z.; Vidotto, M.C.; Guerra, R.L. Reference equations for the performance of healthy adults on field walking tests. *J. Pneumol.* **2011**, *37*, 607–614.
24. Aragão-Santos, J.C.; De Resende-Neto, A.G.; Nogueira, A.C.; Feitosa-Neta, M.L.; Brandão, L.H.; Chaves, L.M.; Da Silva-Grigoletto, M.E. The effects of functional and traditional strength training on different strength parameters of elderly women: A randomized and controlled trial. *J. Sports Med. Phys. Fit.* **2019**, *59*, 380–386. [CrossRef] [PubMed]
25. Gosselink, R.; Troosters, T.; Decramer, M. Peripheral muscle weakness contributes to exercise limitation in COPD. *Am. J. Respir. Crit. Care Med.* **1996**, *153*, 976–980. [CrossRef] [PubMed]
26. Kaymaz, D.; Candemir, İ.Ç.; Ergün, P.; Demir, N.; Taşdemir, F.; Demir, P. Relation between upper-limb muscle strength with exercise capacity, quality of life and dyspnea in patients with severe chronic obstructive pulmonary disease. *Clin. Respir. J.* **2018**, *12*, 1257–1263. [CrossRef] [PubMed]
27. Singh, S.J.; Morgan, M.D.; Scott, S.; Walters, D.; Hardman, A.E. Development of a shuttle walking test of disability in patients with chronic airways obstruction. *Thorax* **1992**, *47*, 1019–1024. [CrossRef] [PubMed]

Journal of
Functional Morphology and Kinesiology

MDPI

Review

Applicability of Field Aerobic Fitness Tests in Soccer: Which One to Choose?

Daniel Bok [1,]* and Carl Foster [2]

1 Faculty of Kinesiology, University of Zagreb, 10000 Zagreb, Croatia
2 Department of Exercise and Sport Science, University of Wisconsin-La Crosse, La Crosse, WI 54601, USA; cfosteruwl@gmail.com
* Correspondence: daniel.bok@kif.hr

Abstract: A desire to make fitness testing cheaper and easier to conduct in a team-sport setting has led to the development of numerous field aerobic fitness tests. This has contributed to a growing confusion among strength and conditioning coaches about which one to use. The main aim of this narrative review was to examine the reliability, validity, sensitivity and usefulness of the commonly used field aerobic fitness tests and to provide practical guidelines for their use in soccer. The University of Montreal track test (UMTT) and Vam Eval test seem the best options for estimation of maximal oxygen uptake (VO_{2max}) while the highest signal-to-noise ratio of the 30-15 intermittent fitness test (30-15IFT) suggests its superior sensitivity to track changes in fitness. The UMTT and 30-15IFT are the best solutions for prescription of long and short high-intensity interval training sessions, respectively. All field tests mostly present with marginal usefulness, but the smallest worthwhile change for UMTT or Vam Eval test, Yo-YoIRT2 and 30-15IFT are smaller than their stage increment making the improvement of only one stage in the test performance already worthwhile. Strength and conditioning coaches are advised to choose the test based on their specific purpose of testing.

Keywords: 30-15 intermittent fitness test; Yo-Yo intermittent recovery test; 20-m shuttle run test; University of Montreal track test; Vam Eval test; maximal aerobic speed; exercise prescription; validity; reliability; sensitivity; usefulness

Citation: Bok, D.; Foster, C. Applicability of Field Aerobic Fitness Tests in Soccer: Which One to Choose? *J. Funct. Morphol. Kinesiol.* **2021**, *6*, 69. https://doi.org/10.3390/jfmk6030069

Academic Editor: Abdul Rashid Aziz

Received: 25 June 2021
Accepted: 16 August 2021
Published: 18 August 2021

1. Introduction

Fitness testing can be considered as a basic professional activity for sport scientists and strength and conditioning coaches [1,2]. It can be conducted for a variety of reasons including assessment of physiological capacities [1,3], talent identification and selection [4], training and performance monitoring [5], evaluation of training program effectiveness [1,6] and training prescription [7]. Due to these multipurpose requirements, one fitness test can hardly be used as an ideal tool able to provide useful information for all aspects of fitness testing. This has led to the development of numerous field aerobic fitness tests generally measuring the same generic fitness quality (i.e., maximal aerobic power), but with better or limited applicability for the specific purpose of testing. Sport-specific fitness tests, for example, appear to show greater ecological validity, but, on the other hand, have limited convergent validity. This makes them appropriate for specific-fitness assessment, but rather poor in assessment of basic fitness capacities (e.g., the maximal oxygen uptake (VO_{2max})) [8]. Similarly, some tests present with limited practical validity and can, therefore, hardly be used to accurately prescribe exercise which is probably the most important part of strength and conditioning coaches' job [3].

Although laboratory incremental exercise test is considered a "gold standard" for testing VO_{2max} [9], field aerobic fitness tests have emerged as time and resource-saving alternatives. During the last four decades, several field tests, including the University of Montreal track test (UMTT) [10], along with its modification the Vam Eval test [11], multistage 20-m shuttle run test (20mSRT) [12], Yo-yo intermittent recovery test level 1 and 2 (Yo-YoIRT1 and 2) [8] and 30-15 intermittent fitness test (30-15IFT) [13] have gained popularity and are widely used in practice for the assessment of aerobic fitness. These tests are different in nature as they include multistage continuous straight-line [10,11], shuttle [12] and intermittent shuttle [8,13] runs to exhaustion. Due to differences in execution, these field tests provide different end-test speeds which are specific to the nature of the effort made during the test. Specifically, introducing changes of direction every 20 m into straight-line running yields higher oxygen uptake (VO_2), heart rate, blood lactate concentration and Rating of Perceived Exertion responses [14,15] which leads to exhaustion at significantly lower end-test speeds during shuttle tests in comparison to incremental straight-line tests [16]. Similarly, omitting inter-effort recoveries while performing the 30-15IFT results in reaching exhaustion at significantly lower end-test speed [17]. For similar levels of aerobic fitness exhaustion will be reached at the lowest running speed in 20mSRT while UMTT or Vam Eval test, Yo-YoIRT and 30-15IFT will have their end-test speeds higher by approximately 2 km/h interval each. This would, for example, make the end-test speed of 30-15IFT approximately 6 km/h greater than the one reached in 20mSRT. Since velocity associated with VO_{2max} (vVO_{2max}) is the preferred method for prescribing exercise intensity for high-intensity interval training (HIIT) [18] this measure should be obtained through a field test which closely mimics the locomotor activity of a certain HIIT format in order to make it usable for prescription. However, not all mentioned field tests are specific to the HIIT formats, so their end-test speeds cannot easily be used for training prescription purposes.

Field tests are very popular as they are less time-consuming and cheaper than tests performed in the laboratory. However, the large number of available field tests has contributed to a growing confusion among coaches about which one to use. All mentioned field tests were nominally created for aerobic fitness assessment. This has led coaches to believe that the tests are basically the same and that the choice can be made solely on preference. However, each test had been developed with a specific purpose and, as such, should be used if and when it has the best metric characteristics for a certain aspect of testing. Therefore, the main aim of this paper is to review the available scientific literature for the purpose of reporting and discussing the reliability, validity, sensitivity and usefulness of the most commonly used field aerobic fitness tests. This will enable the formation of practical guidelines for their proper use in soccer (football).

This paper used a narrative review format. In order to retrieve relevant scientific papers, we searched the Web of Science and PubMed databases using standard search criteria. After accounting for the already retrieved publications, the keywords mentioned in the abstract were used to search for additional scientific papers. Reference lists of retrieved articles and recently published reviews were examined to find additional papers not identified by the keywords-based search. Only full-text articles published in English were included in the review. The searching process included articles retrieved until 1 March 2021.

2. Assessment of Maximal Oxygen Uptake

The general importance of aerobic fitness in soccer is well documented [19,20]. However, even though strength and conditioning coaches seek information about their player's VO_{2max}, recent studies show that matching running performance might not be affected by aerobic fitness capacities [21–23] as much as previously reported [24–27]. Namely, it appears that playing position and game tactics are more important factors in determining how much a player will run during a match than physical fitness [22]. This is further supported by the findings that improvements in aerobic fitness do not necessarily reflect in improvements in high-intensity match running performance in young soccer players [28]. However, aerobically fitter players experience reduced individual running demands during the game which is beneficial in terms of reducing the overall fatigue and injury risk [29], as well as the impairment of technical performance [30]. So, even though aerobic fitness might not be the primary limiting factor for match running performance in soccer players [21–23], since players cover 9 to 12 km in total distance, perform 150 to 350 high-intensity running activities, and execute 50 to 100 accelerations above 2.5 m/s^2 with 300 changes of directions during the match [31], adjusted aerobic conditioning should definitely be implemented in the overall training program. Well-developed aerobic fitness will enable players to perform their technical and tactical requirements with less physiological load [22] and to quickly recover between high-intensity efforts [32] which have been shown to be typical activities for team sports and especially soccer [27,33].

Probably the most important reason for fitness testing is the assessment of physical capacities and abilities [1,3]. Despite recent scientific evidence related to its association with soccer match performance, when it comes to aerobic or cardiorespiratory fitness, many strength and conditioning coaches focus on VO_{2max}. However, it appears that rather than VO_{2max} per se, match running performance is more related to vVO_{2max} [23] and peak incremental test speed [21]. Both variables represent an integrated measure of VO_{2max} and running economy, and can therefore be considered an athlete's peak aerobic locomotor ability [7]. This is further supported by the fact that repeated sprint ability (RSA) [34] shows much larger association with peak incremental test speed than VO_{2max} in team sport players [35]. Therefore, assessing peak aerobic locomotor ability seems to be much more important than VO_{2max} as it provides a more ecologically valid measure of aerobic fitness and can also be used for prescribing exercise.

Apart from being less important than end-test speed for assessment of aerobic fitness in team sports, another reason disputing the assessment of VO_{2max} through test equations is their low prediction accuracy due to the fact that the calculation presumes a standard running economy. While VO_{2max} can be reached in all field tests when executed to exhaustion [12,36,37], the calculation of VO_{2max} using performance data obtained through the test shows different levels of accuracy among the tests. Being the most similar to laboratory incremental exercise tests, the UMTT shows the largest correlation with VO_{2max} ($r = 0.96$) and the lowest standard error of estimate (SEE) of 2.8 mL/kg/min [10] and therefore seems the best option for estimation of VO_{2max}. As originally developed for the assessment of VO_{2max} in limited spaces such as gyms, the 20mSRT also has a high level of criterion-related validity with SEE of 3.5 mL/kg/min in adults [38] and 4.7 and 5.9 mL/kg/min in healthy adults and children, respectively [39]. This indicates a higher validity for adults than for children [40]. On the other hand, both Yo-YoIRTs and 30-15IFT present lower correlation coefficients and limits of agreement with VO_{2max} and, therefore, may not be ideal for estimation of VO_{2max} [8,13,41,42]. However, it seems that criterion-related validity of the tests might be fitness level dependent as higher correlations and lower SEE of Yo-YoIRT1 were reported for recreational [43] and untrained individuals [44].

3. Assessment of Specific Intermittent Endurance

Soccer and most other team sports are intermittent activities with high aerobic demands [20] placed on players due to frequent changes in types of movement [31] and repetition of high-intensity running and sprinting [45]. Even though high-intensity running can be maintained throughout the match [22], the decrease in occurrences of repeated sprint sequences and number of sprints within a sequence [46] suggests the accumulation of fatigue over the course of a match which may negatively impact players' overall physical and technical match performance [47]. As high-intensity running appears to be an important index of match-related physical performance [27], assessing player's ability to repeat such activities and to recover from them quickly seems important. This has led to the development of the Yo-YoIRTs devised with the main purpose of assessing soccer-specific intermittent endurance [8]. Indeed, significant correlations between Yo-YoIRT1 and high-intensity running during the match have been found in young [23,48–50] and senior level [24,26] soccer players. Large and very large significant correlations between Yo-YoIRT1 and high-intensity running (r ranging from 0.56 to 0.76) [8,23,24,26,48–50], very high-intensity running ($r = 0.59$) [50], sprinting ($r = 0.63$ and 0.76) [23,49], total distance covered (r ranging from 0.53 to 0.65) [24,26,48,50] and high-intensity activity (r ranging from 0.56 to 0.77) [23,26,48,49] performed during the match seem to support the ecological validity of the test [44,51]. The same applies for the Yo-YoIRT2 as a very large correlation (r = 0.72) was obtained between Yo-YoIRT2 and peak high-intensity running in a 5-min period during the match [8]. However, the significance of these correlations has lately been brought to question [52–54] as these analyses were performed on pooled data from all the players in a team. This resulted in neglecting the sometimes-substantial differences in physical fitness [21,55] as well as the often-substantial differences in match running performances [21,22,56] between players from different playing positions. Namely, even though significant correlations were found on pooled data, when analyzed according to playing position, the associations between aerobic fitness and match running performances were actually trivial and non-significant, with the only exception of strikers [21]. This suggests that tactical roles dictated by playing positions as well as other contextual factors such as score line, team formation and opponent quality rather than physical fitness are primarily important in determining player's match running performance [21,53]. Additionally, it is also interesting to notice that other field tests, such as UMTT and 20mSRT, which are not initially designed to assess specific intermittent aerobic endurance, also show large to very large correlations with high-intensity running [27,49], very high-intensity running [21,27] and high-intensity activity [49]. In fact, in young soccer players, significant correlations with high-intensity running (r = 0.70 vs. 0.65) and high-intensity activity (r = 0.75 vs. 0.73) were greater for 20mSRT than Yo-YoIRT1, raising doubt to the superiority of Yo-YoIRT1 in terms of ecological validity [49]. Generally, these findings point out that aerobic fitness is not a major limiting factor of match running performance [21,52] and that assessing specific intermittent endurance obviously does not provide an additional benefit in assessing player's physical fitness [3].

However, when choosing the test for the purpose of aerobic fitness assessment of soccer players, the choice of the test used should be based on the player's age, their aerobic fitness level and testing time-point. Namely, it has been shown that high-levels of VO_2 are reached early into the Yo-YoIRT1 and that almost half of the test duration is executed with VO_2 above 95% VO_{2max} [37]. This is quite different from the VO_2 response elicited during continuous tests in which VO_2 kinetics appears to be fairly linear. This means that YoYoIRTs are more metabolically demanding [8,38]. It seems that the difference between findings obtained in the continuous and intermittent tests increases as players improve their VO_{2max}, their anaerobic capacity and the ability to recover quickly following a high-intensity run so that their anaerobic capacity could be expended slowly during the execution of the test. Indeed, the anaerobic contribution to the intermittent test is higher than during a continuous test as blood lactate concentrations and end-test velocities are significantly higher after the intermittent in comparison to the continuous test [17]. This possible differentiation between the 20mSRT and Yo-YoIRT1 as players get fitter is further supported by the almost perfect correlation (r = 0.89) in very young soccer players [49], while slightly lower correlations were observed for adults and elite athletes [43]. Therefore, the literature suggests that 20mSRT should be used with younger and less aerobically fit players as the protocol involves lower starting speeds and smaller increments at the beginning of the test, while Yo-YoIRT1 is more useful for aerobically fitter players and during the in-season period when certain level of conditioning has already been reached [49]. The shorter testing time for Yo-YoIRT1 in young soccer players also makes it a better option for in-season period when time devoted to testing is limited. On the other hand, a very large significant correlation obtained between 30-15IFT and mean sprint time of the RSA test (r = 0.88) [57] and Yo-YoIRT1 (r = 0.75) [58] suggest that even 30-15IFT can be used for evaluation of specific intermittent endurance even though the test was not created for that particular reason [13,42].

4. Performance Monitoring and Assessment of the Training Effects

Very important features of any test are its ability to detect the smallest increase in performance that might be practically significant [59] and its sensitivity to detect training effects after a training program [2,60]. In order for the test to be highly sensitive to changes over time its smallest worthwhile change (SWC) should be greater than the typical error of measurement (TE) [61]. This ensures that the change in the variable really reflects fitness improvement rather than just a variation within the subjects tested. However, the TE alone is not the best indicator of the test sensitivity to training effects, but it is the magnitude (noise) in relation to the usually observed changes (signal) in that test that matters the most [61]. The greater the signal-to-noise ratio, the likely greater sensitivity of the test to detect changes in fitness [61,62].

The TEs are generally lower in 30-15IFT [63–65] and Vam Eval test [66], a modified version of the UMTT, than in 20mSRT [49,67,68] and both Yo-YoIRTs [26,43,48,49,68–77] (Tables 1–5). This is largely due to the fact that TE is dependent on the measurement unit [78] making the speed-related tests less variable than the distance-related ones and therefore rendering a direct comparison between tests inappropriate. The Vam Eval test [66,79] and the 30-15IFT [13,42] both have 0.5 km/h increments presenting much bigger stage steps in comparison to the 40-m shuttle increments which is a minimal detectable change in all distance-related tests. This bigger increment in single stage probably also contributes to a lower variability in test-retest measures. Additionally, the uneven time-dynamic of the speed increments throughout both Yo-YoIRTs and longer time exercised at a single speed stage, requiring maintenance of a very high physiological stress for a longer time which is influenced by motivation, very likely contributes to their higher TEs in comparison to other tests. Indeed, the TEs expressed as CV for Yo-YoIRT1, Yo-YoIRT2 and 20mSRT range between 3.5% and 17.3%, 4.2% and 12.7%, and 2.2% and 6.8%, respectively, while lower values of 3.5% and from 1.5% to 2.5% were found for Vam Eval and 30-15IFT, respectively. Although the number of studies reporting TEs of UMTT and 30-15IFT in soccer players is much lower than the ones reporting TEs for the 20mSRT and Yo-YoIRTs it does appear that the TEs of speed-related tests are more stable than those of distance-related tests. It also can be noticed that the TEs expressed as CVs. within the tests are generally lower in older and fitter players. For example, lower CVs. in older groups of soccer players were reported in most studies in which direct comparisons between age groups were made [70,71], while between-study analysis reveal that recreational [43] and sub-elite [69,70] players generally present higher CVs. compared to their elite counterparts [48,49]. Accordingly, lower TE of 0.23 km/h (CV = 1.3%) for UMTT was reported in moderately trained distance runners [79] in comparison to the TE of 0.57 km/h (CV of 3.5%) found in young soccer players [66] suggesting that greater experience with the mode of running in a test can also contribute to the lower trial-to-trial variability. This is important to acknowledge as both the ability to detect SWC and the sensitivity of the test is influenced by the TE. Therefore, if reliability of the test cannot be directly assessed in a particular group of players, the practitioners are advised to use the TEs from the literature which is derived from the subjects that most closely resemble their athletes.

The "signal" or the usually observed change following a training program in soccer players is also generally greater in the distance-related tests than in the speed-related tests. Namely, the mean change following training programs comprised of different HIIT formats [18] (i.e., long interval HIIT, short interval HIIT, repeated sprint training (RST), sprint interval training (SIT) and small-sided games (SSG)) lasting from 2 to 12 weeks in soccer players averages 2.7% for UMTT or Vam Eval test [80–85] (Table 1), 6.7% for 20mSRT [67,68,86–89] (Table 2), 18.8% for Yo-YoIRT1 [58,68,74,85,90–110] (Table 3), 16.5% for Yo-YoIRT2 [74,77,111–117] (Table 4) and 9.1% for 30-15IFT [58,84,118–121] (Table 5). Although solid conclusions cannot be made due to the differences in duration and experimental designs of the studies it is interesting to notice that training programs comprised of short interval HIIT offered the greatest improvements in UMTT, 20mSRT and 30-15IFT. Namely, performing a combination of short interval HIIT and RST within a 10-week training program yielded an 8.1% or 1.3 km/h increase in UMTT end-test speed [81] while 5 weeks of short interval HIIT program resulted in 20.5% or 365 m increase in 20mSRT total distance covered [87]. Similarly, the greatest improvement in 30-15IFT was also observed following training programs comprised of short interval HIIT as 5.8% (1.3 km/h) and 28.3% (3.6 km/h) increments in end-test speeds were noticed following a 6-week training program in male amateur [84] and semi-professional female soccer players [118], respectively. The fact that short interval HIIT produced the greatest "signal" in these tests is understandable given the fact that short and long interval HIIT have shown the greatest potential to improve VO_{2max} [18,122] and 20mSRT and UMTT both have very large (r = 0.96) [10] and almost perfect (r = 0.84) [40] correlations with VO_{2max}. The 30-15IFT, however, possess only large correlation with VO_{2max} (r = 0.68) [13], but its high specificity to the short interval HIIT sessions probably makes it suitable for capturing the "signal" from this type of interval training programs [42]. On the other hand, a wide range of improvements in Yo-YoIRT1 were noticed after each HIIT program. However, the greatest improvements were obtained after training programs comprised of long interval HIIT and small-sided games (SSG) [91] or a combination of both [58]. The ability of Yo-YoIRT1 to capture such "signals" is in accordance with the soccer-specific nature of the test [8]. Similarly, the greatest improvement in Yo-YoIRT2 was observed following 4 weeks of SIT [116] which is also logical considering the test's capacity to capture larger portion of anaerobic capacity in comparison to Yo-YoIRT1 [8]. Bearing in mind that above presented conclusions could simply be a result of differences in the experimental training interventions we believe that this information could be valuable to coaches when selecting tests for the training program evaluation purposes as tests appear to differentiate between "signals" produced by different training programs. So, it seems as a good practice to select tests based on the training program chosen for evaluation. For example, significant improvement of 17.1% in Yo-YoIRT1 was obtained following 7 weeks of SSG training program even though non-significant decrements of −0.7% in VO_{2max} and 20mSRT were noticed [68]. The SSG training program obviously produced some valuable improvements important for already aerobically well-prepared soccer players which would not be captured if only 20mSRT had been employed [68].

Even though the overview of the studies conducted on soccer players indicate that field tests may differ in their ability to "receive the signal" emitted from different training programs, it is the signal-to-noise ratio that really defines the sensitivity of the test [59,60]. The most accurate measure of sensitivity is the one calculated with the TE and the change in the measure following a training program assessed within the same subject sample, i.e., within the same study. Unfortunately, there are only few intervention studies in which reliability of the tests was assessed prior to the commencement of the training program [67,68,74]. Therefore, for the most studies reported in Tables 1–5, sensitivity was calculated using the TE from other study done on participants with the most similar characteristics. Additionally, the number of studies available for calculation of the signal-to-noise ratio as well as the type and duration of the training programs analyzed differ significantly among the tests and, therefore, direct between-test comparison should be made with caution. Average signal-to-noise ratios were 1, 2.9, 2.7, 2.5 and 5.1 for UMTT or Vam Eval test, 20mSRT, Yo-YoIRT1, Yo-YoIRT2 and 30-15IFT, respectively, suggesting that all tests can be considered sensitive to track adaptations to training. Lower sensitivity of the UMTT or Vam Eval test might partially be due to the fact that reliability measure used for calculation of the ratios was only reported in one study on young soccer players and it turned to be higher (CV = 3.5%) [66] than the one reported on older endurance trained athletes (CV = 1.3%) [79] with more experience in continuous running.

Table 1. Metric characteristics of the University of Montreal Track Test or Vam Eval test extracted from studies conducted on soccer players.

Study	Age and Gender of the Participants	Level of the Participants	Typical Error of Measurement Expressed as Coefficient of Variation	Typical Error of Measurement (Noise)	Smallest Worthwhile Change (0.2× between Subjects SD)	Usefulness of the Test	Training Type and Duration	Initial Level	Usually Observed Change (Signal) Following a Training Program	Usually Observed Change (Signal) Following a Training Program	Signal-to-Noise Ratio
Buchheit et al. (2013) [66] [b]	14.5 ± 1.5 M	Elite	3.5%	0.57 km/h	0.22 km/h	Marginal		≈16.2 km/h			
Los Arcos et al. (2015) [80] [a]	15.5 ± 0.6 M	National, elite			0.18 km/h	Marginal	HIIT l (6 w)	16.8 km/h	1.7%	0.3 km/h	0.5
Dupont et al. (2004) [81] [a]	20.2 ± 0.7 M	National, elite, professional			0.16 km/h	Marginal	HIIT s + RST(10 w)	16.1 km/h	8.1%	1.3 km/h	2.3
Faude et al. (2014) [82] [c]	16.5 ± 0.8 M	High-level, professional conditions			0.2 km/h	Marginal	HIIT s (4 w)	17.8 km/h	−2.8%	−0.5 km/h	−0.8
Faude et al. (2013) [83] [c]	15.9 ± 0.8 M	High-level, professional conditions			0.21 km/h	Marginal	HIIT s (5.5 w)	17.1 km/h	1.5%	0.25 km/h	0.4
Dellal et al. (2012) [84] [b]	26.3 ± 4.7 M	Amateur			n/a	/	HIIT s (6 w)	≈15.8 km/h	6.6%	≈1 km/h	1.9
Wong et al. (2010) [85] [b]	24.6 ± 1.5 M	Elite, professional			0.04 km/h	Marginal	HIIT s (8 w)	15.9 km/h	3.1%	0.5 km/h	0.9
Faude et al. (2014) [82] [c]	15.9 ± 0.8 M	High-level, professional conditions			0.2 km/h	Marginal	SSG (4 w)	17.5 km/h	1.7%	0.3 km/h	0.5
Los Arcos et al. (2015) [80] [a]	15.5 ± 0.6 M	National, elite			0.16 km/h	Marginal	SSG (6 w)	17.0 km/h	−0.6%	−0.1 km/h	−0.2
Dellal et al. (2012) [84] [b]	26.3 ± 4.7 M	Amateur			n/a	/	SSG (6 w)	≈16.1 km/h	5.1%	≈0.8 km/h	1.5

Legend: [a]: University of Montreal Track Test was used in the study, [b]: Vam Eval test was used in the study, [c]: the type of maximal incremental exercise test used in the study was not clearly defined, M: male, HIIT: high-intensity interval training, HIIT s: short format HIIT, HIIT l: long format HIIT, RST: repeated sprint training, SSG: small-sided games, w: weeks. The study from which the TE was taken for calculation of the signal-to-noise ratio is indicated in the brackets.

Table 2. Metric characteristics of the multistage 20-m shuttle run test extracted from studies conducted on soccer players.

Study	Age and Gender of the Participants	Level of the Participants	Typical Error of Measurement Expressed as Coefficient of Variation	Typical Error of Measurement (Noise)	Smallest Worthwhile Change (0.2× between Subjects SD)	Usefulness of the Test	Training Type and Duration	Initial Level	Usually Observed Change (Signal) Following a Training Program	Usually Observed Change (Signal) Following a Training Program	Signal-to-Noise Ratio
Aziz et al. (2005) [67]	27.2 ± 3.3 M	Elite, national team	2.2%	46 m	36 m	Marginal	n/a (5 w)	2.280 m	7.9%	180 m	3.6
Castagna et al. (2010) [49]	14.4 ± 0.1 M	Elite	3.6%	59.5 m	73.4 m	Good		1653 m			
Slettaløkken and Rønnestad (2014) [86]	18–26 M	Semi-professional			49.6 m		HIIT l (6 w)	≈2.433 m	−6.4%	−155 m	−2.9 [67]
Hill-Hass et al. (2009) [68]	14.6 ± 0.9 M	Elite	4.9%	40 m	26.2 m	Marginal	HIIT s + RST (7 w)	2.258 m	3.1%	69 m	0.6
Sanchez-Sanchez et al. (2019) [87]	22.5 ± 2.2 M	Amateur			86.2 m	Good	HIIT s (5 w)	1.770 m	20.5%	362 m	5.7 [49]
Tønnessen et al. (2011) [88]	16.4 ± 0.9 M	Elite			57.6 m	Ok	RST (10 w)	2.360 m	5.7%	144 m	2.6 [67]
Shalfawi et al. (2013) [89]	19.4 ± 4.4 F	Elite			61.6 m	Ok	RST + AgT (10 w)	1.780 m	16.8 %	264 m	4.7 [49]
Hill-Hass et al. (2009) [68]	14.6 ± 0.9 M	Elite	4.9%	40 m	48 m	Ok	SSG (7 w)	2.222 m	−0.7%	−16 m	−0.1

Legend: F: female, M: male, HIIT: high-intensity interval training, HIIT s: short format HIIT, HIIT l: long format HIIT, RST: repeated sprint training, SSG: small-sided games, w: weeks, AgT: agility training. The study from which the TE was taken for calculation of the signal-to-noise ratio is indicated in the brackets.

Table 3. Metric characteristics of the Yo-Yo intermittent recovery test level 1 extracted from studies conducted on soccer players.

Study	Age and Gender of the Participants	Level of the Participants	Typical Error of Measurement Expressed as Coefficient of Variation	Typical Error of Measurement (Noise)	Smallest Worthwhile Change (0.2× between Subjects SD)	Usefulness of the Test	Training Type and Duration	Initial Level	Usually Observed Change (Signal) Following a Training Program	Usually Observed Change (Signal) Following a Training Program	Signal-to-Noise Ratio
Deprez et al. (2014) [69]	12.5 ± 0.6 14.0 ± 0.5 16.2 ± 0.6 M	Sub-elite and Non-elite	U13: 17.3% U15: 16.7% U17: 7.9%	U13: 154 m U15: 171 m U17: 123 m	U13: 70.8 mU15: 88.8 mU17: 95.6 m	Marginal Marginal Marginal		U13: 890 m U15: 1.022 m U17: 1.556 m			
Deprez et al. (2015) [70]	13.9 ± 0.5 16.2 ± 0.6 18.1 ± 0.4 M	High-level	U15: 6.8% U17: 4.3% U19: 4.1%	U15: 137 m U17: 101 m U19: 107 m	U15: 94 m U17: 69.4 m U19: 67.4 m	Marginal Marginal Marginal		U15: 2.024 m U17: 2.404 m U19: 2.547 m			
Castagna et al. (2019) [71]	11.1 ± 0.9 M	2 years' experience	5.1%	51.7 m	90.4 m	Good		1.013 m			
Krustrup et al. (2003) [26]	28 M	Elite	4.9%	91.5 m	14.4 m	Marginal		1.867 m			
Póvoas et al. (2016) [72]	9.7 ± 0.7 F	Regional level competition	10.1%	71.2 m	63.2 m	Ok		705 m			
Póvoas et al. (2016) [73]	9.7 ± 0.7 M	Regional level competition	11.1%	121.9 m	134.4 m	Ok		1.098 m			
Thomas et al. (2006) [43]	24.4 ± 6.0 M	Recreational level	8.7%	107 m	97.6 m	Ok		1.030 m			
Castagna et al. (2010) [49]	14.4 ± 0.1 M	Elite	3.8%	28.9 m	56.6 m	Good		760 m			
Castagna et al. (2009) [48]	14.1 ± 0.2 M	Elite	3.5%	29.5 m	70.4 m	Good		842 m			
Impellizzeri et al. (2008) [90]	17.8 ± 0.6 M	High level			n/a	/	HIIT I (4 w)	≈1.890 m	12%	n/a	1.6 [74]
Özcan et al. (2018) [91]	18.5 ± 1.5 M	Amateur, regional level			71.9 m	Marginal	HIIT I (6 w)	1.057.7 m	89.1%	769 m	10.2 [43]
Ferrari Bravo et al. (2008) [92]	21.1 ± 5.1 M	Professional and amateur			65.8 m	Marginal	HIIT I (7 w)	1.846 m	12.5%	231 m	1.7 [74]
Fanchini et al. (2014) [74]	17 ± 1 M	Professional, 4th national division	7.3%	140 m	66.9 m	Marginal	HIIT I + RST + SSG(11 w)	1.911 m	14.5%	277 m	1.9
Buchheit and Rabbani (2014) [58]	15.4 ± 0.5 M	National level			51.4 m	Marginal	HIIT Ii + SSG (8 w)	1.031 m	35%	360.9 m	4.8 [74]

Table 3. *Cont.*

Study	Age and Gender of the Participants	Level of the Participants	Typical Error of Measurement Expressed as Coefficient of Variation	Typical Error of Measurement (Noise)	Smallest Worthwhile Change (0.2× between Subjects SD)	Usefulness of the Test	Training Type and Duration	Initial Level	Usually Observed Change (Signal) Following a Training Program	Usually Observed Change (Signal) Following a Training Program	Signal-to-Noise Ratio
Arslan et al. (2020) [93]	14.2 ± 0.5 M	Regional level			15 m	Marginal	HIIT s (5 w)	1.240 m	16.4%	244 m	2.2 [74]
Wong et al. (2010) [85]	24.6 ± 1.5 M	Elite, professional			15 m	Marginal	HIIT s (8 w)	1510 m	19.7%	298 m	2.7 [74]
Ouerghi et al. (2014) [94]	22.9 ± 1.7 M	Amateur players, 3rd national division			n/a		HIIT s (12 w)	≈1.440 m	≈70%	1.6 km/h≈1024 m	8 [43]
Hill-Hass et al. (2009) [68]	14.6 ± 0.9 M	Elite	9%	116 m	51.2 m	Marginal	HIIT s + RST (7 w)	1.764 m	21.9%	387 m	2.4
Taylor et al. (2016) [95]	24.1 ± 4.1 M	Semi-professional			54.8 m	Marginal	RST Sl (2 w)	1.830 m	24%	439 m	3.3 [74]
Taylor et al. (2016) [95]	24.1 ± 4.1 M	Semi-professional			120 m	Ok	RST COD (2 w)	1.691 m	31%	524 m	4.2 [74]
Beato et al. (2019) [96]	21 ± 2.4 M	Amateur			73 m	Marginal	RST Sl (2 w)	1.642 m	11%	180 m	1.5 [74]
Beato et al. (2019) [96]	21 ± 2.4 M	Amateur			71.8 m	Marginal	RST COD (2 w)	1.686 m	7.4%	124 m	1 [74]
Soares-Caldeira et al. (2014) [97]	21.4 ± 5.5 M	Professional futsal, regional level			72.6 m	Marginal	RST (4 w)	1.280 m	31.2%	373 m	4.3 [74]
Kavaliauskas et al. (2017) [98]	22 ± 8 M	Semi-professional			81.8 m	Marginal	RST uphill 7% (6 w)	1.468 m	11.9%	175 m	1.6 [74]
Eniseler et al. (2017) [99]	16.9 ± 1.1 M	Elite, national level			50.4 m	Marginal	RST (6 w)	2.306.6 m	7.5%	173.4 m	1 [74]
Ferrari Bravo et al. (2008) [92]	21.1 ± 5.1 M	Professional and amateur			87.8 m	Marginal	RST (7 w)	1.917 m	28.1%	538 m	3.8 [74]
Nedrehagen and Saeterbakken (2015) [100]	19.9 ± 2.5 F 22.0 ± 2.7 M	Semi-professional female and amateur male			37.6 m	Marginal	RST (8 w)	1.455 m	15.3%	222 m	2.1 [74]
Shalfawi et al. (2013) [101]	21.2 ± 2.6 F	Elite			58.6 m	Marginal	RST Sl (8 w)	920 m	27.5%	253 m	3.8 [74]
Shalfawi et al. (2013) [101]	21.2 ± 2.6 F	Elite			54.8 m	Marginal	RST COD (8 w)	1.025 m	9.3%	95 m	1.3 [74]
Beato et al. (2019) [102]	18–21 M	Elite			44.6 m	Marginal	RST Sl (8 w)	2.472 m	5.3%	132 m	0.7 [74]

Table 3. *Cont.*

Study	Age and Gender of the Participants	Level of the Participants	Typical Error of Measurement Expressed as Coefficient of Variation	Typical Error of Measurement (Noise)	Smallest Worthwhile Change (0.2× between Subjects SD)	Usefulness of the Test	Training Type and Duration	Initial Level	Usually Observed Change (Signal) Following a Training Program	Usually Observed Change (Signal) Following a Training Program	Signal-to-Noise Ratio
Beato et al. (2019) [102]	18–21 M	Elite			49.2 m	Marginal	RST COD (8 w)	2.500 m	7.8%	196 m	1.1 [74]
Sanchez-Sanchez et al. (2019) [103]	14.4 ± 0.5 M	Regional level			65.9 m	Marginal	RST COD(8 w)	914 m	8.1%	71 m	1.1 [74]
Sanchez-Sanchez et al. (2019) [103]	14.7 ± 0.5 M	Regional level			66.7 m	Marginal	RST COD (8 w)	1.764 m	2%	34 m	0.3 [74]
Campos-Vazquez et al. (2015) [104]	18.1 ± 0.8 M	Top-level national			60.4 m	Marginal	RST + ST (8 w)	2.297 m	3.5%	80 m	0.5 [74]
Haugen et al. (2014) [105]	17 ± 1 F & M	High-school level			133.8 m	Ok	RST (9 w)	1.583 m	17.4%	275 m	2.4 [74]
Nyberg et al. (2016) [106]	23.5 ± 4.0 M	Semi-professional, 2nd national league			66 m	Marginal	RST (9 w)	1.803 m	11.6%	324 m	1.6 [74]
Hostrup et al. (2019) [107]	24.9 ± 5.4 M	Sub-elite, 2nd amateur league			111.4 m	Marginal	RST (10 w)	1.910 m	1.6%	30 m	0.2 [74]
Macpherson and Weston (2015) [108]	25 ± 4 M	Semi-professional			98.6 m	Marginal	SIT (2 w)	1.523 m	18.1%	275 m	2.5 [74]
Howard & Stavrianeas (2017) [109]	15.1 ± 0.8 M	High-shool level			61.5 m	Marginal	SIT (10 w)	741.6 m	44%	326 m	6 [74]
Arslan et al. (2020) [93]	14.2 ± 0.5 M	Regional level			30.4 m	Marginal	SSG (5 w)	1.284 m	12.8%	188 m	1.8 [74]
Eniseler et al. (2017) [99]	16.9 ± 1.1 M	Elite, national level			77.6 m	Marginal	SSG (6 w)	2.320 m	4.8%	112 m	0.7 [74]
Özcan et al. (2018) [91]	18.4 ± 1.5 M	Amateur, regional level			73.6 m	Marginal	SSG (6 w)	1.235.5 m	63.1%	711 m	7.2 [43]
Hill-Hass et al. (2009) [68]	14.6 ± 0.9 M	Elite	9%	116 m	69 m	Marginal	SSG (7 w)	1.488 m	17.1%	254 m	1.9
Dello Iacono et al. (2019) [110]	18.6 ± 0.6 M	International level			27.6 m	Marginal	SSG (8 w)	1.646 m	20,9%	344 m	2.9 [74]

Legend: F: female, M: male, HIIT: high-intensity interval training, HIIT s: short format HIIT, HIIT l: long format HIIT, SIT: sprint interval training, RST: repeated sprint training, SSG: small-sided games, w: weeks, COD: change of direction, Sl: straight-line, ST: strength training. The study from which the TE was taken for calculation of the signal-to-noise ratio is indicated in the brackets.

Table 4. Metric characteristics of the Yo-Yo intermittent recovery test level 2 extracted from studies conducted on soccer players.

Study	Age and Gender of the Participants	Level of the Participants	Typical Error of Measurement Expressed as Coefficient of Variation	Typical Error of Measurement (Noise)	Smallest Worthwhile Change (0.2× between Subjects SD)	Usefulness of the Test	Training Type and Duration	Initial Level	Usually Observed Change (Signal) Following a Training Program	Usually Observed Change (Signal) Following a Training Program	Signal-to-Noise Ratio
Enright et al. (2018) [75]	18.3 ± 0.2 M	Elite	4.2%	34 m	31,2 m	Ok		920 m			
da Silva et al. (2011) [76]	14 ± 0.8 M	Regional level	11%	49 m	13.6 m	Marginal		445.5 m			
Thomas et al. (2006) [43]	24.4 ± 6.0 M	Recreational level	12.7%	41 m	22 m	Marginal		325 m			
Krustrup et al. (2006) [77]	22–30 17–35 M	Healthy and elite	9.6%	65.5 m	9.2 m	Marginal	Socc T (8 w)	730 m	42%	n/a	4.4
Fanchini et al. (2014) [74]	17 ± 1 M	Professional, 4th national division	7.1%	53.5 m	33.2 m	Marginal	HIIT l + RST + SSG (11 w)	718 m	8.8%	71 m	1.2
Iaia et al. (2017) [111]	17.0 ± 1.0 M	Sub-elite			33.8 m	Ok	RST sh. rest (5 w)	1.000 m	11.4%	111 m	2.7 [75]
Iaia et al. (2017) [111]	17.0 ± 1.0 M	Sub-elite			43.4 m	Good	RST lo. rest (5 w)	1.016 m	6.5%	56 m	1.5 [75]
Sagelv et al. (2019) [112]	16–19 M	High-level national			n/a	/	RST (22 w)	≈890 m	9.1%	/	2.2 [75]
Christensen et al. (2011) [113]	23.4 ± 3.5 M	Elite, 3rd national level			11.2 m	Marginal	SIT + SSG (2 w)	937 m	6.1%	57 m	1.5 [75]
Thomassen et al. (2010) [114]	23.4 ± 0.8 M	Elite			11.2 m	Marginal	SIT + SSG (2 w)	937 m	6.1%	57 m	1.5 [75]
Iaia et al. (2015) [115]	18.5 ± 1 M	Professional, national level			37 m	Ok	SIT (2′ rest) (3 w)	927 m	10.1%	93 m	2.4 [75]
Iaia et al. (2015) [115]	18.5 ± 1 M	Professional, national level			45.2 m	Good	SIT (40″ rest) (3 w)	989 m	3.8%	37 m	0.9 [75]
Mohr and Krustrup (2016) [116]	19 ± 1 M	Sub-elite, university level			13.6 m	Marginal	SIT (4 w)	680 m	49.7%	298 m	7 [74]
Ingebrigtsen et al. (2013) [117]	16.9 ± 0.6 M	Elite			26.6 m	Marginal	SIT (6 w)	559 m	11.3%	63 m	1.6 [74]
Mohr and Krustrup (2016) [116]	19 ± 1 M	Sub-elite, university level			10.4 m	Marginal	SSG (4 w)	693 m	25.8%	165 m	3.6 [74]

Legend: F: female, M: male, HIIT: high-intensity interval training, HIIT s: short format HIIT, HIIT l: long format HIIT, SIT: sprint interval training, RST: repeated sprint training, SSG: small-sided games, Socc T: regular soccer training, w: weeks. The study from which the TE was taken for calculation of the signal-to-noise ratio is indicated in the brackets.

Table 5. Metric characteristics of the 30-15 intermittent fitness test extracted from studies conducted on soccer players.

Study	Age and Gender of the Participants	Level of the Participants	Typical Error of Measurement Expressed as Coefficient of Variation	Typical Error of Measurement (Noise)	Smallest Worthwhile Change (0.2× between Subjects SD)	Usefulness of the Test	Training Type and Duration	Initial Level	Usually Observed Change (Signal) Following a Training Program	Usually Observed Change (Signal) Following a Training Program	Signal-to-Noise Ratio
Čović et al. (2016) [63]	22.8 ± 4.3 F	Elite	1.8%	0.31 km/h	0.2 km/h	Marginal		17.1 km/h			
Thomas et al. (2016) [64]	25.5 ± 4.3 M	Semi-professional	2.5%	1.0 km/h	0.7 km/h	Marginal		n/a			
Valladares-Rodríguez et al. (2017) [65]	24.4 ± 5.6 M 23.3 ± 4.5 F	Professional futsal players	M: 1.5% M F: 1.5% F	M: 0.32 km/h F: 0.21 km/h	M: 0.34 km/hF: 0.26 km/h	Ok Ok		M: 20.2 km/h F: 17.4 km/h			
Buchheit and Rabbani (2014) [58]	15.4 ± 0.5 M	National level			0.22 km/h	Marginal	HIIT l + SSG (8 w)	17.4 km/h	7%	1.2 km/h	4.7 [65]
Dellal et al. (2012) [84]	26.3 ± 4.7 M	Amateur			n/a	/	HIIT s (6 w)	≈19.4 km/h	5.8%	≈1.3 km/	3.9 [65]
Arazi et al. (2017) [118]	23.4 ± 1.3 F	Semi-professional, regional level			0.7 km/h	Marginal	HIIT s (6 w)	12.7 km/h	28.3%	3.6 km/h	11.3 [64]
Paul et al. (2019) [119]	16.2 ± 0.8 M	National level			0.22 km/h	Marginal	HIIT s + SSG (4 w)	17 km/h	8.2%	1.4 km/h	5.5 [65]
Rabbani et al. (2019) [120]	24.1 ± 3.723.2 ± 2.2 M	Semi-professional, 2nd national level			0.22 km/h 0.24 km/h	Marginal	HIIT s + SSG (4 w)	19.5 km/h19.2 km/h	6.9% & 6.2%	1.3 & 1.2 km/h	4.6 [65] 4.1 [65]
Dellal et al. (2012) [84]	26.3 ± 4.7 M	Amateur			n/a	/	SSG (6 w)	≈19.5 km/h	5.1%	≈1 km/h	3.4 [65]
Campos-Vazquez et al. (2017) [121]	27.7 ± 4.3 M	Professional, 2nd national level			0.16 km/h	Marginal	Socc T + M (4 w)	20.1 km/h	5%	1 km/h	3.3 [65]

Legend: F: female, M: male, HIIT: high-intensity interval training, HIIT s: short format HIIT, HIIT l: long format HIIT, SSG: small-sided games, Socc T: regular soccer training, M: match, w: weeks. The study from which the TE was taken for calculation of the signal-to-noise ratio is indicated in the brackets.

It is possible that more experienced soccer players might exhibit better reliability of this test and consequently make it more sensitive to detect long-term adaptations. Namely, it does seem that higher reliability of the 30-15IFT is the main reason for its greater sensitivity as percentage changes after training interventions captured with this test [58,84,119–121] are generally quite similar to the ones captured with UMTT or Vam Eval test [81,84]. This is especially evident in Dellal et al. [84] in which both tests were used for aerobic fitness assessment. However, it is very important to emphasize once again that sensitivity of the test to detect adaptations is mostly influenced by the specificity of the training type imposed on players. For example, even though TE expressed as CV was lower in 20mSRT (4.9%) vs. Yo-YoIRT1 (9%), making the denominator much lower for calculation of signal-to-noise ratio, the adaptations to both the SSG training program and the training program incorporating prolonged short-interval HIIT and RST were better captured with Yo-YoIRT1 [68]. In this case, it appears that both training programs applied to the players were of such volume and activity patterns that it was more similar or specific to the performance of the Yo-YoIRT1 making the players better conditioned to perform on that particular test. Indeed, acute physiological responses to prolonged short-interval HIIT and extensive RST sessions are much more similar to the physiological demand of Yo-YoIRT1 in which half of the test duration is performed with the intensity of ~95% VO_{2peak} [37], therefore creating a better "signal" for that test. Similarly, 11 weeks of long-interval HIIT, SSG, sprint training and technical and tactical drills was better captured with Yo-YoIRT1 than Yo-YoIRT2 in young soccer players, again, mostly due to the large differences in the "signal" [74]. It seems that the overall training program was more focused on the improvement of aerobic capacities and less on anaerobic capacities creating the difference in the adaptation which resulted in different percentage increases in these two tests used. On the other hand, long-interval HIIT, SSG and resistance training intervention lasting for 7 weeks was similarly detected by Yo-YoIRT1 and 30-15IFT [58]. While both tests are intermittent in nature and quite similar in terms of specificity, they presented identical sensitivity probably because the overall training program was comprised of activities that evenly attacked the capacities evaluated by both tests. However, even though the number of studies reporting sensitivity presented in Tables 3 and 5 is significantly different between Yo-YoIRT1 and 30-15IFT, and while direct comparisons should not be made due to the differences in the training programs evaluated, it does seem that the 30-15IFT presents superior overall sensitivity.

Another very important characteristic of the test is its usefulness or the ability to detect the SWC in a measure. Ideally, the TE should be less than half of the SWC and in that case any change in the test greater than the SWC would almost certainly be meaningful [60]. However, the test is rated as *good* whenever the SWC is greater than the TE and *Ok* or *medium* when the TE is equal to the SWC [123]. The TE larger than the SWC makes the test *marginal,* but even with *marginal* test we are still able to detect moderate, large and very large changes in a measure [60,123]. Namely, the changes of 1×, 3×, 6× and 10 × SWC can be considered as small, moderate, large and very large [59]. Determining the magnitude of the SWC is very complex and depends on many factors such as training context, type of adaptations that are being evaluated and the variable itself [60]. For performance variables in team sports the SWC is most often determined as 0.2× between-athletes standard deviation [59,60]. However, using between-athletes SD for calculation makes the SWC susceptible to influence by group homogeneity, i.e., more heterogeneous groups will exhibit larger SWC and may present the test as more useful. Usually, younger and less fit groups of players show more heterogeneity although this is not the general rule as different levels of heterogeneity was found between experimental groups of the same subject sample [95]. On the other hand, great homogeneity was also found in players with significantly different initial fitness status [103]. Anyway, it does appear that all field tests reviewed here present with *marginal* usefulness while the tests were rated as *Ok* and *good* mostly in studies with younger [48,49,68,71–73,88,89] and less fit [43,87,95] or less experienced [71,105] players with a few exceptions noticeable for Yo-YoIRT2 [75,111,115] and 30-15IFT [65]. It is advisable, therefore, that strength and conditioning coaches compare the initial fitness scores of their players with the ones presented in Tables 1–5 and, by choosing the most appropriate TE, estimate the potential usefulness of a particular test for their players. It is also worth noting that the reported SWCs in UMTT or Vam Eval test, Yo-YoIRT2 and 30-15IFT are most often smaller than the test's stage increment making the improvement of only one stage in the test performance already substantial and worthwhile. Indeed, both the UMTT/Vam Eval test and 30-15IFT use 0.5 km/h increments while minimal detectable increment in Yo-YoIRT2 is 40 m making the usually observed SWCs of 0.1–0.2 km/h, 0.2–0.3 km/h and 10–45 m, respectively, almost exclusively outperformed by improvement of just one stage. On the other hand, the usually reported SWCs of 30–70 m and 40 to 135 m for 20mSRT and Yo-YoIRT1, respectively, are much larger than their stage increments of 20 and 40 m rendering one stage increment in the test performance often insufficient for practical significance.

Although VO_{2max} is not related to soccer match performance and should not be a variable of particular interest for strength and conditioning coaches they are still very often interested to find out if the VO_{2max} has increased after a training period. However, if one is still really interested in assessing VO_{2max} and its improvement following a training program, the most logical option would be to use the field test that has the greatest criterion-related validity. The UMTT or the Vam Eval test offer the greatest correlation (r = 0.96) between end-test speed and VO_{2max} and the lowest SEE of 2.81 mL/kg/min amongst the proposed field tests thus appearing as the best candidate for the job [10]. On the other hand, the 20mSRT, Yo-YoIRT1, Yo-YoIRT2 and 30-15IFT all have lower criterion-related validity with correlation coefficients of 0.84 [40], 0.74 [124], 0.47 [124] and 0.68 [13], respectively, which points out their very low capacity to accurately estimate VO_{2max}, especially in top-level athletes. As usually observed changes in VO_{2max} following several weeks of HIIT in soccer players are in range of 5 to 11% or 3 to 6 mL/kg/min [20,122,125,126], even the UMTT or the Vam Eval test with their highest criterion-related validity among the field tests and high reliability (TE of 1.92 mL/kg/min) [10] are not accurate and sensitive enough to capture such small changes induced by a training intervention. Namely, adding the SEE of 2.81 mL/kg/min to the TE of 1.92 mL/kg/min, which is actually the third of the upper range value of the VO_{2max} improvement or the "signal", increases the overall "noise" of the test and renders the test invalid to provide reliable data. Therefore, it is advised that field tests are not used for calculation of VO_{2max} and especially for evaluation of the training effects through the lens of VO_{2max} improvements.

5. Training Prescription

Training prescription is probably the most vital part of strength and conditioning as it involves manipulation of numerous acute training variables in order to reach the desired physiological response [18,127]. For aerobic exercise in particular, prescribing exercise intensity is the key issue, and it becomes especially challenging when prescribing long and short format HIIT [128,129]. These training formats are recognized as optimal for accumulating the most time in the zone >90%VO_{2max} per session and their acute physiological reactions are believed to be the most important for improvement of VO_{2max} [18]. Long format HIIT includes high-intensity intervals lasting from 2 to 6 min performed at 90–105% vVO_{2max}, while short format HIIT includes 10 to 60-s intervals performed at 100–120% vVO_{2max} [18,127]. As this work in the zone > 90%VO_{2max} is performed in the severe intensity domain, usually above the respiratory compensation point, prescribing exercise intensity through percentage of heart rate or VO_{2max}, as is often done for low and moderate-intensity aerobic exercises, is not possible. During long format HIIT the time lag of heart rate response is sometimes as long as the bout itself, so relying on heart rate to control intensity would result in performing very inefficient sessions. The problem with heart rate is even more critical during the short format HIIT as several intervals are needed to reach the desired heart rate zone. This is why vVO_{2max}, or the lowest speed required to elicit VO_{2max}, has emerged as the preferred method for prescribing HIIT [18]. Prescribing exercise intensity using speed for which the cardiovascular response is determined through incremental exercise testing enables better control and more precision in execution of the session.

However, because of the different locomotor nature of the protocol, VO_{2max} is attained at different velocities in each of the discussed field tests. Congruently, end-test velocities represent different physiological qualities, i.e., they are composed of different ratios of aerobic and anaerobic capacities and different degrees of neuromuscular strain which is influenced by the specific nature of the task. Therefore, only those field tests that closely mimic the locomotor activity of the specific HIIT session pose the ability to be used for training prescription. The end-test velocities can hardly be used interchangeably to prescribe HIIT sessions.

The vVO_{2max} required for continuous running is usually obtained through incremental exercise test and, although it differs slightly from the end-test velocity, those two measures are highly correlated [10]. Therefore, this end-test velocity can be used to prescribe long format HIIT because the mode of testing is very similar to the mode of the training session [18]. Namely, the incremental exercise test is performed continuously and in straight-line so no other physiological capacity except for cardiorespiratory fitness and the energetic cost of such mode of running contribute to the task [9]. Performance of the work intervals in long format HIIT sessions rely on the same capacities, so the physiological response during the training sessions will be similar as during testing. As UMTT and Vam Eval test are continuous straight-line field tests their end-test velocities are suitable to prescribe long format HIIT sessions. These tests can also be used to prescribe short format HIIT if the session is performed on the treadmill where by jumping on and off the treadmill accelerations and decelerations can easily be omitted. However, most short format HITT sessions are performed indoors with limited space available requiring introduction of numerous changes-of-directions (COD) and corresponding accelerations and decelerations. These additional actions augment the physiological response of such mode of running [14,16] and in order to accurately individualize such sessions they need to be prescribed based on the test which closely mimics such locomotor activities. Namely, using vVO_{2max} assessed through UMTT to prescribe short interval HIIT sessions performed indoors on a court usually results in very different physiological responses between players [13,42]. This is due to their differences in anaerobic capacities and neuromuscular qualities which are required to change direction and to accelerate and decelerate throughout the session [42]. This has led to the development of 30-15IFT, an intermittent shuttle test, in which the final test speed incorporates aerobic and anaerobic capacities, COD ability and the inter-effort recovery ability in the amount which is required for performance of the short format HIIT [13]. That way all athletes elicit similar physiological reactions while performing short format HIIT at identical relative intensity [13,42]. Namely, the test is highly specific to the training sessions usually performed in intermittent sports, but not to the sports [13,42], which is why it is ideal for training prescription of such sessions.

On the other hand, 20mSRT and Yo-YoIRTs are not specific to either of the HIIT formats and can hardly be used for training prescription. The 20mSRT is a continuous shuttle incremental test, so its final speed does not incorporate inter-effort recoveries making it unsuitable for prescribing short format HIIT. Namely, the variability of the cardiorespiratory response to 10-min intermittent runs was much higher when training prescription was based on the 20mSRT (10.6%) in comparison to the 30-15IFT (2.9%) rendering some subjects unable to finish the session and others below the desired heart rate zone [13]. Similarly, greater anaerobic contribution [16] and poorer running economy [14] during relative shuttle compared to the straight-line running limits the potential of the 20mSRT to be used for prescription of long format HIIT. Namely, the difference between end-test speeds in 20mSRT and UMTT can inform the coach about the COD ability of their athletes, with smaller the difference the better the COD ability [42]. However, as this difference between end-test speeds can be highly variable [130], the cardiorespiratory responses of straight-line running prescribed through the results obtained with the 20mSRT could also appear highly variable [15]. Therefore, it would be very hard to capture the ideal acute cardiorespiratory response during long format HIIT if the 20mSRT is used for training prescription.

As indicated earlier the Yo-YoIRTs are soccer-specific tests with the main purpose of evaluating player's ability to perform intense exercise [8]. Although the tests are very similar to the short format HIIT [131], the protocol design limits their potential to be used for training prescription purposes. At the beginning of the test the speed increments are rather steep and vary in volume while the latter stages have smaller increments which are distance-regulated. Namely, for each speed stage latter in the test the athlete is required to cover the 320-m distance which subsequently shortens time spent at each stage. Additionally, each of the eight shuttles are interspersed with 10-s recovery, making it hard to assess how an athlete would actually cope with the requirement of maintaining the corresponding speed for longer as is necessary during HIIT sessions. Making the test distance-focused is contrary to the concept of HIIT prescription as training sessions are usually time-defined [18]. Therefore, when it comes to training prescription the real question which needs to be answered through testing is whether an athlete can withstand a certain speed level for the duration of an average high-intensity interval. Being a time-defined test the 30-15IFT is, therefore, the best choice to prescribe short format HIIT [13,42].

6. Conclusions

At the beginning aerobic field tests were developed as a cheaper and easier means for the assessment of athletes' fitness compared to laboratory tests. However, different needs have emerged through time and have guided their development. These main goals shaped and determinate the test's main purposes and characteristics. As safer and pacing-free alternative to the 12-min run or the Cooper test, the UMTT was born in 1980 with the main purpose of measuring VO_{2max}. The necessity to assess VO_{2max} indoors with limited space available led to the development of 20mSRT which appeared to be valid and reliable alternative to the UMTT. As strength and conditioning coaches working in soccer were interested in evaluating sport-specific intermittent aerobic ability their requirement resulted in the appearance of the Yo-YoIRTs in the early 1990s. Finally, the inability of all these tests to prescribe short format HIIT, often performed with numerous COD as organized indoors, laid the ground for the birth of 30-15IFT in 2008. All these field tests have their strengths and weaknesses and should be used accordingly, i.e., a test should be selected when it fits best for the particular purpose of the testing (Table 6).

Namely, when it comes to VO_{2max} assessment, the UMTT and Vam Eval test appeared as the best solution even though it must be pointed out that VO_{2max} is not related to soccer match performance and its assessment should not be a priority in soccer players. Additionally, tracking VO_{2max} improvement through time using field tests is not very feasible due to the small magnitude of potential VO_{2max} improvements and an inadequate reliability and criterion-related validity of the tests to give them the necessary sensitivity. The findings presented herein suggest that Yo-YoIRTs are the most often used tests in soccer players. However, the findings obtained with these two, or any other field test for that matter, should not be used to predict on-field match performance as this practice seems to be misleading. The comparison of the signal-to-noise ratios suggested that 30-15IFT is the most sensitive test to track adaptations to training programs. However, this conclusion should be taken with caution as the number of studies reviewed and their methodology differ significantly between the tests reported. Anyway, strength and conditioning coaches are advised to choose the test based on the training program they are about to implement as it appears that the tests differ in their capacities to detect "signals" emitted. While all field tests present with *marginal* usefulness, the usually reported SWCs for UMTT/Vam Eval test, Yo-YoIRT2 and 30-15IFT were smaller than their stage increment making the improvement of only one stage in the test performance already worthwhile. Finally, when it comes to training prescription, UMTT and 30-15IFT should be preferably used for programing long and short HIIT, respectively.

Table 6. Advantages and disadvantages of the reviewed field aerobic fitness tests.

Field Aerobic Fitness Test	Advantages	Disadvantages
UMTT/Vam Eval	Moderate to high reliability High criterion-related validity—best solution for the assessment of VO_{2max} SWC smaller than one stage of the test Best for prescription of long format HIIT	Low to moderate sensitivity Marginal usefulness Athletic track required for testing
20mSRT	Short-distance course required for testing Low end-test running speeds Short testing time High sensitivity Ok to good usefulness	Low to moderate reliability Moderate criterion-related validity for the assessment of VO_{2max} SWC larger than one stage of the test Unsuitable for training prescription
Yo-YoIRT1	Short-distance course required for testing High sensitivity	Low reliability Low criterion-related validity for the assessment of VO_{2max} Marginal usefulness SWC larger than one stage of the testUnsuitable for training prescription
Yo-YoIRT2	Short-distance course required for testing High sensitivity Very short testing time Medium usefulness SWC smaller than one stage of the test Appropriate for players with high aerobic and anaerobic fitness	Low reliability Very low criterion-related validity for the assessment of VO_{2max} Not appropriate for players with low aerobic fitness Unsuitable for training prescription
30-15IFT	Medium-size-distance course required for testing High reliability Excellent sensitivity Medium usefulness SWC smaller than one test stage Best for prescription of short format HIIT	Low criterion-related validity for the assessment of VO_{2max}

Legend: UMTT: University of Montreal Track test, 20mSRT: 20-metre shuttle run test, YoYoIRT1: YoYo intermittent recovery test level 1, YoYoIRT2: YoYo intermittent recovery test level 2, 30-15IFT: 30-15 intermittent fitness test, SWC: smallest worthwhile change, HIIT: high-intensity training.

Author Contributions: Conceptualization, D.B. and C.F.; writing—original draft preparation, D.B.; writing—review and editing, D.B. and C.F. Both authors have read and agreed to the published version of the manuscript.

Funding: This research received no external funding.

Institutional Review Board Statement: Ethical review and approval were not required for this article due to the type of the contribution.

Informed Consent Statement: Not applicable.

Conflicts of Interest: The authors declare no conflict of interest.

References

1. Svensson, M.; Drust, B. Testing soccer players. *J. Sports Sci.* **2005**, *23*, 601–618. [CrossRef]
2. Paul, D.J.; Nassis, G.P. Physical fitness testing in youth soccer: Issues and considerations regarding reliability, validity, and sensitivity. *Pediatr. Exerc. Sci.* **2015**, *27*, 301–313. [CrossRef]
3. Mendez-Villanueva, A.; Buchheit, M. Football-specific fitness testing: Adding value or confirming the evidence? *J. Sports Sci.* **2013**, *31*, 1503–1508. [CrossRef]
4. Carling, C.; Collins, D. Comment on "Football-specific fitness testing: Adding value or confirming the evidence?". *J. Sports Sci.* **2014**, *32*, 1206–1208. [CrossRef]
5. McGuigan, M. *Monitoring Training and Performance in Athletes*; Human Kinetics: Champaign, IL, USA, 2017; pp. 1–12.
6. McGuigan, M. Principles of test selection and administration. In *Essentials of Strength Training and Conditioning*, 4th ed.; Haff, G.G., Triplett, N.T., Eds.; Human Kinetics: Champaign, IL, USA, 2016; pp. 249–258.
7. Buchheit, M.; Laursen, P. Traditional methods of HIIT programming. In *Science and Application of High-Intensity Interval Training: Solutions to the Programming Puzzle*; Laursen, P., Buchheit, M., Eds.; Human Kinetics: Champaign, IL, USA, 2019; pp. 17–31.

8. Bangsbo, J.; Iaia, F.M.; Krustrup, P. The Yo-yo intermittent recovery test: A useful tool for evaluation of physical performance in intermittent sports. *Sports Med.* **2008**, *38*, 37–51. [CrossRef]
9. Bentely, D.J.; Newell, J.; Bishop, D. Incremental exercise test design and analysis: Implications for performance diagnostics in endurance athletes. *Sports Med.* **2007**, *37*, 575–586. [CrossRef] [PubMed]
10. Léger, L.A.; Boucher, R. An indirect continuous running multistage field test: The Université de Montréal track test. *Can. J. Appl. Sport Sci.* **1980**, *5*, 77–84.
11. Mendez-Villanueva, A.; Buchheit, M.; Kuitunen, S.; Poon, T.K.; Simpson, B.; Peltola, E. Is the relationship between sprinting and maximal aerobic speeds in young soccer players affected by maturation? *Ped. Exerc. Sci.* **2010**, *22*, 497–510. [CrossRef]
12. Léger, L.A.; Lambert, J. A maximal multistage 20 m shuttle run test to predict VO_{2max}. *Eur. J. Appl. Physiol. Occup. Physiol.* **1982**, *49*, 1–12. [CrossRef]
13. Buchheit, M. The 30-15 intermittent fitness test: Accuracy for individualizing interval training of young intermittent sport players. *J. Strength Cond. Res.* **2008**, *22*, 365–374. [CrossRef]
14. Buchheit, M.; Haydar, B.; Hader, K.; Ufland, P.; Ahmaidi, S. Assessing running economy during field running with changes of direction: Application to 20 m shuttle runs. *Int. J. Sports Physiol. Perform.* **2011**, *6*, 380–395. [CrossRef]
15. Bekraoui, N.; Boussaidi, L.; Cazorla, G.; Léger, L. Oxygen uptake, heart rate, and lactate responses for continuous forward running and stop-and-go running with and without directional changes. *J. Strength Cond. Res.* **2020**, *34*, 699–703. [CrossRef]
16. Ahmaidi, S.; Collomp, K.; Préfaut, C. The effect of shuttle test protocol and the resulting lactacidaemia on maximal velocity and maximal oxygen uptake during the shuttle exercise test. *Eur. J. Appl. Physiol. Occup. Physiol.* **1992**, *65*, 475–479. [CrossRef]
17. Haydar, B.; Al Haddad, H.; Ahmaidi, S.; Buchheit, M. Assessing inter-effort recovery and change of direction ability with the 30-15 intermittent fitness test. *J. Sports Sci. Med.* **2011**, *10*, 346–354.
18. Buchheit, M.; Laursen, P.B. High-intensity interval training, solutions to the programming puzzle. Part 1: Cardiopulmonary emphasis. *Sports Med.* **2013**, *43*, 313–338. [CrossRef]
19. Hoff, J.; Helgerud, J. Endurance and strength training for soccer players: Physiological considerations. *Sports Med.* **2004**, *34*, 165–180. [CrossRef]
20. Stølen, T.; Chamari, K.; Castagna, C.; Wisløff, U. Physiology of soccer: An update. *Sports Med.* **2005**, *35*, 501–536. [CrossRef]
21. Buchheit, M.; Mendez-Villanueva, A.; Simpson, B.M.; Bourdon, P.C. Match running performance and fitness in youth soccer. *Int. J. Sports Med.* **2010**, *31*, 818–825. [CrossRef]
22. Mendez-Villanueva, A.; Buchheit, M.; Simpson, B.; Bourdon, P.C. Match play intensity distribution in youth soccer. *Int. J. Sports Med.* **2013**, *34*, 101–110. [CrossRef]
23. Rebelo, A.; Brito, J.; Seabra, A.; Oliveira, J.; Krustrup, P. Physical match performance of youth football players in relation to physical capacity. *Eur. J. Sport Sci.* **2014**, *14*, S148–S156. [CrossRef]
24. Krustrup, P.; Mohr, M.; Ellingsgaard, H.; Bangsbo, J. Physical demands during an elite female soccer game: Importance of training status. *Med. Sci. Sports Exerc.* **2005**, *37*, 1242–1248. [CrossRef]
25. Impellizzeri, F.M.; Marcora, S.M.; Castagna, C.; Reilly, T.; Sassi, A.; Iaia, F.M.; Rampinini, E. Physiological and performance effects of generic versus specific aerobic training in soccer players. *Int. J. Sports Med.* **2006**, *27*, 483–492. [CrossRef]
26. Krustrup, P.; Mohr, M.; Amstrup, T.; Rysgaard, T.; Johansen, J.; Steensberg, A.; Pedersen, P.K.; Bangsbo, J. The Yo-Yo intermittent recovery test: Physiological response, reliability, and validity. *Med. Sci. Sports Exerc.* **2003**, *35*, 697–705. [CrossRef]
27. Rampinini, E.; Bishop, D.; Marcora, S.M.; Ferrari Bravo, D.; Sassi, R.; Impellizzeri, F.M. Validity of simple field tests as indicators of match-related physical performance in top-level professional soccer players. *Int. J. Sports Med.* **2007**, *28*, 228–235. [CrossRef] [PubMed]
28. Buchheit, M.; Simpson, B.M.; Mendez-Villanueva, A. Repeated high-speed activities during youth soccer games in relation to changes in maximal sprinting and aerobic speeds. *Int. J. Sports Med.* **2013**, *34*, 40–48. [CrossRef] [PubMed]
29. Carling, C.; Gall, F.L.; Reilly, T.P. Effects of physical efforts on injury in elite soccer. *Int. J. Sports Med.* **2010**, *31*, 180–185. [CrossRef]
30. Rampinini, E.; Impellizzeri, F.M.; Castagna, C.; Azzalin, A.; Ferrari Bravo, D.; Wisløff, U. Effect of match-related fatigue on short-passing ability in young soccer players. *Med. Sci. Sports Exerc.* **2008**, *40*, 934–942. [CrossRef]
31. Taylor, J.B.; Wright, A.A.; Dischiavi, S.L.; Townsend, M.A.; Marmon, A.R. Activity demands during multi-directional team sports: A systematic review. *Sports Med.* **2017**, *7*, 2533–2551. [CrossRef]
32. Ufland, P.; Ahmaidi, S.; Buchheit, M. Repeated-sprint performance, locomotor profile and muscle oxygen uptake recovery: Effect of training background. *Int. J. Sports Med.* **2013**, *34*, 924–930. [CrossRef]
33. Spencer, M.; Bishop, D.; Dawson, B.; Goodman, C. Physiological and metabolic responses of repeated-sprint activities: Specific to field-based team sports. *Sports Med.* **2005**, *35*, 1025–1044. [CrossRef]
34. Girard, O.; Mendez-Villanueva, A.; Bishop, D. Repeated-sprint ability. Part 1: Factors contributing to fatigue. *Sports Med.* **2011**, *41*, 673–694. [CrossRef]
35. Buchheit, M. Repeated-sprint performance in team sport players: Associations with measures of aerobic fitness, metabolic control and locomotor function. *Int. J. Sports Med.* **2012**, *33*, 230–239. [CrossRef] [PubMed]
36. Buchheit, M.; Al Haddad, H.; Millet, G.P.; Lepretre, P.M.; Newton, M.; Ahmaidi, S. Cardiorespiratory and cardiac autonomic responses to 30-15 intermittent fitness test in team sport players. *J. Strength Cond. Res.* **2009**, *23*, 93–100. [CrossRef]
37. Schmitz, B.; Pfeifer, C.; Thorwesten, L.; Krüger, M.; Klose, A.; Brand, S.M. Yo-Yo intermittent recovery level 1 test for estimation of peak oxygen uptake: Use without restriction? *Res. Q. Exerc. Sport.* **2020**, *91*, 478–487. [CrossRef]

38. Ramsbottom, R.; Brewer, J.; Williams, C. A progressive shuttle run test to estimate maximal oxygen uptake. *Br. J. Sports Med.* **1988**, *22*, 141–144. [CrossRef] [PubMed]
39. Léger, L.A.; Mercier, D.; Gadoury, C.; Lambert, J. The multistage 20 metre shuttle run test for aerobic fitness. *J. Sports Sci.* **1988**, *6*, 93–101. [CrossRef]
40. Mayorga-Vega, D.; Aguilar-Soto, P.; Viciana, J. Criterion-related validity of the 20 m shuttle run test for estimating cardiorespiratory fitness: A meta-analysis. *J. Sports Sci. Med.* **2015**, *14*, 536–547.
41. Martínez-Lagunas, V.; Hartmann, U. Validity of the Yo-Yo intermittent recovery test level 1 for direct measurement or indirect estimation of maximal oxygen uptake in female soccer players. *Int. J. Sports Physiol. Perform.* **2014**, *9*, 825–831. [CrossRef]
42. Buchheit, M. The 30-15 intermittent fitness test: 10 year review. *Myorobie J.* **2010**, *1*, 278.
43. Thomas, A.; Dawson, B.; Goodman, C. The Yo-Yo test: Reliability and association with a 20 m shuttle run and VO_{2max}. *Int. J. Sports Physiol. Perform.* **2006**, *1*, 137–149. [CrossRef]
44. Castagna, C.; Krustrup, P.; Póvoas, S. Yo-Yo intermittent tests are a valid tool for aerobic fitness assessment in recreational football. *Eur. J. Appl. Physiol.* **2019**, *120*, 137–147. [CrossRef]
45. Glaister, M. Multiple sprint work: Physiological responses, mechanisms of fatigue and the influence of aerobic fitness. *Sports Med.* **2005**, *35*, 757–777. [CrossRef] [PubMed]
46. Buchheit, M.; Mendez-Villanueva, A.; Simpson, B.M.; Bourdon, P.C. Repeated-sprint sequences during youth soccer matches. *Int. J. Sports Med.* **2010**, *31*, 709–716. [CrossRef]
47. Rampinini, E.; Impellizzeri, F.M.; Castagna, C.; Coutts, A.J.; Wisløff, U. Technical performance during soccer matches of the Italian serie A league: Effect of fatigue and competitive level. *J. Sci. Med. Sport* **2009**, *12*, 227–233. [CrossRef] [PubMed]
48. Castagna, C.; Impellizzeri, F.; Cecchini, E.; Rampinini, E.; Alvarez, J.C.B. Effects of intermittent-endurance fitness on match performance in young male soccer players. *J. Strength Cond. Res.* **2009**, *23*, 1954–1959. [CrossRef]
49. Castagna, C.; Manzi, V.; Impellizzeri, F.; Weston, M.; Alvarez, J.C.B. Relationship between endurance field tests and match performance in young soccer players. *J. Strength Cond. Res.* **2010**, *24*, 3227–3233. [CrossRef]
50. Doncaster, G.; Marwood, S.; Iga, J.; Unnithan, V. Influence of oxygen uptake kinetics on physical performance in youth soccer. *Eur. J. Appl. Physiol.* **2016**, *116*, 1781–1794. [CrossRef]
51. Schmitz, B.; Pfeifer, C.; Kreitz, K.; Borowski, M.; Faldum, A.; Brand, S.M. The Yo-Yo intermittent tests: A systematic review and structured compendium of test results. *Front. Physiol.* **2018**, *9*, 870. [CrossRef]
52. Mendez-Villanueva, A.; Buchheit, M. Physical capacity-match physical performance relationships in soccer: Simply, more complex. *Eur. J. Appl. Physiol.* **2011**, *111*, 2387–2389. [CrossRef]
53. Carling, C. Interpreting physical performance in professional soccer match-play: Should we be more pragmatic in our approach? *Sports Med.* **2013**, *43*, 655–663. [CrossRef]
54. Buchheit, M.; Simpson, B.M. Player-tracking technology: Half-full or half-empty glass? *Int. J. Sports Physiol. Perform.* **2017**, *12*, S2–S35. [CrossRef] [PubMed]
55. Slimani, M.; Nikolaidis, P.T. Anthropometric and physiological characteristics of male soccer players according to their competitive level, playing position and age group: A systematic review. *J. Sports Med. Phys. Fit.* **2019**, *59*, 141–163. [CrossRef] [PubMed]
56. Bradley, P.S.; Sheldon, W.; Wooster, B.; Olsen, P.; Boanas, P.; Krustrup, P. High-intensity running in English FA Premier League soccer matches. *J. Sports Sci.* **2009**, *27*, 159–168. [CrossRef] [PubMed]
57. Buchheit, M. 30-15 intermittent fitness test and repeated sprint ability. *Sci. Sports* **2008**, *23*, 26–28. [CrossRef]
58. Buchheit, M.; Rabbani, A. The 30-15 intermittent fitness test versus the Yo-Yo intermittent recovery test level 1: And sensitivity to training. *Int. J. Sports Physiol. Perform.* **2014**, *9*, 522–524. [CrossRef]
59. Buchheit, M. The numbers will love you back in return—I promise. *Int. J. Sports Physiol. Perform.* **2016**, *11*, 551–554. [CrossRef]
60. Buchheit, M. Monitoring training status with HR measures: Do all roads lead to Rome? *Front. Physiol.* **2014**, *5*, 73. [CrossRef]
61. Lacome, M.; Simpson, B.M.; Buchheit, M. Monitoring training status with player-tracking technology: Still on the road to Rome. Part 1: Traditional practices and new concepts. *Aspetar Sports Med. J.* **2018**, *7*, 54–63.
62. Buchheit, M.; Laursen, P.; Stanley, J.; Plews, D.; Al Haddad, H.; Lacome, M.; Simpson, B.; Saw, A. Response to load. In *Science and Application of High-Intensity Interval Training: Solutions to the Programming Puzzle*; Laursen, P., Buchheit, M., Eds.; Human Kinetics: Champaign, IL, USA, 2019; pp. 179–212.
63. Čović, N.; Jelešković, E.; Alić, H.; Rađo, I.; Kafedžić, E.; Sporiš, G.; McMaster, D.T.; Milanović, Z. Reliability, validity and usefulness of 30-15 intermittent fitness test in female soccer players. *Front. Physiol.* **2016**, *7*, 510. [CrossRef]
64. Thomas, C.; Dos'Santos, T.; Jones, P.A.; Comfort, P. Reliability of the 30–15 intermittent fitness test in semiprofessional soccer players. *Int. J. Sports. Physiol. Perform.* **2016**, *11*, 172–175. [CrossRef]
65. Valladares-Rodríguez, S.; Rey, E.; Mecías-Calvo, M.; Barcala-Furelos, R.; Bores-Cerezal, A.J. Reliability and usefulness of the 30-15 intermittent fitness test in male and female professional futsal players. *J. Hum. Kinet.* **2017**, *60*, 191–198. [CrossRef]
66. Buchheit, M.; Mendez-Villanueva, A. Reliability and stability of anthropometric and performance measures in highly-trained young soccer players: Effect of age and maturation. *J. Sports Sci.* **2013**, *31*, 1332–1343. [CrossRef]
67. Aziz, A.R.; Tan, H.Y.F.; Chuan, T.K. The 20 m multistage shuttle run test: Reliability, sensitivity and its performance correlates in trained soccer players. *Asian J. Exerc. Sports Sci.* **2005**, *2*, 1–7.
68. Hill-Haas, S.V.; Coutts, A.J.; Rowsell, G.J.; Dawson, B.T. Generic versus small-sided game training in soccer. *Int. J. Sports Med.* **2009**, *30*, 636–642. [CrossRef] [PubMed]

69. Deprez, D.; Coutts, A.J.; Lenoir, M.; Fransen, J.; Pion, J.; Philippaerts, R.; Vaeyens, R. Reliability and validity of the Yo-Yo intermittent recovery test level 1 in young soccer players. *J. Sports Sci.* **2014**, *32*, 903–910. [CrossRef]
70. Deprez, D.; Fransen, J.; Lenoir, M.; Philippaerts, R.M.; Vaeyens, R. The Yo-Yo intermittent recovery test level 1 is reliable in young high-level soccer players. *Biol. Sport* **2015**, *32*, 65–70. [CrossRef]
71. Castagna, C.; Krustrup, P.; D'Ottavio, S.; Pollastro, C.; Bernardini, A.; Araújo Póvoas, S.C. Ecological validity and reliability of an age-adapted endurance field test in young male soccer players. *J. Strength Cond. Res.* **2019**, *33*, 3400–3405. [CrossRef] [PubMed]
72. Póvoas, S.C.A.; Castagna, C.; Soares, J.M.C.; Silva, P.; Coelho-e-Silva, M.; Matos, F.; Krustrup, P. Reliability and construct validity of Yo-Yo tests in untrained and soccer-trained schoolgirls aged 9–16. *Pediatr. Exerc. Sci.* **2016**, *28*, 321–330. [CrossRef]
73. Póvoas, S.C.A.; Castagna, C.; Soares, J.M.C.; Silva, P.M.R.; Lopes, M.V.M.F.; Krustrup, P. Reliability and validity of Yo-Yo tests in 9 to 16 year-old football players and matched non-sports active schoolboys. *Eur. J. Sports Sci.* **2016**, *16*, 755–763. [CrossRef]
74. Fanchini, M.; Castagna, C.; Coutts, A.J.; Schena, F.; McCall, A.; Impellizzeri, F.M. Are Yo-Yo intermittent recovery test level 1 and 2 both useful? Reliability, responsiveness and interchangeability in young soccer players. *J. Sports Sci.* **2014**, *32*, 1950–1957. [CrossRef]
75. Enright, K.; Morton, J.; Iga, J.; Lothian, D.; Roberts, S.; Drust, B. Reliability of "in-season" fitness assessments in youth elite soccer players: A working model for practitioners and coaches. *Sci. Med. Footb.* **2018**, *2*, 177–183. [CrossRef]
76. Da Silva, C.D.; Natali, A.J.; de Lima, J.R.P.; Filho, M.G.B.; Garcia, E.S.; Bouzas Marins, J.C. Yo-Yo IR2 test and Margaria test: Validity, reliability and maximum heart rate in young soccer players. *Rev. Bras. Med. Esporte* **2011**, *17*, 344–349.
77. Krustrup, P.; Mohr, M.; Nybo, L.; Jensen, J.M.; Nielsen, J.J.; Bangsbo, J. The Yo-Yo IR2 test: Physiological response, reliability, and application to elite soccer. *Med. Sci. Sports Exerc.* **2006**, *38*, 1666–1673. [CrossRef]
78. Hopkins, W.G. Measures of reliability in sports medicine and science. *Sports Med.* **2000**, *30*, 1–15. [CrossRef]
79. Buchheit, M.; Chivot, A.; Parouty, J.; Mercier, D.; Al Haddad, H.; Laursen, P.B.; Ahmaidi, S. Monitoring endurance running performance using cardiac parasympathetic function. *Eur. J. Appl. Physiol.* **2010**, *108*, 1153–1167. [CrossRef] [PubMed]
80. Los Arcos, A.; Vázquez, J.S.; Martín, J.; Lerga, J.; Sánchez, F.; Villagra, F.; Zuluta, J.J. Effects of small-sided games vs interval training in aerobic fitness and physical enjoyment in young elite soccer players. *PLoS ONE* **2015**, *10*, e0137224.
81. Dupont, G.; Akakpo, K.; Berthoin, S. The effect of in-season, high-intensity interval training in soccer players. *J. Strength Cond. Res.* **2004**, *18*, 584–589. [PubMed]
82. Faude, O.; Steffen, A.; Kellmann, M.; Meyer, T. The effect of short-term interval training during the competitive season on physical fitness and signs of fatigue: A crossover trial in high-level youth football players. *Int. J. Sports Physiol. Perform.* **2014**, *9*, 936–944. [CrossRef]
83. Faude, O.; Schnittker, R.; Schulte-Zurhausen, R.; Müller, F.; Meyer, T. High-intensity interval training vs. high-volume running training during pre-season conditioning in high-level youth football: A cross-over trial. *J. Sports Sci.* **2013**, *31*, 1441–1450. [CrossRef]
84. Dellal, A.; Varliette, C.; Owen, A.; Chirico, E.N.; Pialoux, V. Small-sided games versus interval training in amateur soccer players: Effects on the aerobic capacity and the ability to perform intermittent exercises with changes of direction. *J. Strength Cond. Res.* **2012**, *26*, 2712–2720. [CrossRef] [PubMed]
85. Wong, P.L.; Chaouachi, A.; Chamari, K.; Dellal, A.; Wisloff, U. Effect of preseason concurrent muscular strength and high-intensity interval training in professional soccer players. *J. Strength Cond. Res.* **2010**, *24*, 653–660. [CrossRef]
86. Slettaløkken, G.; Rønnestad, B.R. High-intensity interval training every second week maintains VO_{2max} in soccer players during off-season. *J. Strength Cond. Res.* **2014**, *28*, 1946–1951. [CrossRef]
87. Sanchez-Sanchez, J.; Gonzalo-Skok, O.; Carretero, M.; Pineda, A.; Ramirez-Campillo, R.; Nakamura, F.Y. Effects of concurrent eccentric overload and high-intensity interval training on team sports players' performance. *Kinesiology* **2019**, *51*, 119–126. [CrossRef]
88. Tønnessen, E.; Shalfawi, S.A.I.; Haugen, T.; Enoksen, E. The effect of 40 m repeated sprint training on maximum sprinting speed, repeated sprint speed endurance, vertical jump, and aerobic capacity in young elite male soccer players. *J. Strength Cond. Res.* **2011**, *25*, 2364–2370. [CrossRef]
89. Shalfawi, S.A.I.; Haugen, T.; Jakobsen, T.A.; Enoksen, E.; Tønnessen, E. The effect of combined resisted agility and repeated sprint training vs. strength training on female elite soccer players. *J. Strength Cond. Res.* **2013**, *27*, 2966–2972. [CrossRef]
90. Impellizzeri, F.M.; Rampinini, E.; Maffiuletti, N.A.; Castagna, C.; Bizzini, M.; Wisløff, U. Effects of aerobic training on the exercise-induced decline in short-passing ability in junior soccer players. *Appl. Physiol. Nutr. Metab.* **2008**, *33*, 1192–1198. [CrossRef]
91. Özcan, İ.; Eniseler, N.; Şahan, C. Effects of small-sided games and conventional aerobic interval training on various physiological characteristics and defensive and offensive skills used in soccer. *Kinesiology* **2018**, *50*, 104–111. [CrossRef]
92. Ferrari Bravo, D.; Impellizzeri, F.M.; Rampinini, E.; Castagna, C.; Bishop, D.; Wisloff, U. Sprint vs. interval training in football. *Int. J. Sports Med.* **2008**, *29*, 668–674. [CrossRef]
93. Arslan, E.; Orer, G.E.; Clemente, F.M. Running-based high-intensity interval training vs. small-sided game training programs: Effects on the physical performance, psychophysiological responses and technical skills in young soccer players. *Biol. Sport.* **2020**, *37*, 165–173. [CrossRef]
94. Ouerghi, N.; Khammassi, M.; Boukorraa, S.; Feki, M.; Kaabachi, N.; Bouassida, A. Effects of a high-intensity intermittent training program on aerobic capacity and lipid profile in trained subjects. *J. Sports Med.* **2014**, *5*, 23–248. [CrossRef]

95. Taylor, J.M.; MacPherson, T.W.; McLaren, S.J.; Spears, I.; Weston, M. Two-weeks of repeated-sprint training in soccer: To turn or not to turn? *Int. J. Sports Physiol. Perform.* **2016**, *11*, 998–1004. [CrossRef] [PubMed]
96. Beato, M.; Coratella, G.; Bianchi, M.; Costa, E.; Merlini, M. Short-term repeated-sprint training (straight sprint vs. changes of direction) in soccer players. *J. Hum. Kinet.* **2019**, *70*, 183–190. [CrossRef] [PubMed]
97. Soares-Caldeira, L.F.; de Souza, E.A.; de Freitas, V.H.; de Moraes, S.M.F.; Leicht, A.S.; Nakamura, F.Y. Effects of additional repeated sprint training during preseason on performance, heart rate variability, and stress symptoms in futsal players: A randomized controlled trial. *J. Strength Cond. Res.* **2014**, *28*, 2815–2826. [CrossRef] [PubMed]
98. Kavaliauskas, M.; Kilvington, R.; Babraj, J. Effect of in-season uphill sprinting on physical characteristics in semi-professional soccer players. *J. Sports Med. Phys. Fit.* **2017**, *57*, 165–170.
99. Eniseler, N.; Şahan, C.; Özcan, I.; Dinler, K. High-intensity small-sided games versus repeated sprint training in junior soccer players. *J. Hum. Kinet.* **2017**, *60*, 101–111. [CrossRef]
100. Nedrehagen, E.S.; Saeterbakken, A.H. The effects of in-season repeated sprint training compared to regular soccer training. *J. Hum. Kinet.* **2015**, *49*, 237–244. [CrossRef]
101. Shalfawi, S.A.I.; Young, M.; Tønnesen, E.; Haugen, T.A.; Enoksen, E. The effect of repeated agility training vs. repeated sprint training on elite female soccer players' physical performance. *Kin. Sloven.* **2013**, *19*, 29–42.
102. Beato, M.; Bianchi, M.; Coratella, G.; Merlini, M.; Drust, B. A single session of straight line and change-of-direction sprinting per week does not lead to different fitness improvements in elite young soccer players. *J. Strength Cond. Res.* **2019**. [CrossRef]
103. Sanchez-Sanchez, J.; Ramirez-Campillo, R.; Petisco, C.; Gonzalo Skok, O.; Rodriguez-Fernandez, A.; Miñano, J.; Nakamura, F.Y. Effects of repeated sprints with changes of direction on youth soccer player's performance: Impact of initial fitness level. *J. Strength Cond. Res.* **2019**, *33*, 2753–2759. [CrossRef]
104. Campos-Vazquez, M.A.; Romero-Boza, S.; Toscano-Bendala, F.J.; Leon-Prados, J.A.; Suarez-Arrones, L.J.; Gonzalez-Jurado, J.A. Comparison of the effect of repeated-sprint training combined with two different methods of strength training on young soccer players. *J. Strength Cond. Res.* **2015**, *29*, 744–751. [CrossRef]
105. Haugen, T.; Tonnessen, E.; Leirstein, S.; Hem, E.; Seiler, S. Not quite so fast: Effect of training at 90% sprint speed on maximal and repeated sprint ability in soccer players. *J. Sports Sci.* **2014**, *32*, 1979–1986. [CrossRef] [PubMed]
106. Nyberg, M.; Fiorenza, M.; Lund, A.; Christensen, M.; Rømer, T.; Piil, P.; Hostrup, M.; Christensen, P.M.; Holbek, S.; Ravnholt, T.; et al. Adaptations to speed endurance training in highly trained soccer players. *Med. Sci. Sports Exerc.* **2016**, *48*, 1355–1364. [CrossRef] [PubMed]
107. Hostrup, M.; Gunnarsson, T.P.; Fiorenza, M.; Mørch, K.; Onslev, J.; Pedersen, K.M.; Bangsbo, J. In-season adaptations to intense intermittent training and sprint interval training in sub-elite football players. *Scand. J. Med. Sci. Sports* **2019**, *29*, 669–677. [CrossRef]
108. Macpherson, T.W.; Weston, M. The effect of low-volume sprint interval training on the development and subsequent maintenance of aerobic fitness in soccer players. *Int. J. Sports Physiol. Perform.* **2015**, *10*, 332–338. [CrossRef] [PubMed]
109. Howard, N.; Stavrianeas, S. In-season high-intensity interval training improves conditioning in high school soccer players. *Int. J. Exerc. Sci.* **2017**, *10*, 713–720.
110. Dello Iacono, A.; Beato, M.; Unnithan, V. Comparative effects of game profile-based training and small-sided games on physical performance of elite young soccer players. *J. Strength Cond. Res.* **2019**. [CrossRef] [PubMed]
111. Iaia, M.F.; Fiorenza, M.; Larghi, L.; Alberti, G.; Millet, G.P.; Girard, O. Short or long-rest intervals during repeated-sprint training in soccer? *PLoS ONE* **2017**, *12*, e0171462. [CrossRef] [PubMed]
112. Sagelv, E.H.; Selnæs, I.; Pedersen, S.; Pettersen, S.A.; Randers, M.B.; Welde, B. Effects of linear versus changes of direction repeated sprints on intermittent high intensity running performance in high-level junior football players over an entire season: A randomized trial. *Sports* **2019**, *7*, 189. [CrossRef] [PubMed]
113. Christensen, P.M.; Krustrup, P.; Gunnarsson, T.P.; Kiilerich, K.; Nybo, L.; Bangsbo, J. VO2 kinetics and performance in soccer players after intense training and inactivity. *Med. Sci. Sports Exerc.* **2011**, *43*, 1716–1724. [CrossRef]
114. Thomassen, M.; Christensen, P.M.; Gunnarsson, T.P.; Nybo, L.; Bangsbo, J. Effect of 2-wk intensified training and inactivity on muscle Na+-K+ pump expression, phospholemman (FXYD1) phosphorylation, and performance in soccer players. *J. Appl. Physiol.* **2010**, *108*, 898–905. [CrossRef]
115. Iaia, M.F.; Fiorenza, M.; Perri, E.; Alberti, G.; Millet, G.P.; Bangsbo, J. The effect of two speed endurance training regimes on performance of soccer players. *PLoS ONE* **2015**, *10*, e0138096. [CrossRef]
116. Mohr, M.; Krustrup, P. Comparison between two types of anaerobic speed endurance training in competitive soccer players. *J. Hum. Kinet.* **2016**, *51*, 183–192. [CrossRef]
117. Ingebrigtsen, J.; Shalfawi, S.A.I.; Tønnessen, E.; Krustrup, P.; Holtermann, A. Performance effects of 6 weeks of aerobic production training in junior elite soccer players. *J. Strength Cond. Res.* **2013**, *27*, 1861–1867. [CrossRef]
118. Arazi, H.; Keihaniyan, A.; EatemadyBoroujeni, A.; Oftade, A.; Takhsha, S.; Asadi, A.; Ramirez-Campillo, R. Effects of heart-rate vs speed-based high intensity interval training on aerobic and anaerobic capacity of female soccer players. *Sports* **2017**, *5*, 57. [CrossRef] [PubMed]
119. Paul, D.J.; Marques, J.B.; Nassis, G.P. The effect of a concentrated period of soccer specific fitness training with small-sided games on physical fitness in youth players. *J. Sports Med. Phys. Fit.* **2019**, *59*, 962–968. [CrossRef]

120. Rabbani, A.; Clemente, F.M.; Kargarfard, M.; Jahangiri, S. Combined small-sided game and high-intensity interval training in soccer players: The effect of exercise order. *J. Hum. Kinet.* **2019**, *69*, 249–257. [CrossRef]
121. Campos-Vazquez, M.A.; Toscano-Bendala, F.J.; Mora-Ferrera, J.C.; Suarez-Arrones, L.J. Relationship between internal load indicators and changes on intermittent performance after the preseason in professional soccer players. *J. Strength Cond. Res.* **2017**, *31*, 1477–1485. [CrossRef] [PubMed]
122. Clemente, F.M.; Ramirez-Campillo, R.; Nakamura, F.Y.; Sarmento, H. Effects of high-intensity interval training in men soccer player's physical fitness: A systematic review with meta-analysis of randomized-controlled and non-controlled trials. *J. Sports Sci.* **2021**, *11*, 1–22.
123. Hopkins, W.G. How to interpret changes in athletic performance test. *Sports Sci.* **2004**, *8*, 1–7.
124. Rampinini, E.; Sassi, A.; Azzalin, A.; Castagna, C.; Menaspà, P.; Carlomagno, D.; Impellizzeri, F.M. Physiological determinants of Yo-Yo intermittent recovery tests in male soccer players. *Eur. J. Appl. Physiol.* **2010**, *108*, 401–409. [CrossRef] [PubMed]
125. Helgerud, J.; Engen, L.C.; Wisløff, U.; Hoff, J. Aerobic endurance training improves soccer performance. *Med. Sci. Sports Exerc.* **2001**, *33*, 1925–1931. [CrossRef]
126. Iaia, F.M.; Rampinini, E.; Bangsbo, J. High-intensity training in football. *Int. J. Sports Physiol. Perform.* **2009**, *4*, 291–306. [CrossRef] [PubMed]
127. Buchheit, M.; Laursen, P.B. High-intensity interval training, solutions to the programming puzzle. Part 2: Anaerobic energy, neuromuscular load and practical applications. *Sports Med.* **2013**, *43*, 927–954. [CrossRef] [PubMed]
128. Mann, T.; Lamberts, R.P.; Lambert, M.I. Methods of prescribing relative exercise intensity: Physiological and practical considerations. *Sports Med.* **2013**, *43*, 613–625. [CrossRef] [PubMed]
129. Jamnick, N.A.; Pettitt, R.W.; Granata, C.; Pyne, D.P.; Bishop, D.J. An examination and critique of current methods to determine exercise intensity. *Sports Med.* **2020**, *50*, 1729–1756. [CrossRef]
130. Berthoin, S.; Gerbeaux, M.; Turpin, E.; Guerrin, F.; Lensel-Corbeil, G.; Vandendorpe, F. Comparison of two field tests to estimate maximum aerobic speed. *J. Sports Sci.* **1994**, *12*, 355–362. [CrossRef] [PubMed]
131. Dupont, G.; Defontaine, M.; Bosquet, L.; Blondel, N.; Moalla, W.; Berthoin, S. Yo-Yo intermittent recovery test versus the Université de Montréal Track Test: Relation with a high-intensity intermittent exercise. *J. Sci. Med. Sport* **2010**, *13*, 146–150. [CrossRef]

Journal of
Functional Morphology and Kinesiology

MDPI

Article

Association of Performance in Strength and Plyometric Tests with Change of Direction Performance in Young Female Team-Sport Athletes

Hallvard Nygaard Falch, Eirik Lindset Kristiansen, Markus Estifanos Haugen and Roland van den Tillaar *

Department of Sport Sciences and Physical Education, Nord University, 7600 Levanger, Norway; falch7@hotmail.com (H.N.F.); ek1105@hotmail.com (E.L.K.); haugen@helseogprestasjon.no (M.E.H.)
* Correspondence: roland.v.tillaar@nord.no

Abstract: The change of direction (COD) ability is a task-specific skill dependent on different factors such as the degree of the turn, which has led to differentiating CODs as more force- (>90°) or velocity-oriented (<90°). Considering force and velocity requirements is of importance when designing sport-specific training programs for enhancing COD performance. Thus, 25 female handball and soccer players participated in this study, which investigated the association between three different strength and plyometric exercises and force- and velocity-oriented COD performance. By utilizing the median split analysis, the participants were further divided into a fast ($n = 8$) and a slow ($n = 8$) COD group, to investigate differences in step kinematics between fast and slow performers. The correlational analysis revealed that the bilateral back squat and unilateral quarter squat were significantly associated with several force- and velocity-oriented COD performance ($r = -0.46$ to -0.64), while the association between plyometric and COD performance was limited ($r < 0.44$). The fast COD group revealed higher levels of strength, jump height, peak velocities, higher step frequencies, shorter ground contact times, and greater acceleration and braking power ($d > 1.29$, $p < 0.03$). It was concluded that the observed correlation between strength and COD performance might be due to stronger athletes being able to produce more workload in a shorter time, which was supported by the step kinematics.

Keywords: force; velocity; power; step kinematics

Citation: Falch, H.N.; Kristiansen, E.L.; Haugen, M.E.; van den Tillaar, R. Association of Performance in Strength and Plyometric Tests with Change of Direction Performance in Young Female Team-Sport Athletes. *J. Funct. Morphol. Kinesiol.* **2021**, *6*, 83. https://doi.org/10.3390/jfmk6040083

Academic Editor: Andrea Fusco

Received: 22 September 2021
Accepted: 12 October 2021
Published: 14 October 2021

1. Introduction

In court and field sports, athletes are required to possess numerous physical and tactical skills [1,2]. The physical skills required are distinctive for the different sports, positions on the field, and team tactics [3,4]. However, a minimum threshold for maximal oxygen consumption to endure the match is required [5,6], accompanied by the ability to perform and repeat various high-intensity actions throughout the match [7–9]. Despite the short timeframe of these high-intensity actions relative to total match-time, they are decisive for match outcomes as they often precede match-decisive situations [10,11]. One such important high-intensity action is the physical ability to rapidly perform a change of direction (COD), which frequently occurs in team sports [12–14]. The COD ability can be defined as the athlete's preplanned physical ability to accelerate and decelerate while overcoming inertia before reaccelerating in a new direction [15–17]. Athletes with a well-developed COD ability possess a physical and tactical advantage over their opponent, as, in offensive play, they may have an increased opportunity of bypassing their opponent, creating space, and/or moving into space. Generally, lower-limb strength, power, and reactive strength are leg muscle qualities positive for COD performance [18], due to the importance of rapidly expressing a large amount of force over multiple steps when performing a COD [16,19,20].

Although leg muscle qualities are important factors determining COD performance in general, COD is a task-specific skill [21,22], as improvement in one specific COD task might not transfer to another [18,23]. Factors constraining the specifics of a COD maneuver can be the complexity (i.e., handling a ball) [21], number of COD(s) performed [15], completion time (energy systems utilized), initial velocity approaching the COD, and angle of the turn [19,21,24]. CODs performed at a higher initial velocity will increase braking requirements to manage the turn [19], as shown by more steps for braking [22]. Furthermore, CODs of greater angles require greater redirections of the whole body [25], increases in ground contact time, greater mediolateral force production, and greater knee and hip flexion in the plant step [22,25–28]. Accordingly, CODs of different angles have been suggested to demand different magnitudes and directions of ground reaction forces. This concept is rooted in Newton's laws of motion, as CODs of greater angles (>90°) require greater force productions for braking to change momentum and redirecting the body, as opposed to velocity-oriented CODs (<90°), allowing for velocity maintenance and a transfer of momentum [20,21,24].

Strength and plyometric training are two different training modalities often utilized to improve leg muscle qualities, which may furthermore lead to enhanced COD performance [18,21]. Unfortunately, the distinctiveness of different types of CODs is often neglected when testing COD performance, whereby the COD ability is processed as a general skill, without considering the specific demands of CODs, such as force and velocity magnitudes. Thus, results after conducting training interventions might conflict [21]. To the authors' knowledge, a study by Falch et al. [29] is the only one to specifically examine the association of strength and plyometric exercises with force- and velocity-oriented CODs. The study was conducted on male soccer players and suggested that the plyometric exercises should be COD-specific, while the different strength tests revealed only a 'moderate' insignificant correlation ($r < 0.41$). Furthermore, a review by Falch, Rædergård, and van den Tillaar [21] suggested strength training to be more beneficial for developing force-oriented COD performance, as force-oriented CODs are performed at slower velocities. However, it is unknown how these findings apply to female athletes who are underrepresented in COD research [21]. Females possess, on average, less strength and more fat mass in comparison to males [30], which are important considerations when training for enhancing COD performance. Enhancing relative strength might increase COD performance [31], as acceleration/deceleration of the body is a product of net force [24]. Therefore, strength training might be more beneficial for female athletes for developing overall COD performance due to generally lower levels of relative strength. This assumption is reasonable, as strength training has been found to enhance COD performance in both female volleyball athletes and untrained females when measuring performance changes using a standardized *t*-test [32,33].

Although earlier research suggests plyometric training to be more COD-specific [21,23,29], relative strength might reveal a high association for female athletes in both force- and velocity-oriented CODs, as there might be a "threshold" before further strength gains will not enhance COD performance. Lower-limb strength, commonly dynamically expressed with exercises such as the barbell back squat, might be of even greater importance in force-oriented CODs, as CODs of increased angles require longer ground contact times with greater knee and hip-joint angles [22]. Due to the limited time available to devote to training specific physical abilities such as the COD ability in team sports, it is important to incorporate exercises and training modalities positively affecting overall performance outcomes in a sport-specific context. Thus, the primary objective of the current study was to examine the association between young female court and field sport athletes' performance in different strength and plyometric exercises with performance in a force- and velocity-oriented COD. A secondary objective was to investigate differences in step kinematics and acceleration/deceleration between fast and slow performances in force- and velocity-oriented CODs for further insights into the demands of the force -and velocity-oriented CODs. It was hypothesized that fast COD performers would be better at both accelerating and decelerating in both force- and velocity-oriented CODs due to higher step frequencies and shorter contact times [34]. Such an investigation might provide useful information for strength and conditioning coaches seeking to improve task-specific COD performance in female athletes. Performance in the strength and plyometric exercises was hypothesized to be associated with performance in both force- and velocity-oriented CODs, as all performances are a product of lower-limb force production relative to body mass.

2. Materials and Methods

A within-subject design was conducted during the offseason to investigate the association between performance in strength and plyometric exercises and performance in force- and velocity-oriented CODs in female court and field sport athletes. A between-subject design was used to examine differences between fast and slow COD performers. To avoid a possible learning effect, all athletes had to participate in two familiarization sessions, practicing the different tests of the study. Technical guidance (mainly foot placement and depth) and the study procedure for the familiarization and test day were controlled for by three strength and conditioning professionals.

2.1. Subjects

The subjects of the current study consisted of 25 young female handball (n = 16) and soccer players (n = 9) (age: 19.6 ± 2.8 years, height: 170 ± 7.1 cm, body mass: 68 ± 10.6 kg, body mass index: 23.6 ± 2.3) participating in a minimum of three sessions per week, recruited from two local teams. Both teams competed at the second-highest level in the Norwegian league system of their respective sport. None of the athletes had an injury or any illness in the previous 3 months that could negatively affect the validity of the study. All athletes were informed of the risks and benefits of participation, and a written consent form from the athletes and parents (when under 18 years old) was obtained before the tests. The study was conducted according to the ethical regulations for research, approved by the Norwegian Center for Research Data (project number: 903955) in line with the latest revision of the Declaration of Helsinki. The athletes were instructed to be physically and mentally prepared for performing maximal efforts on the day of testing, which involves the consumption of a light meal 2 h before testing, >7 h of sleep, not consuming alcohol, and avoiding demanding physical training 48 h before testing.

2.2. Procedure

Before testing the performance in the different strength, plyometric, and COD tests, all athletes underwent the standardized warm-up protocol presented by van den Tillaar et al. [35] consisting of submaximal runs and dynamic stretching. Afterward, athletes were randomly assigned by an online randomizer to three different groups, testing maximum performance in the different tests (strength, plyometric, and COD/running). Each group was randomly assigned to start with the strength, plyometric, or COD tests and completed all the tests within the session. Because the athletes already performed a warm-up protocol, only sub-maximal repetitions of the different tests were included as a specific warm-up, leading up to the maximal effort attempts. To investigate the association of force- and velocity-oriented COD performance with strength and plyometric exercises, the athletes were required to perform one velocity- (45°) and one force-oriented (180°) COD task. The COD test was accompanied by a sprint test (large magnitude of concentric force and velocity requirements) and a braking test (large magnitude of eccentric braking force requirements) [36]. Furthermore, the athletes performed three plyometric tests and three strength tests, which were performed bilaterally and unilaterally with the right foot in the vertical and lateral directions. The right foot performed the unilateral movements as the right foot performs the plant step in CODs with a left turn [22].

2.3. COD Tests

The COD tests consisted of a 10 m sprint approaching a 45° or 180° turn and performing the turn before reaccelerating 10 m into the new direction. All CODs were performed with a left turn to ensure that the right foot performed the pivoting step. For the 45° COD, the athlete had to run 10 m before performing the plant step before the 0.8 m line after the 10 m line; she then proceeded by running toward the new direction. For the 180° COD, the athlete had to run 10 m, placing the pivoting foot on the 0.8 m line before reaccelerating into the new direction. Athletes were instructed to complete the test within the shortest amount of time possible. Both total time (10 m + COD + 10 m) and partial time (first and last 10 m) were measured (Figure 1).

In all COD, sprint, or braking tests, the athlete was required to start from a standstill position, with the front foot placed 5 cm behind the first timing gate to prevent a false trigger of a random limb. Time was started after passing the first wireless timing gate (Brower Timing Systems, Salt Lake City, UT, USA, height of 1 m) and stopped when passing the last timing gate (Figures 1 and 2). The tests started on the athletes' own accord, after receiving a signal from a researcher, to limit a reactive component to the different tests. A contact grid, IR-Contactmat-ML6TJP02-870 (Ergotest Innovation, Porsgrunn, Norway,) was utilized to investigate step kinematics in the COD, sprint, and braking tests. The contact grid detected contact and flight times and was placed along the starting line for the COD, sprint, and braking tests. The contact grid covered the whole sprint and braking test area for force-oriented COD, while, for velocity-oriented COD, it covered the first 10 m (Figure 1).

Distance and velocity in the COD, sprint, and braking tests were measured and calculated using a wireless CMP3 distance sensor laser gun positioned 1.8 m behind the athlete (Noptel Oy, Oulu, Finland), with a sampling at 2.56 kHz, which was pointed at the athlete's lower back (approximately center of mass) while running.

Step kinematics were also calculated from the COD test (average step length, average strep frequency, average ground contact and flight times approaching the COD, and ground contact time spent turning in the COD) by Musclelab 10.5.69 (Ergotest innovation A. S, Porsgrunn, Norway), which synchronized the laser gun and the contact mat. All the COD tests (including the sprint and braking test) were conducted on an indoor court surface (Taraflex Sport Evolution M 7.0 mm, Unisport, Finland).

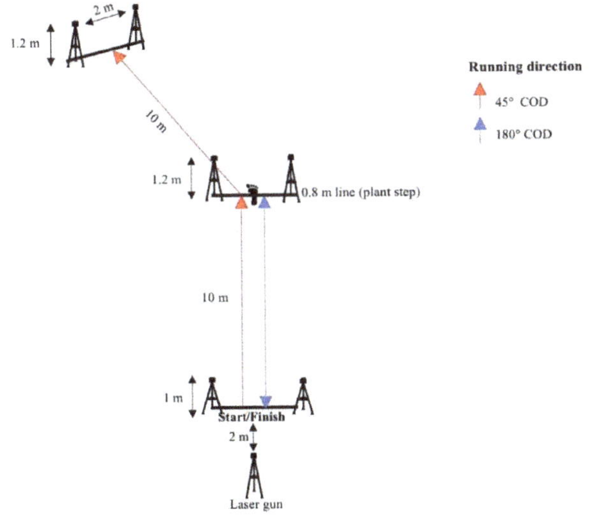

Figure 1. Setup for the COD track.

2.4. Sprint Test

The sprint test consisted of a 30 m straight-line sprint. Athletes were instructed to complete the 30 m sprint within the shortest amount of time possible. The performance variable was the peak velocity, which was measured at 5, 10, 20, and 30 m using a laser gun. Step kinematics for the sprint (average step length, average step frequency, and average ground contact times) were also sampled.

2.5. Maximum Horizontal Braking Test

The maximum horizontal braking test was retrieved from a protocol by Harper et al. [37], using a standardized acceleration–deceleration test, included in the Musclelab v10.5.69 software (Ergotest Innovation A.S, Porsgrunn, Norway). Athletes were instructed to sprint 20 m with maximum effort, before initiating maximum deceleration after passing the 20 m mark (Figure 2). The laser was used to measure the deceleration after passing the 20 m mark. Furthermore, it was used to ensure that the athletes performed a maximal acceleration before deceleration of the laser-measured sprinting velocity at the 20 m marks. If it revealed a 5% decrease in velocity, compared to velocity after 20 m in the straight-line sprint, a reattempt was required after 3 min of rest. Horizontal acceleration power (W/kg), braking power (W/kg), and braking force (N/kg) were calculated from the braking test.

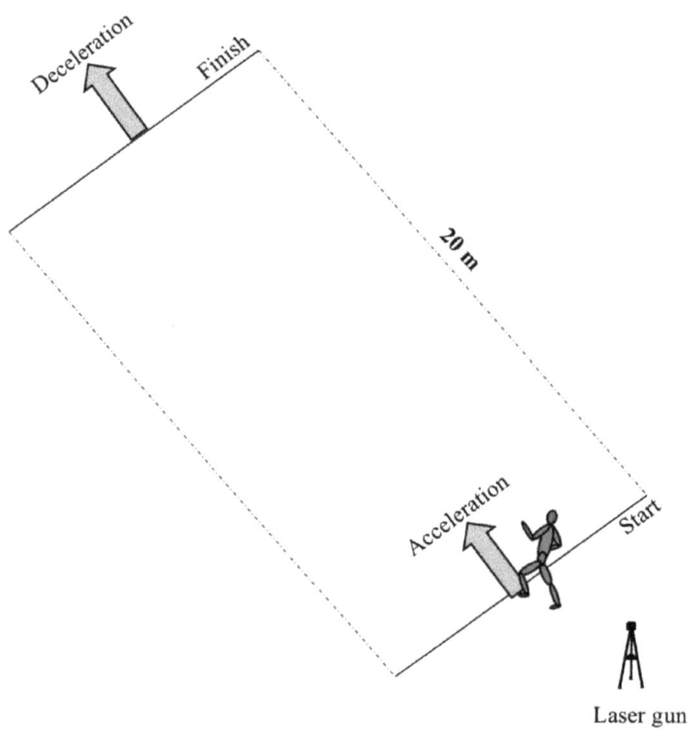

Figure 2. Setup for the maximum horizontal braking test.

2.6. Strength Tests

The strength tests of the current study were similar to an earlier similar study by Falch, Rædergård, and van den Tillaar [29], but in men. It consisted of a bilateral back squat, a unilateral quarter squat performed on a Smith machine, and a lateral barbell squat. Performance was expressed as relative strength (load/body mass). Appropriate depth of the bilateral squat was defined as bending the knee until the trochanter major was in line with the patella. For the quarter and lateral squats, the athlete was required to bend the knee to a 90° flexion in the knee joint (Figure 3). The one-repetition maximum (1-RM) was estimated from the load–velocity relationship [38] using the best-fit regression line of three different data points for each individual athlete. Each data point represents load at a given velocity, whereby the average concentric velocity corresponds to ~1, 0.8, and 0.5 m/s. Thus, the different strength tests required maximal mobilization in the concentric phase of the lift, with three repetitions at each load. The average concentric velocity of the second and third repetitions of each series was used as a data point for calculating 1-RM. This is because, in lighter loads (<80% of 1-RM), the second or third repetition is often the fastest repetition [39]. A linear encoder sampling at 500 Hz (ET-Enc-02, Ergotest Technology AS, Porsgrunn, Norway) was used to measure the concentric velocity in the strength tests. All unilateral tests were performed with the right foot. Performance in the strength tests was estimated by 1-RM/body mass.

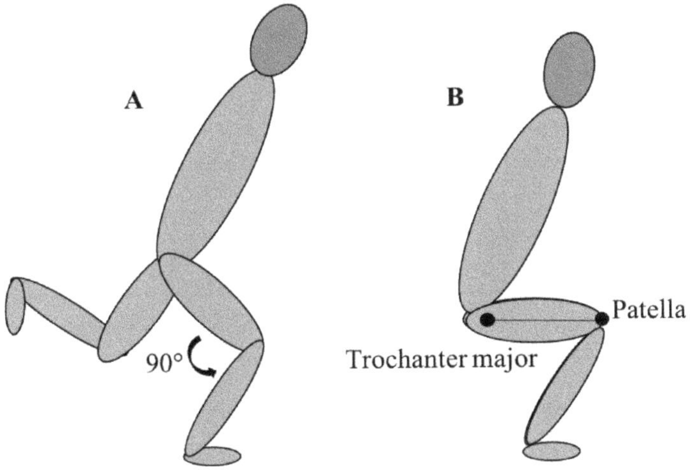

Figure 3. Depth requirements for the unilateral quarter squat, lateral squat (**A**), and bilateral back squat (**B**).

2.7. Plyometric Tests

The plyometric tests were also based on a protocol by Falch, Rædergård, and van den Tillaar [29] consisting of a unilateral vertical countermovement jump for maximal height (cm) and a unilateral lateral countermovement jump for maximal length (cm), herein referred to as a skate-jump and a drop jump with a drop height of 20 cm [40], whereby the reactive strength index (RSI) is the performance variable (jump height/ground contact time). In the unilateral vertical countermovement jump and the drop jump, athletes were instructed to place their hands akimbo, to prevent an arm swing from possibly contributing to performance, limiting the isolated effect of leg power [41]. In the vertical jump, the athlete was furthermore instructed to keep the passive foot in a locked position, to avoid the momentum of the passive foot from contribution to jump height. Reactive strength index (jump height/foot contact) in the drop jump and jump height in the unilateral countermovement jump were determined using a dual force plate (Ergotest Technology AS, Porsgrunn, Norway) sampling at 1000 Hz. The force plate registers contact time and flight time and calculates jump height with the use of flight time according to the following equation: $jump\ height = 1/2 \times 9.81 \times \left(\frac{flight\ time}{2}\right)^2$.

All tests were performed with three approved attempts, with 3 min rest between each attempt. The average of all three attempts was used for further analysis.

2.8. Statistical Analysis

The descriptive statistics are presented as the mean ± standard deviation. To investigate the association between strength and plyometric capabilities with performance in force- and velocity-oriented CODs, correlations between the performance variables were determined utilizing Pearson's correlational coefficient. The relationships between performance variables were based upon r-values defined as small ($0.1 < r < 0.3$), moderate ($0.3 < r < 0.5$), large ($0.5 < r < 0.7$), and very large ($0.7 < r < 0.9$) [42]. Utilizing the median split analysis based on the athlete's average force and velocity COD performances, athletes were furthermore divided into fast and slow COD groups of eight athletes each. The median group ($n = 9$) was excluded from the statistical analysis. Comparisons of the two groups were conducted to investigate the study's secondary objective if fast vs. slow COD performers possessed different step and acceleration/deceleration abilities. Group differences were investigated with multiple independent t-tests (fast vs. slow COD performers). The assumption of homogeneity of variance was controlled for with Levene's test. The assumption of normality was confirmed for all variables using the Kolmogorov–Smirnov test. When assumptions for the independent t-test were violated, the Mann–Whitney U test was conducted. The effect of group differences is presented as Cohen's d, whereby an effect of 0–2 constitutes a very small effect, 0.2–0.5 constitutes a small effect, 0.5–0.8 constitutes a moderate effect, and >0.8 constitutes a large effect. Furthermore, >1.2 was defined as a very large effect and >2 was defined as a huge effect [43]. The level of significance was set at $p < 0.05$, and the confidence interval was set at 95%. All tests were performed in SPSS v.27 (IBM Corp., Armonk, NY, USA).

3. Results

The mean performance in the different strength, plyometric, COD, sprint, and horizontal braking tests for all athletes ($n = 25$) is presented in Table 1.

Table 1. Descriptive statistics of performance in the different tests.

Strength Tests	Mean ± STD
Bilateral squat (1-RM/body mass)	1.28 ± 0.32
Unilateral quarter squat (1-RM/body mass)	0.88 ± 0.28
Lateral squat (1-RM/body mass)	0.57 ± 0.2
Plyometric tests	
Drop jump (reactive strength index)	117 ± 30.8
Unilateral countermovement jump (cm)	11.6 ± 2.5
Skate-jump (cm)	169.1 ± 9.7
CODs (s)	
First 10 m 45°	2.20 ± 0.11
Last 10 m 45°	1.68 ± 0.11
20 m 45°	3.88 ± 0.2
First 10 m 180°	2.40 ± 0.12
Last 10 m 180°	2.77 ± 0.17
20 m 180°	5.17 ± 0.24
Peak velocities in the straight-line sprint (m/s)	
5 m	5.36 ± 0.25
10 m	6.30 ± 0.32
20 m	6.89 ± 0.38
30 m	7.06 ± 0.48
Braking test	
Horizontal acceleration power (W/kg)	8.39 ± 1.22
Horizontal braking power (W/kg)	−10.51 ± 1.48
Horizontal braking force (N/kg)	−3.11 ± 0.47

STD = standard deviation; 1-RM = one-repetition maximum; COD = change of direction.

3.1. Correlations

The bilateral and quarter squat revealed 'moderate to strong' significant correlations with all total-time and first 10 m CODs ($r > -0.43$, $p < 0.04$, Table 2), except for the quarter squat and 20 m 180° COD, where the relationship was nonsignificant ($r = -0.39$, $p > 0.07$). The reactive strength index was only significantly moderately correlated with 10 m 180° COD ($r = -0.44$, $p = 0.03$), while all the other tests revealed insignificant small correlations with COD performance ($r < 0.28$, $p > 0.35$).

Table 2. Correlation of performance in the different sprint, strength, plyometric, braking, and COD tests.

	Bilateral Squat	Quarter Squat	Lateral Squat	RSI	CMJ	Skate-Jump
COD first 10 m 180°	−0.46 *	−0.49 *	0.28	−0.44 *	−0.25	−0.13
COD last 10 m 180°	−0.35	−0.21	0.22	0.03	−0.3	0.15
COD 20 m 180°	−0.48 *	−0.39	0.28	−0.19	−0.28	0.05
COD first 10 m 45°	−0.64 *	−0.62 *	0.19	−0.2	−0.29	−0.01
COD last 10 m 45°	−0.29	−0.17	0.07	−0.02	−0.31	0.02
COD 20 m 45°	−0.5 *	−0.43 *	0.14	−0.13	−0.22	−0.05

CMJ is the unilateral countermovement jump; RSI is the reactive strength index; * indicates a significant correlation at the $p < 0.05$ level. COD = change of direction.

3.2. Fast vs. Slow Performers

When comparing the fast ($n = 8$) vs. slow COD performers ($n = 8$) in both the force- and the velocity-oriented CODs, as well as the different strength and plyometric tests, the fast performers were found to be significantly stronger in the bilateral and unilateral squat ($F < 1.82$, $d > 1.35$, $p < 0.03$). The fast performers also jumped significantly higher in the unilateral countermovement jump ($F < 0.33$, $d > 1.29$, $p < 0.03$, Table 3).

Table 3. Differences in fast vs. slow performers in the different strength and plyometric exercises.

	Force-Oriented CODs		Velocity-Oriented CODs	
	Fast Performers	Slow Performers	Fast Performers	Slow Performers
Strength (load/BM)				
Bilateral squat	1.52 ± 0.31 *	1.05 ± 0.28	1.52 ± 0.23 *	1.1 ± 0.28
Quarter squat	1.06 ± 0.18 *	0.71 ± 0.16	1 ± 0.24 *	0.72 ± 0.17
Lateral squat	0.56 ± 0.17	0.61 ± 0.24	0.61 ± 0.28	0.56 ± 0.17
Plyometric exercises				
Drop jump (RSI)	123.8 ± 28.4	123.3 ± 40.8	112.9 ± 22.8	111 ± 35.9
CMJ (cm)	13.4 ± 1.4 *	10.1 ± 2.4	12.6 ± 1.4 *	10.3 ± 2.2
Skate-jump (cm)	165.5 ± 9.5	169.1 ± 8.3	167.8 ± 9.4	168.3 ± 8

CMJ is the unilateral countermovement jump (right); RSI is the reactive strength index; BM is the body mass; * indicates a significant between fast vs slow performers at the $p < 0.05$ level. COD = change of direction.

3.3. Step Kinematics Differences

The F-statistics from Levene's test revealed unequal between-group variance for step frequency and flight time in the 30 m sprint and 180° COD (Tables 4 and 5). When comparing fast vs. slow performers in the 30 m sprint and both CODs, a huge effect was observed for peak velocities across all the tests and distances, whereby the fast performers revealed significantly higher velocities compared to the slow performers ($d > 3.31$, $p < 0.01$). A large to huge effect was also observed for average ground contact time, flight time, and step frequencies in which the fast performers had shorter ground contact times, shorter flight times, and higher step frequencies ($d > 1.26$, $p < 0.05$). The fast performers also revealed greater horizontal acceleration and braking power in the horizontal braking test ($d > 1.34$, $p < 0.02$), where the effect was very large to huge (Tables 4 and 5). Furthermore, a large nonsignificant effect was observed in the plant step and braking force between the fast and slow performers in the velocity-oriented COD, where the fast performers produced more braking force and shorter ground contact time in the plant step compared to the slow performers ($d > 1.07$, $p > 0.07$, Table 5).

Table 4. Differences in step peak velocities and step kinematics between the fast and slow performers in force-oriented COD.

	Force-Oriented CODs					
	Fast COD Performers	**Slow COD Performers**	**F**	**ES**	**ES**	**CI (95%)**
	Mean ± STD	**Mean ± STD**		***(d)***	**Description**	
Peak velocities (m/s)						
First 5 m (of 30 m sprint)	5.61 ± 0.13	5.08 ± 0.16	0.22	3.61 *	Huge	−0.69, −0.37
First 10 m (of 30 m sprint)	6.62 ± 0.15	5.95 ± 0.2	0.85	3.76 *	Huge	−0.85, −0.47
First 20 m (of 30 m sprint)	7.26 ± 0.17	6.48 ± 0.24	0.7	3.78 *	Huge	−1.00, −0.56
30 m sprint	7.49 ± 0.23	6.56 ± 0.27	1.67	4.14 *	Huge	−1.19, −0.67
180° COD	5.77 ± 0.12	5.21 ± 0.18	2.72	3.78 *	Huge	−0.73, −0.39
Ground contact times (ms)						
30 m sprint, average	154.3 ± 17.9	175.3 ± 12.6	2.02	1.38 *	Very large	4.43, 37.65
180° COD, plant step	1239.7 ± 192.11	1058.7 ± 167.8	0.09	1.01	Large	−37.45, 12.37
180° COD, average	177.4 ± 14.6	202.4 ± 13.3	0.05	1.79 *	Very large	1.00, 39.87
Flight time (ms)						
30 m sprint, average	88.3 ± 5.4	103 ± 10.6	#	1.84 *	Very large	5.11, 24.26
180° COD, average	54.4 ± 9.5	67.6 ± 14.7	0.51	1.09	Large	−0.08, 26.42
Step lengths (m)						
30 m sprint	1.53 ± 0.1	1.52 ± 0.13	0.62	0.11	Very small	−0.14, 0.11
180° COD	1.19 ± 0.08	1.19 ± 0.07	0.64	0.1	Very small	−0.09, 0.08
Step frequencies (n/s)						
30 m sprint	4.22 ± 0.37	3.67 ± 0.14	#	2.18 *	Huge	−0.86, −0.24
180° COD	4.35 ± 0.34	3.74 ± 0.16	#	2.44 *	Huge	−0.91, −0.32
Horizontal braking test						
Acceleration power (W/kg)	9.26 ± 1.03	7.19 ± 0.89	0.15	2.15 *	Huge	−3.90, −1.35
Braking power (W/kg)	−11.58 ± 1.01	−9.73 ± 1.77	0.84	1.34 *	Very large	0.29, 3.38
Braking force (N/kg)	−2.88 ± 0.66	−3.34 ± 0.32	3.86	0.09	Very small	−0.09, 1.01

* Indicates a significant difference at the $p < 0.05$ level; # indicates violated assumptions and the conduction of the Mann–Whitney U test; F is the F-statistic from Levene's test; ES is the effect size; *d* is Cohen's *d*; CI is the confidence interval. STD = standard deviation. COD = change of direction.

Table 5. Differences in performance-related measures between the fast and slow performers in velocity-oriented CODs.

	Velocity-Oriented CODs					
	Fast COD Performers	**Slow COD Performers**	**F**	**ES**	**ES**	**CI (95%)**
	Mean ± STD	**Mean ± STD**		***(d)***	**Description**	
Peak velocities (m/s)						
First 5 m (of 30 m sprint)	5.58 ± 10	5.08 ± 0.16	2.96	3.79 *	Huge	−0.64, −0.35
First 10 m (of 30 m sprint)	6.59 ± 0.1	5.94 ± 0.18	3.34	4.69 *	Huge	−0.81, −0.50
First 20 m (of 30 m sprint)	7.25 ± 0.11	6.44 ± 0.18	3.86	5.48 *	Huge	−0.97, −0.65
30 m sprint	7.53 ± 0.14	6.55 ± 0.25	2.85	6.75 *	Huge	−1.13, −0.43
45° COD	6.01 ± 0.12	5.48 ± 0.2	3.7	3.31 *	Huge	−0.73, −0.36
Ground contact times (ms)						
30 m sprint, averages	156 ± 16.5	176.8 ± 10.1	2.5	1.49 *	Huge	5.20, 34.55
45° COD, plant step	168.2 ± 20.6	186.7 ± 12.3	1.55	1.12	Large	−0.07, 38.78
45° COD, averages	178.9 ± 14.5	195.9 ± 12.4	0.17	1.26 *	Very large	3.34, 35.46
Flight time (ms)						
30 m sprint, average	89.4 ± 8.9	102.3 ± 9.9	0.1	1.38 *	Very large	3.13, 26.5
45° COD, average	56.5 ± 12.4	97.1 ± 36.2	0.01	1.71 *	Very large	1.66, 39.97
Step lengths (m)						
30 m sprint	1.56 ± 0.1	1.54 ± 0.14	0.59	0.15	Very small	−0.15, 0.11
45° COD	1.18 ± 0.15	1.3 ± 0.14	0.55	0.9	Large	−0.04, 0.3
Step frequencies (n/s)						
30 m sprint	4.2 ± 0.35	3.66 ± 0.15	#	2.16 *	Huge	−0.85, −0.24
45° COD	4.33 ± 0.39	3.66 ± 0.26	0.63	2.04 *	Huge	−1.08, −0.25
Horizontal braking test						
Acceleration power (W/kg)	9.25 ± 1	7.21 ± 0.91	0.06	2.15 *	Huge	−3.07, −1.02
Braking power (W/kg)	−11.33 ± 0.91	−9.49 ± 1.65	1.06	1.44 *	Very large	0.42, 3.27
Braking force (N/kg)	−3.27 ± 0.25	−2.81 ± 0.61	4.59	1.07	Large	−0.04, 0.96

* Indicates a significant difference at the $p < 0.05$ level; # indicates violated assumptions and the conduction of the Mann–Whitney U test; F is the F-statistic from Levene's test; ES is the effect size; *d* is Cohen's *d*; CI is the confidence interval. STD = standard deviation. COD = change of direction.

4. Discussion

The primary objective of the current study was to examine the association between performance in strength and plyometric exercises and force- and velocity-oriented COD performance in young female court and field sport athletes. The correlational analysis revealed moderate and strong associations between relative strength in the bilateral squat and quarter squat with several of the COD performances ($r \geq -0.43$, $p \leq 0.04$). This result is in line with earlier research, indicating dynamic lower-limb strength to be important for COD performance in female athletes [44,45]. Furthermore, relative strength has been reported to be associated with COD performance [45–47].

Thus, the bilateral and quarter squat could share similarities with the acceleration/deceleration aspect prior to the plant step in both the force- and the velocity-oriented COD tests due to similar muscle requirements of knee-extensor and hip-flexor strength. Knee-extensor strength has been found to be associated with deceleration abilities [48], by eccentrically absorbing forces [36], which might allow higher velocities when initiating the COD movement. Furthermore, the quadriceps and gluteus maximus, which are agonist muscles in the bilateral and quarter squat, are suggested to contribute largely to acceleration moments when accelerating [49]. On the other hand, the lateral squat was the only strength exercise not revealing any significant correlation with any of the CODs ($r = 0.28$, $p > 0.35$), possibly due to the technical demands of the exercise regarding balance and control [50], inhibiting the athletes' ability to maximize loads at the given velocities. This is logical, as training modalities often predominantly focus on movements in the sagittal plane [51], with exercises such as the bilateral and quarter squat. The balance requirements increase in unilateral movements and are further increased in free-weight exercises. Although it is a unilateral movement, instability in the unilateral quarter squat is reduced by being performed on a Smith machine [52].

Contradictory to earlier research [16,29,53,54], the correlation between plyometrics and COD performance was limited. Balance might also account for this finding, because instability may decrease the ability to express power [55,56]. The countermovement jump and skate jump are unilateral movements that were performed freely, demanding more balance than the unilateral quarter squat performed in the Smith machine when flexing the knee. As such, balance might inhibit a fast pre-stretch, which is desired for the muscles to reach a higher level of active state and subsequently aid in shortening velocities in the jumps [57,58]. Thus, the balance aspect in the eccentric pre-stretch of the jumps could negatively affect performance, which explains the 'small' association with COD performance. Saeterbakken and Fimland [52] indicated that instability reduced the force output of the lower limbs, despite similar muscle activity. As such, it could be speculated that the reduced stability in the unilateral plyometric exercises reduces force output, limiting the correlation with COD.

The drop jump was the only bilateral plyometric exercise, whereby RSI was the only plyometric performance variable significantly correlated with COD performance, revealing a moderate association with the first 10 m 180° COD ($r = -0.44$, $p = 0.03$). This might be due to the relationship between RSI and deceleration ability because the first 10 m 180° COD is a force-oriented COD, whereby performance is more dictated by braking capabilities [22] compared to the 45° COD performances. Drop jumps and high-velocity decelerations share physical similarities, both demanding great eccentric strength and muscle activation in the gastrocnemius to absorb forces [29,36].

The observed correlations in the current study contradict earlier comparisons made in males [29], which suggested plyometric exercises to be more COD-specific. A possible explanation is that strength performance might reveal the greatest association with COD performance in populations with lower levels of relative strength. This is because there might be a threshold to how much strength is beneficial for developing physical abilities underpinning COD performance [59,60]. Furthermore, the bilateral and quarter squat were only significantly correlated with the first 10 m COD and total time to perform the COD test. The small to moderate observed correlation with the last 10 m CODs ($r \leq -0.35$, $p \leq 0.1$), accompanied by the distinctive magnitude of physical ability requirements in force- and velocity-oriented CODs [19,22], suggests the pivoting movement itself to be dependent on different neuromuscular abilities. The pivoting movement requires coordination of the whole body, such as the appropriate inclination angle of the trunk to manipulate the base of support and overcome inertia [21,61]. Without trunk stability and appropriate inclination angles of the trunk, net forces produced while reaccelerating will be limited. As such, transitioning from the weight-acceptance phase to the reacceleration phase might be heavily dependent upon technical demands, limiting the isolated effect of the lower limbs to produce force. Therefore, further investigation into the physical abilities of a priori reaccelerating, differentiating fast and slow performances in force- and velocity-oriented CODs, could increase insights into the associations between COD performance and strength and plyometric performance.

The fast performers in both force- and velocity-oriented CODs revealed similar physical capabilities, indicating that the force- and velocity-oriented CODs consisted of similar demands. The fast COD performers possessed higher levels of acceleration and deceleration abilities. Firstly, the data showed that the fast COD performers attained higher peak velocities in both the force- and the velocity-oriented CODs, as well as sprinting distances over 5, 10, 20, and 30 m, and greater horizontal acceleration power in the horizontal braking test. This finding was expected, as peak velocities have been observed to be associated with COD performance [47,53].

The higher observed acceleration was due to a higher step frequency, as fast performers performed both force- and velocity-oriented CODs with higher step frequencies ($d > 2.04$, $p < 0.01$), while the step length was similar. According to earlier research on sprinting [62], higher step frequencies were expected in the fast performers, as running speed is a result of step length × step frequency [34]. Step frequency, again, is a product of ground contact time and flight time, whereby shorter contact times and/or flight time results in a higher step frequency. Following earlier COD research [63], the ground contact time was observed to differentiate between the groups, whereby the fast performers revealed shorter ground contact times in both force- and velocity-oriented CODs (Tables 4 and 5). Furthermore, the fast performers also revealed shorter flight time in the velocity-oriented COD (Tables 4 and 5). The shorter ground contact and flight times in the fast COD performers accounted for a faster acceleration/deceleration phase. As the step length was the same between the groups, horizontal force production was not indicated to differentiate the fast vs. slow performers. The findings are supported by the horizontal braking test, whereby the fast COD performers revealed greater horizontal braking power, but not horizontal braking force. As the total force production was over the same distance (work), but in a shorter time (work/t), power was higher in the fast performers (Tables 4 and 5). To accelerate/decelerate body weight, athletes need to exert net forces to the ground [21,24], and shorter contact times indicate faster production of net ground reaction forces to change momentum in the acceleration/deceleration phase, according to Newton's laws of motion. The forces required to change the body's momentum are dependent on mass × velocity, and the rate of change in momentum is dependent on the time over which forces are applied (force × time = mass × velocity) [36].

A higher production of power was visible in the strength tests in which the fast performers had a higher 1-RM, which was based on the load–velocity relationship. As such, fast COD performers could, with similar submaximal loads to slow performers, perform at a higher velocity in the different strength tests, thereby performing the same workload over a shorter time, producing more power and, therefore, a higher calculated 1-RM. The findings are in line with earlier research by Barr, Sheppard, Agar-Newman, and Newton [60] who found relative strength in lower-limb exercises to be associated with ground contact time and sprinting velocities ($r = 0.47$ to 0.71) in male athletes accustomed to strength training. Higher power production by the fast COD performers may also explain the higher jump height in the unilateral countermovement jump, as jump height is a result of velocity at takeoff (Table 3). However, the aforementioned balance requirements of the lateral squat and skate jump could have inhibited the fast athletes' ability to express power in these exercises.

Limitations

Strength performances were, for practical reasons, estimated with regression, which can differ from true 1-RM values. To limit differences between estimated and true 1-RM, lifts were performed without a pause to increase ecological validity and sampled from a relatively wide velocity range (±0.5 m/s) [64]. Furthermore, to save testing time and avoid fatigue, performance was only tested for the right foot in the unilateral tests, as the right foot performed the plant step in the COD tests, which does not account for lower-limb asymmetries. However, similar research has observed similar performance in force- and velocity-oriented CODs with a left and right turn [23]. Future research measuring lean body mass and muscle activity in the different tests is warranted, which could provide new useful insights into the nature of the results. Furthermore, testing "stronger" athlete groups would be important to investigate if the correlations for the plyometric and strength exercise with COD performance would change more in the direction of stronger correlations with plyometrics, as found in men [29].

5. Conclusions

In female court and field sport athletes, the bilateral and quarter squat revealed the greatest association with COD performance, possibly due to the demands of the knee-extensor and hip-flexors for rapidly applying force when accelerating/decelerating in a COD. Surprisingly, the association between plyometric and COD performance was limited. These results indicate that stronger athletes have higher levels of power production, positively influencing COD performance. As such, increases in lower-limb strength might positively influence COD performance if the acquired strength gains enhance the ability to produce more workload in a shorter time. Fast vs. slow performers in force- and velocity-oriented CODs were differentiated by higher peak velocities, as well as horizontal acceleration and braking power, which were observable through higher step frequencies and shorter ground contact times. The observed between-group differences in step kinematics might have been a result of differences in relative strength and ability to produce power.

Practical Applications

Relative strength in the bilateral and quarter squat was found to be associated with both force- and velocity-oriented COD performance. According to the results of the current study, female team-sport athletes displaying relative strength of ~<1.5 load/BM in the bilateral back squat and ~<1 load/BM in the unilateral quarter squat might improve COD performance by increasing relative strength in these exercises. However, the practical application does not account for at what point more velocity-specific exercises, such as plyometrics, should be implemented.

Author Contributions: Conceptualization, H.N.F., M.E.H. and R.v.d.T.; methodology, H.N.F. and R.v.d.T.; data curation, H.N.F., M.E.H., E.L.K. and R.v.d.T.; writing—original draft preparation, H.N.F.; writing—review and editing, H.N.F., E.L.K., M.E.H. and R.v.d.T.; supervision, R.v.d.T. All authors read and agreed to the published version of the manuscript.

Funding: This research received no external funding.

Institutional Review Board Statement: The study was conducted according to the ethical regulations for research, approved by the Norwegian Centre for Research Data (project nr: 903955 on 14 September 2020).

Informed Consent Statement: Informed consent was obtained from all subjects involved in the study.

Data Availability Statement: The raw data supporting the conclusions of this article will be made available by the authors, without undue reservation.

Conflicts of Interest: The authors declare no conflict of interest.

References

1. Stølen, T.; Chamari, K.; Castagna, C.; Wisløff, U. Physiology of soccer. *Sports Med.* **2005**, *35*, 501–536. [CrossRef] [PubMed]
2. Carling, C.; Reilly, T.; Williams, A.M. *Performance Assessment for Field Sports*; Routledge: London, UK, 2008.
3. Ostojic, S.M.; Mazic, S.; Dikic, N. Profiling in basketball: Physical and physiological characteristics of elite players. *J. Strength Cond. Res.* **2006**, *20*, 740. [CrossRef]
4. Ziv, G.; Lidor, R. Physical characteristics, physiological attributes, and on-court performances of handball players: A review. *Eur. J. Sport Sci.* **2009**, *9*, 375–386. [CrossRef]
5. Tønnessen, E.; Hem, E.; Leirstein, S.; Haugen, T.; Seiler, S. Maximal aerobic power characteristics of male professional soccer players, 1989–2012. *Int. J. Sports Physiol. Perf.* **2013**, *8*, 323–329. [CrossRef] [PubMed]
6. Reilly, T.; Borrie, A. Physiology applied to field hockey. *Sports Med.* **1992**, *14*, 10–26. [CrossRef] [PubMed]
7. Dalen, T.; Ingebrigtsen, J.; Ettema, G.; Hjelde, G.H.; Wisløff, U. Player load, acceleration, and deceleration during forty-five competitive matches of elite soccer. *J. Strength Cond. Res.* **2016**, *30*, 351–359. [CrossRef]
8. Dalen, T.; Lorås, H.; Hjelde, G.H.; Kjøsnes, T.N.; Wisløff, U. Accelerations–a new approach to quantify physical performance decline in male elite soccer? *Eur. J. Sport Sci.* **2019**, *19*, 1015–1023. [CrossRef]
9. Conte, D.; Favero, T.G.; Lupo, C.; Francioni, F.M.; Capranica, L.; Tessitore, A. Time-motion analysis of Italian elite women's basketball games: Individual and team analyses. *J. Strength Cond. Res.* **2015**, *29*, 144–150. [CrossRef]
10. Faude, O.; Koch, T.; Meyer, T. Straight sprinting is the most frequent action in goal situations in professional football. *J. Sports Sci.* **2012**, *30*, 625–631. [CrossRef]
11. Castagna, C.; D'Ottavio, S.; Abt, G. Activity profile of young soccer players during actual match play. *J. Strength Cond. Res.* **2003**, *17*, 775–780.
12. Bloomfield, J.; Polman, P.; O'Donoghue, P. O-007 Deceleration movements performed during FA Premier League soccer matches. *J. Sports Sci. Med.* **2007**, *10*, 6.
13. Póvoas, S.C.; Seabra, A.F.; Ascensão, A.A.; Magalhães, J.; Soares, J.M.; Rebelo, A.N. Physical and physiological demands of elite team handball. *J. Strength Cond. Res.* **2012**, *26*, 3365–3375. [CrossRef]
14. Duthie, G.M.; Pyne, D.B.; Marsh, D.J.; Hooper, S.L. Sprint patterns in rugby union players during competition. *J. Strength Cond. Res.* **2006**, *20*, 208.
15. Brughelli, M.; Cronin, J.; Levin, G.; Chaouachi, A. Understanding change of direction ability in sport. *Sports Med.* **2008**, *38*, 1045–1063. [CrossRef]
16. Castillo-Rodríguez, A.; Fernández-García, J.C.; Chinchilla-Minguet, J.L.; Carnero, E.Á. Relationship between muscular strength and sprints with changes of direction. *J. Strength Cond. Res.* **2012**, *26*, 725–732. [CrossRef] [PubMed]
17. Sheppard, J.M.; Young, W. Agility literature review: Classifications, training and testing. *J. Sports Sci.* **2006**, *24*, 919–932. [CrossRef] [PubMed]
18. Young, W.; James, R.; Montgomery, I. Is muscle power related to running speed with changes of direction? *J. Sports Med. Phys. Fitn* **2002**, *42*, 282–288.
19. Dos'Santos, T.; Thomas, C.; Comfort, P.; Jones, P.A. The effect of angle and velocity on change of direction biomechanics: An angle-velocity trade-off. *Sports Med.* **2018**, *48*, 2235–2253. [CrossRef]
20. Watts, D. A brief review on the role of maximal strength in change of direction speed. *J. Aust Strength Cond.* **2015**, *23*, 100–108.
21. Falch, H.N.; Rædergård, H.G.; van den Tillaar, R. Effect of different physical training forms on change of direction ability: A systematic review and meta-analysis. *Sports Med.Open* **2019**, *5*, 1–37.
22. Falch, H.N.; Rædergård, H.G.; van den Tillaar, R. Effect of Approach Distance and Change of Direction Angles Upon Step and Joint Kinematics, Peak Muscle Activation, and Change of Direction Performance. *Front. Sports Act. Liv.* **2020**, *2*, 173. [CrossRef]
23. Rædergård, H.G.; Falch, H.N.; van den Tillaar, R. Effects of Strength vs. Plyometric Training on Change of Direction Performance in Experienced Soccer Players. *Sports* **2020**, *8*, 144. [CrossRef]
24. Bourgeois, F.; McGuigan, M.; Gill, N.; Gamble, G. Physical characteristics and performance in change of direction tasks: A brief review and training considerations. *J. Aust. Strength Cond.* **2017**, *25*, 104–117.
25. Havens, K.L.; Sigward, S.M. Whole body mechanics differ among running and cutting maneuvers in skilled athletes. *Gait. Post.* **2015**, *42*, 240–245. [CrossRef]
26. Schot, P.; Dart, J.; Schuh, M. Biomechanical analysis of two change-of-direction maneuvers while running. *J. Orthop. Sports Phys. Ther.* **1995**, *22*, 254–258. [CrossRef] [PubMed]
27. Schreurs, M.J.; Benjaminse, A.; Lemmink, K.A. Sharper angle, higher risk? The effect of cutting angle on knee mechanics in invasion sport athletes. *J. Biomech.* **2017**, *63*, 144–150. [CrossRef] [PubMed]
28. Sigward, S.M.; Cesar, G.M.; Havens, K.L. Predictors of frontal plane knee moments during side-step cutting to 45 and 110 men and women: Implications for ACL injury. *Clin. J. Sports Med.* **2015**, *25*, 529. [CrossRef]
29. Falch, H.N.; Rædergård, H.G.; van den Tillaar, R. Association of strength and plyometric exercises with change of direction performances. *PLoS ONE* **2020**, *15*, e0238580. [CrossRef] [PubMed]
30. Miller, A.E.J.; MacDougall, J.; Tarnopolsky, M.; Sale, D. Gender differences in strength and muscle fiber characteristics. *Eur. J. Appl. Physiol.* **1993**, *66*, 254–262. [CrossRef]
31. Peterson, M.D.; Alvar, B.A.; Rhea, M.R. The contribution of maximal force production to explosive movement among young collegiate athletes. *J. Strength Cond. Res.* **2006**, *20*, 867–873.

32. İnce, İ. Effects of Split Style Olympic Weightlifting Training on Leg Stiffness Vertical Jump Change of Direction and Sprint in Collegiate Volleyball Players. *Univ. J. Edu. Res.* **2019**, *7*, 24–31. [CrossRef]
33. Arazi, H.; Asadi, A.; Roohi, S. Enhancing muscular performance in women: Compound versus complex, traditional resistance and plyometric training alone. *J. Musc. Res.* **2014**, *17*, 1450007. [CrossRef]
34. Murphy, A.J.; Lockie, R.G.; Coutts, A.J. Kinematic determinants of early acceleration in field sport athletes. *J. Sports Sci. Med.* **2003**, *2*, 144.
35. van den Tillaar, R.; Lerberg, E.; von Heimburg, E. Comparison of three types of warm-up upon sprint ability in experienced soccer players. *J. Sport Health Sci.* **2019**, *8*, 574–578. [CrossRef] [PubMed]
36. Hewit, J.; Cronin, J.; Button, C.; Hume, P. Understanding deceleration in sport. *Strength Cond. J.* **2011**, *33*, 47–52. [CrossRef]
37. Harper, D.J.; Morin, J.-B.; Carling, C.; Kiely, J. Measuring maximal horizontal deceleration ability using radar technology: Reliability and sensitivity of kinematic and kinetic variables. *Sports Biomech.* **2020**, 1–17. [CrossRef] [PubMed]
38. Jovanović, M.; Flanagan, E.P. Researched applications of velocity based strength training. *J. Aust. Strength Cond.* **2014**, *22*, 58–69.
39. Rivière, J.R.; Rossi, J.; Jimenez-Reyes, P.; Morin, J.-B.; Samozino, P. Where does the one-repetition maximum exist on the force-velocity relationship in squat? *Int. J. Sports Med.* **2017**, *38*, 1035–1043. [CrossRef]
40. Bogdanis, G.C.; Donti, O.; Papia, A.; Donti, A.; Apostolidis, N.; Sands, W.A. Effect of plyometric training on jumping, sprinting and change of direction speed in child female athletes. *Sports* **2019**, *7*, 116. [CrossRef]
41. Markovic, G.; Dizdar, D.; Jukic, I.; Cardinale, M. Reliability and factorial validity of squat and countermovement jump tests. *J. Strength Cond. Res.* **2004**, *18*, 551–555.
42. Cohen, J. *Statistical Power Analysis for the Behaviour Sciences*; Revised Edition; Academic: New York, NY, USA, 1988; Volume 73, pp. 19–74.
43. Sawilowsky, S.S. New effect size rules of thumb. *J. Mod. Appl. Stat. Met.* **2009**, *8*, 26. [CrossRef]
44. Spiteri, T.; Nimphius, S.; Hart, N.H.; Specos, C.; Sheppard, J.M.; Newton, R.U. Contribution of strength characteristics to change of direction and agility performance in female basketball athletes. *J. Strength Cond. Res.* **2014**, *28*, 2415–2423. [CrossRef]
45. Nimphius, S.; McGuigan, M.R.; Newton, R.U. Relationship between strength, power, speed, and change of direction performance of female softball players. *J. Strength Cond. Res.* **2010**, *24*, 885–895. [CrossRef]
46. Tramel, W.; Lockie, R.G.; Lindsay, K.G.; Dawes, J.J. Associations between absolute and relative lower body strength to measures of power and change of direction speed in Division II female volleyball players. *Sports* **2019**, *7*, 160. [CrossRef]
47. Delaney, J.A.; Scott, T.J.; Ballard, D.A.; Duthie, G.M.; Hickmans, J.A.; Lockie, R.G.; Dascombe, B.J. Contributing factors to change-of-direction ability in professional rugby league players. *J. Strength Cond. Res.* **2015**, *29*, 2688–2696. [CrossRef] [PubMed]
48. Harper, D.J.; Jordan, A.R.; Kiely, J. Relationships between eccentric and concentric knee strength capacities and maximal linear deceleration ability in male academy soccer players. *J. Strength Cond. Res.* **2021**, *35*, 465–472. [CrossRef]
49. Little, T.; Williams, A. *Specificity of Acceleration, Maximum Speed and Agility in Professional Soccer Players*; Routledge: London, UK, 2003.
50. Cross, T. SPEED/STRENGTH EXERCISE: The Lateral Squat. *Strength Cond. J.* **1991**, *13*, 56–64. [CrossRef]
51. Sato, K.; Liebenson, C. The lateral squat. *J. Body Movem. Ther.* **2013**, *17*, 560–562. [CrossRef]
52. Saeterbakken, A.H.; Fimland, M.S. Muscle force output and electromyographic activity in squats with various unstable surfaces. *J. Strength Cond. Res.* **2013**, *27*, 130–136. [CrossRef] [PubMed]
53. Nygaard Falch, H.; Guldteig Rædergård, H.; Van den Tillaar, R. Relationship of performance measures and muscle activity between a 180 change of direction task and different countermovement jumps. *Sports* **2020**, *8*, 47. [CrossRef] [PubMed]
54. Lockie, R.G.; Dawes, J.J.; Jones, M.T. Relationships between linear speed and lower-body power with change-of-direction speed in national collegiate athletic association divisions I and II women soccer athletes. *Sports* **2018**, *6*, 30. [CrossRef]
55. Hammami, R.; Chaouachi, A.; Makhlouf, I.; Granacher, U.; Behm, D.G. Associations between balance and muscle strength, power performance in male youth athletes of different maturity status. *Pediat. Exerc. Sci.* **2016**, *28*, 521–534. [CrossRef]
56. Kean, C.O.; Behm, D.G.; Young, W.B. Fixed foot balance training increases rectus femoris activation during landing and jump height in recreationally active women. *J. Sports Sci. Med.* **2006**, *5*, 138.
57. Pérez-Castilla, A.; Rojas, F.J.; Gómez-Martínez, F.; García-Ramos, A. Vertical jump performance is affected by the velocity and depth of the countermovement. *Sports Biomech.* **2019**, 1–16. [CrossRef]
58. Bobbert, M.F.; Gerritsen, K.G.; Litjens, M.C.; Van Soest, A.J. Why is countermovement jump height greater than squat jump height? *Med. Sci. Sports Exerc.* **1996**, *28*, 1402–1412. [CrossRef] [PubMed]
59. Comfort, P.; Bullock, N.; Pearson, S.J. A comparison of maximal squat strength and 5-, 10-, and 20-meter sprint times, in athletes and recreationally trained men. *J. Strength Cond. Res.* **2012**, *26*, 937–940. [CrossRef] [PubMed]
60. Barr, M.J.; Sheppard, J.M.; Agar-Newman, D.J.; Newton, R.U. Transfer effect of strength and power training to the sprinting kinematics of international rugby players. *J. Strength Cond. Res.* **2014**, *28*, 2585–2596. [CrossRef] [PubMed]
61. Sasaki, S.; Nagano, Y.; Kaneko, S.; Sakurai, T.; Fukubayashi, T. The relationship between performance and trunk movement during change of direction. *J. Sports Sci. Med.* **2011**, *10*, 112. [PubMed]
62. Hewit, J.K.; Cronin, J.B.; Hume, P.A. Kinematic factors affecting fast and slow straight and change-of-direction acceleration times. *J. Strength Cond. Res.* **2013**, *27*, 69–75. [CrossRef] [PubMed]

63. Dos' Santos, T.; Thomas, C.; Jones, P.A.; Comfort, P. Mechanical determinants of faster change of direction speed performance in male athletes. *J. Strength Cond. Res.* **2017**, *31*, 696–705. [CrossRef] [PubMed]
64. Banyard, H.G.; Nosaka, K.; Haff, G.G. Reliability and validity of the load–velocity relationship to predict the 1RM back squat. *J. Strength Cond. Res.* **2017**, *31*, 1897–1904. [CrossRef] [PubMed]

Journal of
Functional Morphology and Kinesiology

MDPI

Article

Drop Jumping on Sand Is Characterized by Lower Power, Higher Rate of Force Development and Larger Knee Joint Range of Motion

George Giatsis *, Vassilios Panoutsakopoulos and Iraklis A. Kollias

Biomechanics Laboratory, Department of Physical Education and Sport Science at Thessaloniki, Aristotle University of Thessaloniki, 54124 Thessaloniki, Greece; bpanouts@phed.auth.gr (V.P.); hkollias@phed.auth.gr (I.A.K.)
* Correspondence: ggiatsis@phed.auth.gr; Tel.: +30-231-099-2208

Abstract: Plyometric training on sand is suggested to result in advanced performance in vertical jumping. However, limited information exists concerning the biomechanics of drop jumps (DJ) on sand. The purpose of the study was to compare the biomechanical parameters of DJs executed on rigid (RIGID) and sand (SAND) surface. Sixteen high level male beach-volleyball players executed DJ from 40 cm on RIGID and SAND. Force- and video-recordings were analyzed to extract the kinetic and kinematic parameters of the DJ. Results of paired-samples *t*-tests revealed that DJ on SAND had significantly ($p < 0.05$) lower jumping height, peak vertical ground reaction force, power, peak leg stiffness and peak ankle flexion angular velocity than RIGID. In addition, DJ on SAND was characterized by significantly ($p < 0.05$) larger rate of force development and knee joint flexion in the downward phase. No differences ($p > 0.05$) were observed for the temporal parameters. The compliance of SAND decreases the efficiency of the mechanisms involved in the optimization of DJ performance. Nevertheless, SAND comprises an exercise surface with less loading during the eccentric phase of the DJ, thus it can be considered as a surface that can offer injury prevention under demands for large energy expenditure.

Keywords: biomechanical analysis; kinetics; kinematics; stretch shortening cycle; vertical jumping; surface stability; balance; impact

Citation: Giatsis, G.; Panoutsakopoulos, V.; Kollias, I.A. Drop Jumping on Sand Is Characterized by Lower Power, Higher Rate of Force Development and Larger Knee Joint Range of Motion. *J. Funct. Morphol. Kinesiol.* **2022**, *7*, 17. https://doi.org/10.3390/jfmk7010017

Academic Editor: Giuseppe Musumeci

Received: 31 December 2021
Accepted: 2 February 2022
Published: 4 February 2022

1. Introduction

Sand surfaces (SAND) are a demanding exercise surface. Research evidence suggests that a higher energy cost is required for running [1–5], walking [6,7], sprinting [8,9] and jumping [10–12] on SAND compared to rigid (RIGID) surfaces. Despite the higher energy cost, training on SAND causes positive adaptations in key strength and conditioning factors such as aerobic endurance, concentric strength of the leg extensor muscles and agility in a variety of sport disciplines [13].

SAND is recommended for training in volleyball and beach-volleyball (BV) players due to the observed favorable performance adaptations, i.e., the improvement in technique, muscle strength, vertical jump height, agility and endurance [14–19]. In particular, BV players improved their jumping ability after the application of training programs on SAND that included jumping exercises utilizing the stretch-shortening cycle (SSC), namely the counter-movement (CMJ) and drop jumps (DJ) [20,21]. However, there is a bias in the literature about the effectiveness of SAND with regard to the facilitation of adaptations in the SSC. The application of training with CMJ and DJ resulted in increased muscle activation of the knee extensor muscles that was interpreted as a positive adaptation in SSC due to SAND [21]. On the opposite, past research findings suggest that, after training on SAND, the improvement in the CMJ jump height was decreased compared to the respective vertical squat jump (SQJ) [10]. In addition, it is suggested that the increased power after the implementation of training on SAND was not a result of an enhanced utilization of SSC [15]. Thus, the effectiveness of SAND for inducing neuromuscular adaptations for the efficient utilization of SSC is questionable [13].

Based on the contemporary knowledge of DJs executed on RIGID, derived from both practice and research, plyometric training enabling the SSC is considered as the best training method to provoke adaptations in key jump performance factors such as force, speed, and power [22]. These adaptations are the reason to include SSC training exercises in volleyball and BV, especially those exercises that require a rapid SSC function [23]. This requirement is fulfilled by executing DJs in a manner that results in an increased power output [24–26]. In detail, the increased mechanical power during vertical jumps is related with the regulation of stiffness that occurs during ground contact [27,28]. However, despite the positive adaptations observed in a variety of jump-related biomechanical variables during vertical jump tests, the implementation of plyometric training on SAND in female volleyball players resulted in a non-significant, yet notable, 3.7% decrease in DJ performance [14].

In general, a good vertical jumping performance is of major importance for BV players since vertical jumping is present in the majority of the skills of the sport [29–32]. However, past research reported differences concerning vertical jumping on SAND compared to RIGID [29,33–36]. In detail, these studies reported lower jumping heights in vertical jumps executed on SAND. Furthermore, it was reported that elite BV players achieved higher jump heights on RIGID comparing to SAND in SQJ and CMJ by 14% and 15.4%, respectively [35,36]. This is due to the compliance of SAND that increases the demands for energy expenditure in order to execute the vertical jump in an explosive manner [11,12]. In addition, lower force application and power production were reported during the propulsion in SQJ and CMJ [10,29,33,35,36]. A previous kinematical analysis in the above-mentioned vertical jump tests revealed that, in order to overcome the constrains imposed by SAND, the ankle joint is extended faster [35,36]. Thus, the knowledge of the biomechanical differences concerning the key kinetic and kinematic parameters of vertical jumping on SAND and RIGID is of importance for designing efficient training programs aiming the optimization of the SSC for BV players.

Despite the current knowledge of the biomechanics of vertical jumping on SAND, there is a gap in the literature concerning the biomechanics of DJs performed on SAND. The effect of SAND is mainly studied in a pre-post study design that examines vertical jumping on RIGID [14,20]. A number of studies has measured vertical jump performance on SAND using jump and reach tests [19,21,29,31]. Other researchers provided information about jumping on SAND using photocell mats [10], accelerometry [15] and inertial measurement devices [32]. Kinetic and kinematic parameters derived from force-plate data have been reported only for SQJ and CMJ [33,35,36]. To the best of our knowledge, there is limited information regarding DJs on SAND since only the magnitude of the ground reaction forces [34] and ground/flight time [11] have being reported. Thus, further insight is needed for the key kinetic factors (i.e., rate of force development, power, stiffness) that can evaluate the effectiveness of the execution of DJs on SAND.

The purpose of the study was to compare the kinetic and kinematic parameters of DJ executed on RIGID and SAND. It was hypothesized that DJ on SAND will result in lower jumping height, force and power output, as well as larger lower extremity joint range of motion compared to RIGID.

2. Materials and Methods

2.1. Participants

The minimal number of participants to achieve an effect size of 0.8, power of 0.9 and $a = 0.05$ for the maximum jump height measurement was found to be 15 according to the estimation made using the G*Power v.3.1.9.7 software [37]. Thus, 16 adult professional male BV players (26.2 ± 5.7 y, 1.87 ± 0.05 m, 83.4 ± 5.8 kg) served as participants in the study. Participation was on a voluntary basis and was allowed after obtaining a signed consent. The inclusion criteria were the participation in an international Federation International de

Volleyball (FIVB) tournament, to be finalists in BV tournaments included in the national championship calendar, to exhibit records of systematic participation in their training and competition program and to have a competitive experience in BV of at least five years. The exclusion criterion was the incident of an injury or locomotor disability in a period of 6 months prior testing. The measurements took place during the competitive season of the national championship. The study was approved by the Institutional Ethics Committee (approval No.: 87/2021).

2.2. Procedure

SAND was simulated by firmly attaching a wooden sand pit on the force-plate. The wooden pit, with dimensions of the bottom and of the top side equal to 46 × 50 cm and 59 × 63 cm, respectively, while its depth was 31 cm, was constructed to contain the sand particles (Figure 1).

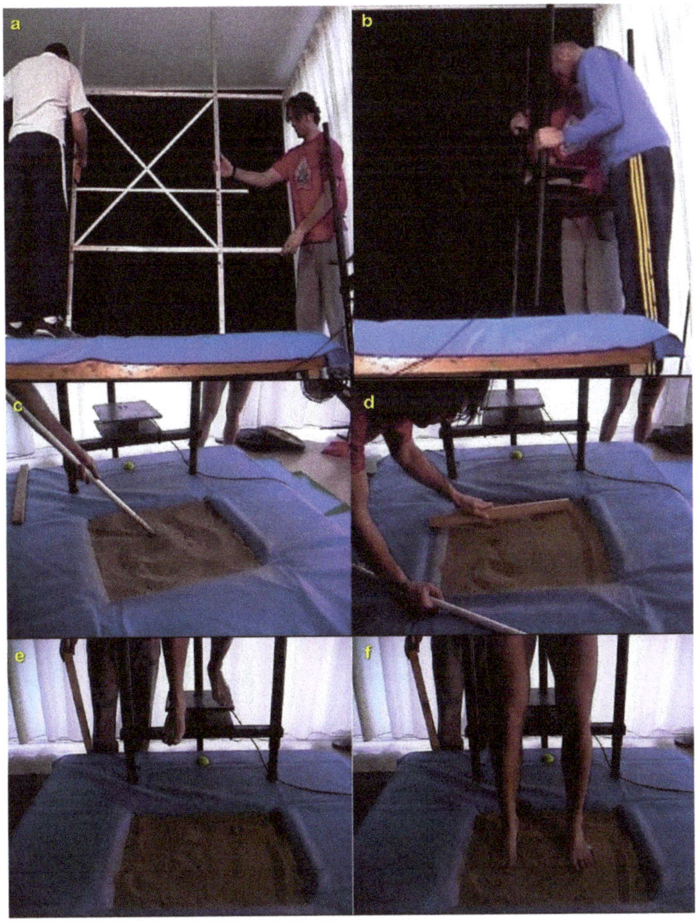

Figure 1. Experimental set-up and procedure for SAND: (**a**) calibration frame placement; (**b**) adjustment of the drop force-plate to the safety platform; (**c**) mixing the sand; (**d**) making the surface even; (**e**) take-off form the drop plate with a roll-off; (**f**) instant of touchdown in the sand pit where the examination for excessive plunging into the SAND was conducted.

The base dimensions of the sandpit were exactly the same as the embedded to the ground force plate in order to be exactly in contact with each other. Before the actual data acquisition, it was established that the participants' mass recorded by the ground force plate was exactly the same with and without the sandpit. The edges of the sandpit were covered with soft materials for safety reasons, i.e., if a faulty landing occurred. Additionally, a canvas sheet was placed away from the sandpit and covered the surrounding safety platform that was 116 × 150 × 31 cm (length, width and height, respectively), which was also used to hold inside the sand particles. The total weight of the wooden pit, including the sand, was 120.12 kg. The sand fulfilled the FIVB requirements for the conduction of official BV tournaments. This was established after checking the physical properties and grain size distribution of the sand as determined from a series of laboratory tests according to the American Society for Testing and Materials (ASTM) and that are described in detail elsewhere [36]. Prior each jump, the sand was mixed throughout its volume with a custom tool that was marked at 31 cm to resemble the sandpit height. This tool was used to evenly spread the sand within the wooden sand pit and to avoid compaction of the sand particles.

Warm-up consisted of cycling on an 817E Monark (Exercise AB, Vansbro, Sweden) Cycle-Ergometer, dynamic stretching exercises and sub-maximal vertical jumps. DJs from 40 cm were executed barefooted in a random order on SAND and on RIGID, with the arms kept on the trunk. The dropping height of 40 cm was selected due to past research suggestions [38]. The drop was performed from a custom made one-dimensional force-plate (1-Dynami, ©: Biomechanics Lab AUTh, Thessaloniki, Greece) that was adjusted and fixed within the safety platform (Figure 1). The instructions given to the participants were to "drop with a roll-off" movement [39] and to "jump as high and as fast as possible" [40]. Participants executed three DJs on each surface. During the experimental DJs on SAND, a Redlake Motionscope PCI 1S camera (Redlake Imaging Corporation, Morgan Hill, CA, USA), operating at 250 fps, was used to visually inspect excessive plunging into the sand that resulted in the annulment of the trial. Finally, only the attempt with maximum jump height (h_{JUMP}) achieved in each condition was selected for further analysis.

2.3. Data Acquisition and Analysis

An AMTI OR6-5-1 force plate (AMTI, Newton, MA, USA) recorded the 3D components of the ground reaction forces (GRF). The sampling frequency for both the force-plates used for data acquisition was set to 500 Hz. The following parameters were calculated [40]:

1. Temporal parameters: total ground contact time (Tc); downward phase duration; time to achieve maximum vertical Ground Reaction Force (tvGRF); time to achieve peak power during the upward phase (tP).
2. Spatial/kinematic parameters: h_{JUMP}; body center of mass (BCM) vertical displacement during the downward and upward phases; BCM vertical velocity.
3. Kinetic parameters: GRF vertical, medio-lateral and anterio-posterior component; rate of force development (RFD); work (W); power (P).

The impact velocity of the BCM on the ground was determined using the take-off data from the drop-force plate [41]. Firstly, the initial velocity of the jump (impact velocity after the drop) was calculated from the time-integral of the net force recorded from the drop force-plate. The flying time of the drop was measured from the synchronous data acquisition from both force-plates. Thus, the BCM velocity at the instant of the landing after the drop phase (U_{IMPACT}) was calculated as shown in Equation (1):

$$U_{IMPACT} = U_{DROP} - g \times t_{FLIGHT} \quad (1)$$

where U_{DROP} is the BCM velocity at the instant of take-off from the drop force-plate, g is the acceleration of gravity and t_{FLIGHT} is the duration of the drop phase.

Afterwards, h_{JUMP} was calculated using the vertical BCM take-off velocity derived from the integration of the net vGRF. RFD was directly extracted as the first time-derivative of the recorded vGRF. Vertical BCM displacement was extracted through the integration of the vertical BCM velocity. Work was calculated by multiplying vertical BCM displacement with net vGRF and power as the time-derivative of work.

Besides the parameters mentioned above, stiffness parameters were also examined. The vertical stiffness was calculated as the ratio of vertical GRF to vertical BCM displacement and leg stiffness as the ratio of vertical GRF to the change of the leg length [42]. To extract the latter, DJs were also video-recorded at 100 fps with a digital video-camera (JVC GR-DVL 9600 EG, Victor Company of Japan Ltd., Yokohama, Japan). The camera was fixed on a tripod placed 7.6 m from the force plate and at a height of 1.2 m, with the camera axis being perpendicular to the plane of motion. A 2.5 m × 2.5 m calibration frame was also recorded to conduct a 2D-DLT analysis for the extraction of the 2D coordinates and the angular kinematics of the lower limb joints [35]. The examined angular kinematic parameters were the ankle, knee and hip range of motion (ROM) and the respective peak angular velocity (ω) of the lower limb joints during the downward and upward phases.

2.4. Statistical Analyses

The Kolmogorov–Smirnov ($p > 0.05$) and the Levene's test ($p < 0.05$) were used to establish the existence of normal distribution and equality of variance of the data, respectively. The results of these tests validated the use of Paired-Samples t-test to check possible significant differences between RIGID and SAND. Effect sizes were estimated after calculating Cohen's d ($\leq$0.49 = small, 0.50–0.79 = medium, $\geq$0.80 = large) [43]. All statistical tests were conducted using the IBM SPSS Statistics v.27.0.1.0 software (International Business Machines Corp., Armonk, NY, USA), with the level of significance set at $a = 0.05$ for all statistical analyses.

3. Results

The results for the spatiotemporal parameters are presented in Table 1. No significant differences ($p > 0.05$) were observed in the examined parameters except for h_{JUMP}, which was lower in SAND (medium effect size).

Table 1. Means ± standard deviations of the comparison for the spatiotemporal parameters of the drop jumps on RIGID and SAND surface ($n = 16$).

Parameter	RIGID	SAND	*MD*	*SE*	*t*	*p*	*d*
Center of Mass displacement (cm)							
Jump height (h_{JUMP})	27.9 ± 4.2	24.4 ± 4.8	3.5	0.9	3.933	0.001 *	0.78
Downward phase	−33.8 ± 12.2	−33.9 ± 8.8	0.1	2.0	0.031	0.976	0.01
Upward phase	39.3 ± 12.6	38.6 ± 12.6	0.6	0.2	0.319	0.754	0.06
Temporal (ms)							
Contact time	408.4 ± 135.5	430.4 ± 121.3	22.0	15.8	1.396	0.183	0.17
Downward time	186.1 ± 72.8	192.0 ± 60.1	5.9	8.7	0.673	0.511	0.09
tvGRF	175.3 ± 82.4	155.3 ± 51.9	19.9	22.1	0.901	0.382	0.29
tP	280.9 ± 126.2	294.8 ± 110.5	13.9	14.2	0.977	0.344	0.12

*: $p < 0.05$; *MD*: mean difference; *SE*: standard error of the mean; h_{JUMP}: jump height; tvGRF: time to achieve maximum vertical Ground Reaction Force; tP: time to achieve maximum power during the upward phase.

The vertical displacement of the BCM was almost identical on either surface for both the downward and upward phase. Significant ($p < 0.05$) differences were observed for the majority of the examined kinetic parameters in the upward but not in the braking phase (Table 2). In detail, larger peak vertical GRF was recorded in RIGID compared to SAND (small effect size). No differences ($p > 0.05$) were observed for the other two components of GRF. RFD was significantly ($p < 0.05$) larger in the downward phase of SAND than in RIGID (large effect size). Power at the downward phase was almost equal between surfaces, but a significantly ($p < 0.05$) larger power output was observed in RIGID compared to SAND in the upward phase (medium effect size). Significantly ($p < 0.05$) lower work was observed both in the downward and upward phase of SAND compared to RIGID (large and medium effect size, respectively). Regarding stiffness, no significant difference ($p > 0.05$) was found for vertical stiffness due to the surface. On the opposite, peak leg stiffness was significantly ($p < 0.05$) lower in SAND than RIGID (medium effect size).

Table 2. Means ± standard deviations of the comparison for the kinetic parameters of the drop jumps on RIGID and SAND surface (n = 16).

Parameter	RIGID	SAND	*MD*	*SE*	*t*	*p*	*d*
Peak Ground Reaction Force (kN)							
Vertical (vGRF; net force)	2.48 ± 0.84	2.14 ± 0.56	0.43	0.15	2.359	0.032 *	0.48
Anterior–Posterior (xGRF)	0.36 ± 0.05	0.37 ± 0.08	0.01	0.03	0.205	0.841	0.15
Mediolateral (yGRF)	0.11 ± 0.04	0.11 ± 0.04	0.01	0.01	0.498	0.627	0.14
Peak Rate of Force Development (kN/s)							
Downward phase	−53.3 ± 14.0	−71.6 ± 25.1	18.3	5.6	3.248	0.005 *	0.90
Upward phase	44.0 ± 11.6	40.1 ± 6.7	3.8	2.6	1.471	0.161	0.41
Peak Power (kW)							
Downward phase	−4.2 ± 1.2	−4.3 ± 1.2	0.1	0.2	0.776	0.289	0.08
Upward phase	3.1 ± 1.0	2.6 ± 0.6	0.5	0.2	2.245	0.040 *	0.61
Peak Work (J)							
Downward phase	−738.4 ± 110.7	−662.6 ±89.2	75.8	21.5	3.518	0.003 *	1.36
Upward phase	778.1 ± 98.6	713.3 ± 86.3	64.8	26.6	2.535	0.023 *	0.70
Stiffness (kN/m)							
Peak Vertical stiffness	11.6 ± 4.0	12.6 ± 3.9	0.8	0.8	1.061	0.305	0.25
Peak Leg stiffness	8.6 ± 4.9	5.1 ± 3.8	3.5	1.5	2.367	0.032 *	0.79
Average Leg stiffness	3.8 ± 2.9	4.0 ± 3.9	0.2	0.9	0.198	0.846	0.06

*: $p < 0.05$; *MD*: mean difference; *SE*: standard error of the mean. Leg stiffness parameters are according to Struzik and Zawadzki [44].

The qualitative examination of the time–history curves of the kinetic parameters revealed almost identical patterns between surfaces (Figure 2). Minor alterations were noted for RFD and vertical stiffness. For the former, a steeper peak was revealed during the initial stage of the downward phase in SAND compared to RIGID (Figure 2b). Vertical stiffness in RIGID exhibited a plateau after reaching its peak value, as for SAND the peak value was of a larger magnitude with respect to the following plateau (Figure 2g).

No significant ($p > 0.05$) differences were observed in the majority of the examined lower limb joint angular kinematical parameters during the downward phase (Table 3). However, a significant difference ($p < 0.05$) was observed for the knee joint range of motion that was larger in SAND compared to RIGID (large effect size). The peak angular velocity of the ankle joint was also significantly different ($p < 0.05$), since it was smaller in SAND than in RIGID (medium effect size). No significant ($p > 0.05$) differences were found in the upward phase.

Table 3. Means ± standard deviations of the comparison for the joint kinematic parameters of the drop jumps on RIGID and SAND surface (n = 16).

Parameter	RIGID	SAND	*MD*	*SE*	*t*	*p*	*d*
Downward phase							
ROM_{ANKLE}	30.30 ± 10.69	37.16 ± 12.51	6.86	3.94	1.742	0.102	0.59
ROM_{KNEE}	42.20 ± 16.84	57.21 ± 14.85	15.01	3.79	3.965	0.001 *	0.95
ROM_{HIP}	23.93 ± 26.37	31.52 ± 18.78	7.59	5.86	1.294	0.215	0.33
ω_{ANKLE}	−6.12 ± 1.74	−5.03 ± 1.76	1.09	0.50	2.168	0.047 *	0.62
ω_{KNEE}	−7.38 ± 1.45	−7.15 ± 0.88	0.23	0.39	0.582	0.569	0.19
ω_{HIP}	−4.05 ± 1.92	−4.23 ± 1.51	0.19	0.37	0.499	0.625	0.10
Upward phase							
ROM_{ANKLE}	64.40 ± 11.66	65.84 ± 11.27	1.44	2.02	0.713	0.487	0.13
ROM_{KNEE}	72.93 ± 17.14	73.65 ± 14.96	0.72	2.90	0.248	0.807	0.05
ROM_{HIP}	66.13 ± 25.18	69.33 ± 18.98	3.20	5.04	0.634	0.536	0.14
ω_{ANKLE}	10.24 ± 1.82	9.62 ± 1.21	0.63	0.36	1.735	0.103	0.40
ω_{KNEE}	10.16 ± 1.29	10.27 ± 0.72	0.11	0.26	0.410	0.687	0.11
ω_{HIP}	8.41 ± 1.13	8.14 ± 0.98	0.27	0.23	1.164	0.263	0.26

*: $p < 0.05$; *MD*: mean difference; *SE*: standard error of the mean; ROM: joint range of motion (in degrees); ω: angular velocity (in rad/s).

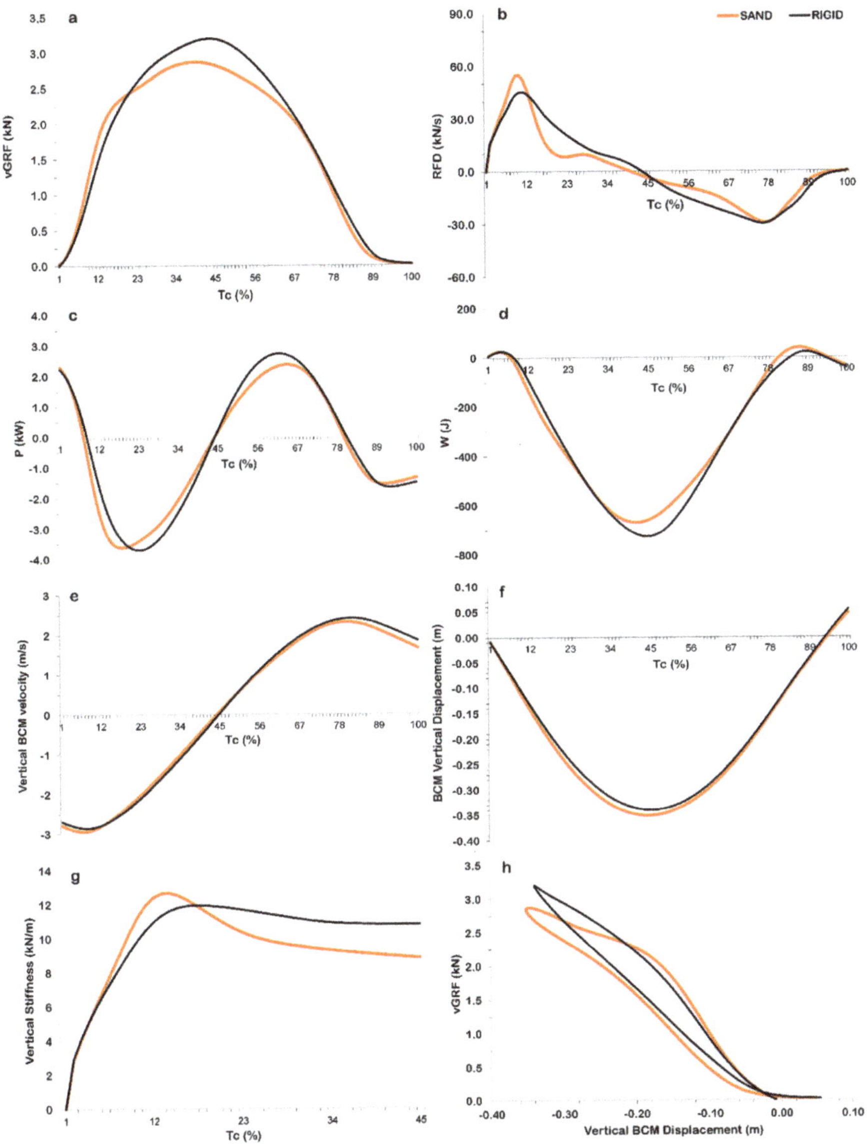

Figure 2. Mean ensemble ($n = 16$) time−history curves for the examined drop jump kinetic parameters on rigid (RIGID) and sand (SAND) surface: (**a**) vertical ground reaction force; (**b**) rate of force development; (**c**) power; (**d**) work; (**e**) body center of mass vertical velocity; (**f**) body center of mass vertical displacement (0 = body center of mass height at the instant of touchdown); (**g**) vertical stiffness; (**h**) vertical stiffness depicted by plotting the vertical body center of mass displacement vs. the vertical ground reaction force. Abbreviations: vGRF: vertical Ground Reaction Force; RFD: Rate of Force Development; P: power; W: work; BCM: body center of mass; Tc: contact time. NOTE: all curves are normalized with respect to Tc; the curves in Figure 2h are depicted for the time period from touchdown to the lowest height of the BCM during the contact with the surface.

The examination of the joint angles at specific instances of the DJ, namely the touchdown, the lowest vertical position of the BCM and the take-off revealed a significant ($p < 0.05$) difference for the knee joint angle at touchdown (Figure 3). In specific, the knee joint was about 13 degrees more extended at touchdown in SAND than in RIGID ($t_{1,15} = -4.202$, $p = 0.001$, $d = 1.23$; large effect size).

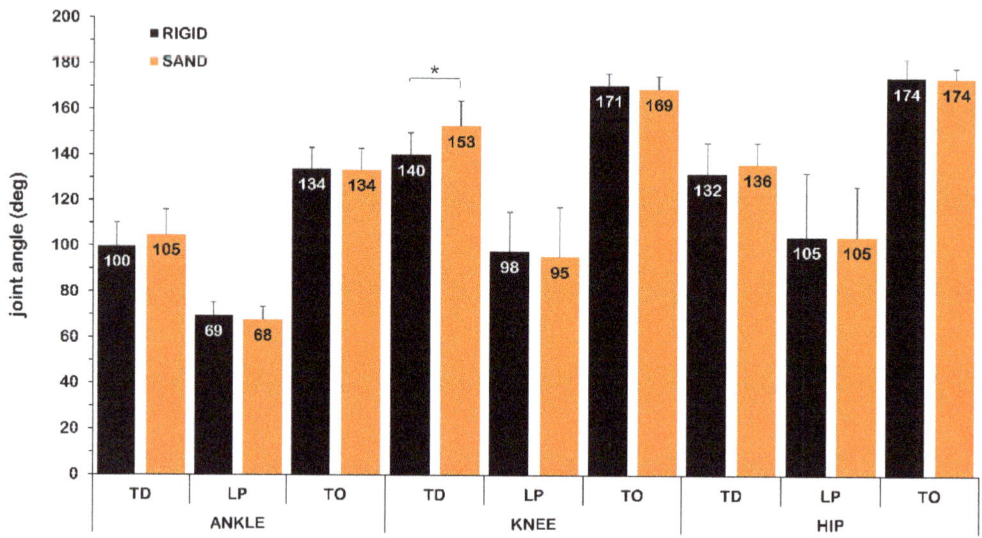

Figure 3. Joint angles at the instants of touchdown (TD), lowest vertical position of the body center of mass (LP) and take-off (TO) of the drop jumps on rigid (RIGID) and sand (SAND) surface ($n = 16$; *: $p < 0.05$).

No significant difference ($p > 0.05$) was revealed for the knee joint at the selected instances of the DJ. In addition, no significant differences ($p > 0.05$) were evident for the ankle and hip joint angles.

The qualitative examination of the time–history curves of the examined angular kinematic parameters also revealed almost identical patterns between surfaces (Figure 4). Minor alterations were noted for the knee joint angle and the angular velocity approximately at the first 30% and at the last 30% of the ground contact phase of the DJ (Figure 4a,c). The deceleration of the body was accompanied with a more rapid knee joint flexion at the downward phase in SAND (Figure 4b). In the upward phase, the knee joint extended to its take-off angular position earlier in SAND than in RIGID (Figure 4d). Regarding leg stiffness (Figure 4f), a similar plateau, in terms of the respective plateau observed for vertical stiffness (Figure 2h), was revealed in the downward phase for SAND.

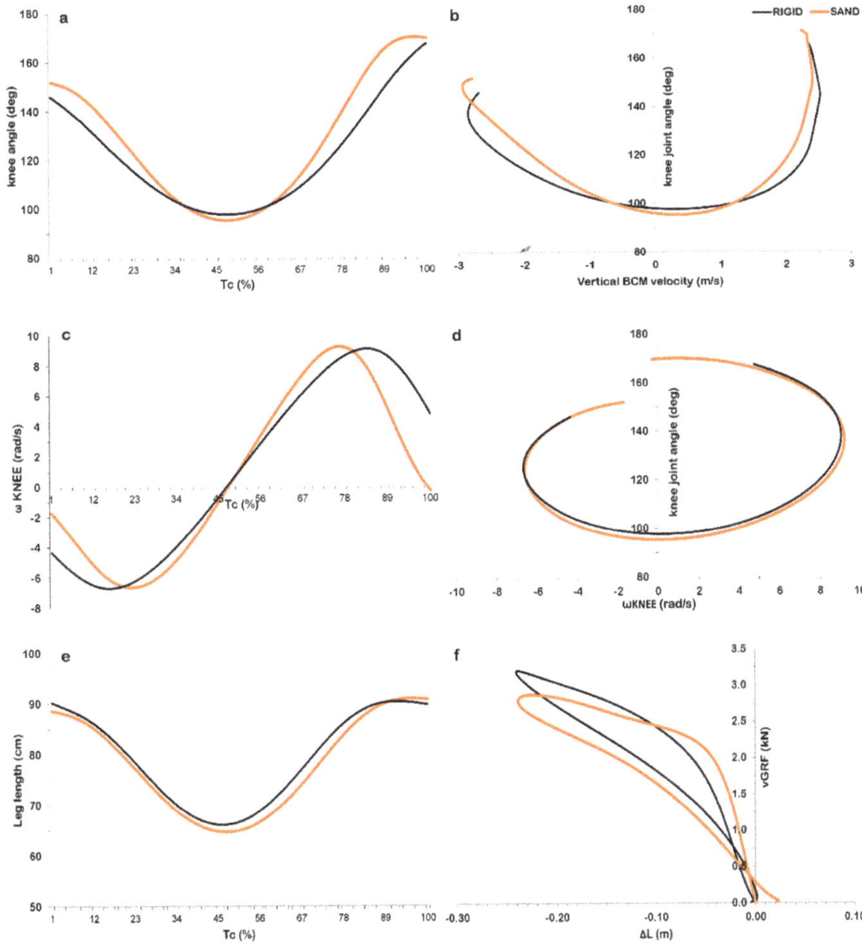

Figure 4. Mean ensemble ($n = 16$) time–history curves for the examined drop jump kinetic parameters on rigid (RIGID) and sand (SAND) surface: (**a**) knee joint angle; (**b**) knee joint angle with respect to the body center of mass vertical velocity; (**c**) angular velocity of the knee joint; (**d**) knee joint angle with respect to the angular velocity of the knee joint; (**e**) leg length; (**f**) change of leg length (0 = leg length at the instant of touchdown from the drop phase) with respect to the vertical Ground Reaction Force. Abbreviations: Tc: contact time; BCM: body center of mass; ωKNEE: angular velocity of the knee joint; ΔL: change of leg length; vGRF: vertical Ground Reaction Force. NOTE: curves depicted in Figure 4a,c,e are normalized with respect to Tc.

4. Discussion

The aim of the study was to compare the biomechanical parameters of DJ executed on SAND and RIGID. Results revealed that DJ on SAND had lower jumping height (−19.8%), force and power output, as well as larger RFD, work, knee ROM and peak ankle angular velocity at the downward phase than RIGID. Thus, the hypothesis of the study was confirmed.

DJ jump height was lower on SAND than RIGID, being in line with past research results concerning the comparison of SQJ and CMJ on different surfaces [11,29,33–36,45,46]. The ground contact time was in considerable agreement with past findings [47,48]. It is commonly agreed among researchers that the lower jumping heights observed in the vertical SQJ and CMJ tests are caused by the lower force and power outputs observed for SAND compared to RIGID [33,35,36]. In the present study, power was significantly lower in the upward phase. Thus, the lower jump height for SAND can be explained by the lower power output, since power is suggested to be a determinant factor for the optimization of DJ performance [40,48–53]. A possible reason for not achieving larger power in SAND can be attributed to the fact that SAND is an unstable surface and inhibits the fast application of force during jumping [29,35,54]. In addition, as depicted in Figure 4, a more rapid knee joint flexion was at the downward phase in SAND. This finding, in combination with the lower angular velocity of the ankle flexion, reveals a different pattern to negotiate the deceleration of the body due to the different stiffness of the surface to execute the jump.

Unlike previous observations in SQJ [35] and CMJ [36], the time to achieve maximum vertical GRF on SAND was not different compared to RIGID. This finding, combined with the larger RFD during the downward phase in SAND indicates that ground contact with the sand was highly unstable. This led the participants to make a strenuous effort to overcome these constraints that were imposed for the execution of the jumping task. However, the medio-lateral and anterio-posterior components of GRF were not different between surfaces. Thus, there is an indirect indication that the balance requirements at the initial phase of the DJ did not differ between surfaces. Nevertheless, it is suggested that the deformation of SAND increases the requirements for dynamic stability [6,12,17,55]. Additionally, SAND comprises a demanding surface to execute explosive movements since its surface is characterized by larger friction compared to other sport surfaces [33]. The interaction with SAND during exercise utilizing the SSC is suggested to absorb large amounts of energy [3,10,11,29,33,34]. In addition, jumping on SAND utilizing the SSC is proposed to lead to lower re-use of the stored elastic energy [10]. These factors eventually result in increased work expenditure. Furthermore, recent research evidence suggests the existence of an additional protective neuromuscular mechanism when "dealing" with landings on harder, less "safe" surfaces, guaranteed even by visual input alone [56].

In the present study, less negative work was done in the downward phase in SAND compared to RIGID. This could be an indirect indicator of a lower rate of energy absorption during the downward phase in SAND [38]. Nevertheless, the lesser negative work could be associated with smaller amounts of elastic energy stored in the series elastic elements and eventually with the lower jumping height in SAND [10,38]. Thus, due to the observed bias, this point has to be further investigated in future studies examining DJ on SAND.

The importance of the knee joint biomechanics as a regulator of DJ performance has been highlighted in past research [57]. resent data revealed that the knee joint extended to its take-off angular position earlier in SAND than RIGID. Thus, less power was generated about the knee joint that could be a cause for the lower DJ performance found in SAND. Power, besides suggested to be a determining factor for DJ performance, is related with changes in vertical stiffness after plyometric training [58]. Furthermore, the mechanical power produced during the upward phase of a DJ is maximized when an optimal leg stiffness occurs in the downward phase [25,27]. In the present study, leg stiffness was lower in SAND. This is not in agreement with past research reporting that equal leg stiffness can be exhibited when performing a DJ on surfaces with different compliance [59]. Leg stiffness is affected by the knee and ankle stiffness [60]. In the present study, both joints showed significant differences concerning their angular kinematics. This finding could be related with the significantly larger peak leg stiffness observed in RIGID. Nevertheless, a notable maintenance of relatively constant values regarding leg stiffness was observed in both RIGID and SAND, confirming past findings [44,59]. In general, lower stiffness during landing is proposed to be related with mechanisms of long-term adaptations caused by eccentric exercise aiming to prevent injuries [61,62]. In conclusion, larger stiffness is related with the inability to resist large eccentric loadings [63], which seems to be the case for RIGID.

Compared to RIGID, DJ on SAND was executed with a more extended knee joint at the instant of impact. In addition, during the downward phase, the knee joint range of motion was larger, while the ankle joint flexed with a slower angular velocity. This combination was reported in the past for DJs executed from higher compared to lower dropping heights [38] and could be considered as a protective mechanism to avoid excessive loading. In addition, in the present study, no differences were observed in the upward phase between the examined surfaces for the lower limb joint angular kinematics. This is not in alignment with past research concerning running or sprinting on SAND, where a backward movement of the feet due to the deformation of the sand at the end of the push off is common [1,8,10]. This might be the result of the relatively long duration of the ground contact time, during which SAND could have been compressed at the downward phase. This could eventually lead the sand surface to dissipate some of its absorptive qualities and thus resemble a more rigid surface [36]. The present finding also cannot support research evidence which suggested that performing SSC on SAND is related to the muscle action during the propulsive phase that resulted by the compensation made in the braking phase in terms of the degradation of elastic energy [10,13]. However, previous studies for the CMJ on SAND reported a significant effect of surface for the knee [45] and ankle joint kinematics [36]. The connection between the two aforementioned joints is the biarticular gastrocnemius muscle, which can affect the ankle range of motion and can cause differences in key biomechanical factors of DJ, such as RFD and knee angular kinematics [64]. It has been proposed that the common adaptation of the neuromuscular system to deal with the differences caused by the instable SAND surface is to exhibit a higher co-contraction of the lower limb muscles, which eventually results in a less optimum flow of energy [1]. However, it was found that muscle activity and muscle–tendon unit mechanical properties of the gastrocnemius muscle increase when jumping from a deformable surface. This led to the conclusion that "internal regulatory mechanisms exist to compensate for differences in surface properties" [65]. Both of the above factors should be further investigated in the future.

Training on SAND was found to be effective for improving sprinting, jumping and balance ability of team sport players [14,20,66–68], and that there is a significant association between specific agility and vertical jump tests on SAND [46]. Additionally, the adaptations of jumping, sprinting and agility were found to be transferred on RIGID [13,15,16]. One reason is the increased motor unit recruitment after the implementation of plyometric training on SAND [21]. In general, performing exercises on SAND is suggested to reduce the musculoskeletal loading in training and rehabilitation programs [69,70]. SAND training is proposed for preseason training due to the decreased muscle soreness, faster recovery and the lower probability of overuse injuries [1,55,71]. Larger training adaptations are expected since heavier training loads can be implemented on SAND [10,56]. However, due to the lower stiffness of SAND, there are limited neuromuscular adaptations since the mechanical stimuli on the musculoskeletal system are reduced. Thus, SAND is less effective for the improvement of explosive movements [10,13,72]. The results of the present biomechanical analysis of DJ on SAND seem to confirm the above notion.

The findings of the present study should be considered given its limitations. The recording of muscle activation patterns during DJ could provide further insight concerning the examination of the regulation of stiffness. Another possible limitation is the fact that only one drop height was selected for analysis; thus, the present results cannot be generalized to interpret DJ on SAND and should be read with caution. Nevertheless, the present study revealed an insight regarding the biomechanics of DJ on SAND that provides information for a widely used jumping modality in training practice and testing environments. It is of importance that the participants in the study were top-level BV players which had extensive training experience in DJ on both SAND and RIGID. Thus, the comparison of DJ biomechanical parameters between the examined surfaces can be considered to be reliable. The present findings are of interest to coaches and researchers, particularly under the perspective of the kinetic and kinematic differences of executing the drop jump on a rigid and on a sand surface.

5. Conclusions

The compliance of the SAND seems to decrease the efficiency of the mechanisms involved in optimizing the DJ performance compared to RIGID. Nevertheless, SAND comprises an exercise surface that imposes a lesser load during the eccentric phase of the DJ compared to RIGID. Therefore, SAND can be used in jumping programs aimed at injury prevention or for rehabilitation programs after an injury in lower extremities. In addition, due to the highly unstable surface of SAND, participants were found to increase knee joint range of motion during the downward phase to fulfill the locomotor requirements to execute the jumping task and to acquire the necessary stability to do so. Finally, due to the higher energy expenditure required on SAND, DJs can be used in the pre- or off-training season not only in beach volleyball, but also in other team and individual sports that include jumping activities in their technique.

Author Contributions: Conceptualization, G.G. and I.A.K.; methodology, G.G. and I.A.K.; software, I.A.K.; validation, G.G., V.P. and I.A.K.; formal analysis, G.G.; investigation, G.G. and V.P.; resources, I.A.K.; data curation, G.G.; writing—original draft preparation, G.G.; writing—review and editing, G.G and V.P.; visualization, G.G. and V.P.; supervision, I.A.K.; project administration, G.G. All authors have read and agreed to the published version of the manuscript.

Funding: This research received no external funding.

Institutional Review Board Statement: The study was conducted in accordance with the Declaration of Helsinki, and approved by the Institutional Ethics Committee of the School of Physical Education and Sport Science at Thessaloniki, Aristotle University of Thessaloniki, Greece (approval No: 87/2021—11 April 2021).

Informed Consent Statement: Informed consent was obtained from all subjects involved in the study.

Data Availability Statement: The data that were acquired and analyzed in the present study are available from the corresponding author upon reasonable request.

Acknowledgments: The authors thank the participating beach-volleyball players for their voluntary participation in the study.

Conflicts of Interest: The authors declare no conflict of interest.

References

1. Gaudino, P.; Gaudino, C.; Alberti, G.; Minetti, A.E. Biomechanics and predicted energetics of sprinting on sand: Hints for soccer training. *J. Sci. Med. Sport* **2013**, *16*, 271–275. [CrossRef]
2. Lejeune, T.M.; Willems, P.A.; Heglund, N.C. Mechanics and energetics of human locomotion on sand. *J. Exp. Biol.* **1998**, *201*, 2071–2080. [CrossRef]
3. Pinnington, H.C.; Dawson, B. The energy cost of running on grass compared to soft dry beach sand. *J. Sci. Med. Sport* **2001**, *4*, 416–430. [CrossRef]
4. Pinnington, H.C.; Dawson, B. Running economy of elite surf iron men and male runners, on soft dry beach sand and grass. *Eur. J. Appl. Physiol.* **2001**, *86*, 62–70. [CrossRef]
5. Zamparo, P.; Perini, R.; Orizio, C.; Sacher, M.; Ferretti, G. The energy cost of walking or running on sand. *Eur. J. Appl. Physiol. Occup. Physiol.* **1992**, *65*, 183–187. [CrossRef]
6. Panebianco, G.P.; Bisi, M.C.; Mangia, A.L.; Fantozzi, S.; Stagni, R. Quantitative characterization of walking on sand inecological conditions: Speed, temporal segmentation, and variability. *Gait Posture* **2021**, *86*, 211–216. [CrossRef] [PubMed]
7. Yun, S.H.; Cho, M.J.; Kwon, J.W. Comparison of forward versus backward walking on spatiotemporal and kinematic parameters on sand: A preliminary study. *J. Biomech.* **2021**, *130*, 110876. [CrossRef] [PubMed]
8. Alcaraz, P.E.; Palao, J.M.; Elvira, J.L.L.; Linthorne, N.P. Effects of a sand running surface on the kinematics of sprinting at maximum velocity. *Biol. Sport* **2011**, *28*, 95–100. [CrossRef]
9. Sanchez-Sanchez, J.; Martinez-Rodriguez, A.; Felipe, J.L.; Hernandez-Martin, A.; Ubago-Guisado, E.; Bangsbo, J.; Gallardo, L.; Garcia-Unanue, J. Effect of natural turf, artificial turf, and sand surfaces on sprint performance. A systematic review and meta-analysis. *Int. J. Environ. Res. Public Health* **2020**, *17*, 9478. [CrossRef]
10. Impellizzeri, F.M.; Rampinini, E.; Castagna, C.; Martino, F.; Fiorini, S.; Wisloff, U. Effect of plyometric training on sand versus grass on muscle soreness and jumping and sprinting ability in soccer players. *Br. J. Sports Med.* **2008**, *42*, 42–46. [CrossRef]
11. Muramatsu, S.; Fukudome, A.; Miyama, M.; Arimoto, M.; Kijima, A. Energy expenditure in maximal jumps on sand. *J. Physiol. Anthropol.* **2006**, *25*, 59–61. [CrossRef]

12. Smith, R. Movement in the sand: Training implications for beach volleyball. *Strength Cond. J.* **2006**, *28*, 19–21. [CrossRef]
13. Binnie, M.J.; Dawson, B.; Pinnington, H.; Landers, G.; Peeling, P. Sand training: A review of current research and practical applications. *J. Sports Sci.* **2014**, *32*, 8–15. [CrossRef] [PubMed]
14. Ahmadi, M.; Nobari, H.; Ramirez-Campillo, R.; Pérez-Gómez, J.; Ribeiro, A.L.d.A.; Martínez-Rodríguez, A. Effects of plyometric jump training in sand or rigid surface on jump-related biomechanical variables and physical fitness in female volleyball players. *Int. J. Environ. Res. Public Health* **2021**, *18*, 13093. [CrossRef]
15. Balasas, D.G.; Christoulas, K.; Stefanidis, P.; Vamvakoudis, E.; Bampouras, T. The effect of beach volleyball training on muscle performance of indoor volleyball players. *J. Sports Med. Phys. Fitness* **2018**, *58*, 1240–1246. [CrossRef]
16. Gortsila, E.; Theos, A.; Nesic, G.; Maridaki, M. Effect of training surface on agility and passing skills of prepubescent female volleyball players. *J. Sports Med. Doping Stud.* **2013**, *3*, 1000128. [CrossRef]
17. Sebastia-Amat, S.; Ardigò, L.P.; Jimenez-Olmedo, J.M.; Pueo, B.; Penichet-Tomas, A. The effect of balance and sand training on postural control in elite beach volleyball players. *Int. J. Environ. Res. Public Health* **2020**, *17*, 8981. [CrossRef] [PubMed]
18. Suresh, T.N.; Jayalakshmi, T.; Begu, R.; Meeran, N.; Sivakumar, V.P.R. Effect of plyometric exercise training on vertical jump height between ground and sand surface in male volleyball players. *Int. J. Pharma. Biol. Sci.* **2017**, *8*, 163–169.
19. Trajkovic, N.; Sporis, G.; Kristicevic, T. Does training on sand during off-season improves physical performance in indoor volleyball players? *Acta Kinesiol.* **2016**, *10*, 107–111.
20. Mirzaei, B.; Asghar Norasteh, A.; Saez de Villarreal, E.; Asadi, A. Effects of six weeks of depth jump vs. countermovement jump training on sand on muscle soreness and performance. *Kinesiology* **2014**, *46*, 97–108.
21. Mirzaei, B.; Norasteh, A.A.; Asadi, A. Neuromuscular adaptations to plyometric training: Depth jump vs. countermovement jump on sand. *Sport Sci. Health* **2013**, *9*, 145–149. [CrossRef]
22. Bobbert, M.F. Drop jumping as a training method for jumping ability. *Sports Med.* **1990**, *9*, 7–22. [CrossRef] [PubMed]
23. Ziv, G.; Lidor, R. Vertical jump in female and male volleyball players: A review of observational and experimental studies. *Scand. J. Med. Sci. Sports* **2010**, *20*, 556–567. [CrossRef]
24. LaStayo, P.C.; Woolf, J.M.; Lewek, M.D.; Snyder-Macler, L.; Reich, T.E.; Lindstedt, S.L. Eccentric muscle contractions: Their contribution to injury, prevention, rehabilitation, and sport. *J. Orthop. Sports Phys. Ther.* **2003**, *33*, 557–571. [CrossRef]
25. Pearson, S.J.; McMahon, J. Lower limb mechanical properties: Determining factors and implications for performance. *Sports Med.* **2012**, *42*, 929–940. [CrossRef] [PubMed]
26. Walshe, A.D.; Wilson, G.J.; Ettema, G.J. Stretch-shorten cycle compared with isometric preload: Contributions to enhanced muscular performance. *J. Appl. Physiol.* **1998**, *84*, 97–106. [CrossRef]
27. Arampatzis, A.; Schade, F.; Walsh, M.; Bruggemann, G.P. Influence of leg stiffness and its effect on myodynamic jumping performance. *J. Electromyogr. Kinesiol.* **2001**, *11*, 355–364. [CrossRef]
28. Bojsen-Moller, J.; Magnusson, S.P.; Rasmussen, L.R.; Kjaer, M.; Aagaard, P. Muscle performance during maximal isometric and dynamic contractions is influenced by the stiffness of the tendinous structures. *J. Appl. Physiol.* **2005**, *99*, 986–994. [CrossRef]
29. Bishop, D.A. Comparison between land and sand-based tests for beach volleyball assessment. *J. Sports Med. Phys. Fitness* **2003**, *43*, 418–423.
30. Giatsis, G.; Pérez-Turpin, J.A.; Hatzimanouil, D. Analysis of time characteristics, jump patters and technical-tactical skills of beach volley men's final in Rio Olympics 2016. *J. Hum. Sport Exerc.* **2020**, *15*, S1013–S1019. [CrossRef]
31. Ricarte Batista, G.; Freire De Araujo, R.; Oliveira Guerra, R. Comparison between vertical jumps of high performance athletes on the Brazilian men's beach volleyball team. *J. Sports Med. Phys. Fitness* **2008**, *48*, 172–176.
32. Schmidt, M.; Meyer, E.; Jaitner, T. Quantifying jump-specific loads in beach volleyball by an inertial measurement device. *Int. J. Sports Sci. Coach.* **2021**, *16*, 391–397. [CrossRef]
33. Bisciotti, G.N.; Ruby, A.; Jaquemod, C. Biomechanics of jumps in the volleyball and in the beach-volley. *Riv. Cult. Sport.* **2001**, *20*, 29–34.
34. Miyama, M.; Nosaka, K. Influence of surface on muscle damage and soreness induced by consecutive drop jumps. *J. Strength Cond. Res.* **2004**, *18*, 206–211. [CrossRef]
35. Giatsis, G.; Kollias, I.; Panoutsakopoulos, V.; Papaiakovou, G. Biomechanical differences in elite beach-volleyball players in vertical squat jump on rigid and sand surface. *Sports Biomech.* **2004**, *3*, 145–158. [CrossRef] [PubMed]
36. Giatsis, G.; Panoutsakopoulos, V.; Kollias, I.A. Biomechanical differences of arm swing countermovement jumps on sand and rigid surface performed by elite beach volleyball players. *J. Sports Sci.* **2018**, *36*, 997–1008. [CrossRef] [PubMed]
37. Faul, F.; Erdfelder, E.; Lang, A.G.; Buchner, A. G* Power 3: A flexible statistical power analysis program for the social, behavioral, and biomedical sciences. *Behav. Res. Methods* **2007**, *39*, 175–191. [CrossRef]
38. Bobbert, M.F.; Huijing, P.A.; van Ingen Schenau, G.J. Drop jumping. II. The influence of dropping height on the biomechanics of drop jumping. *Med. Sci. Sports Exerc.* **1987**, *19*, 339–346. [CrossRef]
39. McKinley, P.; Pedotti, A. Motor strategies in landing from a jump: The role of skill in task execution. *Exp. Brain Res.* **1992**, *90*, 427–440. [CrossRef]
40. Kollias, I.; Panoutsakopoulos, V.; Papaiakovou, G. Comparing jumping ability among athletes of various sports: Vertical drop jumping from 60 centimeters. *J. Strength Cond. Res.* **2004**, *18*, 546–550. [CrossRef]
41. Baca, A. A comparison of methods for analyzing drop jump performance. *Med. Sci. Sports Exerc.* **1999**, *31*, 437–442. [CrossRef] [PubMed]

42. Brughelli, M.; Cronin, J. A review of research on the mechanical stiffness in running and jumping: Methodology and implications. *Scand. J. Med. Sci. Sports* **2008**, *18*, 417–426. [CrossRef]
43. Cohen, J. *Statistical Power Analysis for the Behavioral Sciences*, 2nd ed.; Lawrence Erlbaum Associates Publishers: Hillsdale, NJ, USA, 1988.
44. Struzik, A.; Zawadzki, J. Application of force-length curve for determination of leg stiffness during a vertical jump. *Acta Bioeng. Biomech.* **2016**, *18*, 163–170. [CrossRef] [PubMed]
45. Arianasab, H.; Mohammadipour, F.; Amiri-Khorasani, M. Comparison of knee joint kinematics during a countermovement jump among different sports surfaces in male soccer players. *Sci. Med. Footb.* **2017**, *1*, 74–79. [CrossRef]
46. Buscà, B.; Alique, D.; Salas, C.; Hileno, R.; Pena, J.; Morales, J.; Bantulà, J. Relationship between agility and jump ability in amateur beach volleyball male players. *Int. J. Perform. Anal. Sport* **2015**, *15*, 1102–1113. [CrossRef]
47. Laffaye, G.; Bardy, B.; Taiar, R. Upper-limb motion and drop jump: Effect of expertise. *J. Sports Med. Phys. Fitness* **2006**, *46*, 536–543.
48. Laffaye, G.; Choukou, M.A. Gender bias in the effect of dropping height on jumping performance in volleyball players. *J. Strength Cond. Res.* **2010**, *24*, 2143–2148. [CrossRef]
49. Ferreira, L.C.; Weiss, L.W.; Hammond, K.G.; Schilling, B.K. Structural and functional predictors of drop vertical jump. *J. Strength Cond. Res.* **2010**, *24*, 2456–2467. [CrossRef]
50. Pietraszewski, B.; Rutkowska-Kucharska, A. Relative power of lower limbs in drop jump. *Acta Bioeng. Biomech.* **2011**, *13*, 13–18.
51. Viitasalo, J.T.; Salo, A.; Lahtinen, J. Neuromuscular functioning of athletes and non-athletes in the drop jump. *Eur. J. Appl. Physiol. Occup. Physiol.* **1998**, *78*, 432–440. [CrossRef]
52. Walsh, M.; Arampatzis, A.; Schade, F.; Brüggemann, G.P. The effect of drop jump starting height and contact time on power, work performed, and moment of force. *J. Strength Cond. Res.* **2004**, *18*, 561–566. [CrossRef] [PubMed]
53. Young, W.B.; Wilson, C.J.; Byrne, C. A comparison of drop jump training methods: Effects on leg extensor strength qualities and jumping performance. *Int. J. Sports Med.* **1999**, *20*, 295–303. [CrossRef] [PubMed]
54. Dal Bello, F.; Aedo Muñoz, E.; Gomes Moreira, D.; Brito, C.J.; Miarka, B.; Navarro Cabello, E. Beach and indoor volleyball athletes present similar muscular activation of lower limbs during the countermovement jump. *Hum. Mov.* **2019**, *21*, 43–50. [CrossRef]
55. Binnie, M.J.; Dawson, B.; Pinnington, H.; Landers, G.; Peeling, P. Part 2: Effect of training surface on acute physiological responses after sport-specific training. *J. Strength Cond. Res.* **2013**, *27*, 1057–1066. [CrossRef] [PubMed]
56. De Sire, A.; Marotta, N.; Demeco, A.; Moggio, L.; Paola, P.; Marotta, M.; Iona, T.; Invernizzi, M.; Leigheb, M.; Ammendolia, A. Electromyographic assessment of anterior cruciate ligament injury risk in male tennis players: Which role for visual input? A proof-of-concept study. *Diagnostics* **2021**, *11*, 997. [CrossRef]
57. Horita, T.; Komi, P.; Nicol, C.; Kyrolainen, H. Interaction between pre-landing activities and stiffness regulation of the knee joint musculoskeletal system in the drop jump: Implications to performance. *Eur. J. Appl. Physiol.* **2002**, *88*, 76–84. [CrossRef]
58. Rojano Ortega, D.; Berral-Aguilar, A.J.; Berral de la Rosa, F.J. Kinetics and vertical stiffness of female volleyball players: Effect of low-intensity plyometric training. *Res. Q. Exerc. Sport* **2021**. *(accepted and published on-line)*. [CrossRef]
59. Arampatzis, A.; Stafilidis, S.; Morey-Klapsing, G.; Brüggemann, G.P. Interaction of the human body and surfaces of different stiffness during drop jumps. *Med. Sci. Sports Exerc.* **2004**, *36*, 451–459. [CrossRef]
60. Arampatzis, A.; Bruggemann, G.P.; Morey-Klapsing, G.M. Leg stiffness and mechanical energetic processes during jumping on a sprung surface. *Med. Sci. Sports Exerc.* **2001**, *33*, 923–931. [CrossRef]
61. Butler, R.J.; Crowell III, H.P.; Davis, I.M. Lower extremity stiffness: Implications for performance and injury. *Clin. Biomech.* **2003**, *18*, 511–517. [CrossRef]
62. Hewett, T.E.; Lindenfeld, T.N.; Riccobene, J.V.; Noyes, F.R. The effect of neuromuscular training on the incidence of knee injury in female athletes. *Am. J. Sports Med.* **1999**, *27*, 699–706. [CrossRef]
63. Walshe, A.D.; Wilson, G.J. The influence of musculotendinous stiffness on drop jump performance. *Can. J. Appl. Physiol.* **1997**, *22*, 117–132. [CrossRef] [PubMed]
64. Papaiakovou, G.; Katsikas, F.; Nikodelis, T.; Panoutsakopoulos, V.; Kollias, I. Influence of the ankle joint dorsiflexion on the execution of vertical jumps. In Proceedings of the XXIV International Symposium on Biomechanics in Sports, Salzburg, Austria, 14–18 July 2006; Schwameder, H., Strutzenberger, G., Fastenbauer, V., Lindinger, S., Muller, E., Eds.; University of Salzburg: Salzburg, Austria, 2006; Volume 1, pp. 448–451.
65. Hollville, E.; Rabita, G.; Guilhem, G.; Lecompte, J.; Nordez, A. Effects of surface properties on gastrocnemius medialis and vastus lateralis fascicle mechanics during maximal countermovement jumping. *Front. Physiol.* **2020**, *11*, 917. [CrossRef]
66. Bonavolontà, V.; Carvutto, R.; Di Gioia, A.; De Candia, M. Plyometric training on sand versus grass: Effects on sprinting, jumping, agility and balance in soccer players. *J. Hum. Sport Exerc.* **2021**, *16*, S1104–S1113. [CrossRef]
67. Hammami, M.; Bragazzi, N.L.; Hermassi, S.; Gaamouri, N.; Aouadi, R.; Shephard, R.J.; Chelly, M.S. The effect of a sand surface on physical performance responses of junior male handball players to plyometric training. *BMC Sports Sci. Med. Rehab.* **2020**, *12*, 26. [CrossRef] [PubMed]
68. Pereira, L.A.; Freitas, T.T.; Marín-Cascales, E.; Bishop, C.; McGuigan, M.R.; Loturco, I. Effects of training on sand or hard surfaces on sprint and jump performance of team-sport players: A systematic review with meta-analysis. *Strength Cond. J.* **2021**, *43*, 56–66. [CrossRef]

69. Brown, H.; Dawson, B.; Binnie, M.J.; Pinnington, H.; Sim, M.; Clemons, T.D.; Peeling, P. Sand training: Exercise-induced muscle damage and inflammatory responses to matched-intensity exercise. *Eur. J. Sport Sci.* **2017**, *17*, 741–747. [CrossRef]
70. Richardson, M.C.; Murphy, S.; Macpherson, T.; English, B.; Spears, I.; Chesterton, P. Effect of sand on knee load during a single-leg jump task: Implications for injury prevention and rehabilitation programs. *J. Strength Cond. Res.* **2020**, *34*, 3164–3172. [CrossRef]
71. Binnie, M.J.; Dawson, B.; Arnot, M.A.; Pinnington, H.; Landers, G.; Peeling, P. Effect of sand versus grass training surfaces during an 8-week pre-season conditioning programme in team sport athletes. *J. Sports Sci.* **2014**, *32*, 1001–1012. [CrossRef]
72. Binnie, M.J.; Dawson, B.; Pinnington, H.; Landers, G.; Peeling, P. Effect of surface-specific training on 20 m sprint performance on sand and grass surfaces. *J. Strength Cond. Res.* **2013**, *27*, 3515–3520. [CrossRef]

Journal of
Functional Morphology
and Kinesiology

Article

Exploring Physical Fitness Profile of Male and Female Semiprofessional Basketball Players through Principal Component Analysis—A Case Study

Carlos D. Gómez-Carmona [1,*], David Mancha-Triguero [1,*], José Pino-Ortega [2] and Sergio J. Ibáñez [1]

[1] Research Group in Optimization of Training and Sports Performance (GOERD), Faculty of Sport Sciences, University of Extremadura, Av. de la Universidad, s/n, 10005 Caceres, Spain; sibanez@unex.es
[2] BioVetMed & Sport Sci Research Group, Physical Activity and Sports Department, Sport Science Faculty, University of Murcia, 30720 San Javier, Murcia, Spain; josepinoortega@um.es
* Correspondence: cdgomezcarmona@unex.es (C.D.G.-C.); davidmancha@unex.es (D.M.-T.)

Abstract: Basketball is a sport in continuous evolution, being one of these key aspects of the players' physical fitness that has an impact on the game. Therefore, this study aimed to characterize and identify the physical fitness level and profiles of basketball players according to sex. Total of 26 semi-professional basketball players were assessed (13 male, 13 female) through inertial devices in different previously validated fitness tests. T-test for independent samples and principal component analysis were used to analyze sex-related differences and to identify physical fitness profiles. The results showed differences according to sex in all physical fitness indexes ($p < 0.01$; $d > 1.04$) with higher values in males, except in accelerometer load during small-sided games ($p = 0.17$; $d < 0.20$). Four principal components were identified in male and female basketball players, being two common ([*PC1*] aerobic capacity and in-game physical conditioning, [*PC4* male, *PC3* female] unipodal jump performance) and two different profiles (male: [*PC2*] bipodal jump capacity and acceleration, [*PC3*] curvilinear displacement; female: [*PC2*] bipodal jump capacity and curvilinear displacement, [*PC4*] deceleration). In conclusion, training design must be different and individualized according to different variables, including physical fitness profiles between them. For practical applications, these results will allow knowing the advantages and weaknesses of each athlete to adapt training tasks and game systems based on the skills and capabilities of the players in basketball.

Keywords: team sports; sex-related differences; physical demands; assessment; inertial devices

Citation: Gómez-Carmona, C.D.; Mancha-Triguero, D.; Pino-Ortega, J.; Ibáñez, S.J. Exploring Physical Fitness Profile of Male and Female Semiprofessional Basketball Players through Principal Component Analysis—A Case Study. *J. Funct. Morphol. Kinesiol.* **2021**, *6*, 67. https://doi.org/10.3390/jfmk6030067

Academic Editors: Cristina Cortis, Andrea Fusco and Carl Foster

Received: 16 June 2021
Accepted: 10 August 2021
Published: 13 August 2021

1. Introduction

Basketball is a team sports with dynamic behavior that combines high-intensity actions with specific technical-tactical abilities of the sport, being the level of this abilities and skills more relevant when the competitive level increases [1]. During games, basketball players should be adapted for internal demands that consist of repeated efforts with variable intensity and incomplete rests [2]. Regarding external workload, basketball players covered four-to-six kilometers per game [3,4], realize 400-to-550 changes of direction [5,6], 20–40 accelerations > 3 m/s^2 [7,8], and around 1000 high-intensity actions, two jumps per minute or 45 sprints of few duration [9], resulting in a PlayerLoad of 450–650 a.u. [7,8].

These physical and physiological demands are conditioned by anthropometrical measurements (height, weight, wingspan, etc.) and physical parameters (strength, power, aerobic, and anaerobic capacity, etc.) [10,11], being the physical capacity the most variable aspect throughout a season [6]. It could be influenced by different contextual factors such as sex, age, and competitive level [3,4]. Previous studies have identified sex-related differences with higher demands in male players and an increase difference in relation with the maturity development [12]. In this sense, higher values in male players were found in different capacities/abilities with such as aerobic capacity (related to the body size and age) [13], speed (related to the type II-B muscular fibers), strength (higher size of muscular belly), or agility (related to Q-angle and hip abduction) [11].

All factors influence the playing position and the technical-tactical role in the game. Players with high speed of displacement and agility play further to the basket, while players with higher height, weight, and strength play closer to the basket [6,10]. For this reason, three (guard, forward and center) or five (center, power forward, small forward, point guard, and shooting guard) playing positions have been used traditionally [4,14]. Instead, the playing positions in today's basketball are based on a compendium of physical, technical, and tactical aspects. Recent studies identified between eight and thirteen player's profiles in the National Basketball Association (NBA) [15,16], due to the high specialization of players (each player has a specific role in the game).

For this reason, different assessments need to be realized in different season periods with the aim to control the evolution of the physical fitness and to adapt individually the external and internal workloads during training sessions [17]. To realize the assessment of physical fitness, previously validated field or laboratory tests should be used preferably if they are adapted to sport [18,19]. From the data obtained through these assessment, new mathematical methods could be implemented such as exploratory factor analysis (EFA) to explain many registered variables in a number of extracted factors [20]. Therefore, the purposes of the study were to characterize the physical profiles of male and female basketball players through principal component analysis (PCA), to analyze the differences of physical performance between players' profiles, and to identify the relationship between players' physical profile and the in-game assigned role by the coach.

2. Methods

2.1. Design and Procedures

A cross-sectional study was designed to characterize the physical profiles of basketball players through previous validated field tests, as well as to compare the different profiles obtained and identify the relationship between physical capacities and the playing positions in official games. The physical fitness field tests were performed in two sessions (one for each sex) and at the same time of day (i.e., 9:00 to 11:00 a.m.), under similar environmental conditions (temperature 21.5 ± 0.2 °C; humidity: 42.1 ± 1.2%), and in non-fasting conditions. The order of tests was realized following a previous validated protocol [18]: (1) 6.75-m arc; (2) single leg hop (right and left); (3) abalakov test; (4) 16.25-m RSA; (5) 30–15 IFT; (6) 10 × 15-m 3 vs. 3 small-sided game. The experimental design was detailed in Figure 1.

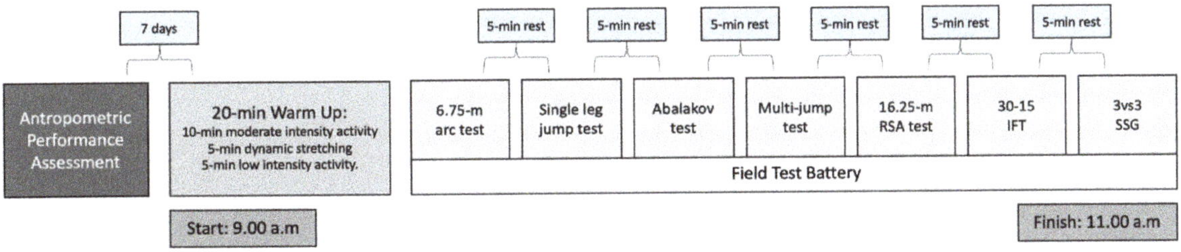

Figure 1. Experimental design of the present study.

Before starting, the basketball players went through the same standardized warm-up that they regularly do before the competition. The warm-up consisted of three work phases with a maximum duration of 20 min [21]. In the first 10-min duration, the players performed moderate activity. In the second phase, the players performed dynamic stretches lasting 5 min. In phase three, players performed low intensity activity for 5 min to prepare for the tests. The tests were performed during training session MD-4 (i.e., four days before the next match day). All tests were realized in their usual indoor training court. During testing, players wore a WIMU PRO^{TM} inertial device (RealTrack Systems, Almeria,

Spain) that register time-motion analysis through ultra-wide band (UWB) radiofrequency technology and microelectromechanical sensors.

2.2. Participants

Twenty-six semi-professional basketball players (n = 13 male, n = 13 female). With age: 18.98 ± 1.84 years (Male: 19.48 ± 1.41 years; Female: 18.49 ± 2.27 years), body mass: 77.13 ± 9.46 kg. (Male: 87.63 ± 7.98 kg; Female: 66.64 ± 10.94 kg), height: 1.82 ± 0.075 m. (Male: 1.91 ± 0.07 m; Female: 1.73 ± 0.08 m), body mass index (BMI): 23.11 ± 1.45 kg/m^2 (Male: 23.98 ± 1.45 kg/m^2; Female: 22.25 ± 3.15 kg/m^2), muscle mass percentage: 75.23 ± 6.98% (Male: 81.31 ± 2.71%; Female: 69.58 ± 4.57%), and fat mass percentage: 20.82 ± 7.29% (Male: 14.48 ± 2.86%; Female: 20.82 ± 7.29%) took part in the present study. Players recruited for the study belonged to the reserve team of a male and female basketball team that play in the first Spanish basketball division. The male basketball team was composed by three guards, six forwards and four centers, while the female basketball team was composed by three guards, five forwards and five centers.

All players met the following inclusion criteria: (a) absence of musculoskeletal injury or health problem in the previous two months, (b) minimum basketball experience of five years, (c) over than two months with high-level monitoring by inertial devices (IMUs), and (d) participated in at least 85% of the training sessions during the two months prior to the study [22]. The study was realized in the first part of the in-season period, where players attended five training sessions and one official game (MD) per week (MD + 1: one recovery session; MD + 2: rest day; MD − 4: strength and on-court conditioning session; MD − 3 and MD − 1: technical-tactical sessions; MD − 2: simulated game session).

2.3. Equipment

Anthropometrical characteristics. Height was registered through a rod stadiometer (SECA, Hamburg, Germany) and body composition through an 8-electrode segmental monitor MC-780MA model (TANITA, Tokyo, Japan).

Time selection and trials' duration. Photocells (ChronoJump, Barcelona, Spain) were used to measure the time to cover each repetition, as well as, to select the duration of each attempt in the timeline of the WIMU PRO^{TM} inertial device (RealTrack Systems, Almería, Spain). Photocells commonly include only two connections: power and communication signal to the software to start and end a timer when the light is interrupted. For this purpose, photocells incorporated an Ant+ transmitter that was connected to the output of the communication signal via a RCA cable (standard communication cable). This process showed almost perfect validity with a bias of 0.006 ± 0.0018 s [23]. The Ant+ transmitter that incorporate a pushbutton was used by the researchers to mark the start and end points on the IMUs timeline in jump tests, 30–15 IFT and small-sided game.

Assessment of players' movements. WIMU PRO^{TM} inertial devices have been used to monitor the workload demands of the players during the assessment. Each device includes tracking (global positioning systems, GPS at 10 Hz; ultra-wide band, UWB at 18 Hz) and microelectromechanical sensors set at 100 Hz (4× accelerometer [2× ±16, 1× ±32 and 1× ±400 g]; 3× gyroscope at 2000°/s; and 1× magnetometer). For time-motion analysis in indoor conditions, a reference system composed of eight UWB antennas was placed around the court following the protocol described in a recent study that showed suitable values of reliability (coefficient of variation, CV < 1%) and validity (mean difference = 0.03 m; magnitude of differences = 0.21% with real measures as reference) [24]. Devices were located at scapulae level using an adjustable harness in each player.

2.4. On-Court Physical Fitness Tests and Registered Variables

Different on-court physical fitness tests were extracted from previous validated field test batteries designed to evaluate the physical performance of male and female basketball players [18,19]. The description of the tests and the variables obtained were mentioned, following the order of realization during the assessment.

6.75-m arc test. This test has been used to assess the ability to complete a curvilinear displacement at the maximum speed as possible [19]. Player must run between the 6.75-m line and 1-m line courtesy from the start line to the end line. The photocells were placed at the start and end line to send the start and end points to the IMUs timeline through Ant+ technology. Ten repetitions were performed (five in each direction) with 1-min rest between repetitions. If the athlete fell or left the running zone, the attempt was repeated. The average of the three best repetitions was selected for analysis. From this test, the average centripetal force generated ($CentF_{AVG}$) in left and right direction in each repetition was obtained [25]. The test CV was 4.2% in males and 5.8% in females.

Single leg jump test. This test has been included to evaluate the power output of each leg independently following Young et al. [26]. Player must performed the takeoff with a single leg. The non-takeoff leg or free leg was flexed at the knee and not permitted to touch the floor. No restrictions or specific instructions were given of the role of free leg during the jumping action, while hands must be placed at the hip as countermovement jump protocol. Left and right takeoff legs were assessed alternatively with 45-s passive rest between jumps (five repetitions with each leg). The average of the three best repetitions was selected for analysis. From this test, the jump height was obtained, that show nearly perfect validity (flight time: CV = 0–13-to-0.29%, Difference = 0.61-to-1.31 ms) and reliability with this device (ICC = 0.96-to-0.97%; SEM = 1.4-to-2.2%; CV = 2.5–3.1%) [27]. The test CV was 9.6% in males and 10.9% in females.

Abalakov test. The bilateral power output and the arms coordination during jump were evaluated following Bosco et al. protocol [28]. The athlete starts from an upright position, with feet shoulder-width apart and arms free. At his discretion, the athlete will flex the legs and then perform an extension of the legs, assisting the arms in the execution of the movement and avoiding the flexion of the trunk. No restrictions were imposed on knee angle during the eccentric phase of the jumps. Subjects were required to maintain straight legs during the flight phase of the jumps. A passive 45-s rest was realized between jumps. From this test, the jump height was obtained. The test CV was 14.4% in males and 16.7% in females.

Multi-jump test: This test assesses the tolerance to fatigue of the lower body [19]. To do this, the player starts on a box with a height of 50 cm. The player jumps down from the box and makes five maximum jumps in a row using the arm swing. From this test, the jump height was obtained. The test CV was 20.5% in males and 23.2% in females.

16.25-m. RSA test. Through this test, the acceleration and deceleration capacities of the athletes were evaluated. The start line of acceleration was placed in the free-throw line, the end line of acceleration and start line of deceleration in the 6.75-m line, and the end line of deceleration in the free-throw line [18]. Players must run as fast as possible from the start line to the end line in acceleration phase and brake as soon as possible into the deceleration phase, without exceeding the end line of the braking zone. The photocells were placed at the start and end line of acceleration zone to send the start and end points to the IMUs timeline through Ant+ technology. Players completed five sprints with 30-s active rest (walking from end line of deceleration zone to start line of acceleration zone) between repetitions. From this test, the average speed ($Speed_{AVG}$) and the maximum deceleration (Dec_{MAX}) were obtained in acceleration and deceleration phase respectively. The test CV was 11.7% in males and 13.6% in females.

30–15 IFT. It is a standardized test in distance and speed to evaluate the aerobic capacity of the players on the court [29]. The baselines (0 and 28 m), the center line (14 m), and four courtesy lines situated at 3 m (2× center line and 1× each baseline) were marked. The test combines 30-s running with 15 s of passive rest. During running time, athletes must be in the zones when it beeps, using the smartphone app for IOS. The start speed was 8 km/h and in each period of 30 s the speed is increased by 0.5 km/h. The test was concluded when the athlete did not reach the zone in two beeps. The last period that the player completes was considered for analysis.

3 vs. 3 small-sided game. 10-min of a 3 vs. 3 small-sided game was played with 3 vs. 3 official rules in a reduced court with dimensions of 10 × 15 meters [18]. To control the official rules, an official referee collaborated in the study. From this game, Player Load by RealTrack Systems (PL_{RT}), total distance covered (Dist), and total distance covered over 16 km/h (Dist > 16 km/h) were registered.

2.5. Procedures

Prior to starting all procedures, the study was approved by the Bioethics Committee of the University of Extremadura (registration number: 232/2019; date of approval: 08/10/2019) because it follows the ethical guidelines of the Declaration of Helsinki (2013). Then, coaches and clubs were contacted to inform about the proposal of the study. Club managers, technical staff, and players signed informed consent. After consent, the selection of testing date was agreed with both teams. Prior to testing, the teams performed two familiarization sessions to know the tests and the high monitoring, reducing the chances of error during the assessment.

Players were cited 30-min prior to the assessment with the aim to place the inertial devices through an anatomical-specific harness at scapulae level and perform the warm-up after the evaluation. At the end of the physical tests, the research team downloaded the data in an laptop and imported them to the SPRO™ software to obtain theand variables. The data were exported from SPRO™ software to an Excel spreandsheet. A database was made in Excel and then introduced in statistical package for further analysis. In addition, researchers made an informative dossier with the results obtained in the different tests in order for the coaching staff to have knowledge about the findings found in order to improve performance or detect possible anomalies.

2.6. Statistical Analysis

Results of physical fitness level of basketball players according to sex are reported as mean and standard deviation (SD). Data normality and homoscedasticity were confirmed through Shapiro–Wilk and Levene tests. The differences in physical fitness level between male and female basketball plaane analyzed by *t*-test for independent samples. The effect sizes were obtained by Cohen's *d* (*d*) and was interpreted as: $d < 0.20$ trivial, $d = 0.20$-to-0.60 low, $d = 0.60$-to-1.20 moderate, $d = 1.20$-to-2.00 high, and $d > 2.00$ very high [30]. The significance level was established at $p < 0.05$.

Then, to identify the physical fitness profile in male and female basketball players, principal component analysis (PCA) was used. Variables were scaled and centered (Z-score). The Kaiser–Meyer–Olkin values (KMO, male = 0.657; female = 0.623) and Barleth Sphericity test confirmed that PCA was suitable ($p < 0.01$). Eigenvalues > 1 were considered for the extraction of prinanipal components. A Varimax-orthogonal rotation method was performed in order to identify the high correlation of components and guarantee that each principal component offered diffanrent information. A threshold of 0.6 in each PC loading was retained for interpretation, extracting the highest anactor loading when a cross-loading was found between the components. Authors did not limit the number of PCs of the model final outcome and PCs selection was based on the guidelines previously described [31]. Data analysis and figures were performed and designed by Statistical Package for the Social Science (SPSS Statistics, version 24, IBM Corporation, Armonk, NY, USA).

3. Results

3.1. Sex-Related Differences in Physical Fitness Level of Basketball Players

The differences in physical fitness level between male and female basketball players are represented in Table 1. Male players obtained better results than female players in all physical fitness variables ($p < 0.01$; $t = 2.65$-to-13.31; $d = 1.04$-to-5.22), except in SSG PL_{RT} ($p = 0.17$; $t = 1.40$; $d < 0.20$ *trivial*). The highest differences were found in 6.75-m arc $CentF_{AVG}$ in both directions and RSA Acc while the lowest differences were found in SSG total distance.

Table 1. Descriptive and inferential analysis of sex-related differences in physical fitness profile of semiprofessional basketball players.

	Male M ± SD	Female M ± SD	t	p	d	Cohen's d Magnitude
6.75-m arc left $CentF_{AVG}$ (N)	467.15 ± 42.91	257.48 ± 37.23	13.31	<0.01	5.22	very high
6.75-m arc right $CentF_{AVG}$ (N)	464.87 ± 50.68	252.17 ± 38.15	12.08	<0.01	4.74	very high
Unipodal jump right leg (cm)	31.67 ± 3.73	20.71 ± 1.30	10.01	<0.01	3.93	very high
Unipodal jump left leg (cm)	33.48 ± 3.45	20.69 ± 1.99	11.57	<0.01	4.54	very high
Abalakov (cm)	40.15 ± 5.30	32.75 ± 3.77	7.12	<0.01	2.79	very high
Multijump (cm)	38.03 ± 6.13	30.45 ± 5.05	3.44	<0.01	1.35	high
RSA Acc (km/h)	26.69 ± 1.21	21.70 ± 0.77	12.53	<0.01	4.92	very high
RSA Dec (m/s^2)	−6.38 ± 0.69	−5.47 ± 0.54	3.76	<0.01	1.48	high
30–15 final players (km/h)	19.88 ± 1.62	17.83 ± 1.55	3.30	<0.01	1.29	high
SSG PL_{RT} (a.u.)	11.01 ± 1.53	10.11 ± 1.74	1.40	0.17	<0.20	trivial
SSG Total Distance (m)	777.66 ± 79.17	704.29 ± 61.06	2.65	<0.01	1.04	moderate
SSG Total Distance > 16 km/h (m)	184.39 ± 41.09	144.29 ± 16.25	3.27	<0.01	1.28	high

Note. M: mean; SD: standard deviation; t: t-value of independent samples t-test; p: significance; d: Cohen's d effect size.

3.2. Physical Fitness Profile of Basketball Players According to Sex

Table 2 and Figure 2 show the principal component analysis in physical fitness test. Four PC were extracted from male and female basketball players that represent an 85.71% and 83.61% of total variance respectively. The PC1 in male players represents 31.01% of total variance and was composed of RSA Dec, 30–15 final players, SSG PL_{RT}, total distance and total distance > 16 km/h, while female players represent 36.00% and are composed of RSA Acc, 30–15 final players, SSG PL_{RT}, total distance and total distance > 16 km/h. The PC2 in male players represents 26.81% and was composed of Abalakov, Multijump, and RSA Acc, while the female players represent 22.91% and were composed of Abalakov, Multijump 6.75-m arc $CentF_{AVG}$ at left and right direction. The PC3 in male players represents 16.15% and was composed of 6.75-m arc $CentF_{AVG}$ at left and right direction, while female players represent 15.32% and were composed of Unipodal jump in both legs. Finally, the PC4 in male players represent 11.74% and was composed of Unipodal jump in both legs, while female players represent 9.37% and were composed of RSA Dec.

Table 2. Principal component analysis by sex with respective eigenvalue, variances and % variance explained.

Sex	Male				Female			
PC	1	2	3	4	1	2	3	4
6.75-m arc left $CentF_{AVG}$ (N)			0.88			0.93		
6.75-m arc right $CentF_{AVG}$ (N)			0.96			0.95		
Unipodal jump right leg (cm)				0.88			0.75	
Unipodal jump left leg (cm)				0.77			0.96	
Abalakov (cm)		0.93				0.51		
Multijump (cm)		0.86				0.74		
RSA Acc (km/h)		0.71			0.90			
RSA Dec (m/s^2)	0.80							−0.90
30–15 final players (km/h)	0.57				0.80			
SSG PL_{RT} (a.u.)	0.76				0.82			
SSG Total Distance (m)	0.89				0.94			
SSG Total Distance > 16 km/h (m)	0.93				0.83			
Eigenvalue	3.72	3.22	1.94	1.41	4.32	2.75	1.84	1.13
Variance	31.01	26.81	16.15	11.74	36.00	22.91	15.32	9.37
%Variance	31.01	57.82	73.97	85.71	36.00	58.91	74.23	83.61

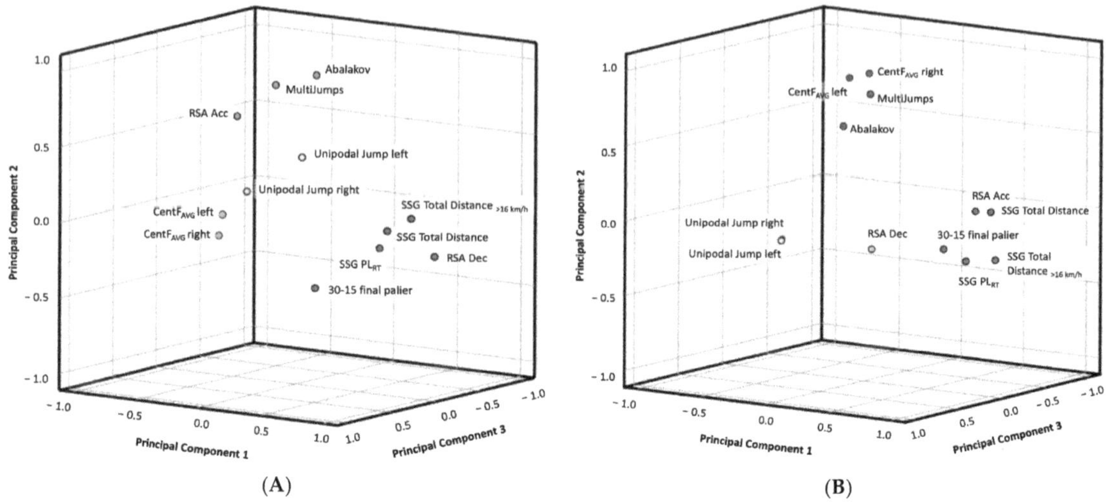

Figure 2. Rotated principal component distribution of physical fitness profile in (**A**) male and (**B**) female semiprofessional basketball players. Each color of the filled circles represents one principal component. In male players, PC1 is represented with red, PC2 with blue, PC3 with green and PC4 with yellow. In female players, PC1 is represented with red, PC2 with purple, PC3 with yellow and PC4 with pink. Red and yellow colors are shown in male and female players due to these components were found in both sexes.

4. Discussion

Basketball performance is determined by physical, technical, and tactical level of the players [6]. Coaches tend to consider mainly the technical and tactical characteristics to determine the playing position during the competition, as well as adapt the collective game to individual profiles in order to optimize performance [4,14]. In contrast, the analysis of physical condition is not usually used to identify specific performance profiles, which is strongly influenced by the sex of the players [32]. Therefore, the purposes of the present study were to identify sex-related differences in the physical profile of semi-professional basketball players, as well as to classify physical profiles based on principal component analysis.

Previously, the literature has shown scientific evidence of the sex-related differences between physical and physiological profiles of basketball players across different ages [10,11,13]. The present research confirms that better results were obtained by male players in (a) curvilinear displacements ($CentF_{AVG}$, N), (b) jump at unipodal, bipodal and repeated efforts (height, cm), (c) acceleration ($Speed_{MAX}$, km/h) and deceleration (Dec_{MAX}, m/s^2), (d) aerobic capacity (30–15 IFT final players, km/h). These higher physical-physiological capacities have impacted in small-sided game demands (total distance and total distance > 16 km/h, m) except in PL_{RT} although the formal and structural elements of the game are similar (10 min of 3 vs. 3 in 10 × 15 meters court). The differences of physical and physiological capacities depend on different factors at anthropometrical (height, weight, wingspan, etc.,) [10,32], morphological (Q-angle, tibiofemoral angle, hip abduction, center of mass displacement) [33], musculoskeletal development (size of muscular belly, bone thickness, type II-B muscular fibers associated) [11], and cardiopulmonary capacity (pulmonary: lung size, respiratory muscle blood flow, cost and work of breathing; cardiovascular: stroke volume, arterial blood pressure and oxygen content, oxygen consumption) [34,35]. For all this, male and female players should be considered as independent populations, so conditioning sessions as well as playing roles during the game need to be modeled based on their specific physical fitness profile.

For reducing the dimensions that explain the physical performance in basketball, mathematical methods are being applied to the sports area as the principal component analysis [20]. PCA is a statistical method for data reduction to explain the most relevant variables of players' behavior. From this analysis, four principal components were extracted in male and female basketball players that explain an elevated percentage of total variance (85.71 and 83.61% respectively). Two principal components were similar in male and female players ((1) aerobic capacity and in-game physical conditioning, (2) single leg jump) and two components were different (male: (3) curvilinear displacements, (4) jump capacity; female: (3) curvilinear displacements and jump capacity, (4) deceleration). Figure 3 represents different examples of basketball technical actions according to the principal components extracted from the physical fitness performance during the tests in both sexes.

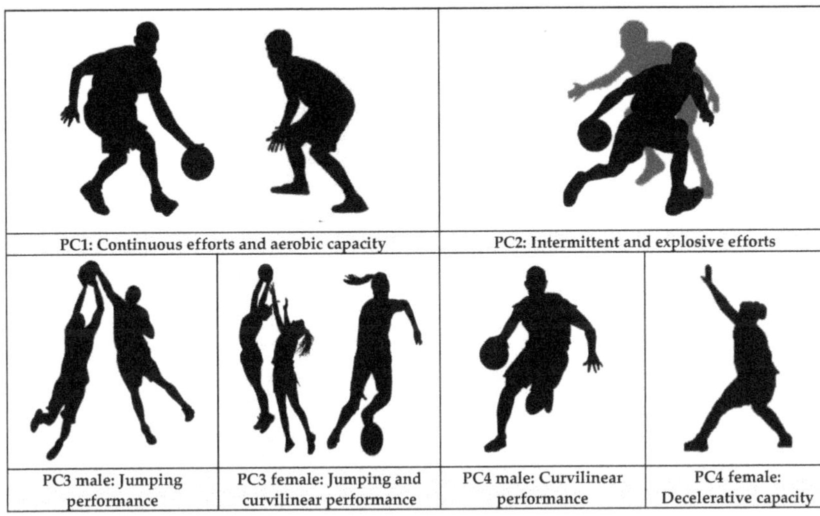

Figure 3. Representation of the principal components extracted according to the sex of the players.

The first component explained 31.01% in male and 36.00% in female of total variance in basketball players. The same variables were found in both sexes that represent aerobic capacity (30–15 final players) and in-game physical-conditioning (total distance, total distance > 16 km/h, PL_{RT}), excepting deceleration in male and acceleration in female players during repeated sprint ability. The game dynamics of basketball requires a high aerobic capacity to repeat high-intensity intermittent efforts during offensive and defensive actions, as well as rapid coast-to-coast transitions during counterattacks [4]. Good values in both variables indicate that the player presents a competitive advantage during the game, being decisive in attack to get better shoot positions and in defense to counteract the actions of the opponent [36,37]. Therefore, the development of aerobic capacity, as well as the integration of physical conditioning in simulated in-game conditions could be useful to improve the fitness level of basketball players, it being determinant for successful.

The other similar component between male and female players is the single leg jump performance that represented the PC4 with 11.74% in males and the PC3 with 15.32% in female players. This ability represents an independent PC due to the prediction of sprint and curvilinear displacement is limited based on single-leg jump at lateral, horizontal, and vertical directions [38]. In both sexes, higher performance in single-leg jump has been associated with the same playing positions (guard and forward) [39,40]. Traditionally, perimeter players were chosen for smaller body size and greater explosiveness than centers, regardless of the evaluation of different physical capabilities [6]. However, players with high values in this variable could be oriented to playing roles out of the paint to take advantage of their single-leg power in the performance of individual technical-tactical actions with short explosive movements (e.g., 1 vs. 1, dribbling, block-outs), without considering the anthropometrical-morphological characteristics.

Performance in curvilinear locomotion, repeated jump ability, and jump with arm-swing also represent key factors in physical fitness in basketball [5,10,11]. These capacities represent two components in male players (jump capacity, PC2: 26.81%; curvilinear locomotion, PC3: 16.15%) and only one component in female players (PC2: 22.91%). Due to the greater specificity of men's basketball, jumping ability and curvilinear movement ability define two different player profiles [15,16]. On the one hand, there are players with greater body size who, from their formative stage, have a specialization in making shots close to the basket and rebound action, while players with greater speed in curvilinear movement present functions related to the outside throw after a race to generate an advantage over the rival [6,10]. However, in women's basketball, due to the lower capacity to perform high intensity actions, the taller players present a multipurpose role, unifying the two roles mentioned in the male sex. These differences may be related to anthropometric and physical characteristics, as well as the different dynamics of the game depending on sex [41]. Therefore, the identification of the players with these specific profiles will require individualized functions and training based on their differentiating characteristics to enhance their performance in competition.

Finally, a main component is observed in female basketball players who are characterized by making a greater number of decelerations and at greater intensity than the rest of the players (PC4, 9.37%). In male players, this profile does not exist (it is integrated in PC1), so it is specific in female basketball. This peculiarity may be due to different aspects related to morphological and musculoskeletal development [6,11,34]. This profile explains the importance of eccentric work in the lower body of female players, where a high number of injuries occur in actions related to decelerations and changes of direction [42]. Instead, due to the musculoskeletal structures of male players (distribution of fibers and muscle belly), the injuries suffered in the lower body are mainly due to overload or fatigue provoked by the high volume of actions in the game and not due to the intensity of them [43]. Therefore, the injury prevention strategies between male and female basketball players should be designed accordingly, where a greater focus is needed on the intensity of actions in female players and on the volume of actions in male players to reduce injury risk.

5. Limitations and Future Research

The present research is the first approach to the identification of physical profiles in basketball based on sex through the principal component analysis, although different limitations should be mentioned. The first of these is related to the size of the sample and its specific competitive level, which means that the data are specific to the study population and cannot be generalized to all basketball players. In addition, the inclusion of new physical condition tests (e.g., agility with and without the ball, anaerobic capacity) to evaluate the physical performance of athletes may lead to the identification of new physical profiles of basketball players. However, the included tests belong to two specific basketball field batteries that are previously validated to evaluate integrally the most important abilities and capabilities in basketball players. Finally, future research that evaluates the physical condition of basketball players through specific tests and classifies the profiles based on principal component analysis will help to understand the physical performance factors throughout the different ages and competitive levels.

6. Conclusions

Sex-related differences were found in physical fitness level with higher values in male players, especially in physical capabilities that depend on power output (curvilinear displacements, unipodal jump, abalakov, and accelerative actions). Four principal components were identified in male and female basketball players with different distribution of physical capabilities. The component that explains the highest total variance in male (31.01%) and female (36.00%) players was represented by aerobic capacity and in-game physical conditioning.

7. Practical Applications

From the conclusions of the present study, different practical applications could be given about the different capabilities of physical fitness in basketball players based on principal component analysis:

1. Because male players presented higher physical fitness values, especially in game actions that depend on power output, it is fundamental to individualize that the training workload depends on the sex as well as the physical characteristics of the players.
2. The comprehension of the different profiles of basketball players in each team based on the physical fitness is fundamental to design individual task oriented in specific physical capabilities to improve sports performance. In addition, the knowledge of physical fitness profile of the players could help the team staff to design playing systems and tactical dispositions adapted to them (e.g., low values in aerobic capacity will entail long-time attacks, low values in changes of speed and curvilinear locomotion will entail more static playing systems, or low values in jump capacity will entail playing systems that end with shoots without rebounds).
3. The integrated work of aerobic capacity through modified game situations seems to be indicated as a fundamental aspect to improve the physical fitness level of basketball players in both sexes, so that the highest total variance was represented by aerobic capacity and in-game physical conditioning.

Author Contributions: Conceptualization, C.D.G.-C., J.P.-O. and S.J.I.; methodology, C.D.G.-C., D.M.-T., J.P.-O. and S.J.I.; software, C.D.G.-C., J.P.-O.; validation, C.D.G.-C., D.M.-T. and S.J.I.; formal analysis, C.D.G.-C., D.M.-T., J.P.-O. and S.J.I.; investigation, C.D.G.-C. and D.M.-T.; resources, S.J.I.; data curation, C.D.G.-C., D.M.-T., J.P.-O. and S.J.I.; writing—original draft preparation, C.D.G.-C. and D.M.-T.; writing—review and editing, J.P.-O. and S.J.I.; visualization, C.D.G.-C., D.M.-T., J.P.-O. and S.J.I.; supervision, J.P.-O. and S.J.I.; project administration, J.P.-O. and S.J.I.; funding acquisition, S.J.I. All authors have read and agreed to the published version of the manuscript.

Funding: The author Carlos D. Gómez Carmona was supported by a grant from the Spanish Ministry of Education, Culture, and Sport (FPU17/00407). This study was co-funded by the Regional Department of Economy and Infrastructure of the Government of Extremadura (Spain) through the European Funds of Regional Development of the European Union (dossier number: GR18170), and the Spanish National Agency of Investigation through the project "Scientific and Technological Support to analyze the Training Workload of Basketball teams according to sex, level of the players and season period" (PID2019-106614GB-I00).

Institutional Review Board Statement: Club managers, technical staff and players were previously informed about the investigation details and signed informed consent. The study was performed based on the ethical guidelines of the Declaration of Helsinki (2013) and approved by the Bioethics Committee of the University (registration number 232/2019).

Informed Consent Statement: Informed consent was obtained from all subjects involved in the study.

Data Availability Statement: The data presented in this study are available on request from the corresponding author. The data are not publicly available due to the Organic Law 3/2018, of 5 December, on the Protection of Personal Data and Guarantee of Digital Rights of the Government of Spain requires that this information must be in custody.

Acknowledgments: The authors would like to thank the study participants, as well as the team staff for their availability and commitment to the present study.

Conflicts of Interest: This work has been developed into the Training Optimization and Sports Performance Research Group (GOERD) of the Sports Science Faculty of the University of Extremadura. The funders had no role in the design of the study; in the collection, analyses, or interpretation of data; in the writing of the manuscript, or in the decision to publish the results. None of the authors has a conflict of interest to declare, and all authors were involved in the study design, data collection, and interpretation, and contributed to the writing of the manuscript.

References

1. Aoki, M.S.; Ronda, L.T.; Marcelino, P.R.; Drago, G.; Carling, C.; Bradley, P.S.; Moreira, A. Monitoring Training Loads in Professional Basketball Players Engaged in a Periodized Training Program. *J. Strength Cond. Res.* **2017**, *31*, 348–358. [CrossRef] [PubMed]
2. Montgomery, P.G.; Pyne, D.B.; Minahan, C.L. The Physical and Physiological Demands of Basketball Training and Competition. *Int. J. Sports Physiol. Perform.* **2010**, *5*, 75–86. [CrossRef] [PubMed]
3. Reina, M.; García-Rubio, J.; Ibáñez, S.J. Training and Competition Load in Female Basketball: A Systematic Review. *Int. J. Environ. Res. Public Health* **2020**, *17*, 2639. [CrossRef]
4. Stojanović, E.; Stojiljković, N.; Scanlan, A.T.; Dalbo, V.J.; Berkelmans, D.M.; Milanović, Z. The Activity Demands and Physiological Responses Encountered During Basketball Match-Play: A Systematic Review. *Sports Med.* **2018**, *48*, 111–135. [CrossRef]
5. Svilar, L.; Castellano, J.; Jukic, I.; Casamichana, D. Positional Differences in Elite Basketball: Selecting Appropriate Training-Load Measures. *Int. J. Sports Physiol. Perform.* **2018**, *13*, 947–952. [CrossRef]
6. Ziv, G.; Lidor, R. Physical Attributes, Physiological Characteristics, on-Court Performances and Nutritional Strategies of Female and Male Basketball Players. *Sports Med.* **2009**, *39*, 547–568. [CrossRef] [PubMed]
7. Vazquez-Guerrero, J.; Reche, X.; Cos, F.; Casamichana, D.; Sampaio, J. Changes in External Load When Modifying Rules of 5-on-5 Scrimmage Situations in Elite Basketball. *J. Strength Cond. Res.* **2018**. [CrossRef] [PubMed]
8. Reina, M.; García-Rubio, J.; Pino-Ortega, J.; Ibáñez, S.J. The Acceleration and Deceleration Profiles of U-18 Women's Basketball Players during Competitive Matches. *Sports* **2019**, *7*, 165. [CrossRef] [PubMed]
9. Delextrat, A.; Badiella, A.; Saavedra, V.; Matthew, D.; Schelling, X.; Torres-Ronda, L. Match Activity Demands of Elite Spanish Female Basketball Players by Playing Position. *Int. J. Perform. Anal. Sport* **2015**, *15*, 687–703. [CrossRef]
10. Cui, Y.; Liu, F.; Bao, D.; Liu, H.; Zhang, S.; Gómez, M.-Á. Key Anthropometric and Physical Determinants for Different Playing Positions During National Basketball Association Draft Combine Test. *Front. Psychol.* **2019**, *10*, 2359. [CrossRef]
11. Mancha-Triguero, D.; García-Rubio, J.; Gamonales, J.M.; Ibáñez, S.J. Strength and Speed Profiles Based on Age and Sex Differences in Young Basketball Players. *Int. J. Environ. Res. Public Health* **2021**, *18*, 643. [CrossRef]
12. Ramos, S.; Volossovitch, A.; Ferreira, A.P.; Fragoso, I.; Massuça, L. Differences in Maturity, Morphological and Physical Attributes between Players Selected to the Primary and Secondary Teams of a Portuguese Basketball Elite Academy. *J. Sports Sci.* **2019**, *37*, 1681–1689. [CrossRef] [PubMed]
13. Mancha-Triguero, D.; García-Rubio, J.; Antúnez, A.; Ibáñez, S.J. Physical and Physiological Profiles of Aerobic and Anaerobic Capacities in Young Basketball Players. *Int. J. Environ. Res. Public Health* **2020**, *17*, 1409. [CrossRef]
14. Ibáñez, S.J.; Mazo, A.; Nascimento, J.; García-Rubio, J. The Relative Age Effect in Under-18 Basketball: Effects on Performance According to Playing Position. *PLoS ONE* **2018**, *13*, e0200408. [CrossRef] [PubMed]
15. Alagappan, M. From 5 to 13 Redefining the Positions in Basketball. 2012. Available online: https://www.math.utk.edu/~{}fernando/Students/GregClark/pdf/Alagappan-Muthu-EOSMarch2012PPT.pdf (accessed on 19 April 2021).
16. Cheng, A. Using Machine Learning to Find the 8 Types of Players in the NBA. Available online: https://medium.com/fastbreak-data/classifying-the-modern-nba-player-with-machine-learning-539da03bb824 (accessed on 19 April 2021).
17. Impellizzeri, F.M.; Marcora, S.M.; Coutts, A.J. Internal and External Training Load: 15 Years On. *Int. J. Sports Physiol. Perform.* **2019**, *14*, 270–273. [CrossRef] [PubMed]
18. Gómez-Carmona, C.D.; Pino-Ortega, J.; Ibáñez, S.J. Design and validity of a field test battery for assessing multi-location external load profile in invasion team sports. *E-Balonmano Com.* **2020**, *16*, 23–48.
19. Mancha-Triguero, D.; García-Rubio, J.; Ibáñez, S.J. Batería de Test de Campo Para Evaluar La Condición Física de Jugadores de Baloncesto: SBAFIT. *E-Balonmano Com* **2019**, *15*, 107–126.
20. Pino-Ortega, J.; Rojas-Valverde, D.; Gómez-Carmona, C.D.; Rico-González, M. Training Design, Performance Analysis and Talent Identification—A Systematic Review about the Most Relevant Variables through the Principal Component Analysis in Soccer, Basketball and Rugby. *Int. J. Environ. Res. Public Health* **2021**, *18*, 2642. [CrossRef]
21. Zarić, I.; Dopsaj, M.; Marković, M. Match Performance in Young Female Basketball Players: Relationship with Laboratory and Field Tests. *Int. J. Perform. Anal. Sport* **2018**, *18*, 90–103. [CrossRef]
22. Chambers, R.; Gabbett, T.J.; Cole, M.H.; Beard, A. The Use of Wearable Microsensors to Quantify Sport-Specific Movements. *Sports Med.* **2015**, *45*, 1065–1081. [CrossRef] [PubMed]
23. Bastida Castillo, A.; Gómez Carmona, C.D.; Pino Ortega, J.; de la Cruz Sánchez, E. Validity of an Inertial System to Measure Sprint Time and Sport Task Time: A Proposal for the Integration of Photocells in an Inertial System. *Int. J. Perform. Anal. Sport* **2017**, *17*, 600–608. [CrossRef]
24. Pino-Ortega, J.; Bastida Castillo, A.; Gómez Carmona, C.D.; Rico-González, M. Validity and Reliability of an Eight Antennae Ultra-Wideband Local Positioning System to Measure Performance in an Indoor Environment. *Sports Biomech.* **2020**. [CrossRef]
25. Granero-Gil, P.; Bastida-Castillo, A.; Rojas-Valverde, D.; Gómez-Carmona, C.D.; de la Cruz Sánchez, E.; Pino-Ortega, J. Accuracy, Inter-Unit Reliability and Comparison between GPS and UWB-Based Tracking Systems for Measuring Centripetal Force during Curvilinear Locomotion. *Proc. Inst. Mech. Eng. Part P J. Sports Eng. Technol.* **2020**, 1–12. [CrossRef]
26. Young, W.B.; MacDonald, C.; Flowers, M.A. Validity of Double- and Single-Leg Vertical Jumps as Tests of Leg Extensor Muscle Function. *J. Strength Cond. Res.* **2001**, *15*, 6–11.
27. Pino-Ortega, J.; García-Rubio, J.; Ibáñez, S.J. Validity and Reliability of the WIMU Inertial Device for the Assessment of the Vertical Jump. *PeerJ* **2018**, *30*, e4709. [CrossRef]

28. Bosco, C.; Komi, P.V.; Tihanyi, J.; Fekete, G.; Apor, P. Mechanical Power Test and Fiber Composition of Human Leg Extensor Muscles. *Eur. J. Appl. Physiol.* **1983**, *51*, 129–135. [CrossRef] [PubMed]
29. Buchheit, M. 30-15 Intermittent Fitness Test | 30-15 IFT by Martin Buchheit. Available online: https://30-15ift.com/ (accessed on 16 January 2020).
30. Cohen, J. Chapter 8. The analysis of variance and covariance. In *Statistical Power Analysis for the Behavioral Sciences*; Routledge Academic: New York, NY, USA, 1988; pp. 273–406.
31. Rojas-Valverde, D.; Pino-Ortega, J.; Gómez-Carmona, C.D.; Rico-González, M. A Systematic Review of Methods and Criteria Standard Proposal for the Use of Principal Component Analysis in Team's Sports Science. *Int. J. Environ. Res. Public Health* **2020**, *17*, 8712. [CrossRef]
32. Mancha-Triguero, D.; Garcia-Rubio, J.; González-Calleja, J.; Ibanez, S.J. Physical Fitness in Basketball Players: A Systematic Review. *J. Sports Med. Phys. Fit.* **2019**, *59*, 1513–1525. [CrossRef]
33. Mendiguchia, J.; Ford, K.R.; Quatman, C.E.; Alentorn-Geli, E.; Hewett, T.E. Sex Differences in Proximal Control of the Knee Joint. *Sports Med.* **2011**, *41*, 541–557. [CrossRef] [PubMed]
34. Wheatley, C.M.; Snyder, E.M.; Johnson, B.D.; Olson, T.P. Sex Differences in Cardiovascular Function during Submaximal Exercise in Humans. *SpringerPlus* **2014**, *3*, 445. [CrossRef] [PubMed]
35. Guenette, J.A.; Witt, J.D.; McKenzie, D.C.; Road, J.D.; Sheel, A.W. Respiratory Mechanics during Exercise in Endurance-Trained Men and Women. *J. Physiol.* **2007**, *581*, 1309–1322. [CrossRef] [PubMed]
36. Gomes, J.H.; Rebello Mendes, R.; de Almeida, M.B.; Zanetti, M.C.; dos S. Leite, G.; Figueira Júnior, A.J. Relationship between Physical Fitness and Game-Related Statistics in Elite Professional Basketball Players: Regular Season vs. Playoffs. *Mot. Rev. Educ. Física* **2017**, *23*, e101626. [CrossRef]
37. Mancha-Triguero, D.; Reina, M.; García-Rubio, J.; Ibáñez, S.J. Does Physical Fitness Influence the Technical-Tactical Performance Indicators in a Professional Female Basketball Team? *RICYDE Rev. Int. Cienc. Deporte* **2021**, *17*, 174–188.
38. Meylan, C.; McMaster, T.; Cronin, J.; Mohammad, N.I.; Rogers, C.; deKlerk, M. Single-Leg Lateral, Horizontal, and Vertical Jump Assessment: Reliability, Interrelationships, and Ability to Predict Sprint and Change-of-Direction Performance. *J. Strength Cond. Res.* **2009**, *23*, 1140–1147. [CrossRef] [PubMed]
39. Delextrat, A.; Cohen, D. Strength, Power, Speed, and Agility of Women Basketball Players According to Playing Position. *J. Strength Cond. Res.* **2009**, *23*, 1974–1981. [CrossRef]
40. Pehar, M.; Sekulic, D.; Sisic, N.; Spasic, M.; Uljevic, O.; Krolo, A.; Milanovic, Z.; Sattler, T. Evaluation of Different Jumping Tests in Defining Position-Specific and Performance-Level Differences in High Level Basketball Players. *Biol. Sport* **2017**, *3*, 263–272. [CrossRef] [PubMed]
41. Sampaio, J.; Godoy, S.I.; Feu, S. Discriminative Power of Basketball Game-Related Statistics by Level of Competition and Sex. *Percept. Mot. Ski.* **2004**, *99*, 1231–1238. [CrossRef]
42. Bahr, R.; Krosshaug, T. Understanding Injury Mechanisms: A Key Component of Preventing Injuries in Sport. *Br. J. Sports Med.* **2005**, *39*, 324–329. [CrossRef]
43. Zuckerman, S.L.; Wegner, A.M.; Roos, K.G.; Djoko, A.; Dompier, T.P.; Kerr, Z.Y. Injuries Sustained in National Collegiate Athletic Association Men's and Women's Basketball, 2009/2010–2014/2015. *Br. J. Sports Med.* **2018**, *52*, 261–268. [CrossRef]

Journal of
Functional Morphology and Kinesiology

MDPI

Article

Pressure. A Qualitative Analysis of the Perception of Concussion and Injury Risk in Retired Professional Rugby Players

Ed Daly [1], Adam White [2], Alexander D. Blackett [3] and Lisa Ryan [1,*]

1 School of Science & Computing, Galway-Mayo Institute of Technology, H91 T8NW Galway, Ireland; ed.daly@gmit.ie
2 Faculty of Health and Life Sciences, Oxford Brookes University, Oxford OX3 0BP, UK; adamwhite@brookes.ac.uk
3 School of Life Sciences and Education, Staffordshire University, Staffordshire ST4 2DE, UK; alexander.blackett@staffs.ac.uk
* Correspondence: lisa.ryan@gmit.ie

Abstract: This study interviewed retired professional rugby union players (≤10 years since retirement) to discuss their careers in the game of rugby union. The primary aim of the study was to document their understanding of concussion knowledge and the analogies they use to describe concussion. In addition, these interviews were used to determine any explicit and implicit pressures of playing professional rugby as described by ex-professional rugby players. Overall, 23 retired professional rugby players were interviewed. The participants had played the game of rugby union (n = 23) at elite professional standard. A semi-structured individual interview design was conducted with participants between June to August 2020. The research team reviewed the transcripts to identify the major themes from the interviews using a reflexive thematic analysis approach. Four major themes were identified: (1) medical and theoretical understanding of concussion, (2) descriptions of concussion and disassociated language, (3) personal concussion experience, and (4) peer influences on concussion within the sport. These were further divided into categories and subcategories. The interviews highlighted that players did not fully understand the ramifications of concussive injury and other injury risk, as it became normalised as part of their sport. This normalisation was supported by trivialising the seriousness of concussions and using dismissive language amongst themselves as players, or with coaching staff. As many of these ex-professional players are currently coaching rugby (48%), these interviews could assist coaches in treating concussion as a significant injury and not downplaying the seriousness of concussion in contact sports.

Keywords: concussion; professional rugby union; long term health; brain injury

Citation: Daly, E.; White, A.; Blackett, A.D.; Ryan, L. Pressure. A Qualitative Analysis of the Perception of Concussion and Injury Risk in Retired Professional Rugby Players. *J. Funct. Morphol. Kinesiol.* **2021**, *6*, 78. https://doi.org/10.3390/jfmk6030078

Academic Editor: Giuseppe Musumeci

Received: 3 September 2021
Accepted: 14 September 2021
Published: 21 September 2021

1. Introduction

Rugby union has been a professional sport (for males) since 1995. It is a collision sport that is highly physical in terms of impact with opposing players, resulting in frequent contact events and collisions throughout the course of a match. The amalgamation of very high physical demands, combined with regular impacts with opposing players, for example, tackling, scrummaging or mauls, means that injury is an inherent risk of the game. In a study examining 12-month match exposure in professional rugby union players, it was proposed that accumulated training workload and recent match exposure influence a player's current injury risk [1]. This study demonstrated that players who had played a high (≥35) or a low (≤15) number of matches in the previous year were more susceptible to injury.

This results in players potentially enduring numerous injuries over the duration of a season in professional rugby (81 per 1000 h (95% CI 63–105), and 3 per 1000 h (95% CI 2–4) during training), and a similar level of injury exposure (ice hockey incidence rates of 79 per 1000 h and rugby league injury incidence rates of 68 per 1000 h) when compared to other collision sports [2]. As such, overall player welfare is the collective responsibility of coaches, team owners and individual players in professional rugby, as it can have a direct impact on players' physical and mental health [3]. At professional levels, compared to rugby at amateur levels, injury risk is significantly higher in terms of physical injury and concussion risk exposure during contact events [4]. In conjunction with these factors, professional players often downplay or underreport concussion incidence and physical injury during their professional rugby careers [5]. Many ex-professional rugby players tend to normalise pain and accept that injury is an integral part of the sport. These beliefs can be reinforced by the environment and become part of the "habitus" of the players involved [6]. The habitus of rugby players includes a subconscious and conscious level of conformity to highly masculine norms and behaviours. These are the accepted behaviours and norms that they *should* display when they are part of a highly masculine environment [7]. By choosing to accept pain and injury as part of professional rugby, this can evolve into a "culture of risk" for the players involved [8]. By extension, this culture of risk can develop into an unquestioned acceptance of physical injury and concussion risk as a normal part of the sport [8,9].

The latent pressure to accept pain and injury risk in sport can become culturally engrained within an athlete's mindset [10]. It has been suggested that many athletes accept pain and consider their ability to tolerate pain a particularly strong personal characteristic [11]. In a group culture, this pressure can be pervasive and become intrinsically linked to the character of the individuals involved and can be exacerbated by coaching or support staff. Under these circumstances, many athletes may consciously or subconsciously perceive that showing physical or mental weakness could tarnish the perception or "superhuman" concept of elite athletes [12]. In professional athletic environments, such as men's professional rugby, a tolerance to injury can be perceived as a characteristic that substantiates their position within a group of fellow professional players. Once concussion and injury risk are accepted as part of professional rugby, a further extension of the culture of risk is to decide not to disclose concussion to medical or coaching staff within a club setting [13].

How can the culture of wanting to play while injured be explained in sport either at amateur or elite level? Research suggests that acceptance of pain can stem from elite athletes trivialising and/or ignoring pain; in these instances, athletes may prefer to self-medicate and disregard medical advice in order to compete [14]. Alternative research noted that rugby players chose to play while knowingly injured because of the "sport ethic" of rugby, where self-sacrifice was a widely established accepted behaviour in the sport, and part of the culture of risk associated with the sport [15]. Furthermore, it can be argued that the culture of risk evident in professional sport is not evident in other occupations [16]. In tandem with this, there is an acceptance that injury is routine and normal in sport [17] and is exacerbated within rugby union by a willingness to participate whilst hurt [18].

The connection between professional athletes and their peers increases pressure to play while injured. The social relations in playing groups generate internal and external pressure to play while injured regardless of the long-term consequences to players' health [19]. The existence of this culture of risk is reinforced by acceptance that injury is part of the game and an aspect that players need to accept to be an elite athlete [8]. This study sought to investigate retired players' understanding of concussion in conjunction with the explicit and implicit sources of pressure experienced in professional rugby union.

2. Methods

2.1. Study Design

A reflexive thematic analysis was utilised for this research study [20]. The primary goal was to examine retired professional rugby union players' understanding of concussion and concussion experiences within the professional game of rugby union. A secondary goal was to examine the latent (implicit and explicit) pressures of being a professional rugby player with regard to concussion incidence and injury risk (n = 23). A semi-structured guide for the interviews was developed to offer an exploratory account of the players' playing background and experience with concussion.

2.2. Eligibility Criteria and Sampling

This study utilised a qualitative research approach, using semi-structured interviews and reflexive thematic analysis. An exponential non-discriminative snowball sampling method was utilised, as the first participants recruited to the study cohort provided other referrals [21]. All included participants were retired professional men's rugby union players who had ended their playing careers within the ten years prior to the commencement of the study (retirement span 2011–2019). The players described their experiences of being involved in professional sport (professional rugby clubs and/or international rugby union representative level) from their subjective perspectives.

Participant Characteristics

The participants associated with this study had all participated in professional rugby union (n = 23). The following nations were represented: Ireland (n = 17), England (n = 1), Scotland (n = 3) and Australia (n = 2). Of the full cohort of players (n = 23), 14 had represented their respective countries at full international test level rugby (61%). The average career span (years) for the cohort was 9.3 (SD 2.7) years in duration and the average age at time of retirement from professional rugby was 30.8 years (SD 2.9).

2.3. Ethics & Procedure

Ethical approval for this study was received via the Research Sub Committee of Galway Mayo Institute of Technology (GMIT; RSC_AC_23062020). Pilot research was gathered from discussions within the research team and exploratory pilot trials of the interview format with potential participants. Data were collected during interviews ranging from twenty-five minutes to seventy minutes in duration with a preliminary conversation with each participant to outline the rationale for the study and how the study would proceed. Each participant was given a participant information sheet with a consent form attached.

2.4. Procedure & Data Collection

Before each one-to-one interview (interviewer (ED) & interviewee), verbal consent was attained, and the interview proceeded by using a standardized series of questions across all interviews. Participants were informed that all information gathered would be treated as confidential and anonymised for the purposes of this qualitative study. Post interview, all transcripts were read and verified by an independent reviewer (LR). The data from the interviews were collected sequentially from the participants and the transcripts were analysed once all interviews were completed. The main purpose of the study was to document the pressures of playing professional rugby union as described by a cohort of ex-professional rugby players. In addition, these interviews were used to ascertain their levels of concussion knowledge and the analogies they used to understand concussion.

2.5. Data Analysis

After data collection from the interviews, each participant was given an opportunity to comment on the recording and offer clarifying comments on their personal interview. The audio recordings of the interviews were transcribed to MS Word format by the lead researcher (ED). After the transcription process, all transcripts were compared to the audio recordings of the interviews. Using this process for syntax correction, final amendments were made to the transcripts to generate the first draft of interview manuscripts (n = 23). Data were analysed thematically according to the Braun and Clarke reflexive thematic analysis approach, following an update to their original thematic analysis approach [20]. For this study, a critical realist framework was utilised to identify the players' descriptions of concussion and their injury experiences in professional rugby [22]. From the 23 interview transcripts, the lead researcher identified eight subcategories, which were subsumed into five categories, resulting in four themes. Interview content and themes were independently assessed and reviewed by the third author to assist and finalise the thematic analysis process [23].

2.6. Researcher Background

The research team nominated ED as the primary interviewer to collect data. This was based on his previous experience as part of a professional rugby organisation and experience in gathering qualitative data. It was determined that the ex-professional players would be more responsive and/or more open in their interview responses with ED as compared to other members of the research team. This could be attributed to the presence of a male interviewer (ED) with prior experience of being part of a professional rugby organisation that would be conducting interviews with male participants [24,25]. This was viewed as a positive aspect with respect to the recruitment of participants for the study.

3. Results

This study examined the pressures associated with being a professional rugby player and the players' understanding of concussion as an injury in professional rugby union. In this study, four major themes were identified by the authorial team. Based on significance, the lead author identified a hierarchical order for these four themes: (1) Medical and theoretical understanding of concussion; (2) Descriptions of symptoms and disassociated language; (3) Concussion experiences, misunderstanding of subconcussive impacts, categories of concussion; and (4) Peer influences on concussion within the sport. Table 1 illustrates the composition of these themes with regard to the categories and subcategories. Throughout the interviews and the process of theme identification, a connecting theme of language emerged across the four major themes.

3.1. Theme 1—Medical & Theoretical Understanding of Concussion

3.1.1. Awareness of the Physiology of Concussion

All players interviewed had a long-term association with the game of rugby. With this association, they had developed a profound connection with the game from an early age, e.g., *"I fell in love rugby around 9–10 years of age"*. This early exposure demonstrated that rugby is a collision sport and has inherent dangers, including concussion risk. The majority of players recalled being coached from an early age to tackle correctly in an attempt to mitigate concussion risk. These skills, when taught correctly, translated into appropriate tackle technique as a professional rugby player: *"I got exposed to good coaching as I was growing up, that's why I love the game and I was lucky enough that I could make a living from it as well"*. The interviewees expressed opinions that direct blows to the head and neck had the highest probability of resulting in concussion while playing the game: *"When the impacts are coming like you know, the brain shakes in the head"*, and "one of mine was where I got chinned like it was wasn't it wasn't a big collision to the head. It was a hip to the chin, and it was almost like you'd see in a boxing match like a punch to the chin", or more clearly expressed as "I got kicked in the head and then there was a loss of consciousness".

This anecdotal evidence is supported by injury surveillance evidence, where the highest incidence of concussion occurs in the tackle and/or the contact areas of the game (i.e., tackles, rucks, mauls, etc.). This belief in improving tackle technique in order to reduce concussion risk in players is regarded as valid by many of the players, in terms such as, *"We can coach our young athletes in in their tackle technique as well, I think will be a would be a big defining factor"*. There was evidence to imply that incorrect technique leads to concussion and other physical injuries to fellow players: *"so many concussions probably around how you tackle and the breakdown area"*, and *"when you look back technically, he wasn't, probably, you know, making tackles the way you should have been"*. Other players believed that certain strength and conditioning protocols could be implemented to reduce concussion risk: *"[i.e., increase the] amount of posterior work around the neck within their program as a prevention strategy"* and

"I think exercise, probably neck strengthening and I don't know much about it, is possibly going to help (prevent concussion)".

As the tackle and/or contact areas of the game are unlikely to change, some players were of the opinion that changes to current tackle laws will do little to mitigate against concussion risk.

"Lowering the tackle height further, which has been discussed, would be a mistake. I think that's going to lead to more concussions, because the reality is for someone who is 6′6″, if you got to defend two players, all of a sudden, you make a last second decision; my head is the thing that's going to take the bang".

3.1.2. Non-Medical Descriptions & Understanding

While many of the players had a clear knowledge of the mechanism of concussion, the language used was frequently framed in non-medical terms, using analogies to describe the effects *"that kind of Deja vu feeling the kind ... where the white stars kind of appear in the corner of your eyes"*. It was apparent that the injury was unclear and unnerving for players:

"you got concussed, but it was a massive brain fog. It's almost like when you wake up from a really good sleep and your alarm was going off. But you're not quite sure exactly where you are at that moment".

This sense of disorientation or uncertainty on a temporary basis had led the players to believe that concussion may not have been as serious as other injuries as they could still continue to play.

"It's a scary feeling when you're in this situation ... ... in front of 50,000 people and I've got absolutely no idea where I am or what's going on. You look around like holy fuck what's going on here? I got this concussion".

Assigning non-medical descriptive language, for example, "head knock" or "head bang", to concussion, may have had the effect of reducing the gravity of concussion in a professional environment. The combination of this type of language and a misunderstanding of concussions created an environment where players could continue to play while symptomatic or choosing to ignore that they had a concussion.

"I had a couple of bangs to the head ... ... at the 2015 World Cup. I did one return to play where I've got head knock, think against South Africa. I managed to play the next week but apart from that I think I had another HIA (Head Injury Assessment) to do ... but I've never been knocked unconscious at all in my career".

Many players chose to ignore "minor" concussions and believed that only being knocked unconscious deemed them to be concussed, which was clearly expressed in the descriptive language for concussions in their careers.

3.2. *Theme 2—Descriptions of Symptoms and Disassociated Language*

Understating the Injury and Using Casual Terminology

This disassociation between the injury and the language used to describe it became the normal manner to describe concussion. This type of language seemed to trivialise the injury and cause a disconnection between declarative and procedural knowledge by the players: *"I would have knocks and seen stars ... knocked out of whack"*. It led to players deeming that they needed to be "knocked out" in order to be diagnosed with concussion: *"I wasn't knocked out or anything, but just a bit wobbly and had gone through the whole protocol"*. It is apparent that players judged their readiness to continue to participate on this metric of being conscious or unconscious and were intentionally dismissive of concussion by using casual language to rationalise the injury. Other descriptions were used where concussions became normalised to the player or were considered an accepted consequence of his occupation. For example, one player described this phenomenon as *"a run of the mill concussion ... ... one was six weeks long"*. This was considered as a standard description which was strongly connected to the previously noted habitus of the environment in which they operated as "tough" professional players [7].

Table 1. Retired players understanding and descriptions of concussion in professional rugby.

Theme	Category	Sub-Category	Terminology	Selected Illustrative Comments from Players
Medical and theoretical understanding of concussion	Awareness of the physiology of concussion Non-medical descriptions	Mechanism of concussion Use of analogies to understand and describe symptoms	Impact, whiplash injury, shaking the brain, chinned Déjà vu, brain fog, deep sleep, being drunk, pressure in the head, disorientated, foggy, deep sleep, blacked out	*"get the impact there going one direction and the impact sends him another direction and it's actually not them hitting the ground. It's the movement of their head at that speed."* **(P6)** *"it's a blinding concussion ... ... like a big charge, but it just went through my whole body and then boom, I fell on the ground".* **(P3)** *"It was like being blackout drunk and not remember anything for hours later".* **(P5)** *"You just feel, you know, your head full of pressure, I had that mild symptoms, a total pain in the back area or between your eyes and then just tension".* **(P21)**
Descriptions of symptoms and disassociated language	Understating the injury and using casual terminology	Dismissive; a non-serious injury	Crack, head knock, bang to the head, run of the mill concussion, Spots, stars, blurry vision, bell rung, dizziness, in a haze	*"that spark sort of thing is when you're going to make that tackle and your head in the wrong spot. If someone really winged you in contact, that's what I'm talking about with those one".* **(P16)** *"To be honest, I'd say, you know, it was one of what do they call having your bell rung or something like that".* **(P18)**
Personal concussion experiences; misunderstanding of sub concussive impacts, categories of concussion	Knock out blows and sub concussive impacts	Knock out blows experience	Numbness, knocked out, panned out, out cold	*"I was knocked out to a point where it's just a real numbness".* **(P3)** *"I got knocked clean out and I just remember being brought in the car after I've been to the hospital".* **(P11)** *"I was out cold in the field, tongue going down my throat".* **(P15)**
		Understanding; concussion and sub concussive impacts	Not fully knocked out, wobbly, gradual onset of symptoms, headaches	*"I know it is a kind of grey area, you know, there's a lot of times, you know, there's a misunderstanding of what I call concussion actually is, I suppose the easiest way to kind of get this across is that I never had a situation where I was out cold where I was like fully unconscious".* **(P17)** *"I took a knock to the head and that was another gradual onset like initial thing was quite painful and the session was finished, very hot day as well. It was horrible because they're terrible headache and you know I didn't really want to talk to anyone or anything like that".* **(P22)**
		Categories of concussion	Minor concussions, strata of concussions, small episodes	*"It's like a minor concussion and you play on, adrenaline gets through, but the next day your neck and your top your head would be sore to touch".* **(P10)** *"The minor ones were, like, suppose any Sunday after a game it will be you're sore everywhere. Like my neck and head would have been sore. And for the three or four days and it was like I could feel the side effects. Vision was slightly blurry, I wasn't myself".* **(P10)**

Table 1. *Cont.*

Theme	Category	Sub-Category	Terminology	Selected Illustrative Comments from Players
		Unacknowledged or hiding concussion symptoms	Shake it off, get on with it, not being right, temporarily not cognitively present	*"You feel like you were there before or yesterday or a week before, and that would be a constant thing where you get knocked all different angles".* **(P10)**
Peer influences on concussion within the sport	Sporting culture, reinforced social norms	Influences on staying in the game	Loyalty, not admitting being injured, badge of honour, pressure from teammates	*"If you were able to stand up and play on and tackle the fellow in front of you and carried ball and everything, you're not coming close enough. You're not concussed enough to go off like, it kind of goes back to the point, that unless you're asleep on the field and can't actually stand up, you're staying, and you play on, was the prevalent attitude".* **(P1)** *"Players are pretty sharp these days as well as you know, it's not just a case of 'man up and get on with it'".* **(P2)** *"The mentality of the whole sport around concussion, there was no talk about them".* **(P8)** *"I understand guys put pressure on themselves to keep playing and toughen up and you know that is rugby for sure"* **(P12)** *"I think that sort of is encouraging, whereas in years gone by, it would be sort of seen as a badge of a badge of honour, whereas now, I think boys are sort of saying 'look, that's not cool'. That's an important thing 'cause obviously there's pressure from your teammates in that sort of environment".* **(P12)**

3.3. Theme 3—Concussion Experiences, Misunderstanding of Subconcussive Impacts, Categories of Concussion

3.3.1. Knock Out Blows and Misunderstanding Subconcussive Impacts

The parameters of concussion definitions as described by the ex-players were predominantly related to whether they were knocked unconscious or whether they remained conscious and able to compete.

"I know it is a kind of grey area, you know, there's a lot of times, you know, there's a misunderstanding of what I call concussion actually is, I suppose the easiest way to kind of get this across is that I never had a situation where I was out cold where I was like fully unconscious".

During the interviews, this led to comments about subconcussive hits, and if they had experienced these types of impacts: "if I got a bang to the head, my memory was wiped for the next 3 or 4 min ... I had no idea where you are, then and all of a sudden the fog starts to lift (and continued to play)". It was noticeable that interviewees did not regard these repeated impacts as subconcussive hits; instead, the lack of clarity about what defined a concussion was a dominant collective comment. An example of such a view was seen when one participated stated:

"You (are) like Christ, what day is it? And it's like that instant for a long period of time and it is a hard one to explain the consciousness of a concussion 'cause sometimes you get knocked out cold, sometimes you're kind of there with it, and sometimes it's your memories aren't that clearly there, so it's obviously a difficult one to explain, but it's just sort of a complete reset of your consciousness at the time. "

3.3.2. Categories of Concussion

An extension of using casual terminology was a categorisation of concussion using euphemistic language, such as one participant's explanation: *"A significant difference from that sparkly glitter stars (subconcussive impact) in your eyes to when you get a full-blown concussion. It's important that you know what they feel like"*. The connection between a "full blown concussion" with unconsciousness and other concussion experiences where the player could continue was interesting to note. In medical terms, while assessing a concussion, medics can categorise concussion symptom severity using various assessment tools at pitch side or in a clinical setting. The players interviewed had their own stratification:

"I always sort of categorized into maybe 3 different areas; extreme is getting knocked out. Then not getting knocked out cold where you don't know what's happening, you're conscious. Then the kind of where you got a big bang to the head, your vision kind of goes; stays with you for a while, it doesn't go away, can't really function all that well."

What became explicit from some players was that they were unaware of the cumulative effects of concussion. These cumulative effects may have short- or long-term effects regardless of what category of concussion they had experienced: *"the small episodes are shaking your brain, you know, even some doctors seem to think that they can be as bad as the other ones as well"*. As concussion frequency escalated in some players, recovery times from concussion increased concurrently.

"I had five (diagnosed) concussions in my career, but we both know the vast majority of players have a lot more than that, one thing that I always took a long time to recover from a concussion, even very minor ones, like I'm talking three weeks, maybe two months".

3.3.3. Unacknowledged or Hiding Concussion Symptomology

Due to the latent pressure associated with a professional rugby environment, for example, contract and financial issues, it was a common trait for players to feel comfortable being dismissive of concussion. Players expressed this in terms of *"I remember multiple times during my career being very, very dazed on the field"* and in a more subtle sense of when they *"weren't cognitively present during the course of the game"*. Many of these occurrences were a clear display of being in an overtly masculine environment and not willing to show any form of weakness as a professional rugby player.

In tandem with these latent pressures, there was evidence to suggest that players are willing to compromise concussion test protocols in an attempt to remain in the game and hide their concussions,

"I was just talking to this friend of mine who just retired last year. The protocols that were in place, we thought were fairly simple. And we thought that we could kind of trick them ... ... answer slower and even if you do have to go off and take the test, you're going to pass".

Some players had other methods to strategize around the Sport Concussion Assessment Tool (SCAT) protocols:

"I remember actively practicing those tests, trying to figure out the months of the year backwards, so I could do it or it's 'apple elbow carpet bubble saddle'. So even when I was concussed, I could go into autopilot and beat those tests ... you can beat the system".

Even when players felt that they did not play whilst concussed, their comments acknowledged that fellow players continued to play while symptomatic: "I've without doubt seen people that have been either hit badly or kind of a bit spaced".

3.4. *Theme 4—Peer Influences on Concussion within the Sport*

Sporting Culture, Reinforced Social Norms

When commenting on the rugby culture that infused their early sporting experiences, it was assumed that playing rugby carried an associated injury risk: "*Even that little bit of stars, that's a concussion, I don't know how many times I got that in my career even all the way up as a kid*". This long-term chronic injury risk was reinforced by culture or familial influences: "*You know what you're signing up for when you're playing*". The inclusive fraternal influence of parents, coaches and peers may not have been openly expressed but it was latently implied:

"rugby was obviously a massive sport where I was from, my father was big into rugby, the school (I attended) was a big rugby school" and "I was told afterwards (experiencing a concussion) that I said, 'please don't take me off, my father kill me' if you take me off'".

To be perceived as physically weak and conceding to concussive injury was a source of humiliation for players. This was evidenced by comments such as:

"it was pretty similar with the delayed (playing while symptomatic) concussions, just an incident but then afterwards you really didn't feel good at all, I mean, it's not something that you flag on the pitch".

The need to keep concussions hidden was deemed as a necessary action to protect their masculinity: "at the time and with an ego, I just tried to ignore it (concussion) as much that I could, so maybe I didn't pay attention to it".

This manifested in a culture where players felt pressure from coaches and peers to be declared fit to play matches because "you're not going to be the one to report yourself; then you're out for the protocols, which could be up to two, three or four weeks". Some players voiced opinions where they did not feel secure in declaring their symptoms to coaches and peers: "like this is the truth, it's very hard to tell because in my early career I didn't even report, I just got back to training and playing". In other instances, players were unsure whether or not they were concussed, and chose to say nothing: "I didn't tell anyone that I was concussed, so I played on, but I didn't know either". The undercurrent of pressure manifesting from peers and coaches was invariably directed back towards the players:

"it is entirely reliant on the honesty of the players. As I said, the players are fully aware of it (pressure to play), they are competitive animals and they're mad to play, particularly as well when you take in other variables like a guy coming out a contract or fighting for his position".

The risk was inherent during competitive matches and, interestingly, was an accepted part of preparation or training sessions leading up to gameday. From examples of concussion incidence discussed by the players, it was described as an accepted aspect of training.

"I got knocked out by my flatmate at the time in a big contact (training) session, two egos colliding, and I ended up coming out the worst. I got knocked out cold. Sort of one of those ones where it's in the highlights of training and people laughing at it because someone got knocked out cold"

If players received a blow to the head and were possibly symptomatic, it was understood that they would return to play: "(I) went off to get stitches and didn't feel right . . . . . . then the one (concussion) after that which was in training that was just in a maul", which was the widely accepted practice.

With the benefit of hindsight after retiring from the game, many participants believed that a cultural shift is required in the game. Most participants were aware of concussion being discussed during their professional careers, but a significant emphasis may not have been placed on disclosing concussions or ongoing symptomology: *"In hindsight now, I should have just told the coaches whatever, I was just not right; yeah (taking) a week out, you know that distinction between a concussion and just having a very hard game, maybe talk a little bit more about"*. According to one participant, in order to implement meaningful change and avoid the current issues of hiding concussion symptoms or how concussion is discussed within the game, *"there needs to be a change and I think it needs to be cultural change; you stop accepting that this (hiding concussion symptoms or concussions) needs to be part of it . . . . . . and getting back on a field that is not necessary."* It is evident that cultural shifts to manage pressure within the game cannot occur in isolation and need to be broad and *momentous, "whereas in years gone by, it was sort of seen as a badge of a badge of honour, where is now. I think boys are sort of saying 'look, that's not cool'. That's an important thing. 'cause obviously pressure from your teammates and that sort of environment"*.

4. Discussion

A key finding from this study was the language that male rugby players used when speaking about concussion and their descriptions of pressure in professional sport. The language associated with concussion was influenced by rugby culture and seemed to be quite dismissive of concussions or concussive symptoms. The language used highlighted the various sources of pressure that professional players experience as elite athletes. Many of the obstacles that players faced when attempting to reveal concussions were latent in nature. This unsaid or latent pressure was underpinned by the understanding that saying nothing was a preferable option for their careers. In a historical sense, the emergence of rugby culture is a legacy from the English public schools system that promoted sportsmanship and endurance with respect to pain tolerance [26]. This historic context partially explains the socialisation of rugby culture, which instilled a tolerance to pain that is generally at odds with what we find in broader society [9].

These pressures to accept pain were expressed and understood from an early age, which continued into their professional careers. Professional rugby is a very competitive environment, where the latent understanding is that players need to display durability, strength of character, physical strength, and a willingness to suffer in silence through physical pain and concussion. Many players did not fully understand the ramifications of concussive injury and other injury risk, as it became normalised as part of their sport [27]. In parallel, many players did not consider the long-term health implications of experiencing multiple concussions over the duration of a professional rugby career. This normalisation was supported by trivialising the seriousness of concussions either via dismissive language amongst themselves as players, or from coaching staff. The pervasive culture saw these risks as acceptable and legitimised the value of this ever-present risk for short-term success in the sport [11].

It was evident from the interviews that players experienced peer pressure from within the playing group. This pressure was intensified by fellow team members and coaching staff encouraging players to compete while injured. This practice may be understood as a "transfer" of the fear of failure from coaches to players as they (coaches) were under pressure to produce results and win games [8]. The latent pressure expressed by fellow players and coaches either directly or indirectly added to the overall burden that was experienced by the players. This was supported by the notion that being able to endure pain and play injured affirmed rugby players' masculinity and affirmed their commitment to the club [19].

Similar comments were expressed by players who were not first choice or "fringe" players. Along with fringe players, there were players who were either recovering from injury or were injured near contract negotiations. These players were continuously uncertain of their futures as they felt excluded from the core playing group or that their "credit in the bank" would be easily forgotten, which led to doubts and issues with self-confidence. It is understandable that the manifestation of these various facets of pressure inevitably led to the players not disclosing concussive symptomology or choosing to continue to play while concussed. The participants in this study opted to dismiss long term physical and mental health primarily for financial gains and to maintain their personal status as professional rugby players. They managed to rationalise and trivialise the seriousness of concussion, either willingly, or unknowingly by a lack of knowledge about the consequences of multiple concussions and subconcussive impacts during their professional careers. Whether these participants used trivialising language to avoid disclosing concussions or did not understand the gravity of their concussive injuries warrants further investigation. The culture of risk expressed by the ex-professional players in this study clearly accepted that injury risk, including concussion, was an accepted part of the game. In this respect, the players accepted the associated implicit and explicit pressures because they wanted to succeed and be prepared for the stringent demands of elite rugby. It is evident from this study that medics and coaches cannot fully rely on players disclosing concussions or concussive symptoms.

In this regard, players used rationalising language to dismiss concussion or underreport concussion incidence. This was supported by self-imposed implicit pressure and explicit pressure applied by peers and coaches. These verbal pressures were expressed through direct and indirect language from peer groups or by coaches within the organisation. It could be argued that much of the language used was due to a lack of understanding of concussion and the mechanisms of being concussed or external pressures on coaching staff to produce results. In these instances, dismissive language or subtle language cues were used to normalise and rationalise concussion as an injury. This study, therefore, provides valuable detail to inform current professional, amateur and young players involved in collision sports. This findings from this study may also influence older retired players in a positive manner by highlighting areas of awareness and education for their long-term brain health. Due to the lack of research in this area, this study is the first of its kind to examine the pressures on professional rugby players, their understanding of concussion and the language they use to describe concussion.

Many of the participants in this study were dismissive of concussion when they played professionally. Now that many of these ex-professional players are currently coaching rugby at amateur, elite, and professional levels (48%), it is interesting to note their current views on concussion. These interviews highlight the continuing need for education of coaching staff at all levels on the signs, symptoms, and recognition of concussion. These data could assist in eliminating outdated beliefs and the recycling of substandard practice associated with concussion and injury risk. These interviews provide valuable detail on the understanding, thoughts and language used by ex-professional players. This knowledge could assist coaches in understanding the importance of treating concussion as a significant injury and not downplaying the seriousness of concussion in contact sports.

4.1. *Reflexivity*

When examining the research process in this study, it is appropriate to acknowledge that the primary researcher attempted to remain critically aware throughout the process. It was notable that the lead researcher built rapport with the participants because of prior experience in professional rugby. Many of the participants offered forthright and authentic replies to the interview questions as a consequence.

4.2. Study Limitations

At the outset of this study, it was anticipated that interviews would be conducted with elite female rugby players; however, this did not materialise. Research into female elite players and their experiences of concussion is an area that warrants further research. This research paper included players who had retired in the previous ten years and may not be reflective of current practice. It is important to acknowledge that the research team involved in this paper have experience in professional rugby union and therefore the interpretation of the findings is through the lens of this previous experience.

5. Conclusions

There exists a disconnect between the language used by players and medical staff when discussing and reporting concussion incidence in the game of rugby. A thorough understanding of how players describe their symptoms is important to enhance recognition of concussion. Future research in this area would require a full discourse analysis of the language that is used by both male and female rugby players at amateur and elite levels of the game. As rugby union is a global sport involving multiple nations where different languages are spoken, discourse analysis can help identify whether these trivialisations of concussion and subconcussive symptoms occur throughout the professional world of rugby. This study also highlighted the fact that many players are dismissive of concussion/subconcussive hits and may not reveal their symptoms to coaches or may downplay their symptoms. Coaches and medical staff should therefore not include players in the decision-making process regarding return to play and should enforce concussion protocols.

Author Contributions: Data analysis, interpretation, article preparation, L.R. Data analysis, interpretation, article preparation, E.D. Article preparation, A.W. & A.D.B. All authors have read and agreed to the published version of the manuscript.

Funding: This research received no external funding.

Institutional Review Board Statement: The study was conducted according to the guidelines of the Declaration of Helsinki, and approved by the Research Sub Committee of Galway Mayo Institute of Technology (GMIT; RSC_AC_23062020).

Informed Consent Statement: Informed consent was obtained from all participants involved in the study.

Conflicts of Interest: The authors declare no conflict of interest.

References

1. Williams, S.; Trewartha, G.; Kemp, S.P.; Brooks, J.H.; Fuller, C.W.; Taylor, A.E.; Cross, M.J.; Shaddick, G.; Stokes, K.A. How Much Rugby is Too Much? A Seven-Season Prospective Cohort Study of Match Exposure and Injury Risk in Professional Rugby Union Players. *Sports Med.* **2017**, *47*, 2395–2402. [CrossRef]
2. Williams, S.; Trewartha, G.; Kemp, S.; Stokes, K. A Meta-Analysis of Injuries in Senior Men's Professional Rugby Union. *Sports Med.* **2013**, *43*, 1043–1055. [CrossRef] [PubMed]
3. Quarrie, K.L.; Raftery, M.; Blackie, J.; Cook, C.; Fuller, C.W.; Gabbett, T.J.; Gray, A.J.; Gill, N.; Hennessy, L.; Kemp, S.; et al. Managing player load in professional rugby union: A review of current knowledge and practices. *Br. J. Sports Med.* **2016**, *51*, 421–427. [CrossRef]
4. Hind, K.; Konerth, N.; Entwistle, I.; Theadom, A.; Lewis, G.; King, D.; Chazot, P.; Hume, P. Cumulative sport-related injuries and longer-term impact in retired male elite-and amateur-level rugby code athletes and non-contact athletes: A retrospective study. *Sports Med.* **2020**, *50*, 2051–2061. [CrossRef] [PubMed]
5. Cunningham, J.; Broglio, S.; Wyse, J.; Mc Hugh, C.; Farrell, G.; Denvir, K.; Wilson, F. Athlete concussion history recall is underestimated: A validation study of self-reported concussion history among current professional rugby union players. *Brain Inj.* **2020**, *35*, 65–71. [CrossRef]
6. Malcolm, D.; Sheard, K. "Pain in the assets": The effects of commercialization and professionalization on the management of injury in English rugby union. *Sociol. Sport J.* **2002**, *19*, 149–169. [CrossRef]

7. Giazitzoglu, A. This Sporting Life: The intersection of hegemonic masculinities, space and emotions among rugby players. *Gend. Work. Organ.* **2020**, *27*, 67–81. [CrossRef]
8. Nixon, H.L. Accepting the Risks of Pain and Injury in Sport: Mediated Cultural Influences on Playing Hurt. *Sociol. Sport J.* **1993**, *10*, 183–196. [CrossRef]
9. Pringle, R.; Markula, P. No Pain Is Sane after All: A Foucauldian Analysis of Masculinities and Men's Rugby Experiences of Fear, Pain, and Pleasure. *Sociol. Sport J.* **2005**, *22*, 472–497. [CrossRef]
10. Curry, T.J. A Little Pain Never Hurt Anyone: Athletic Career Socialization and the Normalization of Sports Injury. *Symb. Interact.* **1993**, *16*, 273–290. [CrossRef]
11. Safai, P. Healing the Body in the "Culture of Risk": Examining the Negotiation of Treatment between Sport Medicine Clinicians and Injured Athletes in Canadian Intercollegiate Sport. *Sociol. Sport J.* **2003**, *20*, 127–146. [CrossRef]
12. Caddick, N.; Ryall, E. The social construction of 'mental toughness'–A fascistoid ideology? *J. Philos. Sport* **2012**, *39*, 137–154. [CrossRef]
13. Fraas, M.R.; Coughlan, G.F.; Hart, E.C.; McCarthy, C. Concussion history and reporting rates in elite Irish rugby union players. *Phys. Ther. Sport* **2014**, *15*, 136–142. [CrossRef]
14. Mayer, J.; Thiel, A. Presenteeism in the elite sports workplace: The willingness to compete hurt among German elite handball and track and field athletes. *Int. Rev. Sociol. Sport* **2018**, *53*, 49–68. [CrossRef]
15. Madrigal, L.; Robbins, J.E.; Gill, D.L.; Wurst, K. A Pilot Study Investigating the Reasons for Playing Through Pain and Injury: Emerging Themes in Men's and Women's Collegiate Rugby. *Sport Psychol.* **2015**, *29*, 310–318. [CrossRef]
16. Hawkins, R.D.; Fuller, C.W. A prospective epidemiological study of injuries in four English professional football clubs. *Br. J. Sports Med.* **1999**, *33*, 196–203. [CrossRef]
17. Messner, M.A. When bodies are weapons: Masculinity and violence in Sport. *Int. Rev. Sociol. Sport* **1990**, *25*, 203–220. [CrossRef]
18. Liston, K.; Reacher, D.; Smith, A.; Waddington, I. Managing Pain and Injury in Non-elite Rugby Union and Rugby League: A Case Study of Players at a British University. *Sport Soc.* **2006**, *9*, 388–402. [CrossRef]
19. Roderick, M.; Waddington, I.; Parker, G. Playing hurt: Managing injuries in English professional football. *Int. Rev. Sociol. Sport* **2000**, *35*, 165–180. [CrossRef]
20. Clarke, V.; Braun, V. Thematic analysis. In *Encyclopaedia of Critical Psychology*; Springer: New York, NY, USA, 2014; pp. 1947–1952.
21. Biernacki, P.; Waldorf, D. Snowball Sampling: Problems and Techniques of Chain Referral Sampling. *Sociol. Methods Res.* **1981**, *10*, 141–163. [CrossRef]
22. Braun, V.; Clarke, V. *Successful Qualitative Research: A Practical Guide for Beginners*; Sage Publishing: Thousand Oaks, CA, USA, 2013.
23. Ellingson, L.L. *Engaging Crystallization in Qualitative Research: An Introduction*; Sage Publishing: Thousand Oaks, CA, USA, 2009.
24. Blackett, A.D.; Evans, A.B.; Piggott, D. "Active" and "passive" coach pathways: Elite athletes' entry routes into high-performance coaching roles. *Int. Sport Coach. J.* **2018**, *5*, 213–226. [CrossRef]
25. Braun, V.; Clarke, V. Reflecting on reflexive thematic analysis. *Qual. Res. Sport Exerc. Health* **2019**, *11*, 589–597. [CrossRef]
26. Bennett, B.; Fyall, G. Power and Control in School Rugby: An Hermeneutic Interpretation of the Pedagogical Intentions of Five Secondary School Rugby Coaches in New Zealand. *J. Glob. Sport Manag.* **2018**, *3*, 284–301. [CrossRef]
27. Albert, E. Dealing with danger: The normalization of risk in cycling. *Int. Rev. Sociol. Sport* **1999**, *34*, 157–171. [CrossRef]

Journal of
Functional Morphology and Kinesiology

MDPI

Review

Indirect Structural Muscle Injuries of Lower Limb: Rehabilitation and Therapeutic Exercise

Stefano Palermi [1], Bruno Massa [1], Marco Vecchiato [2], Fiore Mazza [3], Paolo De Blasiis [4], Alfonso Maria Romano [5], Mariano Giuseppe Di Salvatore [6], Elisabetta Della Valle [1], Domiziano Tarantino [1], Carlo Ruosi [1] and Felice Sirico [1,*]

1 Department of Public Health, University of Naples "Federico II", 80131 Naples, Italy; stefanopalermi8@gmail.com (S.P.); b.massa91@gmail.com (B.M.); elisabetta.dellavalle@unina.it (E.D.V.); domiziano22@gmail.com (D.T.); carlo.ruosi@unina.it (C.R.)
2 Sport and Exercise Medicine Division, Department of Medicine, University of Padova, 35128 Padova, Italy; marcovecchiato.md@gmail.com
3 Rehabilitation Center "Phorma", 84047 Salerno, Italy; fioremazza@hotmail.it
4 Section of Human Anatomy, Department of Mental and Physical Health and Preventive Medicine, University of Campania Luigi Vanvitelli, 80131 Naples, Italy; paolodeblasiis@gmail.com
5 Orthopedics and Sport Medicine Unit, Campolongo Hospital, 84121 Salerno, Italy; alfonso.maria.romano@gmail.com
6 AORN S. Giuseppe Moscati, UOC Ortopedia e Traumatologia, 83100 Avellino, Italy; mariano.disalvatore@gmail.com
* Correspondence: sirico.felice@gmail.com; Tel.: +39-81-7463422

Citation: Palermi, S.; Massa, B.; Vecchiato, M.; Mazza, F.; De Blasiis, P.; Romano, A.M.; Di Salvatore, M.G.; Della Valle, E.; Tarantino, D.; Ruosi, C.; et al. Indirect Structural Muscle Injuries of Lower Limb: Rehabilitation and Therapeutic Exercise. *J. Funct. Morphol. Kinesiol.* **2021**, *6*, 75. https://doi.org/10.3390/jfmk6030075

Academic Editor: Cristina Cortis

Received: 29 July 2021
Accepted: 9 September 2021
Published: 13 September 2021

Abstract: Muscle injuries are the most common trauma in team and individual sports. The muscles most frequently affected are those of the lower limb, and in particular hamstrings, adductors, rectus femoris and calf muscles. Although several scientific studies have tried to propose different rehabilitation protocols, still too often the real rehabilitation process is not based on scientific knowledge, especially in non-elite athletes. Moreover, the growing use of physical and instrumental therapies has made it increasingly difficult to understand what can be truly effective. Therefore, the aim of the present paper is to review proposed therapeutic algorithms for muscle injuries, proposing a concise and practical summary. Following a three-phase rehabilitation protocol, this review aims to describe the conservative treatment of indirect structural muscle injuries, which are the more routinely found and more challenging type. For each phase, until return to training and return to sport are completed, the functional goal, the most appropriate practitioner, and the best possible treatment according to current evidence are expressed. Finally, the last section is focused on the specific exercise rehabilitation for the four main muscle groups with a structured explanatory timetable.

Keywords: muscle injury; rehabilitation; sport medicine

1. Introduction

Muscle injuries (MI) are the most common trauma in team and individual sports and are responsible for most of the time lost in both training and competition [1–3]. In soccer, they account for half of the injuries recorded [1]. In particular, four muscles groups are frequently involved [4,5]: hamstrings are the muscle groups most prone to injury [6–8], followed by adductors, rectus femoris and calf muscles.

The chance to receive an accurate early diagnosis and to have proper rehabilitation is different depending on whether these are professional or amateur athletes [9].

It is a common opinion that rehabilitation protocols designed for muscle injuries should be built mostly on available structure and therapeutic options [10], rather than on scientific knowledge. Clinical experience in the treatment of muscle injuries has proven that a wait-and-see approach is not effective [11]. Several therapeutic options for muscular injuries exist and, even if widely used, many reviews could not find enough evidence for conclusion about any of them. Therefore, unfortunately, common opinion has become to consider that what is popular is what is really effective [9,12,13].

Even though several scientific studies have tried to propose different rehabilitation protocols, to design a particular rehabilitation pathway for each muscular injury based on its grade and/or location is a difficult task; moreover, several commercial physical or instrumental therapies are becoming increasingly used for muscle injury treatment and rehabilitation, even if scientific evidence about their use are discordant [9,14].

In subjects affected by muscle injuries, the diagnostic and rehabilitative approach relies on several factors, such as age, gender, athletic demands, muscular groups involved and type of injury. To this scope, several classifications have been proposed over the years. All these classifications are based on some common criteria such as mechanism of injury (direct or indirect) and degree of lesion of muscle tissue (structural or non structural).

In a real-life scenario, the indirect structural muscle injuries represent the most common type of muscle injuries, and some muscle groups of lower limb are affected predominantly.

Therefore, the aim of the present paper is to review proposed therapeutic algorithms for indirect structural muscle injuries of the lower limb, with a particular insight on hamstrings, adductors, rectus femoris and calf muscles rehabilitation.

2. Muscle Injuries Rehabilitation

There are several classifications of muscle injuries, such as the Munich Muscle Injury Classification, the ISMuLT (Italian Society of Muscles, Ligaments and Tendons) classification, and the British Athletic Classification, that, if used extensively, could improve diagnosis, prognosis and management of muscle injuries [15]. Depending on the mechanism of trauma, according to ISMuLT classification [9], muscle injuries may be distinguished as direct and indirect; indirect ones are in turn classified as non-structural and structural. While direct muscle injuries are often the result of external forces, indirect muscle injuries are stretch-induced injuries caused by a sudden forced lengthening over the viscoelastic limits of muscles occurring during a powerful contraction [9,15]. Indirect structural muscle injuries (commonly referred as "muscle tears") are the more commonly found in everyday clinical practice and represent the biggest challenge in rehabilitation, since these lack a precise therapeutic strategy. Structural muscle injuries classification is shown in Table 1.

Table 1. Indirect structural muscle injuries classification (adapted from [9]).

Severity	Site	Tissue	Relapse
3A: minor partial lesion	P: proximal	MF: myofascial	R0: first lesion
3B: moderate partial lesion	M: medium	MT: muscular belly and myotendinous junction	R1: first relapse
4: subtotal or total lesion and tendon avulsion	D: distal	T: central tendon or free	R2: second relapse
			R3: third relapse

Severity [16], site [17], tissue [3], and relapse [18] are important features to consider when a muscle injury has been diagnosed. Proximal hamstring and quadricep lesions have a worse prognosis, as well as distal calf injuries; moreover, myotendinous junction lesions seem to have a longer recovery period [19]. Therefore, type and location of muscular injuries can influence recovery strategies [3] and proposed exercises should respect the principles of specificity, progression, and individualization, respecting painful symptomatology [14,20,21]. Moreover, location of injury, properly marked, could be useful for a focused therapy. Minor or moderate partial lesions (3A and 3B) are prevalent in sport rehabilitation and their conservative management is more controversial, since (sub)total lesions (4) are generally intended for surgery.

The muscle tissue repair process is completed in a period depending on the severity of the lesion. During this period, different well-defined biological phases are involved (destruction phase, repair/regeneration phase and remodeling phase) [14,22]. Each of these phases must be characterized by a well-defined type of muscular contraction that is consistent with the biological condition observed within the injured area [14,22].

Although rehabilitation is subdivided into a defined number of steps, the duration of each is different, and progression is not time based, but clinical, functional, and imaging criteria based [10,23]. Therefore, the duration of each phase is consistent with the dynamics of the healing processes occurring in the muscle tissue and with the severity of the injury. Each step of this process has a customized duration in accordance with the clinical and imaging criteria required for proceeding from one phase to the next.

Ultrasonography (US) offers dynamic muscle assessment and is fast and relatively inexpensive, allowing serial evaluation of the healing process [9]. However, it should be noted that ultrasonography of skeletal muscles requires a high level of skill on the part of the sports physician. It is recommended to use a 7.5- to 10.0-MHz transducer, starting with a transversal section. A complete scan through the muscle should be performed for the purposes of anatomical orientation. Any apparent abnormalities should be compared with the contralateral side. The transducer pressure should be as light as possible, since compressing the muscle may obscure smaller injuries. The longitudinal section is added in locations where a disturbance of the muscle structure or a gap is suspected. In addition, the use of novel US technique could help in this difficult diagnostic process, such as echo intensity [24]. When clinical and ultrasonography evaluation are discordant, or for muscles not accessible to US examination, in elite athletes, Magnetic Resonance Imaging (MRI) may be required to confirm or exclude minor structural injuries, since this technique is often used as a second-line investigation in musculoskeletal diseases [25,26]. MRI plays only a marginal role in the follow-up and monitoring of structural injuries because the images do not correlate well enough with the clinical evaluation, causing a potential late return to play (RTP) for the athlete.

A physician is responsible for the diagnosis, for overseeing the entire rehabilitation process, and for clinical and ultrasound monitoring; other professionals (physiotherapists, athletic trainers and coaches) control the correct execution of the rehabilitation program, each one for what they are entitled for [9].

Below, we proposed the three-phase rehabilitation protocol, based on ISMuLT [9] and Italian consensus conference [10] recommendations.

0 PHASE (0–72 h post injury)

- Muscle ultrasound allows to detect the structural damage of the skeletal muscle after 36–48 h from injury, because the hemorrhagic collection is maximized after 24 h and decreases after 48 h [9].
- In the immediate post injury period (24–72 h) it is advisable to apply the PRICE (Protection, Rest, Ice, Compression, Elevation) principle [27]. It is widely used, although there are no high quality randomized clinical trials to prove its effectiveness [28–30]. In clinical practice, immediate compression with 15 min cryotherapy cycles, with ice-free phases between, is recommended. Compressive cryotherapy (CC) [31], namely the association between cryotherapy and the application of pressure, deserves separate consideration: CC duration should be 15–20 min, repeated at intervals of 30–60 min for a total of 6 h, so as to substantially limit both the hemorrhage and the myofibril necrosis at the site of injury [32]. It is advisable to apply a compressive bandage and/or compressive cryotherapy within the range of 40–50 mmHg [23].
- A short rest period and/or relative immobilization immediately after the injury is recommended. This rest period optimizes the formation of connective tissue by fibroblasts, thereby reducing the risk of recurrences. Usually crutches are not necessary, while taping can be useful both for immobilization and liquid drainage. However, rest and immobilization should be reduced to only the first postlesion days (3–5 days) [10,14,22,28]. It would be better to have a short immobilization period followed by a progressive load able to favor the correct progression of healing process (POLICE: Protection, Optimal Loading, Ice, Compression, Elevation) [9].

- In the first 72 h postlesion, physical therapies that induce endothermic processes should be avoided for the possible increase in blood extravasation [3,10,33,34].
- After the first 24 h postlesion, it is a good idea to start performing complete lymphatic draining massages and to replace the compression bandage with an elastic bandage [9].
- After the first 24 h postlesion, there is little evidence about the usefulness of pulsed ultrasound therapy (UST) (1 W/cm^2) [13] (often used as cryo-ultrasound, with the adjunct of ice therapy) and low-level laser therapy (LT) (500 mW/cm^2) [9,35–37].

1st PHASE

- Functional goals [9]: treatment of predisposing factors and antagonist muscles; pain-free activity of daily life; pain-free strength training of the injured muscle, at least 50% of theoretical maximum load; recovery of at least 90% of the extensibility deficit of the injured muscle.
- Figure: physician and physiotherapist.
- Location: gym.
- Red Flags [9]: presence of pain when performing strength exercises or low-speed running on the treadmill.
- Image criteria: US check on the 2nd and 4th−5th day after injury [9].

At the beginning of the first phase (second postlesion day), the necrotized parts of the muscle fibers are removed by the macrophages, with an inflammatory process [9]. At the same time, the formation of the scarring connective tissue within the central lesion zone by fibroblasts starts [14,38,39]. Considering that the first 5–7 postlesion days are characterized by a not sufficiently dense and compact scarring, the major risk in this period is that an excessive muscle contraction increases the already existing lesion gap.

- The type of contraction recommended in this first phase is an isometric modality. In fact, during the isometric contraction, there is no myofilaments slippage and, therefore, there is no macro change of the muscle length [22,40]. Between 30 and 50 repetitions of 10–20 s of contraction under the threshold of pain are suggested. According to biomechanics concepts, the internal torque varies along the range of movement (ROM) of each joint. Each joint has specific degrees within the ROM in which the muscle is able to generate the maximum internal force and the anatomical position of muscle–tendon–bone unit give a maximum internal moment arm, generating the maximum torque. To gradually increase mechanical stress on the damaged muscle, it is necessary to proceed along the ROM gradually, by proposing contraction in ROM position where internal force is not able to produce the highest tension of the muscle.
- It is important to correctly perform exercises to recover the extensibility of injured muscle (passive, assisted/active, static or dynamic) [9], and better if with functional schemes. All exercises must be under the threshold of pain. An increased joint range was verified for stretches performed following functional patterns. In case of bi-articular muscles, it is advisable to stretch both insertional areas [41,42].
- Deep massages on the affected area should be avoided [10].
- Elastic bandage is continued until there is liquid collection.
- If there is an excessive hematoma formation within the injured area, it is advisable to proceed to an echo-guided aspiration before the hematoma organization [43].
- It is useful to start an aerobic workout as soon as possible, using non-injured muscles (i.e., upper trunk aerobic workout) [9].
- At the end of each working session, ice massage should be performed for 15–20 min [9].
- The use of electrical stimulation should be encouraged from the first postlesion days to the end of the regeneration phase (up to about the third postlesion week) [10,44–46]. Transcutaneous electric nerve stimulation (TENS) is the form of electrical stimulation most recommended in its two forms: conventional and acupuncture-like; several trials highlight its potential role in inhibition of transmission of pain signals [44]. Neuromuscular electrostimulation (NMES) utilizes high-intensity electrical stimulation to elicit intermittent contraction and relaxation of proximal muscle fibers; it

is widely prescribed for physical rehabilitation and muscle strengthening [44]. It has been demonstrated that these two techniques can stimulate the implantation of muscle resident stem cells inside the injured area, along with the voluntary exercise performed during rehabilitation [47–49].

- There is limited evidence that UST is able both to increase the levels of basic growth factors and to have an antalgic effect [50,51]: it may be recommended after the 0 phase (2 W/cm^2, in continuous modality, 1 MHz) [10].
- Many studies have shown that LT can reduce the inflammatory process of the damaged muscle tissue [52], speed up the tissue regeneration [53], optimize the oxidative metabolism [54] and stimulate cell proliferation [55,56]. Therefore, the use of LT appears to be justified by sufficient evidence, even if not high quality featured [9,10,57].
- Hyperthermia therapy (HT) has proven to be able to stimulate the tissue repair processes, diminish pain symptoms, increase tissue flexibility, and reduce muscular and joint stiffness [58–66]. However, there are poor specific evidence on the HT effectiveness in muscular injuries [9,10].
- Analgesic (paracetamol) can be used in case of pain in the first postlesion days [9,10,67,68], while muscle relaxants, mesenchymal stem cells (MSCs) and platelet-rich plasma (PRP) injections require further evidence-based studies to evaluate their effectiveness [23,69]. The use of nonsteroidal anti-inflammatory drugs (NSAIDs) is controversial [70], and it is not recommended.

2nd PHASE

- Functional goals [9]: absence of pain or feeling of diversity in injured muscle when performing exercises; complete recovery of the extensibility of the injured muscle; recovery of the aerobic sport-specific parameters; complete recovery of the pre-injury weight.
- Figure: physiotherapist and athletic trainer.
- Location: gym and sport-field.
- Clinic criteria [10,23,71]: resolution of swelling, if initially present; absence of pain in response to maximal isometric contraction; absence of pain in response to end-range stretching tests carried out in the active and passive modes; complete range of motion (ROM) of the joints involved in the movement.
- Imaging criteria [10,72,73]: resolution of the lesion gap as observed with US or MRI imaging; the presence of granulation repair tissue within the cicatrix zone (CZ) as revealed by the US. US findings observed during normal healing depend on the nature of the original injury and initial sonographic findings. Minor lesions may increase in echogenicity during the healing process. In these cases, a progressive reduction in intensity or its disappearance is considered normal. More prominent lesions may present as hypoechogenic regions with adjacent fluid collection. Resolution or substantial decrease in the quantity of fluid is to be expected during the normal healing process [74,75].
- Red Flags [9]: extensibility test still positive.

At this stage, the scar area in the CZ is further condensed and reduced in size, and myofibers fill the residual gap of the CZ [14,38,76,77]. During this phase, the granulation tissue gains compactness and elasticity [78]. In this regenerative phase mechanical stimuli should be performed in order to induce an optimal tissue repairing [9].

- There is the introduction of progressively intense concentric exercises. During a concentric contraction, the bulk of the muscle shortens due to the sliding motion of the myofilaments with a relatively constant force proportional to the external load, so the CZ is not subjected to traction and the jagged muscle edges, avoiding diastasis [79]. The concentric contraction should be slow and controlled; they can be manual at the beginning, and subsequently with isotonic equipment [80]. Sixty percent of one repetition maximum (RM) should not be exceeded when performing these exercises in this stage [79,80]. The eccentric phase of the movement must, in all cases, be reduced to the minimum possible intensity [10].

- Keep performing exercises to recover the extensibility of injured muscle [9].
- Proprioceptive exercises should be started [9,81]: balance exercises on stable or unstable different shape surfaces, with or without recurrent destabilization, with or without request for additional cognitive tasks, if possible, with the support of the visive system.
- The practice of massage can be introduced as the completion of tissue healing processes has started [10].
- A 'core stability program' should be introduced in the rehabilitation plan [10,82,83], eventually combined with proprioceptive exercises [9].
- Aerobic exercises can be introduced during this phase [10,23]: the time-progression should be stationary bike, elliptical machine, anti-gravity running and, finally, treadmill running.
- Physical therapies started could be continued in this phase.

3rd PHASE

- Functional goals [9]: consolidation of the strength and extensibility characteristics of the injured muscle; recovery of the sport-specific skills; recovery of the high-intensity sport-specific athletic parameters; working resistance of the injured muscle.
- Figure: physiotherapist and athletic trainer.
- Location: gym and sport-field.
- Clinic criteria [10,23,71]: absence of pain in response to concentric contraction performed at increasing intensity against resistance; absence of pain in response to submaximal eccentric contraction.
- Imaging criteria [10,72,73]: substantial disappearance of the lesion gap on US or MRI examination; presence of compact granulation repair tissue as revealed by US or MRI. Over time small tears may fill with echogenic material, likely representing scar tissue visible at US [84,85]. More extensive scarring results in increased likelihood of recurrent injury [25].
- Red Flags [9]: "different" muscle feeling during or after training.

In this phase, the myofibers intertwining is effectively completed by the interposition of a small amount of scar tissue. There should be proposed strength and extensibility exercises that induce remodeling of the repair tissue based on the sport played [9], depending also on the movement that caused the injury. The remodeling phase may last more than 60 days, depending on the anatomical extent of the injury [9].

- Exercises predominantly based on eccentric contractions of progressively increasing intensity [9,10,23,86–88] could be started after an effective concentric contraction is reached. These should be muscle and location specified [89]. These can be performed even with the use of elastic resistance bandages, where the intensity of the eccentric phase is progressively increased [10,23]. Even if some authors suggest introducing eccentric exercises as soon as possible in the rehabilitation protocol [9], the 3rd phase should be the preferred one for their execution. Moreover, evidence about isoinertial exercises are increasing [90].
- There could be the inclusion of isokinetic exercises [10,28].
- Stretching must be introduced gradually and exercises must not cause the onset of pain. The time of elongation initially is 10–15 s and subsequently up to 1 min, in order to induce a durable, and not just a transient, plastic deformation within the area of structural reorganization [10,23]. For bi-articular muscles, please consider both origin and insertion tendons.
- Running could be improved during this phase, on the condition that dynamometric values of the injured muscle have been reinstated to at least 70% of the preinjury level or that of the opposite limb [10,91], and with the use of GPS monitoring [9].
- Sport-specific exercises can be introduced with caution at the end of the third phase [9,10].
- Even if not supported by strong scientific evidence, physical therapies can be used to avoid muscular fatigue, complications, and re-injury [9]: LT [92], ice water immer-

sion [93], contrast therapy [94], HT [95], TENS [96] and extra-corporeal shock wave therapy (ESWT).

It is important to consider an athlete as "healed" as long as three concepts are respected [9]: progression in the recovery of match intensity; continuous information exchange between coaches, trainers, physiotherapist, athlete and physician; and continuous monitoring of the injured muscle characteristics after trainings and matches.

3. Specific Exercise Rehabilitation

Even though there are so many rehabilitation exercises used, it is the authors' opinion that each muscle injury should be treated differently, trying to individualize it as much as possible. They should follow a well-structured timetable that is appropriate for the specific injury or disorder: as we stated before, the correct progression should be isometric (1st phase), concentric (2nd phase) and eccentric (3rd phase) exercises; proprioceptive, neuromuscular and stretching exercises also have a major role in the rehabilitation process. Below we propose examples of exercises for hamstrings (Table 2), rectus femoris (Table 3), adductors (Table 4), and calf injuries (Table 5), along with their criteria for RTT and secondary prevention programs.

Hamstring

Table 2. Hamstring rehabilitation exercises.

Name	Image	Reference
	Isometric exercises	
(In case of proximal hamstring lesion)	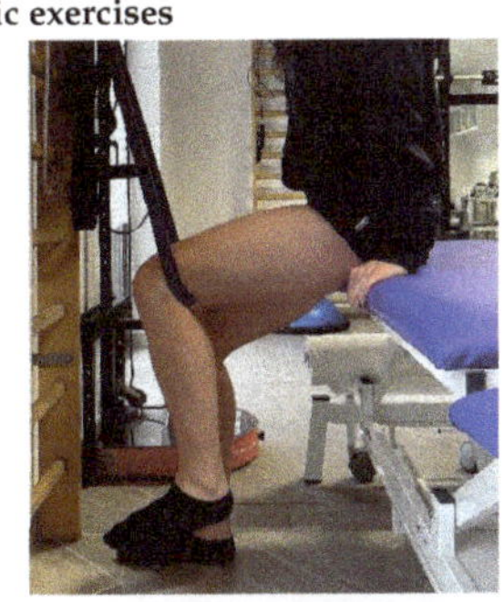	[23]
(In case of medial or distal hamstring lesion)	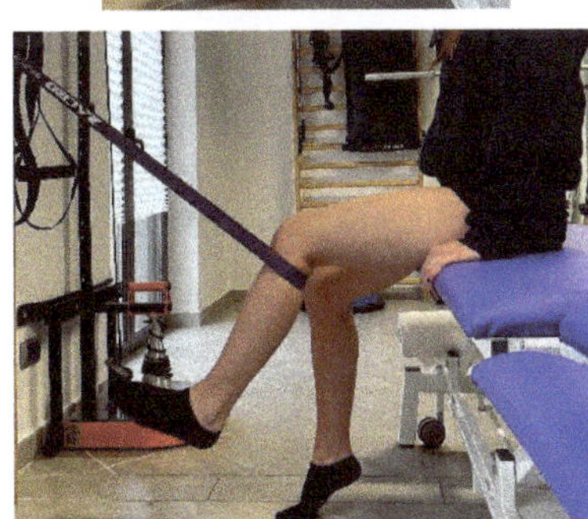	[23]
Isometric exercise at different angles	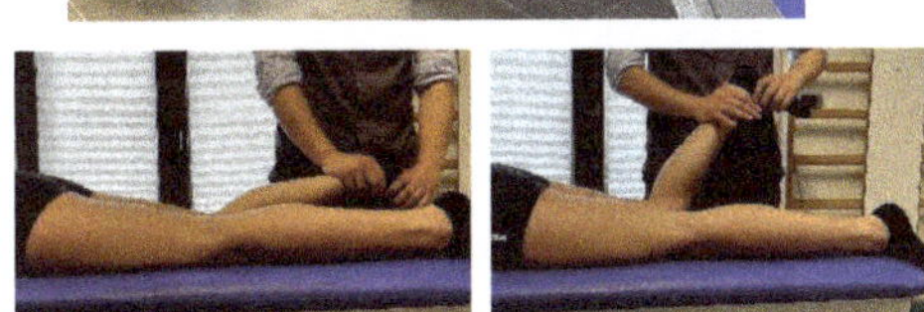	[23]

Table 2. *Cont.*

Name	Image	Reference
	Dynamic exercises	
The extender	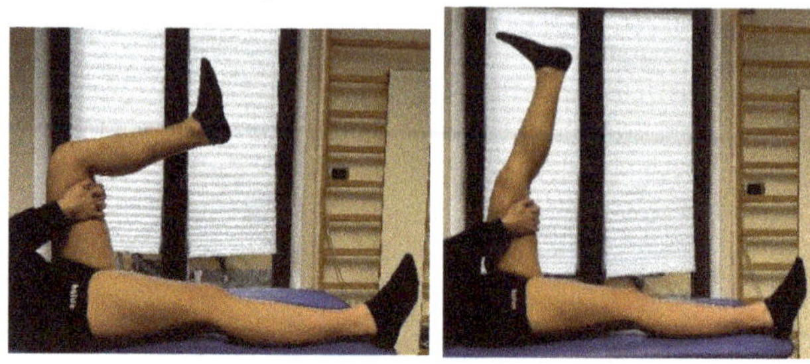	[88,97]
The glider	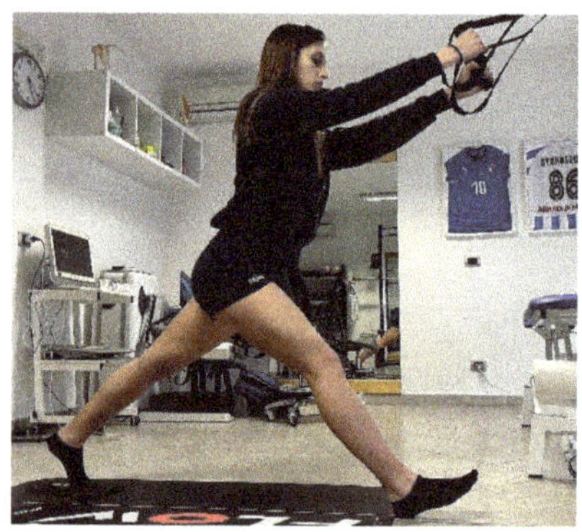	[88,97]
Nordic hamstrings	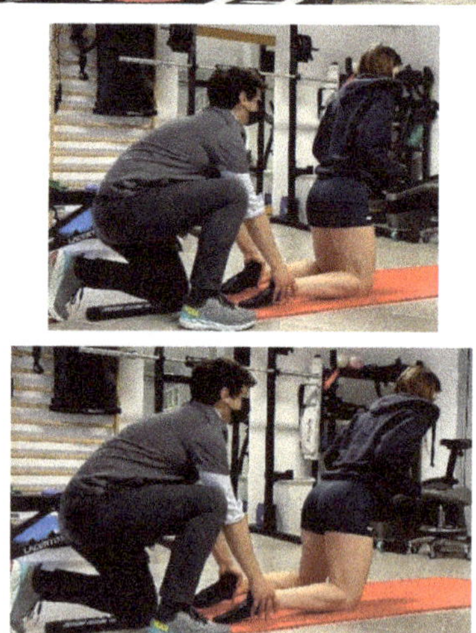	[97]

Table 2. *Cont.*

Name	Image	Reference
	Proprioceptive, neuromuscular and stretching exercises	
Pendulum	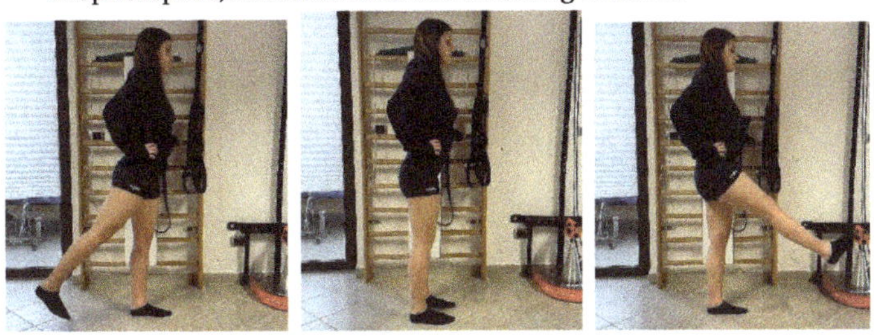	[97]
Stretching Single Leg Raises	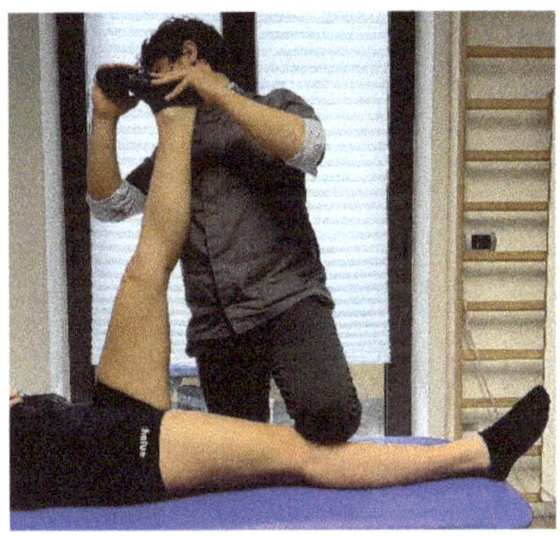	[97]
	Secondary prevention exercises	
Eccentric knee flexor stretch	 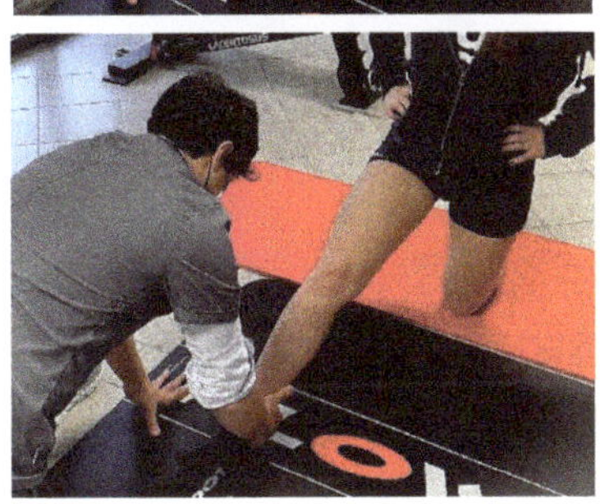	[98]

Table 2. *Cont.*

Name	Image	Reference
	Secondary prevention exercises	
Eccentric hip extensor stretch	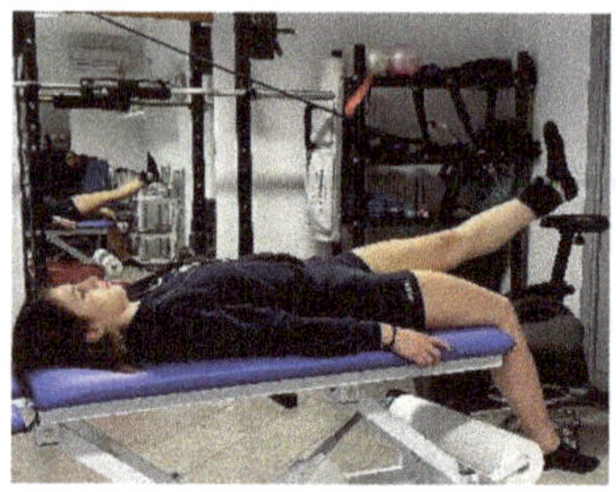 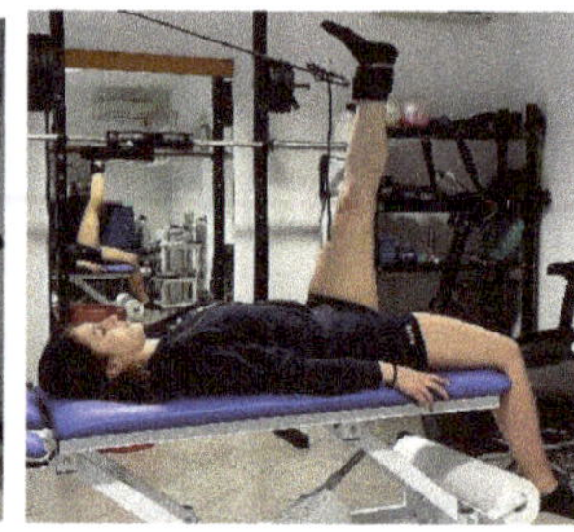	[98]

Quadriceps

Table 3. Quadriceps rehabilitation exercises.

Name	Image	Reference
	Isometric exercises	
(In case of proximal lesion)	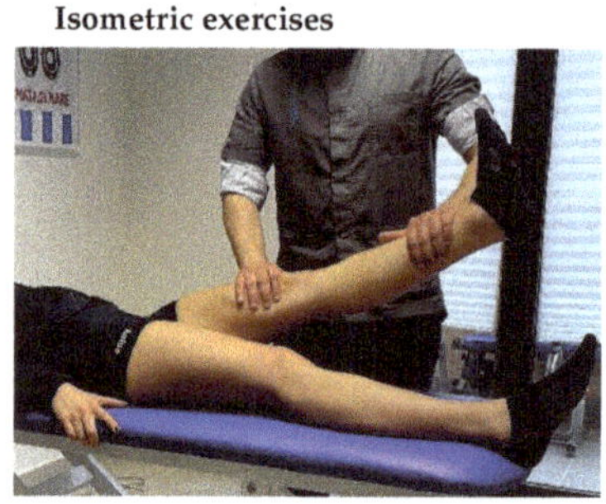	[23]
(In case of medial or distal lesion)	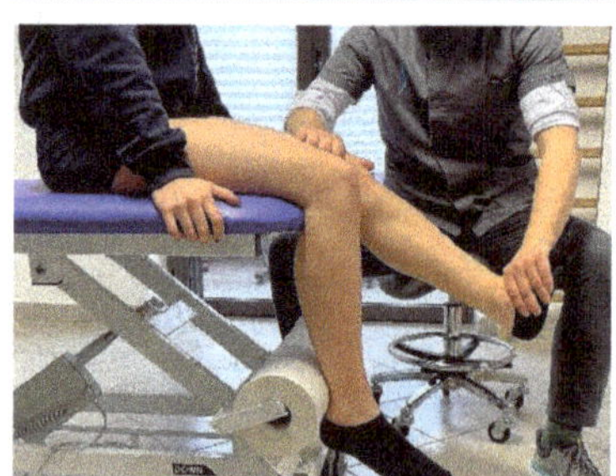	[23]
	Dynamic exercises	
(In case of proximal lesion)	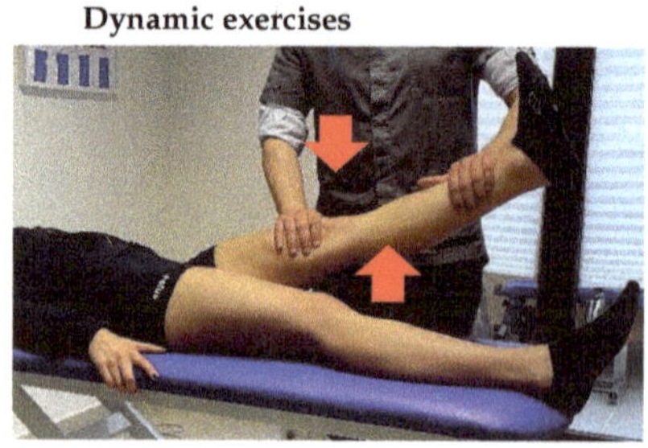	[23]

Table 3. *Cont.*

Name	Image	Reference
	Dynamic exercises	
(In case of medial or distal lesion)	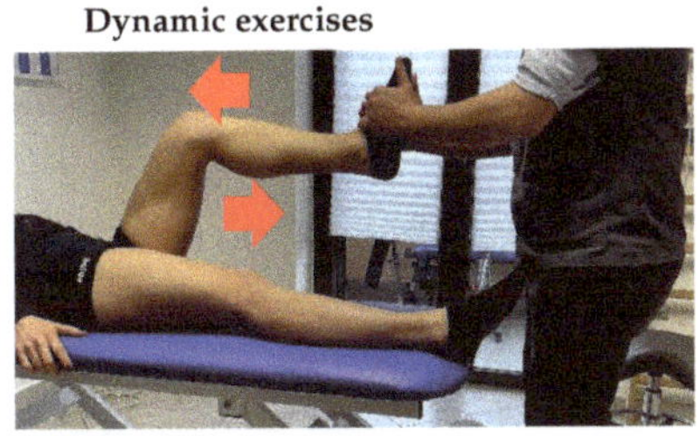	[23]
	Secondary prevention exercises	
Eccentric hip flexor and knee extensor stretch (eccentric load to rectus femoris)		[98]

Adductors

Table 4. Adductors rehabilitation exercises.

Name	Image	Reference
	Isometric exercises	
Isometric exercise with ball	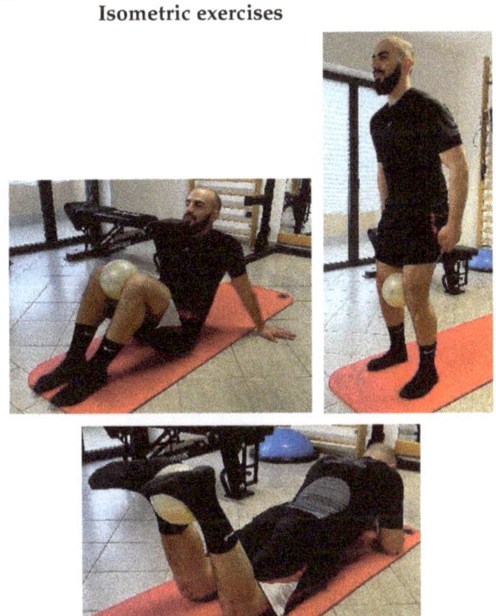	[23]

Table 4. *Cont.*

Name	Image	Reference
	Dynamic exercises	
Manual resisted adduction	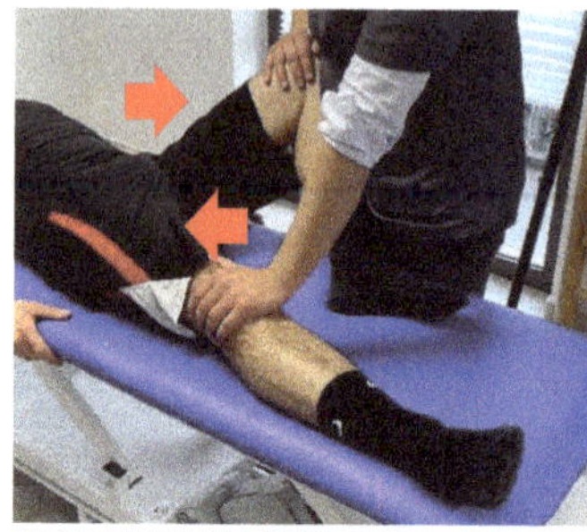 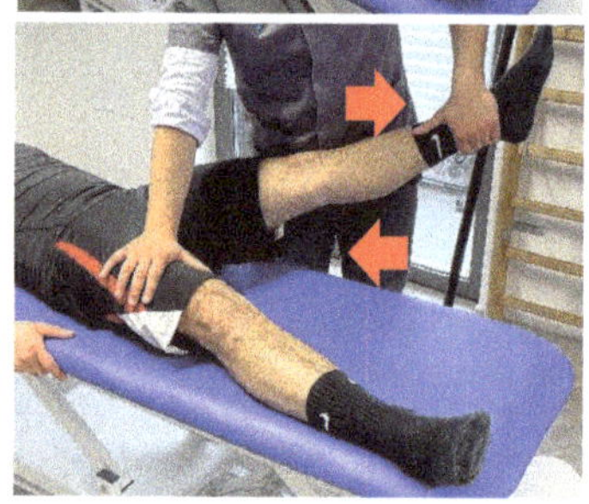	[23]
Adduction with elastic resistance	 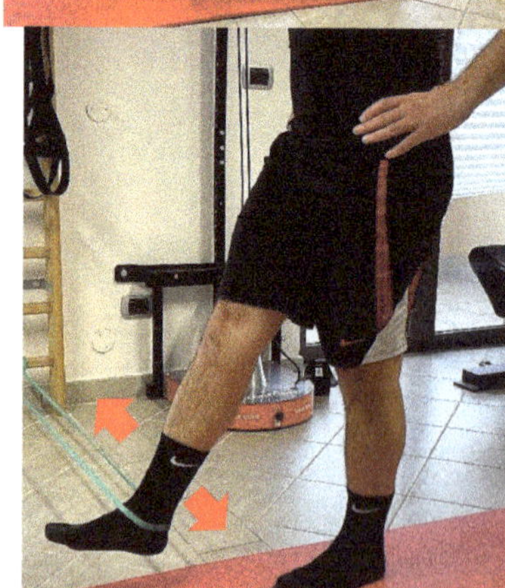	[23]

Table 4. *Cont.*

Name	Image	Reference
	Proprioceptive, neuromuscular, and stretching exercises	
		[23]
	Secondary prevention exercises	
Eccentric side lunge stretch	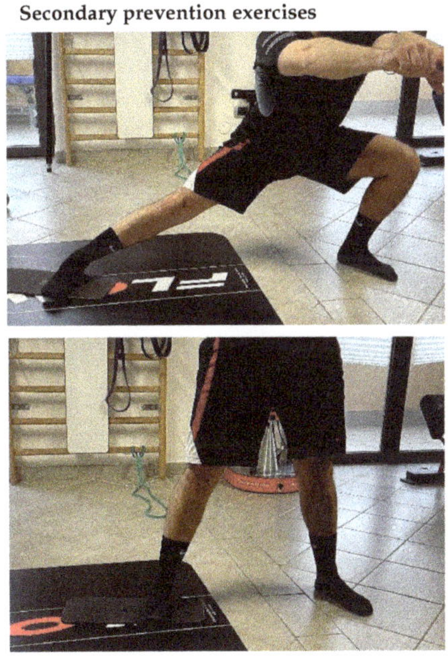	[98]

Table 4. *Cont.*

Name	Image	Reference
	Proprioceptive, neuromuscular, and stretching exercises	
Copenhagen adductor prevention programs	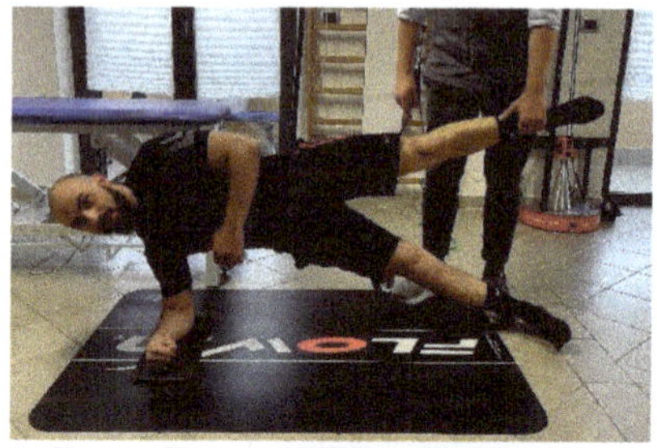 	[99]

Soleus-gastrocnemius

Table 5. Soleus-gastrocnemius rehabilitation exercises.

Name	Image	Reference
	Isometric exercises	
Isometric contraction with manual resistance	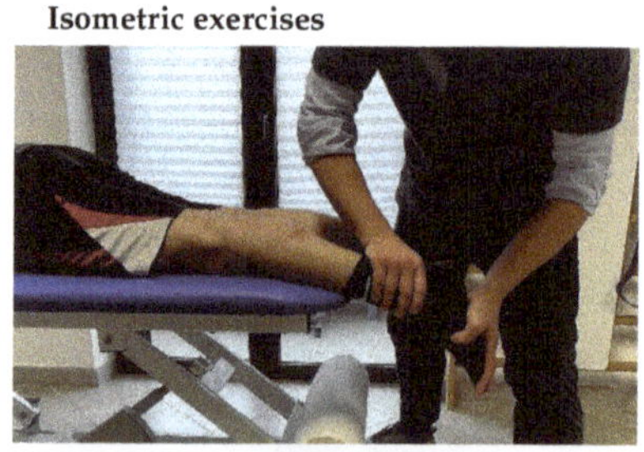 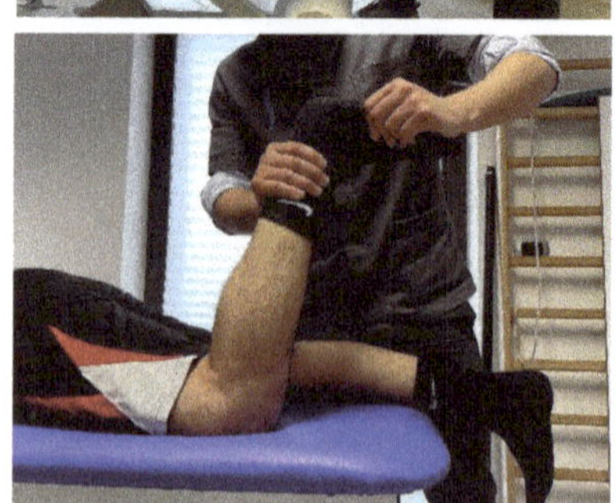	[23]

Table 5. *Cont.*

Name	Image	Reference
	Dynamic exercises	
Concentric/eccentric contraction with manual resistance	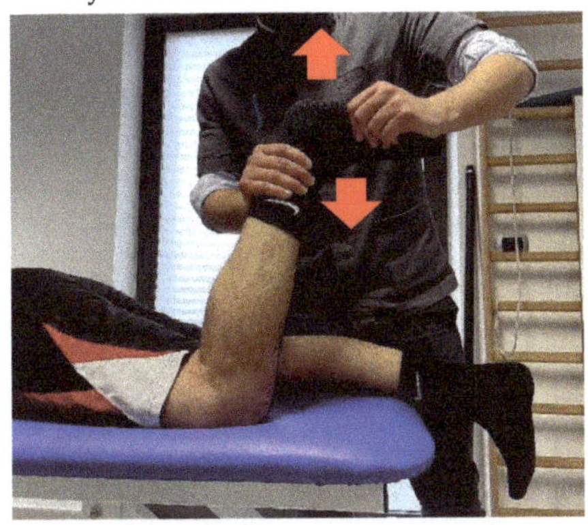	[23]
Concentric/eccentric heel raise	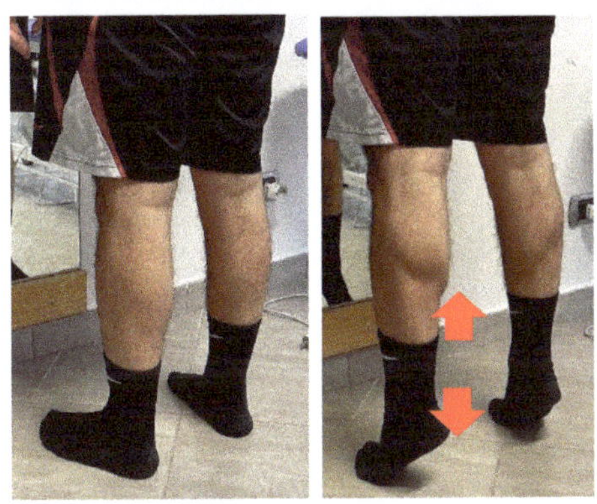	[23]
	Proprioceptive, neuromuscular, and stretching exercises	
	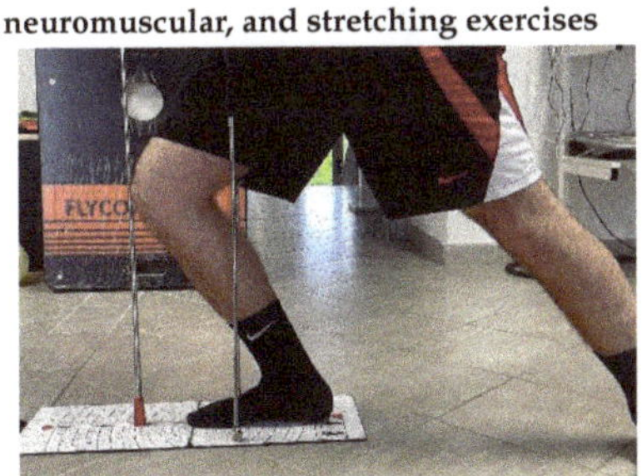	[23]

4. Return to Training (RTT) and Return to Play (RTP)

US examination upon complete RTT and a few days after the RTP is recommended [100]. There are no validated imaging criteria to guide the decision of a safe RTP. To date, no study has suggested US to guide the RTP decision, but a few studies have focused on MRI following hamstring injury [101–104]. Normalization of increased signal intensities on MRI is therefore not required for a successful RTP, since the signal alterations also persist at different weeks after the clinical healing of the injury, suggesting that functional recovery advances structural recovery at imaging [9].

The RTT process should be as individualized as possible, to allow a safe and fast return after a muscle injury. Regarding this point, the Italian consensus conference gave useful advice [101]. General assessment about this process is made up of some key points: absence of clinical symptoms [105–107]; absence of pain or tenderness during muscle palpation [10,107–109]; absence of pain on passive and active stretching [110]; absence of pain on isometric, concentric and eccentric contraction [10]; completion of the prescribed rehabilitation program [108]; MRI and US imaging [111,112]; subjective feelings of the player taken into account [113–115].

It is recommended that the athlete accomplishes a normal of week training of at least four sessions without pain, discomfort, or 'fear'. During this week, performance can be monitored for normality by global positioning system (GPS) and heart rate data [116]; this performance control should be extended to competitions after RTP. The reference value, below which the positive judgement for RTP is postponed, is arbitrarily set at a maximum difference of 10% between preinjury data and the data recorded during the acquisition period following RTT. Furthermore, an evaluation of aerobic capacity is recommended. A VO2 max equal to at least 90% of their preinjury level seems to offer more guarantees for a safer RTP [101].

To define a set of tests to determine the correct timing of RTT is a difficult task. Specific assessment for each muscle group, laboratory tests aimed to assess muscle strength, and functional field tests could be adopted as criteria to define a safe RTT.

Based on the available literature, a list of tests has been defined to each muscle groups and are reported in Table 6.

Table 6. Return to play (RTP) specific tests for muscle groups.

Hamstring	
Specific assessment	• Passive straight leg raise test [110,117,118]; • Dynamic flexibility H test [119]
Laboratory test	• Dynamometric tests (isometric, isotonic and isokinetic tests) [104,120,121]
Field test	• Illinois Agility Test [101,122,123] • Braking test [101] • Backward running [124,125]
Quadriceps	
Specific assessment	• Passive quadriceps stretch test [110,126]
Laboratory test	• Dynamometric tests [107,120,121] • Synchro plates test [121]
Field test	• Illinois Agility Test [101,122,123] • Braking test [101] • Kicking test [10]
Adductors	
Specific assessment	• Pubic stress test [127] • Resisted hip adduction test [120,128] • Squeeze test [129–132] • Adductor passive stretching test [133]
Laboratory test	• Adductor muscles strength assessed by dynamometric tests [104,120,121]
Field test	• Kicking test [10] • Carioca test [134,135]
Soleus-gastrocnemius	
Specific assessment	• Heel-raise test [38,136,137] • Ankle flexibility test [138–140]
Laboratory test	• Dynamometric tests [107,119,120] • Synchro plates test [121] • Drop jump test [141–143]
Field test	• Illinois Agility Test [101,142,143]

The aim of these tests and their specific execution are out of the scope of the present paper, but the rationale behind each test is reported in the references and could be used to guide the RTT and RTP processes.

5. Conclusions

The present paper offers an overview of advice and recommendations on how to set up the rehabilitation process after indirect muscle injuries according to current evidence. The section on specific exercises for the most affected muscles groups adds a practical guide for the practitioner to apply the concepts reported in the guidelines to a real-life scenario. Only with the constant synergistic work between the various professionals involved, who work according to the highest scientific evidence available, the injured athlete can reach the maximum result of the rehabilitation process and return to their sport quickly and safety.

Author Contributions: Conceptualization, S.P. and B.M.; methodology, M.V.; validation, F.M., P.D.B., A.M.R. and M.G.D.S.; data curation, E.D.V.; writing—original draft preparation, S.P. and F.S.; writing—review and editing, D.T. and C.R.; supervision, F.S. All authors have read and agreed to the published version of the manuscript.

Funding: This research received no external funding.

Institutional Review Board Statement: Not applicable.

Informed Consent Statement: Not applicable.

Data Availability Statement: Not applicable.

Conflicts of Interest: The authors declare no conflict of interest.

References

1. Alonso, J.M.; Edouard, P.; Fischetto, G.; Adams, B.; Depiesse, F.; Mountjoy, M. Determination of future prevention strategies in elite track and field: Analysis of Daegu 2011 IAAF Championships injuries and illnesses surveillance. *Br. J. Sports Med.* **2012**, *46*, 505–514. [CrossRef]
2. Freckleton, G.; Pizzari, T. Risk factors for hamstring muscle strain injury in sport: A systematic review and meta-analysis. *Br. J. Sports Med.* **2013**, *47*, 351–358. [CrossRef]
3. Pollock, N.; James, S.L.J.; Lee, J.C.; Chakraverty, R. British athletics muscle injury classification: A new grading system. *Br. J. Sports Med.* **2014**, *48*, 1347–1351. [CrossRef]
4. Hallén, A.; Ekstrand, J. Return to play following muscle injuries in professional footballers. *J. Sports Sci.* **2014**, *32*, 1229–1236. [CrossRef]
5. Waldén, M.; Hägglund, M.; Bengtsson, H.; Ekstrand, J. Perspectives in football medicine. *Unfallchirurg* **2018**, *121*, 470–474. [CrossRef]
6. Ekstrand, J.; Hägglund, M.; Waldén, M. Epidemiology of muscle injuries in professional football (soccer). *Am. J. Sports Med.* **2011**, *39*, 1226–1232. [CrossRef] [PubMed]
7. Hägglund, M.; Waldén, M.; Ekstrand, J. Risk factors for lower extremity muscle injury in professional soccer: The UEFA Injury Study. *Am. J. Sports Med.* **2013**, *41*, 327–335. [CrossRef] [PubMed]
8. Maffulli, N.; Oliva, F.; Frizziero, A.; Nanni, G.; Barazzuol, M.; Giai Via, A.; Ramponi, C.; Brancaccio, P.; Listano, G.; Rizzo, D.; et al. ISMuLT Guidelines for muscle injuries. *Muscles Ligaments Tendons J.* **2013**, *3*, 241–249. [CrossRef]
9. Nanni, G.; Frizziero, A.; Oliva, F.; Maffulli, N. *Gli Infortuni Muscolari-Linee Guida, I.S.Mu.L.T*; Calzetti Mariucci Editore: Torgiano, Italy, 2020.
10. Bisciotti, G.N.; Volpi, P.; Amato, M.; Alberti, G.; Allegra, F.; Aprato, A.; Artina, M.; Auci, A.; Bait, C.; Bastieri, G.M.; et al. Italian consensus conference on guidelines for conservative treatment on lower limb muscle injuries in athlete. *BMJ Open Sport Exerc. Med.* **2018**, *4*, e000323. [CrossRef] [PubMed]
11. Diagnostik und Therapie von Zerrungen und Muskelfaserrissen im Hochleistungssport. In *Manual des Deutschen Fußball-Bundes (DFB)*; Müller-Wohlfahrt, H.W. (Ed.) Deutscher Fußball-Bundes: Frankfurt, Germany, 2006.
12. Tol, J.L.; Hamilton, B.; Best, T.M. Palpating muscles, massaging the evidence? An editorial relating to; Terminology and classification of muscle injuries in sport: The Munich consensus statement. *Br. J. Sports Med.* **2013**, *47*, 340–341. [CrossRef] [PubMed]
13. Hotfiel, T.; Seil, R.; Bily, W.; Bloch, W.; Gokeler, A.; Krifter, R.M.; Mayer, F.; Ueblacker, P.; Weisskopf, L.; Engelhardt, M. Nonoperative treatment of muscle injuries-recommendations from the GOTS expert meeting. *J. Exp. Orthop.* **2018**, *5*, 24. [CrossRef] [PubMed]
14. Järvinen, T.A.; Järvinen, M.; Kalimo, H. Regeneration of injured skeletal muscle after the injury. *Muscles Ligaments Tendons J.* **2013**, *3*, 337–345. [CrossRef] [PubMed]
15. Maffulli, N.; Aicale, R.; Tarantino, D. Classification of Muscle Lesions. In *Muscle and Tendon Injuries*; Canata, G., d'Hooghe, P., Hunt, K., Eds.; Springer: Berlin/Heidelberg, Germany, 2017. [CrossRef]

16. Mueller-Wohlfahrt, H.-W.; Haensel, L.; Mithoefer, K.; Ekstrand, J.; English, B.; McNally, S.; Orchard, J.; van Dijk, C.N.; Kerkhoffs, G.M.; Schamasch, P.; et al. Terminology and classification of muscle injuries in sport: The Munich consensus statement. *Br. J. Sports Med.* **2013**, *47*, 342–350. [CrossRef] [PubMed]
17. Chan, O.; Del Buono, A.; Best, T.M.; Maffulli, N. Acute muscle strain injuries: A proposed new classification system. *Knee Surg Sports Traumatol. Arthrosc.* **2012**, *20*, 2356–2362. [CrossRef] [PubMed]
18. Valle, X.; Alentorn-Geli, E.; Tol, J.L.; Hamilton, B.; Garrett, W.E., Jr.; Pruna, R.; Til, L.; Antoni Gutierrez, J.; Alomar, X.; Balius, R.; et al. Muscle Injuries in Sports: A New Evidence-Informed and Expert Consensus-Based Classification with Clinical Application. *Sports Med.* **2017**, *47*, 1241–1253. [CrossRef] [PubMed]
19. Ekstrand, J.; Askling, C.; Magnusson, H.; Mithoefer, K. Return to play after thigh muscle injury in elite football players: Implementation and validation of the Munich muscle injury classification. *Br. J. Sports Med.* **2013**, *47*, 769–774. [CrossRef]
20. Mendiguchia, J.; Martinez-Ruiz, E.; Edouard, P.; Morin, J.B.; Martinez-Martinez, F.; Idoate, F.; Mendez-Villanueva, A. A Multifactorial, Criteria-based Progressive Algorithm for Hamstring Injury Treatment. *Med. Sci. Sports Exerc.* **2017**, *49*, 1482–1492. [CrossRef]
21. Khan, K.M.; Scott, A. Mechanotherapy: How physical therapists' prescription of exercise promotes tissue repair. *Br. J. Sports Med.* **2009**, *43*, 247–252. [CrossRef]
22. Järvinen, T.A.H.; Järvinen, T.L.N.; Kääriäinen, M.; Kalimo, H.; Järvinen, M. Muscle injuries: Biology and treatment. *Am. J. Sports Med.* **2005**, *33*, 745–764. [CrossRef]
23. Volpi, P.; Bisciotti, G.N. *Muscle Injury in the Athlete*; Springer: Berlin/Heidelberg, Germany, 2019.
24. Wong, V.; Spitz, R.W.; Bell, Z.W.; Viana, R.B.; Chatakondi, R.N.; Abe, T.; Loenneke, J.P. Exercise induced changes in echo intensity within the muscle: A brief review. *J. Ultrasound.* **2020**, *23*, 457–472. [CrossRef]
25. Zappia, M.; Ascione, F.; Di Pietto, F.; Fischetti, M.; Romano, A.M.; Castagna, A.; Brunese, L. Long head biceps tendon instability: Diagnostic performance of known and new MRI diagnostic signs. *Skelet. Radiol.* **2021**, *50*, 1863–1871. [CrossRef]
26. Zappia, M.; Reginelli, A.; Chianca, V.; Carfora, M.; Di Pietto, F.; Iannella, G.; Mariani, P.P.; Di Salvatore, M.; Bartollino, S.; Maggialetti, N. MRI of popliteo-meniscal fasciculi of the knee: A pictorial review. *Acta. Biomed.* **2018**, *19*, 7–17.
27. Bleakley, C.; McDonough, S.; MacAuley, D. The use of ice in the treatment of acute soft-tissue injury: A systematic review of randomized controlled trials. *Am. J. Sports Med.* **2004**, *32*, 251–261. [CrossRef] [PubMed]
28. Järvinen, T.A.H.; Järvinen, T.L.N.; Kääriäinen, M.; Äärimaa, V.; Vaittinen, S.; Kalimo, H.; Järvinen, M. Muscle injuries: Optimising recovery. *Best Pr. Res. Clin. Rheumatol.* **2007**, *21*, 317–331. [CrossRef]
29. Orchard, J.W.; Best, T.M.; Mueller-Wohlfahrt, H.-W.; Hunter, G.; Hamilton, B.H.; Webborn, N.; Jaques, R.; Kenneally, D.; Budgett, R.; Philips, N.; et al. The early management of muscle strains in the elite athlete: Best practice in a world with a limited evidence basis. *Br. J. Sports Med.* **2008**, *42*, 158–159. [CrossRef]
30. Bleakley, C.M.; Glasgow, P.; MacAuley, D.C. PRICE needs updating, should we call the POLICE? *Br. J. Sports Med.* **2012**, *46*, 220–221. [CrossRef]
31. Block, J.E. Cold and compression in the management of musculoskeletal injuries and orthopedic operative procedures: A narrative review. *J. Sport Med.* **2010**, *1*, 105–113. [CrossRef] [PubMed]
32. Schaser, K.-D.; Disch, A.C.; Stover, J.F.; Lauffer, A.; Bail, H.J.; Mittlmeier, T. Prolonged superficial local cryotherapy attenuates microcirculatory impairment, regional inflammation, and muscle necrosis after closed soft tissue injury in rats. *Am. J. Sports Med.* **2007**, *35*, 93–102. [CrossRef] [PubMed]
33. Fyfe, J.J.; Opar, D.A.; Williams, M.D.; Shield, A.J. The role of neuromuscular inhibition in hamstring strain injury recurrence. *J. Electromyogr Kinesiol. Off. J. Int Soc. Electrophys. Kinesiol.* **2013**, *23*, 523–530. [CrossRef]
34. Hägglund, M.; Waldén, M.; Ekstrand, J. Injury recurrence is lower at the highest professional football level than at national and amateur levels: Does sports medicine and sports physiotherapy deliver? *Br. J. Sports Med.* **2016**, *50*, 751–758. [CrossRef]
35. Lim, W.; Lee, S.; Kim, I.; Chung, M.; Kim, M.; Lim, H.; Park, J.; Kim, O.; Choi, H. The anti-inflammatory mechanism of 635 nm light-emitting-diode irradiation compared with existing COX inhibitors. *Lasers Surg. Med.* **2007**, *39*, 614–621. [CrossRef]
36. Ramos, L.; Marcos, R.L.; Torres-Silva, R.; Pallota, R.C.; Magacho, T.; Mafra, F.P.F.; Macedo, M.M.; Carvalho, R.L.D.P.; Bjordal, J.M.; Lopes-Martins, R.A.B. Characterization of Skeletal Muscle Strain Lesion Induced by Stretching in Rats: Effects of Laser Photobiomodulation. *Photomed Laser Surg.* **2018**, *36*, 460–467. [CrossRef]
37. Clijsen, R.; Brunner, A.; Barbero, M.; Clarys, P.; Taeymans, J. Effects of low-level laser therapy on pain in patients with musculoskeletal disorders: A systematic review and meta-analysis. *Eur. J. Phys. Rehabil. Med.* **2017**, *53*, 603–610. [CrossRef] [PubMed]
38. Brooks, S.V. Current topics for teaching skeletal muscle physiology. *Adv. Physiol. Educ.* **2003**, *27*, 171–182. [CrossRef]
39. Delos, D.; Maak, T.G.; Rodeo, S.A. Muscle injuries in athletes: Enhancing recovery through scientific understanding and novel therapies. *Sports Health* **2013**, *5*, 346–352. [CrossRef]
40. Dias da Silva, S.R.; Gonçalves, M. Dynamic and isometric protocols of knee extension: Effect of fatigue on the EMG signal. *Electromyogr. Clin. Neurophysiol.* **2006**, *46*, 35–42.
41. Decoster, L.C.; Cleland, J.; Altieri, C.; Russel, P. The effects of hamstring stretching on range of motion: A systematic literature review. *J. Orthop. Sports Phys. Ther.* **2005**, *35*, 377–387. [CrossRef]

42. Fasen, J.M.; O'Connor, A.M.; Schwartz, S.L.; Watson, J.O.; Plastaras, C.T.; Garvan, C.W.; Bulcao, C.; Johnson, S.C.; Akuthota, V. A randomized controlled trial of hamstring stretching: Comparison of four techniques. *J. Strength Cond. Res.* **2009**, *23*, 660–667. [CrossRef]
43. Smith, T.O.; Hunt, N.J.; Wood, S.J. The physiotherapy management of muscle haematomas. *Phys. Sport Off. J. Assoc. Chart. Physiother. Sport Med.* **2006**, *7*, 201–209. [CrossRef]
44. Moore, S.R.; Shurman, J. Combined neuromuscular electrical stimulation and transcutaneous electrical nerve stimulation for treatment of chronic back pain: A double-blind, repeated measures comparison. *Arch. Phys. Med. Rehabil.* **1997**, *78*, 55–60. [CrossRef]
45. Quintero, A.J.; Wright, V.J.; Fu, F.H.; Huard, J. Stem cells for the treatment of skeletal muscle injury. *Clin. Sports Med.* **2009**, *28*, 1–11. [CrossRef]
46. Di Filippo, E.S.; Mancinelli, R.; Marrone, M.; Doria, C.; Verratti, V.; Toniolo, L.; Dantas, J.L.; Fulle, S.; Pietrangelo, T. Neuromuscular electrical stimulation improves skeletal muscle regeneration through satellite cell fusion with myofibers in healthy elderly subjects. *J. Appl. Physiol.* **2017**, *123*, 501–512. [CrossRef] [PubMed]
47. Ljubicic, V.; Adhihetty, P.J.; Hood, D.A. Application of Animal Models: Chronic Electrical Stimulation-Induced Contractile Activity. *Can. J. Appl. Physiol.* **2005**, *30*, 625–643. [CrossRef] [PubMed]
48. Efthimiadou, A.; Asimakopoulos, B.; Nikolettos, N.; Giatromanolaki, A.; Sivridis, E.; Lialiaris, T.S.; Papachristou, D.N.; Kontoleon, E. The angiogenetic effect of intramuscular administration of b-FGF and a-FGF on cardiac muscle: The influence of exercise on muscle angiogenesis. *J. Sports Sci.* **2006**, *24*, 849–854. [CrossRef] [PubMed]
49. Bellafiore, M.; Sivverini, G.; Palumbo, D.; Macaluso, F.; Bianco, A.; Palma, A.; Farina, F. Increased cx43 and angiogenesis in exercised mouse hearts. *Int. J. Sports Med.* **2007**, *28*, 749–755. [CrossRef] [PubMed]
50. Reher, P.; Doan, N.; Bradnock, B.; Meghji, S.; Harris, M. Effect of ultrasound on the production of IL-8, basic FGF and VEGF. *Cytokine* **1999**, *11*, 416–423. [CrossRef]
51. McBrier, N.M.; Lekan, J.M.; Druhan, L.J.; Devor, S.T.; Merrick, M.A. Therapeutic ultrasound decreases mechano-growth factor messenger ribonucleic acid expression after muscle contusion injury. *Arch. Phys. Med. Rehabil.* **2007**, *88*, 936–940. [CrossRef]
52. Adabbo, M.; Paolillo, F.R.; Bossini, P.S.; Rodrigues, N.C.; Bagnato, V.S.; Parizotto, N.A. Effects of Low-Level Laser Therapy Applied Before Treadmill Training on Recovery of Injured Skeletal Muscle in Wistar Rats. *Photomed. Laser Surg.* **2016**, *34*, 187–193. [CrossRef]
53. Alves, A.N.; Fernandes, K.P.S.; Deana, A.M.; Bussadori, S.K.; Mesquita-Ferrari, R.A. Effects of low-level laser therapy on skeletal muscle repair: A systematic review. *Am. J. Phys. Med. Rehabil.* **2014**, *93*, 1073–1085. [CrossRef]
54. Pastore, D.; Greco, M.; Passarella, S. Specific helium-neon laser sensitivity of the purified cytochrome c oxidase. *Int. J. Radiat. Biol.* **2000**, *76*, 863–870. [CrossRef]
55. Renno, A.C.M.; McDonnell, P.A.; Parizotto, N.A.; Laakso, E.L. The effects of laser irradiation on osteoblast and osteosarcoma cell proliferation and differentiation in vitro. *Photomed. Laser Surg.* **2007**, *25*, 275–280. [CrossRef] [PubMed]
56. Silveira, P.C.L.; da Silva, L.A.; Pinho, C.A.; De Souza, P.S.; Ronsani, M.M.; da Luz Scheffer, D.; Pinho, R.A. Effects of low-level laser therapy (GaAs) in an animal model of muscular damage induced by trauma. *Lasers Med. Sci.* **2013**, *28*, 431–436. [CrossRef]
57. De Freitas, C.E.A.; Bertaglia, R.S.; Vechetti Júnior, I.J.; Mareco, E.A.; Salomão, R.A.; de Paula, T.G.; Nai, G.A.; Carvalho, R.F.; Pacagnelli, F.L.; Dal-Pai-Silva, M. High Final Energy of Low-Level Gallium Arsenide Laser Therapy Enhances Skeletal Muscle Recovery without a Positive Effect on Collagen Remodeling. *Photochem. Photobiol.* **2015**, *91*, 957–965. [CrossRef]
58. Giombini, A.; Casciello, G.; Di Cesare, M.C.; Di Cesare, A.; Dragoni, S.; Sorrenti, D. A controlled study on the effects of hyperthermia at 434 MHz and conventional ultrasound upon muscle injuries in sport. *J. Sports Med. Phys. Fitness* **2001**, *41*, 521–527.
59. Sorrenti, D.; Casciello, G.; Dragoni, S.; Giombini, A. Applicazione della termoterapia endogena nel trattamento delle lesioni muscolari da sport: Studio comparativo tra ipertermia e ultrasuoni. *Med. Dello Sport* **2000**, *53*, 59–67.
60. Goats, G.C. Continuous short-wave (radio-frequency) diathermy. *Br. J. Sports Med.* **1989**, *23*, 123–127. [CrossRef] [PubMed]
61. Lehmann, J.F.; Dundore, D.E.; Esselman, P.C.; Nelp, W.B. Microwave diathermy: Effects on experimental muscle hematoma resolution. *Arch. Phys. Med. Rehabil.* **1983**, *64*, 127–129.
62. Lehmann, J.F.; Brunner, G.D.; Mcmillan, J.A.; Silverman, D.R.; Johnston, V.C. Modification of Heating Patterns Produced by Microwaves at the Frequencies of 2456 and 900 mc. by Physiologic Factors in the Human. *Arch. Phys. Med. Rehabil.* **1964**, *45*, 555–563.
63. Lehmann, J.F.; Brunner, G.D.; Stow, R.W. Pain threshold measurements after therapeutic application of ultrasound, microwaves and infrared. *Arch. Phys. Med. Rehabil.* **1958**, *39*, 560–565.
64. Food and Drug Administration. Physical Medicine Devices; Reclassification of Shortwave Diathermy for All Other Uses, Henceforth To Be Known as Nonthermal Shortwave Therapy. Final order; technical correction. *Fed. Regist.* **2015**, *80*, 61298–61302.
65. D'Agostino, M.C.; Craig, K.; Tibalt, E.; Respizzi, S. Shock wave as biological therapeutic tool: From mechanical stimulation to recovery and healing, through mechanotransduction. *Int. J. Surg.* **2015**, *24*, 147–153. [CrossRef]
66. McGorm, H.; Roberts, L.A.; Coombes, J.S.; Peake, J.M. Turning Up the Heat: An Evaluation of the Evidence for Heating to Promote Exercise Recovery, Muscle Rehabilitation and Adaptation. *Sports Med.* **2018**, *48*, 1311–1328. [CrossRef]
67. Jones, P.; Dalziel, S.R.; Lamdin, R.; Miles-Chan, J.L.; Frampton, C. Oral non-steroidal anti-inflammatory drugs versus other oral analgesic agents for acute soft tissue injury. *Cochrane Database Syst. Rev.* **2015**, *7*, CD007789. [CrossRef]

68. Mackey, A.L.; Rasmussen, L.K.; Kadi, F.; Schjerling, P.; Helmark, I.C.; Ponsot, E.; Aagaard, P.; Durigan, J.L.; Kjaer, M. Activation of satellite cells and the regeneration of human skeletal muscle are expedited by ingestion of nonsteroidal anti-inflammatory medication. *FASEB J. Off. Publ. Fed. Am. Soc. Exp. Biol.* **2016**, *30*, 2266–2281. [CrossRef] [PubMed]
69. Hamid, M.S.A.; Mohamed Ali, M.R.; Yusof, A.; George, J.; Lee, L.P. Platelet-rich plasma injections for the treatment of hamstring injuries: A randomized controlled trial. *Am. J. Sports Med.* **2014**, *42*, 2410–2418. [CrossRef]
70. Almekinders, L.C. Anti-Inflammatory Treatment of Muscular Injuries in Sport. *Sports Med.* **1999**, *28*, 383–388. [CrossRef]
71. Bisciotti, G.N.; Eirale, C. Le lesioni muscolari indotte dall'esercizio: Il delayed muscle soreness. *Med. Dello Sport* **2012**, *65*, 423–435.
72. Megliola, A.; Eutropi, F.; Scorzelli, A.; Gambacorta, D.; De Marchi, A.; De Filippo, M.; Faletti, C.; Ferrari, F.S. Ultrasound and magnetic resonance imaging in sports-related muscle injuries. *Radiol. Med.* **2006**, *111*, 836–845. [CrossRef]
73. Alessandrino, F.; Balconi, G. Complications of muscle injuries. *J. Ultrasound* **2013**, *16*, 215–222. [CrossRef] [PubMed]
74. Takebayashi, S.; Takasawa, H.; Banzai, Y.; Miki, H.; Sasaki, R.; Itoh, Y.; Matsubara, S. Sonographic findings in muscle strain injury: Clinical and MR imaging correlation. *J. Ultrasound Med.* **1995**, *14*, 899–905. [CrossRef]
75. Guermazi, A.; Roemer, F.W.; Robinson, P.; Tol, J.L.; Regatte, R.R.; Crema, M.D. Imaging of Muscle Injuries in Sports Medicine: Sports Imaging Series. *Radiology* **2017**, *282*, 646–663. [CrossRef]
76. Laumonier, T.; Menetrey, J. Muscle injuries and strategies for improving their repair. *J. Exp. Orthop.* **2016**, *3*, 15. [CrossRef]
77. Gordon, T.; English, A.W. Strategies to promote peripheral nerve regeneration: Electrical stimulation and/or exercise. *Eur. J. Neurosci.* **2016**, *43*, 336–350. [CrossRef]
78. Kannus, P.; Parkkari, J.; Järvinen, T.L.N.; Järvinen, T.A.; Järvinen, M. Basic science and clinical studies coincide: Active treatment approach is needed after a sports injury. *Scand. J. Med. Sci. Sports* **2003**, *13*, 150–154. [CrossRef]
79. Chaudhry, S.; Morrissey, D.; Woledge, R.C.; Bader, D.L.; Screen, H.R. Eccentric and concentric loading of the triceps surae: An in vivo study of dynamic muscle and tendon biomechanical parameters. *J. Appl. Biomech.* **2015**, *31*, 69–78. [CrossRef]
80. Chaouachi, A.; Hammami, R.; Kaabi, S.; Chamari, K.; Drinkwater, E.J.; Behm, D.G. Olympic weightlifting and plyometric training with children provides similar or greater performance improvements than traditional resistance training. *J. Strength Cond. Res.* **2014**, *28*, 1483–1496. [CrossRef]
81. Moffa, S.; Perna, A.; Candela, G.; Cattolico, A.; Sellitto, C.; De Blasiis, P.; Guerra, G.; Tafuri, D.; Lucariello, A. Effects of Hoverboard on Balance in Young Soccer Athletes. *J. Funct. Morphol. Kinesiol.* **2020**, *5*, 60. [CrossRef]
82. Sherry, M.A.; Best, T.M. A comparison of 2 rehabilitation programs in the treatment of acute hamstring strains. *J. Orthop. Sports Phys. Ther.* **2004**, *34*, 116–125. [CrossRef]
83. Bliven, K.C.H.; Anderson, B.E. Core stability training for injury prevention. *Sports Health* **2013**, *5*, 514–522. [CrossRef]
84. Bayer, M.L.; Magnusson, S.P.; Kjaer, M. Early versus Delayed Rehabilitation after Acute Muscle Injury. *N. Engl. J. Med.* **2017**, *377*, 1300–1301. [CrossRef]
85. Petersen, J.; Thorborg, K.; Nielsen, M.B.; Skjødt, T.; Bolvig, L.; Bang, N.; Hölmich, P. The diagnostic and prognostic value of ultrasonography in soccer players with acute hamstring injuries. *Am. J. Sports Med.* **2014**, *42*, 399–404. [CrossRef]
86. Mendiguchia, J.; Alentorn-Geli, E.; Idoate, F.; Myer, G.D. Rectus femoris muscle injuries in football: A clinically relevant review of mechanisms of injury, risk factors and preventive strategies. *Br. J. Sports Med.* **2013**, *47*, 359–366. [CrossRef]
87. Tyler, T.F.; Schmitt, B.M.; Nicholas, S.J.; McHugh, M.P. Rehabilitation After Hamstring-Strain Injury Emphasizing Eccentric Strengthening at Long Muscle Lengths: Results of Long-Term Follow-Up. *J. Sport Rehabil.* **2017**, *26*, 131–140. [CrossRef]
88. Askling, C.M.; Tengvar, M.; Thorstensson, A. Acute hamstring injuries in Swedish elite football: A prospective randomised controlled clinical trial comparing two rehabilitation protocols. *Br. J. Sports Med.* **2013**, *47*, 953–959. [CrossRef]
89. Ishøi, L.; Sørensen, C.N.; Kaae, N.M.; Jørgensen, L.B.; Hölmich, P.; Serner, A. Large eccentric strength increase using the Copenhagen Adduction exercise in football: A randomized controlled trial. *Scand. J. Med. Sci. Sports* **2016**, *26*, 1334–1342. [CrossRef]
90. Tesch, P.A.; Fernandez-Gonzalo, R.; Lundberg, T.R. Clinical Applications of Iso-Inertial, Eccentric-Overload (YoYoTM) Resistance Exercise. *Front. Physiol.* **2017**, *8*, 241. [CrossRef]
91. Petersen, J.; Hölmich, P. Evidence based prevention of hamstring injuries in sport. *Br. J. Sports Med.* **2005**, *39*, 319–323. [CrossRef]
92. Lanferdini, F.J.; Bini, R.R.; Baroni, B.M.; Klein, K.D.; Carpes, F.P.; Vaz, M.A. Improvement of Performance and Reduction of Fatigue with Low-Level Laser Therapy in Competitive Cyclists. *Int. J. Sports Physiol. Perform.* **2018**, *13*, 14–22. [CrossRef]
93. Roberts, L.A.; Nosaka, K.; Coombes, J.S.; Peake, J.M. Cold water immersion enhances recovery of submaximal muscle function after resistance exercise. *Am. J. Physiol. Regul. Integr. Comp. Physiol.* **2014**, *307*, R998–R1008. [CrossRef]
94. Bieuzen, F.; Bleakley, C.M.; Costello, J.T. Contrast water therapy and exercise induced muscle damage: A systematic review and meta-analysis. *PLoS ONE* **2013**, *8*, e62356.
95. Duñabeitia, I.; Arrieta, H.; Torres-Unda, J.; Gil, J.; Santos-Concejero, J.; Gil, S.M.; Irazusta, J.; Bidaurrazaga-Letona, I. Effects of a capacitive-resistive electric transfer therapy on physiological and biomechanical parameters in recreational runners: A randomized controlled crossover trial. *Phys. Ther. Sport Off. J. Assoc. Chart. Physiother. Sport Med.* **2018**, *32*, 227–234. [CrossRef]
96. Pérez-Bellmunt, A.; Casasayas, O.; Navarro, R.; Simon, M.; Martin, J.C.; Pérez-Corbella, C.; Blasi, M.; Ortiz, S.; Álvarez, P.; Pacheco, L. Effectiveness of low-frequency electrical stimulation in proprioceptive neuromuscular facilitation techniques in healthy males: A randomized controlled trial. *J. Sports Med. Phys. Fitness* **2019**, *59*, 469–475. [CrossRef]
97. Valle, X.L.; Tol, J.; Hamilton, B.; Rodas, G.; Malliaras, P.; Malliaropoulos, N.; Rizo, V.; Moreno, M.; Jardi, J. Hamstring Muscle Injuries, a Rehabilitation Protocol Purpose. *Asian J. Sports Med.* **2015**, *6*, e25411. [CrossRef]

98. Muscle Injuries Clinical Guide 3.0. 2015. Available online: https://muscletechnetwork.org/wp-content/uploads/2015/04/MUSCLE-INJURIES-CLINICAL-GUIDE-3.0-LAST-VERSION.pdf (accessed on 2 September 2021).
99. Harøy, J.; Clarsen, B.; Wiger, E.G.; Oyen, M.G.; Serner, A.; Thorborg, K.; Holmich, P.; Andersen, T.E.; Bahr, R. The Adductor Strengthening Programme prevents groin problems among male football players: A cluster-randomised controlled trial. *Br. J. Sports Med.* **2019**, *53*, 150–157. [CrossRef]
100. Di Prampero, P.E.; Botter, A.; Osgnach, C. The energy cost of sprint running and the role of metabolic power in setting top performances. *Eur. J. Appl. Physiol.* **2015**, *115*, 451–469. [CrossRef]
101. Bisciotti, G.N.; Volpi, P.; Alberti, G.; Aprato, A.; Artina, M.; Auci, A.; Bait, C.; Belli, A.; Bellistri, G.; Bettinsoli, P.; et al. Italian consensus statement (2020) on return to play after lower limb muscle injury in football (soccer). *BMJ Open Sport Exerc. Med.* **2019**, *5*, e000505. [CrossRef]
102. Silder, A.; Heiderscheit, B.C.; Thelen, D.G.; Enright, T.; Tuite, M.J. MR observations of long-term musculotendon remodeling following a hamstring strain injury. *Skelet. Radiol.* **2008**, *37*, 1101–1109. [CrossRef]
103. Reurink, G.; Almusa, E.; Goudswaard, G.J.; Tol, J.L.; Hamilton, B.; Moen, M.H.; Weir, A.; Verhaar, J.A.; Maas, M. No association between fibrosis on magnetic resonance imaging at return to play and hamstring reinjury risk. *Am. J. Sports Med.* **2015**, *43*, 1228–1234. [CrossRef]
104. Sanfilippo, J.L.; Silder, A.; Sherry, M.A.; Tuite, M.J.; Heiderscheit, B.C. Hamstring strength and morphology progression after return to sport from injury. *Med. Sci. Sports Exerc.* **2013**, *45*, 448–454. [CrossRef]
105. Kvist, J. Rehabilitation following anterior cruciate ligament injury: Current recommendations for sports participation. *Sports Med.* **2004**, *34*, 269–280. [CrossRef]
106. Malliaropoulos, N.; Isinkaye, T.; Tsitas, K.; Maffulli, N. Reinjury after acute posterior thigh muscle injuries in elite track and field athletes. *Am. J. Sports Med.* **2011**, *39*, 304–310. [CrossRef] [PubMed]
107. Delvaux, F.; Rochcongar, P.; Bruyère, O.; Bourlet, G.; Daniel, C.; Diverse, P.; Reginster, J.Y.; Croisier, J.L. Return-to-play criteria after hamstring injury: Actual medicine practice in professional soccer teams. *J. Sports Sci. Med.* **2014**, *13*, 721–723. [CrossRef]
108. Reurink, G.; Goudswaard, G.J.; Tol, J.L.; Almusa, E.; Moen, M.H.; Weir, A.; Verhaar, J.A.N.; Hamilton, B.; Maas, M. MRI observations at return to play of clinically recovered hamstring injuries. *Br. J. Sports Med.* **2014**, *48*, 1370–1376. [CrossRef]
109. Zambaldi, M.; Beasley, I.; Rushton, A. Return to play criteria after hamstring muscle injury in professional football: A Delphi consensus study. *Br. J. Sports Med.* **2017**, *51*, 1221–1226. [CrossRef]
110. Witvrouw, E.; Danneels, L.; Asselman, P.; D'Have, T.; Cambier, D. Muscle flexibility as a risk factor for developing muscle injuries in male professional soccer players. A prospective study. *Am. J. Sports Med.* **2003**, *31*, 41–46. [CrossRef]
111. Connell, D.A.; Schneider-Kolsky, M.E.; Hoving, J.L.; Malara, F.; Buchbinder, R.; Koulouris, G.; Burke, F.; Bass, C. Longitudinal study comparing sonographic and MRI assessments of acute and healing hamstring injuries. *AJR Am. J. Roentgenol.* **2004**, *183*, 975–984. [CrossRef]
112. Slavotinek, J.P. Muscle injury: The role of imaging in prognostic assignment and monitoring of muscle repair. *Semin. Musculoskelet Radiol.* **2010**, *14*, 194–200. [CrossRef]
113. Bauman, J. Returning to play: The mind does matter. *Clin. J. Sport Med. Off. J. Can. Acad. Sport Med.* **2005**, *15*, 432–435. [CrossRef]
114. Glazer, D.D. Development and preliminary validation of the Injury-Psychological Readiness to Return to Sport (I-PRRS) scale. *J. Athl. Train.* **2009**, *44*, 185–189. [CrossRef]
115. Clover, J.; Wall, J. Return-to-play criteria following sports injury. *Clin. Sports Med.* **2010**, *29*, 169–175. [CrossRef]
116. Ravé, G.; Granacher, U.; Boullosa, D.; Hackney, A.C.; Zouhal, H. How to Use Global Positioning Systems (GPS) Data to Monitor Training Load in the "Real World" of Elite Soccer. *Front. Physiol.* **2020**, *11*, 944. [CrossRef]
117. Ridehalgh, C.; Moore, A.; Hough, A. Sciatic nerve excursion during a modified passive straight leg raise test in asymptomatic participants and participants with spinally referred leg pain. *Man. Ther.* **2015**, *20*, 564–569. [CrossRef]
118. Kellis, E.; Ellinoudis, A.; Kofotolis, N. Hamstring Elongation Quantified Using Ultrasonography During the Straight Leg Raise Test in Individuals with Low Back Pain. *PM R* **2015**, *7*, 576–583. [CrossRef]
119. Askling, C.M.; Nilsson, J.; Thorstensson, A. A new hamstring test to complement the common clinical examination before return to sport after injury. *Knee Surg. Sports Traumatol. Arthrosc.* **2010**, *18*, 1798–1803. [CrossRef]
120. Croisier, J.-L.; Forthomme, B.; Namurois, M.-H.; Vanderthommen, M.; Crielaard, J.M. Hamstring muscle strain recurrence and strength performance disorders. *Am. J. Sports Med.* **2002**, *30*, 199–203. [CrossRef]
121. Bisciotti, G.N.; Quaglia, A.; Belli, A.; Carimati, G.; Volpi, P. Return to sports after ACL reconstruction: A new functional test protocol. *Muscles Ligaments Tendons. J.* **2016**, *6*, 499–509. [CrossRef]
122. Hachana, Y.; Chaabène, H.; Nabli, M.A.; Attia, A.; Moualhi, J.; Farhat, N.; Elloumi, M. Test-retest reliability, criterion-related validity, and minimal detectable change of the Illinois agility test in male team sport athletes. *J. Strength Cond. Res.* **2013**, *27*, 2752–2759. [CrossRef]
123. Negra, Y.; Chaabene, H.; Hammami, M.; Amara, S.; Sammoud, S.; Mkaouer, B.; Hachana, Y. Agility in Young Athletes: Is It a Different Ability From Speed and Power? *J. Strength Cond. Res.* **2017**, *31*, 727–735. [CrossRef]
124. Brumitt, J.; Heiderscheit, B.C.; Manske, R.C.; Niemuth, P.E.; Rauh, M.J. Lower extremity functional tests and risk of injury in division iii collegiate athletes. *Int. J. Sports Phys. Ther.* **2013**, *8*, 216–227. [PubMed]
125. Ordway, J.D.; Laubach, L.L.; Vanderburgh, P.M.; Jackson, K.J. The Effects of Backwards Running Training on Forward Running Economy in Trained Males. *J. Strength Cond. Res.* **2016**, *30*, 763–767. [CrossRef] [PubMed]

126. Bouvier, T.; Opplert, J.; Cometti, C.; Babault, N. Acute effects of static stretching on muscle-tendon mechanics of quadriceps and plantar flexor muscles. *Eur. J. Appl. Physiol.* **2017**, *117*, 1309–1315. [CrossRef] [PubMed]
127. Bisciotti, G.N. La tendinopatia degli adduttori nel calciatore: Quando il ritorno alla corsa? *Strength Cond.* **2013**, *5*, 11–16.
128. Engebretsen, A.H.; Myklebust, G.; Holme, I.; Engebretsen, L.; Bahr, R. Intrinsic risk factors for groin injuries among male soccer players: A prospective cohort study. *Am. J. Sports Med.* **2010**, *38*, 2051–2057. [CrossRef]
129. Delahunt, E.; McEntee, B.L.; Kennelly, C.; Green, B.S.; Coughlan, G.F. Intrarater reliability of the adductor squeeze test in gaelic games athletes. *J. Athl. Train.* **2011**, *46*, 241–245. [CrossRef] [PubMed]
130. Delahunt, E.; Kennelly, C.; McEntee, B.L.; Coughlan, G.F.; Green, B.S. The thigh adductor squeeze test: 45° of hip flexion as the optimal test position for eliciting adductor muscle activity and maximum pressure values. *Man Ther.* **2011**, *16*, 476–480. [CrossRef]
131. Hodgson, L.; Hignett, T.; Edwards, K. Normative adductor squeeze tests scores in rugby. *Phys. Ther. Sport Off. J. Assoc. Chart. Physiother. Sport Med.* **2015**, *16*, 93–97. [CrossRef]
132. Nevin, F.; Delahunt, E. Adductor squeeze test values and hip joint range of motion in Gaelic football athletes with longstanding groin pain. *J. Sci. Med. Sport* **2014**, *17*, 155–159. [CrossRef]
133. Atkinson, H.D.E.; Johal, P.; Falworth, M.S.; Ranawat, V.S.; Dala-Ali, B.; Martin, D.K. Adductor tenotomy: Its role in the management of sports-related chronic groin pain. *Arch. Orthop. Trauma Surg.* **2010**, *130*, 965–970. [CrossRef]
134. Kong, D.H.; Yang, S.J.; Ha, J.K.; Jang, S.H.; Seo, J.G.; Kim, J.G. Validation of functional performance tests after anterior cruciate ligament reconstruction. *Knee Surg. Relat. Res.* **2012**, *24*, 40–45. [CrossRef]
135. Jang, S.H.; Kim, J.G.; Ha, J.K.; Wang, B.G.; Yang, S.J. Functional performance tests as indicators of returning to sports after anterior cruciate ligament reconstruction. *Knee* **2014**, *21*, 95–101. [CrossRef] [PubMed]
136. Yocum, A.; McCoy, S.W.; Bjornson, K.F.; Mullens, P.; Burton, G.N. Reliability and validity of the standing heel-rise test. *Phys. Occup. Pediatr.* **2010**, *30*, 190–204. [CrossRef]
137. Möller, M.; Lind, K.; Styf, J.; Karlsson, J. The reliability of isokinetic testing of the ankle joint and a heel-raise test for endurance. *Knee Surg. Sports Traumatol. Arthrosc.* **2005**, *13*, 60–71. [CrossRef]
138. Harris-Love, M.O.; Shrader, J.A.; Davenport, T.E.; Joe, G.; Rakocevic, G.; McElroy, B.; Dalakas, M. Are repeated single-limb heel raises and manual muscle testing associated with peak plantar-flexor force in people with inclusion body myositis? *Phys. Ther.* **2014**, *94*, 543–552. [CrossRef]
139. Baumbach, S.F.; Braunstein, M.; Regauer, M.; Böcker, W.; Polzer, H. Diagnosis of Musculus Gastrocnemius Tightness-Key Factors for the Clinical Examination. *J. Vis. Exp.* **2016**, *7*, 53446. [CrossRef] [PubMed]
140. Baumbach, S.F.; Braunstein, M.; Seeliger, F.; Borgmann, L.; Böcker, W.; Polzer, H. Ankle dorsiflexion: What is normal? Development of a decision pathway for diagnosing impaired ankle dorsiflexion and M. gastrocnemius tightness. *Arch. Orthop. Trauma Surg.* **2016**, *136*, 1203–1211. [CrossRef] [PubMed]
141. Silbernagel, K.G.; Gustavsson, A.; Thomeé, R.; Karlsson, J. Evaluation of lower leg function in patients with Achilles tendinopathy. *Knee Surg. Sports Traumatol. Arthrosc.* **2006**, *14*, 1207–1217. [CrossRef] [PubMed]
142. Hewett, T.; Snyder-Mackler, L.; Spindler, K. The drop-jump screening test: Difference in lower limb control by gender and effect of neuromuscular training in female athletes. *Am. J. Sports Med.* **2007**, *35*, 145–147. [CrossRef] [PubMed]
143. Powell, H.C.; Silbernagel, K.G.; Brorsson, A.; Tranberg, R.; Willy, R.W. Individuals Post Achilles Tendon Rupture Exhibit Asymmetrical Knee and Ankle Kinetics and Loading Rates During a Drop Countermovement Jump. *J. Orthop. Sports Phys. Ther.* **2018**, *48*, 34–43. [CrossRef] [PubMed]

Journal of
Functional Morphology and Kinesiology

MDPI

Article

Influence of COVID-19 Restrictions on Training and Physiological Characteristics in U23 Elite Cyclists

Peter Leo [1,*], Iñigo Mujika [2,3] and Justin Lawley [1]

1 Division of Performance Physiology & Prevention, Department of Sport Science, University of Innsbruck, 6020 Innsbruck, Austria; justin.lawley@uibk.ac.at
2 Department of Physiology, Faculty of Medicine and Nursing, University of the Basque Country, 48940 Leioa, Spain; inigo.mujika@inigomujika.com
3 Exercise Science Laboratory, School of Kinesiology, Faculty of Medicine, Universidad Finis Terrae, Santiago 8320000, Chile
* Correspondence: peter.leo@uibk.ac.at

Abstract: PURPOSE: The COVID-19 pandemic and its associated mobility restrictions caused many athletes to adjust or reduce their usual training load. The aim of this study was to investigate how the COVID-19 restrictions affected training and performance physiology measures in U23 elite cyclists. METHODS: Twelve U23 elite cyclists ($n = 12$) participated in this study (mean ± SD: Age 21.2 ± 1.2 years; height 182.9 ± 4.7 cm; body mass 71.4 ± 6.5 kg). Training characteristics were assessed between 30 days pre, during, and post COVID-19 restrictions, respectively. The physiological assessment in the laboratory was 30 days pre and post COVID-19 restrictions and included maximum oxygen uptake ($\dot{V}O_{2max}$), peak power output for sprint ($Sprint_{Pmax}$), and ramp incremental graded exercise (GXT_{Pmax}), as well as power output at ventilatory threshold (VT) and respiratory compensation point (RCP). RESULTS: Training load characteristics before, during, and after the lockdown remained statistically unchanged ($p > 0.05$) despite large effects (>0.8) with mean reductions of 4.7 to 25.0% during COVID-19 restrictions. There were no significant differences in maximal and submaximal power outputs, as well as relative and absolute $\dot{V}O_{2max}$ between pre and post COVID-19 restrictions ($p > 0.05$) with small to moderate effects. DISCUSSION: These results indicate that COVID-19 restrictions did not negatively affect training characteristics and physiological performance measures in U23 elite cyclists for a period of <30 days. In contrast with recent reports on professional cyclists and other elite level athletes, these findings reveal that as long as athletes are able to maintain and/or slightly adapt their training routine, physiological performance variables remain stable.

Keywords: pandemic; athletes; endurance; detraining; cycling

Citation: Leo, P.; Mujika, I.; Lawley, J. Influence of COVID-19 Restrictions on Training and Physiological Characteristics in U23 Elite Cyclists. *J. Funct. Morphol. Kinesiol.* **2022**, *7*, 1. https://doi.org/10.3390/jfmk7010001

Academic Editors: Cristina Cortis, Andrea Fusco, Carl Foster and Peter Hofmann

Received: 17 November 2021
Accepted: 17 December 2021
Published: 21 December 2021

1. Introduction

Endurance athletes are known for performing high training volumes throughout the year to maintain cardiovascular fitness and to attain peak performance for the competition period [1–3]. The COVID-19 pandemic has been reported to impact the athletes' training, competition, and recovery routines [4–9]. Washif et al. [4] analyzed the training habits of 12,526 athletes of various performance levels (i.e., recreational to professional) from 142 countries and six continents during the COVID-19 confinement. Despite reductions in training session frequency, most athletes focused on maintenance of general endurance and strength rather than exploiting sports-specific training [4,10]. Muriel et al. [11] studied the training and physiological characteristics of 18 male professional cyclists during the 10 weeks prior to the COVID-19 confinement and during the seven-week confinement period: Total training volume significantly decreased by 33.9% during the lockdown. Weekly volumes (hours per week) by standardized training zones declined between 25.8% and 52.2%. There were also large reductions in the best 5-min and 20-min power outputs with declines between 1% and 19% in all the cyclists.

There is increasing evidence that insufficient training stimuli impair major physiological determinants for endurance performance in both the general population [12] and highly trained athletes [13]. Mujika and Padilla [13] reported that training cessation over four weeks causes a rapid decline of 6% to 20% in $\dot{V}O_{2max}$, 10% to 14% in maximal ventilatory volume, 5% to 12% in blood volume and plasma volume, as well as inducing reductions in cardiac dimensions (e.g., left ventricular posterior wall thickness) by 25%.

As the current pandemic situation presents a new challenge for athletes and practitioners, there is limited evidence about the acute effects of a lockdown and mobility restriction period on road cyclists' training characteristics and physiological determinants of performance. Therefore, the aim of the present study was to investigate how COVID-19 restrictions affected pre and post $Sprint_{Pmax}$, GXT_{Pmax}, $\dot{V}O_{2max}$, and submaximal thresholds in U23 elite cyclists. Additionally, training characteristics were compared between 30 days pre, during, and post COVID-19 restrictions. We hypothesized that COVID-19 restrictions would negatively affect physiological performance variables and training characteristics in U23 elite cyclists.

2. Materials and Methods

2.1. Participants

Twelve U23 elite cyclists of a Union Cycliste Internationale (UCI) licensed continental team participated in this study (mean ± SD: Age 21.2 ± 1.2 years; height 182.9 ± 4.7 cm; body mass 71.4 ± 6.5 kg). All methods were approved by the ethics committee under the conditions of the Declaration of Helsinki. All subjects voluntarily participated in the study and provided informed written consent.

2.2. COVID-19 Restrictions

The whole intervention involved 90 days, which were equally divided into 30 days pre, during, and post COVID-19 restrictions. Participants were living either in Austria (n = 6), Germany (n = 4), or Italy (n = 2). COVID-19 restriction guidelines were followed according to each country's own regulations over 30 consecutive days and included mobility restrictions (n = 10) and home confinement (n = 2).

2.3. Training Characteristics

The accumulated training hours, distance covered, and training frequency per week were recorded for the respective periods, mentioned above. All athletes uploaded their training data to an online training platform (Trainingpeaks, Trainingpeaks LLC, Winchester Cir, MA, USA) [14]. Weekly training hours, distance covered, and training frequency were collected and further processed, analyzed, and checked for data spikes in Microsoft Excel (Excel, Microsoft Corporation, Redmond, WA, USA). Intensity ratios, including distance per hour ($km \cdot hour^{-1}$) and distance per session frequency ($km \cdot session^{-1}$), were calculated.

2.4. Physiological Performance Measures

Physiological performance measures were assessed 30 days before and after the COVID-19 restrictions. All laboratory tests were completed on an electromagnetically braked stationary trainer (Cyclus2, RBM electronic-automation GmbG, Leibzig, Deutschland) with the participants' own road bikes (Alto Prestige, KTM Fahrrad GmbH, Mattighofen, Austria). The testing protocol involved a standardized warm-up of 5 min at 100 W, followed by a 15-s sprint test and a laboratory incremental graded exercise test (GXT). The cycling cadence was reduced to 20–30 revolutions per minute (rpm), and the sprint started after a 5-s countdown. After a 10-min recovery phase at 50 W, the GXT was completed. The initial workload for the GXT was set at 150 W and was increased by 20 W per min until volitional exhaustion. The measurements included absolute and relative $Sprint_{Pmax}$ for the 15 s sprint test as well as GXT_{Pmax}. $\dot{V}O_{2max}$ was defined as the highest 30 s average during the GXT [15]. VT and RCP were analyzed from the GXT, according to Beaver et al. [16]. VT was defined as the point where the ventilation rate ($\dot{V}E$)

increased compared to $\dot{V}O2$ ($\dot{V}E/\dot{V}O2$). RCP was defined as the onset of hyperventilation during the GXT, with an increase in $\dot{V}E$ compared to the volume of carbon dioxide ($\dot{V}CO2$) release, known as the $\dot{V}E/\dot{V}CO2$ ratio. Open circuit spiro-ergometry (ZAN600, nSpire Health GmbH, Germany) with a flow sensor (FlowSensor, Type II, nSpire Health GmbG, Oberthulba, Germany) was used to record oxygen uptake ($\dot{V}O2$) and carbon dioxide ($\dot{V}CO2$) release. Continuous recordings of heart rate (HR) were performed via short range telemetry with a 1 Hz sampling rate (V800, Polar Electro Oy, Kempele, Finland).

2.5. Statistical Analysis

All recorded data were initially checked for violations to normality with a Shapiro-Wilk test. Paired samples *t*-tests with Cohen's d effect size were performed between all physiological performance characteristics. Cohen's d was set at 0.2 for small, 0.5 for moderate, and 0.8 for a large effect [17]. A one-way repeated measure analysis of variance (ANOVA) was conducted to compare training characteristics for the 30 days pre, during, and post COVID-19 restrictions. Partial eta square (partial η^2) effect size was set at 0.01 (small), 0.06 (moderate) and 0.14 for a large effect. All statistical analyses (JASP 0.15, JASP Team, Amsterdam, The Netherlands) and graphical illustrations (GraphPad Prism 8, GraphPad Software, San Diego, CA, USA) were conducted with commercially available software packages.

3. Results

The participants' anthropometric and physiological characteristics are presented in Table 1 (mean ± SD).

Table 1. Non-significant differences in anthropometrical and physiological measures between pre and post COVID-19 restrictions; BMI—body mass index; HRmax—maximum heart rate; GXT_{Pmax}—peak power in the laboratory graded incremental exercise test; $\dot{V}O_{2max}$—maximum oxygen uptake.

	Body Mass (kg)	BMI ($kg{\cdot}m^{-2}$)	HR_{max} (bpm)	GXT_{Pmax} (W)	$\dot{V}O_{2max}$ ($mL{\cdot}kg^{-1}{\cdot}min^{-1}$)
PRE	68.8 ± 3.8	20.7 ± 0.7	194 ± 7	471 ± 36	75.4 ± 4.4
POST	70.1 ± 4.4	21.1 ± 0.9	193 ± 5	470 ± 30	75.8 ± 2.8

Training Characteristics

No statistical difference was found in training volume, including completed hours (partial η^2: 0.175), covered distance (partial η^2: 0.227), and session frequency (partial η^2: 0.030) in the 30 days pre, during, and post COVID-19 restrictions ($p > 0.05$). Intensity ratios $km{\cdot}hour^{-1}$ (partial η^2: 0.182) and $km{\cdot}session^{-1}$ (partial η^2: 0.187) also remained unchanged in the 30 days pre, during, and post COVID-19 restrictions. ($p > 0.05$)—see Figure 1.

No statistical differences were found in absolute and relative $Sprint_{Pmax}$ (effect size (ES): 0.451, 0.596, respectively), GXT_{Pmax} (ES: 0.056, 0.198, respectively), power at VT (ES: −0.302, −0.083, respectively) and RCP (ES: −0.522, −0.096, respectively), HRmax (ES: 0.123) as well as absolute and relative $\dot{V}O_{2max}$ (ES: −0.622, −0.071, respectively), between pre and post COVID-19 restrictions—see Figure 2.

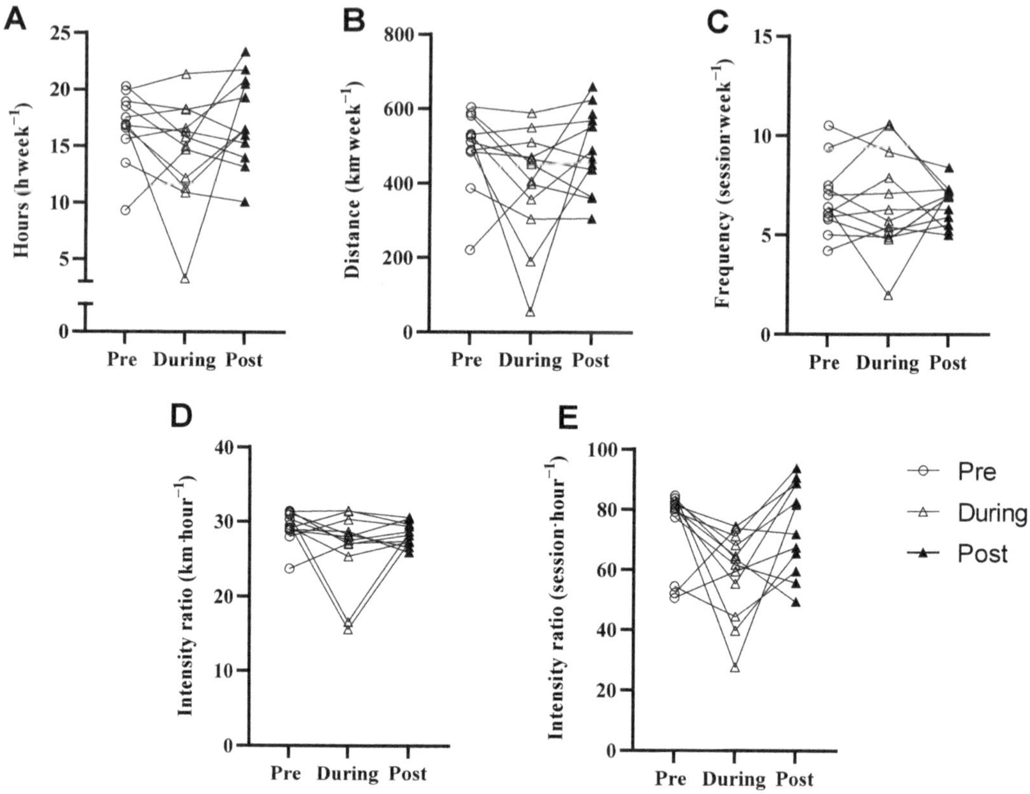

Figure 1. Non-significant differences in training hours (**A**), distance (**B**), frequency (**C**) and intensity ratios (**D**,**E**) between 30 days pre, during, and post COVID-19 restrictions. Physiological Performance Measures.

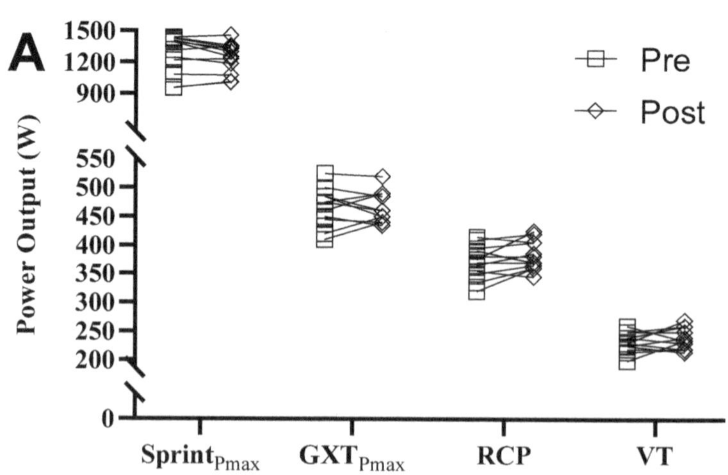

Figure 2. *Cont.*

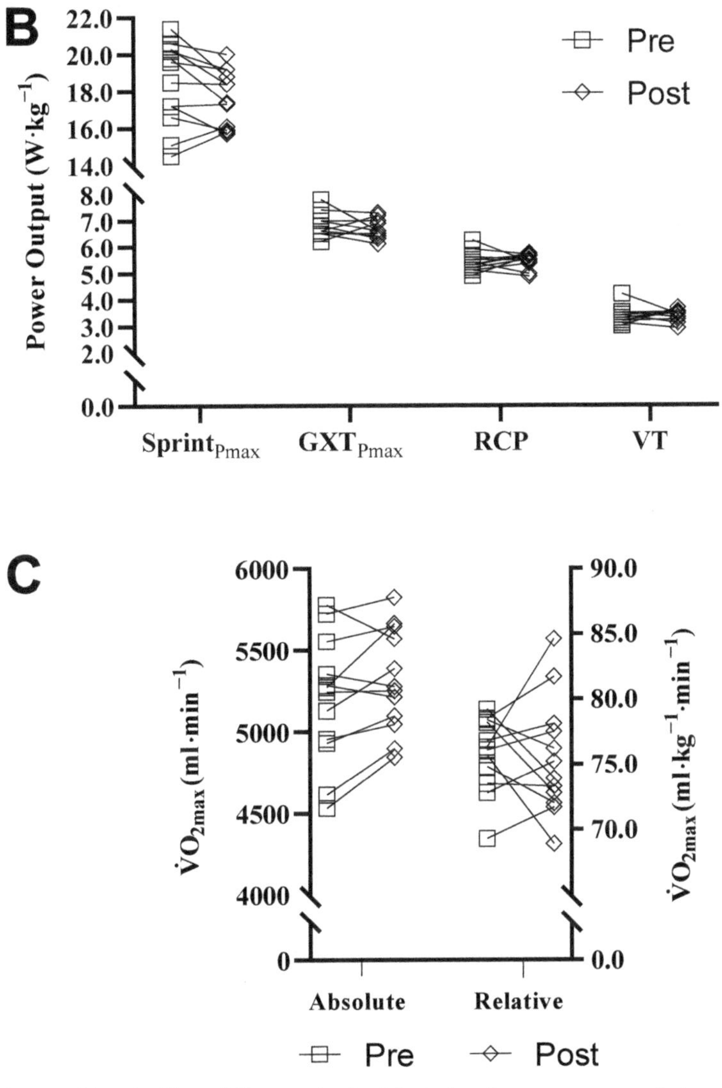

Figure 2. Non-significant differences in physiological performance measures between pre and post COVID-19 restrictions; Absolute (**A**) and relative (**B**) peak power output in $Sprint_{Pmax}$—peak power in the laboratory sprint test; GXT_{Pmax}—laboratory graded incremental exercise test; RCP—respiratory compensation point; VT—ventilatory threshold; $\dot{V}O_{2max}$—maximum oxygen uptake (**C**).

4. Discussion

In contrast to our own initial hypothesis, the present study showed no differences in training characteristics and physiological performance determinants in U23 elite cyclists.

To the authors' knowledge, this is the first study to investigate the effects of COVID-19 restrictions on a U23 elite cycling population. While there has been research conducted on professional cyclists by Muriel et al. [18], their findings were not in line with those of the present study. Muriel et al.'s study [18] reported a significant decline in training characteristics and physiological performance variables during the seven weeks of the COVID-19 confinement in a Spanish professional cycling team. In contrast, the findings of the present paper show that the four-week COVID-19 restrictions did not have a negative effect on training characteristics and physiological performance variables, including $\dot{V}O_{2max}$ and power output at submaximal thresholds. These conflicting findings might lead to the assumption that the COVID-19 restriction policies were different across European countries [19]. The study's participants living in northern European countries could continue their normal training habits (mobility restrictions), while participants living in southern European countries were restricted to indoor training (home confinement). These differences underpin why the subjects in Muriel et al.'s study [18] indicated reduced training volume because they were forced to train indoors (home confinement). Washif et al. [4] undertook a worldwide online survey with 12,526 athletes, including male and female professional and amateur athletes. While most athletes were training more individually to maintain general fitness and health, most athletes reported reduced motivation due to a lack of competitive events. Although professional athletes were coping better with the COVID-19 confinement than amateur athletes, all populations reported reductions in training volume, intensity, and frequency. While these data are based on qualitative assessments (survey), our empirical (quantitative) assessment does not confirm this trend. Despite average reductions in training hours (−15.1% to −18.6%), distance (−23.7% to −25.0%), session frequency (−4.7% to −6.3%), and intensity ratios (−7.2% to −19.1%) compared to pre and post COVID-19 restrictions, the inter-individual variations were too large to result in a significant statistical change. In addition, the small sample size associated with a reduced statistical power increases the likelihood of a false "negative" null hypothesis, known as a Type I error [20]. To better interpret inter-individual differences, effect sizes are provided. "Large" effects (partial $\eta^2 > 0.14$) were found in training characteristics between pre, during, and post COVID-19 restriction periods. The reason for those discrepancies could be due to different COVID-19 restrictions between European countries [19]. Two participants of the present study who experienced home confinement automatically indicated reduced training volume, while the other 10 participants were able to continue their normal training habits. This might explain why collectively the statistical differences remained non-significant. Reductions in training volume during the COVID-19 restrictions were also observed in elite swimmers, which were linked to decreased vagal activity [10].

This study also investigated the change in physiological performance determinants pre and post COVID-19 restrictions. Relative GXT_{Pmax} and $Sprint_{Pmax}$ decreased on average by 0.2 ± 1.3% to 4.3 ± 0.5%, respectively, which was mainly due to a 1.7 ± 0.1% increase in body mass. This change in body mass between pre and post COVID-19 restrictions affected average $\dot{V}O_{2max}$, and power output at RCP and VT when normalized to body mass in the range of −0.5% to −0.9%. Interestingly, average submaximal power outputs at RCP and VT slightly improved, despite small reductions in $\dot{V}O_{2max}$, GXT_{Pmax}, and $Sprint_{Pmax}$. However, statistical significance was not achieved between pre and post COVID-19 restrictions in any of the physiological performance parameters assessed, showing small (<0.2) to moderate (0.6) effects.

Collectively, our data do not support our initial hypothesis that COVID-19 restrictions would negatively affect training and physiological performance characteristics. Despite declining trends in training characteristics when compared to pre and post COVID-19 restrictions, inter-individual variations were too large to reach statistical significance.

5. Limitations

This study is not without its limitations. The findings need to be interpreted and discussed cautiously due to a small sample size and country-specific COVID-19 restrictions.

First, the COVID-19 restrictions were country specific and differently affected the training habits of our participants. For this reason, the most restricted athletes indicated the biggest decline in training characteristics but not in performance. Secondly undertaking research in highly trained athletes is challenging and rare due to team commitments, busy racing schedules and logistical challenges. Additionally, using standardized laboratory testing to inform the present research design and experiment was difficult due to the

COVID-19 situation. Although a bigger sample size of 20–30 athletes would have been beneficial to improve the statistical power, we strongly believe that reporting physiological data in this subgroup of highly trained athletes represents an important contribution to the existing literature. Moreover, qualitative assessments (i.e., survey) were not undertaken in this study, which could have been valuable in explaining inter-individual differences for the training process. In addition, measures of psychological state and well-being would have also been relevant to monitor the participants' coping strategies during COVID-19 restriction periods.

6. Conclusions

The present study demonstrates no statistically significant changes in training characteristics and physiological performance variables during COVID-19 restrictions. These findings need to be interpreted cautiously as country specific COVID-19 regulations and a small sample size with limited statistical power clearly influenced the study's outcome. However, the study's data reveal that if athletes maintain around 75% of their training volume, physiological performance measures remain stable at least for periods of up to 30 days.

Author Contributions: All authors equally contributed to the production of the manuscript. All authors have read and agreed to the published version of the manuscript.

Funding: No funding was received for this project.

Institutional Review Board Statement: The project fulfilled the institutional guidelines for research.

Informed Consent Statement: All participants provided informed written consent.

Data Availability Statement: The datasets generated during and/or analysed during the current study are available from the corresponding author on reasonable request.

Acknowledgments: We would like to thank all participants for their willingness to contribute to this research. We would also like to thank the master students Regina Rathgeb, Jan Pieper and Florian Lenzi for their help in realizing this project.

Conflicts of Interest: The authors declare no conflict of interest.

References

1. Joyner, M.J.; Coyle, E.F. Endurance Exercise Performance: The Physiology of Champions. *J. Physiol.* **2008**, *586*, 35–44. [CrossRef] [PubMed]
2. McKay, A.; Stellingwerff, T.; Smith, E.; Martin, D.T.; Mujika, I.; Goosey-Tolfrey, V.; Sheppard, J.; Burke, L.M. Defining Training and Performance Calibre: A Participant Classification Framework. *Int. J. Sports Physiol. Perform.* **2021**, *17*, 1–10.
3. Mujika, I. Quantification of Training and Competition Loads in Endurance Sports: Methods and Applications. *Int. J. Sports Physiol. Perform.* **2017**, *12*, 229. [CrossRef] [PubMed]
4. Washif, J.; Farooq, A.; Krug, I.; Pyne, D.; Verhagen, E.; Taylor, L.; Wong, D.; Mujika, I.; Cortis, C.; Haddad, M. Training during the COVID-19 Lockdown: Knowledge, Beliefs, and Practices of 12,526 Athletes from 142 Countries and Six Continents. *Sports Med.* **2021**, 1–16. [CrossRef] [PubMed]
5. Washif, J.; Mohd Kassim, S.; Lew, P.; Chong, C.; James, C. Athlete's Perceptions of a "Quarantine" Training Camp During the COVID-19 Lockdown. *Front. Act. Sports Living* **2021**, *2*, 220. [CrossRef] [PubMed]
6. Romdhani, M.; Rae, D.; Nédélec, M.; Ammar, A.; Chtourou, H.; Al Horani, R.; Ben Saad, H.; Bragazzi, N.; Dönmez, G.; Driss, T.; et al. COVID-19 Lockdowns: A Worldwide Survey of Circadian Rhythms and Sleep Quality in 3911 Athletes from 49 Countries, with Data-Driven Recommendations. *Sports Med.* **2021**, 1–16. [CrossRef] [PubMed]
7. Yousfi, N.; Bragazzi, N.; Briki, W.; Zmijewski, P.; Chamari, K. The COVID-19 Pandemic: How to Maintain a Healthy Immune System during the Lockdown—A Multidisciplinary Approach with Special Focus on Athletes. *Biol. Sport* **2020**, *37*, 211–216. [CrossRef] [PubMed]
8. Pillay, L.; van Rensburg, D.; van Rensburg, A.; Ramagole, D.; Holtzhausen, L.; Dijkstra, H.; Cronje, T. Nowhere to Hide: The Significant Impact of Coronavirus Disease 2019 (COVID-19) Measures on Elite and Semi-Elite South African Athletes. *J. Sci. Med. Sport* **2020**, *23*, 670–679. [CrossRef] [PubMed]
9. Facer-Childs, E.; Hoffman, D.; Tran, J.; Drummond, S.; Rajaratnam, S. Sleep and Mental Health in Athletes during COVID-19 Lockdown. *Sleep* **2021**, *44*, 261–270. [CrossRef] [PubMed]

10. Pla, R.; Bosquet, L.; McGibbon, K.; Mujika, I.; Aubry, A. Heart Rate Variability in Elite Swimmers before, during and after COVID-19 Lockdown: A Brief Report on Time Domain Analysis. *Appl. Sci.* **2021**, *11*, 8106. [CrossRef]
11. Muriel, X.; Valenzuela, P.L.; Mateo-March, M.; Pallarés, J.G.; Lucia, A.; Barranco-Gil, D. Physical Demands and Performance Indicators in Male Professional Cyclists during a Grand Tour: WorldTour versus ProTeam Category. *Int. J. Sports Physiol. Perform.* **2021**, *1*, 1–9. [CrossRef] [PubMed]
12. Spiering, B.A.; Mujika, I.; Sharp, M.A.; Foulis, S. Maintaining Physical Performance: The Minimal Dose of Exercise Needed to Preserve Endurance and Strength Over Time. *J. Strength Cond. Res.* **2021**, *35*, 1449–1458. [CrossRef] [PubMed]
13. Mujika, I.; Padilla, S. Detraining: Loss of Training-Induced Physiological and Performance Adaptations. Part I. *Sports Med.* **2000**, *30*, 79–87. [CrossRef] [PubMed]
14. TrainingPeaks. Available online: www.trainingpeaks.com (accessed on 1 November 2021).
15. Scheadler, C.; Garver, M.; Hanson, N. The Gas Sampling Interval Effect on V˙O2peak Is Independent of Exercise Protocol. *Med. Sci. Sports Exerc.* **2017**, *49*, 1911–1916. [CrossRef] [PubMed]
16. Beaver, W.; Wasserman, K.; Whipp, B. A New Method for Detecting Anaerobic Threshold by Gas Exchange. *J. Appl. Physiol.* **1986**, *60*, 2020–2027. [CrossRef] [PubMed]
17. Cohen, J. *Statistical Power Analysis for the Behavioral Sciences*; Academic Press: Cambridge, MA, USA, 2013.
18. Muriel, X.; Courel-Ibáñez, J.; Cerezuela-Espejo, V.; Pallarés, J.G. Training Load and Performance Impairments in Professional Cyclists during COVID-19 Lockdown. *Int. J. Sports Physiol. Perform.* **2020**, *16*, 735–738. [CrossRef] [PubMed]
19. Santamaria, C.; Sermi, F.; Spyratos, S.; Iacus, S.; Annunziato, A.; Tarchi, D.; Vespe, M. Measuring the Impact of COVID-19 Confinement Measures on Human Mobility Using Mobile Positioning Data. A European Regional Analysis. *Saf. Sci.* **2020**, *132*, 104925. [CrossRef] [PubMed]
20. Banerjee, A.; Chitnis, U.; Jadhav, S.; Bhawalkar, J.; Chaudhury, S. Hypothesis Testing, Type I and Type II Errors. *Ind. Psychiatry J.* **2009**, *18*, 127. [CrossRef] [PubMed]

Journal of
Functional Morphology and Kinesiology

MDPI

Case Report

Relationships between Workload, Heart Rate Variability, and Performance in a Recreational Endurance Runner

Daniel Boullosa [1,*], André R. Medeiros [2], Andrew A. Flatt [3], Michael R. Esco [4], Fabio Y. Nakamura [5] and Carl Foster [6]

1 INISA, Federal University of Mato Grosso do Sul, Campo Grande 79070-900, Mato Grosso do Sul, Brazil
2 Federal District Education Secretary, Brasilia 72302-505, Brazil; mdt.ricarte@hotmail.com
3 Biodynamics Laboratory, Department of Health Sciences, Armstrong State University, Savannah, GA 31419, USA; aflatt@georgiasouthern.edu
4 Exercise Physiology Laboratory, Department of Kinesiology, University of Alabama, Tuscaloosa, AL 870312, USA; mresco@ua.edu
5 Research Center in Sports Sciences, Health Sciences and Human Development, CIDESD, University Institute of Maia, ISMAI, 4475-690 Maia, Portugal; fabioy_nakamura@yahoo.com.br
6 Department of Exercise and Sports Science, University of Wisconsin-La Crosse, Lacrosse, WI 54601, USA; cfoster@uwlax.edu
* Correspondence: daniel.boullosa@gmail.com; Tel.: +55-61-98250-2545

Abstract: Background: The association between heart rate variability (HRV), training load (TL), and performance is poorly understood. Methods: A middle-aged recreational female runner was monitored during a competitive 20-wk macrocycle divided into first (M1) and second mesocycle (M2) in which best performances over 10 km and 21 km were recorded. Volume (km), session rating of perceived exertion (sRPE), TL, and monotony (mean TL/SD TL) were the workload parameters recorded. The root mean square of the successive differences in R-R intervals (RMSSD), its coefficient of variation (RMSSDcv), and the RMSSD:RR ratio were the HRV parameters monitored. Results: During M2, RMSSD ($p = 0.006$) and RMSSD:RR ($p = 0.002$) were significantly increased, while RR was significantly reduced ($p = 0.017$). Significant correlations were identified between monotony and volume ($r = 0.552$; $p = 0.012$), RR ($r = 0.447$; $p = 0.048$), and RMSSD:RR ($r = -0.458$; $p = 0.042$). A sudden reduction in RMSSD (from 40.31 to 24.34 ms) was observed the day before the first symptoms of an influenza. Conclusions: The current results confirm the practicality of concurrent HRV and sRPE monitoring in recreational runners, with the RMSSD:RR ratio indicative of specific adaptations. Excessive training volume may be associated to both elevated monotony and reduced RMSSD:RR. Identification of mesocycle patterns is recommended for better individualization of the periodization used.

Keywords: autonomic control of HR; vagal modulations; vagal-sympathetic effect; training monitoring; endurance performance

Citation: Boullosa, D.; Medeiros, A.R.; Flatt, A.A.; Esco, M.R.; Nakamura, F.Y.; Foster, C. Relationships between Workload, Heart Rate Variability, and Performance in a Recreational Endurance Runner. *J. Funct. Morphol. Kinesiol.* **2021**, *6*, 30. https://doi.org/10.3390/jfmk6010030

Academic Editor: Antonino Bianco

Received: 25 February 2021
Accepted: 19 March 2021
Published: 22 March 2021

1. Introduction

Heart rate variability (HRV) is a valid and accessible cardiac-autonomic marker that has been promoted as a technique for monitoring training of recreational runners [1]. The objectives of routine assessment of HRV among athletes include selecting long-term training methods [2], modifying daily exercise prescriptive factors [3], and identifying positive and negative adaptations [4]. There are numerous HRV indices. The square root of mean squared difference of successive R–R intervals (RMSSD) is a robust index of vagal autonomic function that is commonly employed by recreational runners [4]. Indeed, its coefficient of variation (RMSSDcv) and the RMSSD:RR ratio are simple HRV parameters that have been proposed to identify individual responses to endurance training [5,6]. The RMSSD:RR ratio has been proposed to identify vagal saturation as it normalizes vagal modulations by the RR intervals, therefore relating vagal and sympathetic modulations [7].

The associations of endurance running performance with autonomic adaptations and with markers of internal and external training loads separately are well documented. However, there is only one study [8] that has documented a dose–response relationship between heart rate (HR)-derived training impulse (TRIMP), power spectral analyses of HRV, and marathon times. These results suggest a possible sympathetic drift towards the end of the preparation period that positively correlated with performance. However, the weekly autonomic changes associated with weekly training workload indices were not reported in this previous study [8]. This information would be of interest given the previous suggestions on the existence of individual autonomic profiles associated with training cycles and periodization in other endurance sports [5,7]. For instance, a simultaneous reduction in LnRMSSD and LnRMSSD:RR during the final week preceding competition appeared to be indicative of optimal performance in an elite triathlete [7]. In another study with elite rowers [5], different autonomic adaptations to training at different time points of the preparation were also observed. Therefore, identification of how weekly training workload relates to autonomic status would help to better manage training load.

Thus, we present a case report of a recreational female runner who completed a 20-wk competitive macrocycle. Daily HRV and training indices were recorded for subsequently identifying associations between HRV, training workload indices, and running performances. Based on previous studies, we would expect reduced RMSSD and RMSSD:RR prior to better performances.

2. Materials and Methods

The recreational runner is a middle-aged female (50 years; 1.59 m; 50–52 kg; maximum oxygen consumption [VO_2max] = 56 mL·kg^{-1}·min^{-1}) with more than 10 years of endurance training experience, first as a triathlete and more recently (last 7 years) as a road runner. The training history and periodization used have been described elsewhere [9]. Briefly, she completed a 20-week competitive macrocycle, after a 3-month preparatory mesocycle, in which she competed in two 10-km and three 21-km road races in the city of Brasilia, under thermoneutral environmental conditions (<23 °C and <50% relative humidity). Briefly, all the races were completed in early (7:00 a.m.) morning, with similar profiles. Best performances in both distances occurred in the second part of the competitive macrocycle, achieving 99.3% and 97.8% of her best performances for 10 km and 21 km, respectively (recorded 6 years before). Therefore, the competitive macrocycle was divided into first mesocycle (M1 = 10 first weeks) and second mesocycle (M2 = 10 last weeks). She gave her consent for the public use of her data for this case study.

Workload indices included session Rating of Perceived Exertion (sRPE) and its derived indices of training load (sRPE × time in minutes) and monotony (mean weekly training load/SD of weekly training load) [10]. Daily volume in km was recorded with a GPS unit (Forerunner 630, Garmin, Olathe, KS, USA). The typical weekly microcycle (5–7 sessions) included two strength-training sessions plus 20–30 min of submaximal uphill runs on treadmill; one to two running sessions of 'cruise intervals' (at or slightly below the competitive pace), and some intervals at maximum aerobic speed (MAS) in the weeks before competitions. On designated recovery days, she performed 1–2 easy short runs or runs plus walks of 30–60 min, interspersed with some maximum speed progressions over 100-m; and a single long easy run of 70–100 min. The training intensity distribution was "polarized" (75-80/5/15-20) as previously documented [9].

RR intervals were recorded for 2 min (after 1 min of stabilization) every morning, in supine position after awakening, with a validated HR strap (H7, Polar Electro Oy, Kempele, Finland), and exported via Bluetooth to a mobile App (Elite HRV, Asheville, NC, USA). The RR and RMSSD values obtained with the mobile App were subsequently recorded and exported to a custom Excel® spreadsheet, in which weekly RMSSDcv (i.e., [SD of RMSSD/mean RMSSD] × 100) and RMSSD:RR (i.e., mean RMSSD/mean RR) were calculated.

Values are presented as mean ± SD. After normality distribution confirmation, differences between weekly HRV indices in M1 and M2 were performed with a non-paired t test, and effect size (ES) via a Cohen's d. The smallest worthwhile change (SWC) was also calculated as 0.3 × SD of week 1 [11]. The relationships between training workload and HRV indices were performed with a Pearson product correlation coefficient (r). Statistical significance was set at 5%.

3. Results

The evolution of training workload, HRV indices, and running performances (10-km and 21-km running times) over the 20-week macrocycle are presented in Table 1. Differences between M1 and M2 for dependent variables are presented in Table 2. Of note, a sudden reduction in RMSSD (from 40.31 to 24.34 ms) during week 8 was observed the day before the first symptoms of an influenza, which was followed by 2 days of disrupted training.

Table 1. Weekly mean values for HRV, training workload parameters, and competitive performances.

Week	Distance (km)	Training Load (sRPE × Time)	Monotony	RMSSD (ms)	RMSSDcv (%)	RR (s)	RMSSD/RR (ms/s)	Running Performances
1	68	344	2.29	47.75	34.82	1.29	37.06	
2	55	275	1.68	54.73	34.22	1.28	42.85	
3	75	478	2.42	49.09	21.73	1.29	37.95	
4	83	512	3.51	41.29	14.74	1.28	32.19	
5	77	473	2.02	45.46	17.53	1.31	34.59	
6	65	299	1.85	45.64	20.36	1.33	34.23	41:58 (10-km)
7	77	623	1.40	46.47	12.63	1.27	36.56	
8	64	462	1.26	42.60	40.70	1.23	34.66	
9	82	509	2.23	48.48	34.19	1.29	37.60	
10	65	319	1.31	55.48	39.73	1.19	46.67	1:28:25 (21-km)
11	53	262	1.22	50.39	15.26	1.23	40.87	
12	55	381	1.36	59.52	34.60	1.28	46.61	39:56 (10-km)
13	52	363	1.52	52.26	28.21	1.26	41.61	
14	93	531	1.88	56.27	27.12	1.25	45.11	
15	70	522	2.63	54.53	25.64	1.25	43.49	
16	63	395	1.51	61.02	52.34	1.21	50.37	1:26:45 (21-km)
17	55	257	2.13	47.19	16.22	1.21	39.02	
18	84	560	2.44	50.27	12.37	1.24	40.62	
19	47	400	1.50	57.55	23.94	1.25	46.15	
20	61	212	0.77	51.73	33.42	1.19	43.31	1:26:33 (21-km)

Table 2. Comparison of HRV and training workload parameters between M1 and M2.

	M1	M2	*t*-Test (*p*)	Cohen's d
RMSSD (ms)	47.7 (4.60)	54.1 (4.46)	0.006	−1.407
RMSSDcv (%)	27.1 (10.7)	26.9 (11.6)	0.976	0.014
RR (s)	1.28 (0.041)	1.24 (0.025)	0.017	1.207
RMSSD/RR (ms/s)	37.44 (4.34)	43.72 (3.41)	0.002	−1.620
Distance (km)	71.1 (9.06)	63.3 (14.90)	0.174	0.651
Training load (a.u.)	429 (114)	388 (121)	0.444	0.350
Monotony (a.u.)	2.00 (0.67)	1.69 (0.57)	0.296	0.483

There were significant correlations between the km completed each week with training load ($r = 0.738$; $p < 0.00$) and monotony ($r = 0.552$; $p = 0.012$). Conversely, some HRV indices were correlated among them. RMSSD:RR correlated with RMSSD ($r = 0.973$; $p = 0.000$), RMSSDcv ($r = 0.526$; $p = 0.027$), and RR ($r = -0.581$; $p = 0.007$), while RMSSD correlated with RMSSDcv ($r = 0.499$; $p = 0.025$). Further, monotony was correlated with weekly HRV indices: RR ($r = 0.447$; $p = 0.048$) and RMSSD:RR ($r = -0.458$; $p = 0.042$). There were no correlations between training load and HRV parameters. Concurrent weekly changes of monotony and RR, and monotony and RMSSD:RR are shown in Figure 1.

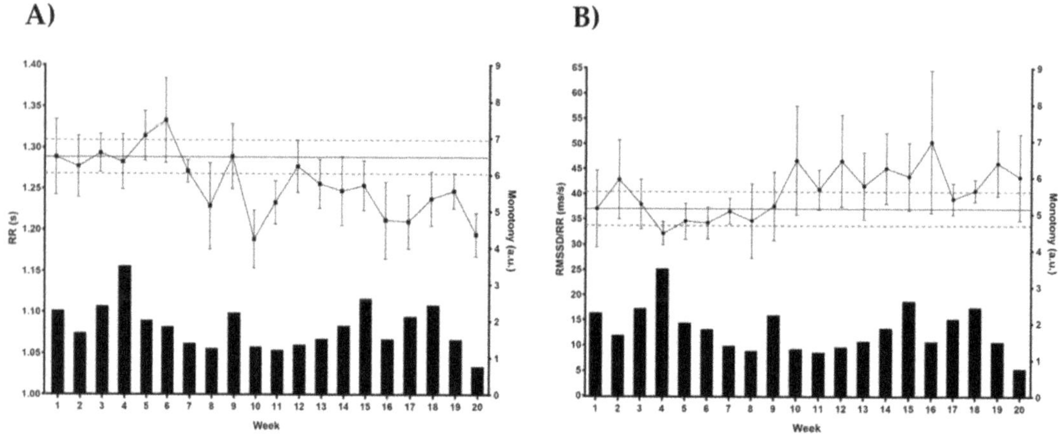

Figure 1. Concurrent weekly changes of monotony and RR intervals (**A**), and monotony and the RMSSD:RR ratio (**B**). The black dots represent the HRV parameters, while the black bars represent the monotony scores.

4. Discussion

The main and novel observation of this case report was the association between monotony, an index of weekly load periodization, and both RR and RMSSD:RR. In addition, our hypothesis was partially confirmed with only reduced RR in M2 being indicative of improved performances. However, a greater RMSSD:RR was observed with better performances in M2, which was strongly correlated to greater RMSSD values in this mesocycle. Thus, simultaneous enhancement of vagal (↑RMSSD) and, probably sympathetic (↓RR and ↑RMSSD:RR) modulations would be suggestive of better adaptations and thus improved performances. Furthermore, these autonomic adaptations would be related to reduced monotony scores. Therefore, simultaneous recording of weekly monotony and RMSSD:RR may be important monitoring tools for endurance runners and other endurance athletes.

The results are in alignment with one study with an elite triathlete [7], suggesting an increased RMSSD associated with positive adaptations and reduced RMSSD:RR associated with improved performances. However, another study with world-champion rowers [5] exhibited consistent substantial reductions in RMSSD:RR prior to outstanding performances. Meanwhile, we did not observe the relationship between reduced RMSSDcv and improved performances observed in another case study with a male recreational athlete [6]. Differences between studies may be attributed to differences between HRV recording characteristics (i.e., position and duration of recordings), periodizations used, and sports demands. However, the increased RMSSD:RR, which would be associated to enhanced vagal modulations and also sympathetic activity, is consistent with previous reports of enhanced performances in samples of recreational runners with greater RMSSD values [4,6], and a possibly increased sympathetic activity near competition [8]. This reinforces the value of the RMSSD:RR ratio to monitor recreational endurance runners, with RR intervals (i.e., a vagal-sympathetic effect index) [12] being a complementary HRV parameter to the most used RMSSD by practitioners. Further studies with samples of recreational runners are needed to corroborate these observations.

The most novel and important observations were the associations identified between monotony with RR ($r = 0.447$; $p = 0.048$) and RMSSD:RR ($r = -0.458$; $p = 0.042$). These correlations are contrary to the desired autonomic adaptations, thus confirming the well-known negative effects of monotony on health and performance of athletes [10]. Monotony was also associated with weekly volume ($r = 0.552$; $p = 0.012$) which, in turn, was associated with training load ($r = 0.738$; $p < 0.00$). This would suggest that volume, a well-known pre-requisite for endurance adaptations, could also favor negative adaptations when associated with high monotony. Of note, the association between training volume and monotony would be mathematically expected in most cases. However, there were no significant differences between training workload parameters between mesocycles, although the ES revealed M1 as the most demanding mesocycle (see Table 2). In this regard, a more detailed analysis reveals that the best performance occurred in week 12 (39:56 in 10 km, which represents 99.3% of her best) with the peak volume achieved, in this case, 3 weeks before competitions instead of 2 weeks as for the other races. Further, this peak volume (i.e., 82 km) was associated to very low monotony scores (i.e., 1.32–1.22) during the 2 weeks of tapering before competition. Therefore, identification of weekly volumes should be accompanied by examination of monotony scores over several weeks, and no single weeks, to identify individual patterns to be replicated in future periodizations. This is an important consideration given the well-known limitation of periodizations to induce peak performances in a purported time. These observations also reinforce the risk associated to high volumes, which are very typical of recreational runners training for performance purposes [1]. In this regard, the approach of the current case report agrees with the recent suggestions on the need for combining both internal and external training load indices for optimized training load monitoring in runners [13]. Therefore, the concurrent use of sRPE and HRV would expand the validity of monotony scores for training monitoring [14], which should be confirmed in further studies.

One interesting finding was the sudden reduction of vagal modulations (i.e., RMSSD = 24.34) because of an influenza. Interestingly, this sudden reduction occurred 1 day before any symptom and served to cancel the programmed training on that day. After 2 days of rest, the runner returned progressively to normal training without any relevant issue to be reported. This is a relevant observation that should be considered by runners exposed to any viral infection (e.g., SARS-CoV), with further studies needed to confirm these observations.

This case study is not without limitations. As this is a single case report, generalization of these results should be considered with caution. Of note, specific characteristics of training and daily activities of the runner may be related to our observations. For instance, the runner followed an "Evolutionary periodization" [9], which accounts for management of both training loads and lifestyle habits, including professional activities, sleeping routines [15], nutritional strategies [16], and incidental physical activity [17] among others. In this regard, as the pre-planned loads were adapted on a daily basis, with consideration of all these factors (including HRV morning data [18]), we do not know if a fixed periodization would result in similar outcomes. In addition, these associations may be different when using other HRV protocols [19], parameters [20], and Apps with different correction algorithms [21–23].

5. Conclusions and Practical Applications

We identified specific autonomic adaptations related to training workload parameters and better performances in a middle-aged recreational female runner. Specifically, an enhanced RMSSD:RR was associated to reduced monotony, a consistent response during the mesocycle of best performances. In addition, a reduction in vagal modulations during the first days of an influenza was also observed. Future studies with runners of different age, sex and levels should confirm these important observations.

Following the current observations, it may be recommended to daily record sRPE, RMSSD, and RR. The subsequent calculation of training load (i.e., sRPE $\times$ time), monotony (i.e., mean weekly training load/SD of weekly training load), and RMSSD:RR, would therefore assist to monitor changes of these parameters on a weekly and mesocycle basis, with respect to changes in training volume (km) and running performance.

Author Contributions: Conceptualization and data curation, D.B.; formal analysis, A.R.M.; Methodology, D.B., A.R.M., A.A.F., M.R.E., F.Y.N. and C.F.; writing—original draft preparation, D.B.; writing—

review and editing, D.B., A.R.M., A.A.F., M.R.E., F.Y.N. and C.F.; All authors have read and agreed to the published version of the manuscript.

Funding: This research received no external funding.

Institutional Review Board Statement: Ethical review and approval were waived for this study, as data analyses and interpretation were conducted as part of the training regime designed and supervised by the first author.

Informed Consent Statement: Informed consent was obtained from the subject involved in the study.

Data Availability Statement: The data that support the findings of this study are available from the corresponding author upon reasonable request.

Acknowledgments: We would like to thank the athlete for her enthusiastic participation and strong adherence to the training periodization used.

Conflicts of Interest: The authors declare no conflict of interest.

References

1. Boullosa, D.; Esteve-Lanao, J.; Casado, A.; Peyré-Tartaruga, L.A.; Da Rosa, R.G.; Del Coso, J. Factors Affecting Training and Physical Performance in Recreational Endurance Runners. *Sports* **2020**, *8*, 35. [CrossRef] [PubMed]
2. Vesterinen, V.; Hakkinen, K.; Laine, T.; Hynynen, E.; Mikkola, J.; Nummela, A.T. Predictors of individual adaptation to high-volume or high-intensity endurance training in recreational endurance runners. *Scand. J. Med. Sci. Sports* **2016**, *26*, 885–893. [CrossRef] [PubMed]
3. Kiviniemi, A.M.; Hautala, A.J.; Kinnunen, H.; Tulppo, M.P. Endurance training guided individually by daily heart rate variability measurements. *Graefe's Arch. Clin. Exp. Ophthalmol.* **2007**, *101*, 743–751. [CrossRef] [PubMed]
4. Buchheit, M.; Chivot, A.; Parouty, J.; Mercier, D.; Al Haddad, H.; Laursen, P.B.; Ahmaidi, S. Monitoring endurance running performance using cardiac parasympathetic function. *Eur. J. Appl. Physiol.* **2010**, *108*, 1153–1167. [CrossRef] [PubMed]
5. Plews, D.J.; Laursen, P.B.; Buchheit, M. Day-to-day heart-rate variability recordings in world-champion rowers: Appreciating unique athlete characteristics. *Int. J. Sports Physiol. Perform.* **2017**, *12*, 697–703. [CrossRef] [PubMed]
6. Flatt, A.A.; Esco, M.R. Endurance Performance Relates to resting heart rate and its variability: A Case Study of a Collegiate Male Cross-country Athlete. *J. Aust. Strength Cond.* **2014**, *22*, 39–45.
7. Stanley, J.; D'Auria, S.; Buchheit, M. Cardiac Parasympathetic Activity and Race Performance: An Elite Triathlete Case Study. *Int. J. Sports Physiol. Perform.* **2015**, *10*, 528–534. [CrossRef]
8. Manzi, V.; Castagna, C.; Padua, E.; Lombardo, M.; D'Ottavio, S.; Massaro, M.; Volterrani, M.; Iellamo, F. Dose-response relationship of autonomic nervous system responses to individualized training impulse in marathon runners. *Am. J. Physiol. Circ. Physiol.* **2009**, *296*, H1733–H1740. [CrossRef]
9. Boullosa, D.; Foster, C. "Evolutionary" based periodization in a recreational runner. *Sport Perform. Sci. Rep.* **2018**, *1*, 36.
10. Foster, C. Monitoring training in athletes with reference to overtraining syndrome. *Med. Sci. Sports Exerc.* **1998**, *30*, 1164–1168. [CrossRef]
11. Hopkins, W.G. How to Interpret Changes in an Athletic Performance Test. *Sportscience* **2004**, *8*, 1–7.
12. Medeiros, A.R.; Michael, S.; Boullosa, D.A. Make it easier! Evaluation of the 'vagal-sympathetic effect' in different conditions with R–R intervals monitoring. *Eur. J. Appl. Physiol.* **2018**, *118*, 1287–1288. [CrossRef]
13. Paquette, M.R.; Napier, C.; Willy, R.W.; Stellingwerff, T. Moving Beyond Weekly "Distance": Optimizing Quantification of Training Load in Runners. *J. Orthop. Sports Phys. Ther.* **2020**, *50*, 564–569. [CrossRef]
14. Foster, C.; Boullosa, D.; McGuigan, M.; Fusco, A.; Cortis, C.; Arney, B.E.; Orton, B.; Dodge, C.; Jaime, S.; Radtke, K.; et al. 25 Years of Session Rating of Perceived Exertion: Historical Perspective and Development. *Int. J. Sports Physiol. Perform.* **2020**, 1–10. [CrossRef]
15. Myles, W.S. Sleep deprivation, physical fatigue, and the perception of exercise intensity. *Med. Sci. Sports Exerc.* **1985**, *17*, 580–584. [CrossRef]
16. Burke, L.M.; Hawley, J.A.; Jeukendrup, A.; Morton, J.P.; Stellingwer, T.; Maughan, R.J. Toward a Common Under-standing of Diet-Exercise Strategies to Manipulate Fuel Availability for Training and Competition Preparation in Endurance Sport. *Int. J. Sport Nutr. Exerc. Metab.* **2018**, *28*, 451–463. [CrossRef]
17. Hautala, A.; Martinmaki, K.; Kiviniemi, A.; Kinnunen, H.; Virtanen, P.; Jaatinen, J.; Tulppo, M. Effects of habitual physical activity on response to endurance training. *J. Sports Sci.* **2012**, *30*, 563–569. [CrossRef]
18. Da Silva, D.F.; Ferraro, Z.M.; Adamo, K.B.; Machado, F.A. Endurance Running Training Individually Guided by HRV in Untrained Women. *J. Strength Cond. Res.* **2019**, *33*, 736–746. [CrossRef]
19. Medeiros, A.R.; Leicht, A.S.; Michael, S.; Boullosa, D. Weekly vagal modulations and their associations with physical fit-ness and physical activity. *Eur. J. Sport Sci.* **2020**. Online Ahead of Print. [CrossRef]
20. Boullosa, D.A.; Abreu, L.; Nakamura, F.Y.; Muñoz, V.E.; Domínguez, E.; Leicht, A.S. Cardiac Autonomic Adaptations in Elite Spanish Soccer Players During Preseason. *Int. J. Sports Physiol. Perform.* **2013**, *8*, 400–409. [CrossRef]

21. Perrotta, A.S.; Jeklin, A.T.; Hives, B.A.; Meanwell, L.E.; Warburton, D.E. Validity of the Elite HRV Smartphone Application for Examining Heart Rate Variability in a Field-Based Setting. *J. Strength Cond. Res.* **2017**, *31*, 2296–2302. [CrossRef] [PubMed]
22. Altini, M.; Amft, O. HRV4Training: Large-scale longitudinal training load analysis in unconstrained free-living settings using a smartphone application. In Proceedings of the 38th Annual International Conference of the IEEE Engineering in Medicine and Biology Society (EMBC), Orlando, FL, USA, 16–20 August 2016; Volume 2016, pp. 2610–2613. [CrossRef]
23. Pereira, R.D.A.; Alves, J.L.D.B.; Silva, J.H.D.C.; Costa, M.D.S.; Silva, A.S. Validity of a Smartphone Application and Chest Strap for Recording RR Intervals at Rest in Athletes. *Int. J. Sports Physiol. Perform.* **2020**, *15*, 896–899. [CrossRef] [PubMed]

Journal of
Functional Morphology and Kinesiology

MDPI

Article

Predictive Validity of the Snatch Pull Force-Velocity Profile to Determine the Snatch One Repetition-Maximum in Male and Female Elite Weightlifters

Ingo Sandau [1,*], Helmi Chaabene [2] and Urs Granacher [2]

1 Research Group Weightlifting, Institute for Applied Training Science, D-04109 Leipzig, Germany
2 Division of Training and Movement Sciences, University of Potsdam, Research Focus Cognition Sciences, D-14469 Potsdam, Germany; chaabanehelmi@hotmail.fr (H.C.); urs.granacher@uni-potsdam.de (U.G.)
* Correspondence: sandau@iat.uni-leipzig.de

Abstract: Background: The prediction of one repetition-maximum (1RM) performance from specific tests is highly relevant for the monitoring of training in weightlifting. Therefore, this study aimed at examining the predictive validity of the theoretical 1RM snatch ($snatch_{th}$) computed from the two-point snatch pull force-velocity relationship (FvR_2) to determine actual snatch 1RM performance in elite weightlifters. Methods: Eight (three female, five male) elite weightlifters carried out a 1RM snatch test followed by a snatch pull test with loads of 80% and 110% of the previously determined 1RM snatch. Barbell kinematics were determined for all lifts using video-tracking. From the snatch pull barbell kinematics, the snatch pull FvR_2 was modeled and the $snatch_{th}$ was calculated. Results: The main findings indicated a non-significant (p = 0.706) and trivial (d = 0.01) mean difference between the actual 1RM snatch performance and the $snatch_{th}$. Both measures showed an extremely large correlation (r = 0.99). The prediction accuracy of the actual 1RM snatch from $snatch_{th}$ was 0.2 ± 1.5 kg (systematic bias ± standard deviation of differences). Conclusions: This study provides a new approach to estimate 1RM snatch performance in elite weightlifters using the snatch pull FvR_2. The results demonstrate that the $snatch_{th}$-model accurately predicts 1RM snatch performance.

Keywords: validation study; performance; monitoring; training

Citation: Sandau, I.; Chaabene, H.; Granacher, U. Predictive Validity of the Snatch Pull Force-Velocity Profile to Determine the Snatch One Repetition-Maximum in Male and Female Elite Weightlifters. *J. Funct. Morphol. Kinesiol.* **2021**, *6*, 35. https://doi.org/10.3390/jfmk6020035

Academic Editor: Cristina Cortis

Received: 27 March 2021
Accepted: 14 April 2021
Published: 16 April 2021

1. Introduction

Systematic performance testing (i.e., monitoring) is a cornerstone to detect intraindividual changes over time for general fitness (e.g., muscle strength, power) and sport-specific performance parameters in elite weightlifters [1]. A widely used and highly reliable test in weightlifting is the one-repetition maximum (1RM) test, for instance, during the snatch [2]. In addition to the 1RM test, performance assessment in weightlifting included vertical jump tests, isometric mid-thigh pull tests, and back/leg extensor tests [3–5].

A prerequisite of a test to be included for performance assessment is that it sufficiently complies with psychometric properties such as validity, reliability, and responsiveness [6]. In terms of test validity, predictive validity is highly relevant for elite sports [7]. In this context, Joffe and Tallent [8] examined the predictive validity of peak power during vertical jumping and peak force during the isometric mid-thigh pull using a multiple linear regression model to determine weightlifting performance (i.e., 1RM snatch, clean and jerk, and total) in highly trained female weightlifters aged 23 years. These authors reported high relative (R^2 = 0.94–0.95) and absolute (standard error of estimate (SEE) = 3.8–9.5 kg) predictive validity. In general, a multiple linear regression model reflects the statistical association between independent (e.g., peak power during vertical jumping, peak force during isometric mid-thigh pull peak force) and dependent (e.g., 1RM snatch) variables in a given sample (i.e., group data). Of note, results from regression models are always specific to the population under investigation [9]. Accordingly, a low predictive value of an estimated outcome can be obtained if established regression equations are translated to different samples (i.e., shrinkage) [10]. This is even more prevalent if the prediction equation has been derived from a rather small sample [10]. In this context, there is preliminary evidence that the prediction of individual performance levels using an individualized biomechanical approach, rather than statistical, provides higher external validity [9].

The individual load-velocity relationship (LvR) (i.e., linear regression model of barbell load vs. barbell velocity) is a good example of such a widely applied biomechanical approach that has the potential to predict 1RM during various strength exercises (e.g., bench press, back squat, deadlift, power clean) [11–13]. In this easy-to-administer approach, the estimated 1RM is the intercept of the extrapolated LvR-regression-line with the previously determined minimal velocity threshold of a maximal lift [11]. In addition to the LvR, the force–velocity relationship (FvR) represents another method to predict 1RM performance [14,15]. A previous study used a commercially available linear encoder (Musclelab, Ergotest, Norway) to compute 1RM bench press performance (systematic bias ± standard deviation of differences (SDD) = 5.4 ± 5.7 kg) from FvR in strength trained adults aged 29 years [15]. However, these authors left the underlying computational algorithm in a black box. In another study, a combined LvR and FvR approach was used to predict 1RM chest press (systematic bias ± SDD = −1.3 ± 1.2 kg) and leg press performance (systematic bias ± SDD = −1.8 ± 2.1 kg) in recreationally active participants aged 24 years [14]. Recently, Sandau, et al. [16] introduced a conceptual biomechanical model to compute the (theoretical) 1RM snatch performance (i.e., $snatch_{th}$) from the two-point snatch pull FvR (FvR_2) in elite male weightlifters aged 27 years. Results indicated high reliability of the $snatch_{th}$ (percentage standard error of measurement (SEM%) = 0.71%, intraclass correlation coefficient (ICC) = 0.99) [16]. However, the predictive validity of the $snatch_{th}$ to determine actual 1RM snatch performance has not yet been investigated.

To the best of the authors' knowledge, no study has examined the accuracy to predict 1RM performance using FvR-profiling with specific weightlifting exercises (i.e., snatch, clean and jerk). Therefore, this study aimed at examining the predictive validity of the $snatch_{th}$ computed from the snatch pull FvR_2 to determine the actual snatch 1RM performance in elite male and female weightlifters. Based on the computational similarities of 1RM prediction from LvR with the $snatch_{th}$ from FvR_2 [16], we hypothesized that the $snatch_{th}$ can be used to accurately predict actual 1RM snatch performance in elite weightlifters.

2. Materials and Methods

2.1. Subjects

Eight elite weightlifters (male = 5; female = 3; age range: 18–29 years), all members of the German national team, volunteered to participate in this study (Table 1). At the time of the study, all weightlifters regularly competed at national and international events and had >6 years of systematic weightlifting training. They were free from any musculoskeletal or neurological diseases or injuries at the time of data collection. This study was conducted according to the latest version of the Declaration of Helsinki, and the experimental protocol was approved by the local Ethics Board of the Institute for Applied Training Science (approval number: ER_2020.28.09_4).

Table 1. Descriptive data in means and standard deviations of the study participants.

	Age (years)	Body Mass (kg)	1RM Snatch (kg)	1RM Clean and Jerk (kg)
Overall	23.3 ± 4.0	79.0 ± 16.6	124.8 ± 37.6	155.5 ± 41.7
Male	22.8 ± 3.3	87.4 ± 15.1	150.8 ± 11.5	184.2 ± 14.6
Female	24.0 ± 5.6	65.0 ± 6.0	81.3 ± 13.1	107.7 ± 12.9

2.2. Data Collection

Data were collected during the preparation period of a macrocycle. Testing was performed on a regular training day at the beginning of the weightlifting training session. Before the tests, an individualized warm-up program was conducted for 15–20 min including cycling on an ergometer at submaximal intensity, mobility exercises with and without the barbell. After the warm-up, the participating weightlifters started with a 1RM snatch test. They were encouraged to reach their maximal snatch performance (i.e., 100% = actual 1RM) using 6–8 loads with 1–2 repetitions per load condition. The rest between load conditions and repetitions was 3 and 2 min, respectively. Weightlifters were instructed to perform every lift at maximal effort. After the 1RM snatch test, a 5 min rest was implemented before the snatch pull test was conducted. During the snatch pull test, weightlifters were encouraged to lift two consecutive repetitions at 80% and 110% of the previously determined 1RM snatch at maximal effort and starting with the 110% load condition. The rest between load conditions was 3 min and between repetitions 2 min. Lifting straps were used during all snatch pull trials.

The snatch and the snatch pull lifts were video recorded (Canon, Legria HF G26) and analyzed using a custom-made real-time barbell tracking software (Realanalyzer, IAT, Leipzig, Germany) [17]. The position of the digital camera followed a routine set-up, with the camera placed 1 m above the floor and positioned next to the athlete at a distance of 5 m. The vertical barbell velocity was computed as the 1st derivative of vertical barbell position data. This system demonstrated excellent absolute (SEM%) and relative (ICC) test-retest reliability for the measurement of maximal barbell velocity (SEM% = 0.72%, ICC > 0.99) and maximal distance (SEM% = 0.45%, ICC > 0.99) [18].

2.3. Data Processing

Maximal vertical barbell velocity was measured during the 1RM snatch (denoted as v_{thres}) and during all snatch pull trials (denoted as v_{max}). If v_{max} values from two consecutive snatch pull lifts differed more than 0.1 $m{\cdot}s^{-1}$ at a specific load condition, a third repetition was conducted. In addition, the distance of vertical acceleration (h_{acc}, i.e., vertical position of the barbell at the instant of v_{max} minus the radius of the barbell plates (0.225 m)) was taken from the measurements during the snatch pull at the 110% load (Figure 1).

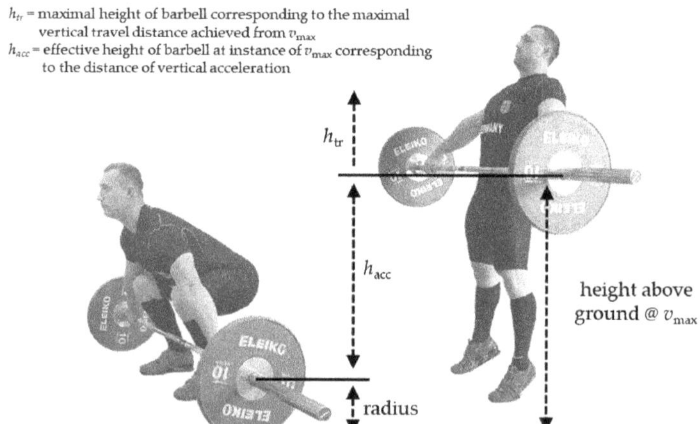

Figure 1. Schematic illustration of the snatch pull exercise from start position (**left**) to position of maximal vertical barbell velocity (v_{max}) (**right**) with vertical barbell kinematic parameters.

Averaged values from two consecutive snatch pull lifts for v_{max} (80% and 110% condition) and h_{acc} (110% condition) were used for further analyses. The snatch pull FvR_2 was modeled using linear regression with mean vertical barbell force ($\overline{F}$) and velocity ($\overline{v}$) from the 80% and 110% load conditions as input parameters. With reference to the approach presented by Samozino et al. [19], the mean vertical barbell force and velocity were computed from vertical barbell kinematics (i.e., v_{max}, h_{acc}).

The mean vertical barbell velocity was computed as follows:

$$\overline{v} = \sqrt{\frac{g \times h_{tr}}{2}} \tag{1}$$

In this equation, h_{tr} stands for the maximal vertical travel distance of the barbell that is achieved from v_{max}, and g stands for the gravitational acceleration:

$$h_{tr} = \frac{{v_{max}}^2}{2g} \tag{2}$$

The mean external vertical force to accelerate the barbell was computed from h_{acc}, h_{tr}, and m (barbell mass) as follows:

$$\overline{F} = m \times g\left(\frac{h_{tr}}{h_{acc}} + 1\right) \tag{3}$$

From the snatch pull FvR_2 regression model, the theoretical maximal mean vertical barbell velocity at zero barbell force (i.e., $\overline{v}_0$; intercept of velocity axis), the theoretical maximal mean vertical barbell force at zero barbell velocity (i.e., $\overline{F}_0$; intercept of force axis), and the theoretical maximal mean vertical barbell power (i.e., $\overline{P}_{max}$) were computed [20]. The $\overline{v}_0$, $\overline{F}_0$, and $\overline{P}_{max}$ were used as typical FvR parameters to evaluate the maximal mechanical capabilities of the neuromuscular system [21]. Based on the individual snatch pull FvR_2 regression model, the $snatch_{th}$ was computed from barbell force at v_{thres} of the 1RM snatch lift [16].

For this purpose, the mean vertical barbell force ($\overline{F}_{thres}$) was computed for v_{thres} from the snatch pull FvR_2 regression model as follows:

$$\overline{F}_{thres} = \frac{\overline{v}_{thres} - \overline{v}_0}{slope_{FvR_2}} \tag{4}$$

where $\overline{v}_{thres}$ was computed from v_{thres} (Equation (1)). The $\overline{F}_{thres}$ is the sum of the gravitational force due to barbell mass and the force from barbell acceleration. To obtain the barbell load at v_{thres} (i.e., $snatch_{th}$), $\overline{F}_{thres}$ has to be divided by the sum of g and the vertical barbell acceleration (a_{thres}) to achieve v_{thres}:

$$snatch_{th} = \frac{\overline{F}_{thres}}{g + a_{thres}} \tag{5}$$

The vertical barbell acceleration to achieve v_{thres} can be expressed as:

$$a_{thres} = \frac{{v_{thres}}^2}{2 \times h_{acc}} \tag{6}$$

To prove the predictive accuracy of the model, the computed $snatch_{th}$ was compared with the actual snatch performance of the 1RM test.

2.4. Statistical Analysis

The level of statistical significance was set for all tests at $p \leq 0.05$. All statistical analyses were conducted using R (version 4.0.2). A list of the applied packages and functions can be found in the Supplementary Materials. The normal distribution of data was assessed and confirmed using the Shapiro–Wilk test. The absence of heteroscedasticity (i.e., the measurement error is related to the magnitude of the measured variable) of the measurements was confirmed using the Breusch–Pagan test. Therefore, no log-transformation of the raw data was necessary. The difference between the 1RM snatch and $snatch_{th}$ was analyzed using a paired-sample t-test alongside effect size (d) and 95% confidence limits (CL). The effect size was interpreted using conventions outlined by Hopkins [22] as small ($|d| > 0.2$), moderate ($|d| > 0.6$), large ($|d| > 1.2$), very large ($|d| > 2.0$), or extremely large ($|d| > 4.0$). An effect size <0.2 was deemed trivial. The correlation between the 1RM snatch and the $snatch_{th}$ was assessed using Pearson product-moment correlation coefficient (r) with 95% CL. Thresholds for the correlation coefficient were considered small ($|r| > 0.1$), medium ($|r| > 0.3$), large ($|r| > 0.5$), very large ($|r| > 0.7$), and extremely large ($|r| > 0.9$) [23]. The absolute and percentage measurement error was computed as SDD and SDD% with 95% CL. Additionally, the Bland–Altman analysis was used to compute the systematic bias (i.e., mean of measurements with 95% CL) and 95% limits of agreement (systematic bias $\pm$ 1.96 $\times$ SDD) with 95% CL. Significant systematic bias was prevalent if the range of the 95% CL of the mean difference did not contain the value 0. Furthermore, a Deming regression was performed to test for constant and proportional bias between the approaches. Significant constant bias was present if the range of the 95% CL of the intercept did not contain the value 0 and significant proportional bias was present if the range of the 95% CL of the slope did not contain the value 1 [24].

3. Results

Descriptive data are displayed in Table 2. The main finding indicated a non-significant ($p = 0.706$) and trivial ($d = 0.01$) mean difference between the actual 1RM snatch performance and the $snatch_{th}$ (Table 3).

Table 2. Descriptive data in means and standard deviations for the actual 1RM snatch, $snatch_{th}$, snatch pull, and snatch pull FvR_2.

1RM Snatch							**$Snatch_{th}$**
actual 1RM (kg)			v_{thre} ($m{\cdot}s^{-1}$)			predicted 1RM (kg)	
113.6 ± 31.1			1.97 ± 0.12			113.4 ± 32.0	
Snatch Pull					**Snatch Pull FvR_2**		
load @80% (kg)	load @110% (kg)	v_{max} @80% ($m{\cdot}s^{-1}$)	v_{max} @110% ($m{\cdot}s^{-1}$)	h_{acc} @110% (m)	$\overline{v}_0$ ($m{\cdot}s^{-1}$)	$\overline{F}_0$ (N)	$\overline{P}_{max}$ (W)
91.3 ± 26.1	123.5 ± 33.7	2.30 ± 0.14	1.83 ± 0.15	0.80 ± 0.08	2.20 ± 0.32	2569.4 ± 613.2	1408.7 ± 377.2

Notes: 1RM = one-repetition maximum of the snatch, v_{thres} = maximal vertical barbell velocity of the 1RM snatch, $snatch_{th}$ = theoretical 1RM snatch computed from the two-point snatch pull force–velocity relationship (FvR_2), load @80/110% = snatch pull barbell load relative to the 1RM snatch, v_{max} = maximal vertical barbell velocity during the snatch pull, h_{acc} = distance of vertical acceleration during the snatch pull, $\overline{v}_0$ = theoretical maximal mean vertical barbell velocity of snatch pull FvR_2, $\overline{F}_0$ = theoretical maximal mean vertical barbell force of snatch pull FvR_2, $\overline{P}_{max}$ = theoretical maximal mean vertical barbell power of snatch pull FvR_2.

Table 3. Comparison between the actual 1RM snatch and the $snatch_{th}$ computed from the snatch pull FvR_2.

$t_{(7)}$	p	d (95% CL)	r (95% CL)	Diff (%)	SDD (95% CL) (kg)	SDD% (%)
0.393	0.706	0.01 (−0.02;0.04)	0.99 (0.99;1.00)	0.18	1.5 (1.0;3.1)	1.3

Notes: t = t-score from the paired-sample t-test, p = p-value from the paired-sample t-test, d = effect size, r = Pearson product-moment correlation coefficient, diff = percentage difference, SDD = standard deviation of differences, SDD% = percentage standard deviation of differences, 95% CL = 95% confidence limits.

Results showed an extremely large correlation between the actual 1RM snatch and $snatch_{th}$ ($r = 0.99$) (Table 3). The outcomes of the Deming regression did not reveal a constant bias (i.e., CL of intercept contained the value 0) or proportional bias (i.e., CL of slope contained the value 1). In addition, the Bland–Altman analysis did not show any significant systematic bias (i.e., CL of the mean differences contained the value 0) (Figure 2). The SDD amounted to 1.5 kg, and the limits of agreement ranged from −2.7 to 3.2 kg.

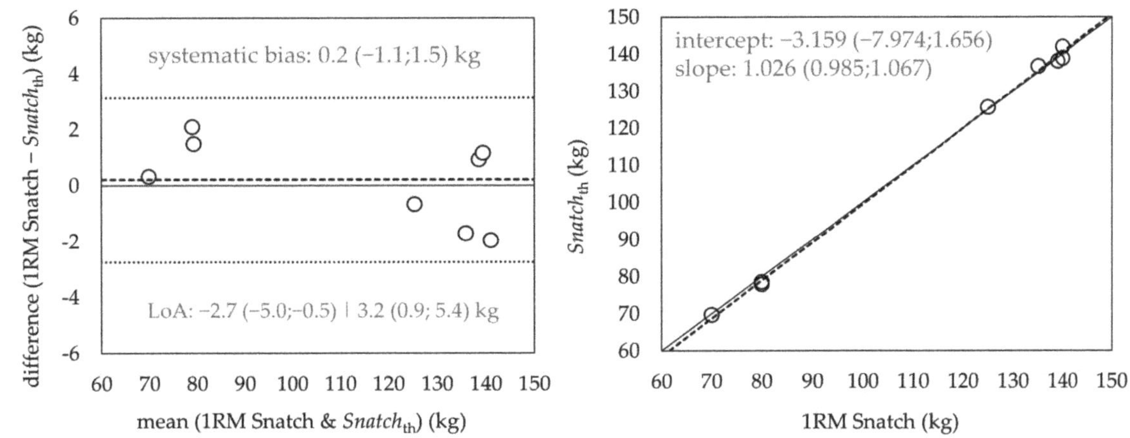

Figure 2. Results of the Bland–Altman analysis (**left**) and Deming regression (**right**) for the comparison between actual 1RM snatch and $snatch_{th}$. The Deming regression plot illustrates the fitted linear model (dashed line) and the identity line (snatch 1RM = $snatch_{th}$, slope = 1) (solid line). The Bland–Altman plot depicts mean differences between actual 1RM snatch and $snatch_{th}$ (dashed line) and 95% limits of agreement (dotted lines). Slope and intercept in the Deming regression plot are reported with 95% CL. The mean difference in the Bland–Altman plot is reported with 95% CL, and limits of agreement (LoA) are reported as systematic bias $\pm$ 1.96 $\times$ SDD with 95% CL. $Snatch_{th}$ = theoretical 1RM of the snatch computed from the snatch pull FvR_2.

4. Discussion

This study aimed to examine the predictive validity of the $snatch_{th}$ computed from the snatch pull FvR_2 to determine the actual 1RM snatch performance of male and female elite weightlifters. It was hypothesized that the $snatch_{th}$ shows high accuracy to predict actual 1RM snatch performance in elite male and female weightlifters. Given that the main findings of this indicated high predictive validity of the $snatch_{th}$ to determine 1RM snatch performance in elite weightlifters, the study hypothesis was confirmed.

The ability of a testing protocol to accurately predict performance in the "real world" setting is a key aspect [7]. In this context, it has been shown that 1RM weightlifting performance can accurately be predicted for the snatch (SEE = 3.8 kg), clean and jerk (SEE = 6.0 kg), and total (SEE = 9.5 kg) using a multiple linear regression model with vertical jump peak power and isometric mid-thigh pull peak force data as predictors [8]. Besides statistical modeling, LvR-profiling is a widely accepted approach to predict 1RM performance that has, in fact, not yet been used for weightlifting competition exercises (i.e., snatch, clean and jerk). However, being similar to the snatch and clean, recent research was conducted on the prediction accuracy of 1RM in the power clean and deadlift exercise using individual LvR profiles [11,12,25]. For instance, Haff, Garcia-Ramos, and James [11] used the individual LvR to predict the 1RM power clean performance in recreationally trained males aged 26 years. These authors reported a measurement error (systematic bias $\pm$ SDD) of 1.4 $\pm$ 7.2 kg. Additionally, the individual LvR was used for the deadlift exercise to predict 1RM performance in resistance trained men aged 24 years [12,25]. Results indicated measurement errors (systematic bias $\pm$ SDD) of 0.6 $\pm$ 8.5 kg [25] and 0.7 $\pm$ 4.7 kg [12].

Of note, the prediction accuracy of the 1RM snatch from the snatch pull FvR_2 model (i.e., $snatch_{th}$) reported in the current study (systematic bias $\pm$ SDD = 0.2 $\pm$ 1.5 kg) is higher compared with the abovementioned studies that used either the multiple linear regression model [8] or the LvR-profiling [11,12,25]. The low measurement error of $snatch_{th}$ can be attributed to three primary reasons. First, the reliability of $snatch_{th}$ has been shown to be very high (SEM% = 0.71%, ICC = 0.99) [16]. This is due to the high reliability of the barbell tracking system used (i.e., Realanalyzer video analysis software) and to the high technical mastery of elite weightlifters. Second, $snatch_{th}$ was computed using the FvR_2-profile of the snatch pull exercise (i.e., test exercise $\neq$ target exercise). During the snatch pull exercise, loads can be used that exceed the 1RM snatch load (i.e., overload) [26]. Therefore, the 1RM prediction using the snatch pull FvR_2 linear regression model is an interpolation (i.e., 1RM condition is located within the measurements). In contrast, all LvR-based 1RM predictions were extrapolations (i.e., 1RM condition is located outside of measurements) given that the test exercise is the same as the target exercise for determining the 1RM. In fact, there is evidence that linear regression-based interpolations are less prone to measurement errors of independent variables than linear regression-based extrapolations [27]. Third, considering the inherent characteristics of the exercises, the snatch pull is a ballistic exercise (i.e., maximal effort lift to achieve maximal barbell velocity), while the deadlift is non-ballistic [28]. It has been shown during LvR-profiling that the reliability of peak and mean barbell velocity is lower in the bench press (i.e., non-ballistic) than in the bench press throw (i.e., ballistic), especially with light loads (e.g., 20% 1RM) [29]. The lower reliability of barbell velocity in a non-ballistic exercise will ultimately affect the slope of the LvR and the prediction of a 1RM load. In theory, the same principle can be applied for the non-ballistic deadlift, explaining the higher random error component (i.e., SDD) for 1RM prediction using LvR in the studies of Benavides-Ubric, Díez-Fernández, Rodríguez-Pérez, Ortega-Becerra and Pareja-Blanco [12] and Jukic, García-Ramos, Malecek, Omcirk and Tufano [25]. In contrast, although the power clean is considered a ballistic exercise [28], very light barbell loads may not be performed at maximal effort due to the catch phase of the power clean (i.e., barbell is "caught" on the shoulders). For the snatch pull, very light barbell loads may limit the lifters' ability to generate maximal force output due to discomfort controlling the barbell at very high velocities [16]. In this case, the limited force output results in a submaximal barbell velocity output that influences the individual LvR slope and the accuracy of the 1RM prediction. In the study of Haff, Garcia-Ramos, and James [11], very light loads (i.e., starting at 30% of 1RM power clean with peak barbell velocities of $3.29\ m\cdot s^{-1}$) were used as initial load stages during LvR-profiling. For the same study, the aforementioned relation (i.e., submaximal barbell velocity output at very light loads) may explain the high random error component reported for the prediction of 1RM power clean using LvR-profiling.

This study presents some limitations that warrant discussion. First, the computed $snatch_{th}$ is based on the concept of threshold velocity (i.e., v_{thres}) for a 1RM snatch [30]. This individual v_{thres} needs to be precisely assessed during 1RM snatch lifts with the same measurement device as used during the snatch pull FvR_2 testing. In addition, since barbell kinematics during weightlifting exercises can differ between barbell sides [17], all measurements need to be done at a standardized barbell side. Second, all lifts for the snatch pull FvR_2 need to be executed at maximal effort. Any submaximal effort will affect the slope of the FvR_2 and thus the model output (i.e., $snatch_{th}$). Third, the size of the sample is rather small. However, the size of the overall population of elite weightlifters is small, resulting in a reduced sample size. Of note, the recruited sample is heterogeneous, including men and women with a wide range of 1RM snatch performances (70–140 kg). Such heterogeneity strengthens the external validity of the computed $snatch_{th}$ from the snatch pull FvR_2. Finally, the presented approach is based on measurements from a custom-made video tracking software (i.e., Realanalyzer) with limited access. However, the computational base can also be used with other measurement units for the assessment of barbell kinematics. However, this would require another study to validate the accuracy of the $snatch_{th}$-model using the new test set up.

5. Conclusions

In conclusion, this study provides a new approach to estimate 1RM snatch performance in elite weightlifters using the snatch pull FvR_2 (i.e., $snatch_{th}$). Our findings demonstrated that the $snatch_{th}$-model accurately predicts 1RM snatch performance with a reduced random error of ± 1.5 kg.

The findings of our study have high practical relevance for weightlifting in particular and strength and conditioning from a broader perspective. More specifically, the snatch pull FvR_2 approach using 2D video analysis is an easy-to-administer test that can regularly be applied during weightlifting training as a valid alternative to the 1RM snatch test to assess individualized progression in weightlifting performance over time. However, when using the approach in practice, the abovementioned limitations need to be considered when interpreting the results. Given the comparable biomechanical structure of the snatch and the clean during the acceleration phase (i.e., pull), the current approach could be adopted for the clean pull exercise to predict the 1RM clean and/or 1RM power clean performance. However, this still needs to be confirmed in future studies.

Supplementary Materials: The following are available online at https://www.mdpi.com/article/10.3390/jfmk6020035/s1, Document S1: Statistics_R.

Author Contributions: Conceptualization, I.S., H.C. and U.G.; methodology, I.S.; validation, I.S., H.C. and U.G.; investigation, I.S.; software, I.S.; writing—original draft preparation, I.S.; writing—review and editing, I.S., H.C. and U.G. All authors have read and agreed to the published version of the manuscript.

Funding: This study is part of the research project (KT 1-17) at the Institute for Applied Training Science and was funded by the German Federal Ministry of the Interior, Building and Community.

Institutional Review Board Statement: The study was conducted according to the guidelines of the Declaration of Helsinki and approved by the Institutional Review Board of the Institute for Applied Trainings Science (approval number: ER_2020.28.09_4).

Informed Consent Statement: Informed consent was obtained from all subjects involved in the study.

Data Availability Statement: All underlying data are in the text, tables and figures.

Acknowledgments: Many thanks to the athletes who participated in this study and the German Weightlifting Federation (BVDG) for supporting this investigation.

Conflicts of Interest: The authors declare no conflict of interest. The funders had no role in the design of the study; in the collection, analyses, or interpretation of data; in the writing of the manuscript, or in the decision to publish the results.

References

1. Bazyler, C.D.; Mizuguchi, S.; Zourdos, M.C.; Sato, K.; Kavanaugh, A.A.; DeWeese, B.H.; Breuel, K.F.; Stone, M.H. Characteristics of a National Level Female Weightlifter Peaking for Competition: A Case Study. *J. Strength Cond. Res.* **2018**, *32*, 3029–3038. [CrossRef]
2. McGuigan, M.R.; Kane, M.K. Reliability of performance of elite Olympic weightlifters. *J. Strength Cond. Res.* **2004**, *18*, 650–653. [CrossRef] [PubMed]
3. Travis, S.K.; Goodin, J.R.; Beckham, G.K.; Bazyler, C.D. Identifying a test to monitor weightlifting performance in competitive male and female weightlifters. *Sports* **2018**, *46*, 46. [CrossRef] [PubMed]
4. Beckham, G.; Mizuguchi, S.; Carter, C.; Sato, K.; Ramsey, M.; Lamont, H.; Hornsby, G.; Haff, G.; Stone, M. Relationships of isometric mid-thigh pull variables to weightlifting performance. *J. Sports Med. Phys. Fit.* **2013**, *53*, 573–581.
5. Shetty, A.B. Quantification of selected segmental strengths in weightlifting. *J. Appl. Sport Sci. Res.* **1990**, *4*, 37–41.
6. Mokkink, L.B.; Terwee, C.B.; Patrick, D.L.; Alonso, J.; Stratford, P.W.; Knol, D.L.; Bouter, L.M.; de Vet, H.C. The COSMIN study reached international consensus on taxonomy, terminology, and definitions of measurement properties for health-related patient-reported outcomes. *J. Clin. Epidemiol.* **2010**, *63*, 737–745. [CrossRef]
7. McGuigan, M. *Monitoring Training and Performance in Athletes*; Human Kinetics: Champaign, IL, USA, 2017.
8. Joffe, S.A.; Tallent, J. Neuromuscular predictors of competition performance in advanced international female weightlifters: A cross-sectional and longitudinal analysis. *J. Sports Sci.* **2020**, *38*, 985–993. [CrossRef] [PubMed]
9. Morin, J.B.; Jiménez-Reyes, P.; Brughelli, M.; Samozino, P. When Jump Height is not a Good Indicator of Lower Limb Maximal Power Output: Theoretical Demonstration, Experimental Evidence and Practical Solutions. *Sports Med.* **2019**, *49*, 999–1006. [CrossRef] [PubMed]
10. Thomas, J.R.; Nelson, J.K.; Silverman, S.J. *Research Methods in Physical Activity*, 7th ed.; Thomas, J.R., Nelson, J.K., Silverman, S.J., Eds.; Human Kinetics: Champaign, IL, USA, 2015; pp. 359–401.
11. Haff, G.G.; Garcia-Ramos, A.; James, L.P. Using Velocity to Predict the Maximum Dynamic Strength in the Power Clean. *Sports* **2020**, *8*, 129. [CrossRef]

12. Benavides-Ubric, A.; Díez-Fernández, D.M.; Rodríguez-Pérez, M.A.; Ortega-Becerra, M.; Pareja-Blanco, F. Analysis of the load-velocity relationship in deadlift exercise. *J. Sports Sci. Med.* **2020**, *19*, 452–459.
13. Caven, E.J.G.; Bryan, T.J.E.; Dingley, A.F.; Drury, B.; Garcia-Ramos, A.; Perez-Castilla, A.; Arede, J.; Fernandes, J.F.T. Group versus individualised minimum velocity thresholds in the prediction of maximal strength in trained female athletes. *Int. J. Environ. Res. Public Health* **2020**, *17*, 7811. [CrossRef] [PubMed]
14. Picerno, P.; Iannetta, D.; Comotto, S.; Donati, M.; Pecoraro, F.; Zok, M.; Tollis, G.; Figura, M.; Varalda, C.; Di Muzio, D.; et al. 1RM prediction: A novel methodology based on the force-velocity and load-velocity relationships. *Eur. J. Appl. Physiol.* **2016**, *116*, 2035–2043. [CrossRef]
15. Bosquet, L.; Porta-Benache, J.; Blais, J. Validity of a commercial linear encoder to estimate bench press 1 RM from the force-velocity relationship. *J. Sports Sci. Med.* **2010**, *9*, 459–463. [PubMed]
16. Sandau, I.; Chaabene, H.; Granacher, U. Validity and reliability of a snatch pull test to model the force-velocity relationship in male elite weightlifters. *J. Strength Cond. Res.* **2021**. [CrossRef]
17. Sandau, I.; Jentsch, H.; Bunk, M. Realanalyzer HD—A real-time barbell tracking software for weightlifting. *EWF Sci. Mag.* **2019**, *5*, 14–23.
18. Sandau, I.; Lippmann, J.; Jentsch, H.; Seidel, I. Reliabilität des Realanalyzer, ein 2-D-Messsystems zur Echtzeiterfassung der Hantelbewegung im Gewichtheben. In Proceedings of the "NeuroMotion" Aufmerksamkeit, Automatisierung, Adaptation, Münster, Germany, 21–23 March 2012; p. 126.
19. Samozino, P.; Morin, J.B.; Hintzy, F.; Belli, A. A simple method for measuring force, velocity and power output during squat jump. *J. Biomech.* **2008**, *41*, 2940–2945. [CrossRef] [PubMed]
20. Samozino, P.; Edouard, P.; Sangnier, S.; Brughelli, M.; Gimenez, P.; Morin, J.B. Force-velocity profile: Imbalance determination and effect on lower limb ballistic performance. *Int. J. Sports Med.* **2014**, *35*, 505–510. [CrossRef]
21. Samozino, P.; Rejc, E.; Di Prampero, P.E.; Belli, A.; Morin, J.B. Optimal force-velocity profile in ballistic movements-altius: Citius or fortius? *Med. Sci. Sports Exerc.* **2012**, *44*, 313–322. [CrossRef]
22. Hopkins, W.G. Spreadsheets for analysis of validity and reliability. *Sportscience* **2015**, *19*, 36–42.
23. Hopkins, W.G.; Marshall, S.W.; Batterham, A.M.; Hanin, J. Progressive statistics for studies in sports medicine and exercise science. *Med. Sci. Sports Exerc.* **2009**, *41*, 3–13. [CrossRef]
24. Payne, R.B. Method comparison: Evaluation of least squares, Deming and Passing/Bablok regression procedures using computer simulation. *Ann. Clin. Biochem.* **1997**, *34 Pt 3*, 319–320. [CrossRef]
25. Jukic, I.; García-Ramos, A.; Malecek, J.; Omcirk, D.; Tufano, J.J. Validity of Load-Velocity Relationship to Predict 1 Repetition Maximum During Deadlifts Performed With and Without Lifting Straps: The Accuracy of Six Prediction Models. *J. Strength Cond. Res.* **2020**. [CrossRef]
26. Suchomel, T.J.; Comfort, P.; Stone, M.H. Weightlifting pulling derivatives: Rationale for implementation and application. *Sports Med.* **2015**, *45*, 823–839. [CrossRef] [PubMed]
27. Rivière, J.R.; Rossi, J.; Jimenez-Reyes, P.; Morin, J.B.; Samozino, P. Where does the one-repetition maximum exist on the force-velocity relationship in squat? *Int. J. Sports Med.* **2017**, *38*, 1035–1043. [CrossRef] [PubMed]
28. French, D.N. Advanced power techniques. In *Developing Power*; McGuigan, M., Ed.; Human Kinetics: Champaign, IL, USA, 2017; pp. 177–198.
29. García-Ramos, A.; Haff, G.G.; Padial, P.; Feriche, B. Reliability of power and velocity variables collected during the traditional and ballistic bench press exercise. *Sports Biomech.* **2018**, *17*, 117–130. [CrossRef]
30. Sandau, I.; Granacher, U. Effects of the barbell load on the acceleration phase during the snatch in elite Olympic weightlifting. *Sports* **2020**, *8*, 59. [CrossRef] [PubMed]

Journal of
Functional Morphology and Kinesiology

MDPI

Article

Evaluation of Strength and Muscle Activation Indicators in Sticking Point Region of National-Level Paralympic Powerlifting Athletes

Felipe J. Aidar [1,2,3,4], Filipe Manuel Clemente [5,6], Dihogo Gama de Matos [3,7], Anderson Carlos Marçal [4], Raphael Fabrício de Souza [1,2], Osvaldo Costa Moreira [8], Paulo Francisco de Almeida-Neto [9], José Vilaça-Alves [10], Nuno Domingos Garrido [10], Jymmys Lopes dos Santos [2], Ian Jeffreys [11], Frederico Ribeiro Neto [12], Victor Machado Reis [9], Breno Guilherme de Araújo Tinoco Cabral [8], Thomas Rosemann [13] and Beat Knechtle [13,14,*]

1 Department of Physical Education, Federal University of Sergipe (UFS), São Cristovão 49100-000, Sergipe, Brazil; fjaidar@gmail.com (F.J.A.); raphaelctba20@hotmail.com (R.F.d.S.)
2 Group of Studies and Research of Performance, Sport, Health and Paralympic Sports-GEPEPS, The Federal University of Sergipe (UFS), São Cristovão 49100-000, Sergipe, Brazil; jymmyslopes@yahoo.com.br
3 Graduate Program of Physiological Science, Federal University of Sergipe (UFS), São Cristovão 49100-000, Sergipe, Brazil; dihogogmc@hotmail.com
4 Graduate Program of Physical Education, Federal University of Sergipe (UFS), São Cristovão 49100-000, Sergipe, Brazil; acmarcal@yahoo.com.br
5 Escola Superior Desporto e Lazer, Instituto Politécnico de Viana do Castelo, Rua Escola Industrial e Comercial de Nun'Álvares, 4900-347 Viana do Castelo, Portugal; filipe.clemente5@gmail.com
6 Instituto de Telecomunicações, Delegação da Covilhã, 1049-001 Lisboa, Portugal
7 Cardiovascular & Physiology of Exercise Laboratory, University of Manitoba, Winnipeg, MB R3T 2N2, Canada
8 Campus of Florestal, University Federal of Viçosa (UFV), Florestal 36570-900, Minas Gerais, Brazil; ocostamoreira@gmail.com (O.C.M.); brenotcabral@gmail.com (B.G.d.A.T.C.)
9 Department of Physical Education, Federal University of Rio Grande do Norte, Natal 59078-970, Rio Grande do Norte, Brazil; paulo220911@hotmail.com (P.F.d.A.-N.); victormachadoreis@gmail.com (V.M.R.)
10 Research Center in Sports Sciences, Health Sciences and Human Development (CIDESD), Trás os Montes and Alto Douro University, 5001-801 Vila Real, Portugal; josevilaca@utad.pt (J.V.-A.); ndgarrido@gmail.com (N.D.G.)
11 Faculty of Life Sciences and Education, University of South Wales, Pontypridd, Wales CF37 5UP, UK; ian.jeffreys@allproperformance.co.uk
12 Paralympic Sports Program, SARAH Rehabilitation Hospital Network, Brasilia 71535-005, Distrito Federal, Brazil; fredribeironeto@sarah.br
13 Institute of Primary Care, University of Zurich, 8091 Zurich, Switzerland; thomas.rosemann@usz.ch
14 Medbase St. Gallen Am Vadianplatz, 9001 St. Gallen, Switzerland
* Correspondence: beat.knechtle@hispeed.ch

Citation: Aidar, F.J.; Clemente, F.M.; Matos, D.G.d.; Marçal, A.C.; de Souza, R.F.; Moreira, O.C.; Almeida-Neto, P.F.d.; Vilaça-Alves, J.; Garrido, N.D.; dos Santos, J.L.; et al. Evaluation of Strength and Muscle Activation Indicators in Sticking Point Region of National-Level Paralympic Powerlifting Athletes. *J. Funct. Morphol. Kinesiol.* **2021**, *6*, 43. https://doi.org/10.3390/jfmk6020043

Academic Editors: Cristina Cortis, Andrea Fusco and Carl Foster

Received: 22 April 2021
Accepted: 12 May 2021
Published: 15 May 2021

Abstract: Background: The sticking region is considered an intervening factor in the performance of the bench press with high loads. Objective: To evaluate the strength indicators in the sticking point region in Powerlifting Paralympic athletes. Methods: Twelve Brazilian Powerlifting Paralympic athletes performed maximum isometric force (MIF), rate of force development (RFD), time at MIF, velocity, dynamic time in sticking, and surface electromyography in several distances from the bar to the chest. Results: For velocity, there was a difference between the pre-sticking and sticking region (1.98 ± 0.32 and 1.30 ± 0.43, $p = 0.039$) and dynamic time between the pre-sticking and the sticking region (0.40 ± 0.16 and 0.97 ± 0.37, $p = 00.021$). In static test for the MIF, differences were found between 5.0 cm and 15.0 cm (CI 95% 784; 1088; $p = 0.010$) and between 10.0 cm and 5.0 cm (CI 95% 527; 768; $p < 0.001$). Regarding the RFD, differences were found (CI 95% 938; 1240; $p = 0.004$) between 5.0 cm and 25.0 cm and between 10.0 cm and 25.0 cm (CI 95% 513; 732; $p < 0.001$). In relation to time, there were differences between 5.0 cm and 15.0 cm (CI 95% 0.330; 0.515; $p < 0.001$), 5.0 cm, and 25.0 cm (CI 95% 0.928; 1.345; $p = 0.001$), 10.0 cm and 15.0 cm ($p < 0.05$) and 15.0 cm and 25.0 cm ($p < 0.05$). No significant differences were observed between the muscles in electromyography, although the triceps showed the highest muscle activation values. Conclusions: The maximum isometric force, rate of force development, time, velocity, and dynamic time had lower values, especially in the initial and intermediate phases in the sticking region.

Keywords: strength indicators; sticking point; Powerlifting Paralympic

1. Introduction

The Powerlifting Paralympic (PP) is part of the sports calendar at the Paralympics, being a strength sport modality, adapted from conventional powerlifting, in which the athlete only performs the bench press. The bench press (BP) is a popular exercise, fairly simple to perform. In accordance with the rules and regulations set by the International Powerlifting Federation (IPF), the lift is performed lying on a bench with the head, shoulders, and buttocks in contact with the bench surface and the feet flat on the floor [1]. The lift begins with arms upright, the elbows stuck out. The barbell is then dropped to the chest or abdomen region until it is forced into straight arms and the elbows are locked out again at the end of the lift.

When performing the bench press at heavy load Madsen and McLaughlin [2] found that in the upward stroke of the bar, there is a point at which the lift slows down or even stops, before accelerating again. Particularly, during a raise at near-maximum loads (>80 percent of 1RM), there is a period where the upward barbell rotation decelerates or even ceases entirely for a brief period [3,4]. This period in which the driving force is less than gravity, leading to the slowdown of the barbell [4], is referred to as the "sticking period" or the "sticking region" [4,5]. This sticking area is also seen as the weakest link in the upward transition—a point at which the lift is likely to collapse [6]. The cause of this sticking region, however, remains unclear. Elliott et al. [5] found that the sticking area was not caused by a rise in the resulting moment arm on the shoulder or elbow or by a reduction in the muscle activity of the prime movers in that area. The leading interpretation model for the sticking zone is based on the assumption that the sticking area is a mechanically disadvantageous region for the development of force [7].

In this context, a possible interpretation of the sticking point is as electromyographic muscle activity (EMG). Few studies have analyzed muscle activity during the sticking point region [4,7]. Van den Tillaar and Ettema [4], concluded that a plausible explanation for the presence of the stuck phase is the reducing potentiation of the contractile elements during the upward step, along with the reduced activity of the pectoral and deltoid muscles during that time. However, in another study by Van den Tillaar, Saeterbakken, and Ettema [7], only the triceps increase in muscle activity during maximal lifting from the sticking to the post-sticking region. These studies did not investigate the effect of several different distances of the bar to the chest upon the muscle activation and sticking region. Investigating the different distances of the bar to the chest and muscle activation during the sticking region of these distances, which is a typical training method to increase maximal strength, would give information regarding whether the sticking region is influenced by the distance.

To our knowledge, no investigation has examined the sticking region and neuromuscular changes within and between the muscles in different distances of the bar to the chest in powerlifting athletes. In addition, possible changes in muscle activation within and between muscles during these distances can help us understand which muscles would be responsible for moving the barbell through the sticking region.

On the other hand, when evaluating the PP, where the legs are extended on the bench during the execution of the movement, situations such as the width of the grip on the bar [8] have been evaluated; however, a hypothesis that has been ignored is that a Paralympian's transfer of force could be impaired making it difficult to maintain strength, power, and velocity and impairing the performance of neuromotor skills needed for lifting [1,9]. Thus, we raised the hypothesis, that due to the lifting position, with the lower limbs on the bench, and the physical deficiencies, which tend to compromise stabilization during the lifting, that the sticking point could be altered in these para-athletes. Therefore, the present study aimed to evaluate the strength indicators and the sticking point region in different distances of the bar to the chest in elite Powerlifting Paralympic athletes.

2. Materials and Methods

The athletes performed the tests of 1RM, velocity maximum (V_{max}), maximum isometric force (MIF), rate of force development (RFD), time at MIF (Time), and surface electromyography (sEMG), distributed over five weeks as shown in Figure 1 (experimental design, weekly schedule of tests). At the beginning of each test day, the athletes performed a previous warm-up for the upper limbs, with three sets of 10 to 20 repetitions in approximately 20 min [10]. Then, a specific warm-up was performed on the bench press with 30% of the load for 1RM, where 10 slow repetitions and 10 fast repetitions were performed to start the procedure. These repetitions in a full range of motion served as a warm-up for the tests. During the test, the athletes received verbal encouragement to make the maximum effort. Each test followed its specific execution protocol, as previously described (Figure 2).

		Day 1	**Day 2**	**Other Days**
Week 1 (Familiarization) Day 1—1MR and velocity test Day 2—MIF, RDF, time, and sEMG test	→	1RM and velocity test Familiarization	sEMG, MIF, RDF, and time test	Rest
Week 2 and 3 MIF, RDF, time, and sEMG test	→	sEMG, MIF, RDF, and time test Distance from the bar to the chest of 5.0; 10.0; 15.0; or 25.0 cm	sEMG, MIF, RDF, and time test Distance from the bar to the chest of 5.0; 10.0; 15.0; or 25.0 cm	Rest
Week 4 and 5 Velocity and image test	→	Velocity and image test Distance from the bar to the chest of 5.0; 10.0; 15.0; or 25.0 cm	Velocity and image test Distance from the bar to the chest of 5.0; 10.0; 15.0; or 25.0 cm	Rest

Figure 1. Experimental design: Weekly test schedule. Legend: 1RM: 1 Repetition Maximum; SEMG: surface electromyography; MIF: Maximum Isocmetric Force; RFD: Rate of Force Development; Time: Time to MIF.

(**a**) Bar 5.0 cm from the chest (**b**) Bar 10 cm from the chest

(**c**) Bar 15 cm from the chest (**d**) Bar 20 cm from the chest

(**e**) photo of the athlete's position on the bench in competition, Brazilian Paralympic Committee

Figure 2. Schematic drawing with the different heights (**a–d**) of the bar to the chest and (**e**) photo of the athlete's position on the bench in competition.

2.1. Participants

The subjects were 12 Powerlifting Paralympic athletes participating in the extension project of the Federal University of Sergipe, Sergipe, Brazil. All participants were Brazilian national level competitors, eligible for the competition [1] and ranked among the ten best in their respective categories. Among the athletes, two athletes were vice-champions of world cups, the third place in a world cup, two were champions, and two third-placed at a national level. In the observed deficiencies, we had four athletes presenting spinal cord injury due to accidents with an injury below the eighth thoracic vertebra; two with sequelae due to polio; four had lower limb malformation (arthrogryposis); and two had cerebral palsy. The characterization of the sample is shown in Table 1, where descriptive statistics based on the mean and standard deviation of each variable were used.

It is noteworthy that the sample size was defined a priori considering the $\eta 2p$ value of 0.766 found in the study by Ettema [7] (when comparing the variable strength at different distances in a male sample, in which the authors verified that the vertical distance of the bar was significant for the performance of the straight supine in healthy men); thus, a sample power of 0.98 (very strong) was estimated for a minimum sample of 12 subjects. For sample power analysis we used the open-source software G* Power (Version 3.0; Berlin,

Germany) considering the *F* statistic with an alpha = 0.05 and a standard beta of 0.80. The analyses were repeated post hoc to expose the strength of the findings of the present study (details in the Section 3).

The athletes participated in the study voluntarily and signed a free and informed consent form, according to the Resolution 466/2012 of the National Research Ethics Commission-CONEP of the National Health Council, following the ethical principles expressed in the Helsinki Declaration (1964, reworded in 2013) of the World Medical Association. This study was approved by the Research Ethics Committee of the Federal University of Sergipe, CAAE: 2.637.882 (date of approval: 7 May 2018).

Table 1. Characterization of subjects.

Variables	Values
Age (years)	26.56 ± 5.55
Body Weight (Kg)	77.89 ± 24.60
Experience (years)	3.47 ± 0.27
Footprint Width (cm)	59.44 ± 17.77
1RM Bench Press Test (Kg)	135.56 ± 30.15 *
Sticking Point (cm)	10.01 ± 3.31 **
1RM/Body Weight	1.81 ± 0.35 ***

* All athletes with loads that keep them in the top 10 of their national categories. ** In the sticking region, the minimum value was 5.0 cm, and the maximum was 15.00 cm. *** Bench press values above 1.4 would be considered elite athletes, according to Ball and Wedman [11].

2.2. *Instruments*

Weighing of the athletes was performed on a digital type Michetti (Michetti, São Paulo, SP, Brazil) platform electronic scale, as this instrument facilitates the weighing of sitting athletes. The scale can hold a maximum weight of 300 kg and has dimensions of 1.50 × 1.50 m. For the bench press exercise, an official straight bench (Eleiko Sport AB, Halmstad, Sweden), approved by the International Paralympic Committee [1], with a total length of 210 cm was used. The IPC-approved powerlifting Olympic bar is serrated and has grooves in its material. They are 220 cm in total length, weighing 20 kg. The Olympic bar shall be marked with the narrowest and widest footprint, according to the International Paralympic Committee [1] official rules 2016–2017, ranging from 42 cm to 81 cm.

2.3. *Load Determination*

The 1RM test was performed, in which each subject started the attempts with a weight that he believed could lift only once using maximum effort. Weight increases were added until the maximum load that could be lifted once was reached. If the practitioner failed to perform a single repetition, 2.4% to 2.5% of the load used in the test was subtracted [12]. The subjects rested for 3–5 min between attempts. The test for determination of 1RM was made one week earlier, at least 48 h in advance so that it could determine the load percentage for the strength and power tests. The results are found in Table 1.

2.4. *Determining Sticking Point Height*

The sticking point occurs during the ascending (concentric) phase of the movement when performing the bench press exercise with a high, maximum, or sub-maximum load [4,5]. Thus, the bench press was performed following the standards of the IPC being performed with a downward motion until it touched the chest and then raising the bar until the elbows were extended. Stopping the bar at the chest tends to increase the occurrence of the sticking point region. It should be noted that isometric assessments in different bench press positions can be used to test the sticking region [4,5].

From the moment the value of 1RM was calculated, athletes rested 10 min [7,10] and then a load that was 2.0% to 2.5% of the load used in the test was added [12] until the subjects failed to raise the bar until the elbows were fully extended. The subjects rested for

3–5 min between attempts. From there, the distance from the bar to the chest where the athletes failed was determined. At this point, the sticking point was determined.

To confirm the height and also to evaluate the velocity in each phase, two phases (eccentric and concentric) were determined, and the concentric phase was divided into three periods, pre-sticking, sticking, and post-sticking. Between the eccentric and the concentric phase, there was a stop of the bar on the chest, which divided the movement and is mandatory in the Paralympic bench press mode [1].

2.5. Data Collection and Analysis

To evaluate the survey in the Bench Press Paralympic (vertical, horizontal, and displacement of the bar in the sagittal plane; a total distance of the bar path in the sagittal plane), dynamic time and the mean velocities of the bar were determined using a digitization software program (Kinovea 0.8.15, www.kinovea.org, accessed on 4 March 2020). The semi-automatic function was used to calculate the kinematic parameters, which was validated in practice [13,14]. For the collection of visual data, a digital video camera (frequency of 60 Hz and quality of 1080 *p*) was used. It was fixed at a height of 0.65 m and a distance of 1.25 m from the bar and aligned perpendicularly to the sagittal plane, for viewing the amplitude of movement. The movements were standardized, and the frames used were with the bar in the highest position (elbows extended) and the bar in the lowest position (against the chest). A 600% zoom was used to correctly identify the points (internal diameter of 2 mm) placed on the bar [13]. Two independent evaluators analyzed the videos, and when the difference was greater than 5%, a third evaluator examined the data.

2.6. Maximum Dynamic Velocity (V_{max}) and Dynamic Time (DTime)

The athletes were evaluated during the competitive phase of the season and were familiar with the testing procedures due to their constant training and testing routines. V_{Max} was defined as 80% of 1RM, using only the weightless bar [15]. A Muscle Lab Encoder (Model PFMA 3010e Muscle Lab System; Ergotest, Langesund, Norway) was used and fixed to the end of the bar. A supine channel was adopted in order not to allow an incline of more than 2.0°, which allowed a dynamic evaluation of high reliability [16].

2.7. Maximum Isometric Strength (MIF), Rate of Force Development Rate (RFD), and Time to MIF

The measure of maximum isometric strength (MIF) (N), rate of force development rate (RFD) ($N \cdot s^{-1}$), and time to MIF (Time) (s) were determined by a Chronojump load cell (Chronojump, BoscoSystem, Madrid, Spain) with a capacity of 500 Kg, the output impedance of 350 ± 3 ohm, insulation resistance over 2000 cc, input impedance 365 ± 5 ohms, 24-bit, and 80 Hz digital–analog converter. The equipment was fixed to the adapted bench, using Spider HMS Simond carabiners (Sigmond Chamonix, Chamonix Mont-Blanc, France) with a breaking load of 21 KN, approved for climbing by the Union Internationaledes Associations d'Alpinisme (UIAA). A steel chain with a breaking load of 2300 kg was used to secure the load cell to the seat. The perpendicular distance between the load cell and the center of the joint was determined [17]. The maximum isometric force (MIF) was measured by the maximum strength generated by the muscles of the upper limbs in the adapted supine exercise [1]. The MIF was determined in Newton (N) conceived by the formula N = (M) × (C), where M = mass in Kg and C = 9.80665, measured between the load cell cable fixation point and the seat of the bench press. For the isometric test, the elbows were positioned at an angle close to 90°, angulation was verified with a device for measuring the amplitude of the angular movement, model FL6010 (Sanny®, São Bernardo do Campo, Brazil). Participants were instructed to perform a single maximum movement looking for elbow extension (as quickly as possible) for evaluation of the MIF; it was determined that the subjects had to maintain the maximum contraction for 5.0 s. The rate of force development (RFD), on the other hand, was determined using the force to time ratio until reaching the maximum force (RFD = ΔStrength/ΔTime) ($N \cdot s^{-1}$). The time to maximum force was determined as the time to reach the MIF. The MIF, RFD, and Time were evaluated with adaptations of the methodology of Van den Tillaar, Saeterbankken, and Ettema [7].

2.8. Surface Electromiography (sEMG)

The electromyographic signals were captured on the dominant side, using double electrodes, positioned parallel to the muscle fibers, 2 cm from the center at the point of greatest muscle area of the following muscles: brachial triceps (long head), anterior deltoid, and in the sternal and clavicular portions of the pectoralis major, on both sides of the body. The ground electrode was positioned over the olecranon. The skin area where the electrodes were placed was previously shaved and cleaned with alcohol. The electrodes (11 mm contact diameter and a 2 cm center-to-center distance) were placed along the presumed direction of the underlying muscle fiber according to the recommendations by SENIAM [18]. For data acquisition, one set was used with one repetition and a maximum load of 100% 1RM. The marker function was used to define the data intervals for each height in the sticking region.

The equipment used was an electromyographic MIOTEC® with 8 channels (MIOTEC, Porto Alegre, RS, Brasil). Data were filtered (second-order Butterworth band-pass filter of 20–500 Hz; notch of 60 Hz). The signal amplitude was calculated through the mean square root (MSR), which was normalized by the percentage of the maximum voluntary isometric contraction (MVIC). MVIC acquisition occurred before the test was performed, and a lift was carried out that remained in an isometric state for 6.0 s. MVIC values were recorded by the equipment and used for normalization. The equipment program issues a report with the values after normalization that was used for analysis in this study, adapted from Golas et al. [19].

2.9. Statistics

Descriptive statistics were performed using measures of central tendency, mean (X) $\pm$ standard deviation (SD), and 95% confidence interval (95% CI). To verify the normality of the variables, the Shapiro–Wilk test was used given the sample size. The data for all variables analyzed were homogeneous and normally distributed. The one-way ANOVA test, with Bonferroni's post hoc correction, was performed to evaluate the differences between phases and distance from the bar to the chest. All statistical analyses were performed using the computerized package Statistical Package for the Social Science (SPSS), version 22.0. The level of significance was set at $p < 0.05$. To check the effect size, (partial eta-squared: $\eta 2p$) adopted values of low effect ($\leq$0.05), medium effect (0.05 to 0.25), high effect (0.25 to 0.50), and very high effect (>0.50) [20].

3. Results

Figure 3 shows the velocity in movement (A) Conventional bench press and (B) Paralympic bench press.

In Table 2 are the results of velocity and dynamic time in each phase of the Paralympic Bench Press, in the eccentric and concentric phases (pre-sticking, sticking, and post-sticking). It is noteworthy that for the results related to velocity (m/s), we made a post hoc sample calculation considering the $\eta 2p$ value of 0.439; thus, a sample power of 0.63 (moderate) was indicated. We did the same procedure for the dynamic time (s) variable considering the $\eta 2p$ value of 0.769; thus, a sample power of 0.96 (very strong) was indicated. For sample power post hoc analysis, we used the open-source software G* Power (Version 3.0; Berlin, Germany) considering the F statistic (ANOVA-one way) with an alpha = 0.05 and a standard beta of 0.80.

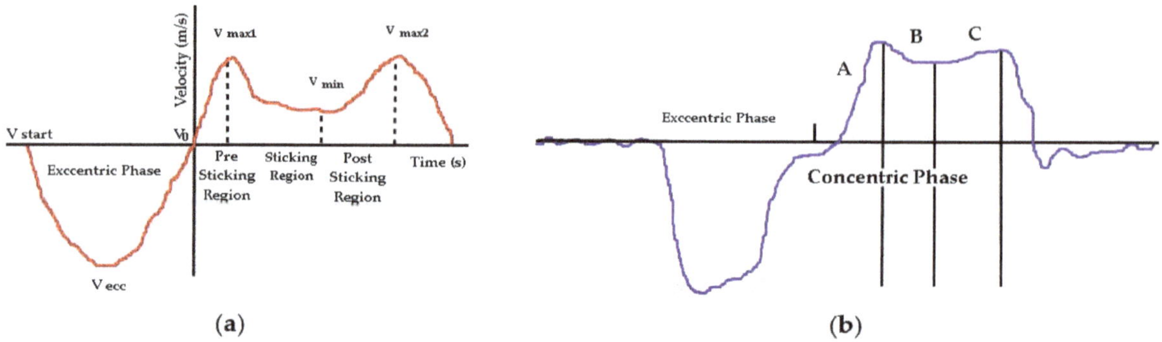

Figure 3. (**a**) Representative velocity–time curve during (sub) maximal bench press lifts (>85% of 1RM) in conventional Bench Press [21–23] and (**b**) representative velocity–time curve during (sub) maximal bench press lifts (95% of 1RM) in Paralympic Power Lifting with the different events and regions. Legend: V: velocity, V_0: at the beginning of the concentric phase, V ecc: Velocity in the eccentric phase, Vmax1: maximum velocity at point 1, Vmax2: maximum velocity at point 2, Vmin: Minimum velocity, (m/s): Meters per second, (s): Second; A: pre-sticking period, B: sticking period; C: post-sticking period.

Table 2. Mean ± SD values and 95% confidence interval (CI 95%) of velocity and time to achieve eccentric, pre-sticking, sticking, and post-sticking in Paralympic Powerlifting.

	Eccentric X ± DP (IC 95%)	Pre-Sticking X ± DP (IC 95%)	Sticking X ± DP (IC 95%)	Post-Sticking X ± DP (IC 95%)	p	η2p
Velocity (m/s)	1.62 ± 0.26 (1.42; 1.83)	1.98 ± 0.32 * (1.73; 2.23)	1.30 ± 0.43 * (0.97; 1.64)	1.86 ± 0.37 (1.58; 2.15)	* 0.039	0.439 #
Dynamic Time (s)	1.81 ± 0.50 a (1.43; 2.19)	0.40 ± 0.16 a,b (0.28; 0.52)	0.97 ± 0.37 a,b (0.68; 1.25)	0.85 ± 0.31 a (0.62; 1.09)	a < 0.001 b = 0.021	0.769 ##

* $p < 0.05$ (ANOVA, one-way); η2p = partial eta-squared; #, high effect (0.25 to 0.50); ##, very high effect (>0.50).

In Figure 4, it was found that in relation to the MIF, the results showed differences between 5.0 cm (CI 95% 492; 802) and 15.0 cm (CI 95% 610; 895). In addition, the findings demonstrate singular statistical significance for the different distances analyzed. (a) There was a significant difference between 5.0 cm and 25.0 cm (95% CI 784; 1088; $p = 0.010$). (b) There was a significant difference between 10.0 cm and 5.0 cm (95% CI 527; 768; $p < 0.001$). (c) Differences were found between 10.0 cm and 15.0 cm ($p = 0.012$), and between 15.0 cm and 25.0 cm. (d) $p = 0.012$; (e) $p = 0.007$).

For the RDF, differences ($p = 0.004$) between 5.0 cm and 25.0 cm (95% CI 938; 1240) were evidenced. There was also a significant difference between 10.0 cm and 25.0 cm (CI 95% 513; 732; $p < 0.001$). There were no significant differences between the other heights of the bar to the chest.

In relation to velocity, differences ($p < 0.05$) were found between 5.0 cm (95% CI 0.614; 0.724) and 10.0 cm (95% CI 0.140; 0.256). Thus, the results pointed out some peculiarities in relation to the analyzed distances: (a) Differences were found between 5.0 cm and 15.0 cm (CI 95% 0.330; 0.515; $p < 0.001$). (b) Differences were found between 5.0 cm and 25.0 cm (95% CI 0.928; 1.345; $p = 0.001$). Additionally, differences were found between 10.0 cm and 15.0 cm. (c) $p = 0.002$; (d) $p = 0.040$. Differences were found between 10.0 cm and 25.0 cm (e) $p < 0.001$; and differences were found between 15.0 cm and 25.0 cm (f) $p = 0.001$.

The surface electromyography (Figure 5) indicated greater activation of the brachial triceps in comparison with the other muscles, mainly at 10.0 cm and 15.0 cm distances. In addition, no significant differences were observed between the studied muscles and between the different distances from the bar to the chest.

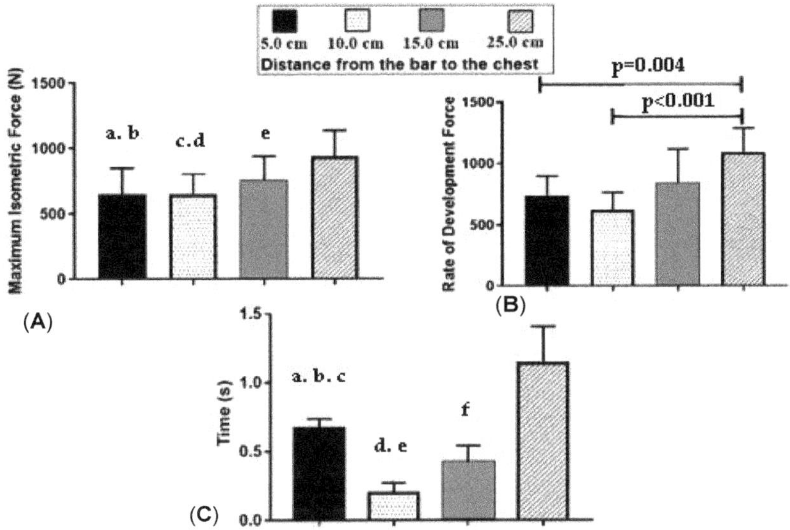

Figure 4. (**A**) Maximum isometric force (MIF), (**B**) rate of force development (RFD), and in relation to (**C**) time, with 5.0; 10.0; 15.0; and 25. cm of the distance from the bar to the chest.

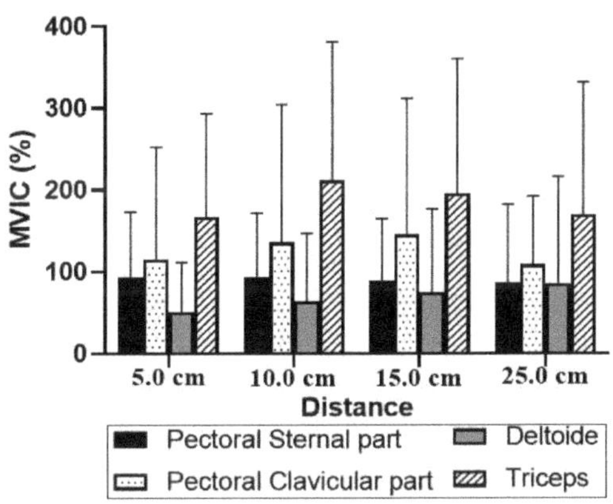

Figure 5. Electromyography of the pectoral, sternal and clavicular parts, deltoid and triceps at different distances from the chest. Legend: MVCI: maximal voluntary isometric contraction (%).

4. Discussion

The objective of the present study was to evaluate the strength indicators in the region of the sticking point in Paralympic weightlifting athletes. In this sense, the main findings were: (i) There was a significant difference for the maximum isometric strength between 5.0 cm and 15 cm, as well as between 5.0 cm and 10.0 cm. (ii) For the rate of force development, differences were found between 5.0 cm and 25.0 cm and between 10.0 cm and 25.0 cm. (iii) For velocity, differences between 5.0 cm and 15.0 cm, between 5.0 cm and 25.0 cm, and between 10.0 cm and 15.0 cm were evidenced. (iv) No significant differences were observed between muscles on electromyography, although the triceps showed the highest values of muscle activation. (v) The maximum isometric force, the rate of force development, and the velocity presented lower values, especially in the initial and intermediate phases in the sticking point region.

The maximum isometric strength showed significant differences between 5.0 and 15.0 cm ($p = 0.001$), between 5.0 and 25.0 cm ($p < 0.001$), between 10.0 and 15.0 cm ($p = 0.012$), and between 15.0 and 25.0 cm ($p = 0.007$). This shows an increase in strength between the initial and final distances—mainly after the sticking region. The study by Van den Tillaar, Saeterbankken, and Ettema [7] investigated the effects of performing the bench press in an isometric way at different moments from the chest bar (0 to 31 cm) and dynamically using 1RM. The study showed a decrease in strength between 4 cm and 13 cm away from the bar on the sternum bone. It also indicated that the FIM increased gradually between the sticking region (4.0 to 13.0 cm), surpassing the force generated by the regular bench press from the end of the sticking point and making it significant from 22.0 cm away from the chest.

Regarding the rate of force development (RFD), our study found significant differences between 5.0 and 25.0 cm ($p = 0.004$) and between 10.0 and 25.0 cm ($p < 0.001$). There were no significant differences between the other heights of the bar to the chest. Corroborating our findings, Drinkwater et al. [24] mention that during the concentric movement of the bench press, there is an initial high-power impulse after contact of the bar with the chest, which tends to be followed immediately by a moment of low power, called sticking point. Thus, a decrease in power and the rate of force development in high-intensity jobs tends to lead to movement failure. If we consider that the RFD would be the capacity to generate strength in less time, this capacity would be a good parameter with which to measure the neuromuscular efficiency of the athletes [25]. The results of our study demonstrate a greater RDF in further distances in the sticking region, which may be a justification for the failure in the region.

In this sense, the sticking point would be more associated with the perspective of the manifestation of strength and conditioning for preventing injuries, in addition to progress in strength adaptations for athletes [26]. Thus, considering that the Paralympic athletes are ranked among the strongest in Brazil, this would justify the increase in RDF to the points 5.0 cm, 10.0 cm, 15.0 cm, and mainly post-sticking point of 25.0 cm.

Regarding velocity, our findings point to significant differences between 5.0 and 10.0 cm ($p < 0.001$); between 5.0 and 15.0 cm ($p = 0.001$); and between 5.0 and 25.0 cm ($p = 0.002$). There were also differences between 10.0 and 15.0 cm ($p = 0.040$) and between 10.0 and 25.0 cm ($p < 0.001$). There was also a difference between 15.0 and 25.0 cm ($p = 0.001$).

In the same direction Gomo and Van Der Tillaar [27] found that the peak velocity and the local minimum velocity occurred with more closed footprints, which was not the target of our study. However, our athletes by competition regulations cannot make wider footprints than 81.0 cm (IPC, 2018), which would be relatively closed footprints. The minimum velocity of the sticking region would occur when the upward movement of the bar decelerated or even stopped completely for a short time during the lift [4]. Our findings indicate that at 5.0 cm we have a high velocity (0.699 m/s), and in intermediate sticking at 10.0 cm velocity tends to fall (0.198 m/s) and then increases by 15 cm (0.423 m/s) and 25.0 cm (1.137 m/s), showing changes in velocity at the various points in the sticking region.

In this sense, this study analyzed the muscle activation (electromyography) of the main muscles involved in the kinetics of this exercise in the sticking region. The results showed greater activation of the triceps compared with the other muscles analyzed. Corroborating this result, other studies [4,28] also indicate greater activation of the triceps, compared with other muscles involved in the bench press. Thus, it is possible to consider that the muscular activation of the triceps contributes to overcoming the sticking region during the lifting of the bar in the concentric phase of the bench press.

However, despite the relevance of the results, this study has some limitations. The sample consisted of national athletes with different disabilities eligible for the modality. In this sense, the findings are for practitioners of Paralympic powerlifting, not evaluating possible differences that could happen in the various disabilities eligible for the sport, since the PP has a unique class, where athletes are not separated by type of disability like in other sports. However, the current findings are still relevant for coaches and researchers for a greater understanding of the sticking point and its effects on sports performance.

5. Conclusions

Thus, it is possible to conclude that in the sticking region, the maximum isometric force, the rate of force development, time, and velocity tend to be impaired to the beginning and after the sticking region and also that the muscular activation of the triceps stands out in all intervals from the sticking point.

The determination of the sticking point becomes important so that coaches can focus on this region, and thus having appropriate and effective training for the point at which the failure normally occurs increases the possibility of improving powerlifting results.

Author Contributions: Conceptualization, F.J.A., F.M.C. and D.G.d.M.; methodology, A.C.M. and R.F.d.S.; software, R.F.d.S.; validation, O.C.M., T.R. and B.K.; formal analysis, P.F.d.A.-N. and B.G.d.A.T.C.; investigation, F.J.A. and A.C.M.; resources, V.M.R.; data curation, F.R.N. and V.M.R.; writing—original draft preparation, J.V.-A. and N.D.G.; writing—review and editing, F.J.A. and J.L.d.S.; visualization, F.M.C., A.C.M. and I.J.; supervision, I.J., T.R. and B.K.; project administration, D.G.d.M., I.J. and V.M.R. All authors have read and agreed to the published version of the manuscript.

Funding: Victor Machado Reis, José Vilaça-Alves and Nuno Domingos Garrido were funded by FCT—Fundação Portuguesa para a Ciência e Tecnologia (UID04045/2020).

Institutional Review Board Statement: This study was approved by the Human Research Ethics Committee of the Federal University of Sergipe (UFS), CAEE: 79909917.0.0000.55.46 (technical advice: 2,637,882, date of approval: 7 May 2018).

Informed Consent Statement: Informed consent was obtained from all subjects involved in the study.

Data Availability Statement: The data that support this study can be obtained from the address www.ufs.br/GPEPS, accessed on 20 February 2021.

Conflicts of Interest: The authors declare no conflict of interest.

References

1. International Paralympic Comitê (IPC). Sports. Available online: https://www.paralympic.org/powerlifting (accessed on 5 December 2018).
2. Madsen, N.; McLaughlin, T. Kinematic factors influencing performance and injury risk in the bench press exercise. *Med. Sci. Sports Exerc.* **1984**, *16*, 376–381. [CrossRef]
3. Kompf, J.; Arandjelović, O. The sticking point in the bench press, the squat, and the deadlift: Similarities and differences, and their significance for research and practice. *Sports Medic.* **2017**, *47*, 631–640. [CrossRef]
4. Van den Tillaar, R.; Ettema, G. The "sticking period" in bench press. *J. Sports Scienc.* **2010**, *28*, 529–535. [CrossRef]
5. Elliott, B.C.; Wilson, G.J.; Kerr, G.K. A biomechanical analysis of the sticking region in the bench press. *Med. Sci. Sports Exerc.* **1989**, *21*, 450–462. [CrossRef]
6. Newton, R. Biomechanics of conditioning exercises. In *Conditioning for Strength and Human Performance*; Chandler, T.J., Brown, L.E., Eds.; Brown Wolters Kluwer/Lippincott Williams & Wilkins: Philadelphia, PA, USA, 2008; pp. 77–93. ISBN 978-1-4511-0084-6.
7. Van Den Tillaar, R.; Saeterbakken, A.H.; Ettema, G. Is the occurrence of the sticking region the result of diminishing potentiation in bench press? *J. Sports Sci.* **2012**, *30*, 591–599. [CrossRef] [PubMed]
8. Dos Santos, M.D.M.; Aidar, F.J.; de Souza, R.F.; Dos Santos, J.L.; da Silva de Mello, A.; Neiva, H.P.; Marinho, D.A.; Marques, M.C. Does the Grip Width Affect the Bench Press Performance of Paralympic Powerlifters? *Int. J. Sports Physiol. Perform.* **2020**, *15*, 1–8. [CrossRef]
9. Ribeiro Neto, F.; Dorneles, J.R.; Costa, R.R.G. Are the flat and arched bench press really similar? *Sports Biomech.* **2020**, 1–2. [CrossRef] [PubMed]
10. Fraga, G.S.; Aidar, F.J.; de Matos, D.G.; Marçal, A.C.; Van den Tillaar, R.; Cabral, B.T.; Reis, V.M. Effects of Ibuprofen Intake in Muscle Damage, Body Temperature and Muscle Power in Paralympic Powerlifting Athletes. *Int. J. Environ. Res. Public Health* **2020**, *17*, 5157. [CrossRef] [PubMed]
11. Ball, R.; Weidman, D. Analysis of USA Powerlifting Federation Data from January 1, 2012–June 11, 2016. *J. Strength Cond. Res.* **2018**, *32*, 1843–1851. [CrossRef] [PubMed]
12. Fleck, S.J.; Kraemer, W.J. *Designing Resistance Training Program*, 4th ed.; Human Kinetics: Champaign, IL, USA, 2014; ISBN 0-7360-8170-4.
13. Puig-Diví, A.; Padullés-Riu, J.M.; Busquets-Faciaben, A.; Padullés-Chando, X.; Escalona-Marfil, C.; Marcos-Ruiz, D. Validity and Reliability of the Kinovea Program in Obtaining Angular and Distance Dimensions. *PLoS ONE* **2017**, *6*, e0216448.
14. Sañudo, B.; Rueda, D.; Pozo-Cruz, B.; de Hoyo, M.; Carrasco, L. Validation of a video analysis software package for quantifying movement velocity in resistance exercises. *J. Strength Cond. Res.* **2016**, *30*, 2934–2941. [CrossRef]

15. González-Badillo, J.J.; Sánchez-Medina, L. Movement velocity as a measure of loading intensity in resistance training. *Int. J. Sports Med.* **2010**, *31*, 347–352. [CrossRef]
16. Gonzalo-Skok, O.; Tous-Fajardo, J.; Arjol-Serrano, J.L.; Mendez-Villanueva, A. Determinants, reliability, and usefulness of a bench press repeated power ability test in young basketball players. *J. Strength Cond. Res.* **2014**, *28*, 126–133. [CrossRef] [PubMed]
17. Sampaio, C.R.S.F.; Aidar, F.J.; Ferreira, A.R.P.; Santos, J.L.D.; Marçal, A.C.; Matos, D.G.; Souza, R.F.; Moreira, O.C.; Guerra, I.; Fernandes Filho, J.; et al. Can Creatine Supplementation Interfere with Muscle Strength and Fatigue in Brazilian National Level Paralympic Powerlifting? *Nutrients* **2020**, *12*, 2492. [CrossRef]
18. Van den Tillaar, R.; Ettema, G. A comparison of muscle activity in concentric and counter movement maximum bench press. *J. Hum. Kinet.* **2013**, *38*, 63–71. [CrossRef]
19. Golas, A.; Maszczyk, A.; Stastny, P.; Wilk, M.; Ficek, K.; Lockie, R.G.; Zajac, A. A new approach to emg analysis of closed-circuit movements such as the flat bench press. *Sports* **2018**, *6*, 27. [CrossRef] [PubMed]
20. Cohen, J. Statistics a power primer. *Psychol. Bull.* **1992**, *112*, 155–159. [CrossRef]
21. Van den Tillaar, R.; Brevik Solheim, J.A.; Bencke, J. Comparions of hamstring muscle activation during high-speed running and various hamstring strengthening exercises. *Int. J. Sports Phys. Ther.* **2017**, *12*, 718–727. [CrossRef] [PubMed]
22. Newton, R.U.; Murphy, A.J.; Humphries, B.J.; Wilson, G.J.; Kreamer, W.J.; Häkkinen, K. Influence of load and stretch shortening cycle on the kinematics, kinetics and muscle activation that occurs during explosive upper-body movements. *Eur. J. Appl. Physiol.* **1997**, *50*, 311–320. [CrossRef]
23. Van den Tillaar, R.; Kwan, K. The Effects of Augmented Eccentric Loading upon Kinematics and Muscle Activation in Bench Press Performance. *J. Funct. Morphol. Kinesiol.* **2020**, *5*, 8. [CrossRef] [PubMed]
24. Drinkwater, E.J.; Galna, B.; McKenna, M.J.; Hunt, P.H.; Pyne, D.B. Validation of an optical encoder during free weight resistance movements and analysis of bench press sticking point power during fatigue. *J. Strength Cond. Res.* **2007**, *21*, 510–517. [PubMed]
25. Corvino, A.; Corvino, F.; Catalano, O.; Sandomenico, F.; Petrillo, A. The Tail and the String Sign: New Sonographic Features of Subcutaneous Melanoma Metastasis. *Ultrasound Med. Biol.* **2017**, *43*, 370–374. [CrossRef] [PubMed]
26. Kompf, J.; Arandjelovic, O. Understanding and overcoming the sticking point in resistance exercise. *Sports Med.* **2016**, *46*, 751–762. [CrossRef] [PubMed]
27. Gomo, O.; Van Den Tillaar, R. The effects of grip width on sticking region in bench press. *J. Sports Sci.* **2016**, *34*, 232–238. [CrossRef]
28. Clark, R.A.; Humphries, B.; Hohmann, E.; Bryant, A.L. The influence of variable range of motion training on neuromuscular performance and control of external loads. *J. Strength Cond. Res.* **2011**, *25*, 704–711. [CrossRef] [PubMed]

Journal of
Functional Morphology and Kinesiology

MDPI

Article

Evaluation of the Post-Training Hypotensor Effect in Paralympic and Conventional Powerlifting

Felipe J. Aidar [1,2,3], Ângelo de Almeida Paz [3], Dihogo de Matos Gama [4], Raphael Fabricio de Souza [2,3], Lúcio Marques Vieira Souza [1,3,5], Jymmys Lopes dos Santos [1,3], Paulo Francisco Almeida-Neto [6], Anderson Carlos Marçal [1,3], Eduardo Borba Neves [7], Osvaldo Costa Moreira [8], Nuno Domingos Garrido [9], Breno Guilherme Araújo Tinôco Cabral [6], Filipe Manuel Clemente [10,11], Victor Machado Reis [9], Pantelis Theo Nikolaidis [12] and Beat Knechtle [13,14,*]

1 Group of Studies and Research of Performance, Sport, Health and Paralympic Sports (GPEPS), Federal University of Sergipe (UFS), São Cristóvão 49100-000, SE, Brazil; fjaidar@gmail.com (F.J.A.); profedf.luciomarkes@gmail.com (L.M.V.S.); jymmys.lopes@gmail.com (J.L.d.S.); acmarcal@yahoo.com.br (A.C.M.)
2 Department of Physical Education, Federal University of Sergipe (UFS), São Cristóvão 49100-000, SE, Brazil; raphaelctba20@hotmail.com
3 Program of Physical Education, Federal University of Sergipe (UFS), São Cristóvão 49100-000, SE, Brazil; angelo-paz@uol.com.br
4 Cardiovascular & Physiology of Exercise Laboratory, University of Manitoba, Winnipeg, MB R3T 2N2, Canada; dihogogmc@hotmail.com
5 Physical Education Course, State University of Minas Gerais (UEMG), Divinopolis 37900-106, MG, Brazil
6 Department of Physical Education, Federal University of Rio Grande do Norte (UFRN), Natal 59078-970, RN, Brazil; paulo220911@hotmail.com (P.F.A.-N.); brenotcabral@gmail.com (B.G.A.T.C.)
7 Graduate Program of Biomedical Engineering, Federal Technological University of Paraná (PPGEB/UTFPR), Curitiba 80230-901, PR, Brazil; eduardoneves@utfpr.edu.br
8 Institute of Biological Sciences and Health, Federal University of Viçosa, Campus Florestal, Viçosa 35690-000, MG, Brazil; ocostamoreira@gmail.com
9 Research Center in Sports Sciences, Health Sciences and Human Development (CIDESD), Trás os Montes and Alto Douro University, 5001-801 Vila Real, Portugal; ndgarrido@gmail.com (N.D.G.); victormachadoreis@gmail.com (V.M.R.)
10 Escola Superior Desporto e Lazer, Instituto Politécnico de Viana do Castelo, Rua Escola Industrial e Comercial de Nun'Álvares, 4900-347 Viana do Castelo, Portugal; filipe.clemente5@gmail.com
11 Instituto de Telecomunicações, Delegação da Covilhã, 1049-001 Lisboa, Portugal
12 School of Health and Caring Sciences, University of West Attica, 12243 Egaleo, Greece; pademil@hotmail.com
13 Institute of Primary Care, University of Zurich, 8091 Zurich, Switzerland
14 Medbase St. Gallen Am Vadianplatz, 9001 St. Gallen, Switzerland
* Correspondence: beat.knechtle@hispeed.ch

Citation: Aidar, F.J.; Paz, Â.d.A.; Gama, D.d.M.; de Souza, R.F.; Vieira Souza, L.M.; Santos, J.L.d.; Almeida-Neto, P.F.; Marçal, A.C.; Neves, E.B.; Moreira, O.C.; et al. Evaluation of the Post-Training Hypotensor Effect in Paralympic and Conventional Powerlifting. *J. Funct. Morphol. Kinesiol.* **2021**, *6*, 92. https://doi.org/10.3390/jfmk6040092

Academic Editor: Peter Hofmann

Received: 12 October 2021
Accepted: 28 October 2021
Published: 1 November 2021

Abstract: High blood pressure (HBP) has been associated with several complications and causes of death. The objective of the study was to analyze the hemodynamic responses in Paralympic bench press powerlifting (PP) and conventional powerlifting (CP) before and after training and up to 60 minutes (min) after training. Ten PP and 10 CP athletes performed five sets of five repetition maximal bench press exercises, and we evaluated systolic, diastolic, and mean blood pressure (SBP, DBP, and MBP, respectively), heart rate (HR), heart pressure product (HPP), and myocardial oxygen volume (MVO_2). The SBP increased after training ($p < 0.001$), and there were differences in the post training and 30, 40, and 60 min later ($p = 0.021$), between 10 and 40 min after training ($p = 0.031$, $\eta^2_p = 0.570$), and between CP and PP ($p = 0.028$, $\eta^2_p = 0.570$). In the MBP, there were differences between before and after ($p = 0.016$) and 40 min later ($p = 0.040$, $\eta^2_p = 0.309$). In the HR, there was a difference between before and after, and 5 and 10 min later ($p = 0.002$), and between after and 10, 20, 30, 40, 50, and 60 min later ($p < 0.001$, $\eta^2_p = 0.767$). In HPP and MVO_2, there were differences between before and after ($p = 0.006$), and between after and 5, 10, 20, 30, 40, 50, and 60 min later ($p < 0.001$, $\eta^2_p = 0.816$). In CP and PP, there is no risk of hemodynamic overload to athletes, considering the results of the HPP, and training promotes a moderate hypotensive effect, with blood pressure adaptation after and 60 min after exercise.

Keywords: blood pressure; hemodynamics; hypotension; resistance training

1. Introduction

High blood pressure (HBP) has been considered a risk factor related to cardiovascular complications and other diseases, such as sudden death syndrome, stroke, acute myocardial infarction, heart failure, peripheral arterial disease, and chronic kidney disease [1–3]. In this regard, blood pressure control with the use of drugs and complementary activities has been widely studied [4,5]. Thus, physical exercise has been used as a form of prevention, control, and nonpharmacological treatment of arterial hypertension, providing an effective and cheaper strategy than pharmacological intervention [4,6]. Therefore, blood pressure reduction below the resting volume after exercise is defined as a hypotensive effect [7,8].

Within the strategies using exercises, studies have indicated that the post-exercise hypotensive effect (PHE) tends to occur regardless of the intensity, and without promoting cardiovascular overload during its practice [9,10]. Among the exercises, resistance exercises have been shown to be effective in controlling and reducing arterial hypertension [2,11,12], providing acute and chronic benefits for hypertensive or nonhypertensive people [6,13]. Among the resistance exercises, we have powerlifting, which is a modality where the one who lifts the most weight wins, and which has presented positive effects on blood pressure levels both for people with no physical disability [11] and for people with disabilities [2].

Powerlifting is characterized as a strength sport where the squat, bench press, and deadlift are performed [14,15], and through the adapted bench press in Paralympic powerlifting (PP) [16]. Studies have shown that powerlifting can be an alternative to prevent and/or control HBP [2,11]; however, to date, there has been no comparison between people with and without disabilities in the powerlifting bench press.

Therefore, our study aimed to analyze the hemodynamic responses generated in PP in comparison with bench press in conventional powerlifting (CP) after training with loads close to five maximum repetitions (5RM) up to 60 min after the end of the training session. In this sense, we raise two hypotheses: (i) powerlifting promotes a hypotensive effect after exercise; (ii) there are hypotensive differences between the conventional and Paralympic powerlifting.

2. Materials and Methods

Blood pressure was checked before the intervention (pre-test) for the collections, with a 10 min rest period [2,17–19], then the training itself was performed and, after the session, a post-test was done.

Figure 1 exemplifies the experimental design of the study.

2.1. Sample

Participants in the study included 10 CP athletes, who trained recreationally, characterized by training at least three times a week but not often participating in official sport competitions, without disabilities, the condition being verified by means of a medical certificate, and 10 PP athletes, all male, participating in the project at the Federal University of Sergipe (UFS) with at least 18 months of training, eligible for the modality and ranked among the top 10 of their categories at the national level, according to the Brazilian Paralympic Committee (CPB) norms. An inclusion criterion to have officially competed at the national level in the last six months was also adopted. Among the eligible deficiencies, four athletes had spinal cord injuries due to accidents with injuries below the eighth thoracic vertebra, two had sequelae due to poliomyelitis, two had malformations in the lower limbs, one was an amputee, and one had cerebral palsy. The characterization of the sample is shown in Table 1.

No fat percentage tests or height tests were performed, given the aforementioned deficiencies, which would make this procedure unfeasible.

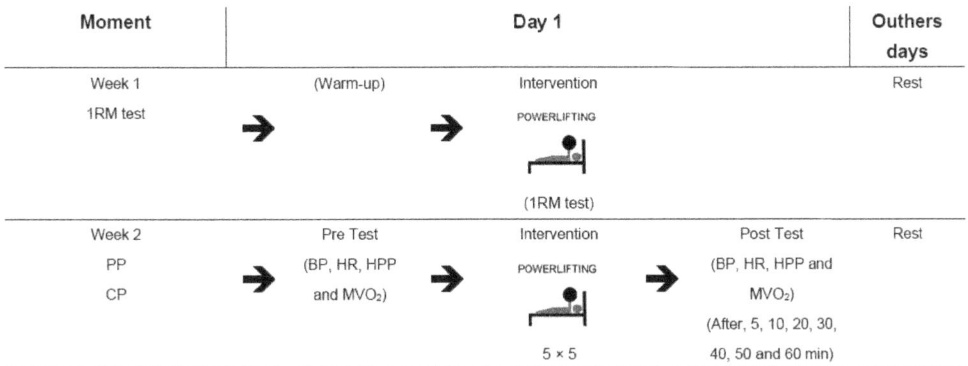

Figure 1. Experimental design—weekly schedule of the tests and intervention. BP: blood pressure, HR: heart rate, HPP: heart pressure product, MVO_2: myocardial oxygen volume, PP: Paralympic powerlifting. CP: conventional powerlifting, 5 × 5: five series of five maximum repetitions.

Table 1. Characterization of the sample.

	PP X ± SD (CI 95%)	CP X ± SD (CI 95%)	*p*	Cohen's d	α	ICC
Age (Years)	26.10 ± 6.95 (21.13–31.07)	23.20 ± 2.62 (21.33–25.07)	0.242	0.55	0.208	0.197
Body Mass (Kg)	76.80 ± 17.42 (64.34–89.26)	77.10 ± 7.68 (71.61–82.59)	0.961	0.02	0.419	0.445
Experience (Years)	2.61 ± 0.46 (2.18–2.99)	2.40 ± 0.16 (2.28–2.51)	0.011 *	0.61	0.18	0.017
1RM (Kg)	123.00 ± 29.93 (101.59–144.41)	92.80 ± 9.60 (85.93–99.67)	0.006 *	1.36	0.020	0.011
1RM/Body Mass	1.64 ± 0.39 ** (1.36–1.92)	1.21 ± 0.06 (1.16–1.25)	0.201	1.54	0.049	0.023

* $p < 0.05$. All PP athletes performed with loads that keep them among the 10 bests in their categories at the national level. ** Values above 1.4 in the bench press would be considered elite athletes, according to Ball e Wedman [20]. Legend: PP: Paralympic powerlifting, CP: conventional powerlifting, ICC: intraclass correlation.

Exclusion criteria included the use of some type of illicit ergogenic resource, some type of symptomatic cardiorespiratory or cardiac disease, metabolic changes, or being involved in any process of rapid weight loss before the competition, because this practice can negatively affect physical performance. The athletes participated in the study voluntarily and signed a free and informed consent form, in accordance with the resolution 466/2012 of the National Research Ethics Commission—CONEP, of the National Health Council, following the ethical principles expressed in the Declaration of Helsinki (1964, reformulated in 1975, 1983, 1989, 1996, 2000, 2008, and 2013) of the World Medical Association. This study was approved by the Research Ethics Committee of the Federal University of Sergipe, CAAE: 2,637,882 (date of approval: 7 May 2018).

2.2. Instruments

The body mass of the athletes was measured with the subjects in a sitting position, if from the PP Group, or standing, if only from the CP Group, using an appropriate Michetti digital electronic scale, Model Mic Wheelchair (Michetti, São Paulo, SP, Brazil). An official 210 cm long straight bench and 220 cm long 20 Kg bar were used herein (Eleiko Sport AB, Halmstad, Sweden); both pieces of equipment were approved by the International Paralympic Committee (IPC) [21].

To assess muscle strength, the entire procedure was performed in the bench press exercise using an official bench (Eleiko Sport AB, Halmstad, Sweden), approved by the International Paralympic Committee (IPC) [21], and an IPC Olympic serrated powerlifting bar that has grooves in its material, is 220 cm in total length, has an internal distance between 131 and 132 cm, with a diameter between 2.8 and 2.9 cm, and an external part of 41.5 cm, with a diameter between 5.0/5.2 cm, weighing 20 kg, in addition to 400 kg of washers from Eleiko (Eleiko Sport AB, Halmstad, Sweden).

2.3. Blood Pressure

Blood pressure (BP), systolic blood pressure (SBP), diastolic blood pressure (DBP), mean blood pressure (MBP) {MBP = DBP + (SBP − DBP)/3}, and Heart Rate (HR) were measured before and immediately after the training session using a noninvasive automated blood pressure (BP) monitor (Microlife 3AC1-1PC, Microlife, Widnau, Switzerland) [22]. The heart pressure product (HPP) was evaluated according to the following equation: HPP = HR × SBP [2,9]. All BP measurements were taken on the left arm, and the fixation of the cuff on the arm occurred with approximately 2.5 cm of distance between its lower extremity and the antecubital fossae, with the subjects in the sitting position [23,24]. The pre-exercise BP did not exceed 160 and 100 mmHg for SBP and DBP, respectively. Initially, the subjects remained comfortably seated for 10 min in a calm, pleasant environment, and the volunteers were also instructed to avoid the Valsalva maneuver throughout the movement, following guidelines from the American College of Sports Medicine [25].

To assess the occurrence of post-exercise hypotension (PEH), BP and HR were also measured at rest, at exercise peak (immediately following the exercise session), and in the sitting position (at rest) at 5, 10, 20, 30, 40, 50 and 60 min after the exercise session. To obtain MVO_2, we used a mathematical function based on a high correlation between the product of cardiac pressure and MVO_2. To estimate myocardial oxygen volume (MVO_2), a mathematical function was used, expressing the result in ml O_2/100 g ventilations per minute (VE/min), as follows: $MVO_2 = (HPP \times 0.0014) - 6.37$ [2,11,26].

2.4. Procedures

The training program took place within two weeks with a seven-day interval for each session, and the bench press exercise was used as the only representative of Paralympic weightlifting. The first training session consisted of five sets of 5RM, and all sessions were performed with ≅ 90% of 1RM at week 2, with 5 min of rest between sets [19,27]. Week 1 served for familiarization and the performance of the 1RM test, where each subject started the attempts with a weight they believed they could lift only once using maximum effort, and weight increments were added until reaching the maximum load that could be lifted all at once. If the practitioner could not perform a single repetition, 2.4% to 2.5% of the load used in the test was subtracted [16,28]. Subjects rested for 3–5 min between trials. All subjects underwent two sessions of the 1RM test, with a 48 to 72 h interval between each session.

2.5. Statistics

Descriptive statistics were performed using measures of central tendency, mean (X) ± Standard Deviation (SD). The statistical treatment was performed using the computerized package Statistical Package for Social Science (SPSS), version 22.0. To verify the normality of the variables, the Shapiro–Wilk test was used, considering the sample size. To assess the performance between the groups, the ANOVA (two-way) and Bonferroni post hoc tests were performed. The significance level adopted was $p < 0.05$. The effect size was determined by the values of "partial eta squared" (η^2_p), considering values of low effect (<0.05), medium effect (0.05–0.25), high effect (0.25–0.50), and very high effect (>0.50) [29–32]. Statistical analyses were performed using the Statistical Package for Social Science (SPSS) version 25.0 software (IBM, North Castle, New York, NY, USA).

3. Results

Our findings are described in Figure 2.

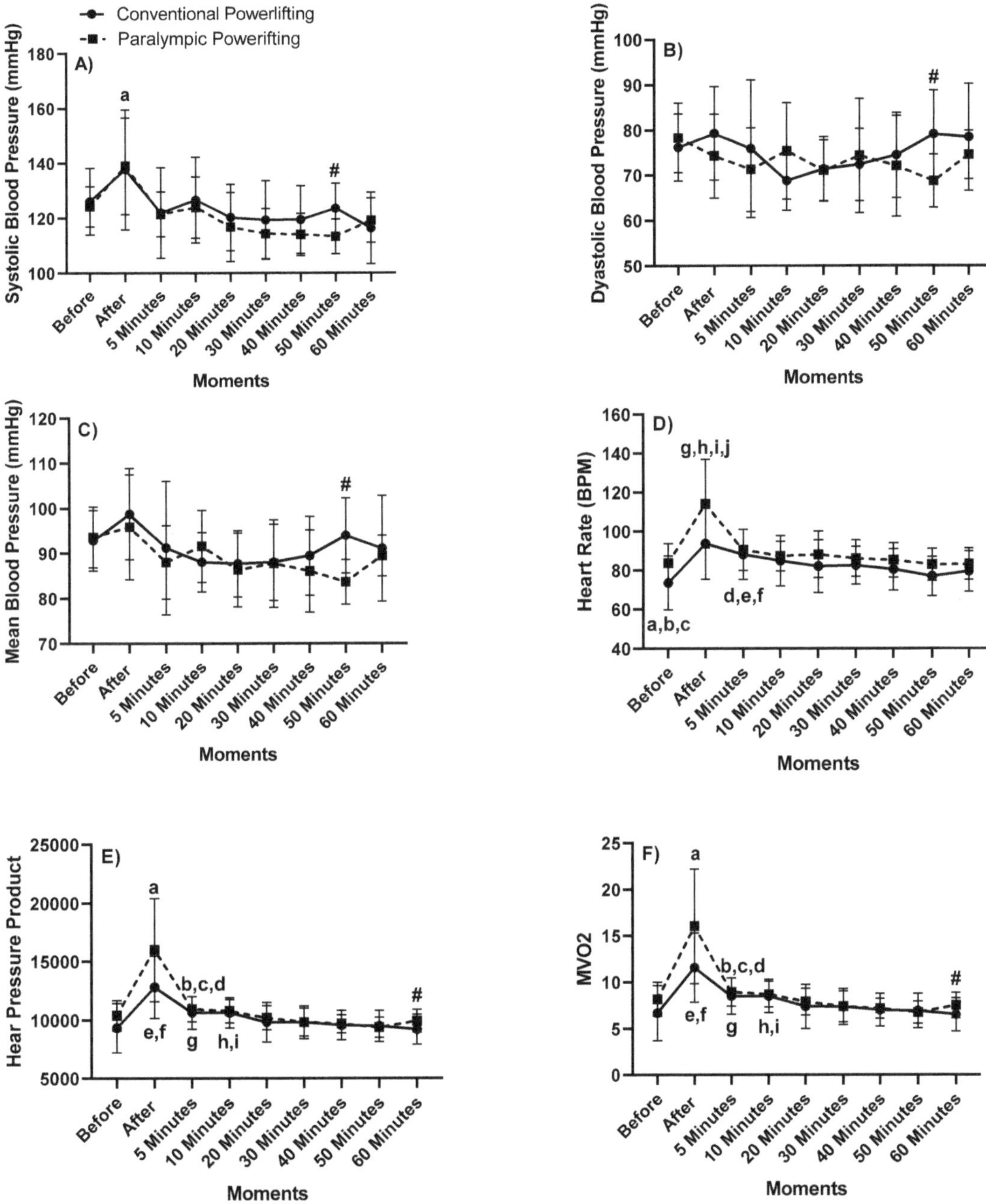

Figure 2. Kinetics of (**A**) systolic blood pressure (SBP), (**B**) diastolic blood pressure (DBP), (**C**) mean blood pressure (MBP), (**D**) heart rate (HR), (**E**) heart pressure product (HPP), and (**F**) MVO_2. Legend: letters "a–j" indicate intraclass differences, and "#" indicates interclass difference ($p < 0.05$).

Concerning SBP, there were significant differences between before and after, with an increase in SBP after 50 min between CP and PP ("a" $p = 0.003$), and in PP between the after and 50 min later ("#" $p = 0.049$; $\eta^2_p = 0.576$; very high effect). The PP kept the SBP lower after and 60 min later, showing a hypotensive effect present even after 50 min.

In DBP, there were no significant differences between the two groups after 50 min ($p = 0.38$; $\eta^2_p = 0.127$; medium effect). In MBP, there were differences between the two groups after 50 min ($p = 0.012$; $\eta^2_p = 0.323$; high effect). In BPD and MBP, there were differences in the 50 min later between groups, with lower values for PP.

In the HR, there were no differences between groups. In Group CP, there were differences between before and after ("a" $p = 0.015$), after 5 min ("b" $p = 0.002$), and in relation to 10 min ("c" $p = 0.003$). There were still differences between after 5 min and 40 min ("d" $p = 0.011$), 50 min ("e" $p = 0.034$), and 60 min ($p = 0.040$). In the PP, there were differences between after and 30 min ("g" $p = 0.021$), 40 min ("h" $p = 0.016$), 50 min ("i" $p = 0.021$), and 60 min later ("j" $p = 0.01$; $\eta^2_p = 0.767$, very high effect). The PP remained with lower HR values than did the CP.

In HPP and MVO2, there were differences between groups after 60 min ("#" $p = 0.047$). In the CP Group, there were differences between after and 30 min ("a" $p = 0.012$) later, and there were still differences between after 5 min and after 40 min ("b" $p = 0.006$), 50 min ("c" $p = 0.003$), and 60 min ("d" $p = 0.009$). In Group PP, there were differences between after and 50 min later ("e" $p = 0.023$) and 60 min later ("f" $p = 0.042$). There were also differences between after 5 min and after 50 min ("g" $p = 0.001$), and there were differences between after 10 min and after 50 min ("h" $p = 0.014$) and after 60 min ("i" $p = 0.038$; $\eta^2_p = 0.819$, very high effect). The PP remained with lower HR values than did the CP.

4. Discussion

The goal of our study was to analyze the hemodynamic responses generated by powerlifting exercise in different groups of CP and PP athletes, evaluating the attentive moments and then up to 60 min after the end of the training session. It is noteworthy that the two training methods were performed with high loads (≅90% of 1RM), intended for the training of CP and PP athletes, and the results did not show any risk of cardiovascular overload (HR after session, 5 × 5 method, ~80 bpm). It is worth mentioning that, in this discussion, we use studies of other modalities and with other athletes, given the lack of research focused on this modality and follow-up, with regard to PP. In this direction, corroborating with our findings, another study also worked with intensity (≅90 of 1RM) with five sets of 5RM (5 × 5), in a single powerlifting session; even so, it was not enough to present hemodynamic overload at peak exercise (160 bpm) [11]. Regarding SBP, there were significant differences between before and after, with an increase in SBP after training in both groups ($p < 0.001$), with higher values for CP, although there were no differences between groups. Relating to 60 min, there were significant differences with hypotensive effect for the PP group ($p = 0.028$) in relation to the other moments and to the CP ($\eta^2_p = 0.570$, very high effect).

Powerlifting training is characterized by small volumes and high intensities reaching close to 100% or more than 1RM [14,15], being this procedure common to CP and PP. Studies have shown that the magnitude of PHE tends to vary due to changes in training intensity [9,33,34]. Corroborating this, one study found no differences in PHE with variations in training volume [35]. Contrary to this, another study indicated that intensity is not the main factor in PHE but rather volume is [36,37], where high volumes tended to lead to a reduction in BP. There are contradictory results regarding the relationship between volume and intensity in the PHE, which also occurs in resistance exercises [9,33,34]. On the other hand, when evaluating the PHE, comparing resistance training with exercises such as bench press and traditional training, it was reported that PHE was presented 20, 30, and 40 min following training [8].

In BPD, our study did not observe significant differences between the two groups during the evaluation. In MBP, there were differences between before and after ($p = 0.016$,

$\eta^2_p = 0.174$, mean effect). In HR, there were no differences. Regarding HPP and MVO_2, there were differences between before and after in both groups ($p = 0.001$, $\eta^2_p = 0.816$, very high effect).

The decrease in blood pressure after the strength training indicates that cardiovascular behavior and PHE tend to be influenced by different mechanisms [9,33]. That is, it could be explained by the cardiac output, systolic volume, occlusion of vessels and arteries, autonomic modulation of HR through sympathetic and parasympathetic nerves, in addition to peripheral vascular resistance during training [2,11].

Another study that observed 40% and 80% of 1RM did not notice a hypotensive difference regarding intensities, and BPD tended to decrease at lower intensities [38]. On the other hand, the type of exercise tends to interfere with pressure levels, and intensities of 80% of 1RM tend to influence PHE more in multi-joint exercises, such as bench press, as was used in our study, compared with single-joint exercises, such as fly machine [10]. The hemodynamic responses are associated with the training load (% of 1RM), in addition to the number of sets, repetitions, and density (rest interval between sets), which are important components for PHE [39]. Contrary to this, PHE is associated with the type of exercise, notably involving large muscle groups, and multiarticular work, and not associated with intensity and volume [40].

When there is a comparison between conventional and Paralympic athletes, for conventional athletes, high intensities (>80% 1RM) tend to reduce the cardiac output mediated by systolic volume [41], and in trained individuals, the hemodynamic responses tend to be improved; thus, the decrease in systolic volume compensates for the increase in HR, caused by the increase in sympathetic activity and reduction in parasympathetic activity in the heart [11]. A study with CP athletes, who performed a training session with five sets of 2RM and 5 min of rest between sets, observed PHE 60 min after training, persisting for up to 24 h after exercise [11]. Furthermore, the reduction in pressure levels after resistance exercises is associated with peripheral vasodilators such as nitric oxide, prostaglandins, adenosine, and potassium that tend to influence PHE [42].

In PP, however, there would be a need to increase blood accumulation in the activated region, promoting increased vasodilation and reduced peripheral vascular resistance, providing PHE [40], and considering that training tends to use multiarticular exercises, which normally involve large muscle groups, increasing the PHE. In a study that evaluated the hypotensive effects in Paralympic athletes, no differences were observed between the two training sessions with different intensities, without significant interaction (training vs. time), and there was an increase in SBP immediately after training at both intensities, followed by a decrease to levels below resting values from 20 to 50 min after resistance training [2]. Other studies have presented similar results between conventional and Paralympic athletes in the bench press exercise of powerlifting [28,43,44].

However, our study has some limitations as the mechanisms that promote hypotension were not investigated in the manuscript. We did not investigate peripheral vascularity, sympathetic activity, stroke volume, beta-adrenergic receptors, or endothelial factors, and the method used to assess blood pressure was through validated devices. This has some limitations compared to invasive methods such as intra-arterial catheterization. Thus, every effort was made to ensure that these measurements were obtained in a consistent, reliable, and accurate manner.

5. Conclusions

We can conclude that both conventional powerlifting and PP do not present a risk of hemodynamic overload to athletes, considering the HPP results, and that powerlifting tends to promote a moderate hypotensive effect, with an adaptation of blood pressure after and 60 min after exercise. Training through high loads and low volume was shown to be important in terms of hypotensive effect from five minutes to 50 min after training, for CP, and was effective for up to 60 min for PP. Thus, it appears that the hypotensive effect in Paralympic athletes tends to remain longer than in CP.

The findings indicate that strength training for Paralympic athletes tends to be safer, with an increased hypotensive effect and even less post-training increase in BP compared to conventional athletes. Perhaps this is explained by the position of Paralympic athletes, with their lower limbs on the bench and not supported on the floor. Another point to consider is that Paralympic athletes tend to have atrophied lower limbs, which are even more hypotonic. Another point that could contribute to this is the fact that Paralympic athletes are wheelchair users or use crutches to get around, which tends to provide additional adaptation in terms of the use of the upper limbs, which could explain the differences found. However, other studies should be carried out evaluating the hypotensive effect of powerlifting in different physical disabilities and involving females.

Author Contributions: Conceptualization, F.J.A., Â.d.A.P., D.d.M.G., and R.F.d.S.; methodology, F.J.A., Â.d.A.P., D.d.M.G., and R.F.d.S.; software, L.M.V.S., J.L.d.S., and P.F.A.-N.; validation, A.C.M., E.B.N.; formal analysis, L.M.V.S., J.L.d.S., and P.F.A.-N.; investigation, F.J.A.; resources, A.C.M., E.B.N.; data curation, O.C.M., N.D.G., and B.G.A.T.C.; writing—original draft preparation, O.C.M., N.D.G., and B.G.A.T.C.; writing—review and editing, F.M.C., V.M.R., P.T.N., and B.K.; visualization, F.M.C., V.M.R., P.T.N., and B.K.; supervision, V.M.R.; project administration, F.M.C. All authors have read and agreed to the published version of the manuscript.

Funding: This research received no external funding.

Institutional Review Board Statement: The study was conducted according to the guidelines of the Declaration of Helsinki, and approved by the Research Ethics Committee of the Federal University of Sergipe, CAAE: 2,637,882 (date of approval: 7 May 2018).

Informed Consent Statement: Informed consent was obtained from all subjects involved in the study.

Data Availability Statement: The data that support this study can be obtained from the address: www.ufs.br/DepartmentofPhysicalEducation (accessed on 12 July 2021).

Conflicts of Interest: The authors declare no conflict of interest.

References

1. Rivera, E.C.; Jiménez, J.M.; Rojas, W.S.; Herrera, A.S. Efeito Agudo do Exercício Sobre a Pressão Arterial: Uma Investigação Metanalítica. *Arq. Bras. Cardiol.* **2016**, *106*, 422–433. (In Portuguese)
2. Paz, Â.A.; Aidar, F.J.; de Matos, D.G.; de Souza, R.F.; da Silva-Grigoletto, M.E.; van den Tillaar, R.; Ramirez-Campillo, R.; Nakamura, F.Y.; Costa, M.D.C.; Nunes-Silva, A.; et al. Comparison of post-exercise hypotension responses in paralympic powerlifting athletes after completing two bench press training intensities. *Medicina* **2020**, *56*, 156. [CrossRef]
3. Lewington, S.; Clarke, R.; Qizilbash, N.; Peto, R.; Collins, R. Prospective Studies Collaboration. Age-specific relevance of usual bloodpressure to vascular mortality: A meta-analysis of individual data for one million adults in 61 prospective studies. *Lancet* **2002**, *360*, 1903–1913. [PubMed]
4. Morgan, S.G.; Lee, A. Cost-related non-adherence to prescribed medicines among older adults: A cross-sectional analysis of a survey in 11 developed countries. *BMJ Open* **2017**, *7*, e014287. [CrossRef] [PubMed]
5. Whittaker, A.C.; Eves, F.F.; Carroll, D.; Roseboom, T.J.; Ginty, A.T.; Painter, R.C.; de Rooij, S.R. Daily stair climbing is associated with decreased risk for the metabolic syndrome. *BMC Public Health* **2021**, *21*, 923. [CrossRef]
6. De Sousa, E.C.; Abrahin, O.; Ferreira, A.L.L.; Rodrigues, R.P.; Alves, E.A.C.; Vieira, R.P. Resistance Training alone reduces systolic and diastolic blood pressure in prehypertensive and ypertensive individuals: Meta-analysis. *Hypertens Res.* **2017**, *40*, 927–931. [CrossRef] [PubMed]
7. Figueiredo, T.; Rhea, M.R.; Peterson, M.; Miranda, H.L.; Bentes, C.M.; Reis, V.M.R.; Simão, R. Influence of number of sets on blood pressure and heart rate variability after a strength training session. *J. Strength Cond. Res.* **2015**, *29*, 1556–1563. [CrossRef]
8. Bentes, C.M.; Costa, P.B.; Corrêa Neto, V.G.; Simão, R.; Paz, G.; Maia, M.F.; Figueiredo, T.; Neto, G.R.; Novaes, J.S.; Miranda, H. Hypotensive Responses of Reciprocal Supersets versus Traditional Resistance Training in Apparently Healthy Men. *Int. J. Exerc. Sci.* **2017**, *10*, 406–416.
9. Cavalcante, P.A.M.; Rica, R.L.; Evangelista, A.L.; Serra, A.J.; Figueira, A., Jr.; Pontes, F.L., Jr.; Kilgore, L.; Baker, J.; Bocalini, D. Effects of exercise intensity on postexercise hypotension after resistance training session in overweight hypertensive patients. *Clin. Interv. Aging* **2015**, *10*, 1487–1495. [CrossRef]
10. Cunha, G.A.; Rios, A.C.S.; Moreno, J.R.; Braga, P.L.; Campbell, C.S.G.; Simões, H.G.; Ruiz Denadai, M.L.D. Hipotensão pós-exercício em hipertensos submetidos ao exercício aeróbico de intensidades variadas e exercício de intensidade constante. *Rev. Bras. Med. Esporte* **2006**, *12*, 313–317. [CrossRef]

11. João, G.A.; Bocalini, D.S.; Rodriguez, D.; Charro, M.A.; Ceschini, F.; Martins, A.; Figueira, A., Jr. Powerlifting sessions promote significant post-exercise hypotension. *Rev. Bras. Med. Esporte* **2017**, *23*, 118–122. [CrossRef]
12. Maior, A.S.; Simão, R.; Martins, M.S.; de Salles, B.F.; Willardson, J.M. Influence of Blood Flow Restriction during Low-Intensity Resistance Exercise on the Postexercise Hypotensive Response. *J. Strength Cond. Res.* **2015**, *29*, 2894–2899. [CrossRef] [PubMed]
13. MacDonald, H.V.; Johnson, B.T.; Huedo-Medina, T.B.; Livingston, J.; Forsyth, K.C.; Kraemer, W.J.; Farinatti, P.T.V.; Pescatello, L.S. Dyniamic resistance training as stand-alone antihypertensive lifestyle therapy: A meta-analysis. *J. Am. Heart Assoc.* **2016**, *5*, e003231. [CrossRef]
14. João, G.A.; Evangelista, A.L.; Gomes, J.H.; Charro, M.A.; Bocalini, D.S.; Cardozo, D.; Seixas da Silva, D.A.d.C.; Simão, R.; Figueira, A., Jr. Effect of 16 weeks of periodized resistance training on gains of powerlifting athletes. *JEPonline* **2014**, *17*, 102–109.
15. Pritchard, H.J.; Morton, R. Powerlifting: Success and failure at the 2012 Oceania and 2013 classic world championships. *J. Aust. Strength Cond.* **2015**, *23*, 67–70.
16. Santos, W.Y.H.D.; Aidar, F.J.; Matos, D.G.; Van den Tillaar, R.; Marçal, A.C.; Lobo, L.F.; Marcucci-Barbosa, L.S.; Machado, S.D.C.; Almeida-Neto, P.F.; Garrido, N.D.; et al. Physiological and Biochemical Evaluation of Different Types of Recovery in National Level Paralympic Powerlifting. *Int. J. Environ. Res. Public Health* **2021**, *18*, 5155. [CrossRef]
17. La Motte, R.H.; Campbell, J.N. Comparison of the responses of warmth and nociceptive C fiber afferents in monkey with human judgments of thermal pain. *J. Neurop.* **1978**, *41*, 509–528. [CrossRef]
18. Bonsu, B.; Terblanche, E. The training and detraining effect of high-intensity interval training on post-exercise hypotensionin young overweight/obese women. *Eur. J. Appl. Physiol.* **2016**, *116*, 77–84. [CrossRef]
19. Fraga, G.S.; Aidar, F.J.; Matos, D.G.; Marçal, A.C.; Santos, J.L.; Souza, R.F.; Carneiro, A.L.; Vasconcelos, A.B.; Da Silva-Grigoletto, M.E.; van den Tillaar, R.; et al. Effects of Ibuprofen Intake in Muscle Damage, Body Temperature and Muscle Power in Paralympic Powerlifting Athletes. *Int. J. Environ. Res. Public Health* **2020**, *17*, 5157. [CrossRef]
20. Ball, R.; Weidman, D. Analysis of USA Powerlifting federation data from January 1, 2012–June 11, 2016. *J. Strength Cond. Res.* **2018**, *32*, 1843–1851. [CrossRef] [PubMed]
21. International Paralympic Comite (IPC). Rules. Official website of IPC Powerlifting. Available online: http://www.paralympic.org/powerlifting/about (accessed on 20 October 2020).
22. Xu, J.; Wu, Y.; Su, H.; Hu, W.; Li, J.; Wang, W.; Liu, X.; Cheng, X. The value of a BP determination method using a novel non-invasive BP device against the invasive catheter measurement. *PLoS ONE* **2014**, *9*, e100287. [CrossRef]
23. Malachias, M.V.B.; Póvoa, R.M.S., Jr.; Nogueira, A.R.; Souza, D.; Costa, L.S.; Magalhães, M.E. 7th Brazilian Guideline of Arterial Hypertension: Chapter 3-Clinical and Complementary Assessment. *Arq. Bras. Cardiol.* **2016**, *107* (Suppl. 3), 14–17. [PubMed]
24. Barroso, W.K.S.; Rodrigues, C.I.S.; Bortolotto, L.A.; Mota-Gomes, M.A.; Brandão, A.A.; Feitosa, A.D.M.; Machado, C.A.; Poli-de-Figueiredo, C.E.; Amodeo, C.; Mion, D., Jr.; et al. Brazilian Guidelines of Hypertension-2020. *Arq. Bras. Cardiol.* **2021**, *116*, 516–658. [CrossRef]
25. American College of Sports Medicine (ACSM). American College of Sports Medicine position stand. Progression models in resistance training for healthy adults. *Med. Sci. Sports Exerc.* **2009**, *41*, 687–708. [CrossRef] [PubMed]
26. Aksentijević, D.; Lewis, H.R.; Shattock, M.J. Is rate–pressure product of any use in the isolated rat heart? Assessing cardiac 'effort' and oxygen consumption in the Langendorff-perfused heart. *Exp. Physiol.* **2016**, *101*, 282–294. [CrossRef] [PubMed]
27. Soares Freitas Sampaio, C.R.; Aidar, F.J.; Ferreira, A.R.P.; Santos, J.L.D.; Marçal, A.C.; Matos, D.G.; Souza, R.F.; Moreira, O.C.; Guerra, I.; Fernandes Filho, J.; et al. Can Creatine Supplementation Interfere with Muscle Strength and Fatigue in Brazilian National Level Paralympic Powerlifting? *Nutrients* **2020**, *12*, 2492. [CrossRef] [PubMed]
28. Aidar, F.J.; Clemente, F.M.; Matos, D.G.; Marçal, A.C.; de Souza, R.F.; Moreira, O.C.; Almeida-Neto, P.F.; Vilaça-Alves, J.; Garrido, N.D.; Dos Santos, J.L.; et al. Evaluation of Strength and Muscle Activation Indicators in Sticking Point Region of National-Level Paralympic Powerlifting Athletes. *J. Funct. Morphol. Kinesiol.* **2021**, *6*, 43. [CrossRef]
29. Cohen, J. *Statistical Power Analysis for the Behavioral Sciences*; Lawrence Erlbaum, Associates: Hillsdale, NJ, USA, 1988.
30. Cohen, J. Statistics a power primer. *Psychol. Bull.* **1992**, *112*, 155–159. [CrossRef]
31. Selya, A.S.; Rose, J.S.; Dierker, L.C.; Hedeker, D.; Mermelstein, R.J. A Practical Guide to Calculating Cohen's f2, a Measure of Local Effect Size, from PROC MIXED. *Front. Psychol.* **2012**, *3*, 111. [CrossRef]
32. Lakens, D. Calculating and reporting effect sizes to facilitate cumulative science: A practical primer for t-tests and ANOVAs. *Front. Psychol.* **2013**, *4*, 863. [CrossRef] [PubMed]
33. Bentes, C.M.; Costa, P.B.; Neto, G.R.; Costa e Silva, G.V.; de Salles, B.F.; Miranda, H.L.; Novaes, J.S. Hypotensive effects and performance responses between diferente resistance training intensities and exercise orders in apparently health women. *Clin. Physiol. Funct. Imaging.* **2015**, *35*, 185–190. [CrossRef] [PubMed]
34. Lizardo, J.H.F.; Simões, H.G. Efeitos de diferentes sessões de exercícios resistidos sobre a hipotensão pos-exercício. *Rev. Braz. Fisioter.* **2005**, *9*, 289–295.
35. Keese, F.; Farinatti, P.; Pescatello, L.; Monteiro, W. A comparison of the immediate effects of resistance, aerobic, and concurrent exercise on postexercise hypotension. *J. Strength Cond. Res.* **2011**, *25*, 1429–1436. [CrossRef]
36. Polito, M.D.; Farinatti, P.T. The effects of muscle mass and number of sets during resistance exercise on postexercise hypotension. *J. Strength Con. Res.* **2009**, *23*, 2351–2357. [CrossRef] [PubMed]
37. Polito, M.D.; Farinatti, P.T.V. Comportamento da pressão arterial após exercícios contra-resistência: Uma revisão sistemática sobre variáveis determinantes e possíveis mecanismo. *Rev. Bras. Med. Esporte* **2006**, *12*, 386–392. [CrossRef]

38. Rezk, C.C.; Marrache, R.C.; Tinucci, T.; Mion, D., Jr.; Forjaz, C.L. Post-resistance exercise hypotension, hemodynamics, and heart rate variability: Influence of exercise intensity. *Eur. J. Appl. Physiol.* **2006**, *98*, 105–112. [CrossRef] [PubMed]
39. Negrão, C.E.; Rodon, M.U.P.B. Exercício físico hipertensão e controle barorreflexo da pressão arterial. *Rev. Bras. Hypertens.* **2001**, *8*, 89–95.
40. Damorim, I.R.; Santos, T.M.; Barros, G.W.P.; Carvalho, P.R.C. Kinetics of Hypotension during 50 Sessions of Resistance and Aerobic Training in Hypertensive Patients: A Randomized Clinical Trial. *Arq. Bras. Cardiol.* **2017**, *108*, 323–330. [CrossRef]
41. Boroujerd, S.S.; Rahimi, R.; Noori, S.R. Effect os high-versus low-intensity resistance training on post-exercise hypotension in male athletes. *Int. Sport Med. J.* **2009**, *10*, 95–100.
42. Goto, C.; Higashi, Y.; Kimura, M.; Noma, K.; Hará, K.; Nakagawa, K.; Kawamura, M.; Chayama, K.; Yoshizumi, M.; Nara, I. Effect of different intensities of exercise on endothelium-dependent vasodilation in humans: Role of endothelium-dependent nitric oxide and oxidative stress. *Circulation* **2003**, *108*, 530–535. [CrossRef]
43. Ribeiro Neto, F.; Dorneles, J.R.; Luna, R.M.; Spina, M.A.; Gonçalves, C.W.; Gomes Costa, R.R. Performance Differences Between the Arched and Flat Bench Press in Beginner and Experienced Paralympic Powerlifters. *J. Strength Cond. Res.* **2020**. [epub ahead]. [CrossRef] [PubMed]
44. García-Ramos, A.; Pérez-Castilla, A.; Villar Macias, F.J.; Latorre-Román, P.Á.; Párraga, J.A.; García-Pinillos, F. Differences in the one-repetition maximum and load-velocity profile between the flat and arched bench press in competitive powerlifters. *Sports Biomech.* **2021**, *20*, 261–273. [CrossRef] [PubMed]

Journal of
Functional Morphology and Kinesiology

MDPI

Article

Impact of BMI, Physical Activity, and Sitting Time Levels on Health-Related Outcomes in a Group of Overweight and Obese Adults with and without Type 2 Diabetes

Roberto Pippi [1,*], Lucia Cugusi [2], Marco Bergamin [3], Vittorio Bini [4], Carmine Giuseppe Fanelli [1], Valentina Bullo [3], Stefano Gobbo [3] and Andrea Di Blasio [5]

1 Healthy Lifestyle Institute, C.U.R.I.A.Mo. (Centro Universitario Ricerca Interdipartimentale Attività Motoria), Department of Medicine and Surgery, University of Perugia, Via G. Bambagioni, 19, 06126 Perugia, Italy; carmine.fanelli@unipg.it

2 Department of Biomedical Sciences, University of Sassari, 07100 Sassari, Italy; lcugusi@uniss.it

3 Department of Medicine, Sport and Exercise Medicine Division, University of Padova, Via Giustiniani 2, 35128 Padova, Italy; marco.bergamin@unipd.it (M.B.); valentina.bullo@unipd.it (V.B.); stefano.gobbo@unipd.it (S.G.)

4 Department of Medicine and Surgery, University of Perugia, Via Gambuli, 1, 06132 Perugia, Italy; vittorio.bini@unipg.it

5 Department of Medicine and Aging Sciences, 'G. D'Annunzio' University of Chieti-Pescara, 66100 Chieti, Italy; andrea.diblasio@unich.it

* Correspondence: roberto.pippi@unipg.it

Abstract: Physical activity level and sedentary behaviors affect health status in people with obesity and type 2 diabetes (DM2); their assessment is mandatory to properly prescribe exercise programs. From January 2011 to February 2014, 293 overweight/obese adults (165 women and 128 men, mean age of 51.9 ± 9.5 years and 54.6 ± 8.3 years, respectively), with and without DM2, participated in a three-month intensive exercise program. Before starting, participants were allocated into three subgroups (overweight, body mass index or BMI = 25–29.9; class 1 of obesity, BMI = 30–34.4; or class 2 (or superior) of obesity, BMI > 35). The international physical activity questionnaire (IPAQ-it) was used to evaluate participants' baseline sitting time (SIT) and physical activity level (PAL). Stratified multiple analyses were performed using four subgroups of SIT level according to Ekelund et al., 2016 (low, 8 h/day of SIT) and three subgroups for PAL (high, moderate, and low). Health-related measures such as anthropometric variables, body composition, hematic parameters, blood pressure values, and functional capacities were studied at the beginning and at the end of the training period. An overall improvement of PAL was observed in the entire sample following the three-month intensive exercise program together with a general improvement in several health-related measures. The BMI group factor influenced the VO_2 max variations, leg press values, triglycerides, and anthropometric variables, while the SIT group factor impacted the sitting time, VO_2 max, glycemic profile, and fat mass. In this study, baseline PAL and SIT did not seem to influence the effects of an exercise intervention. The characteristics of our educational program, which also included a physical exercise protocol, allowed us to obtain positive results.

Keywords: physical activity level; sitting time; body mass index; exercise; obesity; type 2 diabetes

Citation: Pippi, R.; Cugusi, L.; Bergamin, M.; Bini, V.; Fanelli, C.G.; Bullo, V.; Gobbo, S.; Di Blasio, A. Impact of BMI, Physical Activity, and Sitting Time Levels on Health-Related Outcomes in a Group of Overweight and Obese Adults with and without Type 2 Diabetes. *J. Funct. Morphol. Kinesiol.* **2022**, *7*, 12. https://doi.org/10.3390/jfmk7010012

Academic Editors: Cristina Cortis, Andrea Fusco and Carl Foster

Received: 21 December 2021
Accepted: 12 January 2022
Published: 17 January 2022

1. Introduction

It is well established that physical activity level (PAL) and sedentary behaviors affect people's health status [1], especially in populations with obesity and type 2 diabetes (DM2) [2]. The World Health Organization (WHO) has stated that "adults should do at least 150–300 min of moderate-intensity aerobic physical activity; or at least 75–150 min of vigorous-intensity aerobic physical activity; or an equivalent combination of moderate- and vigorous-intensity physical activity (MVPA) throughout the week, for substantial health benefits". Additionally, muscle-strengthening activities, involving all major muscle groups, on two or more days per week at moderate or greater intensity are recommended. People who do not engage in at least 150 min per week of MVPA are defined as inactive [3]. Additionally, sedentariness, defined as "any waking behavior characterized by an energy expenditure $\leq$1.5 METs, while in a sitting, reclining, or lying posture" [4–6], including sitting time (SIT), is a relevant health problem [7] as it is associated with an increase of cardio-metabolic risk [8], obesity, and DM2 [9]. Moreover, literature reports the importance of both SIT and PAL for promoting metabolic health [8,10,11]. In this regard, the WHO recently provided evidence-based public health recommendations for people of all ages on the amount of physical activity, sedentary behavior, and health outcomes [4]. To apply the WHO guidelines, it is crucial to promote interventions aimed at increased levels of physical activity and to monitor trends in both physical activity and SIT, including by low-cost and reliable measuring tools of habitual physical activity, such as questionnaires [12]. In this regard, the International Physical Activity Questionnaire in Short Form (IPAQ-SF) is a validated and most widely used [13] physical activity questionnaire [14].

Previous studies have emphasized the importance of physical exercise type for the optimization of results, recommending prescribing physical exercise for the management of obesity-related comorbidities [15–17] and DM2 [18–23]. Moreover, a part of the literature reports the presence of a negative compensation for spontaneous physical activity, with the inclusion of physical exercise in inactive people, based on the baseline BMI and PAL [24]. The purpose of this study was to evaluate the effects of an intensive exercise program on health-related outcomes (e.g., body mass index or BMI, waist circumference or WC, body composition, muscular strength, and maximal oxygen consumption or VO_2 max) and cardio-metabolic health measures (e.g., blood pressure levels, lipids, and glycemic profile) in a group of overweight and obese adults with and without DM2. We also studied whether, at the end of the exercise period, changes differed among different BMI, PAL, and SIT baseline categories.

2. Materials and Methods

2.1. Participants

From January 2011 to February 2014, a total sample of 293 (Figure 1) overweight/obese adults (165 women and 128 men, mean age 51.9 $\pm$ 9.5 years and 54.6 $\pm$ 8.3 years, respectively) with and without DM2 were recruited at the C.U.R.I.A.Mo. center to follow an intensive and multidisciplinary intervention protocol, as described by De Feo et al. [25].

Inclusion criteria were the presence of all data from clinical, anthropometric, and self-report questionnaires and physical measures collected both before and after the intervention; age between 35 and 70 years, and a BMI $\geq$ 25 kg/m^2. The exclusion criteria were the presence of musculoskeletal disorders or other clinical conditions that could seriously reduce subjects' life expectancy or their ability to participate in the study, particularly any potential contraindications to exercise.

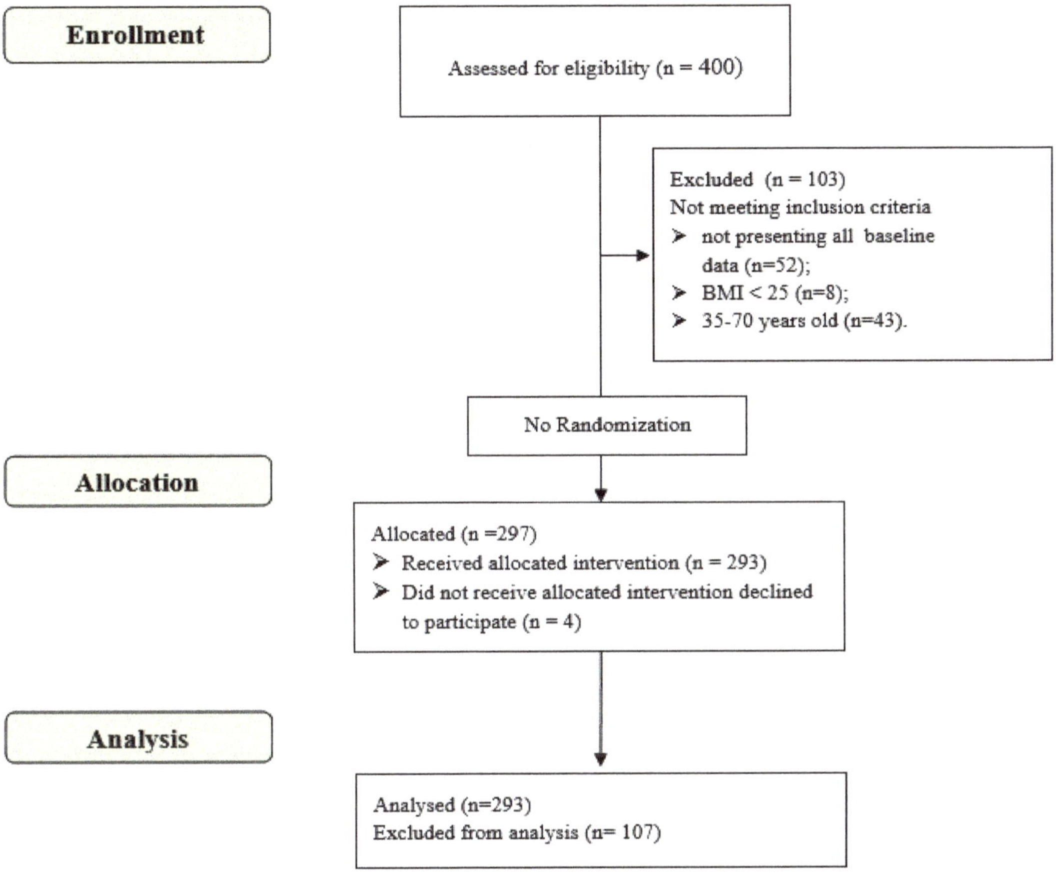

Figure 1. CONSORT flow diagram (adapted from CONSORT 2010 flow diagram).

According to the WHO criteria [26], participants were allocated into three subgroups based on baseline BMI value (Table S1a). The resultant groups were as follows:

- "people with overweight" (or OVER), corresponding to BMI 25–29.9 (n = 63, BMI mean = 28 ± 1.31);
- "people with I degree of obesity" (or I OB), corresponding to BMI 30–34.9 (n = 131, BMI mean = 32.5 ± 1.48);
- "people with II degrees (or superior) of obesity" (or II OB), corresponding to BMI > 35 (n = 99, BMI mean = 38.6 ± 3).

According to IPAQ guidelines [27], participants were further allocated into three resultant subgroups with reference to the baseline level of PAL (Table S1b), as follows:

- "low PAL", (n = 153, mean = 2.2 ± 2.8 MET-h per week);
- "moderate PAL", (n = 108, mean = 20.4 ± 11.8 MET-h per week);
- "high PAL", (n = 32, mean = 71.4 ± 32.7 MET-h per week).

Finally, according to Ekelund et al. (2016) [10], participants were allocated into four categories on baseline levels of SIT (Table S1c). This resulted in the following groups:

- "low SIT", corresponding to <4 h/day (n = 82, mean = 1.5 ± 1.2 h/day);
- "medium SIT", corresponding to 4–5.9 h/day (n = 63, mean = 4.6 ± 0.5 h/day);
- "high SIT", corresponding to 6–8 h/day of sitting time (n = 99, mean = 6.8 ± 0.8 h/day);
- "very high SIT", corresponding to >8 h/day of sitting time (n = 43, mean = 10.4 ± 1.5 h/day).

Please see Table S1a–c for the baseline mean values of all parameters in the entire sample and in the subgroups.

2.2. Intervention

Participants were involved in a three-month physical activity habits intervention, including one individual medical examination conducted by an endocrinologist; one psychological interview focused on lifestyle changes with a psychologist; one individual nutritional, counseling session focused on nutritional habits; and an intensive, gym-based, exercise intervention program. Briefly, as reported by Pippi et al. (2020) [28], the exercise program consisted of 25 bi-weekly small-group practical and counseling sessions (five persons per group), conducted by a certified exercise specialist. Every session lasted 90 min and including a mix of aerobic and strength exercise, administered using the circuit training method.

This approach derived from the original C.U.R.I.A.Mo. clinical model protocol, previously described by De Feo et al. [25]. Briefly, this clinical model utilizes the participation of master trained specialists who work following a multidisciplinary method. It aims to decrease sedentary time by guiding the patient to gradually increase intentional physical activity. The C.U.R.I.A.Mo. project has been registered in the Australian New Zealand Clinical Trials Registry (a Primary Registry in the WHO registry network), with the number: ACTRN12611000255987.

All the participants gave their written informed consent to participate in the study. Using a quasi-experimental study design, individuals were assessed before (T0) and at the end of the multidisciplinary intervention (T1).

2.3. Measures

2.3.1. Clinical and Anthropometric Variable Measures

During the first medical examination, managed by the endocrinologist, clinical variables including systolic (SBP) and diastolic (DBP) blood pressure (through a UM-101 mercury-free sphygmomanometer, A&D Medical, Tokyo, Japan) as well as blood measures such as fasting plasma glucose (GLYC), hemoglobin A1c (HbA1c), total (COL), high-density (HDL) and low-density lipoprotein (LDL) cholesterol, and triglycerides (TRIG), were recorded according to national standards of care [29]. Moreover, anthropometric measures (e.g., weight, BMI, WC) and body composition (fat mass percentage or FM% and muscle mass or MM) were assessed using standard methods with the Tanita body composition analyzer BC-420MA (Tokyo, Japan). Finally, during the medical examination it was also determined whether there were any potential contraindications to exercise.

2.3.2. Physical Performance Measures

Participants' VO_2 max values were assessed with the Rockport fitness walking test [30], and the Brzycki equation was applied to predict the 1-RM value of upper- and lower-body maximal strength [31]. Flexibility was measured using a standard bending test executed from the vertical (VB) and the horizontal position (HB) [32].

2.3.3. Self-Report Questionnaire Measures

PAL and SIT were quantified using the IPAQ short-form questionnaire [33], a validated [34] tool that assesses PA level achieved in the previous week, plus other information about time spent in the sitting position (last question of the questionnaire). According to the IPAQ scoring manual [27], IPAQ data were converted into METs, assigning to each activity the conventionally accepted intensity levels: 3.3 METs for walking, 4 METs for moderate-intensity activity, and 8 METs for vigorous-intensity activity. For example, walking energy expenditure (MET-WALK) was derived by multiplying results from walking minutes × walking days × 3.3, while moderate-intensity activity energy expenditure (MET-MOD) was derived by multiplying moderate-intensity activity minutes × number of days of moderate activity × 4.0. Similarly, vigorous-intensity activity energy expenditure (MET-VIG) was derived by multiplying results from vigorous-intensity activity minutes × number of days of vigorous-intensity activity × 8.0. The final score (the energy expenditure related to total physical activity), calculated as MET-WALK + MET-MOD + MET-VIG, is expressed in MET-minutes per week, subsequently transformed into MET-h per week.

2.4. Data Analysis

Descriptive analyses in terms of means, standard deviations, and/or percentages were carried out for each variable before (T0) the intervention (all the data are available in Table S1a–c, in Supplementary Materials). The percentage of adherence to exercise intervention was calculated as number of sessions performed/total number of sessions × 100.

The univariate analysis of variance (ANOVA) test was run to compare all variables at baseline, across the BMI, PAL, and SIT categories. To evaluate the effects of the exercise program a repeated-measures multivariate analysis of variance was used. Delta (Δ) changes (T1–T0) were computed and studied through univariate ANOVA of all the measures, using BMI, PAL, and SIT categories as a between factor. Post hoc analysis was conducted, using a Bonferroni correction.

p-Values ≤ 0.05 were set as statistically significant. Effect size was measured using partial eta-squares [35]. All the data were digitally archived and the analyses were performed using SPSS® Software, version 25.0 (IBM Corp. Released 2017. IBM SPSS Statistics for Windows, Version 25.0. Armonk, NY, USA: IBM Corp.).

2.5. Sample Size Calculation

A sample size of 293 subjects achieves 94% power to detect a mean of paired differences of 15 min on weekly total time of physical activity with an estimated standard deviation of differences of 74 min and with a significance level (alpha) of 0.05 using a two-sided paired t-test.

The sample size calculation was performed using PASS Software (PASS 16 Power Analysis and Sample Size Software 2018, NCSS, LLC., Kaysville, UT, USA).

3. Results

The percentage of adherence to exercise intervention was 86.9% ± 10.28% for the entire sample. There were no differences between subgroup values.

After the intervention (Table 1), an overall improvement in PAL was observed in the entire sample following the three-month intensive exercise program, with an increase of weekly total time (minutes) dedicated to vigorous physical activity ($p < 0.001$), moderate physical activity ($p = 0.001$), and walking activity ($p = 0.010$). Moreover, a general improvement in several clinical (SBP, $p < 0.001$; DBP, $p < 0.001$; GLYC, $p < 0.001$; TRIG, $p < 0.001$), anthropometric (weight, $p < 0.001$; BMI, $p < 0.001$; WC, $p < 0.001$), and body composition (FM%, $p < 0.001$; MM, $p = 0.048$) variables was observed. Finally, an improvement in physical performance measures (VO_2 max, $p < 0.001$; body strength, $p < 0.001$; flexibility, $p < 0.001$) occurred.

Table 1. Student *t*-test values: mean values of all parameters in the entire sample. Data are presented as means ± SDs, after (T0) and before (T1) the exercise intervention. Statistical significance was set for *p*-values ≤0.05.

Outcomes	T0	T1	*t*	*p*
	Mn ± SD	Mn ± SD		
SBP	**134.31 ± 17.32**	125.06 ± 14.95	−5.547	<0.001
DBP	82.71 ± 8.80	75.37 ± 10.74	−7.255	<0.001
GLYC	120.22 ± 43.44	110.41 ± 35.68	−3.444	<0.001
HbA1c	7.55 ± 10.13	6.25 ± 0.98	−1.435	0.077
COL	200.03 ± 39.52	193.73 ± 39.74	−2.598	0.005
HDL	47.83 ± 9.87	46.75 ± 10.12	−1.584	0.058
LDL	121.30 ± 36.43	122.04 ± 34.68	0.264	0.396
TRIG	154.64 ± 92.81	135.02 ± 81.17	−3.529	<0.001
WEIGHT	95.61 ± 17.25	92.73 ± 16.86	−13.478	<0.001
BMI	33.59 ± 4.49	32.58 ± 4.41	−12.518	<0.001
WC	111.71 ± 12.02	107.45 ± 11.73	−13.565	<0.001
FM%	38.20 ± 7.63	36.50 ± 7.84	−9.090	<0.001
MM	56.11 ± 11.97	55.83 ± 11.60	−1.671	0.048
LAT	39.04 ± 11.28	48.52 ± 12.25	22.897	<0.001
CHEST	27.92 ± 9.95	38.65 ± 12.13	29.754	<0.001
PRESS	155.24 ± 37.43	202.05 ± 45.96	20.290	<0.001
LEXT	31.17 ± 11.07	47.31 ± 14.27	26.586	<0.001
VB	−8.72 ± 9.84	−4.79 ± 10.45	10.623	<0.001
HB	25.80 ± 10.54	29.33 ± 9.34	9.563	<0.001
VO_2 max	19.52 ± 9.30	25.93 ± 7.93	16.409	<0.001
MET-h per week	16.46 ± 24.71	39.80 ± 27.88	12.372	<0.001
VIG_TOT_WEEK_MIN	36.20 ± 126.39	175.54 ± 141.12	14.010	<0.001
VIG_ MET-h per week	4.83 ± 16.85	23.41 ± 18.82	14.010	<0.001
MOD_TOT_WEEK_MIN	73.63 ± 187.45	121.37 ± 208.41	3.071	0.001
MOD_ MET-h per week	4.91 ± 12.50	8.09 ± 13.89	3.071	0.001
WALK_TOT_WEEK_MIN	122.31 ± 175.35	151.01 ± 188.41	2.321	0.010
WALK_ MET-h per week	6.73 ± 9.64	8.31 ± 10.36	2.321	0.010
SIT	5.36 ± 3.17	5.15 ± 2.69	−1.056	0.146

SBP: systolic blood pressure; DBP: diastolic blood pressure; GLYC: fasting plasma glucose; HbA1c: glycosylated hemoglobin; COL: total cholesterol; HDL: high-density lipoprotein; LDL: low-density lipoprotein; TRIG: triglycerides; BMI: body mass index; FM%: fat mass percentage; MM: muscle mass; LAT: lat machine test value; CHEST = chest press test value; PRESS = leg press test value; LEXT = leg extension test value; VB: vertical bending test value; HB: horizontal bending test value; VO_2 max: maximal oxygen consumption value; MET-h per week = weekly total physical activity energy expenditure; VIG_TOT_WEEK_MIN = weekly total time (in minutes) of vigorous physical activity; VIG_ MET-h per week = weekly vigorous physical activity energy expenditure; MOD_TOT_WEEK_MIN = weekly total time (in minutes) of moderate physical activity; MOD_ MET-h per week = weekly moderate physical activity energy expenditure; WALK_TOT_WEEK_MIN = weekly total time (in minutes) of walking activity; WALK_ MET-h per week = weekly walking activity energy expenditure; SIT = daily sitting time.

Please see Table 2a for the results of repeated-measures multivariate analysis of variance to analyze the differences in all variables between T0 and T1, using BMI categories as a between factor.

Table 2. Repeated-measures multivariate analysis of variance values, using BMI (**a**), PAL (**b**) and SIT (**c**).

a. Repeated-measures multivariate analysis of variance to analyze differences in all variables between T0 and T1, using BMI categories as a between factor. Data are presented as means ± SDs. Statistical significance was set for *p*-values ≤0.05.

Outcomes	T0	T1	Time T0 vs. T1			Time * BMI Category			Δ (T1–T0) Post Hoc
	Mn ± SD	Mn ± SD	F	p	Partial η^2	F	p	Partial η^2	
SBP	134.31 ± 17.32	125.06 ± 14.95	31.447	<0.001	0.202	3.252	0.615	0.008	N.S.
DBP	82.71 ± 8.80	75.37 ± 10.74	49.821	<0.001	0.287	3.949	0.022	0.060	N.S.
GLYC	120.22 ± 43.44	110.41 ± 35.68	13.706	<0.001	0.099	0.160	0.852	0.003	N.S.
HbA1c	7.55 ± 10.13	6.25 ± 0.98	2.760	0.099	0.023	1.639	0.199	0.027	N.S.
COL	200.03 ± 39.52	193.73 ± 39.74	6.666	0.011	0.049	1.588	0.208	0.024	N.S.
HDL	47.83 ± 9.87	46.75 ± 10.12	2.809	0.96	0.022	1.238	0.294	0.019	N.S.
LDL	121.30 ± 36.43	122.04 ± 34.68	0.029	0.864	0.000	1.069	0.347	0.018	N.S.
TRIG	154.64 ± 92.82	135.02 ± 81.17	11.871	0.001	0.083	2.285	0.106	0.034	N.S.
WEIGHT	95.61 ± 17.25	92.73 ± 16.86	158.465	<0.001	0.353	123.116	<0.001	0.459	II OB vs. OVER (p < 0.001); II OB vs. I OB (p < 0.001); I OB vs. OVER (p < 0.001)
BMI	33.60 ± 4.49	32.58 ± 4.41	135.345	<0.001	0.318	429.297	<0.001	0.748	II OB vs. OVER (p < 0.001); II OB vs. I OB (p < 0.001); I OB vs. OVER (p < 0.001)
WC	111.71 ± 12.02	107.45 ± 11.73	174.209	<0.001	0.389	127.413	<0.001	0.482	II OB vs. OVER (p < 0.001); II OB vs. I OB (p < 0.001); I OB vs. OVER (p < 0.001)
FM%	38.20 ± 7.63	36.50 ± 7.84	72.460	<0.001	0.201	36.690	<0.001	0.203	II OB vs. OVER (p < 0.001); II OB vs. I OB (p < 0.001); I OB vs. OVER (p < 0.001)
MM	56.11 ± 11.97	55.83 ± 11.60	1.545	0.215	0.005	17.443	<0.001	0.108	II OB vs. OVER (p < 0.001); II OB vs. I OB (p < 0.001); I OB vs. OVER (p < 0.046)
LAT	39.04 ± 11.28	48.53 ± 12.25	453.070	<0.001	0.673	6.993	0.001	0.060	II OB vs. OVER (p < 0.001)
CHEST	27.92 ± 9.95	38.65 ± 12.14	769.477	<0.001	0.778	7.308	<0.001	0.063	II OB vs. OVER (p < 0.001); II OB vs. I OB (p = 0.044)
PRESS	155.24 ± 37.43	202.05 ± 45.96	360.794	<0.001	0.621	12.104	<0.001	0.099	II OB vs. OVER (p < 0.001); II OB vs. I OB (p < 0.001)
LEXT	31.17 ± 11.07	47.31 ± 14.27	621.215	<0.001	0.743	9.548	<0.001	0.082	II OB vs. OVER (p < 0.001); II OB vs. I OB (p = 0.004)
VB	−8.72 ± 9.84	−4.79 ± 10.45	109.377	<0.001	0.279	1.822	0.164	0.013	N.S.
HB	25.80 ± 10.54	29.33 ± 9.34	82.575	<0.001	0.226	3.920	0.021	0.027	II OB vs. I OB (p = 0.019)
VO_2 max	19.52 ± 9.30	25.93 ± 7.93	231.054	<0.001	0.443	14.542	<0.001	0.091	II OB vs. OVER (p < 0.001); II OB vs. I OB (p < 0.001)
MET-h per week	16.46 ± 24.71	39.80 ± 27.88	136.159	<0.001	0.320	1.113	0.330	0.008	N.S.
SIT	5.36 ± 3.17	5.15 ± 2.69	0.714	0.399	0.003	1.704	0.184	0.012	N.S.

b. Repeated-measures multivariate analysis of variance to analyze differences in all variables between T0 and T1, using PAL categories as a between factor. Data are presented as means ± SDs. Statistical significance was set for *p*-values ≤ 0.05.

Outcomes	T0	T1	Time T0 vs. T1			Time * PAL Category			Δ (T1–T0) Post Hoc
	Mn ± SD	Mn ± SD	F	p	Partial η^2	F	p	Partial η^2	
SBP	134.31 ± 17.32	125.06 ± 14.95	18.319	<0.001	0.129	.395	0.674	0.006	N.S.
DBP	82.71 ± 8.80	75.37 ± 10.74	35.135	<0.001	0.221	2.847	0.062	0.044	N.S.
GLYC	120.22 ± 43.44	110.41 ± 35.68	4.264	0.041	0.536	1.170	0.314	0.018	high vs. low (p = 0.041)
HbA1c	7.55 ± 10.13	6.25 ± 0.98	0.972	0.326	0.008	0.599	0.551	0.010	N.S.
COL	200.03 ± 39.52	193.73 ± 39.74	3.272	0.073	0.025	2.132	0.123	0.032	N.S.
HDL	47.83 ± 9.87	46.75 ± 10.12	0.190	0.663	0.002	0.227	0.797	0.004	N.S.
LDL	121.30 ± 36.43	122.04 ± 34.68	0.058	0.811	0.001	1.270	0.285	0.021	N.S.
TRIG	154.64 ± 92.82	135.02 ± 81.17	10.978	0.001	0.078	0.019	0.981	0.001	N.S.
WEIGHT	95.61 ± 17.25	92.73 ± 16.86	127.497	<0.001	0.305	0.269	0.765	0.002	N.S.
BMI	33.60 ± 4.49	32.58 ± 4.41	107.147	<0.001	0.270	3.100	0.047	0.021	moderate vs. low (p = 0.046)
WC	111.71 ± 12.02	107.45 ± 11.73	112.938	<0.001	0.292	0.582	0.560	0.004	N.S.

Table 2. *Cont.*

b. Repeated-measures multivariate analysis of variance to analyze differences in all variables between T0 and T1, using PAL categories as a between factor. Data are presented as means ± SDs. Statistical significance was set for p-values ≤ 0.05.

Outcomes	T0	T1	Time T0 vs. T1			Time * BMI Category			Δ (T1–T0) Post Hoc
	Mn ± SD	Mn ± SD	F	p	Partial η^2	F	p	Partial η^2	
FM%	38.20 ± 7.63	36.50 ± 7.84	63.138	<0.001	0.180	1.120	0.328	0.008	N.S.
MM	56.11 ± 11.97	55.83 ± 11.60	0.374	0.542	0.001	1.141	0.321	0.008	N.S.
LAT	39.04 ± 11.28	48.53 ± 12.25	375.487	<0.001	0.631	3.330	0.038	0.029	high vs. moderate (p = 0.033)
CHEST	27.92 ± 9.95	38.65 ± 12.14	719.275	<0.001	0.767	4.059	0.019	0.036	high vs. moderate (p = 0.017); high vs. low (p = 0.037)
PRESS	155.24 ± 37.43	202.04 ± 45.96	334.875	<0.001	0.604	2.492	0.085	0.022	high vs. moderate (p = 0.014); high vs. low (p = 0.027)
LEXT	31.171 ± 11.07	47.31 ± 14.275	568.074	<0.001	0.725	4.199	0.016	0.038	high vs. moderate (p = 0.017); high vs. low (p = 0.022)
VB	−8.72 ± 9.84	−4.79 ± 10.45	73.522	<0.001	0.206	3.166	0.830	0.001	N.S.
HB	25.80 ± 10.54	29.33 ± 9.34	45.052	<0.001	0.137	2.274	0.105	0.016	N.S.
VO_2 max	19.52 ± 9.30	25.93 ± 7.93	158.397	<0.001	0.353	2.474	0.086	0.017	N.S.
MET-h per week	16.46 ± 24.71	39.80 ± 27.88	35.445	<0.001	0.109	96.579	<0.001	0.400	high vs. moderate (p < 0.001); high vs. low (p < 0.001); moderate vs. low (p < 0.001)
SIT	5.36 ± 3.17	5.15 ± 2.69	0.597	0.440	0.002	8.522	<0.001	0.057	high vs. low (p < 0.001); moderate vs. low (p = 0.017)

c. Repeated-measures multivariate analysis of variance to analyze differences in all variables between T0 and T1, using SIT categories as a between factor. Data are presented as means ± SDs. Statistical significance was set for p values ≤ 0.05.

Outcomes	T0	T1	Time T0 vs. T1			Time * SIT Category			Δ (T1–T0) Post Hoc
	Mn ± SD	Mn ± SD	F	p	Partial η^2	F	p	Partial η^2	
SBP	134.52 ± 17.44	125.31 ± 15.11	21.800	<0.001	0.157	0.055	0.983	0.001	N.S.
DBP	83.1 ± 8.83	75.76 ± 10.71	45.168	<0.001	0.279	0.717	0.544	0.018	N.S.
GLYC	121.75 ± 43.87	111.46 ± 36.03	19.283	<0.001	0.140	0.281	0.839	0.007	N.S.
HbA1c	7.62 ± 10.38	6.26 ± 0.99	1.364	0.245	0.012	0.752	0.524	0.020	N.S.
COL	200.37 ± 38.86	193.88 ± 39.62	7.204	0.008	0.055	0.520	0.669	0.013	N.S.
HDL	47.80 ± 9.98	46.67 ± 10.28	3.451	0.066	0.028	0.850	0.469	0.021	N.S.
LDL	122.248 ± 36.55	122.719 ± 34.40	0.002	0.962	0.024	1.072	0.364	0.029	N.S.
TRIG	154.87 ± 94.23	136.13 ± 82.52	11.146	0.001	0.083	0.770	0.513	0.018	N.S.
WEIGHT	95.793 ± 17.28	92.92 ± 16.92	148.702	<0.001	0.344	3.678	0.013	0.038	low vs. high SIT (p = 0.007)
BMI	33.62 ± 4.50	32.61 ± 4.42	125.817	<0.001	0.308	2.244	0.083	0.023	N.S.
WC	111.78 ± 12.09	107.56 ± 11.78	153.989	<0.001	0.366	1.936	0.124	0.021	N.S.
FM%	38.17 ± 7.59	36.46 ± 7.82	67.703	<0.001	0.194	1.223	0.302	0.013	N.S.
MM	56.24 ± 11.95	55.96 ± 11.59	3.245	0.073	0.011	2.505	0.059	0.026	N.S.
LAT	39.19 ± 11.33	48.65 ± 12.27	430.401	<0.001	0.668	0.735	0.532	0.010	N.S.
CHEST	27.95 ± 9.96	38.64 ± 12.08	752.288	<0.001	0.779	1.322	0.268	0.018	N.S.
PRESS	155.59 ± 37.29	202.34 ± 45.99	340.782	<0.001	0.614	2.106	0.100	0.029	N.S.
LEXT	31.33 ± 11.07	47.45 ± 14.37	572.982	<0.001	0.733	0.799	0.496	0.011	N.S.
VB	−8.66 ± 9.87	−4.71 ± 10.50	108.815	<0.001	0.283	1.061	0.366	0.011	N.S.
HB	25.9 ± 10.62	29.44 ± 9.38	83.961	<0.001	0.233	0.417	0.741	0.005	N.S.
VO_2 max	19.69 ± 9.24	25.95 ± 7.96	223.868	<0.001	0.442	1.304	0.273	0.014	N.S.
MET-h per week	16.61 ± 24.88	39.95 ± 27.97	134.693	<0.001	0.322	6.841	<0.001	0.068	low SIT vs. very high SIT (p = 0.004); low SIT vs. high SIT (p < 0.001); medium SIT vs. high SIT (p = 0.026)
SIT	5.36 ± 3.17	5.15 ± 2.69	6.374	0.012	0.022	271.280	<0.001	0.742	low SIT vs. very high SIT (p < 0.001); low SIT vs. high SIT (p < 0.001); low SIT vs. medium SIT (p < 0.001); medium SIT vs. very high SIT (p < 0.001); medium SIT vs. high SIT (p < 0.001); high SIT vs. very high SIT vs. (p < 0.001); very high SIT vs. high SIT (p < 0.001)

SBP: systolic blood pressure; DBP: diastolic blood pressure; GLYC: fasting plasma glucose; HbA1c: glycosylated hemoglobin; COL: total cholesterol; HDL: high-density lipoprotein; LDL: low-density lipoprotein; TRIG: triglycerides; BMI: body mass index; FM%: fat mass percentage; MM: muscle mass; LAT: lat machine test value; CHEST = chest press test value; PRESS = leg press test value; LEXT = leg extension test value; VB: vertical bending test value; HB: horizontal bending test value; VO_2 max: maximal oxygen consumption value; MET-h per week = weekly total physical activity energy expenditure; SIT = daily sitting time. Between-group comparisons are reported in the last column of the table. Statistically significant differences are then followed by post hoc results (e.g., OVER vs. I OB means that Δ in people with overweight is different from that in the group of people with I degree of obesity). N.S. = not statistically significant. Between-group comparisons are reported in the last column of table. Statistically significant differences are then followed by post hoc results (e.g., high vs. moderate means that Δ in people with high level of physical activity is different from that in people with moderate level of physical activity). N.S. = not statistically significant.

Using PAL categories as a between factor (Table 2b), the entire sample showed a statistically significant improvement in SBP ($p < 0.001$), DBP ($p < 0.001$), GLYC ($p = 0.041$), TRIG ($p = 0.001$), weight ($p < 0.001$), BMI ($p < 0.001$), WC ($p < 0.001$), FM% ($p < 0.001$), physical performance measures ($p < 0.001$), and weekly energy expenditure related to total physical activity ($p < 0.001$).

The PAL group factor impacted BMI ($p = 0.047$), lat ($p = 0.038$) and chest press ($p = 0.019$), leg extension ($p = 0.016$), weekly energy expenditure related to total physical activity ($p < 0.001$), and daily sitting time ($p < 0.001$).

Using SIT categories as a between factor (Table 2c), the entire sample showed a statistically significant improvement in SBP ($p < 0.001$), DBP ($p < 0.001$), GLYC ($p < 0.001$), COL ($p = 0.008$), TRIG ($p = 0.001$), anthropometric and body composition variables ($p < 0.001$), physical performance measures ($p < 0.001$), weekly energy expenditure related to total physical activity ($p < 0.001$), and daily sitting time ($p = 0.012$). The SIT group factor impacted weight ($p = 0.013$), weekly total physical activity energy expenditure ($p < 0.001$), and daily sitting time ($p < 0.001$).

4. Discussion

This study aimed to evaluate the effects of an intensive exercise program of 25 biweekly sessions on the variation of some clinical, anthropometric, body composition, physical performance, and self-report questionnaire variables in a group of overweight and obese adults with and without DM2. We also studied the role played by BMI, PAL, and SIT on the effects of exercise intervention. As expected, we observed an overall improvement of PAL in the entire samples following the three-month intensive exercise program, with a general improvement in several health-related outcomes reviewed; baseline PAL and SIT did not seem to influence all these effects in a sample composed of overweight and obese participants with and without DM2. Similar to the results of other authors [36–40], in our sample, we observed a significative reduction of some clinical and anthropometric variables linked to cardiovascular risk. In fact, as did Cheng et al. [40] and Schwingshackl et al. [41], we observed a significant weight and fat mass reduction, using mixed exercise. With regard to waist circumference, a central adiposity variable useful for identifying specific cardiometabolic risk [42], we observed a significative reduction in the entire sample ($p < 0.001$), even if the post-intervention mean values (107.45 ± 11.73) remained dangerous for health. As also found by Stoner et al. [43], the effects of exercise on LDL, HDL, and HbA1c were inconclusive ($p > 0.005$). As expected, using BMI categories as a between factor, we observed that the previous parameters were influenced by baseline mean values of BMI. In fact, we observed the greatest improvements in subjects with a greater degree of obesity rather than in the overweight group (Table 2a). Particularly, the most important reduction of WC was observed in participants with II degrees (or superior) of obesity rather than in participants with I degree of obesity ($p = 0.006$). In our study, weight and fat mass loss, as well as WC reduction, seemed not to be influenced by PAL and SIT categories as a between factor, with the only exception being the SIT group factor, which impacted weight (low sitting time group vs. high sitting time group, $p = 0.007$). In fact, deepening the weight trend with respect to the SIT groups, we observed a greater reduction in people with low SIT (−3.42 kg) rather than in people with medium (−3.04 kg) and high (−2.85 kg) SIT. These results could be influenced by the SIT trend observed in different SIT groups, although people with low baseline SIT presented a greater increase (+2.49 h per day) rather than those in the medium (−0.39 h per day) and high (−1.50 h per day) groups.

Previous studies of the overweight and obese involved in exercise programs showed improvements in physical measures. Dieli-Conwright et al. [44] found important improvements in estimated VO_2 max (52%) and muscular strength (>30%), as did Hsu et al. [45], who recorded increases in maximal exercise capacity and maximal muscular strength. Balducci et al. [46,47] reported positive changes in VO_2 max, upper- and lower-body strength, and flexibility. In our study, we observed statistical changes in VO_2 max ($p < 0.001$), upper- (lat machine and chest press test, $p < 0.001$) and lower-body strength (leg extension and leg press test, $p < 0.001$) and flexibility (in horizontal bending and vertical bending test, $p < 0.001$) in the entire sample. These results are largely expected and obvious, and the improvements are due to the combined workouts that stimulate the systems more than what happens in activities of daily living. Using BMI categories as a between factor, we noted that baseline BMI values influenced VO_2 max ($p < 0.001$), lower- ($p < 0.001$) and upper-body strength ($p \leq 0.001$), and HB ($p = 0.021$). In these variables, we observed a lower improvement in people with overweight than in the other two groups, according to previous literature concerning obesity's impact on muscular strength [48].

In our sample, baseline PAL (Table 2b) categories appeared to influence the effects of exercise on lat ($p = 0.038$) and chest press ($p = 0.019$) and leg extension ($p = 0.016$) test values. In fact, we observed greater improvement in people with higher baseline PAL than in people in the others two groups. We could postulate that it can be linked to the fact that the highest performing people were able to load more kilograms during strength exercises, thanks to greater resistance to effort. The SIT (Table 2c) category factor did not seem to influence these variables. Such results must encourage us to promote exercise interventions in all people, independently of body weight and SIT. It is essential, however, to tailor exercise programs for the obese by paying attention to different effort perception and motivation in obese people vs. in normal weight people [49,50]. For obese people, their physical condition is reported as a barrier to exercise; difficulties related to the physical condition of obesity may reduce the rhythm of daily activities, such as walking or exercise.

Appropriate management of the health of overweight and obese adults [51] with and without DM2 should include physical activities and exercise prescription [52]. To better tailor exercise prescription [53], assessment of PAL and SIT represents an essential first step. Unfortunately, clinical settings seem to be in increasingly short supply due to scarce time and economic resources, and these constraints often necessitate a simple, low-cost, rapid assessment tool. Even though some authors have explained that self-reported data are often subject to biases and poor agreement between objective and subjective measures of physical activity has been reported [12,34], the IPAQ-SF is a validated tool [34], used in many clinical settings such as ours. In our study, we collected participants' self-reported measures (such as PAL and SIT information) through the IPAQ-SF questionnaire.

High amounts of sedentary time (daily/weekly sitting time) have been associated with a significantly greater risk for metabolic syndrome and DM2 [22,54–57]. As found by Balducci et al. (2019) [58], who reported that an exercise intervention strategy resulted in increased physical activity level and decreased sedentary time, in our study, we observed an improvement in PAL and a decrease in SIT for the entire sample, although the SIT reduction was not statistically significant.

Limitations. This study has some limitations. First, our work did not include a control group. Further, some outcomes were based on self-reported questionnaire measures. To overcome this problem, at least in part, we carefully selected an internationally validated tool (IPAQ). Moreover, we did not use objective measurements (i.e., accelerometry) during exercise sessions. It is also necessary to underline that individuals with musculoskeletal disorders or other clinical conditions that could contraindicate exercise were excluded from the analyses. This aspect may affect the generalizability of our findings. Additionally, in this study, we did not present nutritional data. In future studies, an analysis of eating habits should be carried out before and after the training sessions, given that the nutritional component is relevant in this type of subject. Another limit to underline is the time from data collection to the submission of the manuscript. In the meantime, physical activity programs and new technologies have evolved newer performance and evaluation programs. This could be conditioning the program itself and the result. Finally, we preferred to report the results of this paper using the categories "physical activity level", "sedentary activity time", and "BMI" separately. In real time, these factors could be "mixed" with each other (i.e., we can have a high level of physical activity and a high sedentary time in the same subject or a low level of physical activity and a low sedentary time). This scenario could somehow affect the results.

5. Conclusions

Currently there is an urgent global need to better understand the improvements in health outcomes derived from reducing SIT and implementing PAL, especially in overweight and obese people who represent a worldwide pandemic emergency. Our study results showed an improvement of PAL in the participants following the three-month intensive exercise program, as well as an improvement in several health-related outcomes observed. Our data suggest that baseline PAL and SIT do not seem to influence all the effects observed in a sample composed by overweight and obese participants with and without DM2. These results must encourage us even more to promote exercise interventions in all people, independently from body weight and SIT.

Future investigations that include more objective instruments (i.e., accelerometry) and control groups should be conducted to obtain further evidence through experimental and translational research in order to better inform public health policy, particularly in terms of tailored exercise prescription addressed to people with obesity and/or DM2. Implementations of a supervised exercise intervention—as shown in this study—produced positive results in health-related outcomes in a group of overweight and obese adults with and without DM2. In our opinion, the evidence-based methodology assessments (C.U.R.I.A.Mo. clinical model protocol), including standardized tests to assess physical measures and other variables, are strengths of this study.

Supplementary Materials: The following are available online at https://www.mdpi.com/article/10.3390/jfmk7010012/s1, Table S1a–c. Baseline values: mean values of all parameters in the entire sample and in the subgroups.

Author Contributions: Conceptualization, R.P. and A.D.B.; methodology, R.P., C.G.F. and A.D.B.; software, R.P. and V.B. (Vittorio Bini); validation, R.P. and C.G.F.; formal analysis, R.P. and V.B. (Vittorio Bini); investigation, R.P.; resources, C.G.F.; data curation, R.P. and V.B. (Vittorio Bini); writing—original draft preparation, R.P., L.C., M.B., V.B. (Valentina Bullo), S.G. and A.D.B.; writing—review and editing, R.P., L.C., M.B., V.B. (Vittorio Bini), C.G.F., V.B. (Valentina Bullo), S.G. and A.D.B.; supervision, R.P., C.G.F., M.B., L.C. and A.D.B.; project administration, R.P. and C.G.F.; funding acquisition, C.G.F. All authors have read and agreed to the published version of the manuscript.

Funding: This research received no external funding.

Institutional Review Board Statement: The study was conducted according to the guidelines of the Declaration of Helsinki. The C.U.R.I.A.Mo. project has been approved by local committee CEAS (Comitato Etico Aziende Sanitarie Regione dell' Umbria), with the number: 1/10/1633 and registered in the Australian New Zealand Clinical Trials Registry (a primary registry in the WHO registry network), with the number: ACTRN12611000255987.

Informed Consent Statement: Informed consent was obtained from all subjects involved in the study.

Data Availability Statement: The datasets used and/or analyzed during the current study are available from the corresponding author on reasonable request.

Conflicts of Interest: The authors declare no conflict of interest.

References

1. Whipple, M.O.; Regensteiner, J.G.; Bergouignan, A. Is Being Physically Active Enough to Be Metabolically Healthy? The Key Role of Sedentary Behavior. *Diabetes Care* **2021**, *44*, 17–19. [CrossRef] [PubMed]
2. Nakanishi, S.; Hirukawa, H.; Shimoda, M.; Tatsumi, F.; Kohara, K.; Obata, A.; Okauchi, S.; Katakura, Y.; Sanada, J.; Fushimi, Y.; et al. Impact of physical activity and sedentary time on glycated hemoglobin levels and body composition: Cross-sectional study using outpatient clinical data of Japanese patients with type 2 diabetes. *J. Diabetes Investig.* **2020**, *11*, 633–639. [CrossRef] [PubMed]
3. Thivel, D.; Tremblay, A.; Genin, P.M.; Panahi, S.; Riviere, D.; Duclos, M. Physical Activity, Inactivity, and Sedentary Behaviors: Definitions and Implications in Occupational Health. *Front. Public Health* **2018**, *6*, 288. [CrossRef] [PubMed]
4. World Health Organization. *WHO Guidelines on Physical Activity and Sedentary Behaviour*; WHO: Geneva, Switzerland, 2020.
5. Tremblay, M.S.; Aubert, S.; Barnes, J.D.; Saunders, T.J.; Carson, V.; Latimer-Cheung, A.E.; Chastin, S.F.M.; Altenburg, T.M.; Chinapaw, M.J.M.; On Behalf of SBRN Terminology Consensus Project Participants. Sedentary Behavior Research Network (SBRN)—Terminology Consensus Project process and outcome. *Int. J. Behav. Nutr. Phys. Act.* **2017**, *14*, 75. [CrossRef] [PubMed]
6. Magnon, V.; Dutheil, F.; Auxiette, C. Sedentariness: A Need for a Definition. *Front. Public Health* **2018**, *6*, 372. [CrossRef] [PubMed]
7. Biswas, A.; Oh, P.I.; Faulkner, G.E.; Bajaj, R.R.; Silver, M.A.; Mitchell, M.S.; Alter, D.A. Sedentary time and its association with risk for disease incidence, mortality, and hospitalization in adults: A systematic review and meta-analysis. *Ann. Intern. Med.* **2015**, *162*, 123–132. [CrossRef]
8. Madden, K.M.; Feldman, B.; Chase, J. Sedentary Time and Metabolic Risk in Extremely Active Older Adults. *Diabetes Care* **2020**, *44*, 194–200. [CrossRef]
9. Hu, F.B.; Li, T.Y.; Colditz, G.A.; Willett, W.C.; Manson, J.E. Television Watching and Other Sedentary Behaviors in Relation to Risk of Obesity and Type 2 Diabetes Mellitus in Women. *J. Am. Med. Assoc.* **2003**, *289*, 1785–1791. [CrossRef]
10. Ekelund, U.; Steene-Johannessen, J.; Brown, W.J.; Fagerland, M.W.; Owen, N.; Powell, K.E.; Bauman, A.; Lee, I.-M.; Ding, D.; Lancet Sedentary Behaviour Working Group; et al. Does physical activity attenuate, or even eliminate, the detrimental association of sitting time with mortality? A harmonised meta-analysis of data from more than 1 million men and women. *Lancet* **2016**, *388*, 1302–1310. [CrossRef]
11. Shields, M.; Tremblay, M.S. Sedentary behaviour and obesity. *Health Rep.* **2008**, *19*, 19–30.
12. Lee, P.H.; Macfarlane, D.J.; Lam, T.H.; Stewart, S.M. Validity of the international physical activity questionnaire short form (IPAQ-SF): A systematic review. *Int. J. Behav. Nutr. Phys. Act.* **2011**, *8*, 115. [CrossRef] [PubMed]
13. Craig, C.L.; Marshall, A.L.; Sjöström, M.; Bauman, A.E.; Booth, M.L.; Ainsworth, B.; Pratt, M.; Ekelund, U.; Yngve, A.; Sallis, J.F.; et al. International Physical Activity Questionnaire: 12-Country Reliability and Validity. *Med. Sci. Sports Exerc.* **2003**, *35*, 1381–1395. [CrossRef] [PubMed]
14. Van Poppel, M.N.M.; Chinapaw, M.J.M.; Mokkink, L.B.; Van Mechelen, W.; Terwee, C.B. Physical Activity Questionnaires for Adults: A systematic review of measurement properties. *Sports Med.* **2010**, *40*, 565–600. [CrossRef] [PubMed]
15. Standard Italiani per la Cura dell'Obesità SIO-ADI. Available online: https://www.sio-obesita.org/wp-content/uploads/2017/09/STANDARD-OBESITA-SIO-ADI.pdf (accessed on 2 November 2021).
16. Petridou, A.; Siopi, A.; Mougios, V. Exercise in the management of obesity. *Metabolism* **2019**, *92*, 163–169. [CrossRef] [PubMed]
17. Battista, F.; Ermolao, A.; van Baak, M.A.; Beaulieu, K.; Blundell, J.E.; Busetto, L.; Carraça, E.V.; Encantado, J.; Dicker, D.; Farpour-Lambert, N.; et al. Effect of exercise on cardiometabolic health of adults with overweight or obesity: Focus on blood pressure, insulin resistance, and intrahepatic fat—A systematic review and meta-analysis. *Obes. Rev.* **2021**, *22*, e13269. [CrossRef] [PubMed]
18. Henson, J.; Dunstan, D.W.; Davies, M.; Yates, T. Sedentary behaviour as a new behavioural target in the prevention and treatment of type 2 diabetes. *Diabetes/Metabolism Res. Rev.* **2016**, *32*, 213–220. [CrossRef] [PubMed]
19. Balducci, S.; Sacchetti, M.; Haxhi, J.; Orlando, G.; D'Errico, V.; Fallucca, S.; Menini, S.; Pugliese, G. Physical exercise as therapy for type 2 diabetes mellitus. *Diabetes/Metabolism Res. Rev.* **2014**, *30*, 13–23. [CrossRef]
20. Zanuso, S.; Jimenez, A.; Pugliese, G.; Corigliano, G.; Balducci, S. Exercise for the management of type 2 diabetes: A review of the evidence. *Acta Diabetol.* **2009**, *47*, 15–22. [CrossRef]
21. Gobbo, S.; Bullo, V.; Roma, E.; Duregon, F.; Bocalini, D.S.; Rica, R.L.; Di Blasio, A.; Cugusi, L.; Vendramin, B.; Bergamo, M.; et al. Nordic Walking Promoted Weight Loss in Overweight and Obese People: A Systematic Review for Future Exercise Prescription. *J. Funct. Morphol. Kinesiol.* **2019**, *4*, 36. [CrossRef]
22. Cugusi, L.; Cadeddu, C.; Nocco, S.; Orrù, F.; Bandino, S.; Deidda, M.; Caria, A.; Bassareo, P.P.; Piras, A.; Cabras, S.; et al. Effects of an Aquatic-Based Exercise Program to Improve Cardiometabolic Profile, Quality of Life, and Physical Activity Levels in Men with Type 2 Diabetes Mellitus. *PM&R* **2014**, *7*, 141–148. [CrossRef]
23. Cadeddu, C.; Nocco, S.; Cugusi, L.; Deidda, M.; Bina, A.; Fabio, O.; Bandinu, S.; Cossu, E.; Baroni, M.G.; Mercuro, G. Effects of metformin and exercise training, alone or in association, on cardio-pulmonary performance and quality of life in insulin resistance patients. *Cardiovasc. Diabetol.* **2014**, *13*, 93. [CrossRef]
24. Di Blasio, A.; Di Donato, F.; Di Santo, S.; Bucci, I.; Izzicupo, P.; Di Baldassarre, A.; Gallina, S.; Bergamin, M.; Ripari, P.; Napolitano, G. Aerobic physical exercise and negative compensation of non-exercise physical activity in post-menopause: A pilot study. *J. Sports Med. Phys. Fit.* **2018**, *58*, 1497–1508. [CrossRef]

25. De Feo, P.; Fatone, C.; Burani, P.; Piana, N.; Pazzagli, C.; Battistini, D.; Capezzali, D.; Pippi, R.; Chipi, B.; Mazzeschi, C. An innovative model for changing the lifestyles of persons with obesity and/or Type 2 diabetes mellitus. *J. Endocrinol. Investig.* **2011**, *34*, e349–e354. [CrossRef]
26. World Health Organization. Obesity and Overweight. Available online: http://www.who.int/news-room/fact-sheets/detail/obesity-and-overweight (accessed on 2 November 2021).
27. Guidelines for Data Processing and Analysis of the International Physical Activity Questionnaire (IPAQ). Available online: https://sites.google.com/site/theipaq/scoring-protocol (accessed on 15 June 2021).
28. Pippi, R.; Di Blasio, A.; Aiello, C.; Fanelli, C.; Bullo, V.; Gobbo, S.; Cugusi, L.; Bergamin, M. Effects of a Supervised Nordic Walking Program on Obese Adults with and without Type 2 Diabetes: The CURIA Mo. Centre Experience. *J. Funct. Morphol. Kinesiol.* **2020**, *5*, 62. [CrossRef] [PubMed]
29. Associazione Medici Diabetologi (AMD); Società Italiana di Diabetologia (SID). Standard Italiani Per La Cura Del Diabete Mellito 2018. Available online: http://www.siditalia.it/pdf/Standard%20di%20Cura%20AMD%20-%20SID%202018_protetto2.pdf (accessed on 14 June 2021).
30. Weiglein, L.; Herrick, J.; Kirk, S.; Kirk, E.P. The 1-mile walk test is a valid predictor of VO_{2max} and is a reliable alternative fitness test to the 1.5-mile run in US Air Force males. *Mil. Med.* **2011**, *176*, 669–673. [CrossRef] [PubMed]
31. Brzycki, M. Strength Testing—Predicting a One-Rep Max from Reps-to-Fatigue. *J. Phys. Educ. Recreat. Dance* **1993**, *64*, 88–90. [CrossRef]
32. Mayorga-Vega, D.; Marbán, R.M.; Viciana, J. Criterion-Related Validity of Sit-and-Reach Tests for Estimating Hamstring and Lumbar Extensibility: A Meta-Analysis. *J. Sports Sci. Med.* **2014**, *13*, 1–14.
33. Questionario Internazionale Sull'attività Fisica (IPAQ Italiano) Versione Breve. Available online: https://sites.google.com/site/theipaq/questionnaire_links (accessed on 15 June 2021).
34. Mannocci, A.; Di Thiene, D.; Del Cimmuto, A.; Masala, D.; Boccia, A.; De Vito, E.; La Torre, G. International Physical Activity Questionnaire: Validation and Assessment in an Italian Sample. *Ital. J. Public Health* **2010**, *7*, 369–376. [CrossRef]
35. Cohen, J. *Statistical Power Analysis for the Behavioral Sciences*; Routledge Academic: New York, NY, USA, 1977; p. 283.
36. Shaw, K.A.; Gennat, H.C.; O'Rourke, P.; Del Mar, C. Exercise for overweight or obesity. *Cochrane Database Syst. Rev.* **2006**, *4*, CD003817. [CrossRef]
37. Reginato, E.; Pippi, R.; Aiello, C.; Tomaro, E.S.; Ranucci, C.; Buratta, L.; Bini, V.; Marchesini, G.; De Feo, P.; Fanelli, C. Effect of Short Term Intensive Lifestyle Intervention on Hepatic Steatosis Indexes in Adults with Obesity and/or Type 2 Diabetes. *J. Clin. Med.* **2019**, *8*, 851. [CrossRef]
38. Tomaro, E.S.; Pippi, R.; Reginato, E.; Aiello, C.; Buratta, L.; Mazzeschi, C.; Perrone, C.; Ranucci, C.; Tirimagni, A.; Russo, A.; et al. Intensive lifestyle intervention is particularly advantageous in poorly controlled type 2 diabetes. *Nutr. Metab. Cardiovasc. Dis.* **2017**, *27*, 688–694. [CrossRef]
39. Balducci, S.; Haxhi, J.; Sacchetti, M.; Orlando, G.; Cardelli, P.; Vitale, M.; Mattia, L.; Iacobini, C.; Bollanti, L.; Conti, F.; et al. Relationships of Changes in Physical Activity and Sedentary Behavior with Changes in Physical Fitness and Cardiometabolic Risk Profile in Individuals with Type 2 Diabetes: The Italian Diabetes and Exercise Study 2 (IDES_2). *Diabetes Care* **2021**, *45*, 213–221. [CrossRef] [PubMed]
40. Cheng, C.-C.; Hsu, C.-Y.; Liu, J.-F. Effects of dietary and exercise intervention on weight loss and body composition in obese postmenopausal women: A systematic review and meta-analysis. *Menopause* **2018**, *25*, 772–782. [CrossRef] [PubMed]
41. Schwingshackl, L.; Dias, S.; Strasser, B.; Hoffmann, G. Impact of Different Training Modalities on Anthropometric and Metabolic Characteristics in Overweight/Obese Subjects: A Systematic Review and Network Meta-Analysis. *PLoS ONE* **2013**, *8*, e82853. [CrossRef]
42. Neeland, I.J.; Ross, R.; Després, J.-P.; Matsuzawa, Y.; Yamashita, S.; Shai, I.; Seidell, J.; Magni, P.; Santos, R.D.; Arsenault, B.; et al. Visceral and ectopic fat, atherosclerosis, and cardiometabolic disease: A position statement. *Lancet Diabetes Endocrinol.* **2019**, *7*, 715–725. [CrossRef]
43. Stoner, L.; Rowlands, D.; Morrison, A.; Credeur, D.; Hamlin, M.; Gaffney, K.; Lambrick, D.; Matheson, A. Efficacy of Exercise Intervention for Weight Loss in Overweight and Obese Adolescents: Meta-Analysis and Implications. *Sports Med.* **2016**, *46*, 1737–1751. [CrossRef]
44. Dieli-Conwright, C.M.; Courneya, K.S.; Demark-Wahnefried, W.; Sami, N.; Lee, K.; Sweeney, F.C.; Stewart, C.; Buchanan, T.A.; Spicer, D.; Tripathy, D.; et al. Aerobic and resistance exercise improves physical fitness, bone health, and quality of life in overweight and obese breast cancer survivors: A randomized controlled trial. *Breast Cancer Res.* **2018**, *20*, 124. [CrossRef] [PubMed]
45. Hsu, K.-J.; Chien, K.-Y.; Tsai, S.-C.; Tsai, Y.-S.; Liao, Y.-H.; Chen, J.-J.; Chen, Y.-R.; Chen, C.-N. Effects of Exercise Alone or in Combination with High-Protein Diet on Muscle Function, Aerobic Capacity, and Physical Function in Middle-Aged Obese Adults: A Randomized Controlled Trial. *J. Nutr. Health Aging* **2021**, *25*, 727–734. [CrossRef]
46. Balducci, S.; Zanuso, S.; Cardelli, P.; Salvi, L.; Bazuro, A.; Pugliese, L.; Maccora, C.; Iacobini, C.; Conti, F.G.; Nicolucci, A.; et al. Effect of High- versus Low-Intensity Supervised Aerobic and Resistance Training on Modifiable Cardiovascular Risk Factors in Type 2 Diabetes; The Italian Diabetes and Exercise Study (IDES). *PLoS ONE* **2012**, *7*, e49297. [CrossRef] [PubMed]
47. Balducci, S.; Zanuso, S.; Cardelli, P.; Salvi, L.; Mazzitelli, G.; Bazuro, A.; Iacobini, C.; Nicolucci, A.; Pugliese, G. Changes in Physical Fitness Predict Improvements in Modifiable Cardiovascular Risk Factors Independently of Body Weight Loss in Subjects with Type 2 Diabetes Participating in the Italian Diabetes and Exercise Study (IDES). *Diabetes Care* **2012**, *35*, 1347–1354. [CrossRef]
48. Maffiuletti, N.A.; Ratel, S.; Sartorio, A.; Martin, V. The Impact of Obesity on In Vivo Human Skeletal Muscle Function. *Curr. Obes. Rep.* **2013**, *2*, 251–260. [CrossRef]

49. Di Loreto, C.; Fanelli, C.; Lucidi, P.; Murdolo, G.; De Cicco, A.; Parlanti, N.; Santeusanio, F.; Brunetti, P.; De Feo, P. Validation of a Counseling Strategy to Promote the Adoption and the Maintenance of Physical Activity by Type 2 Diabetic Subjects. *Diabetes Care* **2003**, *26*, 404–408. [CrossRef] [PubMed]
50. de Aguiar, J.B.; Gurgel, L.A.; Santos, A.L.B.; Arruda, S.P.M. Barriers to Physical Exercise and Associated Factors in the Pre- and Postoperative Periods of Bariatric Surgery. *Obes. Surg.* **2021**, *31*, 1696–1704. [CrossRef] [PubMed]
51. Verboven, K.; Hansen, D. Critical Reappraisal of the Role and Importance of Exercise Intervention in the Treatment of Obesity in Adults. *Sports Med.* **2021**, *51*, 379–389. [CrossRef]
52. Minetto, M.A.; Motta, G.; Gorji, N.E.; Lucini, D.; Biolo, G.; Pigozzi, F.; Portincasa, P.; Maffiuletti, N.A. Reproducibility and validity of the Italian version of the International Physical Activity Questionnaire in obese and diabetic patients. *J. Endocrinol. Investig.* **2018**, *41*, 343–349. [CrossRef]
53. De Vos, B.C.; Runhaar, J.; Van Middelkoop, M.; Krul, M.; Bierma-Zeinstra, S.M. Long-term effects of a randomized, controlled, tailor-made weight-loss intervention in primary care on the health and lifestyle of overweight and obese women. *Am. J. Clin. Nutr.* **2016**, *104*, 33–40. [CrossRef]
54. Hamilton, M.T.; Hamilton, D.G.; Zderic, T.W. Sedentary Behavior as a Mediator of Type 2 Diabetes. *Acute Topics Sport Nutr.* **2014**, *60*, 11–26. [CrossRef]
55. Sjöros, T.; Vähä-Ypyä, H.; Laine, S.; Garthwaite, T.; Lahesmaa, M.; Laurila, S.M.; Latva-Rasku, A.; Savolainen, A.; Miikkulainen, A.; Löyttyniemi, E.; et al. Both sedentary time and physical activity are associated with cardiometabolic health in overweight adults in a 1 month accelerometer measurement. *Sci. Rep.* **2020**, *10*, 20578. [CrossRef]
56. Powell, C.; Herring, M.; Dowd, K.; Donnelly, A.; Carson, B.P. The cross-sectional associations between objectively measured sedentary time and cardiometabolic health markers in adults—A systematic review with meta-analysis component. *Obes. Rev.* **2017**, *19*, 381–395. [CrossRef]
57. Skrede, T.; Steene-Johannessen, J.; Anderssen, S.A.; Resaland, G.K.; Ekelund, U. The prospective association between objectively measured sedentary time, moderate-to-vigorous physical activity and cardiometabolic risk factors in youth: A systematic review and meta-analysis. *Obes. Rev.* **2018**, *20*, 55–74. [CrossRef]
58. Balducci, S.; D'Errico, V.; Haxhi, J.; Sacchetti, M.; Orlando, G.; Cardelli, P.; Vitale, M.; Bollanti, L.; Conti, F.; Zanuso, S.; et al. Effect of a Behavioral Intervention Strategy on Sustained Change in Physical Activity and Sedentary Behavior in Patients with Type 2 Diabetes. *J. Am. Med Assoc.* **2019**, *321*, 880–890. [CrossRef] [PubMed]

Journal of
Functional Morphology and Kinesiology

MDPI

Article

Effects of 12 Months of Vitamin D Supplementation on Physical Fitness Levels in Postmenopausal Women with Type 2 Diabetes

Claudio Melibeu Bentes [1,2], Pablo B. Costa [3,*], Monique Resende [2], Claudia Netto [4], Ingrid Dias [5], Anderson Luiz Bezerra da Silveira [1,6], Fabrizio Di Masi [1], Humberto Miranda [5,7], Lucas Monteiro de Carvalho [1,6] and Lizanka Marinheiro [3]

1 Laboratory of Physiology and Human Performance, Department of Physical Education and Sports, Institute of Education, Federal Rural University of Rio de Janeiro, Seropedica 23890-000, RJ, Brazil; claudiomelibeu@ufrrj.br (C.M.B.); andersonsilveira@ufrrj.br (A.L.B.d.S.); fabriziomasi@ufrrj.br (F.D.M.); lucasmonteirodc@gmail.com (L.M.d.C.)
2 Graduate Program, Fernandes Figueira Institute, Oswaldo Cruz Foundation, Rio de Janeiro 22250-020, RJ, Brazil; moniqueres@gmail.com
3 Department of Kinesiology, California State University, Fullerton, CA 92831-3547, USA; lizanka@globo.com
4 Department of Biochemistry, School of Medicine and Surgery, Biomedical Institute, Federal University of the State of Rio de Janeiro (UNIRIO), Rio de Janeiro 20211-040, RJ, Brazil; claucnetto@gmail.com
5 Physical Education Graduate Program, Department of Gymnastics, School of Physical Education and Sports, Federal University of Rio de Janeiro, Rio de Janeiro 21941-599, RJ, Brazil; ingriddias.ufrj@gmail.com (I.D.); humbertomiranda@eefd.ufrj.br (H.M.)
6 Physiological Sciences Multicenter Graduate Program, Department of Physiological Sciences, Health Sciences Center, Institute of Biological and Health Sciences, Federal Rural University of Rio de Janeiro, Seropedica 23890-000, RJ, Brazil
7 Physical Education Graduate Program, Performance, Training, and Physical Exercise Laboratory, Department of Gymnastics, School of Physical Education and Sports, Federal University of Rio de Janeiro, Rio de Janeiro 21941-599, RJ, Brazil
* Correspondence: pcosta@fullerton.edu

Citation: Bentes, C.M.; Costa, P.B.; Resende, M.; Netto, C.; Dias, I.; da Silveira, A.L.B.; Di Masi, F.; Miranda, H.; de Carvalho, L.M.; Marinheiro, L. Effects of 12 Months of Vitamin D Supplementation on Physical Fitness Levels in Postmenopausal Women with Type 2 Diabetes. *J. Funct. Morphol. Kinesiol.* **2021**, *6*, 87. https://doi.org/10.3390/jfmk6040087

Academic Editor: Cristina Cortis

Received: 20 September 2021
Accepted: 21 October 2021
Published: 25 October 2021

Abstract: Introduction: It is common for postmenopausal women to receive a vitamin D supplementation prescription to assist in preventing future falls and to maintain bone health. However, the association between vitamin D supplementation and physical fitness components has not been studied in older women with diabetes. Objective: We examined the influence of 12 months of vitamin D supplementation on the components of physical fitness in postmenopausal women with type 2 diabetes mellitus (T2DM). Methods: Thirty-five postmenopausal women (62.48 ± 7.67 years; 154.6 ± 5.11 cm; 73.93 ± 15.43 kg; 31.13 ± 5.82 BMI) with a diagnosis of T2DM participated in this longitudinal study where participants were supplemented with 1000 IU/day of vitamin D over 12 months. Subjects performed fasting blood samples, anthropometric assessments, body composition, clinical exams, and physical tests at 6-month intervals (P0, P6, and P12). Results and Conclusion: Vitamin D supplementation alone was effective in postmenopausal women in increasing serum vitamin D levels, altering muscle strength levels, promoting improvements in muscle function, as well as preventing and controlling fragility caused by T2DM and aging.

Keywords: cholecalciferol; elderly women; functional capacity; metabolism; athletic performance; hyperglycemia

1. Introduction

The growth of the elderly population is a worldwide phenomenon, and it is estimated that over the next 30 years, the proportion of people over 65 will double from 12% to 22–25% in the US and Canada [1]. Due to the increase in life expectancy, several diseases related to the climacteric period of women now present serious problems for public health [2]. Women tend to report dizziness, tiredness, night sweats, depression, hot flashes, sleep disorders, as well as cardiovascular diseases, osteoporosis, metabolic syndrome, and diabetes, among others [3].

The changes occurring in the transition between menopause and post menopause are physiological, but some may cause symptoms that worsen the quality of life and others that may increase the risk of various diseases [2]. Low levels of female sex steroids, regardless of cause, can lead to endocrine and functional disorders such as sexual dysfunction, loss of libido, altered levels of lipoproteins, and increased risk of obesity, cardiovascular diseases, and cardiometabolic diseases such as type 2 diabetes mellitus (T2D) [3].

T2D is the most prevalent form of diabetes worldwide, representing 90% of global cases, and projection is expected to rise to 300 million cases worldwide by 2025 [4]. Despite the constant campaign to promote lifestyle changes, such as dietary re-education and inclusion of physical activity for better metabolic control, these are difficult measures to achieve and maintain [5]. Similar to T2D, vitamin D deficiency is considered a worldwide epidemic with multiple implications for human health because of its role in various physiological systems [6]. However, several studies have already shown that normal and higher levels of vitamin D may reduce the incidence of non-vertebral fractures and hip fractures [6].

One of the age groups that suffers with comorbidities associated with hypovitaminosis D is postmenopausal women [6]. Several longitudinal studies have already demonstrated an association between hypovitaminosis D and the increased risk of chronic diseases such as cardiovascular diseases [7] and T2D [8]. Low levels of vitamin D and the development of T2D may be associated with the action of insulin resistance, increasing glucose intolerance [9]. In addition, in the postmenopausal period, it is common for women to receive a prescription for vitamin D supplementation as a method of prevention for future falls and maintenance of bone health. However, the association between vitamin D supplementation and physical fitness components has not been studied in the scientific literature in menopausal and diabetic women, although a few studies have investigated the effects of vitamin D supplementation on strength and muscular power gains [10].

In a recent review of the literature, Bentes et al. [11] reported the shortage of studies, and in a short review, the authors describe five longitudinal studies concluding that for vitamin D to have an effect on the postmenopausal age range, daily doses need to be greater than 1000 IU/day. Typically, these studies with vitamin D_3 supplementation used the cholecalciferol form. Consequently, it is essential for longitudinal studies to investigate the effects of vitamin D supplementation on changes in physical fitness components in postmenopausal women with T2D, since this topic requires further investigation as it may represent an important resource in the control and treatment of fragility caused by aging. Therefore, the purpose of the present study was to examine the effects of vitamin D supplementation for 12 months on physical fitness in postmenopausal women with T2D.

2. Methods

2.1. Participants and Research Design

A longitudinal, paired clinical study, with a quasi-experimental characteristic, was performed in a period of 12 months. One hundred and ten women from the gynecology outpatient clinic, who were already postmenopausal and diagnosed with T2D, were recruited to participate in the study. After the invitation, the following exclusion criteria were applied: neurological problems that compromise balance and gait; patients who were supplementing with a dose of more than 1000 u/d of vitamin D (all types) continuously for more than three months; cognitive deficits, deformities in the upper and lower limbs, severe vision problems, dizziness symptoms, severe hearing loss, uncontrolled arterial hypertension, and postural hypotension, use of balance-compromising drugs (sedatives and anticonvulsants); excessive use of alcohol; obesity Grade III, liver diseases and nephropathies, use of glucocorticoids, antiretroviral medication for HIV, hyperparathyroidism, hypercalcemia, lymphomas, granulomatous diseases, current neoplasm. The initial group for the study was 40 participants at the end of the recruitment and consent (Table 1).

Table 1. Sample Characteristics.

	Mean	±	SD	K-S
Age (years)	62.48	±	7.67	0.067
Height (cm)	154.6	±	5.11	0.798
Body mass (kg)	73.93	±	15.43	0.052
Waist circumference (cm)	96.99	±	14.25	0.636
Abdominal circumference (cm)	98.46	±	11.54	0.666
Iliac circumference (cm)	101.42	±	12.24	0.545
Hip circumference (cm)	104.05	±	10.20	0.195
Waist/hip ratio	0.93	±	0.08	0.336
BMI (kg/m^2)	31.13	±	5.82	0.404
Lean body mass (kg)	22.29	±	3.23	0.162
Fat mass (kg)	33.17	±	10.87	0.355
Fat percentage (%)	43.62	±	6.36	0.269
Visceral fat area (cm^2)	124.36	±	31.06	0.248
Resting Metabolic Rate (kcal/d)	1259.54	±	116.20	0.112
Vitamin D status (ng/mL)	27.47	±	8.98	0.073
Fasting glucose (mg/dL)	144.65	±	55.99	0.092

Legend: K–S: Kolmogorov–Smirnov normality test. SD—Standard Deviation.

2.2. Procedures

Participants who met the criteria and who signed the term during the invitation consultation were aware of the start and end dates of data collection and the amount of vitamin D that would be administered, as well as the procedures for the collection, where they should be fasting for 12 h and in light and comfortable clothes. The kits were delivered at the first visit with the amount of supplementation for three months with the daily amount of 1000/ud of vitamin D (cholecalciferol) in pill form and subjects were instructed to return after three months with the empty bottles to be replenished.

All physical tests were performed in random order. At the first visit (P0), the patient underwent blood collection, routine outpatient consultation, body composition analysis, and then received breakfast. After these initial tests, the subjects participated in the physical fitness tests. The routine outpatient consultation consisted of a complete medical examination and analysis of the patient's health status. If necessary, the physician prescribed medication to control the diabetes with oral glucose-regulating drugs. The appointments for replacement of the vitamin D kits and routine laboratory examinations were rescheduled at three-month intervals. However, the procedures for analysis were only repeated with the 6-month intervals (P6 and P12). In addition, they were performed by the same investigators. Thus, the measures for the analysis of the outcome were performed in the following moments: pre-experiment (P0), six months after (P6), and 12 months after (P12) (Figure 1).

2.3. Physical Tests

Handgrip strength was measured using a hydraulic grip strength dynamometer (Jamar Hydraulic Hand Dynamometer Model J00105, Lafayette Instrument Company, La Fayette, IL, USA). Grip strength was measured three times using the dominant hand while the subject was in a seated position, shoulder adducted and neutrally rotated, elbow flexed at 90° with the forearm in neutral position, with a 1 min rest period between attempts. The highest recorded value of three attempts was used for analyses. Relative strength was calculated with the equation: handgrip strength (kg) ÷ body mass (kg), as suggested by Prestes and Tibana [12].

The timed up and go (TUG) test was used to examine functional mobility, muscle function, walking speed, and dynamic balance [13]. This test involves the time taken to rise from a chair, walk 3 m, turn around a marker, walk back to the chair, and sit down [13].

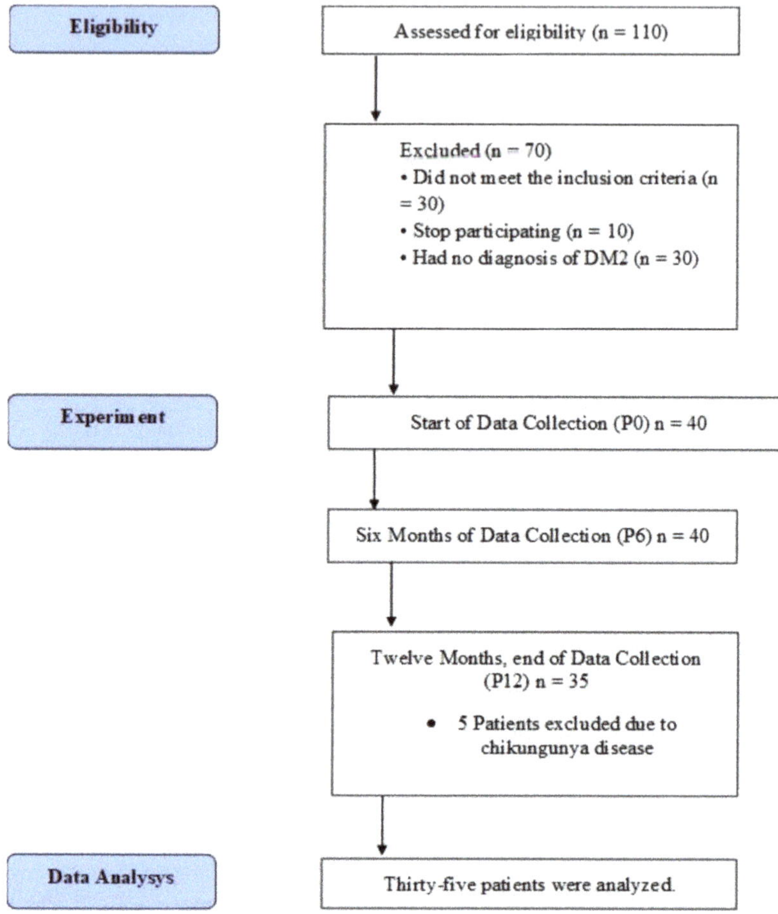

Figure 1. Data Collection Flowchart.

The arm curl test was used to measure the upper body local muscle endurance [14]. Subjects performed seated biceps curls without bending the trunk forward for 30 s with 2.3 kg dumbbells. The score used was the total number of arm curls.

The 30 s chair stand was used to measure lower limb local muscle endurance. The score equals the number of rises from a chair in 30 s with arms folded across the chest [14].

2.4. Anthropometry and Body Composition

Body height (BH) was measured with a stadiometer (Stadiometer Seca 208 Bodymeter), and waist, iliac, abdominal, and hip circumferences were measured with an anthropometric tape. In addition, body mass (BM), fat mass (FM), lean body mass, fat percentage, fat-free mass (FFM), and visceral fat area (VFA) were measured with an octopolar bioimpedance In-Body 720 (Biospace, Seoul, Korea). The validity of this bioimpedance for body composition has been previously documented [15].

2.5. Blood Sample Analyses

Blood samples were collected after fasting for 12 h. The main outcome parameters were fasting plasma glucose and vitamin D (25-hydroxyvitamin D [25-OH D]). The serum fasting glucose was measured by using the enzymatic colorimetric (GOD-PAP) method. The serum vitamin D concentrations were assessed by chemiluminescent assay.

2.6. Statistical Analyses

Statistical analysis was initially performed using the Kolmogorov–Smirnov normality test and the homoscedasticity test (Bartlett criterion). All variables demonstrated normal distribution and homoscedasticity ($p > 0.05$). Repeated measures ANOVA was used to compare the means to verify possible differences in time (P0 vs. P6 vs. P12). In case of significant differences, a Bonferroni post hoc for analysis of multiple comparisons between variables was used (P0 vs. P6; P0 vs. 12; P6 vs. 12). The significance level adopted was $p < 0.05$ and IBM SPSS 24.0 software was used for all statistical analyses.

3. Results

Primary Outcomes

In the comparisons of repeated measurements (P0 vs. P6 vs. P12), the ANOVA results demonstrated patients had significant increases in vitamin D ($p = 0.0001$). In the multiple comparisons, there were gains between the measurements of P0 (27.5 ± 9.0; Limit) vs. P6 (38.8 ± 12.1; Acceptable); $p = 0.0001$ and a percentage increase of 48.39% (11.55 ng/mL), P0 (27.5 ± 9.0; Limit) vs. P12 (38.5 ± 11.3; Acceptable); $p = 0.0001$ and a percentage increase of 32.35% (6.14 ng/mL) (Figure 2).

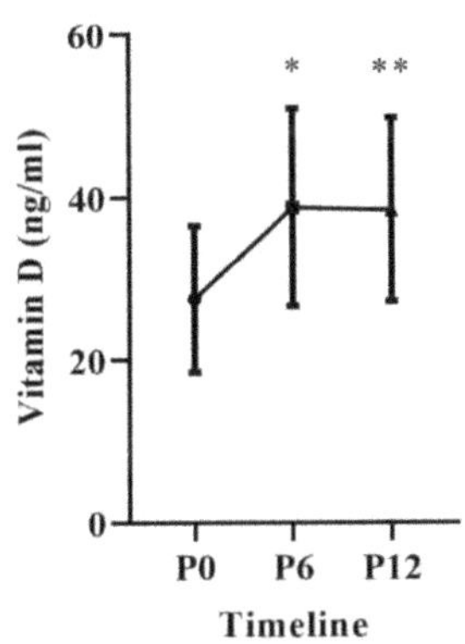

Figure 2. Blood concentration of 25 (OH) D over 12 months of supplementation. * Differences between P0 vs. P6, $p = 0.0001$. ** Differences between P0 vs. P12, $p = 0.0001$.

In repeated measures comparisons (P0 vs. P6 vs. P12), the ANOVA results demonstrated the patients had significant changes in the following variables: handgrip strength ($p = 0.0001$), relative strength ($p = 0.0001$), arm curl test ($p = 0.0002$), and 30 s chair stand ($p = 0.0001$). Additionally, there were no differences in the following variables: timed up and go ($p = 0.107$), sit and reach ($p = 0.625$).

In multiple comparisons, there were significant gains on handgrip strength between P0 (22.10 ± 5.47) vs. P6 (27.73 ± 4.94); $p = 0.0001$, P0 (22.10 ± 5.47) vs. P12 (29.37 ± 5.08); $p = 0.0001$. There were significant gains on relative strength between P0 (0.31 ± 0.08) vs. P6 (0.38 ± 0.09); $p = 0.0001$, P0 (0.31 ± 0.08) vs. P12 (0.41 ± 0.09); $p = 0.0001$. There were significant gains on arm curl test between P0 (12.20 ± 3.88) vs. P6 (15.35 ± 3.89); $p = 0.0001$, P0 (12.20 ± 3.88) vs. P12 (15.94 ± 3.09); $p = 0.0001$. There were significant gains on 30 s chair stand between P0 (8.53 ± 3.19) vs. P6 (11.26 ± 2.60); $p = 0.0001$, P0 (8.53 ± 3.19) vs. P12 (12.20 ± 2.41); $p = 0.0001$. Figure 3.

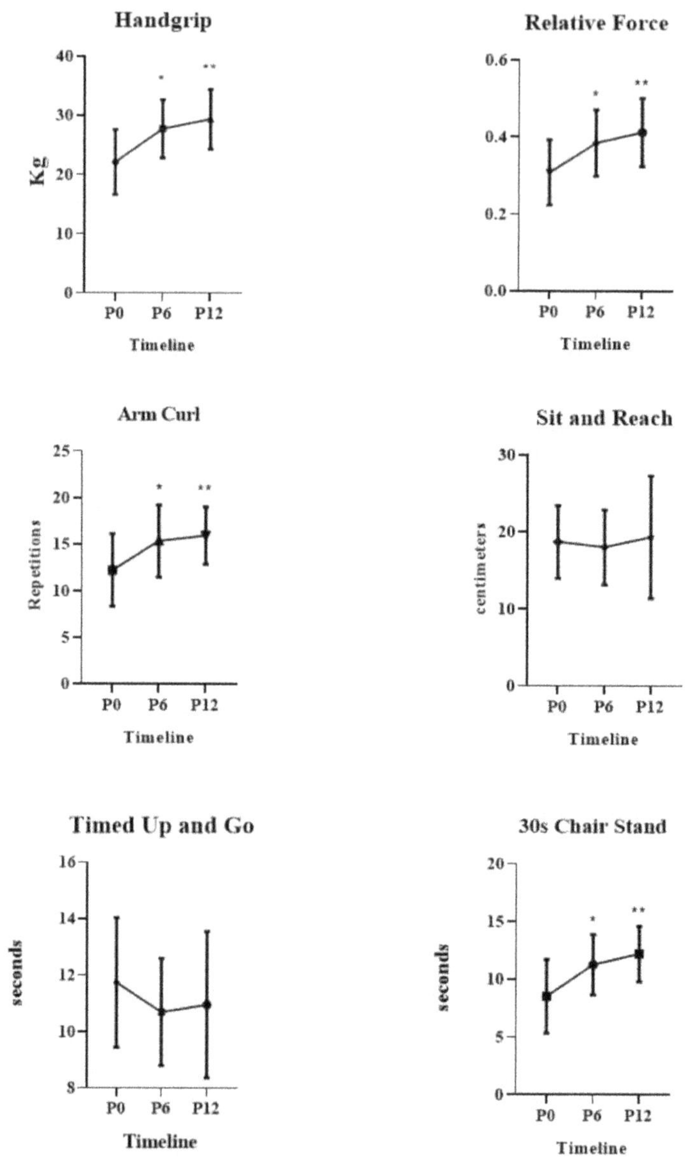

Figure 3. Physical Fitness results over 12-months of Vitamin D supplementation for elderly women. * Differences between P0 vs. P6, $p = 0.0001$, ** differences between P0 vs. P12, $p = 0.0001$.

4. Secondary Outcomes

In repeated measures comparisons (P0 vs. P6 vs. P12), the ANOVA results demonstrated the patients had significant changes on waist/hip ratio (WHR), $p = 0.04$. In multiple comparisons, there were significant changes between P0 (0.93 ± 0.08) vs. P12 (0.88 ± 0.08); $p = 0.041$, Table 2.

Table 2. Secondary Outcomes results.

Variable	Timeline	Mean	±	Standard Deviation	Confidence Interval 95%	
					Lower	Higher
Age (years)	P0	62.64	±	7.64	60.4	65.1
	P6	63.02	±	7.68	60.6	65.6
	P12	63.05	±	7.30	60.6	65.6
Body Mass (Kg)	P0	73.80	±	15.26	69.3	78.6
	P6	74.03	±	14.74	69.5	78.6
	P12	73.02	±	14.17	68.7	77.8
Waist circumference (cm)	P0	96.93	±	14.07	92.9	101.1
	P6	94.46	±	13.41	90.2	98.6
	P12	92.91	±	14.02	88.6	97.6
Abdominal circumference (cm)	P0	98.41	±	11.39	95.2	101.9
	P6	98.28	±	12.15	94.6	102.1
	P12	98.64	±	11.92	94.8	102.5
Iliac circumference (cm)	P0	101.68	±	12.19	98.2	105.6
	P6	101.82	±	11.05	98.4	105.4
	P12	103.80	±	11.56	100.1	107.5
Hip circumference (cm)	P0	104.13	±	10.07	101.1	107.3
	P6	104.60	±	9.13	101.7	107.5
	P12	104.85	±	9.68	101.7	108.0
WHR	P0	0.93	±	0.08	0.9	1.0
	P6	0.90	±	0.08	0.9	0.9
	P12	0.88 *	±	0.08	0.9	0.9
Lean body mass (kg)	P0	31.20	±	5.76	29.4	33.1
	P6	31.07	±	5.72	29.3	32.8
	P12	30.72	±	5.61	29.0	32.7
Bone Mass	P0	2.34	±	0.29	2.3	2.4
	P6	2.32	±	0.29	2.2	2.4
	P12	2.31	±	0.28	2.2	2.4
Lean body mass	P0	22.17	±	3.28	21.2	23.2
	P6	22.53	±	7.53	20.4	25.2
	P12	21.71	±	3.25	20.7	22.8
Fat mass	P0	33.24	±	10.74	30.2	36.5
	P6	33.29	±	10.57	30.0	36.7
	P12	32.80	±	10.25	29.7	36.3
Fat percentage (%)	P0	43.84	±	6.42	41.9	45.9
	P6	44.35	±	6.80	42.3	46.3
	P12	44.05	±	6.56	41.7	46.2
Visceral fat area	P0	124.09	±	31.42	114.7	133.6
	P6	123.26	±	29.43	114.5	132.5
	P12	123.39	±	28.95	114.4	133.5
Rest Metabolic Rate	P0	1255.10	±	118.08	1219.5	1290.5
	P6	1249.73	±	115.25	1215.6	1285.8
	P12	1238.94	±	114.32	1203.1	1277.8

* Differences between P0 vs. P12, $p = 0.041$.

5. Discussion

The main purpose of this study was to examine the influence of vitamin D supplementation for 12 months in postmenopausal women with T2DM on various components of physical fitness. Therefore, to follow the evolution of these outcomes, three time-points were tested over 12 months (P0, P6, and P12). Hypovitaminosis D is common in postmenopausal women. Possible causes of this condition include low sun exposure and reduced capacity to produce vitamin D, reduced renal function, lower absorption of vitamin D from the gastrointestinal tract, and use of multiple medications that may interfere with absorption and metabolism of this vitamin [16,17].

The main results in the present study demonstrated postmenopausal vitamin D supplementation alone, in addition to positively modifying serum levels of this limitrophe to desirable, and promoting significant increases in neuromuscular markers of physical fitness. In addition, the supplementation controlled the chronic loss of lean mass that is natural during aging and a characteristic of patients with T2D who also changed the distribution of body fat, altering indicators such as WHR.

Vitamin D availability in the body is assessed by measuring plasma concentrations of 25 (OH) D [18]. In participants of this study, serum concentrations were within the limitrophe classification in the beginning of data collection, which is below the desirable range. The International Osteoporosis Foundation and Endocrinology Society recommendations based on two meta-analyses suggest concentrations greater than 24 ng/mL to reduce fall rates [19] and above 30 ng/mL to reduce fracture rates [20]

In the present study, patients achieved the desirable classification (above 32 ng/mL) within 6 months and maintained it after 12 months with daily vitamin D supplementation (1000 IU/day). In any case, the population investigated in this clinical study consisted of postmenopausal women with T2DM who have this recommendation because they have risk factors for hypovitaminosis D [21].

Reduced vitamin D blood concentrations are associated with changes in muscle function [12]. Vitamin D receptors (VDR) are found in muscle tissue and are involved in activating muscle protein synthesis, meaning it is an important maintainer and important agent for muscle hypertrophy [22].

One study has shown vitamin D supplementation was able to increase serum VDR concentration by 30% and the amount of muscle fibers by 10% [23]. The main change induced by vitamin D deficiency is fast-twitch type 2 muscle atrophy, which is the first to be recruited during postural balance recovery, a fact that may explain the inverse association between serum levels of 25 (OH) D and falls [19].

The results of the present study support the data reported in the literature that during the follow-up period, as serum 25 (OH) D levels increased, participants had significant increases in muscle function at all times compared to P0. The most plausible explanation for this increase is in the study of Annweiler et al. [24], where the authors state that vitamin D supplementation can alter the oxidative capacity of muscle by improving muscle function.

There were significant increases in all physical fitness variables measured (handgrip, relative strength, arm curl test, 30 s chair stand) at moment P6 and P12. In a recent brief review, Bentes et al. [12] reported that only five studies to date have examined the influence of vitamin D supplementation alone on physical fitness markers; however, only three of the five studies using dosages at or above 1000 IU/day caused significant increases in strength gains, and this was the same dosage used in the present study during 12 months of follow-up.

Corroborating with our results, Zhu et al. [25] examined vitamin D supplementation on muscle strength and mobility in 302 elderly women (70–90 years) with 25 (OH) D insufficiency. The intervention group received doses of 1000 IU/day and the results show that based on baseline values, supplementation was able to increase muscle strength and muscle function in elderly women with vitamin D deficiency. Cangussu et al. [16] evaluated the daily supplementation of 1000 IU in 160 postmenopausal women, and the results demonstrated the intervention group obtained significant gains in vitamin D and consequently, significant increases in lower and upper limb muscle strength in addition to the control of lean mass loss. Similarly, the study by Anek et al. [26] investigated the effects of four weeks of vitamin D supplementation (20,000 IU/week) on markers of bone mass, muscle strength, and balance in 52 postmenopausal women (45–55 years). Results showed that after four weeks there were significant improvements in physical fitness.

In addition to the aforementioned fitness measures, as a secondary outcome, anthropometric measurements and bioimpedance were performed to assess body composition. Only the WHR measurement showed significant improvements. In contrast, during the 12 months of follow-up, there was maintenance of body mass, lean mass, and fat without any significant changes. This can be seen as a fundamental positive factor in the aging process, and an important factor in women with T2DM because of the acceleration in the process of muscle loss due to the pathological feature of the disease [27].

Nevertheless, WHR decreased by 4.08% in 12 months. The explanation lies in the process of transferring body fat. In the postmenopausal process, it is common for women to present changes in the postmenopausal process, it is usual for women to show changes in the abdominal/waist fat accumulation, with decreasing fat layer thickness in the hip and increasing in the visceral region [28]. Unexpectedly, in the present study vitamin D supplementation over the 12 months maintained body mass and body fat but presented significant decreases in WHR.

The strengths of this assay are serum vitamin D (25 (OH) D) measurements with the chemiluminescent assay that is most sensitive for detecting plasma vitamin D [29]. In addition, the functional measures of the present study have practical characteristics and fulfill the participants' daily functions, providing more accurate information on the analysis of variables related to muscle function. The type of population studied (i.e., postmenopausal women with T2DM in an age group under 65 years) demonstrates a strength for the study.

Our most important limitation was not being able to form a control group, as the hospital ethics committee did not allow the formation of a placebo condition in this age category due to the high number of studies already suggesting the benefits for preventing falls and improving bone health. Hence, we highlight the sample size of 40 participants, although the sample size calculation showed a power of 93%, enhancing the external validity of the study. In addition, there could have been a memory bias due to the long period between consultations. However, the investigators performed biweekly calls and quarterly appointments for vitamin replacement and recall of test procedures. Furthermore, the selection bias is associated with admission of patients who were already part of the female endocrinology service which can be assumed to be periodically seen by medical professionals and had constant access to general health care. Another limitation is related to the control of whether the patients used the vitamin D doses, although participants were given biweekly instructions and returned empty bottles indicating total consumption of the product. Lastly, researchers did not have nutrition control of the participants' diets and diet was not monitored. Thus, even though blood tests on P6 and P12 showed increases in serum vitamin D levels, the potential confounding variable of increasing serum vitamin D levels from diet intake alone cannot be overlooked.

6. Conclusions

In conclusion, vitamin D supplementation appeared to be effective in postmenopausal women to increase serum vitamin D levels, hence significantly altering muscle strength levels, and promoting improvements in muscle function, which can help control comorbidities associated with T2D and aging. In addition, serum vitamin D increased significantly, decreasing the WHR, demonstrating that it may be effective in redistributing body fat caused by menopause. Therefore, for 12 months patients maintained body mass, lean and fat mass, and increased muscle strength. Thus, it may be an excellent strategy for women in this age group who do not have any contraindication to supplement with this vitamin. Although the results demonstrate important benefits for postmenopausal diabetic women, the reduced sample size may limit external inferences. Nevertheless, we can consider these important findings as preliminary evidence of potential benefits of vitamin D supplementation for older women with diabetes.

Author Contributions: Conceptualization, C.M.B., M.R., C.N. and H.M.; Data curation, C.M.B. and P.B.C.; Formal analysis, C.M.B. and P.B.C.; Funding acquisition, C.M.B. and C.N.; Investigation, C.M.B. and M.R.; Methodology, C.M.B., I.D., A.L.B.d.S., F.D.M. and H.M.; Project administration, C.M.B., M.R. and L.M.; Supervision, C.N., H.M. and L.M.; Writing—original draft, L.M.d.C.; Writing—review and editing, C.M.B., P.B.C., I.D., A.L.B.d.S., F.D.M. and L.M.d.C. All authors have read and agreed to the published version of the manuscript.

Funding: This work was supported by grants from Coordenação de Aperfeiçoamento de Pessoal de Nível Superior (CAPES—Finance code 001).

Institutional Review Board Statement: The study was conducted according to the guidelines of the Declaration of Helsinki, and approved by the Ethics Committee of Oswaldo Cruz Foundation (protocol code 23081213.6.0000.5269 approved on 4 July 2016).

Informed Consent Statement: Informed consent was obtained from all subjects involved in the study. The research was conducted at the Endocrinology Outpatient Clinic of the National Institute of Women's, Child, and Adolescent Health—Fernandes Figueira, after approval of the project by the Research Ethics Committee of the same institution and after all patients had agreed and signed the term. Informed Consent Form (Number 23081213.6.0000.5269).

Acknowledgments: Lizanka Marinheiro is grateful for the Research and Development Foundation of Rio de Janeiro State (FAPERJ) and to the financial support by the project PPSUS as well as the Coordination of Improvement of Higher Education Personnel (CAPES/Brazil) for the doctoral scholarship (demanda social) conceded to Claudio Melibeu Bentes.

Conflicts of Interest: The authors report no conflict of interest.

References

1. Silverstein, N.M.; Fitzgerald, K.G. Educating a new generation of professionals in aging worldwide. *Gerontol. Geriatr. Educ.* **2017**, *38*, 1–4. [CrossRef] [PubMed]
2. Lobo, R.A.; Davis, S.R.; De Villiers, T.J.; Gompel, A.; Henderson, V.W.; Hodis, H.N.; Lumsden, M.A.; Mack, W.J.; Shapiro, S.; Baber, R. Prevention of diseases after menopause. *Climacteric* **2014**, *17*, 540–556. [CrossRef] [PubMed]
3. Rose, J.; Weiser, T.G.; Hider, P.; Wilson, L.; Gruen, R.L.; Bickler, S.W. Estimated need for surgery worldwide based on prevalence of diseases: A modelling strategy for the WHO Global Health Estimate. *Lancet Glob. Health* **2015**, *3* (Suppl. 2), S13–S20. [CrossRef]
4. Zitkus, B.S. Update on the American Diabetes Association Standards of Medical Care. *Nurse Pract.* **2014**, *39*, 22–32, quiz 32–23. [CrossRef]
5. Colberg, S.R.; Sigal, R.J.; Yardley, J.E.; Riddell, M.C.; Dunstan, D.W.; Dempsey, P.C.; Horton, E.S.; Castorino, K.; Tate, D.F. Physical Activity/Exercise and Diabetes: A Position Statement of the American Diabetes Association. *Diabetes Care* **2016**, *39*, 2065–2079. [CrossRef]
6. Brincat, M.; Gambin, J.; Brincat, M.; Calleja-Agius, J. The role of vitamin D in osteoporosis. *Maturitas* **2015**, *80*, 329–332. [CrossRef]
7. Guo, J.; Cockcroft, J.R.; Elwood, P.C.; Pickering, J.E.; Lovegrove, J.A.; Givens, D.I. Vitamin D intake and risk of CVD and all-cause mortality: Evidence from the Caerphilly Prospective Cohort Study. *Public Health Nutr.* **2017**, *20*, 2744–2753. [CrossRef]
8. Lucato, P.; Solmi, M.; Maggi, S.; Bertocco, A.; Bano, G.; Trevisan, C.; Manzato, E.; Sergi, G.; Schofield, P.; Kouidrat, Y.; et al. Low vitamin D levels increase the risk of type 2 diabetes in older adults: A systematic review and meta-analysis. *Maturitas* **2017**, *100*, 8–15. [CrossRef]
9. Wimalawansa, S.J. Associations of vitamin D with insulin resistance, obesity, type 2 diabetes, and metabolic syndrome. *J. Steroid Biochem. Mol. Biol.* **2018**, *175*, 177–189. [CrossRef]
10. Latham, N.K.; Anderson, C.S.; Reid, I.R. Effects of vitamin D supplementation on strength, physical performance, and falls in older persons: A systematic review. *J. Am. Geriatr. Soc.* **2003**, *51*, 1219–1226. [CrossRef]
11. Bentes, C.M.; Resende, M.; Miranda, H.; Netto, C.C.; Marinheiro, L.P.F. Can Vitamin D supplementation alone effective to increase a physical fitness levels in post-menopausal women with metabolic disorders? Brief Review. *Diabetes Metab. Syndr.* **2018**, *12*, 65–68. [CrossRef]
12. Prestes, J.; Tibana, R.A. Muscular static strength test performance and health: Absolute or relative values. *Rev. Assoc. Médica Bras.* **2013**, *59*, 308–309. [CrossRef]
13. Podsiadlo, D.; Richardson, S. The timed "Up & Go": A test of basic functional mobility for frail elderly persons. *J. Am. Geriatr. Soc.* **1991**, *39*, 142–148. [PubMed]
14. Rikli, R.E.; Jones, C.J. Development and validation of a functional fitness test for community-residing older adults. *J. Aging Phys. Act.* **1999**, *7*, 129–161. [CrossRef]
15. Finn, K.J.; Saint-Maurice, P.F.; Karsai, I.; Ihász, F.; Csányi, T. Agreement between Omron 306 and Biospace InBody 720 bioelectrical impedance analyzers (BIA) in children and adolescents. *Res. Q. Exerc. Sport* **2015**, *86* (Suppl. 1), S58–S65. [CrossRef] [PubMed]

16. Cangussu, L.M.; Nahas-Neto, J.; Orsatti, C.L.; Bueloni-Dias, F.N.; Nahas, E.A. Effect of vitamin D supplementation alone on muscle function in postmenopausal women: A randomized, double-blind, placebo-controlled clinical trial. *Osteoporos. Int.* **2015**, *26*, 2413–2421. [CrossRef]
17. Holick, M.F.; Binkley, N.C.; Bischoff-Ferrari, H.A.; Gordon, C.M.; Hanley, D.A.; Heaney, R.P.; Murad, M.H.; Weaver, C.M. Evaluation, treatment, and prevention of vitamin D deficiency: An Endocrine Society clinical practice guideline. *J. Clin. Endocrinol. Metab.* **2011**, *96*, 1911–1930. [CrossRef]
18. Bischoff-Ferrari, H.A.; Giovannucci, E.; Willett, W.C.; Dietrich, T.; Dawson-Hughes, B. Estimation of optimal serum concentrations of 25-hydroxyvitamin D for multiple health outcomes. *Am. J. Clin. Nutr.* **2006**, *84*, 18–28. [CrossRef]
19. Bischoff-Ferrari, H.A.; Dawson-Hughes, B.; Willett, W.C.; Staehelin, H.B.; Bazemore, M.G.; Zee, R.Y.; Wong, J.B. Effect of Vitamin D on falls: A meta-analysis. *JAMA* **2004**, *291*, 1999–2006. [CrossRef]
20. Bischoff-Ferrari, H.A.; Willett, W.C.; Wong, J.B.; Stuck, A.E.; Staehelin, H.B.; Orav, E.J.; Thoma, A.; Kiel, D.; Henschkowski, J. Prevention of nonvertebral fractures with oral vitamin D and dose dependency: A meta-analysis of randomized controlled trials. *Arch. Intern. Med.* **2009**, *169*, 551–561. [CrossRef] [PubMed]
21. Perez-Lopez, F.R.; Brincat, M.; Erel, C.T.; Tremollieres, F.; Gambacciani, M.; Lambrinoudaki, I.; Moen, M.H.; Schenck-Gustafsson, K.; Vujovic, S.; Rozenberg, S.; et al. EMAS position statement: Vitamin D and postmenopausal health. *Maturitas* **2012**, *71*, 83–88. [CrossRef] [PubMed]
22. Bischoff, H.A.; Borchers, M.; Gudat, F.; Duermueller, U.; Theiler, R.; Stahelin, H.B.; Dick, W. In situ detection of 1,25-dihydroxyvitamin D3 receptor in human skeletal muscle tissue. *Histochem. J.* **2001**, *33*, 19–24. [CrossRef]
23. Ceglia, L.; Niramitmahapanya, S.; da Silva Morais, M.; Rivas, D.A.; Harris, S.S.; Bischoff-Ferrari, H.; Fielding, R.A.; Dawson-Hughes, B. A randomized study on the effect of vitamin D3 supplementation on skeletal muscle morphology and vitamin D receptor concentration in older women. *J. Clin. Endocrinol. Metab.* **2013**, *98*, E1927–E1935. [CrossRef]
24. Annweiler, C.; Beauchet, O.; Berrut, G.; Fantino, B.; Bonnefoy, M.; Herrmann, F.; Schott, A.M. Is there an association between serum 25-hydroxyvitamin D concentration and muscle strength among older women? Results from baseline assessment of the EPIDOS study. *JNHA-J. Nutr. Health Aging* **2009**, *13*, 90–95. [CrossRef]
25. Zhu, K.; Austin, N.; Devine, A.; Bruce, D.; Prince, R.L. A randomized controlled trial of the effects of vitamin D on muscle strength and mobility in older women with vitamin D insufficiency. *J. Am. Geriatr. Soc.* **2010**, *58*, 2063–2068. [CrossRef]
26. Anek, A.; Bunyaratavej, N.; Jittivilai, T. Effects of Short-Term Vitamin D Supplementation on Musculoskeletal and Body Balance for Prevention of Falling in Postmenopausal Women. *J. Med. Assoc. Thai* **2015**, *98* (Suppl. 8), S26–S31.
27. Issa, C.M. Vitamin D and Type 2 Diabetes Mellitus. *Adv. Exp. Med. Biol.* **2017**, *996*, 193–205. [CrossRef] [PubMed]
28. Davis, S.R.; Castelo-Branco, C.; Chedraui, P.; Lumsden, M.; Nappi, R.; Shah, D.; Villaseca, P. Understanding weight gain at menopause. *Climacteric* **2012**, *15*, 419–429. [CrossRef] [PubMed]
29. Simpson, C.A.; Cusano, A.M.; Bihuniak, J.; Walker, J.; Insogna, K.L. Effect of 25(OH) vitamin D reference method procedure (RMP) alignment on clinical measurements obtained with the IDS-iSYS chemiluminescent-based automated analyzer. *J. Steroid Biochem. Mol. Biol.* **2015**, *148*, 41–46. [CrossRef]

Journal of
Functional Morphology and Kinesiology

MDPI

Article

Supervised Versus Unsupervised Pulmonary Rehabilitation in Patients with Pulmonary Embolism: A Valuable Alternative in COVID Era

Vasileios T. Stavrou [1], Michalis Griziotis [1], George D. Vavougios [1,2], Dimitrios G. Raptis [2], Fotini Bardaka [2], Eleni Karetsi [1,2], Athanasios Kyritsis [2], Zoe Daniil [1,2], Konstantinos Tsarouhas [3], Filippos Triposkiadis [3], Konstantinos I. Gourgoulianis [1,2] and Foteini Malli [2,4,*]

1 Laboratory of Cardio-Pulmonary Exercise Testing and Pulmonary Rehabilitation, Department of Respiratory Medicine, Faculty of Medicine, University of Thessaly, 41110 Larissa, Greece; vasileiosstavrou@hotmail.com (V.T.S.); mihgrizio@hotmail.co.uk (M.G.); dantevavougios@hotmail.com (G.D.V.); ekaretsi@uth.gr (E.K.); zdaniil@uth.gr (Z.D.); kgourg@uth.gr (K.I.G.)
2 Department of Respiratory Medicine, Faculty of Medicine, University of Thessaly, 41110 Larissa, Greece; raptisdmed@gmail.com (D.G.R.); bardakafotini@yahoo.gr (F.B.); thanoskyrit@hotmail.com (A.K.)
3 Department of Cardiology, University General Hospital of Larissa, 41110 Larissa, Greece; ktsarouhas14@yahoo.gr (K.T.); ftriposkiadis@gmail.com (F.T.)
4 Faculty of Nursing, University of Thessaly, 41500 Larissa, Greece
* Correspondence: mallifoteini@yahoo.gr

Citation: Stavrou, V.T.; Griziotis, M.; Vavougios, G.D.; Raptis, D.G.; Bardaka, F.; Karetsi, E.; Kyritsis, A.; Daniil, Z.; Tsarouhas, K.; Triposkiadis, F.; et al. Supervised Versus Unsupervised Pulmonary Rehabilitation in Patients with Pulmonary Embolism: A Valuable Alternative in COVID Era. *J. Funct. Morphol. Kinesiol.* **2021**, *6*, 98. https://doi.org/10.3390/jfmk6040098

Academic Editor: Cristina Cortis

Received: 2 November 2021
Accepted: 2 December 2021
Published: 3 December 2021

Publisher's Note: MDPI stays neutral with regard to jurisdictional claims in published maps and institutional affiliations.

Abstract: The aim of our study was to assess the effect of 8 weeks of pulmonary rehabilitation (PR) in patients with pulmonary embolism (PE) during unsupervised PR ($unSPR_{group}$) versus supervised PR (SPR_{group}) on cardiopulmonary exercise testing (CPET) parameters, sleep quality, quality of life and cardiac biomarkers (NT-pro-BNP). Fourteen patients with PE ($unSPR_{group}$, n = 7, vs. SPR_{group}, n = 7) were included in our study (age, 50.7 ± 15.1 years; BMI, 30.0 ± 3.3 kg/m^2). We recorded anthropometric characteristics and questionnaires (Quality of life (SF-36) and Pittsburg sleep quality index (PSQI)), we performed blood sampling for NT-pro-BNP measurement and underwent CPET until exhausting before and after the PR program. All patients were subjected to transthoracic echocardiography prior to PR. The SPR_{group} differed in mean arterial pressure at rest before and after the PR program (87.6 ± 3.3 vs. 95.0 ± 5.5, respectively, p = 0.010). Patients showed increased levels of leg fatigue (rated after CPET) before and after PR (p = 0.043 for SPR_{group}, p = 0.047 for $unSPR_{group}$) while the two groups differed between each other (p = 0.006 for post PR score). Both groups showed increased levels in SF-36 scores (general health; p = 0.032 for SPR_{group}, p = 0.010 for $unSPR_{group}$; physical health; p = 0.009 for SPR_{group}, p = 0.022 for $unSPR_{group}$) and reduced levels in PSQI (cannot get to sleep within 30-min; p = 0.046 for SPR_{group}, p = 0.007 for $unSPR_{group}$; keep up enough enthusiasm to get things done; p = 0.005 for SPR_{group}, p = 0.010 for $unSPR_{group}$) following the PR program. The NT-pro-BNP was not significantly different before and after PR or between groups. PR may present a safe intervention in patients with PE. The PR results are similar in SPR_{group} and $unSPR_{group}$.

Keywords: exercise; pulmonary embolism; pulmonary rehabilitation

1. Introduction

Pulmonary embolism (PE) is an acute and potentially fatal condition in which embolic material, usually a thrombus originating from the deep veins of the legs, blocks the pulmonary circulation resulting in impaired blood flow that may lead to right ventricle dysfunction [1]. PE and deep vein thrombosis are considered to be two manifestations of venous thromboembolism (VTE), which represents the third most common cardiovascular disorder in industrialized countries [2]. PE is difficult to diagnose due to lack of specificity of symptoms and clinical presentation [3]. Patients with history of PE often exhibit functional limitations and decreased quality of life even years after the episode, a condition that is considered as a long-term complication of acute PE and termed "post-PE-syndrome" or "Chronic Thromboembolic Disease" [4].

PE in the setting of COVID-19 is a common complication, frequent in hospitalized patients [5], and is associated with its severity [6]. On the pathophysiological level, the relationship between PE and COVID-19 is bidirectional. Hypercoagulable states and endothelial injury may be induced via virus–host interactions, while subsequent PE may account for persistent hypoxia following the resolution of the acute syndrome [7]. The incidence of PE following COVID-19 varies according to the population studied, the severity of COVID, the thromboprophylaxis dose, the screening protocol for VE, etc. According to a recent meta-analysis, the overall incidence of PE in COVID-19 inpatients is approximately 17%, with increased incidence in patients admitted to ICU (27.9%) versus those hospitalized in wards (7.1%) [8].

The "post-PE-syndrome" is characterized by suboptimal cardiac function but not pulmonary hypertension, altered pulmonary artery flow dynamics, and impaired oxygenation at rest or at exercise, associated with symptoms such as dyspnea, reduced exercise tolerance, or worsening of quality of life, that cannot be explained otherwise [4]. The pathophysiology of the syndrome is poorly understood while its treatment is not specified to measures other than anticoagulation and supportive care. Recent guidelines have outlined that an efficient follow-up strategy after PE should include exercise rehabilitation, although studies addressing the effect of pulmonary rehabilitation programs in these patients are lacking [9]. Pulmonary rehabilitation (PR) includes a supervised program of exercise training and breathing techniques that also addresses issues of health education. PR represents a safe and effective intervention which improves health indicators and quality of life of patients with certain lung diseases such as chronic obstructive pulmonary disease or lung involvement due to other conditions [10]. European Respiratory Society Council and Executive Committee [10] underlies the need to establish specialized rehabilitation programs in order to enhance patient accessibility to this treatment intervention. According to World Health Organization [11], the health indicators related to metabolic profile, physical activity [12], and aerobic and anaerobic capacity are assessed within the cardio-pulmonary exercise testing (CPET). Briefly, CPET is a non-invasive measurement which provides an objective quantitative assessment of metabolic, pulmonary and cardiovascular responses during exercise [13]. Several biomarkers for PE diagnosis, risk stratification and/or risk of recurrence exist but most of them require further validation before being applied in clinical practice. Cardiac troponin T, N-terminal-pro hormone BNP (NT-pro-BNP) and heart-type fatty-acid binding protein, are markers of myocardial strain and injury, which have prognostic value in risk assessment strategies [4].

There is paucity of data concerning the possible role of PR programs in patients with PE. This lack of data extends to unsupervised PR ($unSPR_{group}$), which represents a telemedicine approach that has gained impetus during the COVID-19 pandemic [14]. Telemedicine approaches, including virtual reality applications, have had previous successful implementations in the setting of pulmonary disease rehabilitation. [15]. Home-based unsupervised rehabilitation has been shown to be an effective alternative to formal regimens during the pandemic, ensuring that patients rehabilitation milestones remain on-track following hospitalization [16]. A study from our group has indicated that the efficacy of unsupervised PR in COVID-19 is tangible, and associated with improvements in redox homeostasis, sleep health and anthropometric indices [17]. Considering the overlap between COVID-19 and PE, these studies further demonstrate the rationale and relevance of unsupervised PR in PE.

Currently, there is no study prospectively addressing the efficacy and safety of PR in exercise limitation and quality of life following an episode of PE, despite current guide-lines that suggest that exercise rehabilitation is part of the follow up of these patient group [8]. The effectiveness of different programs of physical activity is not well established but some studies suggest that supervised versus self-selected programs might have similar results [18]. The types of exercise programs in patients with PE have not been addressed in the literature, to this end, we designed this study in order to investigate the effect of 8 weeks of PR in patients with history of an acute episode of PE. Additionally, we aimed to address the results of PR in exercise limitation and quality of life and examine possible differences among patients subjected to supervised versus unsupervised exercise.

2. Materials and Methods

2.1. Study Population

The present research is a pilot study. Patients with a history of PE were prospectively recruited from the PE outpatient clinic (Figure 1) between January 2017 to December 2018. The patients were randomly divided into two groups (using block randomization): unsupervised PR during telerehabilitation ($unSPR_{group}$) and supervised PR (SPR_{group}) in Pulmonary Rehabilitation Center (University of Thessaly). Some of these patients were present in a previous study and belong to the Proceedings of the 9th Conference of Biochemistry and Physiology of Exercise [19]. We included patients with PE diagnosis >6 months prior to enrollment and weekly exercise ≤100-min. Exclusion criteria included contraindications to the performance of CPET (i.e., recent acute myocardial infarction (3–5 days), unstable angina, uncontrolled arrhythmia causing hemodynamic instability, acute endocarditis, acute myocarditis or pericarditis, uncontrolled heart failure, lower extremity thrombosis, pregnancy, presence of severe comorbidity that may interfere with the results of the rehabilitation, i.e., COPD). The study was approved by the Institutional Ethics Committee of the corresponding institution. Verbal and written informed consent were obtained from all participants (No. of Ethical Committee: N° 2800, Scientific Council of University Hospital of Larissa).

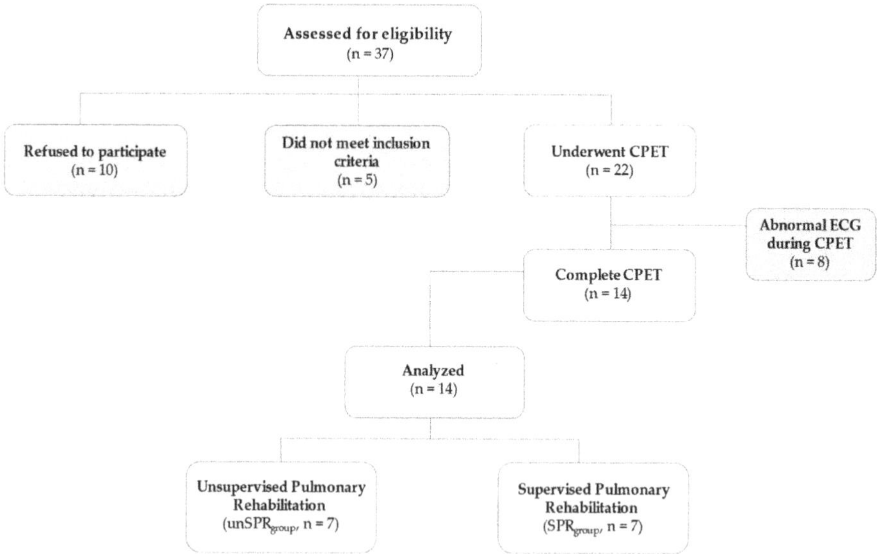

Figure 1. Study flow diagram.

2.2. Procedures

For all patients, we recorded the demographics and characteristics of PE episodes and all subjects underwent echocardiography. Prior to CPET, all participants answered the Pittsburgh Sleep Quality Index (PSQI) [20] and Quality of Life (SF-36) [21] questionnaires while we recorded anthropometric characteristics [22–24], pulmonary function parameters

(FEV_1: forced expiratory volume in 1st sec, FVC: forced vital capacity; Master Screen-CPX, VIASYS HealthCare, Hochberg, Germany) [25]. Blood sampling for NT-pro-BNP measurement was performed 30 min before CPET. BNP measurements were performed in complete blood samples with a commercial analyzer (Triage BNP test; BI-OSITE, San Diego, CA, USA). The same procedure was repeated after 8 weeks.

2.3. Echocardiography

All patients underwent echocardiographic study within 48 h of the CPET and PR program initiation. Two-dimensional echocardiography was performed, with the subjects resting in a left lateral decubitus position, using a Vivid BT08 (General Electric, Miami, FL, USA). Heart images were obtained in the standard parasternal long-axis and short-axis and apical four-and-two chamber planes. Wall thickness was measured from 2D short-axis views at end-diastole, with the greatest measurement within the left ventricular wall defined as the maximal wall thickness. M-mode echocardiograms derived from 2D images in the parasternal long axis were used for the measurement of end-diastolic and systolic dimensions according to the American Society of Echocardiography [26].

2.4. Cardiopulmonary Exercise Testing

CPET was performed on an electronic cycle ergometer (Ergoselect 100, Bitz, Germany) Master Screen-CPX and respiratory and cardiac parameters were recorded (VIASYS Health-Care, Höchberg, Germany). All patients, prior to testing, were familiarized with the test via a 2 min resting stage (1st stage); for a 3 min unloaded cycling as a warm-up (2nd stage); then, after the end of the maximal test (3rd stage), they performed a 5 min unloaded cycling for recovery (4th stage) purposes. In the 3rd stage the ramp work rate increased by 10–15 watts/min until exhaustion was reached. The work rate increment calculated using the Wasserman et al. [27] formula:

$$\text{Work Rate/min}_{\text{(ramp)}} = (VO_{2\text{max}} - VO_{2\text{unloaded}})/100$$

$$VO_{2\text{max}} = (\text{Height}_{\text{(cm)}} - \text{Age}_{\text{(years)}}) \times 20 \text{ (male) or } \times 14 \text{ (female)}$$

$$VO_{2\text{unloaded}} = 150 + (6 \times \text{Weight}_{\text{(kg)}})$$

During testing all patients were instructed to keep a steady speed of 60 ± 5 rpm throughout the four phases of testing. Each trial was terminated when the participant reached symptom-limited maximum exercise, which was confirmed by the presence of respiratory exchange ratio (RER) > 1.10, Heart Rate (HR) $\geq$ 80% of predicted HR_{max}, and/or plateau of oxygen consumption with increasing work load. Moreover, a 12-lead electrocardiogram (ECG) was also employed for HR monitoring, while a pulse oxymeter (MasterScreen, Höchberg, Germany) informed about oxygen saturation (SpO_2). Blood pressure (cuff manometry, Mac, Japan) and Borg CR10 Scales (Leg Fatigue, Dyspnea) point were recorded every 2 min for all phases.

2.5. Pulmonary Rehabilitation Program

The PR program lasted 8 weeks with three sessions per week. The duration of each training session was about 70 min. All sessions were instructed to conduct the PR program either outdoors (walking) or in home (stretching, strength and breathing exercise) without supervision for patients of the $unSPR_{group}$. Patients in the SPR_{group} performed the PR program in the Laboratory of Pulmonary Rehabilitation of the University Hospital of Larissa. Each training session included a warm-up ($unSPR_{group}$ and SPR_{group}: 5 min stretching exercises), the first main set ($unSPR_{group}$: 40 min walking at 60% of VO_2 calculated from heart rate vs. SPR_{group}: 40 min intermittent exercise in cycle ergometer (30 s exercise at 70% of VO_{2max} and 30 s resting)), the second main set of training ($unSPR_{group}$: 10 min (tele)breathing physiotherapy and 10 min multi-joint strength exercises vs. SPR_{group}: 10 min respiratory physiotherapy and 10 min multi-joint strength exercises) and a recovery set ($unSPR_{group}$ and SPR_{group}: 5 min stretching exercises). The set of exercises was analogue for both groups. Minor differences in the two PR programs exist in order to increase safety for injuries. The unSPR$_{group}$ performed exercise-PR unsupervised but according to instructions of a clinical exercise physiologist (VTS) that supervised the SPR$_{group}$.

Adherence to the program of unSPR$_{group}$ was determined via phone calls per week. Each call focused on whether the patients were able to follow the instructions and perform them, troubleshooting and reporting the physiological parameters.

2.6. Statistical Analysis

The Kolmogorov–Smirnov test was utilized to assess normality of distribution of values. A comparison of one group of individuals against themselves (pre-and-post the PR) was performed with the Wilcoxon signed-rank test according to variable distribution. A comparison between the two patient groups was performed with the use Mann–Whitney U-test according to variable distribution. Data are presented as absolute numbers, percentages or mean values and standard deviation (mean $\pm$ SD). For all the statistical analyses the statistical package SPSS 15 (SPSS Inc., Chicago, IL, USA) was used. The level of significance was set at $p < 0.05$.

3. Results

Out of the 37 individuals who were assessed for eligibility, 14 were included in the study (Figure 1). Tables 1 and 2 presents demographical, clinical and echocardiography results of the study subjects. Briefly, the SPR$_{group}$ vs. the unSPR$_{group}$ did not differ significantly in terms of age (49.6 $\pm$ 15.4 vs. 51.9 $\pm$ 16.0, respectively), gender distribution (males 85.71% vs. 71.42%, respectively) and smoking status (smokers: 28.6% vs. 42.9%, respectively). Similarly, SPR$_{group}$ vs. the unSPR$_{group}$ had similar BMI (29.8 $\pm$ 3.9 vs. 30.1 $\pm$ 2.9, kg/m^2, respectively), baseline physical activity (61.7 $\pm$ 24.7 vs. 70.0 $\pm$ 14.1, respectively) (Table 1). The two groups did not differ at baseline in terms of cardiovascular comorbidities, blood pressure measurements and spirometry results (Table 1). Mean Pulmonary Severity Index (PESI) score of the study participants was 52.48 $\pm$ 48.23. All patients were hospitalized during the acute period. Mean PESI score was not significantly different between SPR$_{group}$ and unSPR$_{group}$ (103.25 $\pm$ 50.90 vs. 100.33 $\pm$ 49.13, $p > 0.05$). Ejection fraction was within the normal range in both SPR$_{group}$ and unSPR$_{group}$ (59.2 $\pm$ 2.0 vs. 59.4 $\pm$ 1.3, respectively) (Table 2). Echocardiographic signs of right heart dysfunction (i.e., end-diastolic right ventricular diameter in four chambers view, right ventricular systolic pressure, right atrial area, tricuspid annular plane systolic excursion) were not significantly different between the two study groups (Table 2).

Respiratory parameters and CPET results, before and after the PR for both groups, are presented in Table 3. The SPR$_{group}$ differed in mean arterial pressure (MAP) at rest before and after the PR program (87.6 $\pm$ 3.3 vs. 95.0 $\pm$ 5.5, respectively, $p = 0.010$). MAP levels did not differ significantly before and after the PR program in the unSPR$_{group}$ (85.5 $\pm$ 8.5 vs. 88.6 $\pm$ 9.2, respectively). We observed an increasing trend in $P_{ET}O_2$ in the SPR$_{group}$ after the PR program vs. at baseline that did not reach statistical significance (115.0 $\pm$ 7.0 vs. 108.7 $\pm$ 3.9, respectively, $p = 0.059$) (Table 3). Patients showed differences in leg fatigue before and after PR (SPR$_{group}$: 2.6 $\pm$ 1.4 vs. 3.6 $\pm$ 1.3, respectively, $p = 0.043$ and unSPR$_{group}$: 1.1 $\pm$ 0.7 vs. 1.7 $\pm$ 0.8, respectively, $p = 0.047$). Leg fatigue following the PR program was significantly higher in SPR$_{group}$ when compared to unSPR$_{group}$ (3.6 $\pm$ 1.3 vs. 1.7 $\pm$ 0.8, respectively, $p = 0.006$).

Table 1. Demographical, clinical and spirometry results of the study population.

	SPR_{group}	$UnSPR_{group}$	*p* Value
Age, years	49.6 ± 15.4	51.9 ± 16.0	0.790
Gender, M/F	6/1	5/2	0.552
Smokers, %	28.6	42.9	0.611
Body mass index, kg/m^2	29.8 ± 3.9	30.1 ± 2.9	0.884
Body surface area, m^2	2.0 ± 0.5	2.2 ± 0.3	0.537
Lean body mass, %	60.9 ± 9.0	63.6 ± 6.3	0.531
Total body water, L	44.2 ± 8.9	44.7 ± 7.0	0.904
Alcohol drinking, ml/month	85.0 ± 13.7	83.3 ± 14.4	0.875
Physical Activity, min/week	61.7 ± 24.7	70.0 ± 14.1	0.703
Systolic blood pressure, mmHg	113.6 ± 13.8	117.1 ± 9.9	0.588
Diastolic blood pressure, mmHg	71.4 ± 6.3	72.9 ± 2.7	0.589
Prevalence of CVD and diabetes, %	42.9	57.1	0.626
Prior VTE event, Y/N	1/6	1/6	1.000
Provoked event, Y/N	3/4	6/1	0.109
Under anticoagulant therapy, Y/N	6/1	4/3	0.271
MRC dyspnea scale, 0/I/II	3/4/0	2/4/1	0.690
FEV_1, % of predicted	101.0 ± 8.0	96.9 ± 10.2	0.434
FVC, % of predicted	96.8 ± 7.7	90.9 ± 11.6	0.280

Abbreviations: CDV, cardiovascular disease; FEV_1, forced expiratory volume in 1st sec; FVC, forced vital capacity; M/F, male/female; VTE, venous thromboembolism event; Y/N, yes/no.

Table 2. Echocardiographic characteristics of the study population.

	SPR_{group}	$UnSPR_{group}$	*p* Value
Ejection fraction, %	59.2 ± 2.0	59.4 ± 1.3	0.832
End-diastolic RV diameter (4CH), cm	3.6 ± 0.1	3.6 ± 0.5	0.763
RVSP, mmHg	23.3 ± 4.1	24.2 ± 5.8	0.777
RA area, cm^2	15.3 ± 2.5	15.2 ± 3.1	0.952
TAPSE, mm	20.5 ± 7.8	22.3 ± 4.3	0.626

Abbreviations: RA, right atrial; RV, right ventricle; RVS, right ventricular systolic pressure; TAPSE, tricuspid annular plane systolic excursion.

Table 3. Pulmonary function parameters and cardiopulmonary exercise testing results between groups before and after the pulmonary rehabilitation program. Continuous variables are presented as mean ± standard deviation.

	SPR_{group}			$UnSPR_{group}$			*p* Value between Groups	
	Baseline	**Post-PR**	***p* Value**	**Baseline**	**Post-PR**	***p* Value**	PR_{pre}	PR_{post}
Resting								
VO_2, mL/min	330.6 ± 91.6	349.0 ± 120.8	0.735	336.3 ± 84.5	312.0 ± 53.9	0.533	0.905	0.473
VCO_2, mL/min	257.1 ± 67.1	240.6 ± 101.8	0.095	267.9 ± 102.7	236.1 ± 32.9	0.452	0.821	0.095
$P_{ET}CO_2$, mmHg	29.9 ± 3.7	29.5 ± 2.4	0.832	27.2 ± 4.1	27.1 ± 3.6	0.952	0.227	0.159
$P_{ET}O_2$, mmHg	110.0 ± 5.1	114.0 ± 4.2	0.132	113.8 ± 7.7	112.5 ± 5.4	0.736	0.305	0.568
HR, bpm	81.0 ± 18.2	82.6 ± 18.2	0.875	78.7 ± 10.1	73.9 ± 9.6	0.374	0.776	0.285
MAP, mmHg	87.6 ± 3.3	95.0 ± 5.5	0.010	85.5 ± 8.5	88.6 ± 9.2	0.516	0.549	0.133
Maximal effort								
VO_2, mL/min	1559.1 ± 372.8	1579.3 ± 430.7	0.927	1946.0 ± 640.2	1896.1 ± 390.0	0.863	0.192	0.175
VCO_2, mL/min	1536.6 ± 440.8	1525.5 ± 497.6	0.966	1954.1 ± 672.1	1791.0 ± 453.9	0.604	0.194	0.318
$P_{ET}CO_2$, mmHg	37.5 ± 4.1	36.7 ± 3.5	0.710.	35.5 ± 4.6	35.6 ± 4.4	0.970	0.416	0.615
$P_{ET}O_2$, mmHg	108.7 ± 3.9	115.0 ± 7.0	0.059	112.8 ± 3.8	115.9 ± 4.8	0.199	0.079	0.788
V_E/MVV, %	44.1 ± 10.8	48.1 ± 10.5	0.494	53.3 ± 6.3	53.5 ± 12.7	0.977	0.076	0.408
V_E/VCO_2	28.6 ± 3.0	27.8 ± 3.0	0.279	28.5 ± 2.5	28.2 ± 3.3	0.786	0.357	0.679
HR, bpm	133.3 ± 18.6	133.0 ± 14.2	0.975	149.0 ± 13.3	138.4 ± 12.9	0.158	0.094	0.470
MAP, mmHg	119.5 ± 18.8	117.2 ± 10.6	0.786	123.3 ± 12.5	121.4 ± 9.4	0.757	0.660	0.438
Leg fatigue, Borg Scale	2.6 ± 1.4	3.6 ± 1.3	0.043	1.1 ± 0.7	1.7 ± 0.8	0.047	0.062	0.006
Dyspnea, Borg Scale	1.4 ± 1.0	2.0 ± 1.2	0.337	0.9 ± 1.1	1.3 ± 0.8	0.403	0.317	0.196

Abbreviations: HR, heart rate; MAP, mean arterial pressure; MVV, maximum voluntary volume; $P_{ET}CO_2$, end-tidal carbon dioxide pressure; $P_{ET}O_2$, end-tidal oxygen pressure; VCO_2, carbon dioxide output; V_E, minute ventilation; VO_2, oxygen uptake.

Both patient groups showed statistically significant differences before and after PR in the Quality of Life and Sleep Quality questionnaires. At baseline no differences were observed in both groups for all the subscales of SF-36 (Table 4). After the PR program, we observed that both groups had a higher score in the SF-36 parameters "*physical health*" and *"general health"* versus to their baseline values. In more detail, *"physical health"* increased significantly after the PR program when compared to baseline in the SPR_{group} (92.9 ± 8.1 vs. 71.1 ± 10.7, respectively, p = 0.009) and the $unSPR_{group}$ (93.6 ± 6.9 vs. 76.4 ± 15.7, respectively, p = 0.022). *"General health"* score increased following PR vs. before in the SPR_{group} (81.4 ± 16.8 vs. 63.6 ± 9.9, respectively, p = 0.032) and the $unSPR_{group}$ (72.1 ± 8.6 vs. 51.4 ± 15.7, respectively, p = 0.010). The others parameters of SF-36 were not different before and after PR period (Table 4).

Table 4. Quality of life and sleep quality results between groups before and after pulmonary rehabilitation. Continuous variables are presented as mean ± standard deviation.

		SPR_{group}			$UnSPR_{group}$			*p* Value between Groups	
		Baseline	**Post-PR**	***p* Value**	**Baseline**	**Post-PR**	***p* Value**	**Baseline**	**Post-PR**
Quality of life (SF-36)	Physical Health	71.1 ± 10.7	92.9 ± 8.1	0.009	76.4 ± 15.7	93.6 ± 6.9	0.022	0.923	0.862
	Physical Functioning	78.6 ± 26.7	92.9 ± 18.9	0.271	75.0 ± 25.0	76.4 ± 25.3	0.917	0.801	0.194
	Body Pain	82.5 ± 18.0	87.5 ± 13.4	0.567	80.0 ± 17.0	83.9 ± 13.8	0.643	0.794	0.631
	General Health	63.6 ± 9.9	81.4 ± 16.8	0.032	51.4 ± 15.7	72.1 ± 8.6	0.010	0.109	0.217
	Vitality	67.9 ± 16.0	77.1 ± 13.2	0.260	60.7 ± 13.7	68.6 ± 8.0	0.214	0.387	0.167
	Social Role Functioning	75.0 ± 20.4	92.9 ± 12.2	0.070	85.7 ± 18.3	85.4 ± 14.1	0.968	0.321	0.308
	Emotional Role Functioning	85.7 ± 26.2	98.6 ± 2.4	0.175	85.7 ± 26.2	90.5 ± 16.2	0.690	0.989	0.147
	Mental Health	73.6 ± 24.0	78.3 ± 19.6	0.635	77.1 ± 17.5	81.1 ± 12.2	0.629	0.692	0.749
Sleep quality (PSQI)	Cannot get to sleep within 30 min	2.9 ± 0.4	1.8 ± 0.5	0.046	2.6 ± 0.5	2.0 ± 0.1	0.007	0.779	0.025
	Wake up in the middle of the night or early morning	1.4 ± 0.3	1.4 ± 0.4	0.968	1.3 ± 0.1	1.3 ± 0.2	0.878	0.613	0.694
	Have to get up to use the bathroom	1.1 ± 0.1	1.0 ± 0.5	0.317	1.1 ± 0.4	0.9 ± 0.7	0.986	0.955	0.613
	Cannot breathe comfortably	1.2 ± 0.3	1.1 ± 0.4	0.867	1.1 ± 0.3	1.1 ± 0.6	0.831	0.986	0.978
	Cough or snore loudly	1.0 ± 0.2	0.9 ± 0.6	0.317	0.9 ± 0.3	0.8 ± 0.4	0.326	0.694	0.281
	Feel too cold	1.1 ± 0.3	1.0 ± 0.6	0.325	1.0 ± 0.4	0.9 ± 0.7	0.679	0.732	0.796
	Feel too hot	1.0 ± 0.1	1.0 ± 0.2	0.371	1.1 ± 0.2	0.9 ± 0.3	0.317	0.698	0.789
	Had bad dreams	1.2 ± 0.2	1.1 ± 0.1	0.157	1.1 ± 0.3	1.0 ± 0.2	0.175	0.121	0.779
	Have pain	0.8 ± 0.1	0.7 ± 0.4	0.152	0.8 ± 0.2	0.6 ± 0.8	0.336	0.956	0.232
	… taken medicine to help you sleep	-	-	/	-	-	/	/	/
	… trouble staying awake (driving, eating meals, or social activity)	0.6 ± 0.3	0.6 ± 0.5	0.317	0.5 ± 0.1	0.5 ± 0.6	0.307	0.463	0.397
	… keep up enough enthusiasm to get things done?	1.9 ± 0.4	1.0 ± 0.2	0.005	2.5 ± 0.5	1.3 ± 0.4	0.010	0.779	0.029
	… sleep quality overall	0.3 ± 0.1	0.3 ± 0.2	0.157	0.2 ± 0.2	0.1 ± 0.4	0.317	0.294	0.281

Sleep quality as assessed by PSQI showed differences in both groups before and after PR in parameter *"cannot get to sleep within 30 min"*, (Table 4) and *"keep up enough enthusiasm to get things done"* (Table 4). The PSQI score decreased after PR vs. at baseline ($unSPR_{group}$: 5.7 ± 1.4 vs. 3.9 ± 1.8, respectively, p = 0.035; SPR_{group}: 6.6 ± 1.8 vs. 4.1 ± 1.8, respectively, p = 0.026) compared to the period before PR. The other parameters of PSQI were not different before and after PR period but presented an increasing trend in both groups that did not reach statistical significance.

The NT-pro-BNP levels were not significantly different before and after PR (SPR_{group}: 75.3 ± 10.4 vs. 102.0 ± 45.1 pg/mL, respectively, p = 0.147; $unSPR_{group}$: 76.0 ± 14.4 vs. 104.3 ± 36.5 pg/mL, respectively, p = 0.116, Figure 2) or between groups (Baseline, SPR_{group}: 75.3 ± 10.4 vs. $unSPR_{group}$: 76.0 ± 14.4, pg/mL, p = 0.917; post-PR, SPR_{group}: 102.0 ± 45.1 vs. $unSPR_{group}$: 104.3 ± 36.5, pg/mL, p = 0.919, Figure 2).

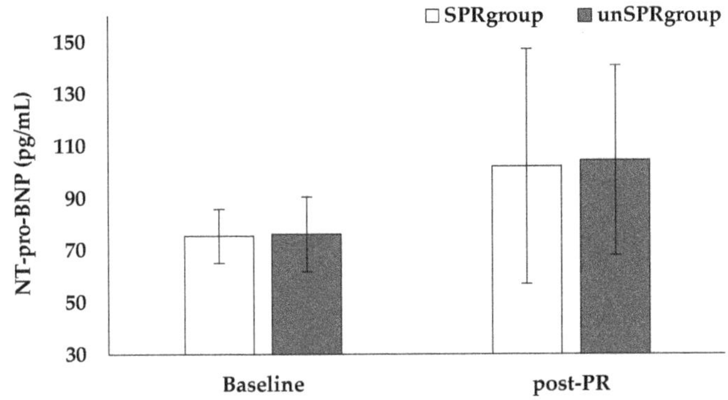

Figure 2. NT-pro-BNP results between groups before and after pulmonary rehabilitation. The bars correspond to the mean value and the outliers correspond to standard deviation.

4. Discussion

The aim of this pilot study was to investigate the effect of 8 weeks of PR in patients with PE. Patients underwent PR program either by SPR_{group} in a Pulmonary Rehabilitation Center or by $unSPR_{group}$ during telerehabilitation and groups were compared to each other. We observed differences in MAP at rest before and after the PR program in the SPR_{group} while both groups showed differences in leg fatigue before and after PR as well as compared to each other. Importantly, the two group of patients showed differences in parameters of quality of life and sleep quality before and after the PR. We did not observe major differences when SPR_{group} was compared to $unSPR_{group}$ with the exception of reduced leg fatigue reported by the SPR_{group} and the parameter *"keep up enough enthusiasm to get things done"* in PSQI which was in favor of the $unSPR_{group}$. However, due to the small number of patients included, no definite conclusions can be drawn concerning differences in the effectiveness of each approach.

4.1. Cardiopulmonary Exercise Testing

CPET provides an objective and quantitative measure of metabolic, pulmonary and cardiovascular responses during exercise which is both non-invasive and safe and can serve as an independent predictor of long-term outcomes [9]. Previous studies have documented reductions in VO_{2max} at various single time points following the PE event [28]. Kahn et al. [29] observed persistent exercise limitations 1-year after PE. Our results showed significant differences in MAP (at rest) in the SPR_{group} before and after PR while both groups showed differences in leg fatigue pre-and-post PR. Additionally, we demonstrated differences in leg fatigue between groups (SPR_{group} versus $unSPR_{group}$). CPET may be limited by leg fatigue that may underlie muscle weakness and leg effort [30]. Most patients tolerate only minimal leg discomfort before stopping the measurement while average healthy individuals may tolerate a greater degree of discomfort, suggesting the subjectiveness of the intensity of leg effort [30]. Leg fatigue may be associated with muscle metabolism impairment and probably increased peripheral oxygen uptake [31]. These findings are of significant clinical relevance and represent the standard of care for the management of patients with cardiovascular diseases such as pulmonary hypertension [32]. According to Kwan et al. [33], the use of exercise training programs of cardiac rehabilitation may benefit patients with a history of acute PE in a manner similar to that of patients suffering acute coronary syndromes.

4.2. Quality of Life and Psychological Aspects

Previous studies have shown that PR may have beneficial effects in patients with cardiovascular diseases, in terms of improved functional capacity and quality of life [34]. Quality of life has become an important outcome aspect of medical care. Patients with

PE may have reduced chronic functional capacity for many years after the event and that may be the main determinant of impaired quality of life [34]. Our results showed difference before and after PR in quality of life, as assessed by the SF-36 questionnaire, in parameters such as "*general health*" and "*physical health*". Reduced functional capacity may relate to persistent dyspnea, while physical activity and exertion were the most common behavior changes in patients. In our study, patients with PE during rehabilitation performed combination exercise and respiratory physiotherapy. This combination may relate to behavior change in patients following PR. Although not statistically significant, we observed a trend for improvement in exercise capacity which may be attributed to improved muscle function and desensitization to dyspnea. Desensitization to dyspnea is often considered a mechanism to explain benefit in the rehabilitation of patients with respiratory diseases and these altered perceptions of dyspnea even without associated physiological changes [34]. It should be noted that the interventions via PR/uns-PR also affect several psychological components. PR for chronic lung diseases has been shown to reduce anxiety and depression in these patients [35]. Telemedicine approaches, either as simple as a follow-up via phone [36] or as intricate as virtual reality [15] have also been shown to confer the same beneficial effect in both anxiety and depression experienced by these patients. A limitation in our study, however, is that we did not specifically assess these parameters, and therefore cannot comment on how rehabilitation may have affected them in this patient population.

4.3. *Sleep Quality*

Sleep disorder breathing and PE are a major health issue in industrialized countries. PE patients have a 2–4 times greater risk of suffering from moderate and severe obstructive sleep apnea syndrome (OSAS) [37]. Previous studies suggest that intermittent hypoxia and fragmentation of sleep may result in blood hypercoagulability, endothelial dysfunction and venous stasis. According to García-Ortega et al. [38], patients with acute PE have increased risk of coexisting moderate and severe OSAS when compared with controls. A polysomnography study showed a lower degree of oxygen saturation in PE patients and higher risk of PE in patients with isolated OSAS while the high hypoxic burden may be related to PE prevalence [39]. Patients with acute PE are at an increased risk of coexisting sleep disorders [38]. Our results showed a difference in sleep quality according to the PSQI questionnaire, in parameters such as "*cannot get to sleep within 30-min*" and "*keep up enough enthusiasm to get things done*" following the PR program. According to Stavrou et al. [40,41], the exercise, in patients with sleep disorders, may reduce the apnea-hypopnea index and improve the sleep quality while the daily physical activity may have a protective role in disease progression.

4.4. *Implication of Rehabilitation Program*

Data concerning rehabilitation after an acute episode of PE are sparse, while the literature lacks data concerning pulmonary rehabilitation program following PE recovery. Many patients have persistent symptoms months after an acute episode of PE and some of those suffer from post-PE syndrome, characterized by exercise limitation and suboptimal right heart function [4]. These patients may benefit the most from PR programs; however, data on this population are lacking. Rehabilitation program may be safe in the PE population in terms of death and serious events [42]. Exercise training after VTE showed differences in VO_{2max} in previous studies with 3-month intervention duration [43]. To our knowledge, this is the first study addressing the effectiveness and feasibility of an unsupervised PR in patients with a PE. The results of our study may help to establish a specialized rehabilitation program with potential important benefits and provide data on the safety, which seems to be a valuable alternative during the pandemic, of exercise programs in pulmonary embolism patient. Moreover, PR could be a highly valuable tool for promoting exercise and symptom recovery following pulmonary embolism and a novel approach concerning the treatment of persistently symptomatic patients with PE may arise.

4.5. Biomarkers

Several biomarkers have been implicated in the diagnosis and risk stratification of PE. Systemic biomarkers of myocardial injury and ventricular dysfunction have been extensively used in everyday clinical practice for the initial risk assessment of patients with acute PE [8]. High levels of BNP may help in the identification of subjects with PE and higher risk of in-hospital complications and death [44]. Additionally, NT-pro-BNP has emerged as a potential biomarker of early diagnosis of Chronic Thromboembolic Pulmonary Hypertension [45]. However, little is known about NT-pro-BNP levels in patients following an acute PE episode, as well as its possible association with exercise limitation. We did not observe differences in NT-pro-BNP levels among patients before and after PR program. We acknowledge that the strength of our results is limited due to the small number of patients included and therefore we suggest that further studies are warranted in order to exert any definite conclusions.

5. Conclusions

In conclusion, pulmonary rehabilitation has beneficial effects on quality of life and sleep quality in patients with pulmonary embolism. It also shows an improving tendency of indicators related to physical capacity. There were no major differences between SPR_{group} and $unSPR_{group}$ (except reduced leg fatigue in the SPR_{group} and improvement in PSQI parameter "*keep up enough enthusiasm to get things done*" in the $unSPR_{group}$). Uns-PR may be a feasible alternative to supervised regimens in the pandemic era, considering its relevance to PE and the increased prevalence of PE due to COVID-19. A larger trial is needed to extend these observations and provide evidences for the long-term effects of pulmonary rehabilitation and confirm the findings.

Author Contributions: V.T.S. and F.M. conceived and designed the experiments; V.T.S., F.B. and E.K. performed the cardiopulmonary exercises testing; V.T.S. analyzed the findings of cardiopulmonary exercises testing; V.T.S. and M.G. performed the pulmonary rehabilitation program; V.T.S., F.B., D.G.R. and A.K. contributed to sample collection; V.T.S. and F.M. performed the experiments and analyzed the data; V.T.S. and M.G. performed the pulmonary rehabilitation programs; K.T. and F.T. performed the echocardiography examinations and analyzed the echocadiographic findings; V.T.S., G.D.V. and F.M. wrote the paper; V.T.S., K.I.G., F.M. and Z.D. edited the paper. All authors have read and agreed to the published version of the manuscript.

Funding: This research received no external funding.

Institutional Review Board Statement: The study was conducted according to the guidelines of the Declaration of Helsinki, and approved by the Institutional Review Board (No. of Ethical Committee: N° 2800, Scientific Council of University Hospital of Larissa).

Informed Consent Statement: Informed consent was obtained from all subjects involved in the study and written informed consent has been obtained from the patients to publish this paper.

Data Availability Statement: The data presented in this study are available on request from the corresponding author.

Acknowledgments: The authors appreciate the friendly cooperation of patients with pulmonary embolism.

Conflicts of Interest: The authors declare no conflict of interest.

References

1. Lavorini, F.; Di Bello, V.; De Rimini, M.L.; Lucignani, G.; Marconi, L.; Palareti, G.; Pesavento, R.; Prisco, D.; Santini, M.; Sverzellati, N.; et al. Diagnosis and treatment of pulmonary embolism: A multidisciplinary approach. *Multidiscip. Respir. Med.* **2013**, *8*, 75. [CrossRef]
2. Douma, R.A.; Kamphuisen, P.; Büller, H.R. Acute pulmonary embolism. Part 1: Epidemiology and diagnosis. *Nat. Rev. Cardiol.* **2010**, *7*, 585–596. [CrossRef]
3. White, R.H. The epidemiology of venous thromboembolism. *Circulation* **2003**, *7*, 107. [CrossRef]
4. Klok, F.; van der Hulle, T.; Exter, P.D.; Lankeit, M.; Huisman, M.; Konstantinides, S. The post-PE syndrome: A new concept for chronic complications of pulmonary embolism. *Blood Rev.* **2014**, *28*, 221–226. [CrossRef] [PubMed]

5. García-Ortega, A.; Oscullo, G.; Calvillo, P.; López-Reyes, R.; Méndez, R.; Gómez-Olivas, J.D.; Bekki, A.; Fonfría, C.; Trilles-Olaso, L.; Zaldívar, E.; et al. Incidence, risk factors, and thrombotic load of pulmonary embolism in patients hospitalized for COVID-19 infection. *J. Infect.* **2021**, *82*, 261–269. [CrossRef] [PubMed]
6. Roncon, L.; Zuin, M.; Barco, S.; Valerio, L.; Zuliani, G.; Zonzin, P.; Konstantinides, S.V. Incidence of acute pulmonary embo-lism in COVID-19 patients: Systematic review and meta-analysis. *Eur. J. Intern. Med.* **2020**, *82*, 29–37. [CrossRef]
7. Poor, H.D. Pulmonary Thrombosis and Thromboembolism in COVID-19. *Chest* **2021**, *160*, 1471–1480. [CrossRef]
8. Jiménez, D.; García-Sanchez, A.; Rali, P.; Muriel, A.; Bikdeli, B.; Ruiz-Artacho, P.; Le Mao, R.; Rodriguez, C.; Hunt, B.J.; Monreal, M.; et al. Incidence of VTE and Bleeding among Hospitalized Patients with Coronavirus Disease 2019: A Systematic Review and Meta-analysis. *Chest* **2021**, *159*, 1182–1196. [CrossRef]
9. Konstantinides, S.V.; Meyer, G.; Becattini, C.; Bueno, H.; Geersing, G.J.; Harjola, V.P.; Huisman, M.V.; Humbert, M.; Jennings, C.S.; Jiménez, D.; et al. 2019 ESC Guidelines for the diagnosis and management of acute pulmonary embolism developed in collaboration with the European Respiratory Society (ERS): The Task Force for the diagnosis and management of acute pulmonary embolism of the European Society of Cardiology (ESC). *Eur. Heart J.* **2019**, *41*, 543–603.
10. Grünig, E.; Eichstaedt, C.; Barberà, J.-A.; Benjamin, N.; Blanco, I.; Bossone, E.; Cittadini, A.; Coghlan, G.; Corris, P.; D'Alto, M.; et al. ERS statement on exercise training and rehabilitation in patients with severe chronic pulmonary hypertension. *Eur. Respir. J.* **2018**, *53*, 1800332. [CrossRef]
11. WHO. Global Reference List of 100 Core Health Indicators (Plus Health-Related SDGs). WHO/HIS/IER/GPM/2018. Available online: https://apps.who.int/iris/handle/10665/259951 (accessed on 1 November 2021).
12. ACSM. *Guidelines for Exercise Testing and Prescription*, 10th ed.; Wolters Kluwer Health: Philadelphia, PA, USA, 2018.
13. Boutou, A.K.; Pitsiou, G.G.; Argyropoulou, P. Non-invasive markers of pulmonary hypertension in interstitial lung disease: Is cardiopulmonary exercise testing the Holy Grail? *Respirology* **2014**, *19*, 621–622. [CrossRef]
14. Werneke, M.W.; Deutscher, D.; Grigsby, D.; Tucker, A.C.; Mioduski, E.J.; Hayes, D. Telerehabilitation during the COVID-19 Pandemic in Outpatient Rehabilitation Settings: A Descriptive Study. *Phys. Ther.* **2021**, *101*, 110. [CrossRef]
15. Rutkowski, S.; Szczegielniak, J.; Szczepańska-Gieracha, J. Evaluation of the Efficacy of Immersive Virtual Reality Therapy as a Method Supporting Pulmonary Rehabilitation: A Randomized Controlled Trial. *J. Clin. Med.* **2021**, *10*, 352. [CrossRef]
16. Li, J.; Xia, W.; Zhan, C.; Liu, S.; Yin, Z.; Wang, J.; Chong, Y.; Zheng, C.; Fang, X.; Cheng, W.; et al. A telerehabilitation programme in post-discharge COVID-19 patients (TERECO): A randomised controlled trial. *Thorax* **2021**, *26*, thoraxjnl-2021-217382. [CrossRef] [PubMed]
17. Stavrou, V.T.; Tourlakopoulos, K.N.; Vavougios, G.D.; Papayianni, E.; Kiribesi, K.; Maggoutas, S.; Nikolaidis, K.; Fradelos, E.C.; Dimeas, I.; Daniil, Z.; et al. Eight Weeks Unsupervised Pulmonary Rehabilitation in Previously Hospitalized of SARS-CoV-2 Infection. *J. Pers. Med.* **2021**, *11*, 806. [CrossRef]
18. Mazzuca, P.; Montesi, L.; Mazzoni, G.; Grazzi, G.; Micheli, M.M.; Piergiovanni, S.; Pazzini, V.; Forlani, G.; Latessa, P.M.; Marchesini, G. Supervised vs. self-selected physical activity for individuals with diabetes and obesity: The Lifestyle Gym program. *Intern. Emerg. Med.* **2016**, *12*, 45–52. [CrossRef] [PubMed]
19. Stavrou, V.; Griziotis, M.; Raptis, D.; Bardaka, F.; Karetsi, E.; Kiritsis, A.; Daniil, Z.; Tsarouhas, K.; Triposkiadis, F.; Gour-goulianis, K.I.; et al. Eight Weeks of Pulmonary Rehabilitation in Patients with Pulmonary Embolism: A Preliminary Report. *Proceedings* **2019**, *25*, 37. [CrossRef]
20. Buysse, D.J.; Reynolds, C.F., III; Monk, T.H.; Berman, S.R.; Kupfer, D.J. The Pittsburgh Sleep Quality Index: A new instrument for psychiatric practice and research. *Psychiatry Res.* **1989**, *28*, 193–213. [CrossRef]
21. Ngo-Metzger, Q.; Sorkin, D.H.; Mangione, C.M.; Gandek, B.; Hays, R.D. Evaluating the SF-36 Health Survey (Version 2) in Older Vietnamese Americans. *J. Aging Health* **2008**, *20*, 420–436. [CrossRef] [PubMed]
22. Mosteller, R.D. Simplified Calculation of Body-Surface Area. *N. Engl. J. Med.* **1987**, *317*, 1098. [CrossRef]
23. Watson, P.E.; Watson, I.D.; Batt, R.D. Total body water volumes for adult males and females estimated from simple anthro-pometric measurements. *Am. J. Clin. Nutr.* **1980**, *33*, 27–39. [CrossRef]
24. Hume, R. Prediction of lean body mass from height and weight. *J. Clin. Pathol.* **1966**, *19*, 389–391. [CrossRef] [PubMed]
25. Miller, M.R.; Hankinson, J.; Brusasco, V.; Burgos, F.; Casaburi, R.; Coates, A.; Crapo, R.; Enright, P.; Van Der Grinten, C.P.M.; Gustafsson, P.; et al. Standardisation of spirometry. *Eur. Respir. J.* **2005**, *26*, 319–338. [CrossRef]
26. Mitchell, C.; Rahko, P.S.; Blauwet, L.A.; Canaday, B.; Finstuen, J.A.; Foster, M.C.; Horton, K.; Ogunyankin, K.O.; Palma, R.A.; Velazquez, E.J. Guidelines for Performing a Comprehensive Transthoracic Echocardiographic Examination in Adults: Recommendations from the American Society of Echocardiography. *J. Am. Soc. Echocardiogr.* **2019**, *32*, 1–64. [CrossRef]
27. Wasserman, K.; Hansen, J.E.; Sue, D.Y.; Stringer, W.W.; Whipp, B. *Principles of Exercise Testing and Interpretation: Including Pathophysiology and Clinical Applications*, 4th ed.; Lippincott Williams and Wilkins: Philadelphia, PA, USA, 2004.
28. Albaghdadi, M.S.; Dudzinski, D.M.; Giordano, N.; Kabrhel, C.; Ghoshhajra, B.; Jaff, M.R.; Weinberg, I.; Baggish, A. Cardiopulmonary Exercise Testing in Patients Following Massive and Submassive Pulmonary Embolism. *J. Am. Hear. Assoc.* **2018**, *7*, e006841. [CrossRef]
29. Kahn, S.R.; Akaberi, A.; Granton, J.T.; Anderson, D.R.; Wells, P.S.; Rodger, M.A.; Solymoss, S.; Kovacs, M.J.; Rudski, L.; Shimony, A.; et al. Quality of life, dyspnea, and functional exercise capacity following a first episode of pulmonary embolism: Results of the ELOPE cohort study. *Am. J. Med.* **2017**, *130*, 990.e9–990.e21. [CrossRef]

30. Stavrou, V.; Boutou, A.K.; Vavougios, G.D.; Pastaka, C.; Gourgoulianis, K.I.; Koutedakis, Y.; Daniil, Z.; Karetsi, E. The use of cardiopulmonary exercise testing in identifying the presence of obstructive sleep apnea syndrome in patients with compatible symptomatology. *Respir. Physiol. Neurobiol.* **2019**, *262*, 26–31. [CrossRef] [PubMed]
31. Vanuxem, D.; Badier, M.; Guillot, C.; Delpierre, S.; Jahjah, F.; Vanuxem, P. Impairment of muscle energy metabolism in patients with sleep apnoea syndrome. *Respir Med.* **1997**, *91*, 551–557. [CrossRef]
32. Mereles, D.; Ehlken, N.; Kreuscher, S.; Ghofrani, S.; Hoeper, M.; Halank, M.; Meyer, F.J.; Karger, G.; Buss, J.; Juenger, J.; et al. Exercise and Respiratory Training Improve Exercise Capacity and Quality of Life in Patients with Severe Chronic Pulmonary Hypertension. *Circulation* **2006**, *114*, 1482–1489. [CrossRef] [PubMed]
33. Kwan, G.; Balady, G.J. Cardiac rehabilitation 2012: Advancing the field through emerging science. *Circulation* **2012**, *125*, 369–373. [CrossRef]
34. Tavoly, M.; Utne, K.K.; Jelsness-Jørgensen, L.-P.; Wik, H.S.; Klok, A.F.; Sandset, P.M.; Ghanima, W. Health-related quality of life after pulmonary embolism: A cross-sectional study. *BMJ Open* **2016**, *6*, e013086. [CrossRef]
35. Paz-Díaz, H.; Montes de Oca, M.; López, J.M.; Celli, B.R. Pulmonary rehabilitation improves depression, anxiety, dyspnea and health status in patients with COPD. *Am. J. Phys. Med. Rehabil.* **2007**, *86*, 30–36. [CrossRef] [PubMed]
36. Kord, Z.; Fereidouni, Z.; Mirzaee, M.S.; Alizadeh, Z.; Behnammoghadam, M.; Rezaei, M.; Abdi, N.; Delfani, F.; Zaj, P. Telenursing home care and COVID-19: A qualitative study. *BMJ Support. Palliat. Care* **2021**, *29*, bmjspcare-2021-003001. [CrossRef]
37. Cooper, C.B. Desensitization to dyspnea in COPD with specificity for exercise training mode. *Int. J. Chronic Obstr. Pulm. Dis.* **2008**, *4*, 33–43. [CrossRef] [PubMed]
38. García-Ortega, A.; Mañas, E.; López-Reyes, R.; Selma, M.J.; García-Sánchez, A.; Oscullo, G.; Jiménez, D.; Martínez-García, M. Ángel Obstructive sleep apnoea and venous thromboembolism: Pathophysiological links and clinical implications. *Eur. Respir. J.* **2019**, *53*, 1800893. [CrossRef] [PubMed]
39. Xie, J.; Li, F.; Wu, X.; Hou, W. Prevalence of pulmonary embolism in patients with obstructive sleep apnea and chronic ob-structive pulmonary disease: The overlap syndrome. *Heart Lung* **2019**, *48*, 261–265. [CrossRef]
40. Stavrou, V.; Bardaka, F.; Karetsi, E.; Seitanidis, G.; Daniil, Z.; Gourgoulianis, K.I. The effect of physical strain on breeders patients with obstructive sleep apnea syndrome. *Respir. Physiol. Neurobiol.* **2019**, *260*, 137–139. [CrossRef]
41. Stavrou, V.; Karetsi, E.; Daniil, Z.; Gourgoulianis, I.K. Four Weeks Exercise in Obstructive Sleep Apnea Syndrome Patient with Type 2 Diabetes Mellitus and without Continuous Positive Airway Pressure Treatment: A Case Report. *Sleep Med. Res.* **2019**, *10*, 54–57. [CrossRef]
42. Amoury, M.; Noack, F.; Kleeberg, K.; Stoevesandt, D.; Lehnigk, B.; Bethge, S.; Heinze, V.; Schlitt, A. Prognosis of patients with pulmonary embolism after rehabilitation. *Vasc. Health Risk Manag.* **2018**, *14*, 183–187. [CrossRef]
43. Lakoski, S.G.; Savage, P.D.; Berkman, A.; Penalosa, L.; Crocker, A.; Ades, P.A.; Kahn, S.R.; Cushman, M. The safety and efficacy of early-initiation exercise training after acute venous thromboembolism: A randomized clinical trial. *J. Thromb. Haemost.* **2015**, *13*, 1238–1244. [CrossRef]
44. Klok, E.; Mos, I.C.M.; Huisman, M.V. Brain-Type Natriuretic Peptide Levels in the Prediction of Adverse Outcome in Patients with Pulmonary Embolism. *Am. J. Respir. Crit. Care Med.* **2008**, *178*, 425–430. [CrossRef] [PubMed]
45. Dentali, F.; Donadini, M.; Gianni, M.; Bertolini, A.; Lonn, E.; Venco, A.; Cattozzo, G.; Ageno, W. Brain natriuretic peptide as a preclinical marker of chronic pulmonary hypertension in patients with pulmonary embolism. *Intern. Emerg. Med.* **2009**, *4*, 123–128. [CrossRef] [PubMed]

Journal of
Functional Morphology and Kinesiology

MDPI

Article

Workload Accomplished in Phase III Cardiac Rehabilitation

Katrina L. Schultz [1], Carl Foster [1,*], Kimberley Radtke [1], Susan Bramwell [1], Cristina Cortis [2], Andrea Fusco [2] and John P. Porcari [1]

[1] Department of Exercise and Sport Science, University of Wisconsin-La Crosse, La Crosse, WI 54601, USA; schultzkatrina82@gmail.com (K.L.S.); kradtke@uwlax.edu (K.R.); sbramwell@uwlax.edu (S.B.); jporcari@uwlax.edu (J.P.P.)

[2] Department of Human Sciences, Society and Health, University of Cassino and Lazio Meridionale, 03043 Cassino, Italy; c.cortis@unicas.it (C.C.); andrea.fusco@unicas.it (A.F.)

* Correspondence: cfosteruwl@gmail.com

Abstract: Exercise training is an important component of clinical exercise programs. Although there are recognized guidelines for the amount of exercise to be accomplished (≥70,000 steps per week or ≥150 min per week at moderate intensity), there is virtually no documentation of how much exercise is actually accomplished in contemporary exercise programs. Having guidelines without evidence of whether they are being met is of limited value. We analyzed both the weekly step count and the session rating of perceived exertion (sRPE) of patients (n = 26) enrolled in a community clinical exercise (e.g., Phase III) program over a 3-week reference period. Step counts averaged 39,818 ± 18,612 per week, with 18% of the steps accomplished in the program and 82% of steps accomplished outside the program. Using the sRPE method, inside the program, the patients averaged 162.4 ± 93.1 min per week, at a sRPE of 12.5 ± 1.9 and a frequency of 1.8 ± 0.7 times per week, for a calculated exercise load of 2042.5 ± 1244.9 AU. Outside the program, the patients averaged 144.9 ± 126.4 min, at a sRPE of 11.8 ± 5.8 and a frequency of 2.4 ± 1.5 times per week, for a calculated exercise load of 1723.9 ± 1526.2 AU. The total exercise load using sRPE was 266.4 ± 170.8 min per week, at a sRPE of 12.6 ± 3.8, and frequency of 4.2 ± 1.1 times per week, for a calculated exercise load of 3359.8 ± 2145.9 AU. There was a non-linear relationship between steps per week and the sRPE derived training load, apparently attributable to the amount of non-walking exercise accomplished in the program. The results suggest that patients in a community clinical exercise program are achieving American College of Sports Medicine guidelines, based on the sRPE method, but are accomplishing less steps than recommended by guidelines.

Keywords: cardiac rehabilitation; step count; sRPE

Citation: Schultz, K.L.; Foster, C.; Radtke, K.; Bramwell, S.; Cortis, C.; Fusco, A.; Porcari, J.P. Workload Accomplished in Phase III Cardiac Rehabilitation. *J. Funct. Morphol. Kinesiol.* **2021**, *6*, 47. https://doi.org/10.3390/jfmk6020047

Academic Editor: Peter Hofmann

Received: 2 April 2021
Accepted: 26 May 2021
Published: 28 May 2021

1. Introduction

Cardiac rehabilitation is of established benefit for decreasing the risk of future cardiovascular events [1–3], delaying/reversing the progression of atherosclerotic disease [4–6], reducing the rate of reinfarction and death [7–9], and accelerating the rate of recovery of functional capacity after clinical episodes [10,11]. The value of cardiac rehabilitation is evident even after accounting for contemporary revascularization and pharmacologic therapy [12]. Rehabilitation exerts its' favorable effect, at least in part, through the very large effect of habitual exercise on both general, and cardiovascular, health [13–16]. Contemporary cardiac rehabilitation is accomplished through monitored exercise, education, and risk factor modification in phase II or phase III cardiac rehabilitation programs. Cardiac rehabilitation, and exercise training in general, decreases the risk of mortality and mitigates metabolic diseases [3], with the effect being somewhat dependent on the dose of exercise [3,6,13–17]. Further, the risk of developing new cardiovascular disease, and of mortality in relation to exercise, has been shown to be dose dependent [13–15,18].

Although there are widely accepted guidelines for exercising patients with known heart disease [19,20] and a rich literature on monitoring exercise training in athletes [21–24], there is virtually no evidence of how much exercise is actually accomplished by patients enrolled in rehabilitation programs. It can be argued that guidelines without evidence of how well they are complied with are of limited value. Accordingly, it seems desirable to have a practical method of quantifying the dose of exercise performed by cardiac rehabilitation patients, particularly in a way that gets beyond exercise mode specificity (e.g., steps per day).

One way to monitor exercise and physical activity is through step counts. Tudor-Locke et al. [18] found that individuals who averaged 5000 steps daily were more obese and had a higher risk of metabolic diseases, compared to individuals averaging 15,000 steps daily. Higher step counts were linked with an absence of obesity and decreased risk of disease in agricultural populations [25]. Recently, Kraus et al. [14] found that in individuals who increased daily steps, the risk of mortality decreased by ~10% for every additional 2000 steps per day. Recording participants' daily step count is a simple method of estimating of how much exercise is done in one session, of how much physical activity is completed, outside of rehabilitation, and a practical way to prescribe and monitor exercise training. Indeed, Hambrecht et al. [6] presented evidence that regression of atherosclerotic lesions was associated with exercise at a level consistent with ~80,000 steps per week. Although step counts do not inherently address the very important aspect of exercise intensity, the strong dose dependency of the benefit of exercise seems to be present even when only step counts are considered [14,18,25]. Despite the presence of guidelines (generally of 7000–10,000 steps per day), to our knowledge there are no systematic data regarding step counts routinely performed during exercise programs primarily designed for secondary prevention.

However, step counts alone do not account for some modes of activity, such as cycling, Nu Step, rowing, arm cranking, and resistance training that are common elements of many rehabilitation programs. It would be useful if there were methods that could account for the multi-modal nature of most cardiac rehabilitation programs, as well as accounting for the net intensity of exercise training. Although accelerometry based methods have become popular, they still require patients to have access to and to wear an accelerometer [16].

The rating of perceived exertion (RPE) is a method of estimating internal training load and has been shown to be an acceptable surrogate of objective markers of exercise intensity such as percentage of heart rate (%HR) reserve and blood lactate accumulation either during an acute bout of exercise [26–28] or during an entire exercise session (thus including exercise duration which causes an upward drift in the workload-RPE relationship), the session RPE (sRPE) [21,22,29,30]. sRPE is easy to use and is accessible to most populations. Based on studies conducted by Foster et al. [22,29,30], Day et al. [31], and Herman et al. [32], sRPE has been shown to be valid in terms of evaluating entire training sessions at intensities ranging from light to vigorous, and in multiple modes of exercise [22]. Historically, sRPE was measured 30-min post exercise to get an accurate judgment of the overall intensity of the exercise training session. However, Christen et al. [33] and Foster et al. [22] have shown that sRPE is very temporally robust. Recent studies by Arney et al. [34,35] have demonstrated the interchangeability of the two most widely used RPE scales, further supporting the utility of sRPE for monitoring training. RPE is also inexpensive and "user-friendly," which makes it ideal for cardiac rehabilitation populations. Further, the ability to use sRPE to "collapse across different modes of exercise" [22,30] makes it attractive for the often-multi-modal nature of cardiac rehabilitation programs. However, despite the generally accepted dose-response nature of responses to training in cardiac patients [6,17], to our knowledge there are no systematic data demonstrating the actual amount of training performed by cardiac patients using the sRPE approach.

Training load can be conceptualized as the amount of work done (duration x intensity) during an exercise bout [22]. Exercise training load has a demonstrated inverse relationship with mortality [3,4,13,15,36]. Total exercise related energy expenditure is also related to weight loss, decreases in abdominal fat, and reduction in waist circumference [17], which are frequent goals of cardiac rehabilitation therapy. Training load can be used to prescribe exercise and to personalize future exercise plans. This can be compared to the American

College of Sports Medicine (ACSM) guidelines for 30 min of moderate intensity five times a week or 10,000 steps a day [14,19]. Accordingly, the purpose of this study was to document the training load accomplished by patients with known cardiovascular disease, or with risk factors likely to cause cardiovascular disease, in a community-based exercise program using both steps/day and the sRPE approach. Such data is uniquely needed relative to understanding how well contemporary programs execute professional society guidelines.

2. Materials and Methods

Thirty-two participants in the phase III La Crosse Exercise and Health Program (LEHP) were recruited. They ranged in age from 35–90, 62% were male and, typical of middle-aged Americans, they generally had elevated values for body mass index (BMI) (23–38). Twenty-two of the patients had experienced a documented cardiovascular event (myocardial infarction, revascularization surgery, angioplasty with stent), and the other 10 were >55 years of age and had risk factors for cardiovascular disease. Most of the patients had been treated for their primary medical problem at one of two local tertiary care hospitals, and had participated in Phase I and II cardiac rehabilitation programs there. They were self-referred to the LEHP as a venue to facilitate continuing exercise therapy, although individual consent from their own physicians was routinely secured. All of the patients were clinically stable, and the only contraindication to participation was clinical instability. Medications were conventional for this population (70% anticoagulants, 60% statins, 10% beta blockers, 12% ACE inhibitors/blockers, 32% anti hyperglycemic medicines) (Table 1). All participants provided written informed consent. The Institutional Review Board for the Protection of Human Subjects of the University of Wisconsin-La Crosse approved the study (Protocol #20-CF-081, 2020). The training program within the LEHP was scheduled to take place 3 times weekly. Although a 2.5 h block of open time was available at the facility, patients were instructed to participate for 1.0–1.5 h each time. Exercise training was intended to be performed primarily with aerobic activities that suited the subject. This could include walking, indoor cycling, arm-leg cycling, or swimming. Some light resistance training was performed by most of the participants. The exact mixture for each participant was highly individualized, primarily because their medical conditions and/or orthopedic history made some activities hard to perform. Most sessions included warm-up and cool-down calisthenics and stretching. We did not have access to maximal exercise tests on any of these patients, so regulation of exercise intensity was accomplished using RPE and the Talk Test.

Table 1. Mean and standard deviation of the steps per week achieved in La Crosse Exercise and Health Program (LEHP) and outside.

Week	LEHP (%)	Outside (%)	Total
1	7354 ± 6934 (17.9)	33,845 ± 15,662 (82.1)	41,199 ± 19,395
2	7483 ± 6342 (18.9)	32,308 ± 14,899 (81.7)	39,502 ± 17,939
3	6945 ± 5232 (17.9)	31,809 ± 15,529 (82.1)	38,753 ± 18,502
Average	7261 ± 6169 (18.2)	32,654 ± 15,360 (82.0)	39,818 ± 18,612

% Represent percentage of steps relative to total amount.

Participants wore an ActiGraph device WGT3X-BT (Pensacola, FL, USA) for 3 weeks. It was worn on the hip, on the non-dominant side during all waking hours, except when bathing. Their daily step count was downloaded at the end of the study period and matched to program, and non-program, step counts based on day of week and time of day. They also tracked their exercise time and sRPE, using the classical 6–20 RPE scale, both during the LEHP, as well as during outside activities [22,29,30]. From this data, workload was calculated to determine the steps per day during the LEHP, during outside activities, the total for the day, and total for the week. Further, to determine training load using the sRPE method, the sRPE x duration in minutes both within the LEHP and in outside activities was computed. Data were collected over 3 consecutive weeks, as a convenient time interval, which is greater than the 2 weekdays and 1 weekend day used in classical epidemiologic studies. Further adding to the representativeness of the data collection period, one of the weeks contained a national holiday (Thanksgiving). The study was conducted during a period of the year when outside environmental temperatures were just above freezing (mid-Fall), although outside snow cover was minimal. One subject was removed due to loss of their ActiGraph device. Five subjects were removed due to either an inadequate amount of data (clear gaps in the recording) or improper recording on the activity monitor log sheet. Thus, a total of twenty-six participants were used in data analysis.

Steps per day and sRPE were presented as descriptive statistics and contrasted to recommended values for exercise training [19,20] and steps per day [14,18]. Statistical comparison of the mean exercise load versus recommendations was not appropriate. Regression statistics were compared between steps per day and the sRPE training load. Descriptive and regression statistics were computed using SPSS.

3. Results

The average number of total steps achieved was around 40,000 per week (Table 1), which is below the conventional benchmarks of 10,000 steps a day (50,000–70,000 steps weekly). About 80% of the steps were accumulated outside the LEHP. This suggests that more total exercise needs to be done to see an optimal decrease in risk of mortality [4,6]. The number of steps accumulated was greater on the days outside of the LEHP, probably because the activity at the LEHP was multi-modal, whereas almost all activity outside of the LEHP was accomplished by walking. The data did not allow fractionation of non-LEHP steps performed as "intentional exercise" versus "activities of daily living" On average, the subjects were exercising 4 days a week with the mean time of 260–270 min a week. This is above the recommended amount based on the ACSM guidelines [19]. The data also showed RPE to be higher in LEHP than during outside exercise (Table 2, Table 3, Table 4).

Table 2. Mean and standard deviation of the time (minutes), rate of perceived exertion (RPE), session RPE (sRPE), and frequency (times per week) for the La Crosse Exercise and Health Program.

Week	Time	RPE	sRPE	Frequency
1	173.5 ± 99.4	12.1 ± 1.9	2108.7 ± 1230.1	1.9 ± 0.8
2	157.9 ± 98.6	12.7 ± 1.9	2035.4 ± 1232.7	1.8 ± 0.7
3	155.8 ± 81.4	12.8 ± 1.9	1983.4 ± 1272.0	1.7 ± 0.6
Average	162.4 ± 93.1	12.5 ± 1.9	2042.5 ± 1244.9	1.8 ± 0.7

Table 3. Mean and standard deviation of the time (minutes), rate of perceived exertion (RPE), session RPE (sRPE), and frequency (times per week) for outside the La Crosse Exercise and Health Program.

Week	Time	RPE	sRPE	Frequency
1	146.9 ± 127.4	11.5 ± 5.7	1700.8 ± 1488.5	2.4 ± 1.4
2	142.1 ± 111.9	12.0 ± 5.8	1729.4 ± 1336.8	2.3 ± 1.5
3	145.7 ± 141.1	11.9 ± 5.8	1741.6 ± 1753.2	2.6 ± 1.5
Average	144.9 ± 126.4	11.8 ± 5.8	1723.9 ± 1526.2	2.4 ± 1.5

Table 4. Mean and standard deviation of the time (minutes), rate of perceived exertion (RPE), session RPE (sRPE), and frequency (times per week) for the La Crosse Exercise and Health Program and outside combined.

Week	Time	RPE	sRPE	Frequency
1	322.4 ± 181.8	12.6 ± 3.8	3617.0 ± 2188.4	4.3 ± 1.1
2	300.0 ± 163.8	12.8 ± 3.9	3273.2 ± 2248.0	4.2 ± 1.1
3	301.5 ± 166.9	12.4 ± 3.8	3189.2 ± 2001.2	4.2 ± 1.2
Average	307.3 ± 170.8	12.6 ± 3.8	3359.8 ± 2145.9	4.2 ± 1.2

Time, RPE, sRPE derived training load and frequency accomplished in the LEHP are represented in Table 2. On average subjects spent about 160–170 min a week exercising with an RPE of about 12.5, which is just below the somewhat hard (RPE = 13) verbal cue widely taken as an idealized intensity for non-athletic individuals [37]. This gives an average workload of 2000–2100 AU (sRPE × minutes). Subjects attended the LEHP just under two times a week, out of the possible three times a week.

Time, RPE, sRPE derived training load and frequency performed outside of the LEHP are shown in Table 3. The average time spent exercising was 140–150 min a week, which is slightly less than the 160–170 min accomplished in LEHP. Average sRPE accomplished is 11.8, which is just above the "light" verbal cue, which is also slightly lower than the sRPE in LEHP. Similarly, the average load is about 1700–1800 AU which is slightly lower than the observed load in LEHP (2000–2100 AU). The frequency of exercising is about 2.4 times/week which is higher than the frequency of LEHP, indicating that participants are exercising more often, but for a shorter time and lower intensity on their own then they are in LEHP.

Table 4 presents the combined time, RPE, sRPE derived training load and frequency from both LEHP and outside exercise. The average frequency was 4.2 ± 1.5 times per week. The average duration was 260–270 min per week indicating that subjects, both in LEHP and outside are exceeding the recommended amount of weekly exercise (>150 min/week) in the ACSM guidelines [19]. Overall RPE averages 12.6 ± 3.84 which is just below the "somewhat hard" verbal cue, indicating that subjects are reaching moderate intensity level [37]. The sRPE Load was 3300–3400 AU per week. This is greater than the ~2000 AU recommended by the ACSM guidelines [19] of RPE = 13 × 30 min per day × 5 days per week.

Figure 1 represents the relationship between total weekly steps and sRPE derived training load, with each week per subject being presented as a data point. The curvilinear line of best fit indicates that as step count increases there is an increase in sRPE derived training load. This graph shows that the majority of weekly averages are less than 60,000 steps and less than 5000 AU load. It also illustrates that a step counts <50,000 per week, there is relatively little change in sRPE based training load, probably because many of the steps were representative of activities of daily living, rather than structured exercise.

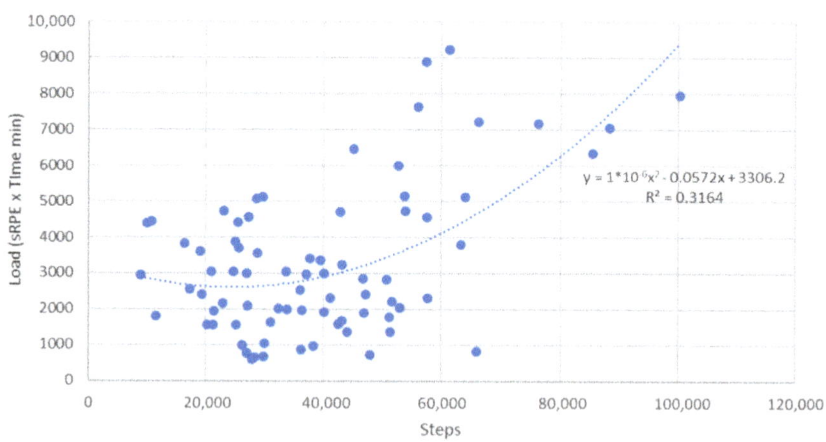

Figure 1. Relationship between total weekly step counts and session Rating of Perceived Exertion (sRPE) training load.

4. Discussion

Despite the unequivocal evidence of exercise-based rehabilitation programs in both rehabilitation and secondary prevention in patients with cardiovascular disease [1–12], physicians are still referring patients to programs in inadequate numbers. Potentially, this non-referral bias may be due to a poor understanding of how well exercise guidelines are implemented. The purpose of this study was to document the number of steps and sRPE derived training load accomplished both inside and outside a community-based Phase III clinical exercise program. These findings suggest that the recommended 10,000 steps/day (~50,000–70,000 steps per week) is not being met, but that the suggested dose of weekly exercise (>150 min/week) being accomplished exceeds the ACSM Guidelines [19]. There was a curvilinear trend between steps and load accomplished (Figure 1) demonstrating that, as step count increases, so does sRPE derived training load, at least after activity of daily living (ADL) steps (~35,000 steps per week) are accounted for. It also suggests that, at high training loads, there is an increased volume of multimodal exercise, which is not reflected by step count.

There were lower average steps observed in LEHP compared to outside, but overall higher time, sRPE, and load accomplished in LEHP. This indicates that these participants are working harder in the LEHP than during outside training. This supports the concept that patients should do more work, and work at higher intensity, in a community-based exercise program than they would otherwise do on their own. This matches general training practice guidelines for cardiac rehabilitation programs.

Currently there are no published data in regard to daily recording in any phase of a cardiac rehabilitation program. Assuming, on average, that participants exercise at a nominal walking pace of 5 kph, a rate requiring ~5.2 kcal per min (11.5 mL/kg $\times$ 90 kg = gross VO_2 = 1.035 L/min, $\times$5 = ~5.2 kcal/min), the total of 266 min of exercise accomplished a week would equate to ~1282 kcal/week. Looking at only LEHP the average time accomplished was 162 min a week, which would equate to ~842 kcal/week. The outside exercise time per week averaged to 144 min, which would equate to ~749 kcal/week. According to Hambrecht et al. [6], 1400 kcal/week need to be accumulated in order to halt progression of coronary atherosclerotic lesions, and 2200 kcal/week needs to be expended to have regression in coronary lesions, although lower amounts of exercise are associated with an improved event free survival over 1-year compared to angioplasty + stenting [38]. The data shows that, on average, the participants are expending about 120 kcal/week less than what is needed to halt atherosclerosis. Assuming that one of the primary goals of exercise-based rehabilitation is secondary prevention, the current observations suggest that more total exercise needs to be accomplished.

An alternative to steps as a metric of exercise training load is the RPE, whether as a momentary index of training intensity, or an expression of the intensity x duration exercise load. Parfitt, Evans, and Eston [37] conducted a study on exercising for 30 min three times a week (90 min/week) for eight weeks at an RPE of 13 which would equal a load of 1170 AU (90 min/week $\times$ 13 RPE). This resulted in improvements in mean arterial pressure, total cholesterol, VO_2max, and body mass index. Our subjects completed on average a workload of 2043 in LEHP and averaged a workload in total of 3360 AU, which is well above the Parfitt et al. [37] workload of 1170 AU, thus these same improvements would be expected to be observed in the current participants. Hambrecht et al. [38] conducted a randomized study that had subjects exercise 20 min daily. Assuming they worked at a similar RPE, their weekly load would be 1750 AU ((20 min $\times$ 7 days) $\times$ 12.5 RPE). He observed a higher event free survival of 88%, versus the 70% event free survival following percutaneous angioplasty with stenting (the nominal gold standard for patients with coronary artery disease). The total exercise load accomplished by our subjects is above that, plausibly accomplished, by Hambrecht et al. [38], suggesting the likelihood of good clinical outcomes. This is consistent with the nearly 50-year history of the LEHP, which has many participants with more than 20 years of participation, and with an experience of only ~1 untoward event per year.

This study does not address the total exercise workload accomplished in phase I and II cardiac rehabilitation, which address the rehabilitative goals of such programs in addition to secondary prevention. Given that one of the goals of the second part of phase II cardiac rehabilitation is secondary prevention, this information needs to be known. This study was conducted on a limited population sample, at a single community-based venue. Extending the results of this study to a larger population, at a larger number of venues, would provide more information about workload accomplished in rehabilitation programs. This is important, as Ades et al. [17] have suggested that the actual exercise load in conventional cardiac rehabilitation programs is inadequate to modify common risk factors that are targets of cardiac rehabilitation therapy.

While the Actigraph is a good device for calculating steps, there is a limitation with using multimodal equipment that may not be recognized by the device. This could be solved by computing MET × min [13] or MET × h [15] from chart records, or Time × RPE [22] from chart records, in established rehabilitation programs, to get a sense of the normal rate of progression of exercise load in Phase I and II rehabilitation programs, which typically include rowing machines, stationary cycling, Nu Step ergometers, and lifting weights. Data collection also happened to occur over the Thanksgiving holiday in Wisconsin. Due to celebration of the holiday and colder weather limiting exercise, this can result in decreased exercise levels that may not have occurred otherwise. Differentiating from outside activity or LEHP activity also provided some limitations. The best evidence is that nominally sedentary individuals accumulate about 5000 steps per day [14,18], but this likely varies greatly with age, season, and occupational demands. Being able to differentiate incidental versus intentional step counts would also be useful in further understanding the workload/health interaction in this population.

It would be beneficial to have this study replicated in phase I and II cardiac rehabilitation to know what workload is occurring at different phases. This information could be used to track if rehabilitation programs are doing enough exercise to accomplish their clinical goals and identify areas that could be improved. Replicating this study with more thorough education on how to use RPE, as well as properly fill out the log sheet, would help with participant error and more accurate data.

5. Conclusions

Our results in a community-based program showed that subjects achieved about 40,000 steps a week, which is less than the recommended 10,000 steps per day. Using an alternative method for calculating load by using sRPE, an average load of 3359 AU was achieved weekly. This information suggests that, while 10,000 steps may not be reached, adequate exercise, via different modalities that do not include steps, is indeed being achieved. Subjects averaged in total exercised 4.2 times weekly, with an average of 1.8 of those occurrences being at LHEP and average of 2.4 being outside of LEHP. The total average time spent exercising was 266 min a week, which is within the recommend amount of weekly exercise according to the ACSM Guidelines [19]. Comparing this study with the works of Hambrecht et al. [6,38] and Parfitt, Evans, and Eston [37], phase III cardiac rehabilitation can improve mean arterial pressure, cholesterol, body mass index, and increase chance of event free survival. However, the step count deficiencies observed here, compared to the regression data from Hambrecht et al. [6] suggest that more exercise could be profitable. Accordingly, community-based programs may benefit from having individuals exercise harder and longer than they otherwise would have on their own.

Author Contributions: Conceptualization, K.L.S., C.F. and K.R.; investigation, K.L.S., C.F. and K.R.; data curation, K.L.S., C.F. and K.R.; writing—original draft preparation, K.L.S., C.F. and K.R.; writing—review and editing, K.L.S., C.F., K.R., S.B., C.C., A.F. and J.P.P.; supervision, S.B. and J.P.P. All authors have read and agreed to the published version of the manuscript.

Funding: This research received no external funding.

Institutional Review Board Statement: The study was conducted according to the guidelines of the Declaration of Helsinki, and approved by the Institutional Review Board for the Protection of Human Subjects of the University of Wisconsin-La Crosse.

Informed Consent Statement: Written informed consent was obtained from all subjects involved in the study.

Data Availability Statement: The data presented in this study are available on request from the corresponding author.

Conflicts of Interest: The authors declare no conflict of interest.

References

1. O'Connor, G.T.; Buring, J.E.; Yusuf, S.; Goldhaber, S.Z.; Olmstead, E.M.; Paffenbarger, R.S.; Hennekens, C.H. An overview of randomized trials of rehabilitation with exercise after myocardial infarction. *Circulation* **1989**, *80*, 234–244. [CrossRef]
2. Oldridge, N.B.; Guyatt, G.H.; Fischer, M.E. Cardiac rehabilitation after myocardial infarction: Combined experience of randomized clinical trials. *JAMA* **1988**, *260*, 945–950. [CrossRef] [PubMed]
3. Eijsvogels, T.M.; Molossi, S.; Lee, D.; Emery, M.S.; Thompson, P.D. Exercise at the extremes. *JACC* **2016**, *67*, 316–329. [CrossRef]
4. Niebauer, J.; Hambrecht, R.; Velich, T.; Hauer, K.; Marburger, C.; Kälberer, B.; Kübler, W. Attenuated progression of coronary artery disease after 6 years of multifactorial risk intervention. *Circulation* **1997**, *96*, 2534–2541. [CrossRef]
5. Haskell, W.; Alderman, E.; Fair, J. Intensive risk factor management in men and women with coronary artery disease: The Stanford Coronary risk intervention project (SCRIP). *Atherosclerosis* **1994**, *109*, 182–189. [CrossRef]
6. Hambrecht, R.; Niebauer, J.; Marburger, C.; Grunze, M.; Kälberer, B.; Hauer, K.; Schuler, G. Various intensities of leisure time physical activity in patients with coronary artery disease: Effects on cardiorespiratory fitness and progression of coronary atherosclerotic lesions. *JACC* **1993**, *22*, 468–477. [CrossRef]
7. Taylor, R.S.; Brown, A.; Ebrahim, S.; Jolliffe, J.; Noorani, H.; Rees, K.; Oldridge, N. Exercise-based rehabilitation for patients with coronary heart disease: Systematic review and meta-analysis of randomized controlled trials. *Am. J. Med.* **2004**, *116*, 682–692. [CrossRef] [PubMed]
8. Lawler, P.R.; Filion, K.B.; Eisenberg, M.J. Efficacy of exercise-based cardiac rehabilitation post–myocardial infarction: A systematic review and meta-analysis of randomized controlled trials. *Am. Heart J.* **2011**, *162*, 571–584. [CrossRef]
9. Hammill, B.G.; Curtis, L.H.; Schulman, K.A.; Whellan, D.J. Relationship between cardiac rehabilitation and long-term risks of death and myocardial infarction among elderly Medicare beneficiaries. *Circulation* **2010**, *121*, 63–70. [CrossRef]
10. Foster, C.; Pollock, M.L.; Anholm, J.D. Work capacity and left ventricular function during rehabilitation after myocardial revascularization surgery. *Circulation* **1984**, *69*, 748–755. [CrossRef] [PubMed]
11. Foster, C.; Oldridge, N.B.; Dion, W. Time course of recovery during cardiac rehabilitation. *J. Cardiopulm. Rehabil.* **1995**, *15*, 209–215. [CrossRef]
12. Rauch, B.; Davod, C.H.; Doherty, P. The prognostic effect of cardiac rehabilitation in the era of acute revascularization and statin therapy: A systematic review and meta-analysis of randomized and non-randomized studies. *Eur. J. Prev. Cardiol.* **2016**, *23*, 1914–1939. [CrossRef] [PubMed]
13. Arem, H.; Moore, S.C.; Patel, A. Leisure time physical activity and mortality. *JAMA Intern. Med.* **2015**, *175*, 959. [CrossRef] [PubMed]
14. Kraus, W.E.; Janz, K.F.; Powell, K.E. Daily step counts for measuring physical activity exposure and its relation to health. *Med. Sci. Sports Exerc.* **2019**, *51*, 1206–1212. [CrossRef] [PubMed]
15. O'Keefe, J.H.; Lavie, C.J.; Guazzi, M. Part 1: Potential dangers of extreme endurance exercise: How much is too much? *Prog. Cardiovasc. Dis.* **2014**, *57*, 396–405. [CrossRef] [PubMed]
16. Ramakishan, R.; Doherty, A.; Smith-Byrne, K. Accelerometer measured physical activity and the incidence of cardiovascular disease. *PLoS Med.* **2021**, *18*, e1003487.
17. Ades, P.A.; Savage, P.D.; Toth, M.J.; Harvey-Berino, J.; Schneider, D.J.; Bunn, J.Y.; Ludlow, M. High-calorie-expenditure exercise. *Circulation* **2009**, *119*, 2671–2678. [CrossRef]
18. Tudor-Loacke, C.; Bassett, D.R. How many steps/day are enough? *Sports Med.* **2004**, *34*, 1–8. [CrossRef] [PubMed]
19. American College of Sports Medicine, Riebe, D.; Ehrman, J.K.; Liguori, G.; Magal, M. *ACSM's Guidelines for Exercise Testing and Prescription*, 10th ed.; Wolters Kluwer: Philadelphia, PA, USA, 2018.
20. Mezzani, A.; Hamm, L.F.; Jones, A.M.; Mcbride, P.E.; Moholdt, T.; Stone, J.A.; Williams, M.A. Aerobic exercise intensity assessment and prescription in cardiac rehabilitation. *J. Cardiopulm. Rehabil. Prev.* **2012**, *32*, 327–350. [CrossRef]
21. Foster, C.; Rodriguez-Marroyo, J.A.; de Koning, J.J. Monitoring training loads: The past, present and future. *Int. J. Sports Physiol. Perf.* **2017**, *12*, s2–s8. [CrossRef]
22. Foster, C.; Boullosa, D.; McGuigan, M. 25 years of session RPE: Historical perspective and development. *Int. J. Sports Physiol. Perf.* **2021**. [CrossRef] [PubMed]
23. Borrensen, J.; Lambert, M.I. Quantifying training load: A comparison of subjective and objective method. *Int. J. Sports Physiol. Perform.* **2008**, *3*, 1–16.

24. Impellizzeri, F.M.; Marcora, S.; Coutts, A.J. Internal and external training: 15 years on. *Int. J. Sports Physiol. Perf.* **2019**, *14*, 270–273. [CrossRef]
25. Bassett, D.R.; Tremblay, M.S.; Esliger, D.W.; Copeland, J.L.; Barnes, J.D.; Huntington, G.E. Physical activity and body mass index in an Old Order Amish community. *Med. Sci. Sports Exerc.* **2007**, *39*, 410–415. [CrossRef] [PubMed]
26. Borg, G. *Borg's Perceived Exertion and Pain Scales*; Human Kinetics: Champaign, IL, USA, 1998.
27. Pollock, M.L.; Foster, C.; Rod, J.L.; Wible, G. Comparison of methods for determining exercise training intensity in cardiac patients and healthy adults. *Adv. Cardiol.* **1982**, *31*, 129–133. [PubMed]
28. Noble, B.J.; Borg, G.V.A.; Jacobs, I.; Ceci, R.; Kaiser, P. A category ratio perceived exertion scale; Relationship to blood and muscle lactates and heart rate. *Med. Sci. Sports Exerc.* **1983**, *15*, 523–528. [CrossRef] [PubMed]
29. Foster, C.; Hector, L.L.; Welsh, R.; Schrager, M.; Green, M.A.; Snyder, A.C. Effects of specific versus cross-training on running performance. *Eur. J. Appl. Physiol. Occup. Physiol.* **1995**, *70*, 367–372. [CrossRef]
30. Foster, C.; Florhaug, J.A.; Franklin, J.; Gottschall, L.; Hrovatin, L.A.; Parker, S.; Dodge, C. A new approach to monitoring exercise training. *J. Streng. Cond. Res.* **2001**, *15*, 109–115.
31. Day, M.L.; Mcguigan, M.R.; Brice, G.; Foster, C. Monitoring exercise intensity during resistance training using the Session RPE scale. *J. Streng. Cond. Res.* **2004**, *18*, 353–358.
32. Herman, L.; Foster, C.; Maher, M.A.; Mikat, R.P.; Porcari, J.P. Validity and reliability of the session RPE method for monitoring exercise training intensity. *S. Afr. J. Sports Med.* **2006**, *18*, 14–17. [CrossRef]
33. Christen, J.; Foster, C.; Porcari, J.P.; Mikat, R.P. Temporal robustness of the session Rating of Perceived Exertion. *Int. J. Sports Physiol. Perform.* **2016**, *11*, 1088–1093. [CrossRef] [PubMed]
34. Arney, B.E.; Glover, R.; Fusco, A.; Cortis, C.; de Koning, J.J.; van Erp, T.; Jaime, S.; Mikat, R.P.; Porcari, J.P.; Foster, C. Comparison of RPE (Rating of Perceived Exertion) scales for Session RPE. *Int. J. Sports Physiol. Perf.* **2019**, *14*, 994–996. [CrossRef] [PubMed]
35. Arney, B.E.; Glover, R.; Fusco, A.; Cortis, C.; de Koning, J.J.; van Erp, T.; Jaime, S.; Mikat, R.P.; Porcari, J.P.; Foster, C. Comparison of rating of perceived exertion scales during incremental and interval exercise. *Kinesiology* **2019**, *51*, 150–157. [CrossRef]
36. Mandsager, K.; Harb, S.; Cremer, P.; Phelan, D.; Nissen, S.E.; Jaber, W. Association of cardiorespiratory fitness with long-term mortality among adults undergoing exercise treadmill testing. *JAMA Netw. Open* **2018**, *1*, 20. [CrossRef] [PubMed]
37. Parfitt, G.; Evans, H.; Eston, R. Perceptually regulated training at RPE13 is pleasant and improves physical health. *Med. Sci. Sports Exerc.* **2012**, *44*, 1613–1618. [CrossRef] [PubMed]
38. Hambrecht, R.; Walther, C.; Möbius-Winkler, S.; Gielen, S.; Linke, A.; Conradi, K.; Schuler, G. Percutaneous coronary angioplasty compared with exercise training in patients with stable coronary artery disease. *Circulation* **2004**, *109*, 1371–1378. [CrossRef]

Journal of
Functional Morphology and Kinesiology

MDPI

Article

Prediction of Exercise Capacity and Training Prescription from the 6-Minute Walk Test and Rating of Perceived Exertion

John P. Porcari [1], Carl Foster [1,*], Maria L. Cress [1], Rachel Larson [1], Hannah Lewis [1], Cristina Cortis [2], Scott Doberstein [1], Marc Donahue [3], Andrea Fusco [2] and Kimberly Radtke [1]

1 Department of Exercise and Sport Science, University of Wisconsin-La Crosse, La Crosse, WI 54601, USA; jporcari007@gmail.com (J.P.P.); mcress3381@gmail.com (M.L.C.); rshaw46@gmail.com (R.L.); brownhann1605@gmail.com (H.L.); sdoberstein@uwlax.edu (S.D.); kradtke@uwlax.edu (K.R.)
2 Department of Human Sciences, Society and Health, University of Cassino and Lazio Meridionale, 03043 Cassino, Italy; c.cortis@unicas.it (C.C.); andrea.fusco@unicas.it (A.F.)
3 Mayo Clinic Health System, La Crosse, WI 54601, USA; donahue.marc@mayo.edu
* Correspondence: cfosteruwl@gmail.com

Abstract: Walking tests, such as the 6-min walk test (6MWT), are popular methods of estimating peak oxygen uptake (VO_2peak) in clinical populations. However, the strength of the distance vs. VO_2peak relationship is not strong, and there are no equations for estimating ventilatory threshold (VT), which is important for training prescription and prognosis. Since the 6MWT is often limited by walking mechanics, prediction equations that include simple additional predictors, such as the terminal rating of perceived exertion (RPE), hold the potential for improving the prediction of VO_2max and VT. Therefore, this study was designed to develop equations for predicting VO_2peak and VT from performance during the 6MWT, on the basis of walking performance and terminal RPE. Clinically stable patients in a cardiac rehabilitation program (N = 63) performed the 6MWT according to the American Thoracic Society guidelines. At the end of each walk, the subject provided their terminal RPE on a 6–20 Borg scale. Each patient also performed a maximal incremental treadmill test with respiratory gas exchange to measure VO_2peak and VT. There was a good correlation between VO_2peak and 6MWT distance (r = 0.80) which was improved by adding the terminal RPE in a multiple regression formula (6MWT + RPE, R^2 = 0.71, standard error of estimate, SEE = 1.3 Metabolic Equivalents (METs). The VT was also well correlated with walking performance, 6MWT distance (r = 0.80), and was improved by the addition of terminal RPE (6MWT + RPE, R^2 = 0.69, SEE = 0.95 METs). The addition of terminal RPE to 6MWT distance improved the prediction of maximal METs and METs at VT, which may have practical applications for exercise prescription.

Keywords: physical fitness; internal load; RPE; performance

Citation: Porcari, J.P.; Foster, C.; Cress, M.L.; Larson, R.; Lewis, H.; Cortis, C.; Doberstein, S.; Donahue, M.; Fusco, A.; Radtke, K. Prediction of Exercise Capacity and Training Prescription from the 6-Minute Walk Test and Rating of Perceived Exertion. *J. Funct. Morphol. Kinesiol.* **2021**, *6*, 52. https://doi.org/10.3390/jfmk6020052

Academic Editor: Peter Hofmann

Received: 18 May 2021
Accepted: 10 June 2021
Published: 14 June 2021

1. Introduction

Exercise capacity is an important quantitative expression of the ability to perform muscular activity. Specifically, the peak oxygen uptake (VO_2peak), used when there is no confirmatory test to prove that a single-measured VO_2peak is equivalent to the classical definition of VO_2max, and the ventilatory threshold (VT) are well correlated with performance in individuals capable of performing prolonged heavy exercise [1–3]. As an integrator of the elements of the Fick equation, VO_2peak is a strong index of overall cardio-pulmonary function [4], and a very strong predictor of survival in clinical populations [5–8]. Similarly, VT, although a physiologically complex concept [9], has come to be recognized as a better measure of exercise capacity than VO_2peak relative to the ability to carry out daily activities [2,3]. It has also been shown to be an important prognostic marker [10,11]. Normally, cardiopulmonary exercise testing (which is technically demanding) [12] is required to measure VO_{2peak} and VT. directly However, cardiopulmonary exercise testing is normally only conducted for the purpose of exploring the differential diagnosis of dyspnea or to determine if there is a multi-organ system explanation of exercise intolerance.

Traditionally VO_2peak and VT are measured in the laboratory, during incremental exercise, on either a treadmill or cycle ergometer, with direct measurement of respiratory gas exchange and/or lactate accumulation [12,13]. However, the technical requirements for performing such evaluations are considerable. A variety of predictive equations for VO_2peak have been developed [14–16]. Although providing reasonable accuracy, they typically require maximal effort exercise testing, which can be challenging and unpleasant for patients and present at least some safety concerns [17]. Accordingly, a number of less technically demanding methods based on real-world ambulatory patterns have arisen [6,7,18–22] which are widely deployed in the fitness and clinical exercise physiology communities. In particular, the 6-min walk test (6MWT) in clinical populations [6,7] and the Rockport 1-mile walking test [21] and the 2-km walking test [22] in fitness populations have shown utility in terms of predicting VO_2peak. Recent data have also suggested that the VT may be predicted from incremental exercise tests based on simple performance measures [23–26].

In some of the waking tests used to predict functional capacity, the analytic strategy is based on starting with a population estimate, with that estimate adjusted for walking performance and with further adjustments made on the basis of terminal heart rate (HR) (e.g., smaller reductions in points on the basis of shorter walking times or lower terminal HR). Some equations [21,22] also adjust on the basis of other variables such as age, sex, height, weight, or body mass index (BMI). This solution is attractive and relatively simple, with independent studies demonstrating good estimates of VO_2peak compared to walking performance alone [23].

However, because of the wide interindividual variability in the HR response and maximal HR, and of the profound effect of many medications on the HR response, it would be desirable to find a method of replacing the HR measure in these prediction equations. Eston et al. [20] have suggested that the progression of the rating of perceived exertion (RPE) during incremental exercise can be used as a tool for predicting VO_2peak. Following this line of thinking, Alamji et al. [23] have suggested that the RPE could also be used to predict VT.

Given that simple walking tests such as the 6MWT are much more assessable to the exercising community than laboratory-based tests, and that most training takes place at intensities just below the VT [3], this study was designed to determine whether data from the 6MWT, with walking performance and terminal RPE as predictor variables, could be used to develop adequate prediction equations for VO_2peak and VT.

2. Materials and Methods

The subjects for the study were 63 adult volunteers. All were participants in either a Phase II cardiac rehabilitation program or a community-based exercise program, designed for the primary and secondary prevention of cardiovascular disease. Diagnoses for patients were conventional: stable angina pectoris (n = 2), post myocardial infarction (n = 22), post revascularization surgery (n = 18), post percutaneous intervention (n = 14), stable heart failure (n = 7), and risk factors for cardiovascular disease (n = 25). The research protocol was approved by the University of Wisconsin-La Crosse Institutional Review Board for the Protection of Human Subjects (Protocol 13-HB-001, approved 13 December 2013). All subjects provided written informed consent before participation.

Each subject performed an incremental treadmill exercise test to fatigue or to clinical signs and symptoms [13] with continuous electrocardiographic (ECG) and hemodynamic monitoring. A modified Balke type protocol was used, with the treadmill speed individually selected to represent comfortable walking during the first 2-min stage, and with subsequent increments in workload provided by 2% increments in treadmill grade every stage. Respiratory metabolism was measured with open-circuit spirometry using a mixing chamber based metabolic system (AEI Industries, Pittsburgh, PA, USA). The pneumotach was calibrated with a 3-L syringe, and the gas analyzers were calibrated using a reference gas (4% CO_2, 16% O_2) and room air. Gas exchange was integrated every 30 s, and the highest 30 s VO_2 was accepted as VO_2peak. The VT was identified using both the v-slope and ventilatory equivalent methods [2]. Because the intent of the study was to predict exercise capacity from 6MWT performance, the VO_2prak and the VO_2 at the VT (VO_2@VT) was expressed as maximal Metabolic Equivalents (METs) (Max METs), and METs at the VT (METs@VT), as we felt that this was more clinically relevant and comparable to estimates of exercise capacity based on performance in standard exercise protocols [13–16]. On a separate day, always within 72 h, each subject performed a 6MWT on a 30-m course with standard conditions and prompting, according to the American Thoracic Society [27]. The distance completed was measured to the nearest meter using laps completed and interpolation between cones placed at 5-m intervals on the walking course. Within 30 s of concluding the 6MWT, the terminal RPE was measured using the classical (6–20) Borg scale [28]. Instructions for using the RPE scale had been discussed with the subject before the start of the 6MWT.

The relationships of 6MWT with Max METs and METs@VT were made using linear regression. Similarly, the relationships of the terminal RPE with Max METs and METs@VT were made using linear regression. Regression equations for predicting Max METs and METs@VT were constructed using multiple linear regression with a stepwise approach, with 6MWT distance entered first and terminal RPE entered second.

3. Results

The characteristics of the subjects are presented in Table 1, with the subjects presented both by sex and as a total group. They were broadly representative of patients in contemporary preventive/rehabilitation programs in terms of age, diagnoses, and medications.

Table 1. Mean and standard deviation of the characteristics of the subjects.

Characteristics	Men (n = 46)	Women (n = 17)	Total (n = 63)
Age (years)	60.2 ± 9.7	55.4 ± 10.1	59.6 ± 10.0
Height (m)	1.78 ± 0.06	1.62 ± 0.06	1.74 ± 0.09
Weight (kg)	94.4 ± 19.1	65.4 ± 12.7	86.8 ± 21.8
6MWT Distance (m)	583 ± 106	600 ± 87	587 ± 101
6MWT RPE	12.1 ± 2.0	12..3 ± 2.3	12.1 ± 2.1
Max METS	7.8 ± 2.3	8.4 ± 2.4	8.1 ± 2.4
METs@VT	5.8 ± 1.4	6.8 ± 2.0	5.9 ± 1.7
Maximal HR	144 ± 24	161 ± 21	149 ± 24

6MWT: 6-min walk test; RPE: rating of perceived exertion; Max METS: maximal Metabolic Equivalents; METs@VT: Metabolic Equivalents at ventilatory threshold; HR: heart rate.

The bivariate relationships between 6MWT distance versus Max METs and METs@VT, and between terminal RPE vs. Max METs and METs@VT, are presented in Figure 1. There was a strong and significant simple correlation between 6MWT distance versus Max METs (r = 0.80) and METs@VT (r = 0.80), and a weak but statistically significant correlation between terminal RPE versus Max METs (r = 0.30) and METs@VT (r = 0.23).

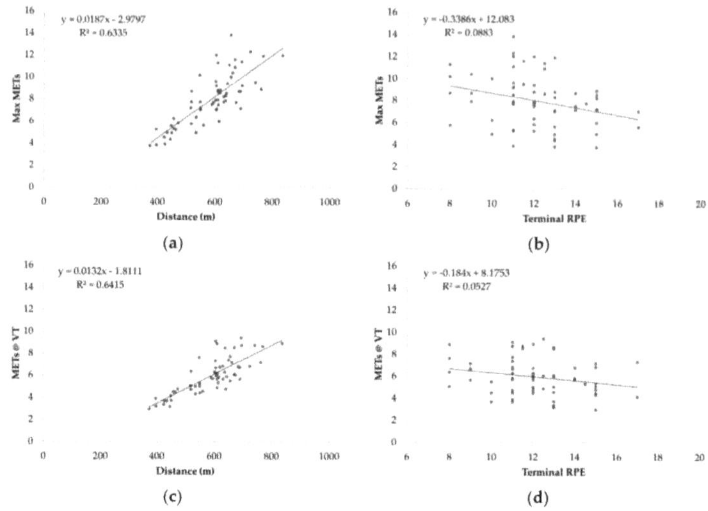

Figure 1. Bivariate relationship between the 6-min walk test (6MWT) distance and the maximal Metabolic Equivalents (Max METs, (**a**)), the terminal rating of perceived exertion (RPE) and Max METs (**b**), the 6MWT distance and Metabolic Equivalents at the ventilatory threshold (METs@VT, (**c**)) and the terminal RPE and METs@VT (**d**).

When the 6MWT distance and terminal RPE were combined in a multiple regression equation to predict Max METs, the R^2 increased from 0.63 to 0.71, with a SEE of 1.3 METs and a standardized residual of 1.0 METs. The prediction equation was:

$$\text{Max METs} = 0.882 + (0.018 * \text{6MWT m}) - (0.308 * \text{RPE}) \quad (1)$$

When the 6MWT distance and terminal RPE were combined in a multiple regression equation to predict METs@VT, the R^2 increased from 0.64 to 0.69, with a SEE of 0.95 METs and a standardized residual of 0.7 METs. The prediction equation was:

$$\text{METs@VT} = 0.140 + (0.013 * \text{6MWT m}) - (0.161 * \text{RPE}) \quad (2)$$

When the derived prediction formulas for Max METs and METs@VT were plotted against the measured Max METs (r = 0.87) and METs@VT (r = 0.85), there was a strong bivariate relationship for both. The residual scores (predicted-observed) revealed a small value for both Max METS (−0.27 ± 1.24 METs) and METs@VT (−0.14 ± 0.92 METs), with most of the outliers at a relatively higher exercise capacity (Figure 2).

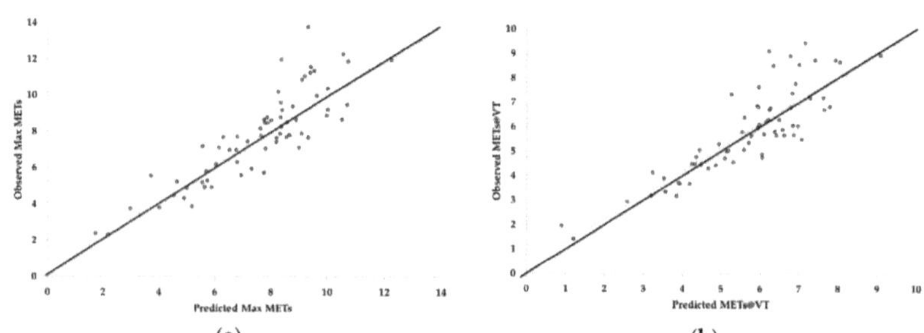

Figure 2. *Cont.*

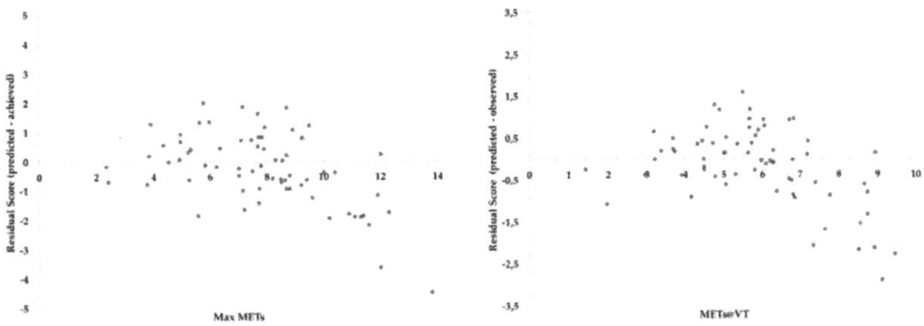

(c) (d)

Figure 2. Bivariate relationship between the 6-min walk test (6MWT) + terminal rating of perceived exertion (RPE) prediction equations and observed values for maximal Metabolic Equivalents (Max METs, **a**), Metabolic Equivalents at the ventilatory threshold (METs@VT, **b**) and the standardized residuals for Max METs (**c**) and METs@VT (**d**).

Tabular presentations of Max METs and METs@VT in relation to 6MWT distance and terminal RPE are presented in Tables 2 and 3. In terms of convenience of use, it is simple to compare 6MWT distance and terminal RPE to derive Max METs and METs@VT.

Table 2. Estimate of maximal Metabolic Equivalents in relation to 6-min walk test (6MWT) distance (m) and terminal rating of perceived exertion (RPE). For convenience of presentation, 6MWT distance is incremented by 50 m. However, as the equation is linear, interpolation of intermediate values is appropriate.

	RPE														
Distance (m)	**6**	**7**	**8**	**9**	**10**	**11**	**12**	**13**	**14**	**15**	**16**	**17**	**18**	**19**	**20**
200	2.63	2.33	2.02	1.71	1.40										
225	3.08	2.78	2.47	2.16	1.85	1.54									
250	3.53	3.23	2.92	2.61	2.30	1.99	1.69								
275	3.98	3.68	3.37	3.06	2.75	2.44	2.14	1.83	1.52						
300	4.43	4.13	3.82	3.51	3.20	2.89	2.59	2.28	1.97	1.66					
325	4.88	4.58	4.27	3.96	3.65	3.34	3.04	2.73	2.42	2.11	1.80				
350	5.33	5.03	4.72	4.41	4.10	3.79	3.49	3.18	2.87	2.56	2.25	1.95	1.64		
375	5.78	5.48	5.17	4.86	4.55	4.24	3.94	3.63	3.32	3.01	2.70	2.40	2.09	1.78	
400	6.23	5.93	5.62	5.31	5.00	4.69	4.39	4.08	3.77	3.46	3.15	2.85	2.54	2.23	1.92
425	6.68	6.38	6.07	5.76	5.45	5.14	4.84	4.53	4.22	3.91	3.60	3.30	2.99	2.68	2.37
450	7.13	6.83	6.52	6.21	5.90	5.59	5.29	4.98	4.67	4.36	4.05	3.75	3.44	3.13	2.82
475	7.58	7.28	6.97	6.66	6.35	6.04	5.74	5.43	5.12	4.81	4.50	4.20	3.89	3.58	3.27
500	8.03	7.73	7.42	7.11	6.80	6.49	6.19	5.88	5.57	5.26	4.95	4.65	4.34	4.03	3.72
525	8.48	8.18	7.87	7.56	7.25	6.94	6.64	6.33	6.02	5.71	5.40	5.10	4.79	4.48	4.17
550	8.93	8.63	8.32	8.01	7.70	7.39	7.09	6.78	6.47	6.16	5.85	5.55	5.24	4.93	4.62
575	9.38	9.08	8.77	8.46	8.15	7.84	7.54	7.23	6.92	6.61	6.30	6.00	5.69	5.38	5.07
600	9.83	9.53	9.22	8.91	8.60	8.29	7.99	7.68	7.37	7.06	6.75	6.45	6.14	5.83	5.52
625	10.28	9.98	9.67	9.36	9.05	8.74	8.44	8.13	7.82	7.51	7.20	6.90	6.59	6.28	5.97
650	10.73	10.43	10.12	9.81	9.50	9.19	8.89	8.58	8.27	7.96	7.65	7.35	7.04	6.73	6.42
675	11.18	10.88	10.57	10.26	9.95	9.64	9.34	9.03	8.72	8.41	8.10	7.80	7.49	7.18	6.87
700	11.63	11.33	11.02	10.71	10.40	10.09	9.79	9.48	9.17	8.86	8.55	8.25	7.94	7.63	7.32

Table 3. Estimate of Metabolic Equivalents at the ventilatory threshold in relation to 6-min walk test (6MWT) distance (m) and terminal rating of perceived exertion (RPE). For convenience of presentation, 6MWT distance is incremented by 50 m. However, as the equation is linear, interpolation of intermediate values is appropriate.

Distance (m)	RPE														
	6	7	8	9	10	11	12	13	14	15	16	17	18	19	20
200	1.83	1.68	1.53	1.38	1.23										
225	2.10	1.94	1.78	1.62	1.46										
250	2.42	2.26	2.10	1.94	1.78	1.62	1.46								
275	2.75	2.59	2.43	2.27	2.11	1.94	1.78	1.62	1.46						
300	3.07	2.91	2.75	2.59	2.43	2.27	2.11	1.95	1.79	1.63	1.46				
325	3.40	3.24	3.08	2.92	2.76	2.59	2.43	2.27	2.11	1.95	1.79	1.63	1.47		
350	3.72	3.56	3.40	3.24	3.08	2.92	2.76	2.60	2.44	2.28	2.11	1.95	1.79	1.63	
375	4.05	3.89	3.73	3.57	3.41	3.24	3.08	2.92	2.76	2.60	2.44	2.28	2.12	1.96	1.80
400	4.37	4.21	4.05	3.89	3.73	3.57	3.41	3.25	3.09	2.93	2.76	2.60	2.44	2.28	2.12
425	4.70	4.54	4.38	4.22	4.06	3.89	3.73	3.57	3.41	3.25	3.09	2.93	2.77	2.61	2.45
450	5.02	4.86	4.70	4.54	4.38	4.22	4.06	3.90	3.74	3.58	3.41	3.25	3.09	2.93	2.77
475	5.35	5.19	5.03	4.87	4.71	4.54	4.38	4.22	4.06	3.90	3.74	3.58	3.42	3.26	3.10
500	5.67	5.51	5.35	5.19	5.03	4.87	4.71	4.55	4.39	4.23	4.06	3.90	3.74	3.58	3.42
525	6.00	5.84	5.68	5.52	5.36	5.19	5.03	4.87	4.71	4.55	4.39	4.23	4.07	3.91	3.75
550	6.32	6.16	6.00	5.84	5.68	5.52	5.36	5.20	5.04	4.88	4.71	4.55	4.39	4.23	4.07
575	6.65	6.49	6.33	6.17	6.01	5.84	5.68	5.52	5.36	5.20	5.04	4.88	4.72	4.56	4.40
600	6.97	6.81	6.65	6.49	6.33	6.17	6.01	5.85	5.69	5.53	5.36	5.20	5.04	4.88	4.72
625	7.30	7.14	6.98	6.82	6.66	6.49	6.33	6.17	6.01	5.85	5.69	5.53	5.37	5.21	5.05
650	7.62	7.46	7.30	7.14	6.98	6.82	6.66	6.50	6.34	6.18	6.01	5.85	5.69	5.53	5.37
675	7.95	7.79	7.63	7.47	7.31	7.14	6.98	6.82	6.66	6.50	6.34	6.18	6.02	5.86	5.70
700	8.27	8.11	7.95	7.79	7.63	7.47	7.31	7.15	6.99	6.83	6.66	6.50	6.34	6.18	6.02

4. Discussion

The main finding of this study was that adding the terminal RPE to 6MWT distance significantly improved the predictability of Max METs. Unique to this study was the ability of 6MWT distance + terminal RPE also to predict the METs@VT, which is a better measure of sustainable exercise capacity [2,9], a powerful emerging measure of prognosis [10,11] and a very useful measure for prescribing exercise [3]. It may also be of particular value in patients with cardiovascular disease who are often on medications that alter the HR response during both exercise testing and training.

The goodness of fit of the equations for predicting Max METs from the 6MWT distance + terminal RPE (R^2 = 0.71) compares favorably with field tests using the 6MWT distance (R^2 = 0.42) [6], the Rockport 1-mile Walking Test (R^2 = 0.78) [21] or R^2 = 0.71) [23], the Cooper 12-min run-walk Test (R^2 = 0.80) [19], or with extrapolation of submaximal RPE (R^2 = 0.85) [20] or (R^2 = 0.71) [23]. The SEE for all of these predictions equations is on the order of 1.0–1.5 METs. Although the correlation of treadmill protocol time vs. Max METs (e.g., VO_2max) is typically higher (R^2 = 0.83–0.94), the SEE is also typically in the range of 1 MET [15,16,29,30]; thus, the simplicity of the 6MWT + terminal RPE is very attractive.

The prediction of METs@VT has more typically been performed using percentages of the maximal power output [25], running speed [23], RPE [23], or as the equivocal stage of the Talk Test [26]. To our knowledge, an approach to estimating the METs@VT in a clinical population with an approach as simple as that used in this study has not previously been reported. Given the importance of the VT in terms of the evaluation of prognosis [10,11] and exercise prescription [3], and the ability of equation-based methods to account for speed and grade equivalents of the METs@VT [31], the ability to predict METs@VT from 6MWT distance + terminal RPE may be of considerable utility.

The workload required to achieve a particular value for the percentage of Max METS (%Max METS), percentage of HR (%HR) reserve, or METs@VT is probably lower than the workload eliciting these markers during exercise testing. Recent research from our laboratory [30] has suggested that "translating" exercise test results to exercise training requires a downregulation of the MET requirement of the workload to ~70–75% of that where a given physiologic response is observed during exercise testing. From that perspective, consider a patient who completes 475 m during the 6MWT, with a terminal RPE of 14. The predicted Max METs would be 5.12 (Table 2), and the METs@VT would be 4.07 (Table 3). Translating to 72% of METs@VT would yield a training intensity of 2.93 METs. Based on the American College of Sports Medicine (ACSM) ambulation equations [31], this workload should be achieved during level-ground ambulation at ~1.13 $m*s^{-1}$ (2.5 mph or 4.1 kph). One would anticipate a RPE during training of ~13 (which has been proposed as the "ideal" intensity for many people) [32] and with comfortable speech still possible [26]. The training load can, of course, be adjusted after the first training bout, but this simple modification of the 6MWT and the simple approach to translating from testing to training [30] suggests a way to define easily the exercise component of the individualized treatment plan for patients in rehabilitation programs.

Another advantage of the 6MWT distance + terminal RPE approach is the ability to make outcome assessments in patients who are mechanically limited to a top walking speed. The 6MWT was originally created for patients with pulmonary disease or heart failure [6,7,26], who often have very limited exercise capacity. In the absence of routine graded exercise testing, and in need of methods for outcome assessment, the 6MWT has more recently been widely used with relatively healthier patients in cardiac rehabilitation programs. These patients often have better exercise capacity and are often constrained by the requirement to only walk during the 6MWT, with a top walking speed unlikely to exceed 2.0 $m \cdot s^{-1}$ (4.5 mph or 7.2 kmh) which requires a VO_2 of ~16 $mL \cdot kg^{-1} \cdot min^{-1}$ (~4.4 METs) [31]. Since common clinical experience shows that many patients can achieve this walking speed more or less comfortably, outcome assessment in rehabilitation programs has been constrained by the use of the 6MWT distance alone. However, in a patient who originally walks 450 m with a RPE of 15, who later walks the same 450 m, but with a RPE of 13, the calculated MaxMETs (4.36 to 4.98 METs) and METs@VT (3.58 to 3.90 METs) improve by 14% and 9%, respectively. Additionally, this follow-up testing can be used to update the exercise proscription.

Although the present study provides meaningful information, some limitations should be acknowledged. First, this study is a small cohort study. Therefore, future researched should consider a larger sample. Second, the present study did not include a large representation of the heart failure population (i.e., potential heart transplant candidates and those with pulmonary hypertension) in which VO_2max might have huge significance, in particular, as most of the times it is hard to obtain cardiopulmonary exercise testing in a heart-failure population, and rough estimates based on 6MWT will be of paramount importance.

5. Conclusions

The results of this study demonstrate that the simple approach of adding the terminal RPE to the 6MWT distance can improve the estimate of Max METs in patients in rehabilitation programs. Additionally, it can be used to make an estimate of the METs@VT, which is important prognostically [10,11] as a better estimate of sustainable exercise capacity [2] and is highly important prescriptively [3]. Lastly, this approach to estimating Max METs and METs@VT may provide a solution to estimating changes in exercise capacity in patients who are mechanically at the limit of their walking speed capacity.

Author Contributions: Conceptualization, J.P.P. and C.F.; investigation, M.L.C., R.L., H.L., M.D. and K.R.; writing—original draft preparation, J.P.P. and C.F.; writing—review and editing: C.C., S.D. and A.F. All authors have read and agreed to the published version of the manuscript.

Funding: This research received no external funding.

Institutional Review Board Statement: The study was conducted according to the guidelines of the Declaration of Helsinki, and approved by the University of Wisconsin-La Crosse Institutional Review Board for the Protection of Human Subjects (Protocol 13-HB-001, approved 13 December 2013).

Informed Consent Statement: Informed consent was obtained from all subjects involved in the study.

Data Availability Statement: The data presented in this study are available on request from the corresponding author.

Conflicts of Interest: The authors declare no conflict of interest.

References

1. Saltin, B.; Astrand, P.O. Maximal oxygen uptake in athletes. *J. Appl. Physiol.* **1967**, *23*, 353–358. [CrossRef] [PubMed]
2. Foster, C.; Cotter, H.M. Blood lactate, respiratory and heart rate markers on the capacity for sustained exercise. In *Physiological Assessment of Human Fitness*, 2nd ed.; Maud, P.J., Foster, C., Eds.; Human Kinetics Press: Champaign, IL, USA, 2005; pp. 63–76.
3. Mezzani, A.; Hamm, L.F.; Jones, A.M.; McBride, P.E.; Moholdt, T.; Stone, J.A.; Urhausen, A.; Williams, M.A. Aerobic exercise intensity assessment and prescription in cardiac rehabilitation. *J. Cardiopulm. Rehabil. Prev.* **2012**, *32*, 327–350. [CrossRef] [PubMed]
4. Levine, B.D. VO2max: What do we know, and what do we still need to know? *J. Physiol.* **2008**, *586*, 25–34. [CrossRef] [PubMed]
5. Myers, J.; Prakash, M.; Froelicher, V.; Do, D.; Partington, S.; Atwood, J.E. Exercise Capacity and Mortality among Men Referred for Exercise Testing. *N. Engl. J. Med.* **2002**, *346*, 793–801. [CrossRef]
6. Ross, R.M.; Murthy, J.N.; Wollak, I.D.; Jackson, A.S. The six minute walk test accurately estimates mean peak oxygen uptake. *BMC Pulm. Med.* **2010**, *10*, 31. [CrossRef]
7. Foreman, D.E.; Fleg, J.K.; Kitzman, D.W.; Brawner, C.A.; Swank, A.M.; McKelvie, R.S.; Clare, R.M.; Ellis, S.J.; Dunlap, M.E.; Bittner, V. 6-min walk test provides prognostic utility comparable to cardiopulmonary exercise testing in ambulatory patients with systolic heart failure. *J. Am. Coll. Cardiol.* **2012**, *60*, 2653–2661. [CrossRef]
8. Keteyian, S.J.; Brawner, C.A.; Savage, P.D.; Ehrman, J.K.; Schairer, J.; Divine, G.; Aldred, H.; Ophaug, K.; Ades, P.A. Peak aerobic capacity predicts prognosis in patients with coronary heart disease. *Am. Heart J.* **2008**, *156*, 292–300. [CrossRef]
9. Poole, D.C.; Rossiter, H.B.; Brooks, G.A.; Gladwell, L.B. The anerobic threshold: 50+ years of controversy. *J. Physiol.* **2020**, *599*, 737–767. [CrossRef]
10. Agostoni, P.; Corrà, U.; Cattadori, G.; Veglia, F.; Battaia, E.; La Gioia, R.; Scardovi, A.B.; Emdin, M.; Metra, M.; Sinagra, G.; et al. Prognostic value of indeterminable anaerobic threshold in heart failure. *Circ. Heart Fail.* **2013**, *6*, 977–987. [CrossRef]
11. Older, P.O.; Levett, D.Z.H. Cardiopulmonary exercise testing and surgery. *Ann. Am. Thorac. Soc.* **2017**, *14*, S74–S83. [CrossRef]
12. Arena, R.A.; Sietima, K.E. Cardiopulmonary exercise testing in the clinical evaluation of patients with heart and lung disease. *Circulation* **2011**, *123*, 668–680. [CrossRef]
13. Foster, C.; Porcari, J.P.; Hipp, M. Clinical exercise testing. In *ACSM's Guidebook for Clinical Exercise Physiologists*; Thompson, W., Ed.; Wolters Kluwer: Baltimore, MD, USA, 2019; pp. 349–364.
14. Bruce, R.A.; Kusumi, F.; Hosmer, D. Maximal oxygen intake and nomographic assessment of functional aerobic impairment in cardiovascular disease. *Am. Heart J.* **1973**, *85*, 546–562. [CrossRef]
15. Foster, C.; Crowe, A.J.; Daines, E.; Dumit, M.; Green, M.A.; Lettau, S.; Thompson, N.N.; Weymier, J. Predicting functional capacity during treadmill testing independent of exercise protocol. *Med. Sci. Sports Exerc.* **1996**, *28*, 752–756. [CrossRef]
16. McConnell, T.R.; Foster, C.; Conlin, N.C.; Thompson, N.N. Prediction of functional capacity during treadmill testing: Effect of handrail support. *J. Cardiopulm. Rehabil.* **1991**, *11*, 255–260. [CrossRef]
17. Myers, J.; Voodi, L.; Umann, T.; Froelicher, V.F. A Survey of Exercise Testing: Methods, Utilization, Interpretation, and Safety in the VAHCS. *J. Cardiopulm. Rehabil.* **2000**, *20*, 251–258. [CrossRef]
18. Balke, B. A simple field test for the assessment of physical fitness. *CARI Rep.* **1963**, *63*, 18–40.
19. Cooper, K.H. A means of assessing maximal oxygen intake: Correlation between field and treadmill testing. *JAMA* **1968**, *203*, 135–138. [CrossRef]
20. Eston, R.; Lambrecht, D.; Shephard, K.; Parfitt, G. Prediction of maximal oxygen uptake from a perceptually regulated sub-maximal graded exercise test. *J. Sports Sci.* **2008**, *26*, 131–139. [CrossRef]
21. Kline, G.M.; Porcari, J.P.; Hintermeister, R.; Freedson, P.S.; Ward, A.; McCarron, R.F.; Ross, J.; Rippe, J.M. Estimation of VO2max from a one-mile track walk, gender, age, and body weight. *Med. Sci. Sports Exerc.* **1987**, *19*, 253–259. [CrossRef]
22. Oja, P.; Laukkanen, R.; Pasanen, M.; Tyry, T.; Vuori, I. A 2-km walking test for assessing the cardiorespiratory fitness of healthy adults. *Int. J. Sports Med.* **1991**, *12*, 356–362. [CrossRef]
23. Alajmi, R.A.; Foster, C.; Porcari, J.P.; Radtke, K.; Doberstein, S. Comparison of non-maximal tests for estimating exercise ca-pacity. *Kinesiology* **2020**, *52*, 10–18. [CrossRef]
24. Condello, G.; Reynolds, E.; Foster, C.; De Koning, J.J.; Casolino, E.; Knutson, M.; Porcari, J.P. A simplified approach for estimating the ventilatory and respiratory compensation thresholds. *J. Sports Sci. Med.* **2014**, *13*, 309–314. [PubMed]
25. De Koning, J.J.; Noordhof, D.A.; Uitslag, T.P.; Galiart, R.E.; Dodge, C.; Foster, C. An approach to estimating gross efficiency during high-intensity exercise. *Int. J. Sports Physiol. Perform.* **2013**, *8*, 682–684. [CrossRef] [PubMed]
26. Foster, C.; Porcari, J.P.; Doro, K.; Ault, S.; Dubiel, J.; Engen, M.; Kolman, D.; Xiong, S. Exercise prescription when there is no exercise test: The Talk Test. *Kinesiology* **2018**, *50*, 333–348.
27. American Thoracic Society. ATS statement: Guidelines for the Six-Minute Walk Test. *Am. J. Respir. Crit. Care Med.* **2002**, *166*, 111–117. [CrossRef]
28. Borg, G.A.V. *Borg's Perceived Exertion and Pain Scales*; Human Kinetics Press: Champaign, IL, USA, 1998.

29. Pollock, M.L.; Bohannon, R.L.; Cooper, K.H.; Ayres, J.J.; Ward, A.; White, S.R.; Linnerud, A.C. A comparative analysis of four protocols for maximal treadmill stress testing. *Am. Heart J.* **1976**, *92*, 39–45. [CrossRef]
30. Pollock, M.L.; Foster, C.; Schmidt, D.H.; Hellman, C.; Linnerud, A.C.; Ward, A. Comparative analysis of three protocols for maximal graded exercise testing of women. *Am. Heart J.* **1982**, *103*, 363–373. [CrossRef]
31. Foster, C.; Anholm, J.; Bok, D.; Boullosa, D.; Condello, G.; Cortis, C.; Fusco, A.; Jaime, S.; De Koning, J.; Lucia, A.; et al. Generalized approach to translating exercise tests and prescribing exercise. *J. Funct. Morphol. Kinesiol.* **2020**, *5*, 63. [CrossRef]
32. Parfitt, G.; Evans, H.; Eston, R. Perceptually regulated training at RPE13 is pleasant and improves physical health. *Med. Sci. Sports Exerc.* **2012**, *44*, 1613–1618. [CrossRef]

Journal of
Functional Morphology and Kinesiology

MDPI

Article

Pattern of the Heart Rate Performance Curve in Subjects with Beta-Blocker Treatment and Healthy Controls

Philipp Birnbaumer [1], Heimo Traninger [2], Matteo C. Sattler [1], Andrea Borenich [3] and Peter Hofmann [1,*]

[1] Institute of Human Movement Science, Sport & Health, University of Graz, 8010 Graz, Austria; philipp.birnbaumer@uni-graz.at (P.B.); matteo.sattler@uni-graz.at (M.C.S.)
[2] ZARG Centre for Outpatient Rehabilitation, 8021 Graz, Austria; heimo.traninger@pro-heart.at
[3] Department of Production and Operations Management, University of Graz, 8010 Graz, Austria; andrea.borenich@uni-graz.at
* Correspondence: peter.hofmann@uni-graz.at; Tel.: +43-316-380-3903

Abstract: (1): Heart rate performance curve (HRPC) in incremental exercise was shown to be not uniform, causing false intensity estimation applying percentages of maximal heart rate (HR_{max}). HRPC variations are mediated by β-adrenergic receptor sensitivity. The aim was to study age and sex dependent differences in HRPC patterns in adults with β-blocker treatment (BB) and healthy controls (C). (2): A total of 535 (102 female) BB individuals were matched 1:1 for age and sex (male 59 ± 11 yrs, female 61 ± 11 yrs) in C. From the maximum incremental cycle ergometer exercise a first and second heart rate (HR) threshold (Th1 and Th2) was determined. Based on the degree of the deflection (kHR), HRPCs were categorized as regular (downward deflection (kHR > 0.1)) and non-regular (upward deflection (kHR < 0.1), linear time course). (3): Logistic regression analysis revealed a higher odds ratio to present a non-regular curve in BB compared to C (females showed three times higher odds). The odds for non-regular HRPC in BB versus C decreased with older age (OR interaction = 0.97, CI = 0.94–0.99). Maximal and submaximal performance and HR variables were significantly lower in BB ($p < 0.05$). $\%HR_{max}$ was significantly lower in BB versus C at Th2 (male: 77.2 ± 7.3% vs. 80.8 ± 5.0%; female: 79.2 ± 5.1% vs. 84.0 ± 4.3%). $\%P_{max}$ at Th2 was similar in BB and C. (4): The HRPC pattern in incremental cycle ergometer exercise is different in individuals receiving β-blocker treatment compared to healthy individuals. The effects were also dependent on age and sex. Relative HR values at Th2 varied substantially depending on treatment. Thus, the percentage of Pmax seems to be a stable and independent indicator for exercise intensity prescription.

Keywords: cardiac rehabilitation; exercise prescription; exercise intensity; heart rate deflection point

Citation: Birnbaumer, P.; Traninger, H.; Sattler, M.C.; Borenich, A.; Hofmann, P. Pattern of the Heart Rate Performance Curve in Subjects with Beta-Blocker Treatment and Healthy Controls. *J. Funct. Morphol. Kinesiol.* **2021**, *6*, 61. https://doi.org/10.3390/jfmk6030061

Academic Editor: Cristina Cortis

Received: 27 May 2021
Accepted: 11 July 2021
Published: 13 July 2021

1. Introduction

To induce desired training effects and to apply safe exercise programs, exercise prescription is commonly based on fixed percentages of maximal heart rate (HR_{max}) or maximal oxygen consumption (VO_{2max}). Usually, intensity ranges, are prescribed between 65% to 85% of HR_{max} or 50% to 75% of VO_{2max} [1]. However, fixed-percentage approaches were shown not to guarantee a uniform load amongst individuals [2,3]. Different metabolic responses, which may vary from over- to under-loading have been prescribed [4]. The inconsistency between intensity domains claims the need for adjustment and new indicators in intensity prescription, proposed by the 2020 position paper from the European Association of Preventive Cardiology (EPAC) [5]. Moreover, individualized prescriptions based on cardiopulmonary exercise tests and individual thresholds such as the first and second ventilatory threshold are recommended [5–7].

In practice, simple approaches are usually favored and most exercise tests measure solely electrocardiogram (ECG) based heart rate (HR). Some problems with the use of $\%HR_{max}$ for exercise prescription have already been addressed to different HR curve patterns [8] (see Figure 1 for different patterns). These authors showed that the increase of

HR during incremental cycle ergometer exercise presented neither a uniform nor a linear pattern which clearly impacts exercise prescription [9]. Most of the young and trained male subjects presented an S-shaped pattern of the heart rate performance curve (HRPC) which was characterized by a distinct flattening of the HR curve at higher intensities (a so-called downward deflection). This course of the HRPC is considered regular [10]. However, in the same study, a significant number of participants showed a linear or even upward deflection [9], which is considered non-regular and was shown to be related to left ventricular function [11]. Recently, we could support these early results by showing this diversity of HRPC patterns in a large cohort of healthy trained and untrained male and female subjects across a wide range of age [12]. Interestingly, age significantly altered the HRPC from regular to non-regular, modulated by maximum exercise performance and sex. These results imply consequences for exercise prescription (e.g., risk of overloading in subjects with non-regular HRPCs) dependent on the above-mentioned variables which has also been critically addressed already by others [2]. Both the downward and upward deflection of the HRPC during incremental exercise is used to determine the so-called HR deflection point (HRDP) [9,13] which is a well-accepted and frequently applied method for threshold determination [14]. Applying the actual standard three Phase-Two threshold model of energy supply, the HRDP corresponds to the second ventilatory or lactate threshold [8,15].

The increase in HR during incremental exercise is strongly related to the increase of plasma catecholamine concentration which drives rate and force of contraction [16]. Pokan et al. [17] showed that the catecholamine response to incremental exercise presented the same pattern than the lactate performance curve in young and trained healthy male subjects, however, the pattern of the HRPC was not related to the catecholamine response in this study (Figure 1). Despite such a similar catecholamine response we could show that the patterns of the HRPC were significantly influenced by a β1-receptor antagonist application [18]. This led to the conclusion that the β1-receptor sensitivity or density are key regulators of the pattern of the HRPC. Reduced β1-receptor sensitivity, either induced by chronic (over) stimulation due to increased sympathetic activity [19] or antagonist treatment, blunts HR increase at rest and submaximal exercise. The reduced sensitivity typically causes inverted HRPCs with upward deflection due to the higher catecholamine effects reaching maximal exercise [18,20] (Figure 1). In a recently published study including 2980 men and 1944 women we could show that the prevalence of HRPC with upward deflection increased from 20 to 70 years (respectively 80 yrs in women) both in men (10–43%) and women (9–30%). This trend might be caused by a decrease in β1-receptor sensitivity with increasing age [21–23]. Therefore, the decline in β1-receptor sensitivity is suggested to be the main factor influencing the HRPC deflection pattern.

Selective or non-selective β-receptor antagonist medication is a standard prescription in cardiology significantly increasing survival when applied long-term [24]. This medication reduces HR and blood pressure at rest and during exertion by a competitive inhibition of the β-receptors counteracting the driving force of stress induced catecholamines [25]. In incremental exercise, β-blocker administration was shown to significantly reduce HR at all workload levels but most at submaximal exercise such as the first and second thresholds in healthy individuals [18,26] as well as patients with cardiovascular disease [27].

As most patients with cardiovascular diseases are on β-adrenoceptor antagonist treatment [28] the above prescribed effects regarding exercise prescription need to be concerned. Individuals receiving a β-blocker are supposed to show a higher number of non-regular HRPCs compared to individuals without such a treatment. Studying the expected differences in the HRPC (and the prevalence of non-regular patterns) due to β-blocker intake is highly relevant for the accuracy of exercise prescription [29]. Therefore, the aim of this study was to investigate age and sex related differences in the distribution of HRPC patterns in adults receiving a β-adrenoceptor antagonist treatment and healthy controls. A secondary aim was to explore age and sex related differences in performance variables (i.e., HR_{max}, P_{max}) between adults receiving beta blocker treatment and healthy controls.

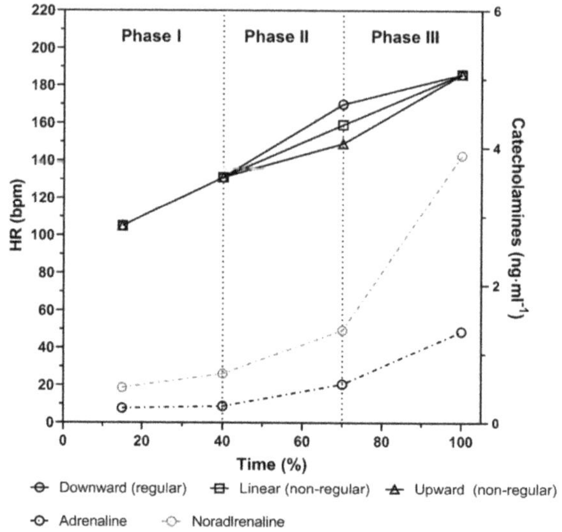

Figure 1. Schematic 3—phase model of the heart rate performance curve (HRPC) during incremental cycle ergometer exercise with downward deflection (regular), linear course or upward deflection (non-regular) (solid black lines) as well as plasma adrenaline and noradrenaline concentrations (dashed and dash-dotted black and grey lines) (modified from Hofmann et al. and Pokan et al. [9,17]).

2. Materials and Methods

2.1. Study Population

In this retrospective, observational study, HR and performance data from maximal cycle ergometer tests from 1070 individuals were analyzed. The groups were composed of 433 male (59 ± 11 yrs) and 102 female (61 ± 11 yrs) individuals treated with β-adrenoceptor antagonists (BB) who were exactly age- and sex-matched 1:1 with 433 male (59 ± 11 yrs) and 102 female (61 ± 11 yrs) healthy individuals in the control group (C). Exercise tests were carried out for performance diagnostics, health preventive or medical reasons between 2004 and 2017 in a cardiology center. The study was approved by the Ethics committee of the local university (GZ. 39/70/63 ex 2016/17). Individuals gave their written consent that their data may be used anonymously for scientific purposes. Only tests with no outliers (non-physiological HR deviation) or missing of HR recordings were included in the analyses. The test protocol was uniform such as to obtain maximal workload within 12–15 min. The protocols for BB and C were independent from age, sex and performance. All ergometer tests in C started at 20 W and workload was increased by 20 W increments every minute up to voluntary exhaustion. In BB, all ergometer tests started at 10 W and workload was increased by 10 W increments every minute up to voluntary exhaustion. In BB, no detailed information about the individual dosing were available for our analysis.

2.2. Assessment of Heart Rate and Performance Data

HR variables were provided as mean values for each single load step (e.g., HR for 20 W, 30 W). Data were analyzed via Vienna CPX-Tool (University of Vienna, Austria) to determine maximal and submaximal HR and performance markers as well as the degree and the direction of the HRPC. To detect a first (Th1) and second threshold (Th2) (equivalent to HRDP) from HR, two regions of interest were defined and consistently applied. Multiple linear regressions were performed for the detection of Th1 between the start of exercise and 66% of P_{max} and for the detection of Th2 between 40% P_{max} and P_{max}. The degree and direction of the deflection of the HRPC ($k_{HR} = (k1 - k2)/(1 + k1 \times k2)$) was calculated from the slopes of two tangents of a second-degree polynomial function fitted between

40% P_{max} and P_{max}. Because Vienna CPX-Tool subtracts the slopes of the tangents (k1 and k2) in a different order compared to earlier analyses [9,11,12] we changed the algebraic sign from negative to positive and vice versa to be consistent with previous studies. Based on the k_{HR} values HRPC´s were categorized as regular HRPC (downward deflection) for $k_{HR} > 0.1$ and as nonregular HRPC (linear course and upward deflection) for $k_{HR} < 0.1$ according to Pokan et al. [11] (Figure 1).

2.3. Statistical Analysis

Descriptive statistics are shown according to their distribution as mean ± standard deviation (SD) or median (interquartile range (IQR)). Data were tested for normal distribution by means of Shapiro–Wilk test. To compare BB and C within male and female individuals, a student *t* test or Mann–Whitney-U test was used. A multivariable logistic regression model was calculated to evaluate the relationship between β-blocker treatment and the presence of non-regular HRPCs (coded as 1 = non-regular, 0 = regular). In this model, we included treatment (0 = C, 1 = BB), sex (0 = male, 1 = female), age (continuous), age^2 and the following interactions of interest as explanatory variables: sex × treatment, age × treatment, age^2 × treatment. We added a quadratic term of age because it better reflected the shape of the relationship. Moreover, a non-linear relationship was also indicated by a significant interaction between age and log(age) ($p = 0.001$). Additionally, two multivariable linear regression models were calculated to evaluate the relationship between β-blocker treatment and HR_{max} and P_{max}, respectively. Here, we included age, sex and the following interactions of interest as explanatory variables: sex × treatment, age × treatment. Age was mean cantered before running the analyses. The assumptions of linear and logistic regression analyses were checked and considered to be met (i.e., after the transformation of age). Please note the main effects of age and sex on HRPC were not addressed in this study. Results are expressed using odds ratios (OR), unstandardized regression coefficients (B) and 95% confidence intervals (CI) for explanatory variables as well as R^2 for the overall model. Finally, to examine age-related associations between β-blocker treatment and submaximal HR and performance variables, data were first categorized into four age groups starting at ≤50 yrs up to >70 yrs. Then, separate 4 × 2 ANOVAs with post hoc Tukey´s multiple comparison tests were applied for males and females. All statistical analyses were performed in SPSS 26 (IBM Corporation, Armonk, NY, USA). Graphical representations were created with Prism 8.0 (GraphPad, San Diego, CA, USA). Statistical significance was considered as $p < 0.05$.

3. Results

The mean age of the study population was 59 ± 11 yrs in male (m) and 61 ± 11 yrs in female (f). Body mass index (BMI) ranged from 18.21 to 44.68 kg/m^2 in males and from 16.36 to 43.25 kg/m^2 in female individuals. In BB, 82% male and 84% female individuals had a selective β1-blocker and 18% and 16%, respectively—a nonselective blocker. Determination of Th1 and Th2 was successful in all cases except for two. P_{max} and performance at Th1 and Th2 as well as HR_{max} and HR at Th1 and Th2 were significantly lower in BB compared to C for both male and female individuals. In addition, relative performance values were significantly different between BB and C in male and female and ranged overall between 37.5–42.1% at Th1 and 69.8–70.8% at Th2. Absolute HR difference between C and BB at Th1 were 6 ± 16 bpm (m) and 8 ± 15 bpm (f) and at Th2 10 ± 16 bpm (m) and 13 ± 15 bpm (f), i.e., HR values at thresholds were lower in BB compared to C. Mean %HR_{max} at Th1 and Th2 was significantly lower in BB compared to C in male and female. However, the difference between BB and C was smaller at Th1 (m: 1.7 ± 9.6%, f: 2.1 ± 8.8%) compared to Th2 (m: 3.6 ± 8.5%, f: 4.7 ± 7.5%) (Table 1).

Table 1. Anthropometrics and characteristics of the incremental cycle ergometer tests in male and female individuals with (BB) and without (C) β-blocker treatment.

	MALES				FEMALES			
Variables	**C** $n = 433$	**BB** $n = 433$	p	**d/r**	**C** $n = 102$	**BB** $n = 102$	p	**d/r**
Age (yrs)	59.0 ±10.9	59.0 ±10.9	1.0 [a]	0.00	60.7 ±10.7	60.7 ±10.7	1.0 [a]	0.00
BMI (kg/m²)	26.2 (4.0; 20.8)	26.8 (4.8; 25.1)	0.003 [b]	0.10	24.6 (5.3; 18.8)	25.7 (5.6; 26.9)	0.116 [b]	0.11
BM (kg)	83 (15; 117)	84 (18; 92)	0.121 [b]	0.05	65 (12; 54)	69 (17; 83)	0.05 [b]	0.14
P_{max} (W)	200 (60; 280)	170 (60; 210)	<0.001 [b]	0.37	140 (40; 160)	110 (40; 130)	<0.001 [b]	0.37
HR_{max} (bpm)	155 (20; 80)	150 (18; 67)	<0.001 [b]	0.18	154.7 ±13.8	148 (16; 61)	<0.001 [b]	0.27
P (W)								
Th1	76.4 (21.8; 105.5)	64.5 (19.1; 79.1)	<0.001 [b]	0.33	50.9 (12.7; 56.4)	46.4 (12.7; 47.3)	<0.001 [b]	0.29
Th2	145.5 (41.8; 210.9)	119.1 ±30.0	<0.001 [b]	0.35	96.9 ±20.3	77.7 (28.4; 95.5)	<0.001 [b]	0.40
HR (bpm)								
Th1	98.1 (15.2; 61.9)	93.4 (14.7; 67.7)	<0.001 [b]	0.24	106.4 ±10.9	97.6 (12.6; 55.3)	<0.001 [b]	0.37
Th2	124.1 (19.3; 78.1)	115.1 (14.6; 77.0)	<0.001 [b]	0.33	129.9 ±14.1	116.9 ±10.2	<0.001 [a]	0.93
P as % P_{max}								
Th1	37.3 (5.0; 9.9)	39.5 (4.2; 14.2)	<0.001 [b]	0.30	39.4 (3.9; 8.3)	42.1 ±2.8	<0.001 [b]	0.47
Th2	69.31 (5.1; 7.6)	71.7 (5.1; 7.1)	0.009 [b]	0.9	72.7 (4.5; 8.3)	68.2 (5.2; 7.1)	0.006 [b]	0.19 *
HR as % HR_{max}								
Th1	63.4 (8.5; 37.6)	62.1 (9.2; 41.5)	0.001 [b]	0.11	68.9 ±5.0	66.8 ±7.0	0.016 [a]	0.34
Th2	80.8 ±5.0	77.2 ±7.3	<0.001 [b]	0.55	84.2 (6.5; 19.7)	79.2 ±5.1	<0.001 [b]	0.45
k_{HR}	−0.4 (1.3; 10.8)	−1.2 (1.32; 10.3)	<0.001 [b]	0.36	0.1 ±1.0	−1.0 ±1.0	<0.001 [a]	−1.00

BMI, body mass index; BM, body mass; P_{max}, maximum power output; HR_{max}, maximum heart rate; P, power output at Th1 and Th2, first and second threshold; HR, heart rate at Th1 and Th2; k_{HR}, degree of the deflection of the heart rate performance curve; d, Cohen's d prescribing the effect size for normal distribution; r, effect size for not normal distributed data ($r = z/\sqrt{n}$); a, parametric t-test; b, Mann–Whitney-Test. * $p < 0.05$.

The number of non-regular HRPCs was higher in BB compared to C (m: 378 vs. 299, $p < 0.001$; f: 90 vs. 50, $p < 0.001$). Whereas the number of regular HRPC was higher in C compared to BB (m: 134 vs. 55, $p < 0.001$; f: 52 vs. 12, $p < 0.001$). The number of non-regular curves for BB and C, dependent on sex and age, are shown in Figure 2.

The logistic regression model (Table A1 in Appendix A), χ^2 (7, N = 1070) =111.07, $p < 0.001$, Nagelkerke´s $R^2 = 14.8$ revealed that the odds of having a non-regular HRPC in an incremental cycle ergometer test were higher among BB than C but the effect was dependent on sex ($OR_{interaction} = 2.77$, 95% CI: 1.24–6.19, $p = 0.013$) and age ($OR_{interaction} = 0.97$, 95% CI = 0.94–0.99, $p = 0.015$). The odds of having a non-regular HRPC due to β-blocker treatment were 2.8 times higher among average-aged females than males (females: OR = 7.65, 95% CI: 3.57–16.39, $p < 0.001$; males: OR = 2.76, 95% CI: 1.79–4.26, $p < 0.001$).

Moreover, the effect of β-blocker treatment on having a non-regular HRPC decreased with older age. For example, the odds of having a non-regular HRPC in BB versus C were OR = 4.04 (95% CI: 2.57–6.34, $p < 0.001$), OR = 2.76 (as above), and OR = 2.12 (95% CI: 1.34–3.34, $p = 0.001$) for 50-years-, 59-years- (mean of the sample) and 70-years-old male individual, respectively (females: $OR_{50\text{-years}} = 11.19$, 95% CI: 5.07–24.72, $p < 0.001$; $OR_{59\text{-years}} = 7.65$, as above, $OR_{70\text{-years}} = 5.86$, 95% CI: 2.76–12.52, $p < 0.001$).

Figure 3 shows the mean values for HR_{max} and P_{max} categorized in four age-groups for male and female individuals.

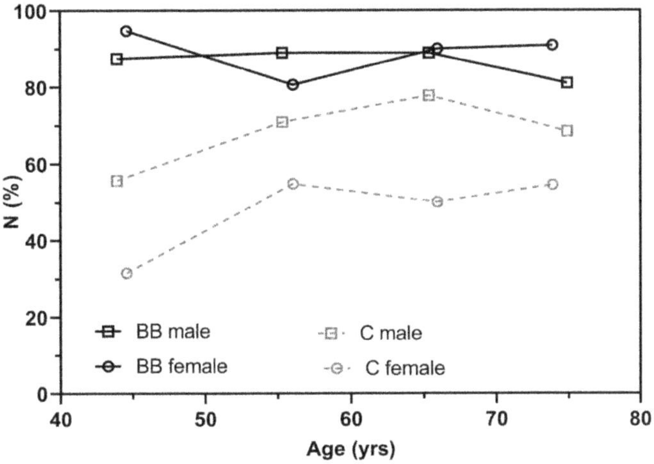

Figure 2. Age related percentage of non-regular HRPCs in male and female individuals with (BB) and without (C) β-blocker treatment.

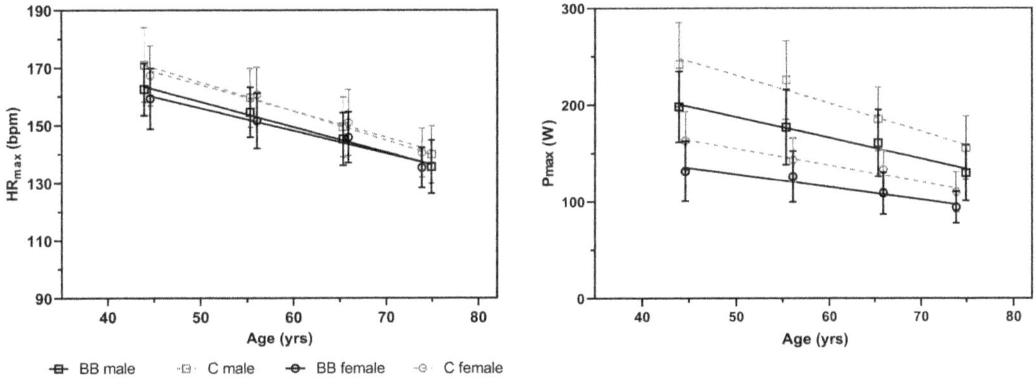

Figure 3. Age related decline of mean maximum heart rate (HR_{max}) and power output (P_{max}) in male and female individuals with (BB) and without (C) β-blocker treatment.

The linear regression model for HR_{max} (Table A2), $F(5, 1064) = 286.01$, $p < 0.001$, $R^2 = 0.57$, revealed that, HR_{max} of the incremental exercise test, of an average-aged, male BB was 5.46 (95% CI: 4.24–6.68, $p < 0.001$) bpm less compared to C. This effect was not dependent on sex but age, indicating a decrease in the difference between BB and C among older age ($B_{interaction} = 0.17$, 95% CI: 0.06–0.27, $p = 0.001$). For example, HR_{max} was, on average, 6.92 (95% 5.40–8.45, $p < 0.001$), 5.46 (as above) and 3.62 (95%CI: 1.97–5.27, $p < 0.001$) bpm lower for 50-years-, 59-years- and 70-years-old male BB, respectively.

The linear regression model for P_{max} (Table A3), $F(5, 1064) = 276.21$, $p < 0.001$, $R^2 = 0.57$, showed that, on average, BB had lower maximum power than C but the magnitude of the difference was dependent on sex and age. When receiving β-blocker treatment, maximum power was reduced by 37.53 W (95% CI: 32.93–42.13, $p < 0.001$) and 23.34 W ($-37.53 + 14.19$; 95% CI: 13.84–32.84, $p < 0.001$) for an average-aged male and female individual, respectively. These differences were decreasing with older age ($B_{interaction} = 0.83$, 95% CI: 0.45–1.21, $p < 0.001$). For example, 50-years-, 59-years- and 70-years-old male BB showed, on average, 44.88 W (95% CI: 39.14–50.63, $p < 0.001$), 37.53 W (as above) and 28.31 W (95% CI: 22.09–34.54, $p < 0.001$) lower maximum power compared to C (females: $B_{50\text{-}years} = 30.69$, 95% CI:

20.38–41.01, $p < 0.001$; $B_{59\text{-years}} = 23.34$ W, as above; $B_{70\text{-years}} = 14.12$ W, 95% CI: 4.00–24.25, $p = 0.006$).

ANOVAs revealed significantly lower absolute power output at Th1 and Th2 in male BB compared to C in all four age groups (Table 2). However, relative performance values varied less between BB and C and were not significantly different for Th1 and in two age groups for Th2. Absolute HR values were significantly lower at Th2 in BB, and difference decreased from about 13 bpm in the age group < 50 yrs to 7 bpm in the age group > 70 yrs. $\%HR_{max}$ was not statistically different between male BB and C, but values were consistently 2.5 to 4.4% lower in BB. In female individuals, no significant differences at all were found between BB and C, although absolute submaximal HR and performance values were consistently lower in BB. Comparable to male individuals, the HR difference decreased from the youngest to the oldest age group from about 12 to 4 bpm at Th1 and 18 to 8 bpm at Th2. $\%P_{max}$ at Th1 and Th2 were comparable to the male sample and varied less between groups.

Table 2. Characteristics of the incremental cycle ergometer test in male and female individuals with (BB) and without (C) β-blocker treatment in four representative age groups shown as mean (±SD).

MALES *n* = 433		**≤50 yrs**	**51–60 yrs**	**61–70 yrs**	**>70 yrs**
n		176	155	117	73
P Th1 (W)	C	87.8 ± 15.6	83.1 ± 15.0	69.9 ± 11.1	60.7 ± 11.9
	BB	75.21 ± 14.0 *	68.5 ± 13.7 *	62.5 ± 11.4 *	52.3 ± 9.3 *
P Th2 (W)	C	169.7 ± 31.6	158.0 ± 30.2	129.2 ± 22.8	109.7 ± 22.5
	BB	139.9 ± 26.9 *	125.0 ±2 7.4 *	113.3 ± 25.1 *	90.7 ± 20.2 *
$\%P_{max}$ Th1 (%)	C	36.4 ± 2.5	36.9 ± 2.4	37.9 ± 2.5	39.2 ± 2.3
	BB	38.0 ± 2.4	39.0 ± 2.9	39.3 ± 2.9	40.9 ± 3.2
$\%P_{max}$ Th2 (%)	C	70.2 ± 2.6	69.9 ± 2.5	69.7 ± 2.5	70.7 ± 2.5
	BB	70.5 ± 2.5	70.7 ± 2.5 *	70.6 ± 2.6 *	70.0 ± 2.5
HR Th1 (bpm)	C	106.5 ± 11.2	100.2 ± 11.1	96.3 ± 10.6	94.2 ± 11.3
	BB	99.0 ± 11.4	94.8 ± 10.5	90.3 ± 9.3	90.7 ± 10.2
HR Th2 (bpm)	C	139.1 ± 13.8	128.6 ± 11.4	119.4 ± 10.2	114.6 ± 10.0
	BB	126.1 ± 11.2 *	118.9 ± 8.9 *	111.6 ± 8.8 *	107.7 ± 9.6 *
$\%HR_{max}$ Th1 (%)	C	62.3 ± 5.4	62.8 ± 6.2	64.5 ± 6.9	67.5 ± 7.8
	BB	60.9 ± 6.2	61.4 ± 6.6	61.2 ± 10.2	67 ± 7.4
$\%HR_{max}$ Th2 (%)	C	81.3 ± 4.8	80.5 ± 4.6	79.9 ± 5.3	82.0 ± 5.4
	BB	77.6 ± 5.2	77.0 ± 5.0	75.7 ± 11.1	79.5 ± 7.4
k_{HR}	C	−0.1 ± 1.1	−0.3 ± 1.0	−0.5 ± 1.2	−0.3 ± 1.4
	BB	−1.0 ± 1.0	−1.1 ± 1.1	−1.2 ± 1.0	−0.9 ± 1.5
FEMALES *n*= 102		**≤50 yrs**	**51–60 yrs**	**61–70 yrs**	**>70 yrs**
n		19	31	30	22
P Th1 (W)	C	62.6 ± 9.6	56.1 ± 9.4	52.7 ± 7.2	44.7 ± 8.7
	BB	53.3 ± 11.4	51.3 ± 8.9	45.7 ± 7.8	41.5 ± 6.2
P Th2 (W)	C	115.5 ± 19.5	101.0 ± 16.2	94.7 ± 14.3	78.1 ± 15.9
	BB	90.8 ± 20.6	89.0 ± 19.4	75.4 ± 15.3	65.7 ± 11.3
$\%P_{max}$ Th1 (%)	C	38.6 ± 2.3	39.2 ± 2.1	39.7 ± 2.2	40.6 ± 1.2
	BB	40.9 ± 2.8	41.1 ± 2.6	42.3 ± 2.3	44.4 ± 2.0
$\%P_{max}$ Th2 (%)	C	71.0 ± 2.5	70.5 ± 2.4	71.0 ± 2.3	70.8 ± 2.3
	BB	69.1 ± 2.3	70.7 ± 2.6	69.4 ± 2.4	69.9 ± 2.4
HR Th1 (bpm)	C	112.8 ± 11.2	109.6 ± 10.4	103.5 ± 9.9	100.1 ± 7.4
	BB	101.2 ± 10.2	101.5 ± 13.5	94.7 ± 6.9	96.7 ± 8.2
HR Th2 (bpm)	C	142.0 ± 13.5	134.1 ± 11.9	126.4 ± 11.1	118.1 ± 9.2
	BB	123.7 ±8.1	121.3 ± 10.6	113.1 ± 7.2	110.2 ± 8.3
$\%HR_{max}$ Th1 (%)	C	67.5 ± 5.2	68.2 ± 4.7	68.6 ± 5.1	71.3 ± 4.5
	BB	67.5 ± 5.2	68.2 ± 4.7	68.6 ± 5.1	71.3 ± 4.5
$\%HR_{max}$ Th2 (%)	C	84.8 ± 4.4	83.4 ± 4.1	83.8 ± 4.4	84.0 ± 4.4
	BB	77.9 ± 6.5	80.0 ± 4.5	77.6 ± 4.3	81.4 ± 4.5
k_{HR}	C	0.41 ± 1.12	−0.07 ± 0.86	0.12 ± 1.06	0.05 ± 1.07
	BB	−0.98 ± 0.90	−1.06 ± 0.98	−1.05 ± 0.99	−1.02 ± 0.97

n, Number of tests; P, power output at Th1 and Th2, first and second threshold; $\%P_{max}$ Th1 and Th2, P at Th1 as a percentage of maximum power output; HR Th1 and Th2, heart rate at Th1 and Th2; $\%HR_{max}$ Th1 and Th2, HR at Th1 and Th2 as a percentage of maximum heart rate; k_{HR}, degree of the deflection of the heart rate performance curve; * $p < 0.05$.

4. Discussion

Our study showed that individuals receiving a β-blocker treatment had usually more than two times higher odds of having a non-regular HRPC (including both linear time course and upward deflection), compared to healthy individuals. However, the odds were dependent on sex and age, indicating higher odds for females and decreasing odds among older individuals. Maximum heart rate as well as maximum power was reduced in BB compared to C, but the effect decreased with higher age.

Overall, we found 87% compared to 69% non-regular curves in male BB compared to C and 88% compared to 49% in female individuals. The number of non-regular curves in male and female BB was high independent of age, but increased with age in C. The matched healthy group was a sub-sample of a previous investigation, and results are in line with our earlier findings [12] and Hofmann et al. [9], who presented a significantly lower number of only 14% of non-regular curves in a group of 227 young, healthy and trained male subjects. Regarding the number of non-regular curves in BB, no comparative studies were found. However, β-blocker administration was already shown to change the direction of the HRPC from a regular downward deflection to non-regular curve patterns in healthy individuals [18,26]. In particular, Hofmann et al. [18] even showed, that these changes were significantly related to the degree of the deflection in participants randomly receiving placebo or selective β1-adrenoreceptor antagonist. The "more regular" the curves were in placebo conditions, the greater was the change in β-blocker treatment, whereas non-regular patterns were not affected. Therefore, the higher number of non-regular curves in the present study is consistent with previous literature.

Differences in receptor sensitivity were thought to be a cause for varying responses of the HRPC. In subjects with a regular HRPC, β1-adrenergic receptors were suggested to be sensitive to catecholamines at low intensity levels (Phase 2), leading to a proportional HR increase. At high intensity above Th2 (Phase 3), receptors saturate, and HR increase is damped. This is suggested to cause the flattening of the HRPC, resulting in a regular HRPC [18]. Contrary in subjects with non-regular curves the receptor sensitivity was suggested to be lower, or receptors are blocked due to β-blocker administration. This leads to a blunted HR increase between Th1 and Th2 (Phase 2) (see Figure 1), were catecholamine levels continuously increase but are still moderate. Above Th2 (Phase 3) HR increases disproportionally due to exponentially increasing catecholamine levels [18]. Therefore, the considerably higher number of non-regular HRPCs in the BB group in our study is caused by the β-blocker treatment, especially in the younger subjects. As ageing is associated with a decreased receptor sensitivity [21] and women were shown to have a higher sensitivity compared to men [30], changes in the odds among older individuals and sex can be addressed to changes in the receptor sensitivity. In our previous study we already showed, that women show less non-regular curves compared to men and that the number of non-regular HRPC´s increase with increasing age in healthy individuals [12]. Therefore, the HRPC pattern from BB and C are thought to become more similar with increasing age due to receptor sensitivity decreases with age.

These alterations in the HRPC also influenced the submaximal HR markers, where absolute and $\%HR_{max}$ at Th1 and Th2 were found significantly lower in BB compared to C. This was already shown for patients, with and without β-blocker treatment [27], as well as for healthy individuals who randomly received β-blocker or placebo [18,26,31]. The reduced HR at maximal and submaximal intensities in individuals receiving a β-blocker can be addressed to the negative chronotropic effect as desired [25,32]. In our study, the HR differences between BB and C were found between 6 to 8 bpm at Th1 and 10 to 13 bpm at Th2. These values are markedly smaller compared to values of other studies which showed larger differences (Th1: 19 to 26 bpm, Th2: 22 to 37 bpm) [18,26,27,31]. The smaller differences between groups in our study could be due to differences in dosing. However, no information about the dosing was available in the current study. However, a minimum dosage to avoid side-effects during long-term treatment might be explanatory. A further reason might be the higher mean age of our study population. Decreasing maximum

HR with increasing age [33] reduces the amplitude of HR from rest to maximal exercise, theoretically effecting the absolute difference between groups. Furthermore, aging was associated with reduced β-adrenergic receptor sensitivity [21] which possibly leads to smaller differences in the HR response to exercise between individuals with and without β-blocker with increasing age. Both age dependent reasons can be supported by our data, were the difference in HR at maximal and submaximal values is shown to decrease between groups with increasing age (Table 2). In terms of HR_{max}, values in BB were approximately 5 bpm lower compared to C. The difference was greater in younger compared to older age. Hence, increasing age reduces the difference between individuals with and without β-blocker treatment. Considering that age reduces the receptor sensitivity, older people seem to more likely respond like individuals on β-blocker treatment. However, comparable studies showed larger differences in HR_{max} of about 19 to 38 bpm between patients [27] and healthy individuals with and without β-blocker treatment [18,26,27,31].

Regarding relative values, we found significantly smaller $\%HR_{max}$ at Th1 and Th2 in BB. Values at Th2 were below the common upper limit of 85% HR_{max} in BB in 93% and 86%, respectively in C in 79% and 56% of all cases in male and female individuals. Therefore, both groups are overestimated with respect to the upper limit of exercise intensity if based on such a standard prescription. Furthermore, individuals with β-blocker treatment presented a lower $\%HR_{max}$ at Th2 (as well as Th1) compared to healthy individuals. Although the differences between BB and C were small, we can show for the first time a different heart rate response during an incremental cycle ergometer exercise in most of the individuals receiving a β-blocker therapy.

Overall, performance in male and female BB was normal with respect to age predicted P_{max} and C were slightly better trained. Interestingly, $\%P_{max}$ at Th1 and Th2 was found very similar in BB and C, both in females and males. The percentage ranged between 37.5 and 42.1% at Th1 and between 69.8 and 70.8% at Th2. These values are highly comparable to results from young healthy subjects were a percentage of 38.8 and 72.0% was shown for the first and second lactate threshold [29]. Based on these findings, intensity prescription by means of $\%P_{max}$ seems to be more generalizable compared to relative heart rate values [29]. A more individualized prescription and increased consideration of exercise intensity prescription including $\%P_{max}$ was recommended by EPAC [5]. Therefore, exercise prescription via fixed percentage of HR_{max} should be avoided and individual thresholds or at least percentages of P_{max} are recommended, especially in individuals on β-blocker treatment in order to avoid overloading [7].

The detection of a first threshold of HR within a fixed region of interest revealed a mean $\%HR_{max}$ at Th1 of healthy individuals comparable to HR values of the first ventilatory threshold from the literature [26]. Only a few studies examined the determination of a first heart rate threshold/turn point by mathematical models and showed no significant difference compared to a gold standard detection (e.g., AT-Wassermann). Therefore, the detection of changes in response patterns of HR is suggested to be adequate and promising for the detection of a first threshold in a three-phase model [34,35]. Although this method is yet not very common, this might be of interest for practice.

The present study is not without limitations. One limitation is the lack of any medical diagnosis in C as well as in BB. Due to the advanced age, the likelihood that also C have undiagnosed health conditions is high. Therefore, we cannot exclude that the deflection of the HRPC has been influenced by the consequences of any cardiovascular disease or events [36]. This also refers to BB, where a medical diagnosis is obvious. Furthermore, we do not have any information regarding the β-blocker dosing, which does not allow to discuss any dose-response effects of β-blocker administration on the deflection of the HRPC. However, the presented data reflect the practical situation in rehabilitation and secondary prevention and did not underly any controlled study setting, possibly explaining the smaller differences between groups in this study. Further, also limiting is that performance in BB was lower compared to C, although individuals were matched 1:1 for age and sex. Based on our earlier findings, this might influence the pattern of the

HRPC due to the fact that individuals with lower performance were shown to present a higher number of non-regular curves [12]. Regarding exercise prescription, the prescription via $\%VO_{2max}$ is an even more common and accepted model beside $\%HR_{max}$. Due to the proportional relationship of VO_2 and performance, exercise prescription based on $\%VO_{2max}$ might be more accurate compared to $\%HR_{max}$ although not measured in our study. However, Hofmann et al. [18] showed, that compared to $\%HR_{max}$, $\%VO_{2max}$ was not affected by a selective β-blocker. In the future, investigations regarding the dosing, the influence of selective and non-selective β-blocker application, including spirometric data, are necessary to better understand the implications of β-blocker treatment for exercise prescription. Nevertheless, our study prescribes relevant information regarding age and sex-dependent influences of β-blocker treatment on HR changes during a standardized incremental exercise and its consequences for exercise prescription.

5. Conclusions

The HR increase to incremental cycle ergometer exercise is neither uniform nor linear and differs considerably between individuals. BB showed higher odds of having a non-regular HRPC and lower HR_{max} and P_{max} compared to C. However, the effects were modified by age and sex. Relative HR values at the Th2 were not constant but varied substantially with β-blocker treatment, questioning the validity of fixed values for exercise prescription. Therefore, a generalization in terms of exercise prescription via fixed percentage of HR_{max} is not recommended and misclassifications and overestimation of upper limits are very likely in individuals with β-blocker treatment. Interestingly $\%P_{max}$ seems to be a good trade off to provide valid estimations about exercise intensity, even in patients with β-blocker treatment and supports the request in the actual guidelines to apply such an easy tool [5].

Author Contributions: Conceptualization, P.H.; H.T. and P.B.; methodology, P.H., M.C.S. and P.B.; formal analysis, M.C.S., A.B. and P.B.; investigation, P.B.; resources, P.H. and H.T.; data curation, A.B. and P.B.; writing—original draft preparation, P.B.; writing—review and editing, P.H., M.C.S.; A.B. and H.T.; visualization, P.B.; supervision, P.H.; project administration, P.H. All authors have read and agreed to the published version of the manuscript.

Funding: Open Access Funding by the University of Graz.

Institutional Review Board Statement: The study was conducted according to the guidelines of the Declaration of Helsinki, and approved by the Ethics Committee of the University of Graz (GZ. 39/70/63 ex 2016/17).

Informed Consent Statement: Informed consent was obtained from all subjects involved in the study.

Data Availability Statement: The data presented in this study are available on request from the corresponding author. The data are not publicly available due to hospital confidentiality.

Conflicts of Interest: The authors declare no conflict of interest.

Appendix A

Table A1. Results of the main logistic regression model to evaluate sex and age-related differences in heart rate performance curves (HRPC) dependent on medication.

	B (SE)	OR (95% CI)	*p*
β-blocker	1.015 (0.221)	2.761 (1.788, 4.261)	<0.001
Age	0.022 (0.009)	1.022 (1.005, 1.040)	0.01
Age^2	−0.002 (0.001)	0.998 (0.997, 0.999)	0.001
Female	−0.914 (0.228)	0.401 (0.256, 0.628)	<0.001
Female × β-blocker	1.019 (0.410)	2.771 (1.239, 6.194)	0.013
Age × β-blocker	−0.34 (0.014)	0.966 (0.940, 0.993)	0.015
Age^2 × β-blocker	0.001 (0.001)	1.001 (0.999, 1.003)	0.331

B (SE), unstandardized regression coefficients (standard error); OR, odds ration; CI, 95% confidence interval; logistic regression model: χ^2 (7, N = 1070) = 111.07, $p < 0.001$, Nagelkerke's $R^2 = 14.8$.

Table A2. Results of the linear regression model with maximum heart rate as the outcome.

Model	B (95% CI)	*p*
Constant	156.070 (155.206, 156.933)	<0.001
β-blocker	−5.457 (−6.678, −4.236)	<0.001
Age	−1.017 (−1.088, −0.945)	<0.001
Female	0.448 (−1.533, 2.429)	0.657
Female × β-blocker	−1.690 (−4.492, 1.111)	0.237
Age × β-blocker	0.165 (0.064, 0.266)	0.001

B, unstandardized regression coefficients; CI, 95% confidence interval; linear regression model: $F(5, 1064) = 286.01$, $p < 0.001$, $R^2 = 0.57$.

Table A3. Results of the linear regression model with maximum power as the outcome.

	B (95% CI)	*p*
Constant	206.659 (203.407, 209.912)	<0.001
β-blocker	−37.528 (−42.128, −32.929)	<0.001
Age	−2.717 (−2.988, −2.447)	<0.001
Female	−64.954 (−72.416, −57.493)	<0.001
Female × β-blocker	14.189 (3.637, 24.742)	0.008
Age × β-blocker	0.829 (0.447, 1.210)	<0.001

B, unstandardized regression coefficients; CI, 95% confidence interval; linear regression model: $F(5, 1064) = 276.21$, $p < 0.001$, $R^2 = 0.57$.

References

1. Lavie, C.J.; Thomas, R.J.; Squires, R.W.; Allison, T.G.; Milani, R.V. Exercise Training and Cardiac Rehabilitation in Primary and Secondary Prevention of Coronary Heart Disease. *Mayo Clin. Proc.* **2009**, *84*, 373–383. [CrossRef]
2. Iannetta, D.; Inglis, E.C.; Mattu, A.T.; Fontana, F.Y.; Pogliaghi, S.; Keir, D.A.; Murias, J.M. A Critical Evaluation of Current Methods for Exercise Prescription in Women and Men. *Med. Sci. Sports Exerc.* **2019**, *1*, 466–473. [CrossRef]
3. Hansen, D.; Bonné, K.; Alders, T.; Hermans, A.; Copermans, K.; Swinnen, H.; Maris, V.; Jansegers, T.; Mathijs, W.; Haenen, L.; et al. Exercise Training Intensity Determination in Cardiovascular Rehabilitation: Should the Guidelines Be Reconsidered? *Eur. J. Prev. Cardiol.* **2019**, *26*, 1921–1928. [CrossRef] [PubMed]
4. Scharhag-Rosenberger, F.; Meyer, T.; Gässler, N.; Faude, O.; Kindermann, W. Exercise at given Percentages of VO2max: Heterogeneous Metabolic Responses between Individuals. *J. Sci. Med. Sport* **2010**, *13*, 74–79. [CrossRef]
5. Ambrosetti, M.; Abreu, A.; Corrà, U.; Davos, C.H.; Hansen, D.; Frederix, I.; Iliou, M.C.; Pedretti, R.F.; Schmid, J.-P.; Vigorito, C.; et al. Secondary Prevention through Comprehensive Cardiovascular Rehabilitation: From Knowledge to Implementation. 2020 Update. A Position Paper from the Secondary Prevention and Rehabilitation Section of the European Association of Preventive Cardiology. *Eur. J. Prev. Cardiol.* **2020**, *28*, 460–495. [CrossRef] [PubMed]
6. ACSM. *ACSM's Guidelines for Exercise Testing and Prescription*, 10th ed.; Lippincot Williams & Wilkins: Philadelphia, PA, USA, 2016.
7. Li, W.G.; Huang, Z.; Zhang, X.A. Exercise Prescription in Cardiac Rehabilitation Needs to Be More Accurate. *Eur. J. Prev. Cardiol.* **2020**, *30*, 2047487320936021.
8. Hofmann, P.; Tschakert, G. Special Needs to Prescribe Exercise Intensity for Scientific Studies. *Cardiol. Res. Pract.* **2010**, *2011*, 209302. [CrossRef]
9. Hofmann, P.; Pokan, R.; von Duvillard, S.P.; Seibert, F.J.; Zweiker, R.; Schmid, P. Heart Rate Performance Curve during Incremental Cycle Ergometer Exercise in Healthy Young Male Subjects. *Med. Sci. Sports Exerc.* **1997**, *29*, 762–768. [CrossRef]
10. Brooke, J.D.; Hamley, E.J. The Heart-Rate–Physical Work Curve Analysis for the Prediction of Exhausting Work Ability. *Med. Sci. Sports* **1972**, *4*, 23–26.
11. Pokan, R.; Hofmann, P.; Preidler, K.; Leitner, H.; Dusleag, J.; Eber, B.; Schwaberger, G.; Füger, G.F.; Klein, W. Correlation between Inflection of Heart Rate/Work Performance Curve and Myocardial Function in Exhausting Cycle Ergometer Exercise. *Eur. J. Appl. Physiol. Occup. Physiol.* **1993**, *67*, 385–388. [CrossRef] [PubMed]
12. Birnbaumer, P.; Traninger, H.; Borenich, A.; Falgenhauer, M.; Modre-Osprian, R.; Harpf, H.; Hofmann, P. Heart Rate Performance Curve Is Dependent on Age, Sex, and Performance. *Front. Public Health* **2020**, *8*, 98. [CrossRef] [PubMed]
13. Conconi, F.; Ferrari, M.; Ziglio, P.G.; Droghetti, P.; Codeca, L. Determination of the Anaerobic Threshold by a Noninvasive Field Test in Runners. *J. Appl. Physiol.* **1982**, *52*, 869–873. [CrossRef] [PubMed]
14. Bodner, M.E.; Rhodes, E.C. A Review of the Concept of the Heart Rate Deflection Point. *Sports Med.* **2000**, *30*, 31–46. [CrossRef]
15. Binder, R.K.; Wonisch, M.; Corra, U.; Cohen-Solal, A.; Vanhees, L.; Saner, H.; Schmid, J.-P. Methodological Approach to the First and Second Lactate Threshold in Incremental Cardiopulmonary Exercise Testing. *Eur. J. Cardiovasc. Prev. Rehabil.* **2008**, *15*, 726–734. [CrossRef]

16. Zouhal, H.; Jacob, C.; Delamarche, P.; Gratas-Delamarche, A. Catecholamines and the Effects of Exercise, Training and Gender. *Sports Med. Sports Med.* **2008**, *38*, 401–423. [CrossRef] [PubMed]
17. Pokan, R.; Hofmann, P.; Lehmann, M.; Leitner, H.; Eber, B.; Gasser, R.; Schwaberger, G.; Schmid, P.; Keul, J.; Klein, W. Heart Rate Deflection Related to Lactate Performance Curve and Plasma Catecholamine Response during Incremental Cycle Ergometer Exercise. *Eur. J. Appl. Physiol. Occup. Physiol.* **1995**, *70*, 175–179. [CrossRef]
18. Hofmann, P.; Wonisch, M.; Pokan, R.; Schwaberger, G.; Smekal, G.; von Duvillard, S. Beta1-Adrenoceptor Mediated Origin of the Heart Rate Performance Curve Deflection. *Med. Sci. Sports Exerc.* **2005**, *37*, 1704–1709. [CrossRef]
19. Brodde, O.E.; Bruck, H.; Leineweber, K. Cardiac Adrenoceptors: Physiological and Pathophysiological Relevance. *J. Pharmacol. Sci.* **2006**, *100*, 323–337. [CrossRef] [PubMed]
20. Chwalbińska-Moneta, J. Threshold Increases in Plasma Growth Hormone in Relation to Plasma Catecholamine and Blood Lactate Concentrations during Progressive Exercise in Endurance-Trained Athletes. *Eur. J. Appl. Physiol. Occup. Physiol.* **1996**, *73*, 117–120. [CrossRef]
21. Christou, D.D.; Seals, D.R. Decreased Maximal Heart Rate with Aging Is Related to Reduced β-Adrenergic Responsiveness but Is Largely Explained by a Reduction in Intrinsic Heart Rate. *J. Appl. Physiol.* **2008**, *105*, 24–29. [CrossRef] [PubMed]
22. Stratton, J.R.; Cerqueira, M.D.; Schwartz, R.S.; Levy, W.C.; Veith, R.C.; Kahn, S.E.; Abrass, I.B. Differences in Cardiovascular Responses to Isoproterenol in Relation to Age and Exercise Training in Healthy Men. *Circulation* **1992**, *86*, 504–512. [CrossRef] [PubMed]
23. Fleg, J.L.; Schulman, S.; O'Connor, F.; Becker, L.C.; Gerstenblith, G.; Clulow, J.F.; Renlund, D.G.; Lakatta, E.G. Effects of Acute Beta-Adrenergic Receptor Blockade on Age-Associated Changes in Cardiovascular Performance during Dynamic Exercise. *Circulation* **1994**, *90*, 2333–2341. [CrossRef]
24. Zeitouni, M.; Kerneis, M.; Lattuca, B.; Guedeney, P.; Cayla, G.; Collet, J.P.; Montalescot, G.; Silvain, J. Do Patients Need Lifelong β-Blockers after an Uncomplicated Myocardial Infarction? *Am. J. Cardiovasc. Drugs* **2019**, *19*, 431–438. [CrossRef]
25. Ladage, D.; Schwinger, R.H.G.; Brixius, K. Cardio-Selective Beta-Blocker: Pharmacological Evidence and Their Influence on Exercise Capacity. *Cardiovasc. Ther.* **2013**, *31*, 76–83. [CrossRef]
26. Wonisch, M.; Hofmann, P.; Fruhwald, F.; Kraxner, W.; Hödl, R.; Pokan, R.; Klein, W. Influence of beta-blocker use on percentage of target heart rate exercise prescription. *Eur. J. Cardiovasc. Prev. Rehabil.* **2003**, *10*, 296–301. [CrossRef]
27. Díaz-Buschmann, I.; Jaureguizar, K.V.; Calero, M.J.; Aquino, R.S. Programming Exercise Intensity in Patients on Beta-Blocker Treatment: The Importance of Choosing an Appropriate Method. *Eur. J. Prev. Cardiol.* **2014**, *21*, 1474–1480. [CrossRef]
28. Sauer, G. β-Blocker in Der Sekundärprävention von Herz-Kreislauf-Erkrankungen. *Z. Kardiol.* **2003**, *92*, 24–30. [CrossRef]
29. Hofmann, P.; Von Duvillard, S.P.; Seibert, F.J.; Pokan, R.; Wonisch, M.; Lemura, L.M.; Schwaberger, G. %HRmax Target Heart Rate Is Dependent on Heart Rate Performance Curve Deflection. *Med. Sci. Sports Exerc.* **2001**, *33*, 1726–1731. [CrossRef] [PubMed]
30. Turner, M.J.; Mier, C.M.; Spina, R.J.; Schechtman, K.B.; Ehsani, A.A. Effects of Age and Gender on the Cardiovascular Responses to Isoproterenol. *J. Gerontol. A Biol. Sci. Med. Sci.* **1999**, *54*, 393–400. [CrossRef] [PubMed]
31. Mitchell, B.L.; Kade, D.; Gaynor, P.; Simon, S.; Roger, G.E. Physiological and Perceived Exertion Responses during Exercise: Effect of β-Blockade. *Med. Sci. Sports Exerc.* **2019**, *51*, 782–791. [CrossRef]
32. Brawner, C.A.; Ehrman, J.K.; Schairer, J.R.; Cao, J.J.; Keteyian, S.J. Predicting Maximum Heart Rate among Patients with Coronary Heart Disease Receiving Beta-Adrenergic Blockade Therapy. *Am. Heart J.* **2004**, *148*, 910–914. [CrossRef]
33. Zhu, N.; Suarez-Lopez, J.R.; Sidney, S.; Sternfeld, B.; Schreiner, P.J.; Carnethon, M.R.; Lewis, C.E.; Crow, R.S.; Bouchard, C.; Haskell, W.L.; et al. Longitudinal Examination of Age-Predicted Symptom-Limited Exercise Maximum HR. *Med. Sci. Sports Exerc.* **2010**, *42*, 1519–1527. [CrossRef] [PubMed]
34. Higa, M.N.; Silva, E.; Neves, V.F.C.; Catai, A.M.; Gallo, L.; Silva de Sá, M.F. Comparison of Anaerobic Threshold Determined by Visual and Mathematical Methods in Healthy Women. *Braz. J. Med. Biol. Res.* **2007**, *40*, 501–508. [CrossRef] [PubMed]
35. Zamunér, A.R.; Catai, A.M.; Martins, L.E.B.; Sakabe, D.I.; Da Silva, E. Identification and Agreement of First Turn Point by Mathematical Analysis Applied to Heart Rate, Carbon Dioxide Output and Electromyography. *Braz. J. Phys. Ther.* **2013**, *17*, 614–622. [CrossRef]
36. Pokan, R.; Hofmann, P.; Von Duvillard, S.P.; Schumacher, M.; Gasser, R.; Zweiker, R.; Fruhwald, F.M.; Eber, B.; Smekal, G.; Bachl, N.; et al. Parasympathetic Receptor Blockade and the Heart Rate Performance Curve. *Med. Sci. Sports Exerc.* **1998**, *30*, 229–233. [CrossRef]

Journal of
Functional Morphology and Kinesiology

MDPI

Article

Effects of Activity Tracker-Based Counselling and Live-Web Exercise on Breast Cancer Survivors during Italy COVID-19 Lockdown

Andrea Di Blasio [1,*], Teresa Morano [1], Federica Lancia [1], Gianluca Viscioni [1], Angelo Di Iorio [1], Simona Grossi [2], Ettore Cianchetti [2], Lucia Cugusi [3], Stefano Gobbo [4], Marco Bergamin [4], Anna D'Eugenio [1], Laura Masini [1], Massimo Rinaldi [1], Maria Teresa Scognamiglio [2], Anastasios Vamvakis [5] and Giorgio Napolitano [1]

1 Department of Medicine and Ageing Sciences, "G. d'Annunzio" University of Chieti-Pescara, Via Polacchi L. 11, 66100 Chieti, Italy; moranoteresa@gmail.com (T.M.); lanciafederica@gmail.com (F.L.); gianluca.viscioni@gmail.com (G.V.); a.diiorio@unich.it (A.D.I.); annadeugenio.medicinaintegrata@gmail.com (A.D.); lauramasini.medicinaintegrata@gmail.com (L.M.); drmassimorinaldi@gmail.com (M.R.); giorgio.napolitano@unich.it (G.N.)
2 Eusoma Breast Centre, "G. Bernabeo" Hospital, ASL02 Lanciano-Vasto-Chieti, c.da S. Liberata, 66026 Ortona, Italy; sgrossi@unich.it (S.G.); ettore.cianchetti@gmail.com (E.C.); mt.scognamiglio66@gmail.com (M.T.S.)
3 Department of Biomedical Sciences, University of Sassari, Viale San Pietro 43/B, 07100 Sassari, Italy; lucia.cugusi@uniss.it
4 Department of Medicine, University of Padova, Via Giustiniani 2, 35128 Padova, Italy; stefano.gobbo@unipd.it (S.G.); marco.bergamin@unipd.it (M.B.)
5 Department of Internal Medicine, Papageorgiou Hospital, Aristotle University of Thessaloniki, 56403 Thessaloniki, Greece; tvamvakis@yahoo.gr
* Correspondence: andrea.diblasio@unich.it

Citation: Di Blasio, A.; Morano, T.; Lancia, F.; Viscioni, G.; Di Iorio, A.; Grossi, S.; Cianchetti, E.; Cugusi, L.; Gobbo, S.; Bergamin, M.; et al. Effects of Activity Tracker-Based Counselling and Live-Web Exercise on Breast Cancer Survivors during Italy COVID-19 Lockdown. *J. Funct. Morphol. Kinesiol.* **2021**, *6*, 50. https://doi.org/10.3390/jfmk6020050

Academic Editors: Cristina Cortis, Andrea Fusco and Carl Foster

Received: 26 April 2021
Accepted: 7 June 2021
Published: 9 June 2021

Abstract: Background: To prevent and fight the increase of daily sedentary time and to promote and stimulate the positive effects of physical activity and exercise on health, both traditional interventions and new strategies are important for breast cancer survivors (BCS). The research goal was to compare the effects of weekly personal feedback, based on objectively measured physical activity, on the trends of both daily sedentary time and on the physical activity of BCS (E^- group) with those of an intervention also including online supervised physical exercise sessions (E^+ group), during the Italy COVID-19 lockdown. Methods: The Italian COVID-19 emergency allowed the possibility to also observe the effects of social and personal limitations. A total of 51 BCS were studied over an 18-week period and had an objective registration of day-to-day sedentary time, physical activity, and sleep. Both subsamples received weekly or fortnight personal feedback. Data were analysed considering four key periods, according to the COVID-19 emergency steps. Results: Statistical analysis showed an additive effect for sedentary time and a multiplicative effect both for light-to vigorous and light-intensity physical activities. The E^- group had a high overall sedentary time and a different trend of light-to vigorous and light-intensity physical activities, with a reduction from the 1st to the 2nd periods (national and personal restrictions), showing a significant rise just at the end of the national restrictions. Conclusions: The use of an activity tracker and its accompanying app, with the reception of weekly tailored advice and supervised online physical exercise sessions, can elicit proper physical activity recomposition in BCS in the COVID-19 era.

Keywords: light-intensity physical activity; Polar Loop 2; sedentary time; breast cancer

1. Introduction

A diagnosis of cancer and both pharmacological and non-pharmacological treatments of breast cancer could have negative effects on daily physical activity (reducing it), while sedentary time is increased [1–6]. Indeed, according to a study by De Groef et al. [5], 2 years after surgery, all activity domains were still significantly lower compared to preoperative values. After the first 12 months, the only significant improvement was seen in the occupational domain, while Gal et al. [6] found that breast cancer survivors, with and without systemic treatment, were less likely to spend time in physical activity compared to the general population, until 3 years post-diagnosis. The contemporaneous increase of sedentary time and reduction of physical activity needs particular and early attention due to its negative consequences on psychophysical health [7], as well as through its characteristic pro-inflammatory pattern, which is considered the starting point of the most common chronic non-communicable diseases [8], including breast cancer onset and recurrence. To prevent and fight the increase of daily sedentary time and to promote and stimulate the positive effects of physical activity and exercise on health [9,10], both traditional interventions, based on in-person ambulatory counselling and supervised adapted physical exercise sessions [9–12], and new strategies are important to reach as many women as possible, according to personal differences that are linked with psychological, familiar, working, and environmental differences. Thanks to advances in technology, increasing literature supports the importance of the use of activity trackers to improve the daily physical activity of breast cancer survivors, both alone [13–17] and integrated into supervised exercise programmes [18]. Indeed, they can stimulate people to be more physically active and less sedentary, as they provide insights into physical activity variables promoting "self-knowledge, and health, through numbers" [19]. Therefore, the study by Wu et al. [18] furnished important results, underlying the need to combine technology with both human feedback and interventions. Indeed, breast cancer survivors participating in a combined 12-week in-person physical exercise programme and physical activity promotion, using activity trackers, underlined that the use of an activity tracker and its accompanying app raised lifestyle awareness. Therefore, patients need personalized advice and a more realistic representation of total daily physical activity, together with more integration between the interventions concerning their recovery. This is of particular importance in the COVID-19 era. Worldwide, during the COVID-19 pandemic lockdown, the "stay-at-home" measures adopted to counteract the spread of the virus would dramatically reduce the physical activity levels of the general population [20,21], including cancer patients, suffering from psychophysical constriction characterized by the contemporaneous increase of sedentary time and reduction of daily physical activity, while the fear of the virus increases, together with psychosocial and emotional disorders, sleep disruption, and consequently sedentary time [22–27]. The described situation is particularly dangerous for cancer survivors as leads to poor psychophysical health, due to the fact that poor physical activity is linked to some of the side effects of cancer treatments, such as poor sleep, reinforcing the negative loop [28,29]. To prevent and counteract the negative consequences of the "stay-at-home" measures it was widely suggested to support personalized and supervised physical activity programs, with the option to group-play physical activity programs (e.g., exergames) [25]

Therefore, in order to optimize the recovery of breast cancer survivors, the original research goal was to compare the effects of weekly personal feedback, based on objectively measured physical activity, on daily sedentary time, and on physical activity of breast cancer survivors with those of an intervention also including online supervised physical exercise sessions. In this case, the consequences of the COVID-19 emergency (i.e., the government restrictions to counteract the spread of the virus) occurred during the execution of the study, allowing us to also observe the interaction of the treatments with the phases of the first Italian lockdown. Therefore, the final research goal was to verify whether merging weekly personalized feedback and online supervised physical exercise sessions confers major benefits on daily sedentary time and on physical activity of breast cancer survivors than just weekly personalized feedback, in the presence of a personal confinement and of its progressive regression. Our hypothesis was that the absence of supervised workouts, even in the presence of tailored personal suggestions, has lower power in the maintenance/improvement of daily sedentary time and physical activity characteristics of breast cancer survivors, in the presence of a personal confinement and, also, during its progressive regression.

2. Materials and Methods

2.1. Participants

The Integrative Medicine Clinic of both ASL02 of Lanciano-Vasto-Chieti (Italy) and Department of Medicine and Ageing Sciences of the "G. d'Annunzio" University of Chieti-Pescara (Italy) at "G. Bernabeo" Hospital (Ortona, Italy) recruited study participants. A total of 51 breast cancer survivors (50.98 ± 6.28 years), among those who had visits from 1 October 2019 to 12 January 2020, matching both inclusion and exclusion criteria were selected for this study. The inclusion criteria for this study were age between 30–60 years, 6–48 months after breast surgery, actual hormone therapy, and participation in the "Angel Project", which is described in the following section. The exclusion criteria for this study were actual chemotherapy, actual radiotherapy, actual diseases limiting motion, actual chronic use of hypnotic pills, actual pharmacological treatment for anxiety and/or depression or no interest in participating in live online physical exercise sessions. The term "actual" relates to a period starting from the date of the basal evaluation of each participant and continuing until the end of the study. The Ethics Committee of Chieti-Pescara approved this study (# 312/2015), and participants gave their written informed consent.

2.2. Study Design

As displayed in Figure S1, the Integrative Medicine Clinic, which was activated on 15 November 2017, furnishes integrative support for breast cancer survivors during the follow-up phase, including evaluations, behavioural counselling, and interventions. The clinic's role was significant in directing patients regarding physical activity, sleep, body composition and nutrition; acupuncture; analysis and control of blood, salivary, metabolic, immune and endocrine parameters; psychotherapy, mindfulness, and both adapted and supervised physical exercise. Patients participating in the "Angel Project" were requested to continuously wear a scientifically validated commercial accelerometer (i.e., Polar Loop 2 (Kempele, Finland)) [30–32] to record and remotely control daily physical activity, sedentary time, and sleep characteristics through the use of a dedicated website (i.e., Polar Flow (Kempele, Finald)), to receive personalized weekly feedback from the Integrative Medicine Clinic for 18 consecutive weeks. In the same period, they received personalized qualitative dietary suggestions, with a fortnight frequency. As the objective of the project was to educate persons to progressively self-evaluate their lifestyle (i.e., nutrition, daily physical activity, sedentary time, and sleep characteristics) during the first 18 weeks, as well as for an additional 12 weeks, each participant in the project had their data from the past week sent to the Integrative Medicine Clinic for interpretation and feedback, after having uploaded the activity tracker's data to the website. The feedback, which was inherent to sedentary time, physical activity, and sleep characteristics, listed their positive and negative points and how the latter needed to be improved in order to receive feedback concerning self-interpretation. With a fortnight frequency, each participant did the same, with feedback concerning the qualitative characteristics of their nutrition and receiving the feedback about their appropriateness. During the whole experimental period, each participant was followed by a different researcher in each field, remaining the same until the end of the 30-week period. Each researcher followed a maximum of 30 persons. In each field, participants were randomly assigned to one of two available researchers during the recruitment process (in an alternate manner). In each field, researchers had the same cultural background and formation, and to properly set their own work, they had the possibility to see the interaction of the assigned participant with the other researcher.

All selected participants started the project in the same day, 3 February 2020, and finished the first phase on 8 June 2020. Due to the COVID-19 emergency, which was characterized by a national quarantine in Italy from 8 March 2020 to 4 May 2020, the first phase of observation period had the following characteristics: (i) From 3 February 2020 to 8 March 2020 the lifestyles of participants were not influenced by government restrictions; (ii) from 8 March 2020 to 4 May 2020 the "stay-at-home" measures, which were nationally adopted to counteract the spread of the virus, dramatically influenced the lifestyles of participants, dramatically restricting the possibility to go outside the home, work, and provide for primary necessities; (iii) from 4 May 2020 to 1 June 2020, due to the progressive reduction of the "stay-at-home" measures, participants had the possibility to progressively recover normal habits and movements outside the home. Indeed, all the shops were reopened on 18 May 2020, and all sports centres and gyms were reopened on 25 May 2020.

2.3. Recording and Control of Daily Physical Activity, Sedentary, and Sleep Time

To participate in the "Angel Project", participants, after medical examinations including points 1 and 2 of the Integrative Medicine Clinic procedures (Figure S1), were requested to buy a scientifically validated commercial triaxial accelerometer, the Polar Loop 2 (Kempele, Finland) [30–32], to have a personal device to be continuously followed for 30 weeks and to continue to use it also after the end of their participation in the project to control and improve proper daily physical activity, sedentary and sleep characteristics. After having bought the device, an in-person appointment with the assigned researcher (i.e., a sport science specialist well-versed in physical exercise for breast cancer survivors, with more than 5 years of experience in the field of female physical activity analysis and counselling) was scheduled to explain the functioning of both the device and its connected webpage, as well as to create a personal account on it. At the end of each week, each person uploaded weekly data from the device to the webpage in order to give the assigned researcher an opportunity to analyse data and furnish, within 24 h, personalised feedback, including focus on both the positive and negative points and on and how the latter needed to be improved. The day of the data upload, before uploading, each participant recorded their morning body weight in light clothing immediately after waking up in a fasting condition and after voiding and reported it on the website. After the first period, through a new in-person appointment, each researcher furnished the operative instructions for the next period to the assigned participants. Each participant wore the device for the whole day, on the non-dominant wrist and in an adherent way. The webpage integrated the information gathered by the three-axis accelerometer with gender, age, stature, weight, and handedness of the user. As a result, qualitative, quantitative, and distributive information about both daily physical activity and sleep were obtained [30–32]. From the recorded data, we focused our attention on time spent in sedentary activities, and in light-, moderate-, and vigorous-intensity physical activities. Sedentary activities relate to those activities requiring an engagement $\leq$1.5 METs while in a sitting, reclining or lying posture and awake [33]. Light-intensity physical activities relate to those activities requiring a metabolic engagement >1.5 METs and <3 METs. Moderate-intensity physical activities relate to those activities requiring metabolic engagement $\geq$3 METs and $\leq$6 METs, while vigorous-intensity physical activities relate to those requiring a metabolic engagement >6 METs and $\leq$9 METs [34]. The device, combined with the webpage, furnished information about sleep characteristics. Daily nap periods were considered sedentary time, while nocturnal sleeping results are not discussed in this manuscript.

2.4. Dietary Habits

According to the results of basal evaluations and for the first 18 weeks, each participant received online personalized qualitative nutritional suggestions, also taking into account the symptoms and habits of the past weeks, according to the following subsequent principles: Support of organ functions, reduction of proinflammatory nutrients, reduction of nutrients eliciting increases in insulin and growth factors, and increase of nutrients stimulating the immune system. The counsellor was a nutritionist with more than 5 years of experience in the field of nutrition for breast cancer survivors. Feedback was received every 2 weeks. With a fortnight frequency, after the first 18 weeks, for 12 weeks, each participant sent information concerning the qualitative characteristics of proper nutrition and receiving the feedback about their appropriateness.

2.5. Live Online Physical Exercise Sessions

Live online physical exercise sessions were offered to project participants three times a week. Each workout session lasted 50 min, was conducted on the same days and hours, and was composed of a maximum of 10 women to allow exercise supervision. Twice a week, the workout included 10 min of a standing analytic warm-up, 25 min of circuit training (including two sets of seven standing and three lying down adapted calisthenic exercises), and 15 min of stretching and relaxation executed in a lying position. Once a week, the middle workout session included 10 min of a standing analytic warm-up, 25 min of standing aerobic-based exercise, and 15 min of stretching and relaxation executed in a lying position. The intensities of both calisthenics and aerobic-based exercises were assigned and controlled through the Borg 15-point RPE scale [35]. In both cases, the assigned intensity was 12–13 of the used scale. Each researcher recorded the attendance of each participant at the end of each workout. Each live online physical exercise session was conducted by a sports science specialist well-versed in physical exercise for breast cancer survivors, with more than 5 years of experience in this specific field.

2.6. Statistical Analysis

From 3 February 2020 to 1 June 2020, we obtained 17 weeks of continuous valid data for all of the 51 women participating in the project. The recorded weeks were gathered in four periods: 1st period (i.e., the period not including government restrictions), from 3 February 2020 to 8 March 2020; 2nd period (i.e., the first month of government restrictions), from 11 March 2020 to 7 April 2020; 3rd period (i.e., the second month of government restrictions), from 8 April 2020 to 3 May 2020; and 4th period (i.e., the first month of progressive reduction of government restrictions), from 4 May 2020 to 1 June 2020. Among the 51 recruited women, the 24 women who were able to participate in the two live online exercise sessions were placed in the E^{+} group, receiving both workouts and weekly personal counselling concerning sedentary time and physical activity. The 27 women wanting to participate but not having the ability to attend the two live online exercise sessions due to time and/or day incompatibility were placed in the E^{-} group, receiving just weekly personal advice.

The analysis of variance and chi-square test were used to verify whether subsamples differed for age, time from surgery, chemotherapy (y/n), radiation therapy (y/n), and pharmacological treatments ancillary to hormonal therapy. Basal differences of time spent in sedentary, light, moderate, and vigorous physical activities, and their variations, as well as those of body weight, according to the four periods of the study, were evaluated with linear mixed models (LMMs). As both sedentary time and physical activity could vary across time and persons, we assessed the differences among the tow exercise interventions analyzing the 1st period, as the basal (run in time), with LMMs. Mixed models increase the repeated measures precision of the estimate and provide easier handling of missing data compared to those with the ANOVA statistic. When applicable, each table contains the chosen LMM estimates and parameters, which are described at their bottom. Data, when applicable, are presented as means $\pm$ standard deviations; $p \leq 0.05$ was considered significant. Data were analysed using the SAS 9.4 software (SAS Institute Inc., Cary, NC, USA).

3. Results

3.1. Basal Characteristics of the Sample

Table 1 shows basal characteristics of the sample, including breast cancer survivors with or without chemotherapy, radiation therapy, and pharmacological therapy to lower blood pressure and plasma lipids, in addition to hormone therapy, which are present in each person in different combinations. Table 1 also shows if subsamples (i.e., the E^- and E^+ groups) differ in the reported characteristics; no differences were found, except for time spent in light and vigorous-intensity physical activities: E+ spent low time in light-intensity and more time in vigorous-intensity physical activities than E-. Adherence to exercise sessions in the E^+ group was 94.37 ± 5.23%. Adherence of all participants with regard to the uploading of data was 100%.

Table 1. Basal characteristics of the sample and basal differences among subsamples.

	N = 51	E^- (*n* = 27)	E^+ (*n* = 24)	E^- vs. E^+ *p*
Age (years)	50.98 ± 6.17	50.62 ± 3.71	51.37 ± 8.18	0.67
Time from surgery (months)	13.68 ± 7.03	14.14 ± 6.72	13.16 ± 7.46	0.62
Chemotherapy (y/n)	22/29	12/15	10/14	0.49
Radiation therapy (y/n)	33/18	18/9	15/9	0.28
Blood pressure-lowering drugs (y/n)	37/14	6/21	8/16	0.28
Lipid-lowering drugs (y/n)	29/22	9/18	7/17	0.49
Sedentary time (min)	457.62 ± 101.36	465.67 ± 97.36	448.56 ± 105.36	0.53
Light-intensity physical activities (min)	327.42 ± 90.00	351.77 ± 87.77	300.01 ± 84.77	0.02
Moderate-intensity physical activities (min)	62.15 ± 41.96	60.53 ± 37.96	63.98 ± 46.15	0.81
Vigorous-intensity physical activities (min)	10.51 ± 16.81	6.08 ± 10.78	15.48 ± 20.62	0.04

Statistical significances concerning age and time from surgery are inherent to the analysis of variance. Statistical significances concerning chemotherapy, radiation therapy, blood pressure-lowering drugs, and lipid-lowering drugs are inherent to the Chi-square test, while statistical significances concerning other variables are inherent to the LMMs.

3.2. Sedentary Time

Table 2 shows results concerning statistical analysis on daily sedentary time. Model A, the unconditional means model, showed that the total daily sedentary time was on average 468.25 ± 14.64 min, and the amount of variance within each person over time was 3547.98 ± 170.41 min ($\delta^2{}_e$), whereas the amount of variation between participants, regardless of time, was 10,730.01 ± 2185.46 min ($\delta^2{}_0$). The unconditional growth model, using the 2nd period as a reference (i.e., the first 4 weeks of government restrictions), showed that, in the 1st, 3rd, and 4th periods, participants spent less time in sedentary activities. Of the total variance, pseudo-R^2 ($R^2{}_{y,y1}$) demonstrated that 2% could be attributable to the different periods of the study. The personal level covariate model showed that the E^- group spent more time in sedentary activities (49.56 ± 28.39 min) than the E^+ group (i.e., it underlines the presence of an additive effect). Taking into account the interaction between exercise intervention and the four periods, statistical analysis did not show a multiplicative effect. Descripting the result, the E^- group, compared to the E^+ group, increased its sedentary time from the 1st to the 2nd periods, while the same behaviour was shown for both subsamples during the 3rd and 4th periods (Figure S2). Of the total variance, pseudo-R^2 ($R^2{}_{y,y1}$) demonstrated that 6% could be attributable to the interaction between time and intervention.

Table 2. Mixed model analyses of variations in sedentary time.

			Model A Unconditional Means Model	Model B Unconditional Growth Model	Model C Personal Level Covariate
Initial status	Intercept	γ_{00}	468.25 ± 14.64 ***	482.50 ± 14.31 ***	456.18 ± 20.66 ***
	Intervention	γ_{01}			49.56 ± 28.39 *
Rate of change	Intercept (time)	γ_{10}-1		−24.90 ± 5.60 ***	−8.18 ± 8.03
		γ_{10}-2		Reference	Reference
		γ_{10}-3		−10.92 ± 5.60 *	−17.88 ± 8.05 *
		γ_{10}-4		−26.59 ± 7.56 ***	−35.05 ± 10.78 **
Interaction	Time * intervention				
		γ_{11}-1			−31.53 ± 11.04 **
		γ_{11}-2			Reference
		γ_{11}-3			13.09 ± 11.06
		γ_{11}-4			15.89 ± 14.82
Level 1	Within-person	δ^2_e	3547.98 ± 170.41 ***	2417.51 ± 119.84 ***	2415.17 ± 119.91 ***
Level 2	In initial status	δ^2_0	10,730.01 ± 2185.46 ***	9582.94 ± 2015.14 ***	9730.47 ± 2064.95 ***
	In rate of change	δ^2_1		35.68 ± 8.09 **	31.89 ± 7.39 ***
	Covariance	δ_{01}		−81.03 ± 91.87	−99.58 ± 89.64
		ρ	0.75		
		$R^2_{y,y1}$		0.02	0.06
		R^2_e			0.30
		R^2_0			0.01
		R^2_1			0.13
		AIC	10,309	10,063	10,025
		BIC	10,313	10,071	10,032

Note: * $0.05 < p < 0.01$, ** $0.01 \leq p < 0.001$, *** $p \leq 0.001$. Abbreviations: γ_{00} = intercept of the average trajectory; γ_{01} = intercept of the intervention trajectory; γ_{10}-1 = intercept time effect of the trajectory for the run-in period; γ_{10}-2 = reference time, 2nd period; γ_{10}-3 = intercept of the trajectory for the 3rd period; γ_{10}-4 = intercept of the trajectory for the 4th period; γ_{11}-1 = slope of the trajectory for the interaction between intervention and run-in phase/period; γ_{11}-2 = reference; γ_{11}-3 = slope of the trajectory for the interaction between intervention and 3rd phase/period; γ_{11}-4 = slope of the trajectory for the interaction between intervention and 4th phase/period; δ^2_e = within-person variance components; δ^2_0 = in initial status variance components; δ^2_1 = in rate of change variance components; δ_{01} = covariance estimate; ρ = intraclass coefficient correlation; $R^2_{y,y1}$ = percentage of total variability associated linearly with time; R^2_e = pseudo-R^2 statistic assesses the proportion of within-person variation "explained by time"; R^2_0 = pseudo-R^2 statistic assesses the percentage variation in initial status; R^2_1 = pseudo-R^2 statistic assesses the percentage variation in rate of change; AIC = Akaike information criterion; BIC = Bayesian information criterion.

3.3. Time Spent in Light- to Vigorous-Intensity Physical Activities

Table 3 shows results concerning the statistical analysis on time spent practicing light- to vigorous-intensity physical activities. Model A, the unconditional means model, showed that daily time spent practicing light-intensity physical activities was on average 371.32 ± 11.98 min and the amount of variance within each person over time was 4411.31 ± 211.87 min (δ^2_e), whereas the amount of variation between participants, regardless of time, was 7071.39 ± 1463.34 min (δ^2_0). The unconditional growth model, using the 2nd period as a reference (i.e., the first 4 weeks of government restrictions), showed that, in the 1st, 3rd, and 4th periods, participants spent more time practicing light- to vigorous-intensity physical activities. Of the total variance, pseudo-R^2 ($R^2_{y,y1}$) demonstrated that 3% could be attributable to the different periods of the study. The personal level covariate model, considering the effect of intervention and the interaction with time, showed that the E^- group reduced its light- to vigorous-intensity physical activities from the 1st to the 4th periods (Figure S3). Of the total variance, pseudo-R^2 ($R^2_{y,y1}$) demonstrated that 7% could be attributable to the interaction between time and intervention.

Table 3. Mixed model analyses of variation of light- to vigorous-intensity physical activities.

			Model A Unconditional Means Model	Model B Unconditional Growth Model	Model C Personal Level Covariate
Initial status	Intercept	γ_{00}	371.32 ± 11.98 ***	341.75 ± 12.23 ***	338.78 ± 17.97 ***
	Intervention	γ_{01}			5.86 ± 24.70
Rate of change	Intercept (time)	γ_{10}-1		54.13 ± 5.90 ***	33.30 ± 11.63 **
		γ_{10}-2		Reference	Reference
		γ_{10}-3		24.69 ± 5.88 ***	−16.89 ± 11.61
		γ_{10}-4		54.78 ± 8.02 ***	−35.86 ± 15.73 *
Interaction	Time * intervention				
		γ_{11}-1			19.67 ± 9.53 *
		γ_{11}-2			Reference
		γ_{11}-3			5.27 ± 9.94
		γ_{11}-4			47.55 ± 11.76 ***
Level 1	Within-person	δ^2_e	4411.31 ± 211.87 ***	2568.43 ± 127.38 ***	2561.44 ± 127.20 ***
Level 2	In initial status	δ^2_0	7071.39 ± 1463.34 ***	10,265 ± 2159.66 ***	9984.46 ± 2120.10 ***
	In rate of change	δ^2_1		43.94 ± 9.85 ***	39.55 ± 9.01 ***
	Covariance	δ_{01}		−368.56 ± 118.64 **	−325.05 ± 110.97 **
		ρ	0.64		
		$R^2_{y,y1}$		0.03	0.07
		R^2_e			0.37
		R^2_0			0.04
		R^2_1			0.10
		AIC	10,478	10,100	10,069
		BIC	10,482	10,115	10,077

Note: * $0.05 < p < 0.01$, ** $0.01 \leq p < 0.001$, *** $p \leq 0.001$. Abbreviations: γ_{00} = intercept of the average trajectory; γ_{01} = intercept of the intervention trajectory; γ_{10}-1 = intercept time effect of the trajectory for the run-in period; γ_{10}-2 = reference time, 2nd period; γ_{10}-3 = intercept of the trajectory for the 3rd period; γ_{10}-4 = intercept of the trajectory for the 4th period; γ_{11}-1 = slope of the trajectory for the interaction between intervention and run-in phase/period; γ_{11}-2 = reference; γ_{11}-3 = slope of the trajectory for the interaction between intervention and 3rd phase/period; γ_{11}-4 = slope of the trajectory for the interaction between intervention and 4th phase/period; δ^2_e = within-person variance components; δ^2_0 = in initial status variance components; δ^2_1 = in rate of change variance components; δ_{01} = covariance estimate; ρ = intraclass coefficient correlation; $R^2_{y,y1}$ = percentage of total variability associated linearly with time; R^2_e = pseudo-R^2 statistic assesses the proportion of within-person variation "explained by time"; R^2_0 = pseudo-R^2 statistic assesses the percentage variation in initial status; R^2_1 = pseudo-R^2 statistic assesses the percentage variation in rate of change; AIC = Akaike information criterion; BIC = Bayesian information criterion.

3.4. Time Spent in Light-Intensity Physical Activities

Table 4 shows results concerning statistical analysis on time spent practicing light-intensity physical activities. Model A, the unconditional means model, showed that daily time spent practicing light-intensity physical activities was on average 310.27 ± 9.39 min and the amount of variance within each person over time was 2528.76 ± 121.45 min (δ^2_e), whereas the amount of variation between participants, regardless of time, was 4355.68 ± 899.26 min (δ^2_0). The unconditional growth model, using the 2nd period as a reference (i.e., the first 4 weeks of government restrictions), showed that, in the 1st, 3rd, and 4th periods, participants spent more time practicing light-intensity physical activities. Of the total variance, pseudo-R^2 ($R^2_{y,y1}$) demonstrated that 3% could be attributable to the different periods of the study. The personal level covariate model, considering the effect of intervention and the interaction with time, shows that the E^- group reduced its light-intensity physical activity time from the 1st to the 4th periods (Figure S4). Of the total variance, pseudo-R^2 ($R^2_{y,y1}$) demonstrated that 7% could be attributable to the interaction between time and intervention.

Table 4. Mixed model analyses of variations in light-intensity physical activities.

			Model A Unconditional Means Model	Model B Unconditional Growth Model	Model C Personal Level Covariate
Initial status	Intercept	γ_{00}	310.27 ± 9.39 ***	292.06 ± 9.64 ***	279.85 ± 13.97 ***
	Intervention	γ_{01}			23.21 ± 19.20
Rate of change	Intercept (time)	γ_{10}-1		34.70 ± 4.58 ***	20.05 ± 6.51 **
		γ_{10}-2		Reference	Reference
		γ_{10}-3		13.72 ± 4.61 **	20.63 ± 6.58 **
		γ_{10}-4		31.81 ± 6.10 ***	45.81 ± 8.61 ***
Interaction	Time * intervention				
		γ_{11}-1			27.76 ± 8.95 **
		γ_{11}-2			Reference
		γ_{11}-3			−13.11 ± 9.04
		γ_{11}-4			−26.57 ± 11.83 *
Level 1	Within-person	δ^2_e	2528.76 ± 121.45 ***	1740.58 ± 86.28 ***	1736.73 ± 86.20 ***
Level 2	In initial status	δ^2_0	4355.68 ± 899.26 ***	6965.27 ± 1460.77 ***	6331.01 ± 1345.51 ***
	In rate of change	δ^2_1		19.84 ± 4.67 ***	16.82 ± 4.08 ***
	Covariance	δ_{01}		−235.36 ± 69.32 ***	−188.39 ± 60.84 **
		ρ	0.63		
		$R^2_{y,y1}$		0.03	0.07
		R^2_e			0.28
		R^2_0			0.11
		R^2_1			0.17
		AIC	9971	9727	9688
		BIC	9975	9735	9696

Note: * $0.05 < p < 0.01$, ** $0.01 \leq p < 0.001$, *** $p \leq 0.001$. Abbreviations: γ_{00} = intercept of the average trajectory; γ_{01} = intercept of the intervention trajectory; γ_{10}-1 = intercept time effect of the trajectory for the run-in period; γ_{10}-2 = reference time, 2nd period; γ_{10}-3 = intercept of the trajectory for the 3rd period; γ_{10}-4 = intercept of the trajectory for the 4th period; γ_{11}-1 = slope of the trajectory for the interaction between intervention and run-in phase/period; γ_{11}-2 = reference; γ_{11}-3 = slope of the trajectory for the interaction between intervention and 3rd phase/period; γ_{11}-4 = slope of the trajectory for the interaction between intervention and 4th phase/period; δ^2_e = within-person variance components; δ^2_0 = in initial status variance components; δ^2_1 = in rate of change variance components; δ_{01} = covariance estimate; ρ = intraclass coefficient correlation; $R^2_{y,y1}$ = percentage of total variability associated linearly with time; R^2_e = pseudo-R^2 statistic assesses the proportion of within-person variation "explained by time"; R^2_0 = pseudo-R^2 statistic assesses the percentage variation in initial status; R^2_1 = pseudo-R^2 statistic assesses the percentage variation in rate of change; AIC = Akaike information criterion; BIC = Bayesian information criterion.

3.5. Time Spent in Moderate-Intensity Physical Activities

Table 5 shows results concerning the statistical analysis on time spent practicing moderate-intensity physical activities. Model A, the unconditional means model, showed that daily time spent practicing moderate-intensity physical activities was on average 53.09 ± 5.98 min and the amount of variance within each person over time was 716.61 ± 34.42 min (δ^2_e), whereas the amount of variation between participants, regardless of time, was 1781.34 ± 364.24 min (δ^2_0). The unconditional growth model, using the 2nd period as a reference (i.e., the first 4 weeks of government restrictions), showed that, in the 1st, 3rd, and 4th periods, participants spent more time practicing moderate-intensity physical activities. Of the total variance, pseudo-R^2 ($R^2_{y,y1}$) demonstrated that 3% could be attributable to the different periods of the study. Taking into account the interaction between exercise intervention and the four periods, statistical analysis did not show a multiplicative but just a time effect. Describing the result, the E^- group, compared to the E^+ group, reduced its moderate-intensity physical activities from the 1st to the 2nd periods, while the same behaviour was shown for both subsamples during the 3rd and 4th periods (Figure S5). Of the total variance, pseudo-R^2 ($R^2_{y,y1}$) demonstrated that 4% could be attributable to the interaction between time and intervention.

Table 5. Mixed model analyses of variations in moderate-intensity physical activities.

			Model A Unconditional Means Model	**Model B Unconditional Growth Model**	**Model C Personal Level Covariate**
Initial status	Intercept	γ_{00}	53.09 ± 5.98 ***	47.99 ± 5.36 ***	55.43 ± 7.84 ***
	Intervention	γ_{01}			−14.28 ± 10.77
Rate of change	Intercept (time)	γ_{10}-1		14.74 ± 2.38 ***	8.99 ± 3.45 *
		γ_{10}-2		Reference	Reference
		γ_{10}-3		9.62 ± 2.33 ***	11.55 ± 3.38 ***
		γ_{10}-4		20.50 ± 3.31 ***	23.29 ± 4.80 ***
Interaction	Time * intervention				
		γ_{11}-1			10.91 ± 4.74 *
		γ_{11}-2			Reference
		γ_{11}-3			−3.71 ± 4.65
		γ_{11}-4			−5.39 ± 6.59
Level 1	Within-person	δ^2_e	716.61 ± 34.42 ***	345.29 ± 17.14 ***	344.68 ± 17.13 ***
Level 2	In initial status	δ^2_0	1781.34 ± 364.24 ***	1200.33 ± 254.36 ***	1223.10 ± 261.47 ***
	In rate of change	δ^2_1		11.36 ± 2.43 ***	11.09 ± 2.38 ***
	Covariance	δ_{01}		−12.05 ± 17.93	−13.31 ± 17.99
		P	0.71		
		$R^2_{y,y1}$		0.03	0.04
		R^2_e			0.48
		R^2_0			0.03
		R^2_1			0.01
		AIC	8833	8316	8289
		BIC	8837	8323	8297

Note: * $0.05 < p < 0.01$, *** $p \leq 0.001$. Abbreviations: γ_{00} = intercept of the average trajectory; γ_{01} = intercept of the intervention trajectory; γ_{10}-1 = intercept time effect of the trajectory for the run-in period; γ_{10}-2 = reference time, 2nd period; γ_{10}-3 = intercept of the trajectory for the 3rd period; γ_{10}-4 = intercept of the trajectory for the 4th period; γ_{11}-1 = slope of the trajectory for the interaction between intervention and run-in phase/period; γ_{11}-2 = reference; γ_{11}-3 = slope of the trajectory for the interaction between intervention and 3rd phase/period; γ_{11}-4 = slope of the trajectory for the interaction between intervention and 4th phase/period; δ^2_e = within-person variance components; δ^2_0 = in initial status variance components; δ^2_1 = in rate of change variance components; δ_{01} = covariance estimate; ρ = intraclass coefficient correlation; $R^2_{y,y1}$ = percentage of total variability associated linearly with time; R^2_e = pseudo-R^2 statistic assesses the proportion of within-person variation "explained by time"; R^2_0 = pseudo-R^2 statistic assesses the percentage variation in initial status; R^2_1 = pseudo-R^2 statistic assesses the percentage variation in rate of change; AIC = Akaike information criterion; BIC = Bayesian information criterion.

3.6. Time Spent in Vigorous-Intensity Physical Activities

Table 6 shows results concerning the statistical analysis on time spent practicing vigorous-intensity physical activities. Model A, the unconditional means model, showed that daily time spent practicing vigorous-intensity physical activities was on average 7.96 ± 1.34 min and the amount of variance within each person over time was 93.03 ± 4.46 min (δ^2_e), whereas the amount of variation between participants, regardless of time, was 86.49 ± 18.33 min (δ^2_0). The unconditional growth model, using the 2nd period as a reference (i.e., the first 4 weeks of government restrictions), showed that, in the 1st, 3rd, and 4th periods, participants spent more time practicing vigorous-intensity physical activities. Of the total variance, pseudo-R^2 ($R^2_{y,y1}$) demonstrated that 2% could be attributable to the different periods of the study. Taking into account the interaction between exercise intervention and the four periods, statistical analysis did not show a multiplicative but just a time effect. Describing the result, the E^- group, compared to the E^+ group, reduced its vigorous-intensity physical activities from the 1st to the 2nd periods, while the same behaviour was shown for both subsamples during the 3rd and 4th periods (Figure S6). Of the total variance, pseudo-R^2 ($R^2_{y,y1}$) demonstrated that 6% could be attributable to the interaction between time and intervention.

Table 6. Mixed model analyses of variations in vigorous-intensity physical activities.

			Model A Unconditional Means Model	Model B Unconditional Growth Model	Model C Personal Level Covariate
Initial status	Intercept	γ_{00}	7.96 ± 1.34 ***	5.80 ± 1.40 ***	7.36 ± 1.98 ***
	Intervention	γ_{01}			−2.79 ± 2.73
Rate of change	Intercept (time)	γ_{10}-1		4.00 ± 0.90 ***	6.98 ± 1.30 *
		γ_{10}-2		Reference	Reference
		γ_{10}-3		1.97 ± 0.91 *	1.87 ± 1.31
		γ_{10}-4		3.53 ± 1.20 **	5.37 ± 1.73 **
Interaction	Time * intervention				
		γ_{11}-1			5.62 ± 1.79 **
		γ_{11}-2			Reference
		γ_{11}-3			0.17 ± 1.80
		γ_{11}-4			−3.50 ± 2.38
Level 1	Within-person	δ^2_e	93.03 ± 4.46 ***	69.34 ± 3.44 ***	68.21 ± 3.39 ***
Level 2	In initial status	δ^2_0	86.49 ± 18.33 ***	142.82 ± 31.53 ***	128.28 ± 28.90 **
	In rate of change	δ^2_1		0.73 ± 0.17 ***	0.71 ± 0.17 ***
	Covariance	δ_{01}		−6.36 ± 1.99 **	−5.65 ± 1.87
		ρ	0.48		
		$R^2_{y,y1}$		0.02	0.06
		R^2_e			0.24
		R^2_0			0.11
		R^2_1			0.03
		AIC	6913	6747	6713
		BIC	6917	6755	6721

Note: * $0.05 < p < 0.01$, ** $0.01 \leq p < 0.001$, *** $p \leq 0.001$. Abbreviations: γ_{00} = intercept of the average trajectory; γ_{01} = intercept of the intervention trajectory; γ_{10}-1 = intercept time effect of the trajectory for the run-in period; γ_{10}-2 = reference time, 2nd period; γ_{10}-3 = intercept of the trajectory for the 3rd period; γ_{10}-4 = intercept of the trajectory for the 4th period; γ_{11}-1 = slope of the trajectory for the interaction between intervention and run-in phase/period; γ_{11}-2 = reference; γ_{11}-3 = slope of the trajectory for the interaction between intervention and 3rd phase/period; γ_{11}-4 = slope of the trajectory for the interaction between intervention and 4th phase/period; δ^2_e = within-person variance components; δ^2_0 = in initial status variance components; δ^2_1 = in rate of change variance components; δ_{01} = covariance estimate; ρ = intraclass coefficient correlation; $R^2_{y,y1}$ = percentage of total variability associated linearly with time; R^2_e = pseudo-R^2 statistic assesses the proportion of within-person variation "explained by time"; R^2_0 = pseudo-R^2 statistic assesses the percentage variation in initial status; R^2_1 = pseudo-R^2 statistic assesses the percentage variation in rate of change; AIC = Akaike information criterion; BIC = Bayesian information criterion.

3.7. Body Weight

When the same statistical analysis was repeated for body weight, no significant effects were shown for intervention and for its interaction with time. On the contrary, a significant time effect was shown ($F_{(3,147)} = 27.62$; $p < 0.001$) (Figure 1). Specifically, using the 2nd period as a reference (i.e., the first 4 weeks of government restrictions), our data showed that body weight significantly increased from the 1st (Est. = −0.33; St. Error = 0.09; df = 150; t = −3.72; $p < 0.001$) to 3rd (Est. = 0.25; St. Error = 0.08; df = 150; t = 2.87; $p < 0.004$) periods and then declined in the 4th period (i.e., the first month of progressive reduction of government restrictions) (Est. = −0.28; St. Error = 0.12; df = 150; t = −2.25; $p = 0.02$).

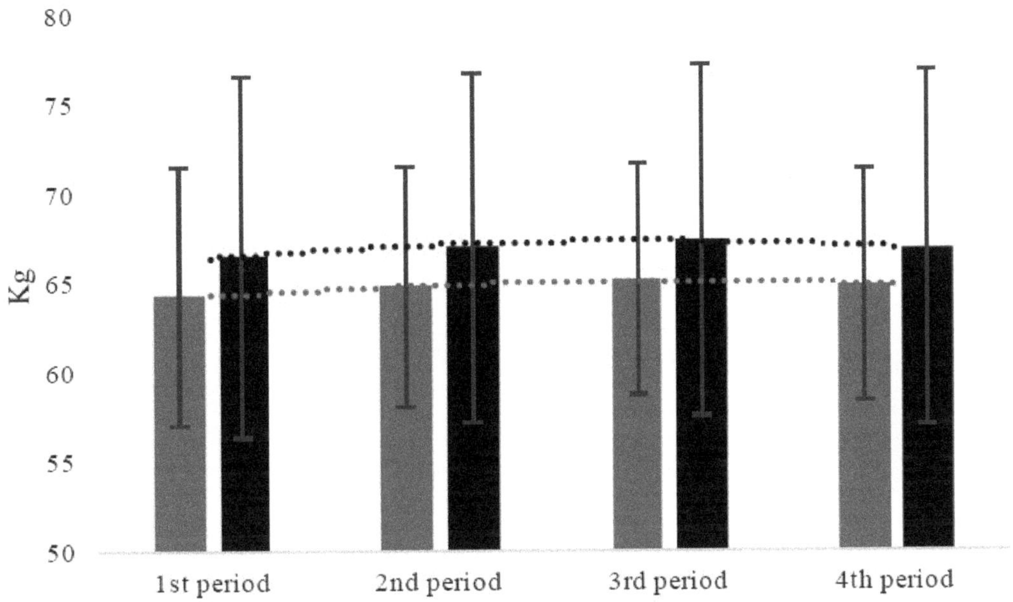

Figure 1. Variation of body weight according to the time of the study and type of intervention, with just a significant time effect, according to the linear mixed models analysis. Note: Grey columns, non-exercising women; black columns, exercising women; 1st period (i.e., the period not including government restrictions), from 3 February 2020 to 8 March 2020; 2nd period (i.e., the first month of government restrictions), from 11 March 2020 to 7 April 2020; 3rd period (i.e., the second month of government restrictions), from 8 April 2020 to 3 May 2020; 4th period (i.e., the first month of progressive reduction of government restrictions), from 4 May 2020 to 1 June 2020.

4. Discussion

Concerning sedentary time and physical activity, the important result of our study is that, notwithstanding the presence of weekly personal advice and the use of technology, the absence of participation in a supervised physical exercise programme did not allow for effective improvement in the daily movement of breast cancer survivors in the presence of a situation simultaneously limiting geographic mobility and socialization, such as the COVID-19 emergency. Our results also showed that the use of technology allows the promotion of "self-knowledge, and health, through numbers" [19], as each participant had the possibility, to see, through the webpage, proper sedentary time and physical activity trends; through the activity tracker, the instantaneous amount of daily physical activity; and had the possibility to be alerted when 1 h of continuous sedentary time was reached. These results could be explained through those of Wu et al. [18], even though their study was not conducted in a situation such as that created by the COVID-19 emergency. Indeed, even though their participants reported that the activity tracker and its accompanying app functioned as a motivational tool and created more awareness of physical activity and sedentary behaviour, they underlined that tailored and personalized advice were particularly important, as well as the role of the physiotherapist giving them every 2 weeks. Indeed, all the women had undergone cancer treatment and experienced disease- and treatment-related side effects, as well as fatigue and a decreased physical fitness level, and according to participants, it was sometimes frustrating to read messages generated by a system that did not consider the side effects of the treatment. Translating these results in our case, including women participating in online supervised exercise sessions and women who did not (all receiving personalized advice every week), we probably have a similar situation—the presence of limitations concerning geographic mobility and socialization, creating a negative self-reinforcing loop including fear, anxiety, sedentary time, and treatment-related side effects, such as fatigue and a decreased physical fitness level, does not allow properly applying the tailored advice, stagnating the person in a poor psychophysical condition that is characterized by the increase of sedentary time and the decrease of daily physical activity [12,24,27]. On the contrary, online supervised exercise sessions could be used to safely maintain/improve the daily physical activity level of breast cancer survivors in the presence of personal restrictions, since thanks to its characteristics, physical exercise properly reduces/prevents some negative side effects of treatment, such as fatigue and pain, negatively affecting daily movements, and promotes psychological health, which is undoubtedly linked with physical health and daily physical activity [12,19,36]. Nevertheless, statistical analysis did not show a significant interaction of intervention with time, both for moderate- and vigorous-intensity physical activities, although the E^+ group had a mean adherence to exercise sessions of $94.37 \pm 5.23\%$. This stimulates two hypotheses. The first is that, while the E^+ group partially replaced daily moderate- to vigorous-intensity spontaneous physical activities with live online workouts, being more functional for health, due to their characteristics (i.e., proper total and continuous duration, proper modulation of both intensity and recovery, and proper activity selection), the E^- group did not and tried to maintain its daily routines. The second hypothesis is that the activity tracker simply underestimated some of the physical exercises that were performed on the spot, recognizing it as light-intensity physical activity or sedentary time [37]. Indeed, when we focused the attention on the analysis of the trend of time spent on light- to vigorous-intensity physical activities and on that of light-intensity physical activities, we observed the presence of a multiplicative effect, with the E^+ group showing a better trend. Therefore, it is conceivable to speculate that, in the presence of personal limitations, breast cancer survivors participating in online supervised exercise sessions, physical activity monitoring and counselling programmes, as presented, benefit from a tailored theoretical and practical intervention positively affecting their daily physical activity. On the contrary, when the online supervised exercise sessions are not present, breast cancer survivors benefit from just a theoretical intervention that, according to our results, is not enough to support proper daily physical activity recomposition (i.e., sedentary time reduction with a contemporaneous increase of light-, moderate- and/or vigorous-intensity physical activities). We are also in accordance with Newton et al. [36] reporting that, in the era of COVID-19, it is necessary that exercise oncology programmes adapt to the changing environment, as patients with cancer and survivors risk to regress to a sedentary lifestyle, resulting in a decline of health and their quality of life, particularly those undergoing treatment or suffering adverse effects of treatment. To do this, according to Newton et al. [36], the key elements are online exercise led by an exercise professional that has to be able to create a tailored lesson and a personal interaction. Interpreting our results, the willful condition of breast cancer survivors is particularly important since, according to Gardner et al. [38], if self-control is not diminished, habit formation alone may not be sufficient for the maintenance of behaviour change. The statement by Gardner et al. [38] is also supported by the observations of the unconditional growth models of sedentary time and physical activity variables, notwithstanding the increase of sedentary time and the decrease of light-, moderate-, and vigorous-intensity physical activities from the 1st to 2nd periods. Moving on from the 2nd period, it is possible to observe, in the E^- group, the inversion of the trends reaching the same values of the 1st period (Tables 2–6, Figures S2–S6). Therefore, if habit alone had been sufficient for the maintenance of behaviour change, we would have had to observe a stagnation of the variables and not a return to the starting point.

When body weight was analysed, just a time effect was found, showing its increase from the 1st to the 3rd periods and then its decrease, with a mean variation from one period to another widely lower than 1 kg. The absence of a significant multiplicative effect in body weight, notwithstanding the presence of a significant additive effect in sedentary time and multiplicative effect in both physical activity and physical exercise participation, allows us to hypothesize that the subsamples differently managed quantity of food. Indeed, the presence of different trends in sedentary time and physical activity, with a same trend in body weight, allowed speculating that the E^- group probably adopted a more quantitatively restricted diet to compensate for inadequate daily movements and the inability to better address it, notwithstanding personal advice, without live online exercise sessions. Unfortunately, the absence of objective data concerning the nutritional habits of study participants, including both qualitative and quantitative information, does not allow going beyond this hypothesis, as we furnished fortnight personal qualitative nutritional counselling, without knowing how effectively they applied them and with what quantities.

Study limitations included: (i) The absence of objective data concerning the nutritional habits of study participants, including both qualitative and quantitative information; (ii) with regard to sample characteristics, indeed, we had a sample of breast cancer survivors having the possibility to buy an activity tracker, even if it is a low-cost device, to use it in conjunction with its web app to have an Internet connection and both ability and possibility to routinely connect for workout and/or nutritional suggestions. This implies that our results are not generalizable to the whole population but are applicable just to a similar population, probably with a middle-high socioeconomic status and probably under the "healthy worker effect" [39], as the volition to participate in the project means that participants have the volition to improve their daily physical activity; (iii) the absence of data concerning psychological fields, which certainly could close the circle and better illustrate the trend of all components of behaviour, allowing us to better identify the causes, mediators, and correlates of the trend of each area; (iv) the interaction between on spot exercises, including callisthenic exercises, with just a wrist accelerometer that represents a partial study limitation, as it is possible to obtain an underestimation of the intensity due to body position, notwithstanding that the body was moderately to vigorously engaged. Lastly, another study limitation concerns the low effect size of our results, expressed as ($R^2_{y,y1}$), meaning that the models considered could explain only a little percentage of the total variance. Therefore, the presence of day-to-day data, coming from 17 consecutive weeks, concerning sedentary time and physical activity variables and the COVID-19 emergency represented the strength of the study. Indeed, the latter, through the government restrictions, allowed us to observe the personal response to, until now, a unique situation in the technological era, furnishing the possibility to translate them in a similar population (i.e., cancer survivors requesting the same pharmacological treatments) and in a similar situation (i.e., pandemic emergency and personal restrictions due to immune deficiency not allowing social activities).

5. Conclusions

Our results suggest that the use of an activity tracker, its accompanying app, and the reception of weekly tailored advice concerning the improvement of sedentary time and physical activity are not enough to elicit proper physical activity recomposition in breast cancer survivors in the COVID-19 era. On the contrary, using them in addition to online supervised physical exercise sessions seems able to counteract the negative effects of COVID-19 personal restrictions on sedentary time and physical activity. Therefore, the COVID-19 pandemic emergency and its related government restrictions have been shown to not negatively influence the sedentary time and daily physical activity of breast cancer survivors prone to change, recovering their behaviour when restrictions were reduced. According to our opinion, our results could be translated into situations similar to that of COVID-19, including patients with breast cancer needing particular attention and confinement, in order to optimize health through movement, remembering that, in the field of physical activity, merging technology with the live and tailored approach seems the optimal combination to favor health through movement, as it furnishes the possibility to move without discomfort and, as a consequence, to continue to move, to increase the fitness level and explore new opportunities of movement and health.

Supplementary Materials: The following are available online at https://www.mdpi.com/article/10.3390/jfmk6020050/s1.

Author Contributions: A.D.B., T.M., G.V., M.B. and G.N. conceived and designed the study; A.D.B., T.M., F.L., G.N., S.G. (Simona Grossi), E.C., A.D., L.M., M.R. and M.T.S. conducted the study according to the proper specialization; A.D.B., A.D.I., L.C. and S.G. (Stefano Gobbo) analysed the data and interpreted the results; A.D.B., A.D.I. and A.V. prepared the tables and figures and drafted the manuscript; A.D.B., T.M., F.L., G.V., A.D.I., S.G. (Simona Grossi), E.C., L.C., S.G. (Stefano Gobbo), M.B., A.D., L.M., M.R., M.T.S., A.V. and G.N. edited and revised the manuscript. All authors have read and agreed to the published version of the manuscript.

Funding: The authors report no involvement in the research by the sponsor that could have influenced the outcome of this work.

Institutional Review Board Statement: The study was conducted according to the guidelines of the Declaration of Helsinki, and approved by the Ethics Committee of Chieti-Pescara (# 312/2015).

Informed Consent Statement: Informed consent was obtained from all subjects involved in the study.

Data Availability Statement: The data presented in this study are available on request from the corresponding author.

Conflicts of Interest: The authors declare no conflict of interest.

References

1. Agrawal, S. Late effects of cancer treatment in breast cancer survivors. *S. Asian J. Cancer* **2014**, *3*, 112–115. [CrossRef]
2. Wang, X.S.; Woodruff, J.F. Cancer-related and treatment-related fatigue. *Gynecol. Oncol.* **2015**, *136*, 446–452. [CrossRef]
3. Duregon, F.; Vendramin, B.; Bullo, V.; Gobbo, S.; Cugusi, L.; Di Blasio, A.; Neunhaeuserer, D.; Zaccaria, M.; Bergamin, M.; Ermolao, A. Effects of exercise on cancer patients suffering chemotherapy-induced peripheral neuropathy undergoing treatment: A systematic review. *Crit. Rev. Oncol.* **2018**, *121*, 90–100. [CrossRef]
4. Fassier, P.; Zelek, L.; Partula, V.; Srour, B.; Bachmann, P.; Touillaud, M.; Druesne-Pecollo, N.; Galan, P.; Cohen, P.; Hoarau, H.; et al. Variations of physical activity and sedentary behavior between before and after cancer diagnosis: Results from the prospective population-based NutriNet-Santé cohort. *Medicine* **2016**, *95*, e4629. [CrossRef] [PubMed]
5. De Groef, A.; Geraerts, I.; Demeyer, H.; Van der Gucht, E.; Dams, L.; de Kinkelder, C.; Dukers-van Althuis, S.; Van Kampen, M.; Devoogdt, N. Physical activity levels after treatment for breast cancer: Two-year follow-up. *Breast* **2018**, *40*, 23–28. [CrossRef] [PubMed]
6. Gal, R.; Monninkhof, E.M.; Peeters, P.H.M.; Van Gils, C.H.; Van Den Bongard, D.H.J.G.; Wendel-Vos, G.C.W.; Zuithoff, N.P.A.; Verkooijen, H.M.; May, A.M. Physical activity levels of women with breast cancer during and after treatment, a comparison with the Dutch female population. *Acta Oncol.* **2019**, *58*, 673–681. [CrossRef] [PubMed]
7. Welch, W.A.; Ehlers, D.; Gavin, K.L.; Aguinaga, S.; Cottrell, A.; Nielsen, A.; Solk, P.; McAuley, E.; Phillips, S.M. Effects of reallocating sedentary time with physical activity on quality of life indicators in breast cancer survivors. *Psychooncology* **2019**, *28*, 1430–1437. [CrossRef]
8. Mahumud, R.A.; Alam, K.; Dunn, J.; Gow, J. The burden of chronic diseases among Australian cancer patients: Evidence from a longitudinal exploration, 2007–2017. *PLoS ONE* **2020**, *15*, e0228744. [CrossRef] [PubMed]
9. Campbell, K.L.; Winters-Stone, K.M.; Wiskemann, J.; May, A.M.; Schwartz, A.L.; Courneya, K.S.; Zucker, D.S.; Matthews, C.E.; Ligibel, J.A.; Gerber, L.H.; et al. Exercise guidelines for cancer survivors: Consensus statement from International Multidisciplinary Roundtable. *Med. Sci. Sports Exerc.* **2019**, *51*, 2375–2390. [CrossRef]
10. Di Blasio, A.; Morano, T.; Cianchetti, E.; Gallina, S.; Bucci, I.; Di Santo, S.; Tinari, C.; Di Donato, F.; Izzicupo, P.; Di Baldassarre, A.; et al. Psychophysical health status of breast cancer survivors and effects of 12 weeks of aerobic training. *Complement Ther. Clin. Pract.* **2017**, *27*, 19–26. [CrossRef] [PubMed]
11. Di Blasio, A.; Morano, T.; Napolitano, G.; Bucci, I.; Di Santo, S.; Gallina, S.; Cugusi, L.; Di Donato, F.; D'Arielli, A.; Cianchetti, E. Nordic walking and the Isa method for breast cancer survivors: Effects on upper limb circumferences and total body extracellular water—A pilot study. *Breast Care* **2016**, *11*, 428–431. [CrossRef]
12. Di Blasio, A.; Morano, T.; Bucci, I.; Di Santo, S.; D'Arielli, A.; Castro, C.G.; Cugusi, L.; Cianchetti, E.; Napolitano, G. Physical exercises for breast cancer survivors: Effects of 10 weeks of training on upper limb circumferences. *J. Phys. Ther. Sci.* **2016**, *28*, 2778–2784. [CrossRef]
13. Hartman, S.J.; Nelson, S.H.; Weiner, L.S. Patterns of Fitbit use and activity levels throughout a physical activity intervention: Exploratory analysis from a randomized controlled trial. *JMIR Mhealth Uhealth* **2018**, *6*, e29. [CrossRef]
14. Valle, C.G.; Deal, A.M.; Tate, D.F. Preventing weight gain in African American breast cancer survivors using smart scales and activity trackers: A randomized controlled pilot study. *J. Cancer Surviv.* **2017**, *11*, 133–148. [CrossRef]

15. Lloyd, G.R.; Oza, S.; Kozey-Keadle, S.; Pellegrini, C.A.; Conroy, D.E.; Penedo, F.J.; Spring, B.J.; Phillips, S.M. Breast cancer survivors' beliefs and preferences regarding technology-supported sedentary behavior reduction interventions. *AIMS Public Health* **2016**, *3*, 592–614. [CrossRef] [PubMed]
16. Nguyen, N.H.; Hadgraft, N.T.; Moore, M.M.; Rosenberg, D.E.; Lynch, C.; Reeves, M.M.; Lynch, B.M. A qualitative evaluation of breast cancer survivors' acceptance of and preferences for consumer wearable technology activity trackers. *Support. Care Cancer* **2017**, *25*, 3375–3384. [CrossRef]
17. Rosenberg, D.; Kadokura, E.; Bouldin, E.; Miyawaki, C.; Higano, C.; Hartzler, A. Acceptability of Fitbit for physical activity tracking within clinical care among men with prostate cancer. In Proceedings of the AMIA Annual Symposium, Chicago, IL, USA, 12–16 November 2016; pp. 1050–1059.
18. Wu, H.S.; Gal, R.; van Sleeuwen, N.C.; Brombacher, A.C.; IJsselsteijn, W.A.; May, A.M.; Monninkhof, E.M. Breast cancer survivors' experience with an activity tracker integrated into a supervised exercise program: Qualitative study. *JMIR Mhealth Uhealth* **2020**, *7*, e10820. [CrossRef] [PubMed]
19. Lee, V.R. What's happening in the "quantified self" movement? ICLS 2014 Proceedings. In Proceedings of the International Conference of the Learning Sciences, Boulder, CO, USA, 23–27 June 2014; pp. 1032–1036.
20. Meyer, J.; McDowell, C.; Lansing, J.; Brower, C.; Smith, L.; Tully, M.; Herring, M. Changes in physical activity and sedentary behavior in response to COVID-19 and their associations with mental health in 3052 US adults. *Int. J. Environ. Res. Public Health* **2020**, *17*, 6469. [CrossRef] [PubMed]
21. Castañeda-Babarro, A.; Arbillaga-Etxarri, A.; Gutiérrez-Santamaría, B.; Coca, A. Physical activity change during COVID-19 confinement. *Int. J. Environ. Res. Public Health* **2020**, *17*, 6878. [CrossRef]
22. Ammar, A.; Brach, M.; Trabelsi, K.; Chtourou, H.; Boukhris, O.; Masmoudi, L.; Bouaziz, B.; Bentlage, E.; How, D.; Ahmed, M.; et al. Effects of COVID-19 Home Confinement on Eating Behaviour and Physical Activity: Results of the ECLB-COVID19 International Online Survey. *Nutrients* **2020**, *12*, 1583. [CrossRef]
23. Trabelsi, K.; Ammar, A.; Masmoudi, L.; Boukhris, O.; Chtourou, H.; Bouaziz, B.; Brach, M.; Bentlage, E.; How, D.; Ahmed, M.; et al. Globally altered sleep patterns and physical activity levels by confinement in 5056 individuals: ECLB COVID-19 international online survey. *Biol. Sport* **2021**, *38*, 495–506.
24. Ammar, A.; Trabelsi, K.; Brach, M.; Chtourou, H.; Boukhris, O.; Masmoudi, L.; Bouaziz, B.; Bentlage, E.; How, D.; Ahmed, M.; et al. Effects of home confinement on mental health and lifestyle behaviours during the COVID-19 outbreak: Insight from the ECLB-COVID19 multicenter study. *Biol. Sport* **2021**, *38*, 9–21. [CrossRef] [PubMed]
25. Bentlage, E.; Ammar, A.; How, D.; Ahmed, M.; Trabelsi, K.; Chtourou, H.; Brach, M. Practical Recommendations for Maintaining Active Lifestyle during the COVID-19 Pandemic: A Systematic Literature Review. *Int. J. Environ. Res. Public Health* **2020**, *17*, 6265. [CrossRef]
26. Swainston, J.; Chapman, B.; Grunfeld, E.A.; Derakshan, N. COVID-19 lockdown and its adverse impact on psychological health in breast cancer. *Front. Psychol.* **2020**, *11*, 2033. [CrossRef]
27. Vanni, G.; Materazzo, M.; Pellicciaro, M.; Ingallinella, S.; Rho, M.; Santori, F.; Cotesta, M.; Caspi, J.; Makarova, A.; Pistolese, C.A.; et al. Breast cancer and COVID-19: The effect of fear on patients' decision-making process. *In Vivo* **2020**, *34*, 1651–1659. [CrossRef] [PubMed]
28. Buman, M.P.; Youngstedt, S.D. Physical activity, sleep, and biobehavioral synergies for health. In *Sleep and Affect*; Elsevier: Amsterdam, The Nehterlands, 2015; pp. 321–337.
29. A Creasy, S.; E Crane, T.; O Garcia, D.; A Thomson, C.; Kohler, L.N.; Wertheim, B.C.; Baker, L.D.; Coday, M.; Hale, L.; Womack, C.R.; et al. Higher amounts of sedentary time are associated with short sleep duration and poor sleep quality in postmenopausal women. *Sleep* **2019**, *42*, zsz093. [CrossRef]
30. Hardcastle, S.J.; Galliott, M.; Lynch, B.M.; Nguyen, N.H.; Cohen, P.A.; Mohan, G.R.; Johansen, N.J.; Saunders, C. Acceptability and utility of, and preference for wearable activity tracker amongst non-metropolitan cancer survivors. *PLoS ONE* **2018**, *13*, e0210039. [CrossRef]
31. Wright, S.P.; Brown, T.S.H.; Collier, S.R.; Sandberg, K. How consumer physical activity monitors could transform human physiology research. *Am. J. Physiol. Regul. Integr. Comp. Physiol.* **2017**, *312*, R358–R367. [CrossRef]
32. Brooke, S.M.; An, H.-S.; Kang, S.-K.; Noble, J.M.; Berg, K.E.; Lee, J.-M. Concurrent validity of wearable activity tracker under free-living conditions. *J. Strength Cond. Res.* **2017**, *31*, 1097–1106. [CrossRef] [PubMed]
33. Sedentary Behaviour Research Network. Letter to the Editor: Standardized use of the terms "sedentary" and "sedentary behaviours". *Appl. Physiol. Nutr. Metab.* **2012**, *37*, 540–542. [CrossRef]
34. Haskell, W.L.; Lee, I.M.; Pate, R.R.; Powell, K.E.; Blair, S.N.; Franklin, B.A.; Macera, C.A.; Heath, G.W.; Thompson, P.D.; Bauman, A. Physical activity and public health: Updated recommendation for adults from the American College of Sports Medicine and the American Heart Association. *Med. Sci. Sports Exerc.* **2007**, *39*, 1423–1434. [CrossRef] [PubMed]
35. Borg, G. *Borg's Perceived Exertion and Pain Scales*; Human Kinetics: Champaign, IL, USA, 1988; p. 104.
36. Newton, R.U.; Hart, N.H.; Clay, T. Keeping patients with cancer exercising in the age of COVID-19. *JCO Oncol. Pract.* **2020**, *16*, 656–664. [CrossRef] [PubMed]
37. Morris, C.E.; Wessel, P.A.; Tinius, R.A.; Schafer, M.A.; Maples, J. Validity of activity trackers in estimating energy expenditure during high-intensity functional training. *Res. Q. Exerc. Sport* **2019**, *90*, 377–384. [CrossRef]

38. Gardner, B.; Lally, P.; Rebar, A. Does habit weaken the relationship between intention and behavior? Revisiting the habit-intention interaction hypothesis. *Soc. Personal. Psychol. Compass* **2020**, *14*, e12553. [CrossRef]
39. Shah, D. Healthy worker effect phenomenon. *Indian J. Occup. Environ. Med.* **2009**, *13*, 77–79. [CrossRef]

Journal of
Functional Morphology and Kinesiology

MDPI

Article

Detection of the Anaerobic Threshold in Endurance Sports: Validation of a New Method Using Correlation Properties of Heart Rate Variability

Bruce Rogers [1,*], David Giles [2], Nick Draper [3], Laurent Mourot [4,5] and Thomas Gronwald [6]

1 College of Medicine, University of Central Florida, 6850 Lake Nona Boulevard, Orlando, FL 32827-7408, USA
2 Lattice Training Ltd., Chesterfield S41 9AT, UK; dave@latticetraining.com
3 School of Health Sciences, College of Education, Health and Human Development, University of Canterbury, 8041 Christchurch, New Zealand; nick.draper@canterbury.ac.nz
4 EA3920 Prognostic Factors and Regulatory Factors of Cardiac and Vascular Pathologies, Exercise Performance Health Innovation (EPHI) Platform, University of Bourgogne Franche-Comté, 25000 Besançon, France; laurent.mourot@univ-fcomte.fr
5 Division for Physical Education, National Research Tomsk Polytechnic University, Lenin Ave, 30, 634050 Tomsk Oblast, Russia
6 Department of Performance, Neuroscience, Therapy and Health, Faculty of Health Sciences, MSH Medical School Hamburg, University of Applied Sciences and Medical University, Am Kaiserkai 1, 20457 Hamburg, Germany; thomas.gronwald@medicalschool-hamburg.de
* Correspondence: bjrmd@knights.ucf.edu

Abstract: Past attempts to define an anaerobic threshold (AnT) have relied upon gas exchange kinetics, lactate testing and field-based evaluations. DFA a1, an index of heart rate (HR) variability (HRV) fractal correlation properties, has been shown to decrease with exercise intensity. The intent of this study is to investigate whether the AnT derived from gas exchange is associated with the transition from a correlated to uncorrelated random HRV pattern signified by a DFA a1 value of 0.5. HRV and gas exchange data were obtained from 15 participants during an incremental treadmill run. Comparison of the HR reached at the second ventilatory threshold (VT2) was made to the HR reached at a DFA a1 value of 0.5 (HRVT2). Based on Bland–Altman analysis and linear regression, there was strong agreement between VT2 and HRVT2 measured by HR ($r = 0.78$, $p < 0.001$). Mean VT2 was reached at a HR of 174 (±12) bpm compared to mean HRVT2 at a HR of 171 (±16) bpm. In summary, the HR associated with a DFA a1 value of 0.5 on an incremental treadmill ramp was closely related to that of the HR at the VT2 derived from gas exchange analysis. A distinct numerical value of DFA a1 representing an uncorrelated, random interbeat pattern appears to be associated with the VT2 and shows potential as a noninvasive marker for training intensity distribution and performance status.

Keywords: autonomic nervous system; HRV; intensity distribution; endurance training

Citation: Rogers, B.; Giles, D.; Draper, N.; Mourot, L.; Gronwald, T. Detection of the Anaerobic Threshold in Endurance Sports: Validation of a New Method Using Correlation Properties of Heart Rate Variability. *J. Funct. Morphol. Kinesiol.* **2021**, *6*, 38. https://doi.org/10.3390/jfmk6020038

Academic Editors: Cristina Cortis, Andrea Fusco and Carl Foster

Received: 31 March 2021
Accepted: 23 April 2021
Published: 26 April 2021

1. Introduction

The identification of physiologic indicators representing breakpoints involved in endurance exercise intensity is of vital importance for both performance monitoring and exercise intensity distribution [1,2]. In the classic three-zone model, the intensity boundaries are defined by either certain gas exchange parameters or blood lactate determination [3,4]. The lowest intensity zone is felt to be delimited by the first ventilatory (VT1) or lactate threshold (LT1) and is described as representing an aerobic threshold (AeT). The highest intensity zone, encompassing work rates above the second ventilatory (VT2) or lactate thresholds (LT2), is described as an anaerobic threshold (AnT) and is felt to be unsustainable for long durations [5]. Although the VT2 appears to be characterized as the first systematic increase in the ventilatory equivalent of CO_2 or the first decrease in the expiratory fraction of CO_2 during heavy exercise, the LT2 has had several alternate definitions [4,6]. These include the exercise intensity associated with reaching a fixed lactate of 4 mmol/l, various calculations and exponential computations models and a maximal lactate steady state (MLSS) [4,6]. To complicate matters further, many of these concepts can yield different results depending on incremental ramp progression, stage protocol or even expert visual interpretation [2,6,7]. In addition, both gas exchange and lactate testing require specialized equipment, operators, can be invasive and are costly. An initially promising approach to AnT measurement, the identification of deoxyhemoglobin breakpoints of locomotor muscle [8,9], has unfortunately been hindered by the abandonment of commercial products due to poor financial outcomes. In an effort to identify the AnT noninvasively, field-based tests have been devised with the functional threshold power (FTP) evaluation being a popular example [10]. However, results have been shown to vary depending on warmup protocol and test procedures [11–13]. In addition, the FTP is dependent on motivation, individual pacing strategy [14] and by its definition, physically exhausting. Another maximal effort test, critical power (CP) has also received attention as a means of AnT delineation [15]. However, discrepancy between the CP, FTP, VT2 and MLSS [5,14,16] has been debated and concordance is unclear. Therefore, agreement of performance based tests of the AnT can be variable, and each has a potential detrimental impact on an athlete's training intensity distribution strategy. Given these considerations, a search for alternative, objective, noninvasive methods for determining the AnT seems reasonable.

Previous analysis of the dynamic change in heart rate (HR) variability (HRV) during exercise has also shown potential for demarcation of threshold boundaries, particularly the AeT [17]. The mechanism affecting HRV during exercise is felt to be related to alteration in autonomic balance to the cardiac sinoatrial pacemaker center. As intensity rises, there is a withdrawal of parasympathetic influence and augmentation of sympathetic stimulation [18]. Conventional HRV indexes such as standard deviation of normal-to-normal RR intervals (SDNN), high-frequency (HF) power and standard deviation 1 from Poincaré plot analysis (SD1) can be used to identify the AeT during incremental exercise ramps by observing at what exercise intensity a nadir HRV value occurs [19–21]. Since these particular indexes reach their lowest value at the AeT, they are not felt to be helpful for AnT delineation given the loss of any further dynamic range. In contrast, a nonlinear index of fractal correlation properties called alpha1 of Detrended Fluctuation Analysis (DFA a1) possesses a much wider dynamic range [22,23]. This index is based on both the fractal nature and "correlation pattern" of a series of cardiac beat to beat intervals over time [17]. Fractal behavior in relation to HRV can be described as degrees of self-similarity between RR interval fluctuations over different time scales and allows a distinction of the fractal character of the physiological signal between fractional Brownian motion (fBm: DFA a1 > 1.0) and fractional Gaussian noise (fGn: DFA a1 < 1.0) [24,25]. Analogously, this conception allows the determination of different alterations of complexity in HR time series, either toward rigid order with increasing correlation properties or toward disorder with decreasing correlation properties [26]. At low exercise intensity, DFA a1 values are usually in a well-correlated fractal range near or above 1.0. As intensity rises, DFA a1 passes 0.75 at the AeT [27] and continues to drop through the 0.5 range associated with uncorrelated random behavior of interbeat pattern, finally to drop below 0.5 representing an anticorrelated range at the very highest work rates, which could be seen as a protective feedback and stabilizing mechanism where interactions and/or coordination of subsystems fail before the whole system fails [23,26,28]. Advantages of DFA a1 for intensity monitoring also revolve around its dimensionless nature, which makes calibration to other internal load parameters such as gas exchange or lactate unnecessary. In other words, a HR or VO_2 associated with a DFA a1 value of 0.75 corresponds to an intensity near the AeT/VT1 in a wide spectrum of individuals and was referred to as the HRVT (heart rate variability threshold) [27].

Since DFA a1 still possesses reasonable dynamic range at intensities above the AeT, the question arises whether another numerical threshold exists corresponding to the AnT.

From a mathematical standpoint, a DFA a1 value of 0.5 is of interest as a breakpoint for a high-intensity physiologic process. A value of 0.5 denotes a loss of correlation and fractal patterns in the beat to beat cardiac time series. To better understand the concept of "correlation properties", a comparison to a random walk can be used [25,28,29]. During a hypothetical random walk, at each advancing step, the walker can choose to go either left or right. If the choice the walker makes is not random but based on the pattern (series of left or right choices) of what went beforehand, the pattern is described as being correlated (DFA a1 about 1.0), since the future pattern is based on the past sequence. However, if each new step is taken with equal, random chances of left or right, an "uncorrelated" pattern exists (DFA a1 about 0.5). Additionally, from an observational perspective, DFA a1 values well below 0.5 occur at the termination of maximal exercise ramps, leading to the suspicion that the 0.5 value may have significance as a high-intensity boundary near the AnT [23]. Therefore, the intent of this study is to investigate whether reaching a DFA a1 of 0.5 during an incremental exercise ramp is associated with the VT2, a ventilatory marker of the AnT. Since both artifact correction and recording device bias could be an issue with DFA a1 precision [30], high quality ECG tracings will be utilized for HRV interpretation. In addition, given practical considerations for training prescription in sports and rehabilitation, only HR measurements will be compared given the excellent agreement between the HR and VO_2 relationship during exercise [31].

2. Methods

2.1. Participants

Seventeen male recreational runners aged 19 to 52, without previous medical history, current medications or physical difficulties were tested. Participants were informed of the possible testing risks and institutionally approved consent was given. Approval for the research was granted by the University of Derby, UK (LSREC_1415_02) and followed the principles of the Declaration of Helsinki. Runners did not consume caffeine, alcohol or any stimulant for the 24 h before testing. There was no tobacco usage. Background data for each participant including, age, body weight, and training volume in hours per week are presented in Table 1 and were also published in an earlier work [27]. Testing was performed in the afternoon and at least 3 h after food (with no set diet). No intense activity was performed the day prior to the test. Two participants with excessive cardiac ectopy (frequent atrial premature beats and atrial trigeminy) during testing were excluded from HRV analysis.

Table 1. Age, training volume (TV), bodyweight (BW), maximal oxygen uptake (VO_{2MAX}), Oxynet-derived HR at VT2 and HR at HRVT2 for all participants.

Nr.	Age [yrs]	TV [hrs/wk]	BW [kg]	VO_{2MAX} [mL/kg/min]	VT2 [bpm]	HRVT2 [bpm]
1	19	3–6	82	58	179	180
2	19	3–6	82	57	183	183
3	20	3–6	82	47	194	187
4	22	1–3	73	45	170	188
5	23	>6	77	71	148	160
6	24	3–6	69	64	166	144
7	24	>6	65	54	177	173
8	24	3–6	76	47	182	176
9	25	>6	78	54	169	170
10	26	>6	69	72	192	194
11	30	1–3	92	46	160	143
12	30	>6	73	74	172	161
13	32	1–3	65	49	186	182
14	36	>6	75	57	180	171
15	50	3–6	94	41	159	150
Mean (SD)	27 (±8)	-	77 (±8)	56 (±10)	174 (±12)	171 (±16)

2.2. Exercise Protocol

A motorized treadmill (Woodway, Birmingham, UK) was used for an incremental maximal cardiopulmonary exercise test for all runners. The treadmill was set for the Bruce protocol with increases in speed and inclination from 2.7 km/h at ten percent grade, increasing by 1.3 km/h and two percent grade every 3 min until exhaustion. A fan was used for cooling. Room temperature was approximately 24 °C for all tests.

Gas Exchange Testing and Calculation of the AnTGas exchange kinetics were recorded continuously using a breath to breath metabolic cart (Metalyzer; Cortex Biophysik GmbH, Leipzig, Germany), with a Polar H7 (Polar Electro Oy, Kempele, Finland) wirelessly paired to the Metalyzer cart for the purpose of HR recording concurrent with gas exchange data. Heart rate corresponded to each reported breath to breath data point. Assessment of the VO_2 over time relationship was performed to determine any significant plateau of the VO_2 curve for estimation of VO_2 linearity. VT2 associated HR was determined by Oxynet [32,33], a convolutional neural network previously shown to have excellent agreement (average mean absolute error = 6.1%, $r = 0.99$) with manually derived results especially in individuals with medium to high aerobic fitness levels. Raw gas exchange data were uploaded to the Oxynet web app (http://oxynetresearch.promfacility.eu (accessed on 20 February 2021)) followed by a download of results.

2.3. RR Measurements and Calculation of DFA a1 Derived Threshold

Each participant's ECG/RR times series was recorded with a 3-lead ECG (MP36; Biopac Systems Ltd., Essen, Germany) with a sampling rate of 1000 Hz. Electrodes were placed in the CM5 distribution after appropriate skin cleaning. MP36 test data were saved as acq files. ECG data for each participant were imported into Kubios HRV Software (Version 3.4.3, Biosignal Analysis and Medical Imaging Group, Department of Physics, University of Kuopio, Kuopio, Finland). Kubios preprocessing settings were set to the default values including the RR detrending method which was kept at "Smoothn priors" (Lambda = 500). For DFA a1 estimation, the root mean square fluctuation of the integrated and detrended data is measured in observation windows of different sizes. The data are then plotted against the size of the window on a log–log scale [34]. The scaling exponent represents the slope of the line, which relates (log) fluctuation to (log) window size. DFA a1 window width was set to $4 \leq n \leq 16$ beats. Visual inspection of the entire test recording was done to determine sample quality, noise, arrhythmia, and missing beat artifact. As mentioned above, two participants with excessive atrial ectopy were excluded from analysis. The RR series of the included participants was then corrected by the Kubios "automatic method" [35] and relevant HRV results exported as text files for further analysis. Percent artifact occurring during threshold interpretation segments was below 5%.

The following procedure was used to indicate at what level of running intensity HR the DFA a1 would cross a value of 0.5: DFA a1 was calculated from the incremental exercise test RR series using 2 min time windows with a recalculation every 5 s throughout the test. Two-minute time windowing was chosen based on the minimal required RR interval calculations by Chen et al. [36]. Plotting of DFA a1 vs. HR was then performed. Inspection of the DFA a1 relationship with HR generally showed a reverse sigmoidal curve, with a stable area above 1.0 at low work rates, a rapid, near linear drop reaching below 0.5 at higher intensity, then flattening without major change. Linear regression was done on the subset of data consisting of the rapid, near linear decline from values close to 1.0 (correlated) to approximately 0.5 (uncorrelated) or below if the values continued in a straight fashion. The HR where DFA a1 reached 0.5 was calculated based on the regression equation from that linear section (Figure 1).

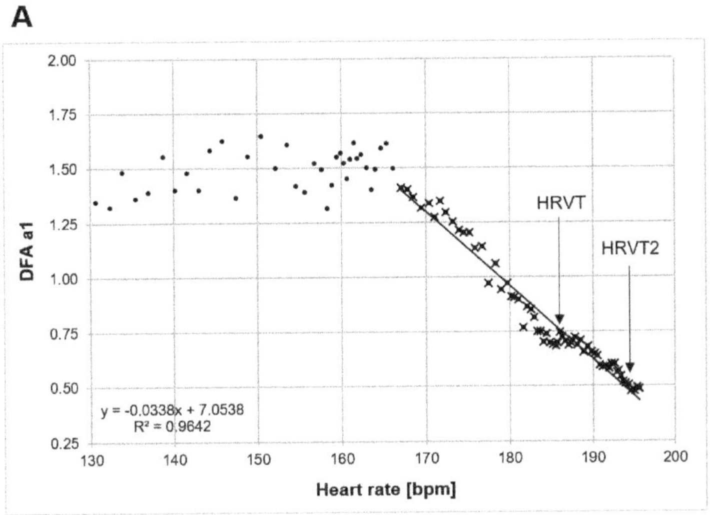

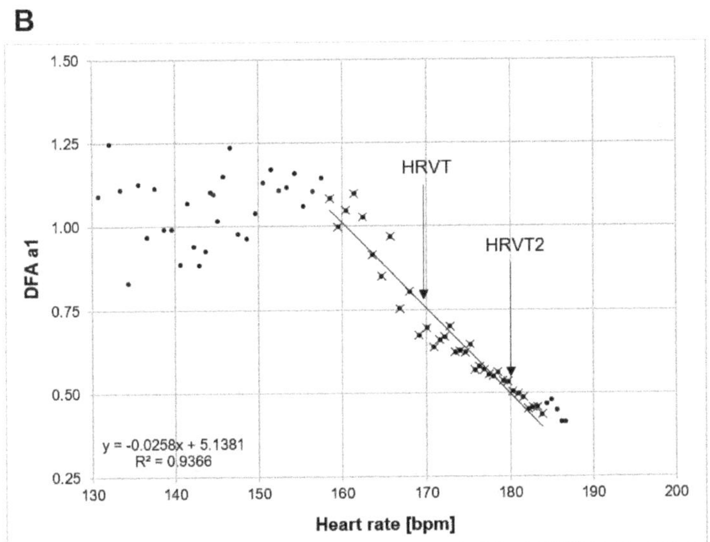

Figure 1. Individual examples of calculation methods for both HRVT and HRVT2. A line is drawn through the straight section of the plot of DFA a1 vs. HR. HRVT is at the intersection of DFA a1 = 0.75 and HRVT2 is at the intersection of DFA a1 = 0.5. (**A**) Participant with a VO_{2MAX} of 72 mL/kg/min, VT2 at 192 bpm and HRVT2 at 194 bpm. (**B**) Participant with a VO_{2MAX} of 58 mL/kg/min, VT2 at 179 bpm and HRVT2 at 180 bpm. Points with X are used for linear regression.

3. Statistics

Statistical analyses were performed for the variables HR at VT2 derived from gas exchange testing and HR at DFA a1 of 0.5 (HRVT2). Standard statistical methods were used for the calculation of means and standard deviations (SD). Normal distribution of data was checked by Shapiro–Wilk's test. The agreement against the VT2 HR was assessed using linear regression, Pearson's r correlation coefficient, coefficient of determination (R^2), standard error of estimate (SEE) and Bland–Altman plots with limits of agreement [37]. The size of Pearson's r correlations were evaluated as follows: $0.3 \leq r < 0.5$ low; $0.6 \leq r < 0.8$ moderate and $r \geq 0.8$

high [38]. The paired t-test was used for comparison of VT2 HR vs. HRVT2 HR. For all tests, the statistical significance was accepted as $p \leq 0.05$. Analysis was performed using Microsoft Excel 365 with Real Statistics Resource Pack software (Release 6.8) and Analyse-it software (Version 5.66).

4. Results

VO_{2MAX} varied considerably among participants, ranging between 41 and 74 mL/kg/min. VT1 was reached at HRs between 108 and 183 bpm [27]. Oxynet-derived HR at VT2 showed a mean value of 174 (±12) bpm compared with a mean HRVT2 HR of 171 (±16) bpm ($p = 0.18$) for all participants (Table 1). Regression analysis for VT2 HR vs. HRVT2 HR showed significant correlation ($p < 0.001$) with Pearson's r = 0.78, $R^2 = 0.60$ and SEE = 10.5 bpm (Figure 2). Bland–Altman evaluation of VT2 HR vs. HRVT2 HR is shown in Figure 3 with a mean bias of −4 (±10) bpm and LOA from −24 to +16 bpm.

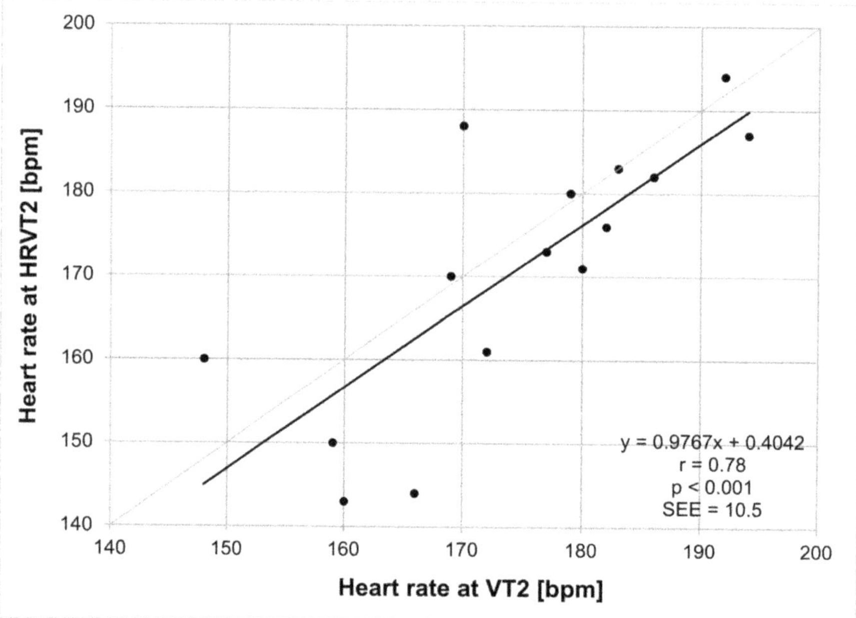

Figure 2. Regression plot analysis VT2 HR vs. HRVT2 HR for all participants. Bisection line in light gray. SEE: standard error of estimate; r: Pearson's correlation coefficient.

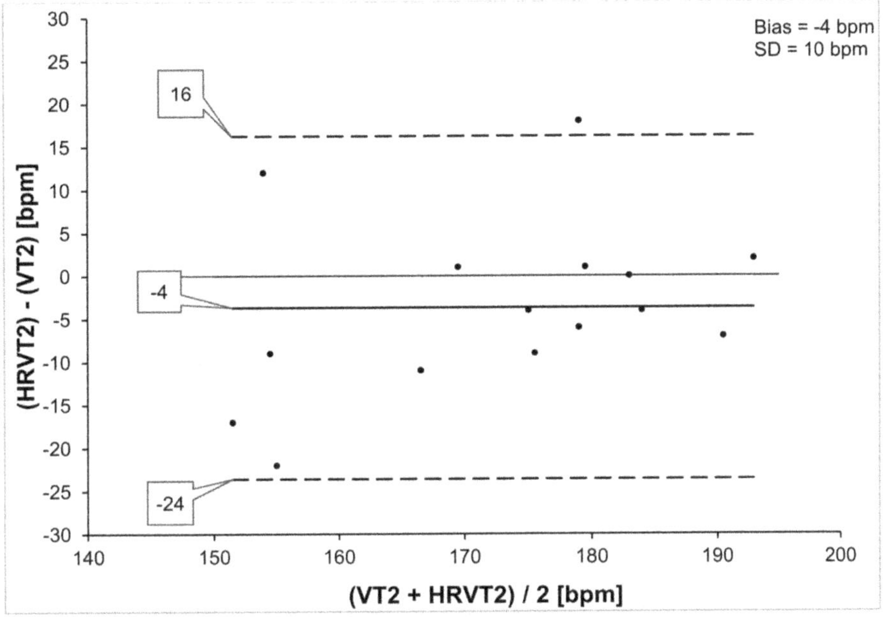

Figure 3. Bland–Altman analysis VT2 HR vs. HRVT2 HR for all participants. Center line represents the mean difference between each paired value, the top and bottom lines are 1.96 standard deviations from the mean difference.

5. Discussion

The purpose of this study was to examine whether DFA a1, a HRV index of fractal correlation properties would exhibit an uncorrelated pattern at the AnT derived by gas exchange data, a boundary separating sustainable from unsustainable exercise intensity [5]. In a previous investigation, it was shown that a DFA a1 value of 0.75, which represents a midpoint between well-correlated and uncorrelated patterns, was associated with the AeT as measured by VT1 [27]. In the present report, the HR reached at a DFA a1 of 0.5 was closely related to the HR reached at the AnT as measured by the VT2 during a treadmill running ramp. Multiple prior studies have shown similar behavior during incremental exercise with DFA a1, declining past the 0.75 mark with mild to moderate intensity then surpassing 0.5 during the highest work rates attained [23]. Prior to this report no attempt has been made to determine if the AnT is associated with a particular DFA a1 value using a method validation comparison. Strengths of this study include RR recording done by a research-grade ECG device containing few artifacts and the inclusion of recreational runners with a wide age range and performance spectrum. In addition, VT2 was computed by a validated neural network system utilizing the raw gas exchange data, eliminating any observer error or bias [32,33].

The results obtained here show good agreement between the HR derived from the HRVT2 and the HR associated with VT2 obtained through Oxynet analysis. This was supported by comparison of mean HR parameters by paired t testing, Pearson's correlation coefficient and Bland–Altman analysis. Although participants had variable differences in HR concordance, the results are clinically meaningful in the context of the reported agreement of other surrogate markers. Both the mean difference of −4 bpm and the LOA seen here are of similar magnitudes to that of a comparison of the MLSS and FTP [12] as well as the muscle oxygen desaturation breakpoint association to the MLSS [8].

The question as to why a DFA a1 of 0.5 could be the area of interest for an anaerobic breakpoint should be discussed. Prior studies have shown DFA a1 to drop past the uncorrelated value of 0.5 to an anticorrelated range at near maximal attained work rates [23].

Recently, a study by Naranjo-Orellana et al. [39] showed that during constant-intensity exercise performed over a 5 min span at the VT2, a DFA a1 value of 0.48 (±0.11) was seen, very similar to our results. Therefore, from an observational perspective, it is not unreasonable to look for the AnT to occur near the 0.5 value. Perhaps most importantly, a DFA a1 of 0.5 specifically represents a transition from an uncorrelated random to an anticorrelated pattern in HR time series [28]. Viewpoints regarding the significance of correlation properties can revolve around practical aspects (empirically validated breakpoints such as 0.75) but can also be considered from a network physiology standpoint [26]. The later concept entails the notion that fractal correlation properties of HRV depend upon "organismic demand", a model of multiple neuromuscular, biochemical, peripheral and central nervous system inputs [17]. In this framework, hypothetical acute physiologic responses and cardiocirculatory advantages may lie behind the changes seen in correlation patterns due to increasing exercise intensity and/or overall organismic demands. Depending on the internal load situation, the correlation properties of HR time series change to best suit the current and perhaps even the anticipated requirements as an optimization and/or stabilization strategy [28,40,41]. Therefore, the anticorrelated behavior during very high-intensity exercise could be interpreted in the sense of progressive segregation and centralization or "mechanization" of a complex (open) biological system as proposed by von Bertalanffy [42,43] and could indicate a maximum energy flux at the cost of cardiovascular self-regulation, which may reduce the adaptability to further perturbations and ultimately endangers the integrity of the overall system [44–46]. Thus, every fluctuation is corrected immediately in the opposite direction by a dominant attractor [29], e.g., performance attracter, as stated by Gronwald et al. [17], which results in an anticorrelated signal pattern [28]. This organismic regulatory withdrawal may also be interpreted as a loss of systemic integrity in the sense of a hazardous situation for homeostasis [47], which may only be tolerated for a short period of time.

6. Limitations and Future Directions

A potential problem with using HRV-related indexes to determine a physiologic threshold boundary involves the quality and precision of the RR time series [30,48]. Since the rate of missed beat artifact rises with increasing exercise intensity [49], excess amounts of artifact correction can affect DFA a1 resulting in bias and erroneous estimation of threshold values. Common methods of artifact correction may produce a positive proportional bias in DFA a1, predominantly affecting values 0.5 and below, especially at artifact rates above 5% [30]. No participant in the cohort examined here exceeded that limit. In addition, it is also possible that recording device bias may occur leading to results that differ from those of high resolution ECG monitoring. In view of these issues, further validation of this approach is recommended with commonly used consumer monitoring devices and typical artifact levels. However, if these issues can be resolved, analysis of DFA a1 over the course of an exercise ramp may provide both aerobic and anerobic threshold boundaries for the purpose of endurance sport training intensity distribution. Since all participants in this study were male, evaluation in females is important for widespread future usage. It is also unclear what the effects of athlete status (elite, recreational, inactive), ramp protocol (short vs. long, slope), exercise type (cycling vs. running vs. XC skiing), food intake, caffeine and recent high-intensity exercise would have on the HRVT2. Future use of this approach in the study of exercise training interventions may also be of interest. Though no data currently exist, following DFA a1 over the course of an intervention protocol could be of interest as a surrogate marker for AnT-related performance improvement. An intriguing thought centers on the methods used to obtain each AnT-related metric in this study, one from a convolutional neural network of gas exchange parameters and the other from a relatively simple mathematical relationship of HRV. Although at this time Oxynet relies purely on respiratory parameters, it could be of interest to determine whether adding DFA a1 HRVT-related measurements would improve final accuracy. Lastly, as a DFA a1 reaching

0.5 is sufficient for boundary determination, exercise efforts to exhaustion such as the FTP or CP can be avoided, both for health-related and exercise intensity distribution purposes.

7. Conclusions

Nonlinear heart rate variability analysis during an incremental treadmill run demonstrated that the heart rate reached at the second ventilatory threshold was closely associated with that of the heart rate associated with a DFA a1 of 0.5 in a population of recreational athletes. This DFA a1 value represents a distinct mathematical breakpoint in the cardiac interbeat series, from a correlated pattern seen with light to moderate exercise intensity to an uncorrelated, random pattern of heart rate time series occurring at the point of an unsustainable work rate. Although promising, additional study and verification in females, other exercise modalities, recording devices and disease states are recommended. Since this method may not require testing to exhaustion, application to athletes avoiding maximal stress during a given training cycle and to those unable/unsuitable to undergo maximal intensity exercise may be possible. In combination with evidence of a DFA a1 of 0.75 representing the aerobic threshold boundary, a comprehensive solution for training boundary demarcation using only heart rate variability may soon be achieved.

Author Contributions: B.R. and T.G. conceived the study. D.G. and N.D. performed the physiologic testing. B.R. wrote the first draft of the article. B.R. performed the data analysis. All authors (B.R., D.G., N.D., L.M. and T.G.) revised it critically for important intellectual content, final approval of the version to be published, and accountability for all aspects of the work. All authors have read and agreed to the published version of the manuscript.

Funding: This research received no external funding.

Institutional Review Board Statement: The studies involving human participants were reviewed and approved by University of Derby, United Kingdom (LSREC_1415_02). The study was conducted according to the guidelines of the Declaration of Helsinki.

Informed Consent Statement: The patients/participants provided their written informed consent to participate in this study.

Data Availability Statement: The raw data supporting the conclusions of this article will be made available by the authors, without undue reservation.

Conflicts of Interest: D.G. was employed by company Lattice Training. The remaining authors declare that the research was conducted in the absence of any commercial or financial relationships that could be construed as a potential conflict of interest.

References

1. Stöggl, T.L.; Sperlich, B. Editorial: Training Intensity, Volume and Recovery Distribution Among Elite and Recreational Endurance Athletes. *Front. Physiol.* **2019**, *10*, 592. [CrossRef]
2. Jamnick, N.A.; Pettitt, R.W.; Granata, C.; Pyne, D.B.; Bishop, D.J. An Examination and Critique of Current Methods to Determine Exercise Intensity. *Sports Med.* **2020**, *50*, 1729–1756. [CrossRef] [PubMed]
3. Binder, R.K.; Wonisch, M.; Corra, U.; Cohen-Solal, A.; Vanhees, L.; Saner, H.; Schmid, J.-P. Methodological approach to the first and second lactate threshold in incremental cardiopulmonary exercise testing. *Eur. J. Cardiovasc. Prev. Rehabil.* **2008**, *15*, 726–734. [CrossRef] [PubMed]
4. Meyer, T.; Lucía, A.; Earnest, C.P.; Kindermann, W. A Conceptual Framework for Performance Diagnosis and Training Prescription from Submaximal Gas Exchange Parameters—Theory and Application. *Int. J. Sports Med.* **2004**, *26*, S38–S48. [CrossRef] [PubMed]
5. Keir, D.A.; Fontana, F.Y.; Robertson, T.C.; Murias, J.M.; Paterson, D.H.; Kowalchuk, J.M.; Pogliaghi, S. Exercise Intensity Thresholds: Identifying the Boundaries of Sustainable Performance. *Med. Sci. Sports Exerc.* **2015**, *47*, 1932–1940. [CrossRef] [PubMed]
6. Faude, O.; Kindermann, W.; Meyer, T. Lactate threshold concepts: How valid are they? *Sports Med.* **2009**, *39*, 469–490. [CrossRef]
7. Caen, K.; Pogliaghi, S.; Lievens, M.; Vermeire, K.; Bourgois, J.G.; Boone, J. Ramp vs. step tests: Valid alternatives to determine the maximal lactate steady-state intensity? *Eur. J. Appl. Physiol.* **2021**, 1–9. [CrossRef]
8. Bellotti, C.; Calabria, E.; Capelli, C.; Pogliaghi, S. Determination of maximal lactate steady state in healthy adults: Can NIRS help? *Med. Sci. Sports Exerc.* **2013**, *45*, 1208–1216. [CrossRef] [PubMed]
9. Fontana, F.Y.; Keir, D.A.; Bellotti, C.; de Roia, G.F.; Murias, J.M.; Pogliaghi, S. Determination of respiratory point compensation in healthy adults: Can non-invasive near-infrared spectroscopy help? *J. Sci. Med. Sport* **2015**, *18*, 590–595. [CrossRef]

10. Allen, H.; Coggan, A. *Training and Racing with a Power Meter*; Velo Press: Boulder, CO, USA, 2010.
11. Lillo-Beviá, J.R.; Courel-Ibáñez, J.; Cerezuela-Espejo, V.; Morán-Navarro, R.; Martínez-Cava, A.; Pallarés, J.G. Is the Functional Threshold Power a Valid Metric to Estimate the Maximal Lactate Steady State in Cyclists? *J. Strength Cond. Res.* **2019**. [CrossRef]
12. Klitzke Borszcz, F.; Ferreira Tramontin, A.; Pereira Costa, V. Is the Functional Threshold Power Interchangeable with the Maximal Lactate Steady State in Trained Cyclists? *Int. J. Sports Physiol. Perform.* **2019**, *14*, 1029–1035. [CrossRef]
13. Inglis, E.C.; Iannetta, D.; Passfield, L.; Murias, J.M. Maximal Lactate Steady State Versus the 20-Minute Functional Threshold Power Test in Well-Trained Individuals: "Watts" the Big Deal? *Int. J. Sports Physiol. Perform.* **2020**, *15*, 541–547. [CrossRef]
14. Valenzuela, P.L.; Morales, J.S.; Foster, C.; Lucía, A.; de La Villa, P. Is the Functional Threshold Power a Valid Surrogate of the Lactate Threshold? *Int. J. Sports Physiol. Perform.* **2018**, *13*, 1293–1298. [CrossRef] [PubMed]
15. Jones, A.M.; Burnley, M.; Black, M.I.; Poole, D.C.; Vanhatalo, A. The maximal metabolic steady state: Redefining the 'gold standard'. *Physiol. Rep.* **2019**, *7*, e14098. [CrossRef] [PubMed]
16. Karsten, B.; Petrigna, L.; Klose, A.; Bianco, A.; Townsend, N.; Triska, C. Relationship between the Critical Power Test and a 20-min Functional Threshold Power Test in Cycling. *Front. Physiol.* **2021**, *11*, 613151. [CrossRef] [PubMed]
17. Gronwald, T.; Rogers, B.; Hoos, O. Fractal Correlation Properties of Heart Rate Variability: A New Biomarker for Intensity Distribution in Endurance Exercise and Training Prescription? *Front. Physiol.* **2020**, *11*, 550572. [CrossRef]
18. White, D.W.; Raven, P.B. Autonomic neural control of heart rate during dynamic exercise: Revisited. *J. Physiol.* **2014**, *592*, 2491–2500. [CrossRef] [PubMed]
19. Tulppo, M.P.; Makikallio, T.H.; Takala, T.E.; Seppanen, T.; Huikuri, H.V. Quantitative beat-to-beat analysis of heart rate dynamics during exercise. *Am. J. Physiol. Circ. Physiol.* **1996**, *271*, H244–H252. [CrossRef]
20. Cottin, F.; Médigue, C.; Lopes, P.; Lepretre, P.-M.; Heubert, R.; Billat, V. Ventilatory Thresholds Assessment from Heart Rate Variability during an Incremental Exhaustive Running Test. *Int. J. Sports Med.* **2006**, *28*, 287–294. [CrossRef]
21. Karapetian, G.K.; Engels, H.J.; Gretebeck, R.J. Use of heart rate variability to estimate LT and VT. *Int. J. Sports Med.* **2008**, *29*, 652–657. [CrossRef]
22. Gronwald, T.; Hoos, O.; Ludyga, S.; Hottenrott, K. Non-linear dynamics of heart rate variability during incremental cycling exercise. *Res. Sports Med.* **2019**, *27*, 88–98. [CrossRef]
23. Gronwald, T.; Hoos, O. Correlation properties of heart rate variability during endurance exercise: A systematic review. *Ann. Noninvasive Electrocardiol.* **2020**, *25*, e12697. [CrossRef]
24. Ivanov, P.C.; Amaral, L.A.N.; Goldberger, A.L.; Havlin, S.; Rosenblum, M.G.; Stanley, H.E.; Struzik, Z.R. From 1/f noise to multifractal cascades in heartbeat dynamics. *Chaos Interdiscip. J. Nonlinear Sci.* **2001**, *11*, 641–652. [CrossRef]
25. Eke, A.; Herman, P.; Kocsis, L.; Kozak, L.R. Fractal characterization of complexity in temporal physiological signals. *Physiol. Meas.* **2002**, *23*, R1–R38. [CrossRef]
26. Balagué, N.; Hristovski, R.; Almarcha, M.D.C.; Garcia-Retortillo, S.; Ivanov, P.C. Network Physiology of Exercise: Vision and Perspectives. *Front. Physiol.* **2020**, *11*, 611550. [CrossRef] [PubMed]
27. Rogers, B.; Giles, D.; Draper, N.; Hoos, O.; Gronwald, T. A New Detection Method Defining the Aerobic Threshold for Endurance Exercise and Training Prescription Based on Fractal Correlation Properties of Heart Rate Variability. *Front. Physiol.* **2021**, *11*, 596567. [CrossRef]
28. Hardstone, R.; Poil, S.S.; Schiavone, G.; Jansen, R.; Nikulin, V.V.; Mansvelder, H.D.; Linkenkaer-Hansen, K. Detrended fluctuation analysis: A scale-free view on neuronal oscillations. *Front. Physiol.* **2012**, *3*, 450. [CrossRef] [PubMed]
29. Karasik, R.; Sapir, N.; Ashkenazy, Y.; Ivanov, P.C.; Dvir, I.; Lavie, P.; Havlin, S. Correlation differences in heartbeat fluctuations during rest and exercise. *Phys. Rev. E* **2002**, *66*, 062902. [CrossRef] [PubMed]
30. Rogers, B.; Giles, D.; Draper, N.; Mourot, L.; Gronwald, T. Influence of Artefact Correction and Recording Device Type on the Practical Application of a Non-Linear Heart Rate Variability Biomarker for Aerobic Threshold Determination. *Sensors* **2021**, *21*, 821. [CrossRef]
31. Reis, V.M.; Tillaar, R.V.D.; Marques, M.C. Higher Precision of Heart Rate Compared with VO2 to Predict Exercise Intensity in Endurance-Trained Runners. *J. Sports Sci. Med.* **2011**, *10*, 164–168.
32. Zignoli, A.; Fornasiero, A.; Stella, F.; Pellegrini, B.; Schena, F.; Biral, F.; Laursen, P.B. Expert-level classification of ventilatory thresholds from cardiopulmonary exercising test data with recurrent neural networks. *Eur. J. Sport Sci.* **2019**, *19*, 1221–1229. [CrossRef]
33. Zignoli, A.; Fornasiero, A.; Rota, P.; Muollo, V.; Peyré-Tartaruga, L.; Low, D.; Fontana, F.; Besson, D.; Pühringer, M.; Ring-Dimitriou, S.; et al. Oxynet: A collective intelligence that detects ventilatory thresholds in cardiopulmonary exercise tests. *Eur. J. Sport Sci.* **2021**, *31*, 1–11. [CrossRef] [PubMed]
34. Tarvainen, M.P.; Niskanen, J.-P.; Lipponen, J.A.; Ranta-Aho, P.O.; Karjalainen, P.A. Kubios HRV—Heart rate variability analysis software. *Comput. Methods Programs Biomed.* **2014**, *113*, 210–220. [CrossRef] [PubMed]
35. Lipponen, J.A.; Tarvainen, M.P. A robust algorithm for heart rate variability time series artefact correction using novel beat classification. *J. Med. Eng. Technol.* **2019**, *43*, 173–181. [CrossRef]
36. Chen, Z.; Ivanov, P.C.; Hu, K.; Stanley, H.E. Effect of nonstationarities on detrended fluctuation analysis. *Phys. Rev. E* **2002**, *65*, 041107. [CrossRef] [PubMed]
37. Bland, J.M.; Altman, D.G. Measuring agreement in method comparison studies. *Stat. Methods Med. Res.* **1999**, *8*, 135–160. [CrossRef]

38. Chan, Y.H. Biostatistics 104: Correlational analysis. *Singap. Med. J.* **2003**, *44*, 614–619.
39. Naranjo-Orellana, J.; Nieto-Jiménez, C.; Ruso-Álvarez, J. Non-linear heart rate dynamics during and after three controlled exercise intensities in healthy men. *Physiol. Int.* **2021**, *107*, 501–512. [CrossRef]
40. Goldberger, A.L.; Amaral, L.A.; Hausdorff, J.M.; Ivanov, P.C.; Peng, C.K.; Stanley, H.E. Fractal dynamics in physiology: Alterations with disease and aging. *Proc. Natl. Acad. Sci. USA* **2002**, *99*, 2466–2472. [CrossRef]
41. Fossion, R.; Rivera, A.L.; Estañol, B. A physicist's view of homeostasis: How time series of continuous monitoring reflect the function of physiological variables in regulatory mechanisms. *Physiol. Meas.* **2018**, *39*, 084007. [CrossRef]
42. von Bertalanffy, L. An outline on General System Theory. *Br. J. Philos. Sci.* **1950**, *2*, 134–165. [CrossRef]
43. von Bertalanffy, L. The Theory of Open Systems in Physics and Biology. *Science* **1950**, *111*, 23–29. [CrossRef]
44. Rosenwinkel, E.T.; Bloomfield, D.M.; Arwady, M.A.; Goldsmith, R.L. Exercise and autonomic function in health and cardiovascular disease. *Cardiol. Clin.* **2001**, *19*, 369–387. [CrossRef]
45. Kannankeril, P.J.; Goldberger, J.J. Parasympathetic effects on cardiac electrophysiology during exercise and recovery. *Am. J. Physiol. Heart Circ. Physiol.* **2002**, *282*, H2091–H2098. [CrossRef]
46. Billman, G.E. Cardiac autonomic neural remodeling and susceptibility to sudden cardiac death: Effect of endurance exercise training. *Am. J. Physiol. Circ. Physiol.* **2009**, *297*, H1171–H1193. [CrossRef]
47. Seely, A.J.; Macklem, P.T. Complex systems and the technology of variability analysis. *Crit. Care* **2004**, *8*, R367–R384. [CrossRef]
48. Cassirame, J.; Chevrolat, S.; Mourot, L. Effects of R-R time series accuracy on heart rate variability indexes. *Mov. Sport Sci. Sci. Mot.* **2019**, *106*, 27–35. [CrossRef]
49. Giles, D.A.; Draper, N. Heart Rate Variability During Exercise: A Comparison of Artefact Correction Methods. *J. Strength Cond. Res.* **2018**, *32*, 726–735. [CrossRef] [PubMed]

Journal of
Functional Morphology and Kinesiology

MDPI

Article

High-Intensity Functional Training Guided by Individualized Heart Rate Variability Results in Similar Health and Fitness Improvements as Predetermined Training with Less Effort

Justin A. DeBlauw [1,*], Nicholas B. Drake [2], Brady K. Kurtz [1], Derek A. Crawford [3], Michael J. Carper [1], Amanda Wakeman [3] and Katie M. Heinrich [1]

1 Department of Kinesiology, Kansas State University, Manhattan, KS 66506, USA; bkkurtz@ksu.edu (B.K.K.); mcarper@pittstate.edu (M.J.C.); kmhphd@ksu.edu (K.M.H.)
2 Department of Health, Human Performance, and Recreation, Pittsburg State University, Manhattan, KS 66506, USA; nichbinderdrake@gmail.com
3 School of Nutrition, Kinesiology, and Psychological Science, University of Central Missouri, Warrensburg, MO 64093, USA; dcrawford@ucmo.edu (D.A.C.); wakeman@ucmo.edu (A.W.)
* Correspondence: jdeblauw@ksu.edu

Citation: DeBlauw, J.A.; Drake, N.B.; Kurtz, B.K.; Crawford, D.A.; Carper, M.J.; Wakeman, A.; Heinrich, K.M. High-Intensity Functional Training Guided by Individualized Heart Rate Variability Results in Similar Health and Fitness Improvements as Predetermined Training with Less Effort. *J. Funct. Morphol. Kinesiol.* **2021**, *6*, 102. https://doi.org/10.3390/jfmk6040102

Academic Editors: Cristina Cortis, Andrea Fusco and Carl Foster

Received: 12 October 2021
Accepted: 8 December 2021
Published: 13 December 2021

Abstract: Heart rate variability (HRV) may be useful for prescribing high-intensity functional training (HIFT) exercise programs. This study aimed to compare effects of HRV-guided and predetermined HIFT on cardiovascular function, body composition, and performance. Methods: Recreationally-active adults (n = 55) were randomly assigned to predetermined HIFT (n = 29, age = 24.1 ± 4.1 years) or HRV-guided HIFT (n = 26, age = 23.7 ± 4.5) groups. Both groups completed 11 weeks of daily HRV recordings, 6 weeks of HIFT (5 d·week-1), and pre- and post-test body composition and fitness assessments. Meaningful changes in resting HRV were used to modulate (i.e., reduce) HRV-guided participants' exercise intensity. Linear mixed models were used with Bonferroni post hoc adjustment for analysis. Results: All participants significantly improved resting heart rate, lean mass, fat mass, strength, and work capacity. However, no significant between-groups differences were observed for cardiovascular function, body composition, or fitness changes. The HRV-guided group spent significantly fewer training days at high intensity (mean difference = −13.56 ± 0.83 days; $p < 0.001$). Conclusion: HRV-guided HIFT produced similar improvements in cardiovascular function, body composition, and fitness as predetermined HIFT, despite fewer days at high intensity. HRV shows promise for prescribing individualized exercise intensity during HIFT.

Keywords: autonomic nervous system; exercise intensity; training prescription

1. Introduction

Exercise training programs relying on predetermined volume and intensity often result in heterogeneous fitness outcomes across individuals [1]. To maximize training potential, employing an individualized training program is the most practical applied strategy [2]. An important factor in individualizing training and reducing the risk of maladaptation, is the ability to effectively monitor responses to training stressors [3]. Training stress is often described as the input variable that is manipulated to elicit a desired physiological response and is categorized as either external (e.g., speed, repetitions) or internal (e.g., heart rate, lactate) load [4,5].

A promising, non-invasive tool to monitor internal load to optimize training outcomes is heart rate variability (HRV) [6,7]. HRV is assessed by measuring the time intervals between successive heartbeats, since an increase or decrease in these intervals reflects changes in cardiac autonomic regulation [8]. HRV is a valid tool to assess individual variation in adaptation, fatigue, and overtraining during training programs [7,9,10]. Daily HRV measurements are often used to adjust training prescriptions for endurance activities such as running [9,11], cross country skiing [12], and cycling [6,10]. Endurance training programs utilizing HRV-guided individualization improve VO2peak, peak power in runners [11], and 40 min time trial performance in cyclists [10]. Additionally, resistance training frequency can be increased when using HRV to determine recovery intervals [13].

While these findings are promising, their focus on single modality endurance training regimens does not reflect the complexity of high-level sport training or current trends in exercise programs. High-intensity functional training (HIFT), a "Top 10 Fitness Trend" in 2018, is comprised of functional, multi-joint aerobic and muscle strengthening exercises performed at relative high effort or intensity [14]. HIFT combines components of aerobic, weightlifting, and body weight exercises into training sessions in constantly variable patterns across multiple time domains, creating a unique stimulus virtually every day [15]. This uniqueness of HIFT creates difficulty when attempting to quantify training loads with external markers [15,16]. However, HIFT is inherently individually modified as the exercises, intensity levels, and/or time domains can be adapted as needed for each individual [17]. Thus, HIFT programs are ideally situated to benefit from implementing HRV-guided training prescriptions.

To the best of our knowledge, no study has investigated the efficacy of HIFT exercise programs when guided by daily HRV. The purpose of the current investigation was to determine the effects of HRV-guided HIFT training compared to predetermined HIFT training on cardiovascular function, body composition, and performance outcomes in recreationally active participants. We hypothesized that HRV-guided HIFT (i.e., prescribing training volume and intensity of HIFT in response to daily HRV status) would result in reduced training volume at high intensity and improved fitness outcomes compared to predetermined HIFT training.

2. Materials and Methods

2.1. Experimental Design

This study was an 11 week, two-site prospective randomized controlled trial intervention, designed to determine the efficacy of HRV as means to modulate HIFT. Participants were randomly assigned at each site to either an experimental (HRV-guided) or control group (predetermined), with groups balanced for sex. After assignment, both groups completed 14 days of resting HRV measurements, which served as baseline values. Following the baseline period, participants continued taking morning HRV readings and began the exercise intervention which consisted of two three-week training blocks interspersed with pre- and post-intervention testing weeks; a mid-point week was used to recalibrate HRV metrics. Table 1 illustrates the study timeline. During training weeks, participants completed 60-min HIFT sessions on five consecutive days (Monday–Friday) followed by two days of recovery (Saturday and Sunday). Participants were asked to participate in 30 total training sessions with multiple training times available during the training intervention so as to maintain an appropriate participant-to-researcher ratio and accommodate schedules. The HRV-guided group had exercise intensity and volume modulated based on their morning HRV values, while the predetermined group completed training as prescribed. Performance measurements were assessed with participants attending two laboratory sessions during testing weeks, with 48 h of rest in-between.

Table 1. Study timeline from baseline to post-testing.

Study Duration 11 Weeks						
Weeks 1–2	**Random Assignment**	**Week 3**	**Weeks 4–6**	**Week 7**	**Weeks 8–10**	**Week 11**
Baseline HRV & randomization	HRV-guided	Pre-testing VO_2max, strength & body composition	HRV-modulated training	Mid-point recalibration of HRV SWC windows	HRV-modulated training	Post-testing VO_2max, strength & body composition
	Predetermined		Predetermined training		Predetermined training	

2.2. Participants

Fifty-five recreationally active men and women aged 18–35 years were recruited for this study. Participant baseline characteristics sorted by group and sex are presented in Table 2. All participants had been exercising regularly, while not pursuing any specific health or fitness-related goal (e.g., weight loss or competition preparation) for at least six months at the time of enrollment for this study. All participants reported no physical or health limitations for vigorous exercise, as determined by a medical health history questionnaire and physical activity readiness questionnaire [18]. In addition, no participants indicated a health condition or medication that would alter cardiac rhythms. Written informed consent was obtained from all participants prior to study commencement. The study was performed in accordance with the Declaration of Helsinki, and two university institutional review boards approved all procedures (IRB #9131).

Table 2. Participant descriptives by group * sex.

	Men (HRV-Guided) (n = 12)	**Men (Predetermined) (n = 14)**	**Female (HRV-Guided) (n = 12)**	**Female (Predetermined) (n = 17)**
Age (years)	25.0 ± 5.1	23.3 ± 2.8	22.4 ± 3.4	24.6 ± 4.8
Weight (kg)	83.4 ± 10.8	89.8 ± 15.5	72.5 ± 21.9	71.8 ± 9.6
Height (cm)	181 ± 8	182 ± 6	164 ± 5	165 ± 4

2.3. Procedures

2.3.1. Heart Rate Variability

All participants took daily morning HRV readings using a commercially available smartphone application for both iOS and Android (Amsterdam, The Netherlands; see http://www.hrv4training.com/ (1 June 2021)). The HRV4Training software utilizes photoplethysmography to determine the variability in R-R intervals from continuous heart rate data [19,20]. To maintain HRV reliability, participants were instructed to use the application in the morning upon waking, after excretion from the urinary bladder and resting for five minutes. To perform readings, participants placed their index finger over the smartphone camera for one-minute while in the supine position [21]. The HRV4Training application has a built-in methodology for signal filtering, processing, interpolation, artifact correction, and R-R peak detection which can be found in the reference for the application development [19]. For day-to-day monitoring of individual recovery (i.e., sympathovagal balance) HRV was measured as the root mean squared of successive differences (RMSSD). Due to the lack of normality, the RMSSD was transformed using the natural logarithm (LnRMSSD), which was then multiplied by two so that LnRMSSD (HRVdaily) could be viewed on a scale of approximately one to ten for ease of interpretation and to reflect the application display [22].

2.3.2. Resting Heart Rate

Participant resting heart rate (rHR) was collected daily simultaneously with morning HRV readings using photoplethysmography via the HRV4Training smartphone application.

2.3.3. Coefficient of Variance of Heart Rate Variability

Participant coefficient of variation in HRV (CV of HRV), the amount of day-to-day variability in HRV scores, was collected simultaneously with morning HRV readings using photoplethysmography via the HRV4Training smartphone application [23].

2.3.4. Body Composition

Body composition was measured for all participants at pre- and post-testing. Participant height was measured to the nearest 0.1 cm with a Charder stadiometer (Model HM 200P; Taichung City, Taiwan) at both sites. Weight was measured to the nearest 0.1 kg via a

Tanita (Tanita TBF-140, Tokyo, Japan) at site 1 and Tanita TBF310 bioelectrical impedance scale (Arlington Heights, IL, USA) at site 2. Body fat percentage (BF%), fat mass (FM), and lean mass (LM) were measured using a dual energy x-ray (DEXA; Discovery A QDR, Hologic, Inc., Marlborough, MA, USA) at site 1 and a Tanita TBF310 bioelectrical impedance scale at site 2.

2.3.5. Aerobic Capacity

Aerobic capacity was determined as maximal oxygen consumption (VO_2max) via the Bruce treadmill test protocol [24]. Site 1 used a predictive-regression equation based upon time to exhaustion [25] to determine aerobic capacity; the standard error of the estimate for males was ±3.55 mL/kg/min and ±2.70 mL/kg/min for females. Site 2 completed the Bruce treadmill test protocol, followed by a maximal oxygen consumption validation to ensure there was no further increase in oxygen consumption with increasing workload [26]. Expired gas fractions were assessed through breath-by-breath data recording, and measurements were analyzed at 15 s intervals (ParvoMedics TrueOne 2400 Metabolic, Salt Lake City, UT, USA). The gas calibration and metabolic cart flow were calibrated before each testing session using a 3 L syringe and following manufacturer instructions. Heart rate was recorded continuously using a Polar H7 chest strap heart rate monitor (Polar Electro OY, Kempele, Finland).

2.3.6. Physical Work Capacity

Physical work capacity was measured through a 10 min workout in which participants completed as many rounds as possible of the following: 12 goblet squats (20 kg kettlebell for men, 12 kg kettlebell for women), 12 burpees, and 24 calories on a rowing ergometer (Model D, PM5 Monitor, Concept 2 Inc., Morrisville, VM, USA).

2.3.7. Muscular Strength

Maximal strength was determined by the one-repetition maximum (1RM) protocol for the barbell back squat, barbell overhead (OH) press and barbell deadlift in kilograms [27], in accordance with previous research methodology [15]. Individual sum totals for 1RM for back squat, OH press, and deadlift were designated as each participant's CrossFit Total (CFT) [28]. Each lift was supervised and verified by certified exercise professionals, who were also research assistants, and participant rest times were controlled with a minimum of three minutes and a maximum of five minutes between maximal attempts [29].

2.3.8. High-Intensity Exercise Training Program

The high-intensity exercise program employed in this study was HIFT, utilizing a popular, community-based HIFT template [30]. All training sessions were performed indoors as group exercise and supervised by a research assistant holding a CrossFit® Level 1 certificate. Training sessions for site 1 were held at a community HIFT facility while sessions for site 2 were held within the Functional Intensity Training Lab at Kansas State University. The training protocol for this program has been previously described by Crawford et al. [15], and specific details of the structure and components for each daily training session can be found in Table A1. All training sessions lasted approximately 1 h and consisted of an instructor-led warm-up, movement preparation period, daily workout, and cool-down. A total of 30 training sessions were programmed, and an adherence rate of 80% was required for participant inclusion in data analysis. Participants remained in free-living conditions and were asked to not engage in any additional exercise training outside of the study.

2.4. Modulation for High-Intensity Exercise Training Program

A 14 day baseline period was used to establish individual baseline HRV values. Individual seven-day rolling averages (Ln rMMSD7day) were calculated to determine and track shifts in resting HRV in response to the training. The Ln rMMSD7day was used as it has been demonstrated to be superior in predicting training stress rather than single-day HRV values [7]. Smallest worthwhile change (SWC) windows were set to monitor meaningful changes from baseline HRV. Previous investigations have established the SWC in resting HRV as ±0.5 standard deviation from an individual's mean Ln rMSSD [2,7,11,31]. For this study, two SWC change windows were calculated as ±0.5 standard deviation (SWC1) and ±1 standard deviation (SWC2) from the individual's mean Ln rMMSD, in order to modulate training stress during the exercise intervention.

Each HRV-guided participant was prescribed reduced training volume and intensity when their rolling seven-day average of HRVdaily (HRV7day as indexed by Ln rMMSD7day) differed meaningfully from baseline values such that it fell within a SWC window [32]. When a participant's Ln rMMSD7day was within the SWC1 no training modifications were prescribed. If participant's Ln rMMSD7day fell between SWC1 and SWC2, their scheduled workout was reduced 25% in volume (i.e., repetitions) and external load (i.e., absolute weight). If the participant's Ln rMMSD7day exceeded the SWC2, they completed a low-intensity (i.e., >50% HRR) active recovery session (e.g., walking and light stretching activities) for a fixed duration of 20 min. A detailed description of the modified and light training session is provided in Table A2. The HRV values obtained during the baseline period were used for the first block of training. After the pre-intervention testing and three weeks of training (15 training sessions) HRV means and both SWC monitoring windows were recalculated for the second training period, as previous findings have demonstrated how changes in fitness may alter resting HRV [33–35], and the dose of completing 15 HIFT sessions should be sufficient to elicit fitness improvements [16,36,37]. The Predetermined group completed all training sessions without intensity modulation. Site 1 assessed Edward's training load to evaluate the efficacy of the modulated training, which is available in previously published research [32].

2.5. Statistical Analyses

Data were analyzed using the R statistical computing environment and language [38] via the Jamovi graphical user interface [39]. Data were checked for normality using the Shapiro-Wilk test and visual inspection of the corresponding Q-Q plots of residuals. Relationships between fixed effects (i.e., group and timepoint) and outcome metrics (i.e., cardiovascular, body composition, and performance) data were assessed using linear mixed-effects models via the GAMLj General Analysis for Linear Models module [40]. Individual participants were input as random factors within the models and lean body mass was used as a covariate, due to significant correlations identified with outcomes metrics. An alpha level of 0.05 was used for all statistical inferences. Post hoc assessments were adjusted using the Bonferroni correction. Effect sizes (ES) were calculated for within and between group changes. ES were classified as 0.2 "small", 0.5 "medium", and 0.8+ "large" [41].

3. Results

Baseline and post-test values for each training group are shown in Table 3. The HRV-guided training resulted in similar changes in cardiovascular function, body composition, and performance as the predetermined training (Table 3 & Figure 1). The greatest percentage changes were for predetermined BF% (15.7% decrease) and FM (15.1% decrease), and HRV-guided squat (14.2% increase) and deadlift (12.6% increase). The HRV-guided group completed significantly fewer days at high intensity (DHI) than the predetermined group, as shown in Table 4. Participants displayed a high training and daily HRV monitoring adherence (Table 4).

Table 3. Within and between group comparisons of pre- and post-test changes in key outcomes.

	HRV-Guided				Predetermined				Between Group
	Pre	Post	% Change	ES	Pre	Post	% Change	ES	ES
Cardiovascular function									
Resting heart rate (bpm)	73.6 ± 9.8	69.3 ± 9.0	−5.84	0.46	74.6 ± 14.6	72.7 ± 11.4	−2.55	0.15	0.33
Heart rate variability (ms)	8.4 ± 1.1	8.6 ± 1.1	2.38	0.14	8.7 ± 1.2	8.7 ± 1.2	0	0.01	0.09
CV of HRV (ms)	10.1 ± 3.9	9.0 ± 3.8	−10.89	0.28	8.7 ± 3.3	9.5 ± 3.1	9.20	−0.24	0.14
Body composition									
Body fat %	31.8 ± 11.1	29.2 ± 9.7	−8.18	0.63 *	31.8 ± 8.3	26.8 ± 8.1	−15.73	0.61 *	0.27
Lean mass (kg)	54.5 ± 13.5	54.8 ± 13.3	0.55	0.02	52.6 ± 11.2	54.0 ± 11.5	2.66	−0.12	0.06
Fat mass (kg)	23.9 ± 8.8	23.5 ± 8.7	−1.67	0.05	23.9 ± 8.8	20.3 ± 8.5	−15.06	0.42	0.37
Fitness outcomes									
VO_2max (mL * kg * min)	42.1 ± 6.8	43.0 ± 7.5	2.14	0.13	44.4 ± 6.4	44.2 ± 8.0	−0.45	0.03	0.15
Work capacity (reps)	131 ± 36	147 ± 35	12.21	0.45	127 ± 24	145 ± 26	14.17	−0.70 *	0.06
Squat (kg)	90.2 ± 44.5	103 ± 45.0	14.19	0.29	87.6 ± 33.2	99.1 ± 31.5	13.13	−0.36	0.10
Press (kg)	41.6 ± 18.9	45.3 ± 21.4	8.89	0.18	41.5 ± 16.2	45.5 ± 16.4	9.64	−0.25	0.01
Deadlift (kg)	103 ± 46	116 ± 47	12.62	0.27	107 ± 34	121 ± 47	13.08	−0.34	0.11
CrossFit total (kg)	232 ± 109	259 ± 108	11.63	0.25	237 ± 82	266 ± 85	12.24	−0.35	0.07

Values are presented as mean ± SD. * moderate effect size. VO_2max, maximal oxygen consumptions; ES, effect size.

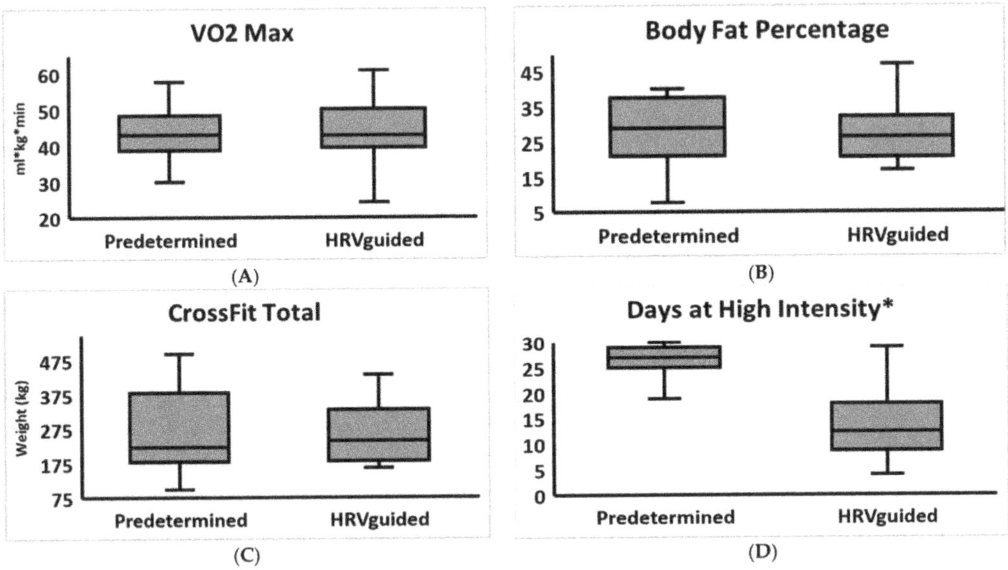

Figure 1. Changes in primary outcome metrics (**A**) VO_2max, (**B**) body fat percentage, (**C**) CrossFit total, and (**D**) days at High Intensity sorted by group. * Indicates a statistically significant difference.

Table 4. Study intervention metrics by group.

	HRV-Guided		Predetermined		Between Group
	Mean	95% CI	Mean	95% CI	ES
Days at high intensity (days)	12.9 ± 5.6	11.7; 14.1	26.5 ± 2.6	25.4; 14.1	3.12 **
Training adherence (days completed)	86.8 ± 9.5	84.3; 89.0	89.8 ± 7.6	87.5; 92.0	0.35
HRV compliance (% days recorded)	95.1 ± 4.8	93.8; 96.7	94.6 ± 5.9	92.9; 95.6	0.09

Values are presented as mean ± SD. ** large effect size. CI, confidence interval; ES, effect size.

3.1. Effects on Cardiovascular Function

A significant main effect for time was observed for HR (F = 4.89; mean difference = −3.25 ± 1.47 bpm; 95% CI = −6.14, −0.37; $p = 0.035$) with a reduction in resting HR being observed across both conditions from pre- to post-test. No other main effects on cardiovascular function were observed.

3.2. Effects on Body Composition

Significant main effects for time were observed for LM (F = 16.43; mean difference = 1.19 ± 0.29 kg; 95% CI = 0.61, 1.77; $p < 0.001$) and FM (F = 4.39; mean difference = −0.62 ± 0.3 kg; 95% CI = −1.12, −0.36; $p = 0.045$), with all individuals improving LM and FM at post-test. No other main effects on body composition were observed.

3.3. Effects on Performance Outcomes

Significant main effects for time were observed for work capacity (F = 14.92; mean difference = 16.87 ± 4.37; 95% CI = 8.31, 25.44; $p < 0.001$), squat (F = 29.16; mean difference = 7.98 ± 1.48 kg; 95% CI = −5.08, 10.87; $p < 0.001$), OH Press (F = 10.52; mean difference = 2.62 ± 0.81 kg; 95% CI = 1.04, 4.20, $p < 0.003$), deadlift (F = 22.09; mean difference = 10.37 ± 2.21 kg; 95% CI = 6.05, 14.70; $p < 0.001$), and CFT (F = 20.68; mean difference = 21.79 ± 4.18 kg; 95% CI = 13.61, 29.88; $p < 0.001$) where both groups improved at post-test. No other main effects on performance outcomes were observed.

3.4. Effects on Intervention Metrics

A significant main effect for group was observed for DHI (F = 270.46; mean difference = −13.56 ± 0.83 days; 95% CI = −15.20, −11.99; $p < 0.001$) with the HRV-guided group training fewer DHI. Training adherence to the 30 prescribed training sessions for the Predetermined group was 26.3–27.6 sessions and for the HRV-guided group was 25.3–26.7 sessions.

4. Discussion

This study tested the effects of HRV-guided and predetermined HIFT on health and fitness outcomes in recreationally active participants. Our results support our first hypothesis, as HRV-guided prescription resulted in fewer DHI compared to a predetermined prescription. This is demonstrated by the HRV-guided group completing 17 of 30 days as modulated, lower intensity training days. Our second hypothesis that HRV-guided prescription would elicit greater improvements in fitness outcomes than the predetermined group was not supported by the data. This is evident through lack of significant differences between groups for changes in all primary outcome fitness measures. Collectively, these findings are of interest as they demonstrate that HRV-guided training results in similar improvements across fitness outcomes while spending fewer training sessions at high intensity compared to a predetermined prescription.

Our finding that HRV-guided training did not result in greater changes in aerobic or work capacity than predetermined training in a 9-week HIFT program was similar to previous aerobic exercise investigations where the HRV-guided group displayed increases in aerobic capacity with no significant difference between groups [6,10,11]. Additionally, neither a small or moderate effect size was observed between groups as previously reported by Vesterninen et al. [42] and Nuuttila et al. [43], respectfully. This finding is not atypical as Hautala et al., [44] has shown that aerobic capacity adaptations are not universal and may be driven by intrinsic factors that predispose individuals to favorable adaptations based on training mode.

We observed no significant differences between groups on improvements in maximal squat, OH press, deadlift, and CrossFit total. The lack of observed group differences is similar to the findings of De Oliveira et al. [13] on maximal strength in young resistance-trained men undergoing HRV-guided training. However, De Oliveria et al. [13] used HRV to augment training frequency, while we used HRV to modulate training intensity. Our findings extend those of De Olivieria et al. [13] and suggest that HRV is a practical tool to individualize the prescription of training frequency and intensity. This enhances the practitioner/coach's ability to determine when and how much stress to apply in training.

Our participants showed an increase in overall strength following HIFT participation regardless of group. The finding that HIFT is a valid program structure for improving strength is supported by the findings of Heinrich etl al. [45] and Buckley et al. [46] in which HIFT participants displayed increases in bench press, back squat, OH press and deadlift 1RM. It is possible that the observed changes in strength were a result of our participants being classified as "novice", or as a result of an effective training paradigm. In order to determine the cause, future investigations need to apply this intervention across different experience classifications of HIFT participants. These findings demonstrate that HIFT 5 days/week-1 is an effective methodology for improving muscular strength.

Morning rHR significantly decreased for HRV-guided and Predetermined groups from pre- to post-test, whereas no significant changes were observed in HRV or the CV of HRV. Our findings conflict with those of Kliszczewicz et al. [47] who did not observe improvements in rHR after 15 weeks of HIFT, although they also did not find change in HRV. The lack of observed change in HRV may be a function of the nature of HIFT, as Schneider et al. [48] observed a decrease and no change in HRV following a microcycle of strength training and high-intensity interval training, respectively. Although non-significant, changes were found; we did observe a trend for increases in HRV suggesting an increase in parasympathetic activity. Previously, it has been demonstrated that increases in parasympathetic activity are associated with improved fitness characteristics as well as reduced homeostatic perturbations in response to subsequent stressors [7,49,50].

Of note we observed similar fitness improvements in both groups despite the HRV-guided group spending significantly less time training at high-intensity, namely, 13 less days. This is consistent with the findings of Vesterinen et al. [42] in which HRV-guided recreational endurance runners spent less time training at moderate and high intensity. Since an individual's HRV response or ability to maintain homeostatic balance can vary due to training history, exercise modality and exercise intensity, a predetermined training prescription may under- or over-estimate the necessary recovery time required [13,51–53]. The use of an HRV-guided training prescription may aid practitioners/coaches in optimizing the timing of training stress application.

In addition, participant body composition improved in both of our training groups. This finding contrasts with those of Nuuttila et al. [43] in which no changes in body weight or fat percentage were observed after 11 weeks of HRV-guided running. The changes we observed may be attributed to the high levels of body fat of our participants at >22% versus < 13% for Nuuttila et al.'s [43] participants. Favorable changes in body composition were also found by Feito et al. [54] in both men and women, following 16 weeks of HIFT. As in Feito et al.'s [54] study, our participants were not engaged in physical activity specifically targeting changes in body composition prior to the study, thus allowing for a significant change in body composition from pre- to post-test as a result of the training intervention.

A limitation of this study is that HRV measurements were taken by the individual participants and not within a lab setting, which impairs the standardization process. We are unable to say with certainty that all HRV measurements were protocol adherent throughout the study. A degree of inherent trust must be allotted to participants to strictly adhere to the measurement protocols, and while this increases the external validity of our findings it may have affected our internal validity. We were unable to record and quantify participant internal load (e.g., rHR) or external load (e.g., total training volume) during each training session as was done by De Oliveira et al. [13], to demonstrate the difference in the total work completed by both groups. Quantifying the total work completed by participants within each group would provide additional support to the reduced training load completed by the HRV-guided group. Due to the two-site design, different measures of body composition were used at each site, which may contribute to an increased variability in this outcome metric. Participants were instructed to refrain from engaging in any additional exercise outside of the intervention, yet we were unable to ensure these instructions were adhered to throughout the study period. Finally, we were unable to determine the contribution of muscular hypertrophy to the strength gains observed, as muscle cross-sectional area was not assessed.

A key strength of our study is that we were able to demonstrate how a commercially available smartphone application, with a low individual cost, can be an effective tool for modulating the individual prescription of exercise intensity. Additionally, we demonstrated high adherence to daily HRV recordings and the exercise protocols by participants. This demonstrates that daily HRV recordings are manageable for participants over a period of 11 weeks. Our participants displayed a high level of HRV compliance (>90%) and training adherence (>80%) over the 11 week intervention. Finally, this is the first study to modulate HIFT training prescription based on individual HRV.

5. Conclusions

In conclusion, modulating HIFT exercise intensity by individual HRV status, among recreationally active participants, resulted in similar fitness improvements as predetermined HIFT for aerobic capacity, strength, cardiovascular adaptations, and body composition, despite spending fewer days training at high intensity. Practically, our findings suggest that the use of a rolling average of HRV is an effective tool for modulating daily training intensity, with a focus on individual prescription. Coaches and practitioners can use HRV as a tool to effectively individualize exercise prescription for HIFT participation, although additional research is needed to examine the effects for well-trained participants.

Author Contributions: Conceptualization, J.A.D., D.A.C., N.B.D., B.K.K., M.J.C. and K.M.H.; methodology, J.A.D., D.A.C., N.B.D., B.K.K., M.J.C. and K.M.H.; software, J.A.D., D.A.C., N.B.D. and B.K.K.; validation, J.A.D. and D.A.C.; formal analysis, J.A.D. and D.A.C.; investigation, J.A.D., D.A.C., N.B.D. and B.K.K.; resources, D.A.C. and K.M.H.; data curation, J.A.D. and D.A.C.; writing—original draft preparation, J.A.D.; writing—review and editing, J.A.D., D.A.C., N.B.D., A.W. and K.M.H.; visualization, J.A.D.; supervision, D.A.C. and K.M.H.; project administration, D.A.C. and K.M.H.; funding acquisition, J.A.D., B.K.K. and K.M.H. All authors have read and agreed to the published version of the manuscript.

Funding: This work was supported by the Mindlin foundation (48-0771751) and the Kansas State University Office of Undergraduate Research and Inquiry.

Institutional Review Board Statement: The study was conducted according to the guidelines of the Declaration of Helsinki and approved by the Institutional Review Boards of Kansas State University (IRB #9131; 14 March 2018) and Pittsburg State University (26 July 2017).

Informed Consent Statement: Informed consent was obtained from all subjects involved in the study.

Data Availability Statement: The data presented in this study are available on request from the corresponding author.

Conflicts of Interest: The authors declare no conflict of interest. The funders had no role in the design of the study; in the collection, analyses, or interpretation of data; in the writing of the manuscript, or in the decision to publish the results.

Appendix A

Table A1. Detailed description of daily high-intensity functional training intervention.

Day	Structure	Structured Daily Workout
1	M	Two-mile Run (no time cap)
2	GW	[8 Push Press (135/95 lbs.) + 8 Pull-Ups] × 5 rounds for time
3	MGW	[12 Goblet Squats (45/25 lbs.) + 12 Burpees + 24 Calorie Row] AMRAP in 10 min
4	MG	[400-m Run + 25 Box Jumps (18/12")] × 3 rounds for time
5	W	Deadlift 5-5-5-5-5 working up to target 85% of 1RM
6	G	Kipping Pull-Up practice for 20 min
7	WM	[10 Thrusters (135/95 lbs.) + 100 Double Unders] × 4 rounds for time
8	GWM	[6 Handstand Push-Ups + 12 Deadlifts (185/135 lbs.) + 500 m Row] AMRAP in 12 min
9	GW	[15 Ring Rows + 20 Wall Balls (20/14 lbs.)] × 4 rounds for time
10	M	8 km Partner Row (no time cap)
11	W	Front Squat 1-1-1-1-1-1-1-1-1-1 working up to target a 1RM
12	MG	[400-m Run + 20 Push-Ups] × 5 rounds for time
13	WMG	[5 Cleans (135/95 lbs.) + 10 Pull-Ups + 15 Double Unders] AMRAP in 15 min
14	WM	[10/20–8/16–6/12–4/8–2/4 repetitions of Power Clean/Calorie Row] for time
15	G	Handstand Push-Up Practive for 20 min
16	W	Squat 3-3-3-3-3-3-3 working up to target 90% 1RM
17	MG	[800-m Run + 25 Sit-Ups] × 3 rounds for time
18	MGW	[50 Double Unders + 5 Box Jumps (18/12") + 15 Ball Slams (20/14 lbs.)] AMRAP in 15 min
19	GW	[6 Strict Pull-Ups + 6 Front Squats (50% Squat 1RM)] × 4 rounds for time
20	M	Two-mile Run (no time cap)
21	M	Tabata Double Unders × 2
22	GW	[Maximum repetitions Handstand Push-Ups + 6 Deadlifts (75% 1RM)] × 5 rounds for time
23	GWM	[20 Sit-Ups + 16 Dumbbell Clean and Jerk (45/20 lbs.)]
24	WM	[30 Kettlebell Swings (45/20 lbs.) + 400 m Run] × 5 rounds for time
25	G	Strict Pull-Up Practice (Loaded) for 25 min
26	G	Muscle Up Practice for 25 min
27	WM	[6 Squats (50% 1RM) + 50 Double Unders] × 4 rounds for time
28	WMG	[12 Goblet Squats (45/25 lbs.) + 12 Burpees + 24 Calorie Row] AMRAP in 10 min
29	MG	[400-m Run + 10 Handstand Push-Ups] × 5 rounds for time
30	W	Clean 1-1-1-1-1-1-1-1-1-1 working up to target 1RM

M = monostructural (i.e., a single cardiovascular exercise modality) exercise, G = gymnastics exercise, W = weightlifting exercise, and AMRAP = "as many rounds as possible". Daily workouts were scaled to match individual capabilities on an as-needed basis. All scaling options were in accordance with outlined CrossFit scaling practices as per Glassman (2016) (p. 75). Table is adapted from Crawford et al. (2019).

Table A2. Detailed description the modified and light training.

Day	Modified Training	Light Training
1	1.5 m	WALK 20 min
2	25% Volume Reduction. 115/65#	Barbell Press/DB/PVC Pending Strength Levels, DEAD hang stretch or lat banded distraction
3	NO MODULATION	NO MODULATION
4	300 m/Walk 100 m & 18 Jumps 16/12" Step up	Walk 400 m, 2/Leg/Rd Sampson Stretch
5	5RM Load with 2 RIR "5 × 3" at 85%	5 × 3 at 40% 1RM
6	3 Strict Pull ups, Kipping practice (no kipping pull-ups)	Shoulder-strengthing exercise, light lat pull down machine (RPE LIGHT)
7	8 Thrusters, 150 Single Unders. 115/65	Barbell, DB, PVC thrusters and hopping in place, no rope, 20 min
8	8 min: 4 pushups, 8 deadlifts 135/95, 375 row, RPE 13–17	20 min: Pushups, BB/DB/PVC deadlifts, 400 m walking
9	3 RFT, 16/10	Ring Rows, childs pose, walking, 20 min
10	6 K partner row, not for time or 3 k row if solo	Walking 20 min
11	85% of Back Squat, 10 × 1, 2 RIR	KB/Goblet Squat/Barbell Squat, Walking
12	200 m run, walk back, 15 pushups	20 min: 400 m walk, knee pushups
13	5 Deadlifts, ring rows, 15 single unders	Med Ball Cleans, childs pose, shoulder exercise, walking, 20 min

Table A2. *Cont.*

Day	Modified Training	Light Training
14	4 rounds, 2 min break b/t: 115/65 6 power cleans, 12 cal row	Med ball cleans, walking 20 min RPE 6–13
15	HS holds	push-ups, walking
16	7 × 3 2 RIR, 75% target	40% 7 × 3
17	600 m Run, 200 m walk. 20 Situps	20 min Walk
18	12 min	20 min Walk
19	3 RFT	20 min: Goblet Squats, Lat Pulldowns
20	1.5 m Run	20 min Walk
21	1.5 Tabata Rounds	20 min Walk
22	4 RFT Max Push ups 2 RIR, 6 Deadlift 50%	20 min: Push ups, Handstand holds, deadlifts, hamstring curls
23	15 min	20 min Stretching
24	4 RFT 28 KB Swings, 200 M run, 200 walk	20 min Walk
25	30 Pull-Ups	Lat Pull-Downs, Ring Row
26	30 Pull-Ups, 20 dips	Lat Pull-Downs, Ring Row, Press
27	3 RFT	20 min: Air Squats, Box Jumps, walking
28	Testing Day, no modifications	Testing Day, no modifications
29	4 RFT, 200 m run, 200 m walk. 9 push ups	20 min Walk
30	7 × 1 front squat, 2 RIR	20 min: Goblet squat, front rack mobility, barbell front squat

References

1. Bouchard, C.; Rankinen, T. Individual differences in response to regular physical activity. *Med. Sci. Sports Exerc.* **2001**, *33*, S446–S451. [CrossRef]
2. Kiviniemi, A.M.; Hautala, A.J.; Kinnunen, H.; Nissilä, J.; Virtanen, P.; Karjalainen, J.; Tulppo, M.P. Daily Exercise Prescription on the Basis of HR Variability among Men and Women. *Med. Sci. Sports Exerc.* **2010**, *42*, 1355–1363. [CrossRef]
3. Halson, S.L. Monitoring Training Load to Understand Fatigue in Athletes. *Sports Med.* **2014**, *44*, 139–147. [CrossRef] [PubMed]
4. Impellizzeri, F.M.; Marcora, S.M.; Coutts, A.J. Internal and external training load: 15 years on. *Int. J. Sports Physiol. Perform.* **2019**, *14*, 270–273. [CrossRef]
5. Bourdon, P.C.; Cardinale, M.; Murray, A.; Gastin, P.; Kellmann, M.; Varley, M.C.; Gabbett, T.J.; Coutts, A.J.; Burgess, D.J.; Gregson, W.; et al. Monitoring athlete training loads: Consensus statement. *Int. J. Sports Physiol. Perform.* **2017**, *12*, 161–170. [CrossRef]
6. Javaloyes, A.; Sarabia, J.M.; Lamberts, R.P.; Plews, D.; Moya-Ramon, M. Training Prescription Guided by Heart Rate Variability Vs. Block Periodization in Well-Trained Cyclists. *J. Strength Cond. Res.* **2020**, *34*, 1511–1518. [CrossRef]
7. Plews, D.J.; Laursen, P.B.; Kilding, A.E.; Buchheit, M. Evaluating training adaptation with heart-rate measures: A methodological comparison. *Int. J. Sports Physiol. Perform.* **2013**, *8*, 688–691. [CrossRef]
8. Makivic, B.; Djordjevic, M.; Willis, M. Heart Rate Variability (HRV) as a Tool for Diagnostic and Monitoring Performance in Sport and Physical Activities. *J. Exerc. Physiol. Online* **2013**, *16*, 103–131.
9. Vesterinen, V.; Nummela, A.; Heikura, I.; Laine, T.; Hynynen, E.; Botella, J.; Häkkinen, K. Individual Endurance Training Prescription with Heart Rate Variability. *Med. Sci. Sports Exerc.* **2016**, *48*, 1347–1354. [CrossRef]
10. Javaloyes, A.; Sarabia, J.M.; Lamberts, R.P.; Moya-Ramon, M. Training prescription guided by heart-rate variability in cycling. *Int. J. Sports Physiol. Perform.* **2019**, *14*, 23–32. [CrossRef]
11. Kiviniemi, A.M.; Hautala, A.J.; Kinnunen, H.; Tulppo, M.P. Endurance training guided individually by daily heart rate variability measurements. *Eur. J. Appl. Physiol.* **2007**, *101*, 743–751. [CrossRef] [PubMed]
12. Schmitt, L.; Regnard, J.; Parmentier, A.L.; Mauny, F.; Mourot, L.; Coulmy, N.; Millet, G.P. Typology of " Fatigue " by Heart Rate Variability Analysis in Elite Nordic-skiers. *Int. J. Sports Med.* **2015**, *36*, 999–1007. [CrossRef] [PubMed]
13. De Oliveira, R.M.; Ugrinowitsch, C.; Kingsley, J.D.; Da Silva, D.G.; Bittencourt, D.; Caruso, F.R.; Borghi-Silva, A.; Libardi, C.A. Effect of individualized resistance training prescription with heart rate variability on individual muscle hypertrophy and strength responses. *Eur. J. Sport Sci.* **2019**, *19*, 1092–1100. [CrossRef]
14. Feito, Y.; Heinrich, K.; Butcher, S.; Poston, W. High-Intensity Functional Training (HIFT): Definition and Research Implications for Improved Fitness. *Sports* **2018**, *6*, 76. [CrossRef]
15. Crawford, D.A.; Drake, N.B.; Carper, M.J.; DeBlauw, J.A.; Heinrich, K.M. Are Changes in Physical Work Capacity Induced by High-Intensity Functional Training Related to Changes in Associated Physiologic Measures? *Sports* **2018**, *6*, 26. [CrossRef] [PubMed]
16. Crawford, D.A.; Drake, N.B.; Carper, M.J.; DeBlauw, J.A.; Heinrich, K.M. Validity, Reliability, and Application of the Session-RPE Method for Quantifying Training Loads during High Intensity Functional Training. *Sports* **2018**, *6*, 84. [CrossRef]

17. Heinrich, K.M.; Crawford, D.A.; Langford, C.R.; Kehler, A.; Andrews, V. High-Intensity Functional Training Shows Promise for Improving Physical Functioning and Activity in Community-Dwelling Older Adults: A Pilot Study. *J. Geriatr. Phys. Ther.* **2021**, *44*, 9–17. [CrossRef] [PubMed]
18. Thomas, S.; Reading, J.; Shephard, R.J. Revision of the Physical Activity Readiness Questionnaire (PAR-Q). *Can. J. Sport Sci.* **1992**, *17*, 338–345. [PubMed]
19. Plews, D.J.; Scott, B.; Altini, M.; Wood, M.; Kilding, A.E.; Laursen, P.B. Comparison of Heart-Rate-Variability Recording With Smartphone Photoplethysmography, Polar H7 Chest Strap, and Electrocardiography. *Int. J. Sports Physiol. Perform.* **2017**, *12*, 1324–1328. [CrossRef] [PubMed]
20. Esco, M.R.; Flatt, A.A.; Nakamura, F.Y. Agreement between a Smartphone Pulse Sensor Application and Electrocardiography for Determining lnRMSSD. *J. Strength Cond. Res.* **2017**, *31*, 380–385. [CrossRef]
21. Esco, M.R.; Flatt, A.A. Ultra-short-term heart rate variability indexes at rest and post-exercise in athletes: Evaluating the agreement with accepted recommendations. *J. Sports Sci. Med.* **2014**, *13*, 535–541.
22. Williams, S.; Booton, T.; Watson, M.; Rowland, D.; Altini, M. Heart rate variability is a moderating factor in the workload-injury relationship of competitive crossfit™ athletes. *J. Sports Sci. Med.* **2017**, *16*, 443–449. [PubMed]
23. Altini, M. Coefficeint of Variation (CV): What Is It and How Can You Use It? Available online: https://www.hrv4training.com/blog/coefficient-of-variation-cv-what-is-it-and-how-can-you-use-it (accessed on 1 June 2021).
24. Bruce, R.A.; Kusumi, F.; Hosmer, D. Maximal oxygen intake and nomographic assessment of functional aerobic impairment in cardiovascular disease. *Am. Heart J.* **1973**, *85*, 546–562. [CrossRef]
25. Foster, C.; Jackson, A.S.; Pollock, M.L.; Taylor, M.M.; Hare, J.; Sennett, S.M.; Rod, J.L.; Sarwar, M.; Schmidt, D.H. Generalized equations for predicting functional capacity from treadmill performance. *Am. Heart J.* **1984**, *107*, 1229–1234. [CrossRef]
26. Poole, D.C.; Jones, A.M. Measurement of the maximum oxygen uptake Vo2max: Vo2peak is no longer acceptable. *J. Appl. Physiol.* **2017**, *122*, 997–1002. [CrossRef] [PubMed]
27. Butcher, S.; Neyedly, T.; Horvey, K.; Benko, C. Do physiological measures predict selected CrossFit® benchmark performance? *Open Access J. Sports Med.* **2015**, *6*, 241. [CrossRef] [PubMed]
28. Rippetoe, M. The CrossFit Total. Available online: http://journal.crossfit.com/2006/12/the-crossfit-total-by-mark-rip.tpl.1/05/2018 (accessed on 1 May 2018).
29. Schoenfeld, B.J.; Pope, Z.K.; Benik, F.M.; Hester, G.M.; Sellers, J.; Nooner, J.L.; Schnaiter, J.A.; Bond-Williams, K.E.; Carter, A.S.; Ross, C.L.; et al. Longer interset rest periods enhance muscle strength and hypertrophy in resistance-trained men. *J. Strength Cond. Res.* **2016**, *30*, 1805–1812. [CrossRef] [PubMed]
30. Glassman, G. The CrossFit Level 1 Training Guide. *CrossFit J.* **2016**, 15–25. Available online: http://library.crossfit.com/free/pdf/CFJ_English_Level1_TrainingGuide.pdf (accessed on 11 October 2021).
31. Plews, D.J.; Laursen, P.B.; Kilding, A.E.; Buchheit, M. Heart rate variability in elite triathletes, is variation in variability the key to effective training? A case comparison. *Eur. J. Appl. Physiol.* **2012**, *112*, 3729–3741. [CrossRef] [PubMed]
32. Crawford, D.A.; Heinrich, K.M.; Drake, N.B.; DeBlauw, J.; Carper, M.J. Heart rate variability mediates motivation and fatigue throughout a high-intensity exercise program. *Appl. Physiol. Nutr. Metab.* **2020**, *45*, 193–202. [CrossRef]
33. Buchheit, M.; Chivot, A.; Parouty, J.; Mercier, D.; Al Haddad, H.; Laursen, P.B.; Ahmaidi, S. Monitoring endurance running performance using cardiac parasympathetic function. *Eur. J. Appl. Physiol.* **2010**, *108*, 1153–1167. [CrossRef] [PubMed]
34. Boutcher, S.H.; Park, Y.; Dunn, S.L.; Boutcher, Y.N. The relationship between cardiac autonomic function and maximal oxygen uptake response to high-intensity intermittent-exercise training. *J. Sports Sci.* **2013**, *31*, 1024–1029. [CrossRef]
35. Hynynen, E.; Vesterinen, V.; Rusko, H.; Nummela, A. Effects of moderate and heavy endurance exercise on nocturnal HRV. *Int. J. Sports Med.* **2010**, *31*, 428–432. [CrossRef] [PubMed]
36. Drake, N.; Smeed, J.; Carper, M.J.; Crawford, D.A. Effects of Short-Term CrossFit™ Training: A Magnitude-Based Approach. *J. Exerc. Physiol. Online* **2017**, *20*, 111–133.
37. Heinrich, K.M.; Becker, C.; Carlisle, T.; Gilmore, K.; Hauser, J.; Frye, J.; Harms, C.A. High-intensity functional training improves functional movement and body composition among cancer survivors: A pilot study. *Eur. J. Cancer Care* **2015**, *24*, 812–817. [CrossRef]
38. R Core Team. R: A Language and Environment for Statistical Computing. (Version 4.0) (Computer Software). 2020. Available online: https://cran.r-project.org (accessed on 24 August 2020).
39. The Jamovi Project. Jamovi. (Version 1.6) (Computer Software). 2021. Available online: https://www.jamovi.org (accessed on 11 October 2021).
40. Gallucci, M. GAMLj: General Analyses for Linear Models. (Jamovi Module). 2019. Available online: https://gamlj.github.io/ (accessed on 11 October 2021).
41. Cohen, J. *Statistical Power Analysis for the Behavioral Sciences*, 2nd ed.; Routledge: Hillsdale, NJ, USA, 1988; ISBN 978-0805802832.
42. Vesterinen, V.; Hakkinen, K.; Laine, T.; Hynynen, E.; Mikkola, J.; Nummela, A. Predictors of individual adaptation to high-volume or high-intensity endurance training in recreational endurance runners. *Scand. J. Med. Sci. Sports* **2016**, *26*, 885–893. [CrossRef] [PubMed]
43. Nuuttila, O.P.; Nikander, A.; Polomoshnov, D.; Laukkanen, J.A.; Häkkinen, K. Effects of HRV-Guided vs. Predetermined Block Training on Performance, HRV and Serum Hormones. *Int. J. Sports Med.* **2017**, *38*, 909–920. [CrossRef]

44. Hautala, A.J.; Kiviniemi, A.M.; Mäkikallio, T.H.; Kinnunen, H.; Nissilä, S.; Huikuri, H.V.; Tulppo, M.P. Individual differences in the responses to endurance and resistance training. *Eur. J. Appl. Physiol.* **2006**, *96*, 535–542. [CrossRef] [PubMed]
45. Heinrich, K.M.; Spencer, V.; Fehl, N.; Poston, W.S.C. Mission essential fitness: Comparison of functional circuit training to traditional Army physical training for active duty military. *Mil. Med.* **2012**, *177*, 1125–1130. [CrossRef] [PubMed]
46. Buckley, S.; Knapp, K.; Lackie, A.; Lewry, C.; Horvey, K.; Benko, C.; Trinh, J.; Butcher, S. Multimodal high-intensity interval training increases muscle function and metabolic performance in females. *Appl. Physiol. Nutr. Metab.* **2015**, *40*, 1157–1162. [CrossRef] [PubMed]
47. Kliszczewicz, B.; Feito, Y.; Bycura, D.; Brown, D.; Price, B. Vagal Response To 15-Weeks of High- Intensity Functional Training: A Pilot Study. *J. Sport Hum. Perform.* **2016**, *4*, 1–10.
48. Schneider, C.; Wiewelhove, T.; Raeder, C.; Flatt, A.A.; Hoos, O.; Hottenrott, L.; Schumbera, O.; Kellmann, M.; Meyer, T.; Pfeiffer, M.; et al. Heart rate variability monitoring during strength and high-intensity interval training overload microcycles. *Front. Physiol.* **2019**, *10*, 582. [CrossRef]
49. Borresen, J.; Lambert, M.I. Autonomic control of heart rate during and after exercise: Measurements and implications for monitoring training status. *Sports Med.* **2008**, *38*, 633–646. [CrossRef]
50. Kiviniemi, A.M.; Tulppo, M.P.; Hautala, A.J.; Vanninen, E.; Uusitalo, A.L.T. Altered relationship between R-R interval and R-R interval variability in endurance athletes with overtraining syndrome. *Scand. J. Med. Sci. Sport.* **2014**, *24*, 77–85. [CrossRef] [PubMed]
51. Buchheit, M. Monitoring training status with HR measures: Do all roads lead to Rome? *Front. Physiol.* **2014**, *5*, 73. [CrossRef] [PubMed]
52. Stanley, J.; Peake, J.M.; Buchheit, M. Cardiac Parasympathetic Reactivation Following Exercise: Implications for Training Prescription. *Sports Med.* **2013**, *43*, 1259–1277. [CrossRef] [PubMed]
53. Schmitt, L.; Regnard, J.; Millet, G.P. Monitoring fatigue status with HRV measures in elite athletes: An avenue beyond RMSSD? *Front. Physiol.* **2015**, *6*, 2013–2015. [CrossRef]
54. Feito, Y.; Hoffstetter, W.; Serafini, P.; Mangine, G.T. Changes in body composition, bone metabolism, strength, and skill-specific performance resulting from 16-weeks of HIFT. *PLoS ONE* **2018**, *16*, e0198324. [CrossRef] [PubMed]

Journal of
Functional Morphology and Kinesiology

MDPI

Article

Functional Translation of Exercise Responses from Exercise Testing to Exercise Training: The Test of a Model

Tristan Tyrrell [1], Jessica Pavlock [1], Susan Bramwell [1], Cristina Cortis [2], Scott T. Doberstein [1], Andrea Fusco [2], John P. Porcari [1] and Carl Foster [1,*]

1 Department of Exercise and Sport Science, University of Wisconsin-La Crosse, La Crosse, WI 54601, USA; tyrrell2697@uwlax.edu (T.T.); pavlock0615@uwlax.edu (J.P.); sbramwell@uwlax.edu (S.B.); sdoberstein@uwlax.edu (S.T.D.); jporcari007@gmail.com (J.P.P.)

2 Department of Human Sciences, Society and Health, University of Cassino and Lazio Meridionale, 03043 Cassino, Italy; c.cortis@unicas.it (C.C.); andrea.fusco@unicas.it (A.F.)

* Correspondence: cfosteruwl@gmail.com

Abstract: Exercise prescription based on exercise test results is complicated by the need to downregulate the absolute training intensity to account for cardiovascular drift in order to achieve a desired internal training load. We tested a recently developed generalized model to perform this downregulation using metabolic equivalents (METs) during exercise testing and training. A total of 20 healthy volunteers performed an exercise test to define the METs at 60, 70, and 80% of the heart rate (HR) reserve and then performed randomly ordered 30 min training bouts at absolute intensities predicted by the model to achieve these levels of training intensity. The training HR at 60 and 70% HR reserve, but not 80%, was significantly less than predicted from the exercise test, although the differences were small. None of the ratings of perceived exertion (RPE) values during training were significantly different than predicted. There was a strong overall correlation between predicted and observed HR ($r = 0.88$) and RPE ($r = 0.52$), with 92% of HR values within ±10 bpm and 74% of RPE values within ±1 au. We conclude that the generalized functional translation model is generally adequate to allow the generation of early absolute training loads that lead to desired internal training loads.

Keywords: exercise prescription; target heart rate; RPE; METs

Citation: Tyrrell, T.; Pavlock, J.; Bramwell, S.; Cortis, C.; Doberstein, S.T.; Fusco, A.; Porcari, J.P.; Foster, C. Functional Translation of Exercise Responses from Exercise Testing to Exercise Training: The Test of a Model. *J. Funct. Morphol. Kinesiol.* **2021**, *6*, 66. https://doi.org/10.3390/jfmk6030066

Academic Editor: Peter Hofmann

Received: 29 June 2021
Accepted: 26 July 2021
Published: 28 July 2021

1. Introduction

Exercise is a lifestyle factor that is beneficial to health [1–4]. It is beneficial on a dose–response basis, up to amounts of training of several times professional society guidelines [1,2] and only associated with health problems at levels consistent with heavy athletic training in middle-aged and older individuals [1,2,5]. In patients with chronic diseases, exercise forms a cornerstone of the treatment scheme of rehabilitation programs being of value both in terms of accelerating the rate of recovery [6], secondary prevention [7–9], and is even of value after contemporary medical care (percutaneous interventions and statins) is accounted for [10]. However, exercise training can present a significant risk of clinical events [11], particularly in the setting of unaccustomed heavy exercise in sedentary individuals [12] or in the setting of myocardial ischemia during exercise training [13]. Further, as with other behavioral interventions, new exercise programs have a disappointing pattern of compliance [14]. Although high-intensity exercise has recently become popular [15] and has well-documented efficacy [16], even in clinical populations [17–19], it is associated with at least a potentially increased risk of complications and reductions in program enjoyment and compliance [20].

In the setting of an increased risk based on age, risk factors, history of disease or symptoms, professional societies recommend graded exercise testing before beginning exercise programs, both to rule out clinically occult disease and to guide the exercise prescription [21]. The dominant model for subsequent exercise prescription is based on a percentage of the maximal exercise capacity (metabolic equivalents (METs)), heart rate (HR) or HR reserve (%HRR) [21], rating of perceived exertion (RPE) [22], or ventilatory threshold (VT) [23]. Recently, the talk test, which can be taken as a surrogate of VT, has been effectively used to prescribe exercise training [24,25] and to avoid ischemia during training [26].

Unfortunately, the same exercise workload during graded exercise testing (GXT) that produces a given HR, RPE, VT, or talk test response during incremental exercise, does not produce an equivalent response during sustained exercise training. The effect of cardiovascular drift dictates that the response to sustained exercise is often significantly larger than during brief stages at that same workload during a GXT. Potentially, this leads to training sessions that are harder than desired [27], which may, in turn, have adverse effects on safety and compliance. Accordingly, the exercise training workload must be reduced, downregulated, or "translated" in order to achieve desirable results during exercise training. Previous work from our laboratory has demonstrated a solution for translating exercise responses from GXT to ambulation and cycle ergometry [27,28], to arm–leg ergometry [29], to recreational activities [30]. A recent report from our laboratory demonstrated a potentially viable strategy to generalize the process of translating exercise test responses based on computing the MET cost during GXT and during training [31]. If this generalized model were shown to be accurate, then the process of translating either maximal or submaximal GXT responses to workloads useable on the first day of training would be more effective and safer. Accordingly, the intent of this study was to provide a systematic test of this generalized model on the basis of both %HRR and RPE responses.

2. Materials and Methods

The subjects for this study were 20 healthy, young adult volunteers. Although all were physically active, none were systematically trained athletes. The study protocol was approved by the Institutional Review Board for the Protection of Human Subjects at the University of Wisconsin-La Crosse (protocol # 45CFR46, approved 13 April 2021). All subjects provided written informed consent prior to participation. Characteristics of the subjects are presented in Table 1. The overall experimental approach was to start with MaxMETs, which is a normally measured variable during exercise testing, and then apply our predictive model for converting MaxMETs into the workload for exercise training [31] to test the degree to which exercise training responses (HR and RPE) during training fell within the desired range.

Table 1. Mean and standard deviation of the characteristics of the subjects.

Characteristics	Males (n = 10)	Females (n = 10)
Age (years)	23.5 ± 2.2	22.3 ± 0.9
Height (cm)	183.1 ± 7.3	167.6 ± 6.7
Weight (kg)	82.3 ± 18.4	70.0 ± 13.2
VO_2max (ml·kg^{-1}·min^{-1})	53.3 ± 6.1	40.3 ± 5.3
HRmax (bpm)	189 ± 7	185 ± 5
RPEmax	19.9 ± 0.2	19.9 ± 0.2

VO_2max: maximal oxygen uptake; HRmax: maximal heart rate; RPEmax: maximal rating of perceived exertion.

Each subject performed a GXT, to volitional fatigue, using a modified Bruce treadmill protocol [32]. During the test, HR and RPE (6–20 scale) [33] were measured at the end of each 1 min stage and blood pressure was measured every 3 min. During the test, respiratory metabolism was measured using open-circuit spirometry using a mixing chamber-based metabolic system (AEI Technologies, Pittsburgh, PA, USA). Integration of gas exchange values was performed every 30 s and VO_2max was accepted as the highest 30 s value obtained during the test. The VT was computed using both the v-slope and ventilatory equivalent methods [34]. The HR reserve (HRR) was computed based on the observed maximal HR at the end of the test, and resting HR obtained in the standing position just before the test. The individual exercise time vs. HR curves were examined to determine the moment that the HR achieved reference values of 60% HRR, 70% HRR, and 80%HRR, which are surrogates for easy, moderate, and hard exercise. The RPE at these moments was also noted. The speed and grade of the treadmill belt at these moments were used to calculate the MET requirement based on conventional equations [21] and are referred to as GXT METs. The training workload was computed from the generalized functional translation model [31] as 72% of the GXT METs at each of the % HRR targets. The speed and grade required to achieve this MET level were computed by backward solving of the same equations [21]. From the solution, treadmill speeds between 1.79 and 2.23 $m \cdot s^{-1}$ (4–5 mph, 6.4–8 kmh) were avoided since solely walking or solely running is inconvenient in this speed range.

Each subject then performed three, randomly ordered 30 min training bouts (5 min warm-up at 1.33 $m \cdot s^{-1}$, 0% grade, 20 min at the targeted workload, and 5 min cool-down at 1.33 $m^{\cdot -1}$). HR and RPE were measured at 5 min intervals, and the HR and RPE at 15 min, 20 min, and 25 min were averaged and accepted as the HR and RPE response to the training workloads predicted by the generalized functional translation model.

The HR and RPE achieved during steady-state training at the intended easy, moderate, and hard workloads were compared to that calculated from responses during the GXT using repeated measures Analysis of variance (ANOVA). A p-value of <0.05 was accepted as statistically significant. When justified by ANOVA, pairwise comparisons were made using Tukey's test. Correlations are calculated using the Pearson product–moment correlation.

3. Results

The predicted and achieved HR and RPE during the last 15 min of the easy, moderate, and hard training bouts are presented in Table 2. There were small but significant differences between predicted and achieved HR at the 60% HRR (easy) and 70%HRR (moderate) levels of intensity. There was no significant difference between predicted and achieved HR at the 80%HRR intensity. There were no significant differences between predicted and achieved RPE at any of the intensity levels.

Table 2. Mean and standard deviation of the comparison between the predicted and achieved values of heart rate (HR) and rate of perceived exertion (RPE) at the 60%, 70%, and 80% of the heart rate reserve (% HRR) intensities.

% HRR	Predicted HR	Achieved HR	Predicted RPE	Achieved RPE
60%	135.1 ± 4.3	130.2 ± 7.6 *	11.8 ± 1.2	11.2 ± 1.5
70%	147.9 ± 4.6	142.8 ± 8.9 *	13.0 ± 1.4	12.4 ± 1.7
80%	160.4 ± 5.4	157.3 ± 9.7	14.1 ± 1.1	13.8 ± 1.3

* = significantly ($p < 0.05$) different from predicted HR.

A scatterplot of predicted vs. achieved HR for the combined results of the three intensities bouts is presented in Figure 1. There did not appear to be a significant bias in the pattern of responses, and the correlations between predicted and observed HR responses for easy (r = 0.52), moderate (r = 0.74), hard (r = 0.64), and combined (r = 0.88) were uniformly strong. Overall, 92% of observed HR values were within ±10 bpm of the predicted values.

A scatterplot of predicted vs. achieved RPE for the combined results of the three intensities bouts is presented in Figure 2. There did not appear to be a significant bias in the pattern of responses, and the correlations between predicted and observed RPE responses for easy (r = 0.66), moderate (r = 0.81), hard (r = 0.61), and combined (r = 0.88) were uniformly strong. Overall, 74% of the observations were within the ±1 RPE unit of the predicted values.

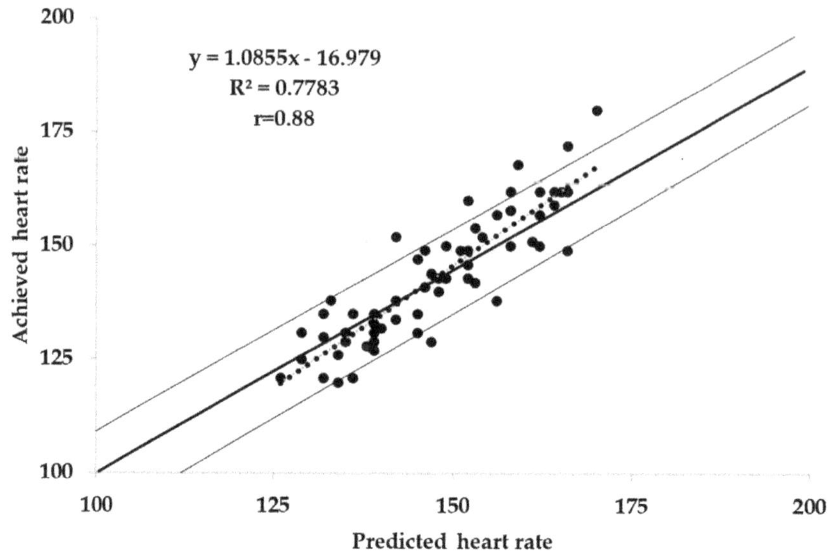

Figure 1. Predicted vs. achieved heart rate during the three intensities bouts.

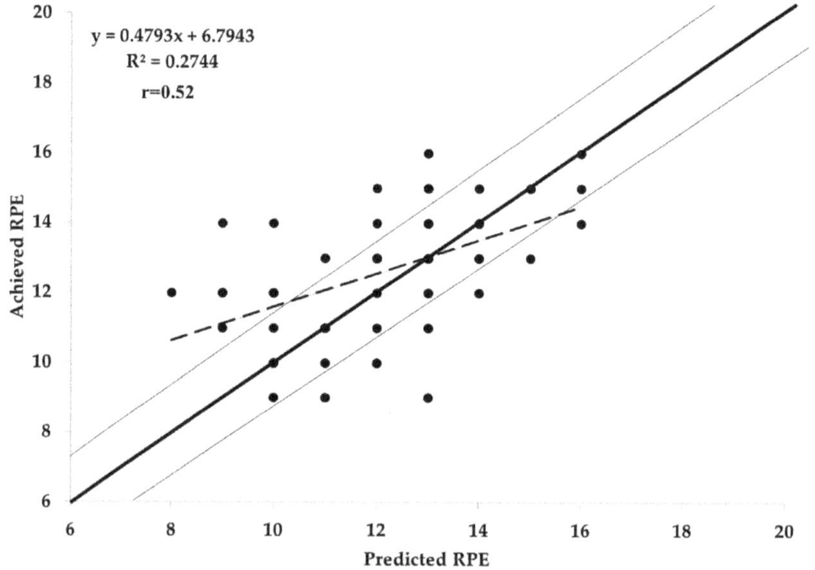

Figure 2. Predicted vs. achieved rating of perceived exertion (RPE) during the three intensities bouts.

4. Discussion

The main finding of this study was that the generalized model for translating GXT responses into training workloads [31] appears to be accurate over the range of training intensities commonly prescribed in fitness and rehabilitation programs. While there was a bias for the predicted HR at the 60% (135 vs. 131) and 70% (148 vs. 143) HRR workloads to be slightly higher than observed, 92% of observed values were within ±10 bpm. Similarly, 74% of RPE values were within the ±1 RPE unit. Thus, the results support our earlier findings with more specific exercise translation approaches [27–30] but using a generalized approach.

While there was a small but significant overprediction of HR responses at the easy and moderate training loads, the error was small, and in a direction that would be acceptable clinically. One of the biggest concerns during exercise prescription is that workloads during the beginning days and weeks of a training program will be too heavy. This may result in reduced enjoyment [20] and may bias toward an increased likelihood of untoward events [11–13]. Thus, the tendency for the "translated" training load to be a little easier than predicted is a very acceptable error.

The primary limitation to the current results is attributable to a limitation of subject selection by the COVID-19 pandemic. In an idealized test of the generalized model, we would have selected fully sedentary individuals, or even patients from a rehabilitation program, as these are the individuals who actively need a strategy for translating GXT results into training prescriptions. However, restrictions on laboratory use dictated that younger and more active students served as the subjects. However, as none of the subjects was systematically training for sports competition, and as their VO_2max spanned a considerable range of fitness, we feel that the experimental test of the generalized functional translation model remains valid.

The generalized functional translation model tested in this study was based on the results of a maximal GXT, as had been the results of the progenitor studies [27–30]. Contemporary practice in both fitness and rehabilitation communities is not to have maximal GXT results available. Where a preliminary GXT is performed, it is often submaximal in nature, limited to an RPE of 15 (hard). However, the strong relationship between the progression of RPE and relative exercise intensity [35,36] and the strong relationship between the talk test and relative exercise intensity [24,25,36] suggest that picking target values for RPE or the talk test from a submaximal GXT and then applying the generalized functional translation model is likely to yield the same prescriptive result as a maximal effort GXT [25].

5. Conclusions

The results of this study suggest that a generalized model for translating GXT results to exercise training loads, based on calculated MET values [31], yields both HR and RPE responses during training bouts that are close to predicted values. Therefore, the results suggest a prescriptive strategy based on 70–75% of the MET requirement at that level of internal training load during a GXT is likely to make the beginning portions of an exercise training program both more pleasant and safer while still being effective.

6. Practical Applications

As an example of the use of the functional translation model, which is validated in the present data, consider the following: A sedentary, although healthy, person performs a GXT (without handrail support) using a Balke type treadmill protocol (a constant speed with grade increments every 2 min) with the results in Table 3.

Table 3. Example of a subject's speed, grade, metabolic equivalents (METs), heart rate (HR), and rating of perceived exertion (RPE) response values during a graded exercise testing using a Balke type treadmill protocol.

Speed ($m{\cdot}s^{-1}$)	Grade (%)	METs [21]	HR	RPE
0	0	1	70	6
1.34	0	3.3	95	8
1.34	2	4.1	115	10
1.34	4	4.9	133	12
1.34	6	5.8	150	14
1.34	8	6.6	155	16
1.34	10	7.4	160	18.5

The target HR (THR) is calculated at 70% HRR as follows:

$$THR = [(160 - 70) \times 0.7] + 70$$

$$THR = [90 \times 0.7] + 70 = 133 \text{ bpm}$$

METs are calculated per standard equations [21] as

$$VO_2 = [(\text{speed } (\text{m}\cdot\text{min}^{-1}) \times 0.1] + [\text{speed } (\text{m}\cdot\text{min}^{-1}) \times 1.8 \times \text{grade}/100] + 3.5$$

For example, at 4% grade (e.g., 70%HRR),

$$VO_2 = [80.4 \times 0.1] + [80.4 \times 1.8 \times 0.04] + 3.5$$

$$VO_2 = 8.4 + 5.8 + 3.5 = 17.7 \text{ mL}\cdot\text{min}^{-1}\cdot\text{kg}^{-1} = 5.1 \text{ METs}$$

Applying the generalized functional translation model, the training intensity becomes

$$5.1 \times 0.72 = 3.7 \text{ METs} = 13.0 \text{ mL}\cdot\text{min}^{-1}\cdot\text{kg}^{-1}$$

Solving for grade at a walking speed of 1.34 $\text{m}\cdot\text{s}^{-1}$,

$$13.0 = [80.4 \times 0.1] + [80.4 \times 1.8 \times \text{grade}] + 3.5$$

$$13.0 = 8.0 + [144.7 \times \text{grade}] + 3.5$$

$$(13.0 - 8.0 - 3.5)/144.7 = \text{grade}$$

$$1.5/144.7 = \text{grade} = 0.01 = 1\%$$

Thus, with the tabled GXT results, one would expect ~70% HRR with an RPE of 11–12 at a walking speed of 1.34 $\text{m}\cdot\text{s}^{-1}$ (3.0 mph or 4.8 kmh) at a grade of 1%. While this may not provide the specific results desired, it does provide a reasonable candidate for how to prescribe the first workouts in the training facility.

Author Contributions: Conceptualization, J.P.P. and C.F.; investigation, T.T., J.P. and C.F.; data curation, T.T., J.P., J.P.P. and C.F.; writing—original draft preparation, T.T., J.P. and C.F.; writing—review and editing, T.T., J.P., S.B., C.C., S.T.D., A.F., J.P.P. and C.F.; supervision, S.B. and J.P.P. All authors have read and agreed to the published version of the manuscript.

Funding: This research received no external funding.

Institutional Review Board Statement: The study was conducted according to the guidelines of the Declaration of Helsinki and approved by the Institutional Review Board for the Protection of Human Subjects at the University of Wisconsin-La Crosse (protocol # 45CFR46, approved 13 April 2021).

Informed Consent Statement: Informed consent was obtained from all subjects involved in the study.

Data Availability Statement: The data presented in this study are available on request from the corresponding author.

Conflicts of Interest: The authors declare no conflict of interest.

References

1. Arem, H.; Mozre, S.C.; Patel, A.; Hartge, P.; De Gonzalez, A.B.; Visvanathan, K.; Campbell, P.T.; Freedman, M.; Weiderpass, E.; Adami, H.O.; et al. Leisure time physical activity and mortality: A detailed pooled analysis of the dose-response relationship. *JAMA Intern. Med.* **2015**, *175*, 959–967. [CrossRef]
2. O'Keefe, J.; Lavie, C.J.; Guazzi, M. Potential danger of extreme endurance exercise: How much is too much? *Prog. Cardiovas. Dis.* **2015**, *57*, 396–405. [CrossRef]
3. Kraus, W.E.; Janz, K.F.; Powell, K.E.; Campbell, W.W.; Jakicic, J.M.; Troiano, R.P.; Sprow, K.; Torres, A.; Piercy, K.L. Daily step counts for measuring physical activity exposure and its relation to health. *Med. Sci. Sports Exerc.* **2019**, *51*, 1206–1212. [CrossRef]
4. Lee, I.M.; Shiroma, E.J.; Lobelo, F.; Puslza, P.; Blair, S.N.; Katzmarek, P.J. Effect of physical inactivity on major non-communicable diseases worldwide: An analysis of burden of disease and life expectany. *Lancet* **2012**, *380*, 21–27. [CrossRef]

5. Paffenbarger, R.S.J.; Hyde, R.T.; Wing, A.L.; Hsieh, C.C. Physical activity and all-cause mortality and longevity of college alumni. *N. Engl. J. Med.* **1986**, *314*, 605–613. [CrossRef]
6. Foster, C.; Pollock, M.L.; Anholm, J.D.; Squires, R.W.; Ward, A.; Dymond, D.S.; Rod, J.L.; Saichek, R.P.; Schmidt, D.H. Work capacity and left ventricular function during rehabilitation from myocardial revascularization surgery. *Circulation* **1984**, *69*, 748–755. [CrossRef]
7. Neibauer, J.; Hambrecht, R.; Velicht, T.; Hauer, K.; Marburger, C.; Kälberer, B.; Weiss, C.; von Hodenberg, E.; Schlierf, G.; Schuler, G.; et al. Attenuate progression of coronary artery disease after 6-years of multi-factorial risk intervention. *Circulation* **1997**, *96*, 2534–2541. [CrossRef]
8. Haskell, W.; Alderman, E.; Fair, J. Intensive risk factor management in men and women with coronary artery disease. The Stanford coronary risk intervention project (SCRIP). *Atherosclerosis* **1984**, *109*, 182–189. [CrossRef]
9. O'Connor, G.T.; Buring, J.E.; Yuset, S.; O'Connor, G.T.; Buring, J.E.; Yusuf, S.; Goldhaber, S.Z.; Olmstead, E.M.; Paffenbarger, R.S., Jr.; Hennekens, C.H. An overview of randomized trials of rehabilitation with exercise after myocardial infarction. *Circulation* **1989**, *80*, 234–244. [CrossRef]
10. Rauch, B.; Davod, C.O.; Doherty, P.; Saure, D.; Metzendorf, M.I.; Salzwedel, A.; Voeller, H.; Jensen, K.; Schmid, J.P. The prognostic effect of cardiac rehabilitation in the era of acute revascularization and statin therapy: A systematic review and meta analysis of randomized and non-randomized trails. *Eur. J. Prev. Cardiol.* **2016**, *23*, 1914–1939. [CrossRef]
11. Foster, C.; Porcari, J.P.; Battista, R.A.; Udermann, B.; Wright, G.; Lucia, A. The risk of exercise training. *Am. J. Lifestyle Med.* **2008**, *2*, 279–284. [CrossRef]
12. Franklin, B.; Billezke, S. Putting the benefits and risks of aerobic exercise in perspective. *Curr. Sports Med. Repts.* **2012**, *12*, 201–208. [CrossRef]
13. Hassock, K.H.; Hartwig, R. Cardiac arrest associated with supervised cardiac rehabilitation. *J. Cardiopulm. Rehabil.* **1982**, *2*, 402–408.
14. Oldridge, N.B. Compliance and experience in primary and secondary prevention of coronary heard disease: A review. *Prev. Med.* **1982**, *12*, 52–70.
15. Burgomaster, K.A.; Hughes, S.C.; Heigenhauser, G.I.F.; Bradwell, S.N.; Gibala, M.J. Six sessions of sprint interval training increases muscle oxidative potential and cycle endurance capacity in humans. *J. Appl. Physiol.* **2005**, *98*, 1985–1998. [CrossRef] [PubMed]
16. Gibala, M.J.; Little, J.P.; Macdonald, M.J.; Hawley, J.A. Physiological adaptation to low volume, high intensity interval training in health and disease. *J. Physiol.* **2012**, *590*, 1077–1084. [CrossRef]
17. Meyer, K.; Foster, C.; Georgakopoulous, N.; Hajric, R.; Westbrook, S.; Ellestad, A.; Tilman, K.; Fitzgerald, D.; Young, H.; Weinstein, H.; et al. Comparison of left ventricular function during interval and steady state exercise in patients with chronic heart failure. *Am. J. Cardiol.* **1998**, *82*, 1382–1387. [CrossRef]
18. Babraj, J.A.; Volland, N.B.J.; Keast, C.; Guppy, F.M.; Cottrell, G.; Tiernan, J.A. Extremely short duration, high intensity interval training substantially improves insulin action in young healthy males. *BMC Endoc. Dis.* **2009**, *9*, 1–8. [CrossRef]
19. Wisloff, U.; Stoylen, A.; Loeuuecker, J.P.; Bruvold, M.; Rognmo, Ø.; Haram, P.M.; Tjønna, A.E.; Helgerud, J.; Slørdahl, S.A.; Lee, S.J.; et al. Superior cardiovascular effect of aerobic interval training versus moderate continuous training in heart failure patients A randomized study. *Circulation* **2007**, *115*, 3086–3094. [CrossRef]
20. Foster, C.; Farland, C.V.; Guidotti, F.; Harbin, M.; Roberts, B.; Schuette, J.; Tuuri, A.; Doberstein, S.T.; Porcari, J.P. The effects of high intensity training vs steady state training on aerobic and anaerobic capacity. *J. Sports Sci. Med.* **2015**, *14*, 747–755.
21. ACSM. *ACSM's Guidelines for Exercise Testing and Prescription*, 10th ed.; Wolters Kluwer: Philadelphia, PA, USA, 2017.
22. Parfitt, G.; Evans, H.; Eston, R. Perceptually regulated training at RPE13 is pleasant and improves physical health. *Med. Sci. Sports Exerc.* **2012**, *44*, 1613–1618. [CrossRef]
23. Mezzani, A.; Hamm, L.F.; Jones, A.M.; McBride, P.E.; Moholdt, T.; Stone, J.A.; Urhausen, A.; Williams, M.A. Aerobic exercise intensity assessment and prescription in cardiac rehabilitation. *J. Cardiopulm. Rehab. Prev.* **2012**, *32*, 327–350. [CrossRef]
24. Foster, C.; Porcari, J.P.; Doro, K.; Dubiel, J.; Engen, M.; Kolman, D.; Xiong, S. Exercise prescription when there is no exercise test: The Talk Test. *Kinesiology* **2018**, *50*, 333–348.
25. Porcari, J.P.; Falck, K.; Turek, J.; Sweeney, S.; Wargowski, A.; Doberstein, S.T.; Foster, C. Comparison of %HRR and Talk Test for exercise prescription. *Kinesiology* **2018**, *50*, 3–10. [CrossRef]
26. Cannon, C.; Foster, C.; Porcari, J.P.; Skemp-Arlt, K.M.; Fater, D.C.W.; Backes, R. The Talk Test as a measure of exertional ischemia. *Am. J. Med. Sport* **2004**, *6*, 52–57.
27. Foster, C.; Porcari, J.P.; Gibson, M.; Wright, G.; Greany, J.; Talati, N.; Recalde, P. Translation of submaximal exercise test responses to exercise prescription using the Talk Test. *J. Str. Cond. Res.* **2009**, *23*, 2425–2429. [CrossRef]
28. Foster, C.; Lemberger, K.; Thompson, N.N.; Sennett, S.M.; Hare, J.; Pollock, M.L.; Pels, A.E., III; Schmidt, D.H. Functional translation of exercise responses from graded exercise testing to exercise training. *Am. Heart J.* **1986**, *112*, 1309–1316. [CrossRef]
29. Foster, C.; Thompson, N.N.; Bales, S. Functional translation of exercise responses using combined arm and leg ergometery. *Cardiology* **1991**, *78*, 150–155. [CrossRef] [PubMed]
30. Foster, C.; Thompson, N.N. Functional translation of exercise test responses to recreational activities. *J. Cardiopulm. Rehabil.* **1991**, *11*, 373–377. [CrossRef]
31. Foster, C.; Anholm, J.D.; Bok, D.; Boullosa, D.; Condello, G.; Cortis, C.; Fusco, A.; Jaime, S.J.; de Koning, J.J.; Lucia, A.; et al. Generalized approach to translating exercise tests and prescribing exercise. *J. Funct. Morph. Kin.* **2020**, *5*, 63. [CrossRef]

32. Foster, C.; Porcari, J.P.; Hipp, M. Clinical exercise testing. In *ACSM's Guidebook for Clinical Exercise Physiologists*; Thompson, W., Ed.; Wolters Kluwer: Baltimore, MD, USA, 2019; pp. 349–364.
33. Borg, G.A.V. *Borg's Perceived Exertion and Pain Scales*; Human Kinetics Press: Champaign, IL, USA, 1998.
34. Foster, C.; Cotter, H.M. Blood lactate, respiratory and heart rate markers on the capacity for sustained exercise. In *Physiological Assessment of Human Fitness*, 2nd ed.; Maud, P.J., Foster, C., Eds.; Human Kinetics Press: Champaign, IL, USA, 2005; pp. 63–76.
35. Eston, R.; Lambrecht, D.; Shephard, K.; Parfitt, G. Prediction of maximal oxygen uptake from a perceptually regulated submaximal graded exercise test. *J. Sports Sci.* **2008**, *26*, 131–139. [CrossRef] [PubMed]
36. Alamji, R.A.; Foster, C.; Porcari, J.P.; Radtke, K.; Doberstein, S. Comparison of non-maximal tests for estimating exercise capacity. *Kinesiology* **2020**, *52*, 10–18.

Journal of
Functional Morphology and Kinesiology

MDPI

Article

Physiological and Psychological Responses to Three Distinct Exercise Training Regimens Performed in an Outdoor Setting: Acute and Delayed Response

Stefano Benítez-Flores [1,*], Carlos A. Magallanes [1], Cristine Lima Alberton [2] and Todd A. Astorino [3]

1 Department of Physical Education and Health, Higher Institute of Physical Education, University of the Republic, Montevideo 11600, Uruguay; camagallanes@gmail.com
2 Neuromuscular Assessment Laboratory, Physical Education School, Federal University of Pelotas, Pelotas 625-96055630, Brazil; tinialberton@yahoo.com.br
3 Department of Kinesiology, California State University San Marcos, San Marcos, CA 92096-0001, USA; astorino@csusm.edu
* Correspondence: stefanobenitez@gmail.com; Tel.: +598-92199656

Abstract: The aim of this study was to compare the acute responses to three time-matched exercise regimens. Ten trained adults (age, maximum oxygen consumption (VO_{2max}), and body mass index (BMI) = 25.9 ± 5.6 yr, 50.9 ± 5.4 $mL \cdot kg^{-1} \cdot min^{-1}$, and 22.1 ± 1.8 $kg \cdot m^{-2}$) completed sprint interval training (SIT) requiring 14 × 5 s efforts with 35 s of recovery, high-intensity interval training (HIIT) consisting of 18 × 15 s efforts at ~90% of peak heart rate (HR_{peak}) with 15 s of recovery, and vigorous continuous training (CT) consisting of 8.75 min at ~85 %HR_{peak}, in randomized order. Heart rate, blood lactate concentration, rating of perceived exertion, affective valence, and enjoyment were monitored. Moreover, indices of neuromuscular function, autonomic balance, diet, mental stress, incidental physical activity (PA), and sleep were measured 24 h after each session to analyze the magnitude of recovery. Both HIIT and CT exhibited a greater %HR_{peak} and time ≥ 90 %HR_{peak} than SIT ($p < 0.05$). Blood lactate and rating of perceived exertion were higher in response to SIT and HIIT vs. CT ($p < 0.05$); however, there were no differences in enjoyment ($p > 0.05$). No differences were exhibited in any variable assessed along 24 h post-exercise between conditions ($p > 0.05$). These data suggest that HIIT and CT accumulate the longest duration at near maximal intensities, which is considered a key factor to enhance VO_{2max}.

Keywords: high-intensity interval training; sprint interval training; continuous training; fatigue; recovery

Citation: Benítez-Flores, S.; Magallanes, C.A.; Alberton, C.L.; Astorino, T.A. Physiological and Psychological Responses to Three Distinct Exercise Training Regimens Performed in an Outdoor Setting: Acute and Delayed Response. *J. Funct. Morphol. Kinesiol.* **2021**, *6*, 44. https://doi.org/10.3390/jfmk6020044

Academic Editors: Cristina Cortis, Andrea Fusco and Carl Foster

Received: 9 April 2021
Accepted: 19 May 2021
Published: 24 May 2021

1. Introduction

For many decades, the efficacy of interval training in human wellness has consistently been shown [1]. There are two distinct categories of interval training, consisting of high-intensity interval training (HIIT), which involves submaximal efforts near the workload associated with maximal heart rate or oxygen uptake, and sprint interval training (SIT), which involves supramaximal bouts requiring "all-out" effort [2]. Both training modalities induce similar physiological adaptations to continuous training (CT) with a lower training volume [1]. Despite the many well-documented benefits of HIIT and SIT on various indices of cardiometabolic health [1], many interval regimens require extremely demanding efforts that may elicit negative perceptions (e.g., Tabata protocol and Wingate-based SIT) [3,4] that could be impractical for non-athletic populations.

Since the level of self-efficacy, motivation, and enjoyment are predictors of physical activity participation (PA) [5], it has been suggested that reducing bout duration elicits more positive affective responses [4]. This is a great advantage of modified SIT which has been shown to significantly improve the cardiometabolic profile in healthy and unhealthy adults [2,6–8]. In addition, modified SIT requires lower total volume [6,8] which could mitigate the lack of time indicated as the main barrier to engaging in regular PA [5]. In fact, current guidelines recommend small blocks of vigorous PA given that they induce several health benefits [9]. However, most interval training regimens are conducted in laboratory environments using expensive technology [10] that limit the external validity of these findings to real-world settings such as the home or workplace where minimal equipment is available.

Chronic adaptations to exercise training are caused by the accumulation of acute cardiovascular and metabolic responses [11] which highlights the importance of assessing the change in variables including heart rate (HR) and blood lactate concentration to exercise. However, only a few studies have compared differences in these variables in response to interval training regimens performed outside of a laboratory. Warr-di Piero et al. [12] recruited a sample of heterogeneous athletes to perform four different regimens of HIIT with bout durations equal to 10, 50, 90, and 130 s. The results showed greater blood lactate and rating of perceived exertion in response to longer bouts (i.e., 90 and 130 s) compared to the shorter efforts. This result is in agreement with data reported by Cipryan et al. [13] acquired in a laboratory. In healthy subjects, Eigendorf et al. [14] evaluated physiological responses to SIT, HIIT, and CT having identical mean power output. This study showed no significant difference between conditions in oxygen consumption (VO_2), respiratory exchange ratio, or plasma ammonia concentration. Nevertheless, the exercise duration was substantial (~75 min), reducing its applicability in untrained populations.

Besides the acute responses to exercise, latent physiological responses also influence training adaptations. In addition, the magnitude of delayed-onset muscle soreness (DOMS) in the days following exercise may impair adherence to PA, particularly in sedentary adults [15]. Nevertheless, a recent study by Farias-Junior et al. [16] in overweight men showed no alterations in numerous markers of muscle damage and inflammation 48 h after HIIT and CT matched for volume. Yet, to the best of our knowledge, there are no data concerning changes in variables including sleep, mental stress, or neuromuscular function in response to time-matched exercise regimens completed using accessible equipment in field conditions. At present, it has been suggested that heart rate variability (HRV), jump performance, and psychometric questionnaires are reliable and practical non-invasive measurements to elucidate the overall recovery status from prior exercise [17].

Thus, the aim of this study was to compare acute physiological and psychological responses to three different field-based exercise protocols with identical total duration yet different structure, and additionally, to observe changes in various markers of recovery for 24 h post-exercise employing affordable tools. Our hypothesis is that no differences in HR values and recovery status will be evident between conditions due to the similar exercise workload.

2. Materials and Methods

2.1. Subjects

Twelve healthy adults (range 20–40 yr; 8 males, 4 females) participated in the study during November/December 2019 (Table 1). The inclusion criteria were: (1) absence of musculoskeletal injuries and cardiometabolic risk factors; (2) highly physically active according to the short IPAQ; (3) not consuming any nutritional supplements, drugs, or tobacco products; (4) not competing professionally in any sport; and (5) previous experience in intense exercise training. Participants were instructed to abstain from PA and alcohol consumption for 48 h before all sessions and also to avoid stimulating drinks (e.g., coffee, mate, etc.) in the morning/afternoon of each session. In addition, they were asked not to change lifestyle habits (e.g., work, sleep, food, etc.) throughout the experiment. Prior to involvement, all procedures, potential risks, and benefits were fully explained to participants and subsequently, they provided their informed consent. This study was carried out in accordance with the principles stipulated in the Declaration of Helsinki of 1975, revised in 2013.

Table 1. Participants' characteristics.

Variable	
Age (yr)	25.9 ± 5.6
Height (cm)	173.2 ± 7.6
Weight (kg)	66.3 ± 7.3
BMI (kg·m^{-2})	22.1 ± 1.8
Body fat mass (%)	19.0 ± 7.6
Body skeletal muscle mass (%)	38.5 ± 5.8
HR_{rest} (beat·min^{-1})	55 ± 5
HR_{peak} SRT (beat·min^{-1})	200 ± 9
VO_{2max} SRT (mL·kg^{-1}·min^{-1})	50.9 ± 5.4

BMI = body mass index; HR_{peak} = peak heart rate; HR_{rest} = rest heart rate; VO_{2max} = maximum oxygen consumption; SRT = shuttle run test. Data are mean ± SD.

2.2. Study Design

This study adopted a randomized crossover design that consisted of one session to measure various physical and morphological variables followed by 3 time-matched sessions of endurance training (SIT, HIIT, and CT). The design was developed with the aim of being integrated in real-world circumstances, using low-cost tools that are easy to employ. All procedures took place on a 400 m outdoor public track. Every session consisted of different efforts ("all-out", intermittent submaximal, and continuous submaximal) with a similar internal load (i.e., HR) and completion of ~9 min of exercise per session, and was separated by 7 days. All sessions were held on the same days (Monday and Tuesday), time of day (0800–1100 a.m. and 200–500 p.m.), and season (Spring), with similar environmental conditions (18–25 °C temperature, 40–50% humidity, and 11–20 km·h^{-1} wind). Dietary consumption was monitored for 24 h before the first exercise session using diet recalls, and we requested that participants replicate the same food intake 24 h before the other sessions. Physiological and psychological responses were monitored during all sessions. In addition, in the 24 h after each session, various indicators of residual neuromuscular/metabolic fatigue were monitored to determine the level of recovery/stress. Additionally, incidental PA and sleep were controlled as possible confounding variables (Figure 1).

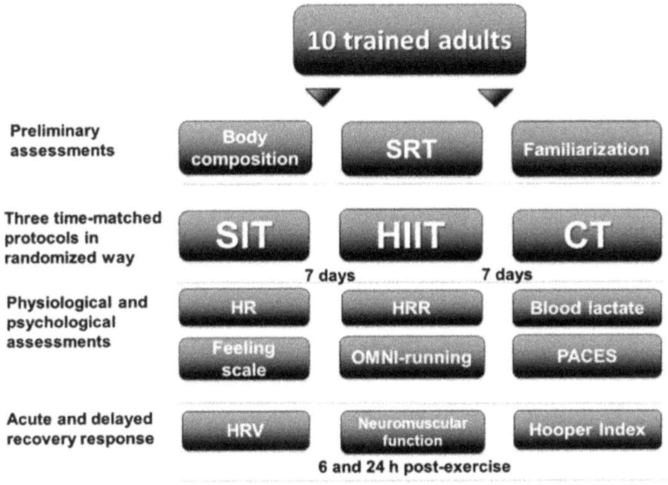

Figure 1. Study design. SRT = shuttle run test; SIT = sprint interval training with 5 s efforts; HIIT = high-intensity interval training with 15 s efforts; CT = continuous training; HR = heart rate; HRR = heart rate recovery; OMNI-running = OMNI-running rating of perceived exertion scale; PACES = physical activity enjoyment scale; HRV = heart rate variability.

2.3. Procedures

2.3.1. Preliminary Assessments

In the first session, the participants completed the short IPAQ followed by anthropometric measurements. Subsequently, the 20-m shuttle run test (SRT) was performed to estimate VO_{2max}. Finally, they were familiarized with the exercise protocols.

Body Composition

The following anthropometric measurements were recorded: height (cm), weight (kg), body fat (%), and body skeletal muscle mass (%), utilizing a digital body composition bioimpedance sensor HBF-514C, OMRON (Kyoto, Japan).

20-m Shuttle Run Test

The SRT is a practical approach to assess cardiorespiratory fitness in non-laboratory settings without need for sophisticated equipment [18]. The test consists of running for as long as possible between 2 lines separated by 20 m with a rhythm imposed by audio. The initial speed was equal to 8.5 km$\cdot$h^{-1} and was increased by 0.5 km$\cdot$h^{-1} every minute. The speed obtained in the last stage is considered as the maximal aerobic speed. The end of the test is determined when the distance of 20 m cannot be covered in two consecutive efforts. The HR was monitored during the test using a telemetric system Firstbeat Technologies Ltd. (Jyväskylä, Finland). Peak heart rate (HR_{peak}) was registered and VO_{2max} was estimated according to the formula proposed by Leger and Gadoury [18] (VO_{2max} = (6 $\times$ maximal aerobic speed) $-$ 27.4).

Familiarization

Participants were familiarized with the psychological scales and neuromuscular function tests by completing 2–3 repetitions of each test. Additionally, they were familiarized with the three exercise protocols as they completed 1 min of CT at ~85% HR_{max} (estimated according to: 220 minus age), two to three 15 s bouts of HIIT at ~90% HR_{max} (same as CT), and two 5 s bouts of SIT.

Exercise Protocols

The warm-up included 3 min of running at a self-selected pace. Total time for every mode was ~14 min. The CT and HIIT bouts were prescribed at intensities equal to 85 and 90% of the HR_{peak}. During each session, HR was instantaneously monitored and feedback was provided so that the subjects maintained the target intensity. Previously, our group identified that HR was equal to ~90% HR_{max} in response to a very similar modified SIT protocol as the one used in the current study [19]. During SIT, participants were instructed to run as fast as they could for 5 s. After every repetition of HIIT and SIT, passive pauses were used to facilitate recovery [11], and then participants were warned by a sound signal to run in the opposite direction. The characteristics of these protocols are described in Table 2.

2.3.2. Physiological and Psychological Assessments in the Three Exercise Protocols

External and Internal Load

Throughout all exercise regimens, the distance achieved was quantified by placing landmarks every 5 m on the track, and an investigator individually followed each participant to record the distance run.

Internal load of the sessions was determined using HR data collected with chest straps Firstbeat Technologies Ltd. (Jyväskylä, Finland) that were later exported to the Firstbeat Sports software version 4.7.3.1, Firstbeat Technologies Ltd. (Jyväskylä, Finland). The intensity descriptors selected were maximal (HR_{max}), mean (HR_{mean}), minimum (HR_{min}), percent peak HR (%HR_{peak}), time $\geq$ 70 %HR_{peak} (i.e., between 70 and 80% of the HR_{peak}), time $\geq$ 80 %HR_{peak} (i.e., between 80 and 90% of the HR_{peak}), and time $\geq$ 90 %HR_{peak} (i.e., between 90 and100% of the HR_{peak}).

HR Recovery (HRR)

The relative HRR was defined as the difference between HR registered at the end of exercise (HR_{end}) and after 30, 60, 90, and 120 s of recovery (i.e., HRΔ30s, HRΔ60s, HRΔ90s, and HRΔ120s). This variable was recorded as the subjects walked at ~4 km·h^{-1}, similar to a previous study [20].

Blood Lactate Concentration

Fingertip blood samples (15 μL of blood) were collected pre- and 4 min post-exercise using disposable lancets and placed in reagent strips for subsequent analysis with a portable lactate analyzer Accutrend, Roche Diagnostics (Basel, Switzerland).

Affective Valence, Rating of Perceived Exertion, and Enjoyment

The Feeling scale was employed pre- and immediately post-exercise to measure the affective valence (pleasure and displeasure), ranging from −5 (very bad) to +5 (very good) [21]. To determine perceived exertion pre- and immediately post-exercise, the OMNI-running rating of perceived exertion scale (OMNI-running) was used [22]. This scale was validated for adults during walking/running exercise and is easier to interpret and apply than the classic Borg scale (6–20). The physical activity enjoyment scale (PACES) is traditionally used to characterize the level of enjoyment induced by exercise. In the present study, a valid and reliable Spanish short version was completed immediately post-exercise [23].

Table 2. HR response for the training protocols.

Variable	SIT	HIIT	CT
Total exercise duration (min)	13.75	13.75	13.75
Exercise duration (min)	8.75	8.75	8.75
Number of efforts	14	18	1
Work/recovery (s)	5/35	15/15	None
Intensity	all-out	~90 %HR_{peak}	~85 %HR_{peak}
Distance (m)	428.8 ± 27.4 *	1318.4 ± 112.9 †	1784.5 ± 192.6
HR_{max} (beat·min^{-1})	177 ± 10	185 ± 10 †	182 ± 8
HR_{mean} (beat·min^{-1})	166 ± 10 *	175 ± 8	173 ± 8
HR_{min} (beat·min^{-1})	146 ± 15	148 ± 18	134 ± 17
time ≥ 70 %HR_{peak} (s)	114 ± 100	30 ± 32	18 ± 29
time ≥ 80 %HR_{peak} (s)	384 ± 99	300 ± 141	414 ± 143
time ≥ 90 %HR_{peak} (s)	30 ± 58 ‡	180 ± 141	90 ± 156
HR_{end} (beat·min^{-1})	168 ± 12	182 ± 10†	174 ± 8
HRΔ30s (beat·min^{-1})	14 ± 8	15 ± 5	18 ± 11
HRΔ60s (beat·min^{-1})	27 ± 9	30 ± 10	36 ± 13
HRΔ90s (beat·min^{-1})	36 ± 11 *	49 ± 7	50 ± 11
HRΔ120s (beat·min^{-1})	46 ± 13 ‡	59 ± 6	53 ± 10

SIT = sprint interval training with 5 s efforts; HIIT = high-intensity interval training with 15 s efforts; CT = continuous training; HR = heart rate. Data are mean ± SD. * $p < 0.05$ vs. HIIT and CT. † $p < 0.05$ vs. SIT and CT. ‡ $p < 0.05$ vs. HIIT.

2.3.3. Acute and Delayed Recovery Response

Heart Rate Variability (HRV)

To assess HRV, the participants were oriented comfortably in the supine position and requested to breathe normally and avoid any kind of movements throughout data acquisition. The subjects remained in this position for 2 min pre- and post-exercise, and 6 and 24 h post-exercise. For the analysis, Firstbeat Sports software version 4.7.3.1, Firstbeat Technologies Ltd. (Jyväskylä, Finland) was adopted, and only the second minute of recording was analyzed because the first minute is considered a stabilization period. This type of recording has been used to assess autonomic regulation accurately in a field environment [24]. The variable selected was the root mean square of successive differences between R-R intervals (RMSSD), established as the strongest indicator to monitor autonomic balance [17].

Neuromuscular Function

To observe the neuromuscular function, the PUSH band version 2.0 PUSH Inc. (Toronto, ON, Canada) was employed with waist belt secured properly. The PUSH is a portable device based on a triaxial accelerometer and gyroscope that register samples at 1000 Hz, transforming them into one 200 Hz signal. This device was created to track the velocity of movement during a variety of strength exercises and shows adequate validity and reliability [25]. Two tests were chosen to assess the impact of the different exercise regimens on neuromuscular function. First, the countermovement jump (CMJ) was performed, given that it is a validated technique to determine lower-body power [26]. Second, the reactive strength index (RSI) was evaluated through completion of 10 continuous rebound jumps to assess the capacity of reactive strength. The RSI 10/5 stiffness was considered, including only the best 5 jumps [27]. The CMJ and RSI tests were performed before, 4 min post-exercise, and 6 and 24 h post-exercise. In the case of the CMJ, the average of 2 repetitions separated by 1 min recovery was analyzed, while in the case of RSI, only 1 repetition was analyzed.

Hooper Index

The Hooper index is an ecological method to assess general parameters of recovery and wellness for athletes [28]. This psychometric questionnaire is based on ratings relative to fatigue, stress level, muscle soreness, and sleep quality, scored on a seven-point Likert scale with 1-point increments scores of 1–7, with 1 and 7 representing very, very low (very, very good in the case of sleep) and very, very high (very, very bad in the case of sleep), respectively. The Hooper index was administered pre-, immediately post-exercise, and 6 and 24 h post-exercise in all conditions.

Incidental PA

Incidental PA was monitored by accelerometry GT3X, ActiGraph, LCC. (Pensacola, FL, USA) 48 h before and 24 h after completion of each exercise condition. Subjects were instructed to wear the device on their non-dominant wrist, and to only remove it for any aquatic activity. Analysis days were included when $\geq$ 10 h·day of recording was attained. The daily energy expenditure (EE) and incidental PA patterns were calculated, applying an epoch of 1 s and a frequency of 100 Hz, with the Freedson's algorithm for cutoff points: (1) light, <1951 counts·min; (2) moderate, 1952–5724 counts·min; (3) vigorous, 5725–9498 counts·min; and (4) very vigorous, >9499 counts·min [29]. All estimations were completed in manufacturer software ActiLife version 6.13.4, ActiGraph, LCC. (Pensacola, FL, USA).

Sleep

Sleep indicators were calculated 48 h before (i.e., 2 nights) and 24 h after (i.e., 1 night) for each condition, utilizing accelerometry GT3X, ActiGraph, LCC. (Pensacola, FL, USA) with an epoch of 60 s based on the Cole and Kripke algorithm [30]. All estimations were performed using manufacturer ActiLife version 6.13.4, ActiGraph, LCC. (Pensacola, FL, USA).

2.4. Statistical Analysis

Sample size was defined with the following input parameters in G*Power version 3.1.9.7, Düsseldorf University https://www.psychologie.hhu.de/arbeitsgruppen/allgemeine-psychologie-und-arbeitspsychologie/gpower (Düsseldorf, Germany): (1) F test for one group and three measurements; (2) effect size of 0.45; (3) a-value of 0.05; (4) statistical power of 0.80; and (5) correlation between measures of 0.5. The calculated sample size was 10.

Data are presented as mean ± SD. Normality was assessed by means of standard distribution measures, visual inspection of Q–Q plots, box plots, and the Shapiro–Wilk test. Variables with a non-normal distribution were log-transformed for analysis. Where normalization was not possible, non-parametric methods were used. One-way repeated measures ANOVA was conducted to examine differences in select variables (i.e., distance, HR, HRR, PACES, etc.) during the three sessions. In addition, this technique was used to compare changes in incidental PA and sleep (before 48 h and after 24 h). A series of 2 (time equal to pre- and post-) × 3 (condition equal to SIT, HIIT, and CT) two-way repeated measures ANOVA was performed to assess differences in blood lactate. Additionally, the same analysis with a 4 × 3 model was used to assess differences across time (pre-, post-, post-6 h, and post-24 h) and condition (SIT, HIIT and CT) for the evaluation of HRV and RSI. Mauchly's sphericity was tested and if sphericity could not be assumed the Greenhouse–Geisser correction was used. When required, pairwise comparisons were conducted using Bonferroni's corrections. In the case of non-normally distributed variables (OMNI-running, Feeling scale, CMJ, and Hooper index) an analysis with non-parametric techniques was applied using the Friedman rank sum test with post-hoc Nemenyi. Effect sizes were calculated using n_p^2 in order to examine the magnitude of the differences between the three sessions. For non-parametric variables, the Kendall's W coefficient (k) was used as the measure of the Friedman test effect size. Cohen's d (for normal variables) and $r = z/\sqrt{N}$ (for non-normal variables) were calculated for ES analyses representing ≤ 0.20 as a small effect, 0.50 as a medium effect, and $\geq$0.80 as a large effect. Parametric statistics were performed with IBM SPSS version 23.0 (Armonk, NY, USA) and non-parametric statistics with R (R Core Team, 2018). In all cases, the alpha level was set at $p < 0.05$. All graphics were made with GraphPad Prism version 6.01 (San Diego, CA, USA).

3. Results

One female participant withdrew from the study due to time constraints. Additionally, data from one female participant were not considered since in one session, the HR signal was not detected. All other participants were able to complete the three training conditions.

3.1. External and Internal Load

3.1.1. HR

The completed distance was significantly different ($p < 0.001$; $n_p^2 = 0.98$) across all conditions ($p < 0.001$) (Table 2). Significant differences were detected in HR_{max} ($p < 0.001$; $n_p^2 = 0.57$) as it was higher in HIIT vs. SIT ($p = 0.050$; $d = 0.80$) and CT ($p = 0.027$; $d = 0.33$). In addition, the results show that HR_{mean} and $\%HR_{peak}$ ($p < 0.001$; $n_p^2 = 0.60$) were lower in SIT than HIIT ($p = 0.003$; $d = 0.99$) and CT ($p = 0.029$; $d = 0.77$) (Figure 2A), although there was no difference in the HR_{min} ($p = 0.105$; $n_p^2 = 0.22$). For time $\geq$ 70 $\%HR_{peak}$, a significant difference was found ($p = 0.032$; $k = 0.34$) although post-hoc analyses showed no differences between means ($p > 0.05$). For time $\geq$ 80 $\%HR_{peak}$, no significant difference was shown ($p = 0.196$; $k = 0.16$) but for time $\geq$ 90 $\%HR_{peak}$, a significant difference between conditions occurred ($p = 0.017$; $k = 0.40$) as it was lower in SIT than HIIT ($p = 0.011$; $r = 0.80$) (Table 2).

3.1.2. HRR and Blood Lactate

The results show a significant difference in HR_{end} ($p \leq 0.001$; $n_p^2 = 0.57$) that was higher in response to HIIT than SIT ($p = 0.002$; $d = 1.26$) and CT ($p = 0.031$; $d = 0.88$). A significant difference was observed in HRΔ90s ($p = 0.002$; $n_p^2 = 0.51$) that was lower in SIT vs. HIIT ($p = 0.031$; $d = 1.41$) and CT ($p = 0.002$; $d = 1.27$). Moreover, a significant difference was identified in HRΔ120s ($p = 0.007$; $n_p^2 = 0.43$) that was lower in response to SIT vs. HIIT ($p = 0.018$; $d = 1.28$). No differences were observed in HRΔ30s ($p = 0.517$; $n_p^2 = 0.10$) or HRΔ60s ($p = 0.070$; $n_p^2 = 0.26$) across conditions (Table 2).

Significant condition ($p < 0.001$; $n_p^2 = 0.65$), time ($p < 0.001$; $n_p^2 = 0.92$), and condition × time interaction ($p < 0.001$; $n_p^2 = 0.65$) effects were identified for blood lactate. Post-hoc outcomes revealed higher post-exercise values vs. pre- in SIT ($p < 0.001$; $d = 3.71$), HIIT ($p < 0.001$;

d = 3.21), and CT ($p < 0.001$; d = 2.10). Additionally, higher post-exercise values occurred in SIT (11.5 ± 3.4 mmol L^{-1}; $p < 0.001$; d = 2.26) and HIIT (10.6 ± 3.7 mmol L^{-1}; p = 0.005; d = 1.80) than CT (5.4 ± 1.7 mmol L^{-1}). These data are shown in Figure 2B.

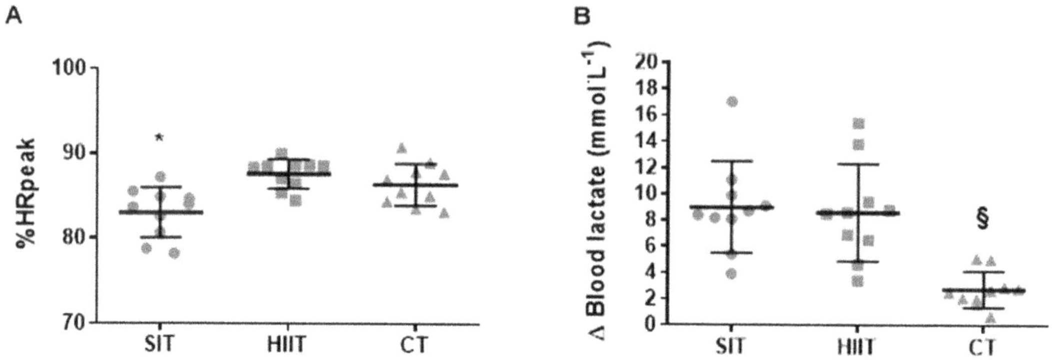

Figure 2. Individual response of (**A**) $\%HR_{peak}$ through the training protocols and individual (**B**) Δ of blood lactate through the training protocols. SIT = sprint interval training with 5 s efforts; HIIT = high-intensity interval training with 15 s efforts; CT = continuous training. * $p < 0.05$ differences vs. HIIT and CT. § $p < 0.05$ differences vs. HIIT and SIT.

3.1.3. Affective Valence, Perceived Exertion, and Enjoyment

Significant main effects of condition (p = 0.030; k = 0.37), time (p = 0.020; k = 0.54), and a training × time interaction ($p < 0.001$; k = 0.54) were identified for the Feeling scale. Post-hoc analysis identified differences in affective valence during HIIT (p = 0.009; r = 0.89) and lower post-exercise values in HIIT versus CT (p = 0.020; r = 0.80) (Figure 3A).

There were significant main effects of condition (p = 0.007; k = 0.50), time (p = 0.002; k = 1), and training × time interaction ($p < 0.001$; k = 0.85), indicating a different pattern in response to the three regimens for OMNI-running. Pairwise comparisons showed differences in pre- vs. post- for SIT (p = 0.002; r = 0.89) and HIIT ($p < 0.001$; r = 0.88). No inter-conditions differences were found at any time ($p > 0.05$) (Figure 3B).

No differences in PACES were observed between conditions (p = 0.734; n_p^2 = 0.034) (Figure 3C).

3.2. *Acute and Delayed Recovery Response*

3.2.1. HRV

There were significant main effects of time ($p < 0.001$; n_p^2 = 0.91) and training × time (p = 0.011; n_p^2 = 0.26) for lnRMSSD. Pairwise comparisons revealed decrements in post-exercise measures for SIT ($p < 0.001$; d = 4.31), HIIT ($p < 0.001$; d = 4.63), and CT ($p < 0.001$; d = 2.71) and lower values than the other time points ($p < 0.01$). No differences between conditions were found at any time point ($p > 0.05$) and there was no main effect of condition (p = 0.875; n_p^2 = 0.01) (Figure 4).

3.2.2. CMJ and RSI

Significant main effects of time (p = 0.002; k = 0.49) and training × time (p = 0.002; k = 0.26) interaction were noted, yet post-hoc tests showed no differences between means. No main effect of training was observed (p = 0.120; k = 0.21) (Figure 5A). The results show no main effects of training, time, or training × time interaction ($p > 0.05$; $n_p^2 \geq 0.22$) for RSI (Figure 5B).

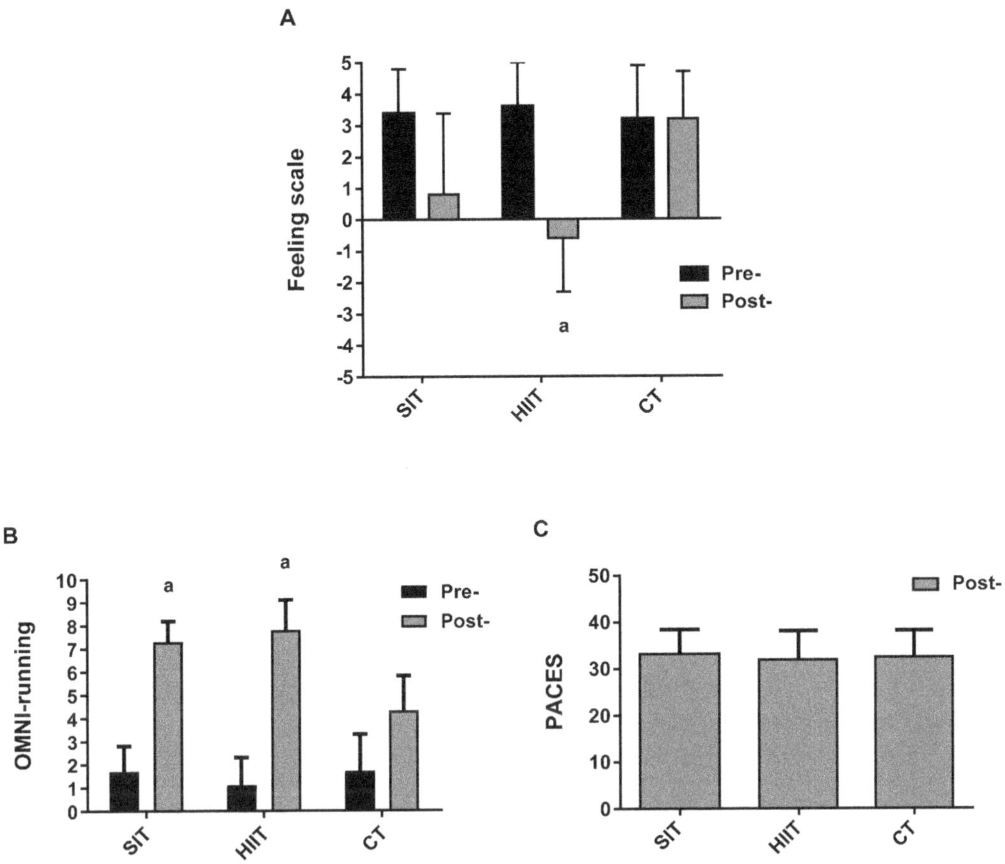

Figure 3. Group response for (**A**) Feeling scale; (**B**) OMNI-running, and (**C**) PACES through the training protocols. SIT = sprint interval training with 5 s efforts; HIIT = high-intensity interval training with 15 s efforts; CT = continuous training. [a] $p < 0.05$ differences vs. pre-training.

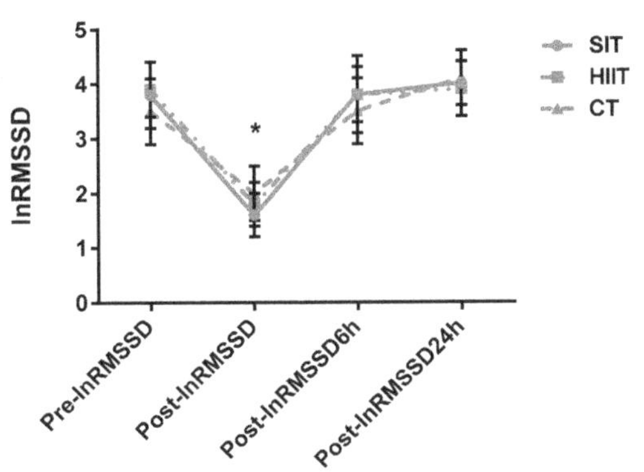

Figure 4. Group response comparisons of the lnRMSSD at the different time points. SIT = sprint interval training with 5 s efforts; HIIT = high-intensity interval training with 15 s efforts; CT = continuous training. * $p < 0.05$ differences vs. pre-training for all protocols.

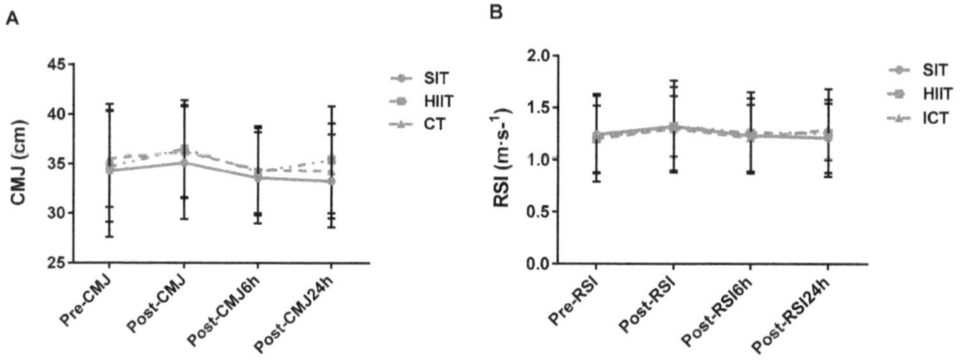

Figure 5. Group response comparisons of the (**A**) CMJ and (**B**) RSI at the different time points. SIT = sprint interval training with 5 s efforts; HIIT = high-intensity interval training with 15 s efforts; CT = continuous training.

3.2.3. Hooper Index

Regarding fatigue, significant training ($p = 0.007$; $k = 0.50$), time ($p < 0.001$; $k = 0.63$), and training×time interaction ($p < 0.001$; $k = 0.50$) effects were identified. Paired comparisons showed a difference between pre- vs. post-exercise for SIT ($p = 0.005$; $r = 0.89$) and HIIT ($p = 0.010$; $r = 0.84$). Regarding stress, significant time ($p = 0.020$; $k = 0.34$) and training×time interaction ($p = 0.004$; $k = 0.25$) effects were detected, without a main effect of training ($p = 0.140$; $k = 0.19$). Neither paired differences were found ($p > 0.05$). Regarding muscle soreness, significant time ($p = 0.002$; $k = 0.51$) and training×time interaction ($p = 0.002$; $k = 0.27$) effects were observed, without a main effect of training ($p = 0.070$; $k = 0.27$). Neither paired differences were noted ($p > 0.05$). Regarding sleep quality, no effect was detected ($p > 0.05$; $k \geq 14$).

3.2.4. Incidental PA and sleep

No differences were detected before and after for all parameters of incidental PA ($p > 0.05$; $n_p{}^2 \geq 0.26$) or sleep ($p > 0.05$; $n_p{}^2 \geq 0.024$) (Table 3).

Table 3. Incidental PA and sleep for the training protocols.

	SIT	HIIT	CT	SIT	HIIT	CT
Variable	**Before 48 h**			**After 24 h**		
INCIDENTAL PA						
Kcal·day^{-1}	717.5 ± 185	842.4 ± 300.1	849.7 ± 239.2	1138 ± 351.5	1303.4 ± 418.3	1218.9 ± 458.6
%sedentary domain	74.6 ± 5.5	70.6 ± 9.9	68.1 ± 7.2	69.6 ± 5.7	65.8 ± 6.3	68.3 ± 6.4
%light domain	13.6 ± 3	15.4 ± 4.7	17.3 ± 3.1	15 ± 4.9	16.2 ± 3	15.1 ± 3.7
%MVPA domain	11.8 ± 3	14 ± 5.8	14.6 ± 4.9	15.5 ± 1.2	18 ± 4.5	16.6 ± 3.2
SLEEP						
Efficiency	88.1 ± 5.9	89.2 ± 4.3	90 ± 3.3	89.8 ± 5.1	91.6 ± 4.4	89.6 ± 5.3
TST	365 ± 76.4	373.9 ± 106.2	366.4 ± 68.5	360.1 ± 90.3	317.3 ± 82.2	349.2 ± 85.6
WASO	49.8 ± 26.6	43.9 ± 16.5	38.6 ± 18.9	43 ± 24.3	27.8 ± 11.5	37.7 ± 17.8
N°awakenings	17.6 ± 6.6	19.5 ± 6.5	16.8 ± 7.8	18.8 ± 11.4	15.1 ± 6	17.7 ± 8.6

SIT = sprint interval training with 5 s efforts; HIIT = high-intensity interval training with 15 s efforts; CT = continuous training; MVPA = moderate vigorous physical activity; WASO = wake after sleep onset; TST = total sleep time. Data are mean ± SD.

4. Discussion

This is the first study that compared acute and delayed responses of time-matched interval training regimens utilizing brief efforts vs. continuous exercise in real-world settings. Our results show no significant differences between HIIT and CT in HR_{mean} and time ≥ 90 %HR_{peak}, yet these outcomes are lower in response to SIT. Second, blood lactate concentration and perception of effort were higher in SIT and HIIT than CT. Third, no differences between protocols were detected in acute fatigue and general recovery status over 24 h, suggesting that brief bouts of interval exercise or vigorous exercise do not interfere with subsequent recovery.

Data from many recent reviews demonstrate that low-volume interval training is effective to enhance cardiorespiratory fitness, glycemic regulation, and body fat [2,8]. However, to date there is still a lack of results elucidating responses to different interval training protocols in relation to CT, since most studies compare regimens having dissimilar exercise load [31]. Obtaining a better understanding of acute responses to these regimens is important, as different iterations of peak workload, bout number, intensity, duration, mean workload, and intensity and duration of recovery may elicit specific acute disturbances of homeostasis that in turn promote specific physiological adaptations [32].

It has been suggested that for exercise training to improve central and peripheral factors associated with O_2 transport and utilization, participants should exercise at intensities near VO_{2max} and spend at least several minutes at this target intensity [11]. In elite cyclists, Almquist et al. [33] compared acute responses to brief (30 s) vs. long intervals (5 min) having the same volume and work:rest ratio. Their results show that brief intervals led to a 14% higher mean power output and 153% longer duration above 90% HR_{peak}. Our data show that 15 s bouts of HIIT elicited the highest HR throughout the session in the form of a greater $\%HR_{peak}$ (+5%) and longest duration running at intensities $\geq 90\ \%HR_{peak}$ (+500%) (Figure 2 and Table 2). Furthermore, HIIT and CT revealed similar HR_{min} and HR_{max} values, suggesting that both sessions elicit the same cardiorespiratory intensity despite different running velocity. These findings coincide with the information presented by Tschakert et al. [34] and propose that reduced durations of effort have a greater ability to stabilize cardiac function.

During intense, brief efforts of exercise, including HIIT and SIT, the contribution of phosphagen and glycolytic metabolism is high. When recoveries are too brief and/or VO_2 kinetics are slowed, oxymyoglobin availability is attenuated [35]. Such a situation of "partial hypoxia" may lead to a decline in PCr concentrations and to an increase in anaerobic glycolysis towards ATP supply. Our data support this idea, as SIT and HIIT elicit similar blood lactate values that are higher compared to CT (Figure 2) with a large effect size (≥ 1.80), despite equal exercise duration. Our outcomes are supported by those obtained in active men and women performing lab-based cycling [36]. A likely mechanism explaining this result is the higher recruitment of type IIx glycolytic muscle fibers required by both interval regimens. Similar findings were observed by Eigendorf et al. [14], who compared three exercise approaches at 50% maximal power output (SIT 6 s work × 24 s rest: 33 min; HIIT 30 s work × 30 s rest: 38 min; CT: 45 min). Furthermore, our protocols showed a lesser blood lactate accumulation than a remarkable number of previous designs that applied SIT or HIIT [11]. Overall, reduced durations of effort, regardless of the exercise modality used (i.e., cycling or running), seem to rely more on oxidative metabolism with less dependence on glycolytic metabolism, attenuating the residual fatigue [12,13,19,34,37].

The pleasure:displeasure and enjoyment experienced during exercise could be a predictor of future engagement in exercise programs [5]. When exercise intensity surpasses the workload associated with the anaerobic threshold, affective valence declines regardless of total work completed [38]. In this regard, some studies have compared the psychological responses to interval training vs. CT matched for internal and external parameters. For example, Jung et al. [39] documented that HIIT (10 bouts of 1 min at ~100% × 1 min at ~20% maximal power output) was more enjoyable and preferable than CT (20 min at 80% maximal power output) and CT (40 min at 40% maximal power output) despite a more aversive response. In contrast, Oliveira et al. [40] observed a lower affective valence in response to HIIT vs. CT at the same average intensity (85% of respiratory compensation point). Similarly, Saanijoki et al. [41] indicated that perceived exertion and arousal was more negative after Wingate-based SIT vs. CT. However, Oliveira et al. [40] and Saanijoki et al. [41] employed relatively high-volume regimens of HIIT (2 min) and

SIT (30 s) which induce marked disruptions of homeostasis. Another study that exhibited more positive affective valence did require very short efforts (5 s) [4]. Interestingly, we found that 15 s bouts of HIIT show the most aversive response (Figure 3) and a significant post-effort sensation of fatigue (effect size = 0.84). Despite these differences, no difference in enjoyment was reported, supporting previous data [36,40]. Alternatively, it could be that the PACES scale is cognitive, and affective valence is not, and is more reflective of an more instantaneous perception of exercise. The opponent-process theory states that pleasant affective states are easier to achieve after adverse physical stimulus, as a kind of reward mechanism [42]. This may be linked to the release of chemical modulatory neurotransmitters associated with pleasure and decreased anxiety [40]. Moreover, the intermittent nature of interval training induces a "rebound effect" that generates a better balance of pleasure [39]. However, we did not detect differences in affective valence pre-exercise, given that 62% of the change in psychological variables in response to SIT and HIIT was explained by baseline values [43]. Therefore, it can be argued that the format of interval training protocols implemented in our study (HIIT and SIT) does not harm tolerance or perceptual response. However, chronic studies are necessary to further explore this hypothesis. In this sense, present findings indicate that self-paced HIIT can induce greater physical enjoyment than CT in active young adults [44].

In order to optimize responses to exercise training, it is important to monitor fatigue, fitness, or performance adaptations [17]. Heart rate variability is a non-invasive metric that can be applied daily in large groups to assess the magnitude of physiological recovery using cardiac autonomic balance [17]. It is evident that interval training sessions should be conducted when vagal HRV indicators are high, when they return to baseline, or even when they exceed values from baseline assessments [11]. We observed that lnRMSSD, which represents the vagal influence on autonomic control, is similar over the 24 h after completion of SIT, HIIT, and CT. In fact, differences only were noted for all conditions immediately post-exercise (Figure 4). Although greater anaerobic metabolism occurs during SIT and HIIT, this phenomenon was not reflected in sympathovagal balance. Likewise, no differences were observed in any variable between protocols 6 and 24 h post-exercise, which is attributed to the low-time commitment of these sessions leading to a relatively rapid recovery.

One interesting finding was the lack of difference in CMJ and RSI post-exercise (Figure 5). These tests can be used to assess potential reductions in neuromuscular function and onset of residual fatigue in field conditions. The results suggest that low-volume, intense exercise sessions having distinct stimuli and structure (i.e., "all-out", intermittent submaximal, and continuous submaximal) do not impair the capacity of subsequent muscular contraction. Similarly, our group reported that a ~12 min session of 5 s sprints did not mitigate subsequent performance in a vertical jump test in healthy active males [19]. Previously, others demonstrated that the small training workload of HIIT and CT did not increase markers of muscle damage and inflammation 48 h post-exercise [16]. Thus, low-volume regimens as completed in the present study may be useful to implement in active adults who choose concurrent training or exercise twice daily, as these bouts do not seem to reduce subsequent neuromuscular function.

Our results from the Hooper index show no significant changes in the general recovery time-course patterns for any regimen. Previously, an association between changes in lnRMSSD and Hooper index has been reported in sub-elite athletes during subsequent days to the competition [45], suggesting that this approach could be an easy way to quantify global psychophysiological stress. Overall, affordable and practical tools such as HRV, jump performance, and psychometric questionnaires could help practitioners and coaches monitor acute and delayed metabolic responses and onset of functional overreaching after different exercise protocols. Additionally, there were no differences from pre- to post-exercise in multiple indicators of sleep quality and daily PA, which is likely due to the low-volume of our regimens. These variables are paramount to consider since they can significantly influence internal and external responses allowing return to homeostasis.

This study has some limitations that must be considered. First, HR during interval exercise may have a faster kinetic response than VO_2 and other physiological mechanisms, including acidosis; thus, it may not be the most robust marker for monitoring exercise intensity [17]. However, due to the development of wearable technology, HR can be an useful indicator of internal load, since the perception of effort may not be accurate in inactive adults. Second, variables related to the recovery status were only monitored for a single day after exercise, so we do not know how these variables would change if assessed over an extended period. However, our participants regularly engaged in team sports, resistance training, and endurance exercise, which may have conditioned their psychological or physiological responses. Third, our sample was small, physically trained, and included predominantly men, so these findings cannot be generalized to higher volume protocols, women, or inactive populations.

5. Conclusions

Our findings suggest that brief bouts of HIIT and CT with a low-time commitment (~14 min) elicit a substantial cardiorespiratory demand ≥ 90 %HR_{peak} that is superior to that induced by SIT. Moreover, our data support those from laboratory-based protocols requiring cycling or running and demonstrate that track-based HIIT consisting of repeated brief efforts can elicit the intense characteristics of interval training while being more accessible to the general population. As a high contribution of anaerobic metabolism attendant with HIIT and SIT could generate more aversive psychological responses, it is advisable to start a training program with a few brief efforts (i.e., 4–6) and then gradually increase bout duration or number. Additionally, it is necessary to examine these training models under unsupervised circumstances, seeking to progress on real-world conditions that are truly scalable in the long-term. Finally, strength and conditioning coaches have to take into account that when endurance exercise sessions are reduced (i.e., < 15 min), regardless of the type of stimulus (SIT, HIIT, or CT), the recovery process can be completed quickly (~6 h).

Author Contributions: Conceptualization, S.B.-F.; methodology, S.B.-F.; investigation, S.B.-F. and C.A.M.; formal analysis, S.B.-F., C.A.M., C.L.A., and T.A.A.; writing—original draft preparation, S.B.-F., C.A.M., C.L.A., and T.A.A., writing—review and editing, S.B.-F., C.A.M., C.L.A., and T.A.A. All authors have read and agreed to the published version of the manuscript.

Funding: This research was founded by Proyectos I + D, Comisión Sectorial de Investigación Científica-Universidad de la República, Uruguay 2018, grant number 192.

Institutional Review Board Statement: The study was conducted according to the guidelines of the Declaration of Helsinki, and approved by the Ethics Committee of Instituto Superior de Educación Física, Universidad de la República, Uruguay (2/2020 and date of final approval 4 November 2020).

Informed Consent Statement: Informed consent was obtained from all subjects involved in the study.

Data Availability Statement: Data pertaining to this research study are available from the corresponding author upon reasonable request.

Acknowledgments: We want to thank all the participants, as well as Mariana Gómez, Pablo Pérez-Ifrán, Maximiliano Rial, and Andrés Parodi for their assistance during data collection.

Conflicts of Interest: The authors declare no conflict of interest.

References

1. Batacan, R.B.; Duncan, M.J.; Dalbo, V.J.; Tucker, P.S.; Fenning, A.S. Effects of high-intensity interval training on cardiometabolic health: A systematic review and meta-analysis of intervention studies. *Br. J. Sports Med.* **2017**, *51*, 494–503. [CrossRef]
2. Gibala, M.J.; Little, J.P. Physiological basis of brief vigorous exercise to improve health. *J. Physiol.* **2020**, *598*, 61–69. [CrossRef]
3. Foster, C.; Farland, C.V.; Guidotti, F.; Harbin, M.; Roberts, B.; Schuette, J.; Tuuri, A.; Doberstein, S.T.; Porcari, J.P. The effects of high intensity interval training vs steady state training on aerobic and anaerobic capacity. *J. Sports Sci. Med.* **2015**, *14*, 747–755.
4. Townsend, L.K.; Islam, H.; Dunn, E.; Eys, M.; Robertson-Wilson, J.; Hazell, T.J. Modified sprint interval training protocols. Part II. Psychological responses. *Appl. Physiol. Nutr. Metab.* **2017**, *42*, 347–353. [CrossRef]

5. Trost, S.G.; Owen, N.; Bauman, A.E.; Sallis, J.F.; Brown, W. Correlates of adults' participation in physical activity: Review and update. *Med. Sci. Sports Exerc.* **2002**, *34*, 1996–2001. [CrossRef]
6. Benítez-Flores, S.; Medeiros, A.R.; Voltarelli, F.A.; Iglesias-Soler, E.; Doma, K.; Simões, H.G.; Rosa, T.S.; Boullosa, D.A. Combined effects of very short "all out" efforts during sprint and resistance training on physical and physiological adaptations after 2 weeks of training. *Eur. J. Appl. Physiol.* **2019**, *119*, 1337–1351. [CrossRef]
7. Schaun, G.Z.; Pinto, S.S.; Silva, M.R.; Dolinski, D.B.; Alberton, C.L. Whole-body high-intensity interval training induce similar cardiorespiratory adaptations compared with traditional high-intensity interval training and moderate-intensity continuous training in healthy men. *J. Strength Cond. Res.* **2018**, *32*, 2730–2742. [CrossRef]
8. Vollaard, N.B.J.; Metcalfe, R.S. Research into the Health Benefits of Sprint Interval Training Should Focus on Protocols with Fewer and Shorter Sprints. *Sports Med.* **2017**, *47*, 2443–2451. [CrossRef]
9. Stamatakis, E.; Johnson, N.A.; Powell, L.; Hamer, M.; Rangul, V.; Holtermann, A. Short and sporadic bouts in the 2018 US physical activity guidelines: Is high-intensity incidental physical activity the new HIIT? *Br. J. Sports Med.* **2019**, *53*, 1137–1139. [CrossRef]
10. Gray, S.R.; Ferguson, C.; Birch, K.; Forrest, L.J.; Gill, J.M. High-intensity interval training: Key data needed to bridge the gap from laboratory to public health policy. *Br. J. Sports Med.* **2016**, *50*, 1231–1232. [CrossRef]
11. Buchheit, M.; Laursen, P.B. High-intensity interval training, solutions to the programming puzzle: Part I: Cardiopulmonary emphasis. *Sports Med.* **2013**, *43*, 313–338. [CrossRef]
12. Warr-di Piero, D.; Valverde-Esteve, T.; Redondo-Castán, J.C.; Pablos-Abella, C.; Sánchez-Alarcos Díaz-Pintado, J.V. Effects of work-interval duration and sport specificity on blood lactate concentration, heart rate and perceptual responses during high intensity interval training. *PLoS ONE* **2018**, *13*, e0200690. [CrossRef]
13. Cipryan, L.; Tschakert, G.; Hofmann, P. Acute and post-exercise physiological responses to high-intensity interval training in endurance and sprint athletes. *J. Sports Sci. Med.* **2017**, *16*, 219–229.
14. Eigendorf, J.; Maassen, M.; Apitius, D.; Maassen, N. Energy metabolism in continuous, high-Intensity, and sprint interval training protocols with matched mean intensity. *J. Strength Cond. Res.* **2019**. [CrossRef]
15. Cheung, K.; Hume, P.; Maxwell, L. Delayed onset muscle soreness: Treatment strategies and performance factors. *Sports Med.* **2003**, *33*, 145–164. [CrossRef] [PubMed]
16. Farias-Junior, L.F.; Browne, R.A.V.; Freire, Y.A.; Oliveira-Dantas, F.F.; Lemos, T.; Galvão-Coelho, N.L.; Hardcastle, S.J.; Okano, A.H.; Aoki, M.S.; Costa, E.C. Psychological responses, muscle damage, inflammation, and delayed onset muscle soreness to high-intensity interval and moderate-intensity continuous exercise in overweight men. *Physiol. Behav.* **2019**, *199*, 200–209. [CrossRef]
17. Buchheit, M. Monitoring training status with HR measures: Do all roads lead to Rome? *Front. Physiol.* **2014**, *5*, 73. [CrossRef]
18. Léger, L.; Gadoury, C. Validity of the 20 m shuttle run test with 1 min stages to predict VO2max in adults. *Can. J. Sport Sci.* **1989**, *14*, 21–26. [PubMed]
19. Benitez-Flores, S.; De Sousa, A.F.; Da Cunha Totó, E.C.; Rosa, T.S.; Del Rosso, S.; Foster, C.; Boullosa, D.A. Shorter sprints elicit greater cardiorespiratory and mechanical responses with less fatigue during time-matched sprint interval training (SIT) sessions. *J. Electromyogr. Kinesiol.* **2018**, *50*, 137–148. [CrossRef]
20. Boullosa, D.A.; Barros, E.S.; del Rosso, S.; Nakamura, F.Y.; Leicht, A.S. Reliability of heart rate measures during walking before and after running maximal efforts. *Int. J. Sports Med.* **2014**, *35*, 999–1005. [CrossRef]
21. Hardy, C.J.; Rejeski, W.J. Not what, but how one feels: The measurement of affect during exercise. *J. Sport Exerc. Psychol.* **1989**, *11*, 304–317. [CrossRef]
22. Utter, A.C.; Robertson, R.J.; Green, J.M.; Suminski, R.R.; McAnulty, S.R.; Nieman, D.C. Validation of the Adult OMNI Scale of perceived exertion for walking/running exercise. *Med. Sci. Sports Exerc.* **2004**, *36*, 1776–1780. [CrossRef]
23. Fernández García, E.; Sánchez Bañuelos, F.; Salinero Martín, J.J. Validation and adaptation of the PACES scale of enjoyment of the practice of physical activity for Spanish adolescent girls. *Psicothema* **2008**, *20*, 890–895. [PubMed]
24. Esco, M.R.; Williford, H.N.; Flatt, A.A.; Freeborn, T.J.; Nakamura, F.Y. Ultra-shortened time-domain HRV parameters at rest and following exercise in athletes: An alternative to frequency computation of sympathovagal balance. *Eur. J. Appl. Physiol.* **2018**, *118*, 175–184. [CrossRef]
25. Montalvo, S.; Gonzalez, M.P.; Dietze-Hermosa, M.S.; Eggleston, J.D.; Dorgo, S. Common vertical jump and reactive strength index measuring devices: A validity and reliability analysis. *J. Strength Cond. Res.* **2021**. [CrossRef] [PubMed]
26. Markovic, G.; Dizdar, D.; Jukic, I.; Cardinale, M. Reliability and factorial validity of squat and countermovement jump tests. *J. Strength Cond. Res.* **2004**, *18*, 551–555. [CrossRef] [PubMed]
27. Comyns, T.M.; Flanagan, E.P.; Fleming, S.; Fitzgerald, E.; Harper, D.J. Interday reliability and usefulness of a reactive strength index derived from 2 maximal rebound jump tests. *Int. J. Sports Physiol. Perform.* **2019**, *14*, 1200–1204. [CrossRef]
28. Hooper, S.L.; Mackinnon, L.T.; Howard, A.; Gordon, R.D.; Bachmann, A.W. Markers for monitoring overtraining and recovery. *Med. Sci. Sports Exerc.* **1995**, *27*, 106–112. [CrossRef]
29. Freedson, P.S.; Melanson, E.; Sirard, J. Calibration of the Computer Science and Applications, Inc. accelerometer. *Med. Sci. Sports Exerc.* **1998**, *30*, 777–781. [CrossRef]
30. Cole, R.J.; Kripke, D.F.; Gruen, W.; Mullaney, D.J.; Gillin, J.C. Automatic sleep/wake identification from wrist activity. *Sleep* **1992**, *15*, 461–469. [CrossRef] [PubMed]

31. Andreato, L.V.; Esteves, J.V.; Coimbra, D.R.; Moraes, A.J.P.; de Carvalho, T. The influence of high-intensity interval training on anthropometric variables of adults with overweight or obesity: A systematic review and network meta-analysis. *Obes. Rev.* **2019**, *20*, 142–155. [CrossRef]
32. Tschakert, G.; Hofmann, P. High-intensity intermittent exercise: Methodological and physiological aspects. *Int. J. Sports Physiol. Perform.* **2013**, *8*, 600–610. [CrossRef] [PubMed]
33. Almquist, N.W.; Nygaard, H.; Vegge, G.; Hammarström, D.; Ellefsen, S.; Rønnestad, B.R. Systemic and muscular responses to effort-matched short intervals and long intervals in elite cyclists. *Scand. J. Med. Sci. Sports* **2020**, *30*, 1140–1150. [CrossRef]
34. Tschakert, G.; Kroepfl, J.; Mueller, A.; Moser, O.; Groeschl, W.; Hofmann, P. How to regulate the acute physiological response to "aerobic" high-intensity interval exercise. *J. Sports Sci. Med.* **2015**, *14*, 29–36. [PubMed]
35. Glaister, M. Multiple sprint work: Physiological responses, mechanisms of fatigue and the influence of aerobic fitness. *Sports Med.* **2005**, *35*, 757–777. [CrossRef]
36. Olney, N.; Wertz, T.; LaPorta, Z.; Mora, A.; Serbas, J.; Astorino, T.A. Comparison of acute physiological and psychological responses between moderate-intensity continuous exercise and three regimes of high-intensity interval training. *J. Strength Cond. Res.* **2018**, *32*, 2130–2138. [CrossRef] [PubMed]
37. Islam, H.; Townsend, L.K.; Hazell, T.J. Modified sprint interval training protocols. Part I. Physiological responses. *Appl. Physiol. Nutr. Metab.* **2017**, *42*, 339–346. [CrossRef]
38. Kilpatrick, M.; Kraemer, R.; Bartholomew, J.; Acevedo, E.; Jarreau, D. Affective responses to exercise are dependent on intensity rather than total work. *Med. Sci. Sports Exerc.* **2007**, *39*, 1417–1422. [CrossRef]
39. Jung, M.E.; Bourne, J.E.; Little, J.P. Where does HIT fit? An examination of the affective response to high-intensity intervals in comparison to continuous moderate- and continuous vigorous-intensity exercise in the exercise intensity-affect continuum. *PLoS ONE* **2014**, *9*, e114541. [CrossRef]
40. Oliveira, B.R.; Slama, F.A.; Deslandes, A.C.; Furtado, E.S.; Santos, T.M. Continuous and high-intensity interval training: Which promotes higher pleasure? *PLoS ONE* **2013**, *8*, e79965. [CrossRef] [PubMed]
41. Saanijoki, T.; Nummenmaa, L.; Koivumäki, M.; Löyttyniemi, E.; Kalliokoski, K.K.; Hannukainen, J.C. Affective adaptation to repeated SIT and MICT protocols in insulin-resistant subjects. *Med. Sci. Sports Exerc.* **2018**, *50*, 18–27. [CrossRef] [PubMed]
42. Solomon, R.L. The opponent-process theory of acquired motivation: The costs of pleasure and the benefits of pain. *Am. Psychol.* **1980**, *35*, 691–712. [CrossRef] [PubMed]
43. Astorino, T.A.; Vella, C.A. Predictors of change in affect in response to high intensity interval exercise (HIIE) and sprint interval exercise (SIE). *Physiol. Behav.* **2018**, *196*, 211–217. [CrossRef] [PubMed]
44. Soylu, Y.; Arslan, E.; Sogut, M.; Kilit, B.; Clemente, F. Effects of self-paced high-intensity interval training and moderate intensity continuous training on the physical performance and psychophysiological responses in recreationally active young adults. *Biol. Sport* **2021**, *38*, 555–562. [CrossRef]
45. Rabbani, A.; Clemente, F.M.; Kargarfard, M.; Chamari, K. Match fatigue time-course assessment over four days: Usefulness of the hooper index and heart rate variability in professional soccer players. *Front. Physiol.* **2019**, *10*, 109. [CrossRef]

Journal of
Functional Morphology and Kinesiology

Article

Electromyographic Analysis of the Lumbar Extensor Muscles during Dynamic Exercise on a Home Exercise Device

John M. Mayer [1,*], Brian E. Udermann [2] and Joe L. Verna [1,3]

1 The Vert Mooney Research Foundation (DBA US Spine & Sport Foundation), San Diego, CA 92111, USA; joev@spineandsport.com
2 Department of Exercise and Sport Science, College of Science and Health, University of Wisconsin—LaCrosse, LaCrosse, WI 54601, USA; budermann@uwlax.edu
3 Spine & Sport Physical Therapy Inc., San Diego, CA 92111, USA
* Correspondence: johnmayer@usssf.com; Tel.: +1-858-583-4494

Abstract: Resistance exercise with devices offering mechanisms to isolate the lumbar spine is effective to improve muscle strength and clinical outcomes. However, previously assessed devices with these mechanisms are not conducive for home exercise programs. The purpose of this study was to assess the surface electromyographic (EMG) activity of the lumbar extensor muscles during dynamic exercise on a home back extension exercise device. Ten adults (5 F, 5 M) performed dynamic lumbar extension exercise on a home device at three loads: 1.00 × body weight (BW), 1.25 × BW and 1.50 × BW. Surface EMG activity from the L3/4 paraspinal region was collected. The effect of exercise load, phase of movement, and position in the range of motion on lumbar extensor EMG activity (normalized to % maximum voluntary isometric contraction) was assessed. Lumbar extensor EMG activity significantly increased from 1.00 BW to 1.50 BW loads ($p = 0.0006$), eccentric to concentric phases ($p < 0.0001$), and flexion to extension positions ($p < 0.0001$). Exercise using a home back extension exercise device progressively activates the lumbar extensor muscles. This device can be used for home-based resistance exercise programs in community-dwelling adults without contraindications.

Keywords: low back pain; electromyography; exercise training; home exercise program; virtual care

Citation: Mayer, J.M.; Udermann, B.E.; Verna, J.L. Electromyographic Analysis of the Lumbar Extensor Muscles during Dynamic Exercise on a Home Exercise Device. *J. Funct. Morphol. Kinesiol.* **2022**, 7, 26. https://doi.org/10.3390/jfmk7010026

Academic Editors: Cristina Cortis, Andrea Fusco and Carl Foster

Received: 11 January 2022
Accepted: 25 February 2022
Published: 1 March 2022

1. Introduction

Low back pain is common, costly, disabling, and greatly impacts the quality of life of adults around the world [1,2]. Hundreds of treatment options are available for low back pain, many of which have only modest results [2]. Physical exercise is typically recommended as an effective prevention and treatment strategy for low back pain [2]. Among the options for therapeutic exercise, progressive resistance exercise using back extension machines that isolate the lumbar spine has a relatively large body of evidence to support safety, and ability to improve physical function and reduce disability [3].

The premise behind use of these machines is that isolating the lumbar spine through various restraint mechanisms forces the lumbar extensor muscles (e.g., multifidus, erector spinae) to be the primary producers of torque during compound trunk extension, thereby limiting the input of the more powerful gluteal and hamstring muscles [4,5]. This strategy has been shown to effectively activate the lumbar extensors and other trunk extensor muscles [6,7]. Progressive resistance exercise training on these devices has been shown to result in large strength gains in healthy adults [4,5], and relieve symptoms and restore functional capacity in individuals with chronic low back pain [3,8–10]. Existing back extension machines that stabilize the pelvis in the seated position and allow for gradual loading of lumbar extensors are intended for in-clinic use, which requires face-to-face interactions with a therapist or trainer. The computerized lumbar extension dynamometer (MedX, Altamonte Springs, FL, USA) is an example of such a machine. Despite its benefits, it is relatively costly, is not portable, has a large footprint [3], and is not intended for home use. Therefore, the development of portable, cost-effective, and efficacious back extension exercise devices is needed to foster home exercise programs.

The demand for implementation of virtual solutions to deliver exercise programs for the management of low back pain is increasing, and the COVID-19 pandemic has provided an impetus to explore such solutions. A recent systematic review found that telehealth is safe and effective for the management of non-acute low back pain [11]. A recent observational study found that virtual care focusing on exercise for managing low back pain resulted in similar improvements in function and pain reduction compared to in-clinic programs [12]. Moreover, another recent observational study found benefits of virtual physical therapy in terms of patient satisfaction and improving access to care [13]. However, the overall body of evidence on virtual care is minimal and no standard exists for home-based exercise delivery, particularly for programs focusing on progressive resistance exercise training. Thus, implementation of progressive resistance exercise programs delivered virtually has been limited for managing low back pain.

A smaller, portable, and less costly back extension exercise device (MedX Home Back Device, Converge Medical Technology LLC, Austin, TX, USA) was developed as an alternative to provide isolated progressive resistance exercise for the lumbar extensors. The device is intended for home use and is not difficult to implement, administer, and complete exercise sessions. It utilizes body weight and external loads (metal plates—assessed prototype; resistance bands or hydraulics: current version) to apply resistive loading during back extension exercise in the seated position. It also incorporates similar mechanisms to isolate the lumbar spine as the computerized lumbar extension dynamometer (Figures 1 and 2). However, the ability of the home back extension exercise device to effectively activate the lumbar extensor muscles and provide progressive loads has not been explored. Therefore, the purpose of this study was to assess the surface electromyographic (EMG) activity of the lumbar extensor muscles during full range of motion, dynamic exercise on a home back extension exercise device at three exercise loads.

(a) (b)

Figure 1. Illustration of pelvic restraint mechanisms and movement patterns on the home back extension exercise device. (**a**) Start position (lumbar flexion). (**b**) Finish position (lumbar extension).

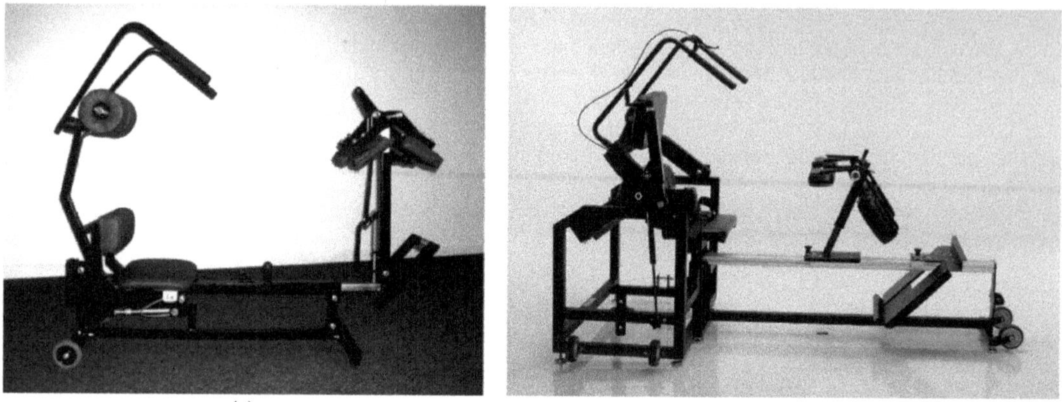

(a) (b)

Figure 2. Home Back Extension Exercise Device (MedX Home Back Device, Converge Medical Technology LLC., Austin, TX, USA). (**a**) Prototype assessed in study. (**b**) Current version.

2. Materials and Methods

2.1. Study Design

An observational study with repeated measures was conducted at a clinical facility. Participants reported to the facility on one occasion during which physical performance measures and surface EMG data were collected multiple times.

2.2. Participants

A convenience sample of participants were recruited by word of mouth and posted flyers to include an equal number of males and females. The study's protocol was reviewed and approved by Biomed IRB (San Diego, CA, USA) and each participant provided written informed consent. Inclusion criteria for participation were [7]: 18–45 years of age; good general health; ability to provide written informed consent. Exclusion criteria were: History of significant clinical low back pain; history of lumbar spine pathology, deformity, or surgery; knee or hip disorders contraindicating use of the exercise testing device's pelvic restraint mechanisms; cardiovascular or other orthopedic contraindications to resistance exercise; a "yes" response for any item on the physical activity readiness questionnaire at screening [14]; history of high blood pressure; resting blood pressure and heart rate measurements outside of the normal range at screening; current participation in a resistance exercise program for the back musculature; pregnant females.

2.3. Sample Size Calculation

Sample size was estimated using G*Power 3.1 [15], based on previous work assessing lumbar extensor muscle surface EMG activity during exercise [7,16,17] and the following parameters: 25% increase in mean value from exercise at 1.00 $\times$ body weight (BW) to 1.50 $\times$ BW with a standard deviation of approximately 50% of the mean value (effect size = 0.50), repeated measures, power = 0.80, alpha = 0.05. Based on these parameters, a sample size of n = 10 was adequate.

2.4. Procedures

2.4.1. Participant Selection and Screening

Candidates who responded to recruitment efforts contacted the investigator by telephone. To confirm eligibility, a standardized telephone screening questionnaire was used. Candidates who were eligible according to the telephone screen were referred to the study site to complete additional screening procedures. After providing informed consent, candidates completed a health history questionnaire and the physical activity readiness questionnaire [14]. Next, resting blood pressure and heart rate were recorded from eligible candidates. Finally, females were administered a urine pregnancy test. At this time, eligible candidates were invited to participate in the study.

2.4.2. Assessment of Isometric Lumbar Extension Strength

Immediately following the screening procedures, height and body weight were recorded from eligible participants. Next, isometric lumbar extension strength was assessed, which was used to normalize EMG data as a percentage of Maximum Voluntary Isometric Contraction (%MVIC). For the strength test, the participant was seated in an upright position on a computerized lumbar extension dynamometer (MedX Corp.) with its pelvic restraint mechanisms engaged (e.g., lap belt, femur restraint pad, pelvic restraint pad, footboard). In the seated testing position, the participant was upright with the hips flexed at approximately 70–80 degrees and in slight internal rotation, and the knees flexed at approximately 20 degrees. The participant performed two trials of strength tests on the device—submaximal test and maximal test. After establishing the test position, the participant performed light dynamic exercise and submaximal isometric strength tests in the sagittal plane for familiarization to the isometric testing and dynamic exercise procedures. For the submaximal isometric strength tests, the participant pushed against the thoracic pad on the device while using moderate effort. Submaximal testing was performed one time at three positions—72, 36, and 0 degrees of lumbar flexion, which represents the full range of motion in the sagittal plane allowed by the testing device. After a 15-min rest, the participant performed the actual tests to determine maximum voluntary isometric lumbar extension torque at the same three positions that were used for the submaximal testing. At each position, the participant gradually built up force against the thoracic pad (using the trunk extensor muscles) and pushed as hard as possible for approximately one second. A monitor provided visual feedback of performance and the investigator provided verbal encouragement for the participant to generate maximum force. Isometric strength (torque) was recorded electronically by the device in foot-pounds (ft-lb) and converted to Newton-meters (N-m). This isometric strength testing protocol has been validated and described in detail [7,18].

2.4.3. Assessment of Dynamic Lumbar Extension Exercise

Following the strength test and a 15-min rest, the participants completed 1 set of full range of motion dynamic exercise on the prototype version of the home back extension exercise device (Figure 2A) at 3 exercise loads (3 sets total—1 set at each of 3 exercise loads), with a 3-min rest between each set. The order of exercise at the 3 loads was balanced across participants. The exercise loads for the 3 sets were 1.00 times body weight (BW), 1.25 BW, and 1.50 BW. Loads greater than body weight were accommodated by metal plates that were attached to the device.

Each set consisted of three repetitions using a slow movement exercise protocol (i.e., 10 s concentric, 10 s eccentric). For each set of dynamic exercise, the participant was positioned at full flexion on the device and completed the concentric phase by extending their low back against the thoracic pad until reaching full pain-free extension. Upon reaching full extension, the participant completed the eccentric phase by slowly returning to the starting position. Three repetitions were completed for each set. To standardize the tempo of the movement, a metronome was used and was set at 60 beats per minute. Information about adverse events (e.g., muscle soreness) was gathered through verbal subjective reports from the participants during and after lumbar extension strength tests and dynamic exercises on the study visit, and during four subsequent days until symptoms resolved.

2.4.4. Instrumentation and EMG Processing

Surface EMG signals were collected from the right and left lumbar paraspinal region at the L3–4 level during the isometric strength test and each set of dynamic exercise utilizing techniques adapted from previous work [7,16]. First, the skin was palpated to establish landmarks for the regions of interest and was scrubbed with an alcohol pad. Two round (1.5 cm) self-adhesive, disposable silver/silver chloride, pre-gelled, snap surface electrodes were placed on the skin surface. The location of electrode placement was 1 cm above and below the L3–4 interspace over the central portion of the paraspinal muscle belly bilaterally, which was approximately 2–3 cm from the midline of the spine in the sagittal plane. Active electrodes were used according to the manufacturer's recommendations (Noraxon USA Inc., Scottsdale, AZ, USA) [19,20], and the inter-electrode distance was 2 cm.

Electromyographic signals were collected with a 1000 Hz sampling rate. Characteristics of the differential amplifier were as follows according to the manufacturer's recommendations [19,20]: bandpass filter: high pass cutoff—10 Hz, low pass cutoff—500 Hz; common mode rejection: minimum 85 dB at 1000 Hz; input impedance: >10 Megaohm. Prior to processing, raw EMG data were visually inspected in order to detect noise (e.g., mechanical or movement artifacts, electrical signals from other sources, such as electrocardiograms and power lines). Raw EMG data were rectified, smoothed (via root mean square (RMS) technique with 50 ms interval), filtered (median 5 filtering technique), and normalized (based on MVIC values obtained from the strength test). Myoresearch v.2.1 software (Noraxon USA Inc.) was used for EMG data processing and analysis. The reliability of EMG data collected in a similar manner was shown to be acceptable [7].

2.5. Outcome Measures

The primary outcome measure was lumbar extensor muscle surface EMG activity expressed as the normalized value in %MVIC. Data obtained from the second repetition of exercise were used for analysis. The second repetition of exercise was arbitrarily selected to minimize potential artifacts in muscle activity due to acceleration in the concentric phase to start the exercise set (i.e., first repetition at its start point) and deceleration at the eccentric phase to end the exercise set (i.e., third repetition at its end point). Thus, the second repetition was the most likely repetition to be performed in the desired smooth, controlled fashion without artifacts. Data were normalized using a previously published method [7], using the following equation:

$$\%\text{MVIC} = (\text{Raw EMG (mV/sec) dynamic exercise}/\text{Raw EMG (mV/sec) peak isometric contraction}) \times 100\%$$

2.6. Data Analysis

Descriptive data (group means and standard deviations) were calculated for normalized EMG (in %MVIC) by exercise load (1.00 BW, 1.25 BW, 1.50 BW), phase of movement (concentric, eccentric), position in the dynamic range of motion (flexion, mid, extension). Position in the dynamic range of motion was categorized as approximately 49–72 degrees for flexion, 25–48 degrees for mid, and 0–24 degrees for extension, which was approximately equivalent to 3.33-s intervals within each movement phase (concentric, eccentric). Lumbar extensor muscle surface EMG activity was evaluated for the effect of exercise load (1.00 BW, 1.25 BW, 1.50 BW), phase of movement (concentric, eccentric), position in range of motion (flexion, mid, extension) using analysis of variance (ANOVA) with repeated measures. Post hoc, pairwise comparisons were conducted using Tukey's procedure, as needed. Statistical significance was set at alpha = 0.05. Stata 7.0 (Stata Corp., College Station, TX, USA) statistical package was utilized for all analyses.

3. Results

Participant characteristics and lumbar extension peak isometric torque are shown in Table 1. Peak torque values were generally within normal limits using normative data established by the manufacturer [21]. No serious adverse events were reported following isometric exercise testing on the lumbar dynamometer and dynamic exercise on the home back extension exercise device. 30% (3/10) of the participants reported delayed onset muscle soreness (DOMS) in the low back region that peaked 24–36 h after exercise, disappeared within 96 h, and did not affect physical function. All participants completed three repetitions of dynamic exercise at the three assigned loads and there was no indication that the loads were near maximal effort. 80% (8/10) of the participants displayed a progressive increase in lumbar extensor muscle surface EMG activity as exercise load was increased from 1.00 BW to 1.50 BW.

Table 1. Participant demographic characteristics and lumbar extension torque values.

Variable	Total (n = 10)		Female (n = 5)		Male (n = 5)	
	Mean	SD	Mean	SD	Mean	SD
Age (year)	33.0	8.4	29.0	9.5	37.0	5.5
Body Height (cm)	174.0	6.9	170.7	8.1	177.3	3.3
Body Weight (kg)	76.2	18.4	62.3	9.8	90.2	13.4
Peak IM torque (N-m)	392.0	190.5	254.2	95.3	529.9	157.8

Key: Peak IM torque (N-m) = Peak lumbar extension isometric torque in Newton-meters (N-m) assessed on a dynamometer.

Normalized lumbar extensor muscle surface EMG activity by exercise load, phase of movement, and position in range of motion is shown in Table 2 and Figure 3. There was a significant effect of exercise load on lumbar extensor muscle surface EMG activity [F (2, 9) = 7.77, $p = 0.0006$]. Post-hoc analysis revealed that EMG activity at an exercise load of 1.50 BW was significantly greater than 1.00 BW ($p < 0.05$). There was a significant effect of phase of movement on lumbar extensor muscle surface EMG activity [F (1, 9) = 31.33, $p < 0.0001$], indicating that EMG activity was greater for the concentric phase compared to the eccentric phase. There was a significant effect of position in range of motion on lumbar extensor muscle surface EMG activity [F (2, 9) = 30.55, $p < 0.0001$]. Post hoc analysis revealed that EMG activity at the extension position was significantly greater than the mid position and flexion position ($p < 0.05$), and that EMG activity was greater at the mid position than the flexion position ($p < 0.05$).

Table 2. Normalized lumbar extensor muscle surface EMG activity (in % MVIC) during dynamic exercise on a home back extension exercise device depicted by exercise load, phase of movement, and position in range of motion.

	Exercise Load					
	1.00 BW		1.25 BW		1.50 BW	
	Mean	SD	Mean	SD	Mean	SD
Full Repetition	34.9	16.0	42.1	11.8	47.1	9.8
Concentric Phase:						
Full Concentric Phase	41.2	17.9	50.5	16.2	52.8	10.3
Flexion Position	33.0	16.1	36.6	16.6	37.7	11.3
Mid Position	40.1	19.9	50.0	16.3	50.8	8.7
Extension Position	50.4	21.9	65.0	20.7	69.9	16.0
Eccentric Phase:						
Full Eccentric Phase	28.7	16.0	33.6	12.7	41.4	13.8
Flexion Position	23.3	14.9	27.4	12.7	31.2	13.0
Mid Position	27.3	16.8	30.4	13.4	38.5	12.3
Extension Position	35.3	18.5	42.9	14.4	54.7	27.8

Key: Values are in % Maximum Voluntary Isometric Contraction (MVIC), BW = Body Weight, SD = Standard Deviation.

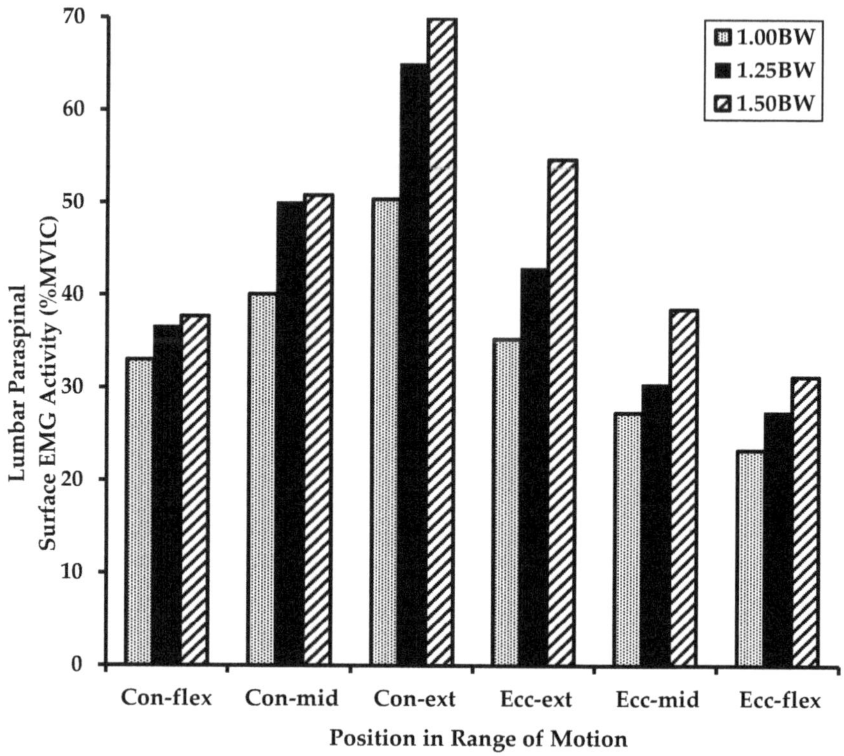

Figure 3. Graph of normalized lumbar extensor muscle surface EMG activity (in % MVIC) during dynamic exercise on a home back extension exercise device depicted by exercise load, phase of movement, and position in range of motion. Key: Mean values in % Maximum Voluntary Isometric Contraction (MVIC), BW = Bodyweight, Con = Concentric Phase, Ecc = Eccentric Phase, flex = flexion position in range of motion, mid = mid position, ext = extension position.

4. Discussion

The findings of this study indicate that full range of motion lumbar extension exercise on a home back extension exercise device effectively activates the lumbar extensor muscles in a progressive manner. Supported by the findings of no adverse events, the study suggests that a home exercise device is well-tolerated and safe for lumbar extensor exercise training in adults without contraindications to resistance exercise. While DOMS was experienced by some participants, it is a typical response following unfamiliar lumbar extension exercise—whether isometric exercise tests or dynamic exercise training [3]. Nevertheless, educating clients and patients on expectations regarding likelihood of muscle soreness and stiffness is important.

The observed range of mean values for normalized lumbar extensor surface EMG activation levels suggests that the home back extension exercise device can provide gradual progressive resistance for the lumbar extensors. Furthermore, the mean value of the lowest observed activation level is likely low enough for patients during early phases of therapeutic exercise programs. Since the exercise loads for this study were arbitrarily selected, the observed lumbar extensor EMG activation levels do not necessarily represent the maximum attainable levels during exercise on the device. Given the progressive nature of the observed activation levels, it is possible that higher activation levels can be attained with additional external loads. Whether these activation levels provide the overload stimulus necessary for lumbar extension strength gains is unknown.

One explanation for the lack of progressive resistance for the lumbar extensors despite higher exercise loads in 20% of participants is that activation of the lumbar extensor muscles during compound trunk extension is variable [22,23]. Other trunk extensor muscles, such as the glutes and hamstrings, may be recruited at varying levels during compound trunk extension at different loads, which is consistent with previous research on other exercise devices [22,23]. The specific biomechanical strategies for individuals to generate force on the home back extension exercise device are unknown.

As expected, lumbar extensor muscle activation levels during exercise on the home back extension exercise device were higher during the concentric phase than the eccentric phase. A possible explanation for the wide variation of activation levels among the positions in the range of motion (i.e., nearly 100% greater in the extended position) is that isotonic exercise was performed (versus variable resistance exercise). While it is unknown if similar variations in lumbar extensor muscle activation exist during exercise on the computerized lumbar dynamometer and other trunk extension movements, this finding is consistent with the flexion-relaxation phenomena of the posterior lumbar musculature [24]. Therefore, slow, controlled movements emphasizing both concentric and eccentric phases throughout the full pain-free range of motion (particularly extension) while gradually increasing resistance over time is recommended for safety and optimal activation of the lumbar extensor muscles [3]. This recommendation is generally consistent with the guidelines of the American College of Sports Medicine for resistance exercise training of major muscle groups [25].

Many of the design characteristics to isolate the lumbar spine of the prototype home back extension exercise device that was tested in this study are similar to those of the computerized lumbar dynamometer, such as a footboard, femur restraint, and pelvic restraint pad (Figures 1 and 2). However, the prototype device does not incorporate a lap belt. Without restraint from a lap belt, participants were able to hike (extend) their hips during the terminal extension phase of exercise, which may have permitted the hamstrings and gluteal muscles to elicit force production at the expense of the lumbar extensors [4,7].

Based on the findings of this study and other prototype testing, a new version of the device (Figure 2B) was developed and is currently available (Figure 2B). The current version has similar overall design features and addresses shortcomings of the prototype. For example, the current version uses hydraulic mechanisms to apply external loads and has a lap belt to enhance pelvic stabilization. These changes could help accommodate a wider range of exercise loads (both lower and higher) and improve isolation of the lumbar spine. There is no reason to expect that the improvements in the new version would negatively impact the ability to apply progressive resistance compared to the prototype tested.

4.1. Limitations

This study has some limitations that impact its generalizability. For example, the sample size was small and consisted of individuals without a history of clinical low back pain. The exercise loads were arbitrarily selected and are not representative of the full range of loads possible with the device. Future research is warranted to assess loading conditions at different ranges, such as those lower than body weight, in healthy individuals and those with low back pain. Also, the study did not assess longer-term exercise training programs. Moreover, future research would be useful in healthy individuals and patients with low back pain to compare the effectiveness of the home back extension exercise device with other exercise devices, such as the Variable Angle Roman Chair [26], on the ability to activate the lumbar extensor muscles, optimize strength gains, and enhance clinical outcomes.

4.2. Pragmatic Applications

The home back extension exercise device was able to safely administer progressive loads for the lumbar extensor muscles. Thus, trainers and clinicians can incorporate the device for delivering lumbar extensor exercise training programs in community-dwelling adults without contraindications to resistance exercise. The safety of lumbar extensor strengthening exercises has been documented and can be enhanced by starting the program at a low load and gradually applying progressive resistance at subsequent sessions [3]. Implementation of this device for home use outside of clinical settings does not preclude adequate supervision, which is needed to monitor safety, encourage proper movement, and improve adherence. Recent research suggested that education is needed to enhance acceptance of telehealth physical therapy by patients with chronic low back pain [27], and ongoing supervision provides an opportunity to do so. While numerous approaches for supervision are possible, supervision of a home exercise program using this device could be accomplished through an initial on-site orientation in the home setting followed by periodic virtual sessions hosted by a qualified professional. The Official Disability Guidelines (ODG) Medical Treatment Guidelines generally recommend lumbar extension exercise equipment (e.g., MedX lumbar extension machine) for the management of chronic low back pain [28]. The ODG indicates this modality may be an option for first-line treatment when implemented within a supervised physical therapy program, such as face-to-face in the clinic or virtually via telehealth [26]. Implementation of the home back extension exercise device appears to be appropriate for this purpose. Lumbar strengthening exercises and other exercises are recommended for the management of chronic low back pain [3,29]. However, subclassification of specific individuals for the management of low back pain to receive lumbar strengthening exercises or other exercises (e.g., motor control) has not been validated through research and is beyond the scope of this study. Thus, the role of lumbar strengthening exercises within an exercise program depends on client/patient preferences, functional goals, and trainer/clinician experiences [2,3]. If the goal is to strengthen the lumbar extensor muscles, then exercises that apply progressive loads to the lumbar extensor muscles should be implemented, such as the home back extension exercises assessed in this study.

5. Conclusions

The findings of this study indicate that dynamic exercise on a home back extension exercise device is safe and provides a mechanism to progressively activate the lumbar extensor muscles. This device can be used for progressive resistance exercise training programs for community-dwelling adults without contraindications to resistance exercise.

Author Contributions: Conceptualization, J.M.M. and J.L.V.; methodology, J.M.M. and J.L.V.; software, J.M.M.; validation, J.M.M., J.L.V. and B.E.U.; formal analysis, J.M.M., J.L.V. and B.E.U.; interpretation of data, J.M.M., J.L.V. and B.E.U.; investigation, J.M.M. and J.L.V.; resources, J.M.M. and J.L.V.; data curation, J.M.M. and J.L.V.; writing—original draft preparation, J.M.M.; writing—review and editing, J.M.M., J.L.V. and B.E.U.; visualization, J.M.M.; supervision, J.M.M.; project administration, J.M.M. and J.L.V.; funding acquisition, J.M.M. All authors have read and agreed to the published version of the manuscript.

Funding: This research was funded by a grant from MedX Corp. (Altamonte Springs, FL, USA). Converge Medical Technology LLC (Austin, TX, USA) provided funding to support post-study procedures related to the material discussed in this manuscript. The protocol number (SSF00306).

Institutional Review Board Statement: The study was conducted according to the guidelines of the Declaration of Helsinki, and approved by the Institutional Review Board of Biomed IRB (San Diego, CA, USA) (protocol number: SSF00306, date of approval: 9 August 2006).

Informed Consent Statement: Informed consent was obtained from all subjects involved in the study.

Data Availability Statement: The data presented in this study are available on request from the corresponding author.

Conflicts of Interest: J.M.M. and J.L.V. are officers and directors of US Spine & Sport Foundation which received funding from MedX Corp. to support this research. J.M.M. was financially compensated by Converge Medical Technology LLC to support post-study procedures related to the material discussed in this manuscript. B.E.U. has no conflicts of interest related to the material discussed in this manuscript. The funders had a role in the decision to publish the results. The funders had no role in the design of the study, in the collection, analyses, or interpretation of data, or in the writing of the manuscript.

References

1. Hartvigsen, J.; Hancock, M.J.; Kongsted, A.; Louw, Q.; Ferreira, M.L.; Genevay, S.; Hoy, D.; Karppinen, J.; Pransky, G.; Sieper, J.; et al. Underwood, M. What low back pain is and why we need to pay attention. *Lancet* **2018**, *391*, 2356–2367. [CrossRef]
2. Foster, N.E.; Anema, J.R.; Cherkin, D.; Chou, R.; Cohen, S.P.; Gross, D.P.; Ferreira, P.H.; Fritz, J.M.; Koes, B.W.; Peul, W.; et al. Prevention and treatment of low back pain: Evidence, challenges, and promising directions. *Lancet* **2018**, *391*, 2368–2383. [CrossRef]
3. Mayer, J.M.; Mooney, V.; Dagenais, S. Evidence-informed management of chronic low back pain with lumbar extensor strengthening exercises. *Spine J.* **2008**, *8*, 96–113. [CrossRef] [PubMed]
4. Graves, J.E.; Webb, D.C.; Pollock, M.L.; Leggett, S.H.; Carpenter, D.M.; Foster, D.N.; Cirulli, J. Pelvic stabilization during resistance training: Its effect on the development of lumbar extension strength. *Arch. Phys. Med. Rehabil.* **1994**, *75*, 210–215. [CrossRef]
5. Pollock, M.L.; Leggett, S.H.; Graves, J.E.; Jones, A.; Fulton, M.; Cirulli, J. Effect of resistance training on lumbar extension strength. *Am. J. Sports Med.* **1989**, *17*, 624–629. [CrossRef]
6. Udermann, B.E.; Graves, J.E.; Donelson, R.; Iriso, J.; Boucher, J. Effect of pelvic restraint on hamstring, gluteal, and lumbar muscle emg activation. *Arch. Phys. Med. Rehabil.* **1999**, *80*, 1176–1179. [CrossRef]
7. San Juan, J.G.; Yaggie, J.A.; Levy, S.S.; Mooney, V.; Udermann, B.E.; Mayer, J.M. Effects of pelvic stabilization on lumbar muscle activity during dynamic exercise. *J. Strength Cond. Res.* **2005**, *19*, 903–907. [CrossRef]
8. Leggett, S.; Mooney, V.; Matheson, L.; Nelson, B.; Dreisinger, T.; Van Zytveld, J.; Vie, L. Restorative exercise for clinical low back pain: A prospective two-center study with 1-year follow-up. *Spine* **1999**, *24*, 889–898. [CrossRef]
9. Mooney, V.; Gulick, J.; Perlman, M.; Levy, D.; Pozos, R.; Leggett, S.; Resnick, D. Relationships between myoelectric activity, strength, and MRI of the lumbar extensor muscles in back pain patients and normal subjects. *J. Spinal Disord.* **1997**, *10*, 348–356. [CrossRef]
10. Risch, S.; Norvell, N.; Pollock, M.; Risch, E.; Langer, H.; Fulton, M.; Graves, J.E.; Leggett, S. Lumbar strengthening in chronic low back pain patients: Physiological and psychosocial benefits. *Spine* **1993**, *18*, 232–238. [CrossRef]
11. Corso, M.; Cancelliere, C.; Mior, S.; Salmi, L.R.; Cedraschi, C.; Nordin, M.; Taylor-Vaisey, A.; Côté, P. Are non-pharmacological interventions delivered through synchronous telehealth as effective and safe as in-person interventions for the management of patients with non-acute musculoskeletal conditions? A systematic rapid review. *Arch. Phys. Med. Rehabil.* **2022**, *103*, 145–154.e11. [CrossRef] [PubMed]
12. Raiszadeh, K.; Tapicer, J.; Taitano, L.; Wu, J.; Shahidi, B. In-clinic versus web-based multidisciplinary exercise-based rehabilitation for treatment of low back pain: Prospective clinical trial in an integrated practice unit model. *J. Med. Internet Res.* **2021**, *23*, e22548. [CrossRef] [PubMed]
13. Grundstein, M.J.; Fisher, C.; Titmuss, M.; Cioppa-Mosca, J. The role of virtual physical therapy in a post-pandemic world: Pearls, pitfalls, challenges, and adaptations. *Phys. Ther.* **2021**, *101*, pzab145. [CrossRef] [PubMed]
14. Shephard, R.J. PAR-Q, Canadian home fitness test, and exercise screening alternatives. *Sports Med.* **1988**, *5*, 1985–1995. [CrossRef] [PubMed]
15. Faul, F.; Erdfelder, E.; Lang, A.G.; Buchner, A. G*Power 3: A flexible statistical power analysis program for the social, behavioral, and biomedical sciences. *Behav. Res. Methods* **2007**, *39*, 175–191. [CrossRef]
16. Mayer, J.M.; Graves, J.E.; Robertson, V.L.; Pierra, E.A.; Verna, J.L.; Ploutz-Snyder, L.L. Electromyographic activity of the lumbar extensor muscles: Effect of angle and hand position during Roman chair exercise. *Arch. Phys. Med. Rehabil.* **1999**, *80*, 751–755. [CrossRef]
17. Mayer, J.M.; Verna, J.L.; Manini, T.; Mooney, V.; Graves, J.E. Electromyographic activity of the trunk extensor muscles: Effect of varying hip position and lumbar posture during Roman chair exercise. *Arch. Phys. Med. Rehabil.* **2002**, *83*, 1543–1546. [CrossRef] [PubMed]
18. Graves, J.E.; Pollock, M.L.; Carpenter, D.M.; Leggett, S.H.; Jones, A.; MacMillan, M.; Fulton, M. Quantitative assessment of full range-of-motion isometric lumbar extension strength. *Spine* **1990**, *15*, 289–294. [CrossRef] [PubMed]
19. Konrad, P. *The ABC of EMG. A Practical Introduction to Kinesiological Electromyography*; Noraxon USA Inc.: Scottsdale, AZ, USA, 2005; pp. 1–60.
20. Noraxon USA Inc. *MyoSystem 1400. Operation & Technical Manual*; Noraxon USA, Inc.: Scottsdale, AZ, USA, 2010; pp. 1–29.
21. MedX Corp. *Lumbar Extension Machine Instructional Manual*; MedX Corp.: Altamonte Springs, FL, USA, 2004; pp. 1–67.
22. Clark, B.C.; Manini, T.M.; Ploutz-Snyder, L.L. Derecruitment of the lumbar musculature with fatiguing trunk extension exercise. *Spine* **2003**, *28*, 282–287. [CrossRef]

23. Mayer, J.M.; Graves, J.E.; Clark, B.C.; Formikell, M.; Ploutz-Snyder, L.L. The use of magnetic resonance imaging to evaluate lumbar muscle activity during trunk extension exercise at varying intensities. *Spine* **2005**, *30*, 2556–2563. [CrossRef] [PubMed]
24. Shirado, O.; Ito, T.; Kaneda, K.; Strax, T.E. Flexion-relaxation phenomenon in the back muscles. a comparative study between healthy subjects and patients with chronic low back pain. *Am. J. Phys. Med. Rehabil.* **1995**, *74*, 139–144. [PubMed]
25. American College of Sports Medicine. *ACSM's Guidelines for Exercise Testing and Prescription*, 11th ed.; Wolters Kluwer Health: Philadelphia, PA, USA, 2021.
26. Verna, J.L.; Mayer, J.M.; Mooney, V.; Pierra, E.A.; Robertson, V.L.; Graves, J.E. Back extension endurance and strength: The effect of variable-angle roman chair exercise training. *Spine* **2002**, *27*, 1772–1777. [CrossRef]
27. Fritz, J.M.; Lane, E.; Minick, K.I.; Bardsley, T.; Brennan, G.; Hunter, S.J.; McGee, T.; Rassu, F.S.; Wegener, S.T.; Skolasky, R.L. Perceptions of telehealth physical therapy among patients with chronic low back pain. *Telemed. Rep.* **2021**, *2*, 258–263. [CrossRef]
28. Official Disability Guidelines (ODG). *Medical Treatment Guidelines: Lumbar Extension Exercise Equipment for Low Back Conditions; and MedX Lumbar Extension Machine for Low Back Conditions*; MCG Health: Austin, TX, USA, 2021; Available online: https://www.mcg.com/odg/ (accessed on 23 December 2021).
29. George, S.Z.; Fritz, J.M.; Silfies, S.P.; Schneider, M.J.; Beneciuk, J.M.; Lentz, T.A.; Gilliam, J.R.; Hendren, S.; Norman, K.S. Interventions for the management of acute and chronic low back pain: Revision 2021. *J. Orthop. Sports. Phys. Ther.* **2021**, *51*, CPG1–CPG60. [CrossRef] [PubMed]

Journal of
Functional Morphology and Kinesiology

MDPI

Review

The Efficacy of Flywheel Inertia Training to Enhance Hamstring Strength

Joey O' Brien [1,2,*], Declan Browne [1], Des Earls [1] and Clare Lodge [1]

[1] HealthCore, Department of Science and Health, Institute of Technology Carlow, R93 V960 Carlow, Ireland; Declanbrowne@itcarlow.ie (D.B.); desearls@itcarlow.ie (D.E.); clarelodge@itcarlow.ie (C.L.)
[2] High Performance Unit, WIT Arena, X91 P20H Waterford, Ireland
* Correspondence: C00232530@itcarlow.ie

Abstract: The purpose of this narrative review is to examine the efficacy of flywheel inertia training to increase hamstring strength. Hamstring strain injury is common in many sports, and baseline strength deficits have been associated with a higher risk of hamstring strain injury. As a result, strength and conditioning professionals actively seek additional techniques to improve hamstring strength with the aim of minimising the incidence of hamstring strain injury. One method of strength training gaining popularity in hamstring strength development is flywheel inertia training. In this review, we provide a brief overview of flywheel inertia training and its supposed adaptions. Next, we discuss important determinants of flywheel inertia training such as familiarisation, volume prescription, inertia load, technique and specific exercise used. Thereafter, we investigate its effects on hamstring strength, fascicle length and hamstring strain injury reduction. This article proposes that hamstring specific flywheel inertia training can be utilised for strength development, but due to the low number of studies and contrary evidence, more research is needed before a definite conclusion can be made. In addition, as with any training modality, careful consideration should be given to flywheel inertia training determinants. This review provides general recommendations of flywheel inertia training determinants that have value when integrating flywheel inertia training into a hamstring strengthening program.

Keywords: flywheel inertia training; hamstring strength; eccentric; inertia load

Citation: O' Brien, J.; Browne, D.; Earls, D.; Lodge, C. The Efficacy of Flywheel Inertia Training to Enhance Hamstring Strength. *J. Funct. Morphol. Kinesiol.* **2022**, *7*, 14. https://doi.org/10.3390/jfmk7010014

Academic Editors: Cristina Cortis, Andrea Fusco and Carl Foster

Received: 28 November 2021
Accepted: 19 January 2022
Published: 20 January 2022

1. Introduction

Hamstring strain injuries (HSI) are among the most common types of injuries seen in the area of sports medicine in elite athletes, with a prevalence of 6 to 25%, depending on the sport [1–3]. It is the single most frequent injury in professional football [4], rugby union [5] and track and field, especially among sprinters and jumpers [6]. Elkstrand et al., in a prospective study, [3] established that HSI accounts for 37% of muscle injuries in professional soccer players and 25% of athlete match absences. HSI has a high re-injury rate [7], with many taking place within two weeks of the original injury [8]; a rationale for this may be due to inadequate rehabilitation programs and/or a premature return to sport [9]. The function of the hamstring is complicated, as it acts as a hip extensor, knee flexor, and external rotator of the hip and knee, depending on leg positioning and connection to the ground [4]. There are two main types of acute hamstring strain mechanisms, which are best characterised by their injury settings [10]. The most prevalent type of HSI occurs during high-speed running, followed by HSI as a result of activities that cause rapid hamstring lengthening such as high kicking, sliding tackle, and sagittal split [10]. There are two views in the literature about the mechanism of hamstring injuries sustained during high-speed running. One is that the hamstring is most susceptible to damage during active lengthening, which occurs during the running gait cycle's late swing phase [11]. However, it has also been suggested that hamstring injury occurs during the initial stance phase as the body is driven forward above the touchdown point, due to the presence of large forces acting in opposing directions [12]. Following recent reviews [13,14] it now seems to be most accepted that HSI during sprinting are most likely to occur due to excessive muscle strain caused by eccentric contraction during the late swing phase of the running gait cycle [15].

HSI's can result in adverse effects on an individual's performance, thereby hampering a team's success [16]; therefore, raising the awareness of common risk factors is essential. Older age and previous injury history have been commonly associated with a higher risk of future HSI [17,18] and have been previously described as non-modifiable risk factors [19]. A topical literature review [20] highlighted numerous modifiable risk factors, including flexibility, fatigue, high speed running loads, sprint running, lumbo-pelvic hip control, insufficient/inadequate warm-up, strength and intra-limb and inter-limb asymmetry and biceps femoris fascicle length. To date, hamstring strength is the most researched, with various types of muscular strength deficits being associated with a higher risk of in HSI numerous studies [21–25]. As a result, strength and conditioning professionals actively seek additional techniques to improve hamstring strength to minimise the incidence of HSI with non-traditional methods of strength training such as Flywheel Inertial Training (FIT) gaining popularity.

When FIT was first developed, it was intended to help astronauts cope with the neuromuscular dysfunctions and contemporaneous muscle atrophy that result due to the lack of gravity experienced during long-term space travel [26,27]. Many studies have since described the mechanical advantages of flywheel devices and attempted to clarify the neural and physiological mechanisms, structural adaptations and training effects induced by flywheel exercise [28]. However, few have specifically investigated it as a hamstring strength training modality. FIT is based on the flywheel principle, whereby the inertia generated by rotating flywheels provides resistance [29]. The exerted force unwinds a strap attached to the device's shaft during the concentric phase, causing the flywheel to rotate; when the concentric phase is completed, the strap rewinds and the user must resist the device by performing an eccentric muscular action [30]. This technique can result in brief moments of eccentric overload if performed correctly [31]. This eccentric overload may induce positive gains in hamstring strength. Currently, there is a scarcity of research investigating FIT to increase hamstring strength, specifically eccentric hamstring strength. Therefore, the main aim of this review is to investigate the efficacy of FIT to increase hamstring strength while also investigating determinants of hamstring specific FIT to provide practical guidelines to be used in hamstring specific FIT.

2. Materials and Methods

The following search strategies were used to locate relevant articles online. Databases searched included EBSCO host, Web for Science, PubMed, Pub Med, SPORTDiscus and Google Scholar. Relevant key terms in the search included training flywheel inertia training OR flywheel device OR rotational inertia device OR inertia load AND hamstring strength AND eccentric strength. Additionally, articles cited in the reference lists of acknowledged journals were manually searched and examined.

2.1. Familiarisation

To fully benefit from FIT, a certain amount of coordination is required [32], thus experience with these devices will have an impact on the results gained, especially if the exercises entail movements in a closed kinetic chain where the displacement is done against gravity [33]. Familiarisation with FIT could alter how the concentric and eccentric phases of an exercise are performed, which may, in turn, alter the eccentric overload stimulus experienced [34]. For example, Tous-Fajardo et al. [32] reported that participants with experience in flywheel training achieved a higher eccentric overload when compared to participants with no previous experience. Twenty male field sport athletes took part in the study. Ten participants had previous experience (>5 sessions) in the flywheel leg curl exercise, while ten with similar physical characteristics and training history had only experienced <2 familiarisation sessions. Athletes with more FIT experience had considerably higher eccentric peak forces than concentric peak forces (<0.05); additionally, athletes of the same ability who were familiar with FIT generated greater eccentric and concentric peak force. It appears that a certain amount of coordination is required to apply braking forces at the end of the eccentric action, which is warranted to elicit an eccentric overload; fine-tuning this strategy appears to require adequate familiarisation [32].

Piqueras-Sanchiz et al. [35] reported high (0.7–0.89) to very high (>0.9) reliability during both concentric and eccentric power production over varying inertial loads. The authors tested both males and females across four testing sessions in the flywheel leg curl exercise using varying inertial loads (0.083, 0.132, 0.182, 0.266 and 0.350 kg·m^2). An increase in power reliability was established across all testing sessions with very high reliability (>0.9) scores for all inertial loads in session four. Both males and females presented the highest reliability scores in the fourth session, prompting authors to conclude that to achieve reliable and stable outputs, a familiarisation process with the flywheel leg curl exercise is required with inexperienced subjects, recommending two to four sessions depending on the training experience. Less experienced subjects may need more extended familiarisation periods.

Similar research [36] conducted during the flywheel squat exercise observed comparable reliability scores (0.79–0.93). The authors in that study recommended performing at least three sessions to obtain a stable measure during the flywheel squat exercise. A more recent [33] study examined the differences in kinetic and kinematic profiles between two different FIT devices. A half-squat incremental test was performed by 39 healthy males on two different FIT devices: one a horizontal cylinder and the other a vertical cone-shaped axis. The study reported differences in biomechanical output between athletes with different levels of experience in the use of FIT; peak force during the ECC phase was delayed with respect to the start of the next CON phase in the less experienced group. This delay may affect the amount of force that can be produced in the ECC phase. This agrees with previous research [32], which established that to apply maximal force towards the end of the ECC stages of movement, a certain amount of co-ordination is required. Research suggests that this co-ordination is best achieved by FIT experience. The differences described in this study [33] were consistent in both FIT devices analysed.

The research definitively shows that a familiarisation period is warranted to obtain reliable, stable measures in flywheel training (two to four sessions). It may be conceivable that less complex exercises such as a flywheel leg curl may require shorter familiarisation periods when compared to more complex exercises such as flywheel Romanian deadlifts, but more research is required to make a final determination. It is crucial to understand how FIT variables such as power output production and eccentric overload are affected by participant familiarisation.

2.2. Inertial Load

An important determinant that can affect FIT performance is inertial loading. Depending on the inertial load used, different mechanical responses have been observed, which appears to be an essential factor in optimising training results [36]. Tous-Fajardo et al. [32] was the first study to investigate the effects of inertial load in a hamstring specific leg curl exercise. Twenty male amateur soccer and rugby players performed one testing session, which consisted of two sets of six-coupled concentric–eccentric actions. Each set used two different inertial loads (0.11 and 0.22 kg·m^2). The larger inertia showed greater peak force, although it should be noted that the inertial loads used would be classed as low inertial loads in comparison to more recent similar research.

Piqueras-Sanchiz et al. [35] also investigated the effect of inertial load on power output in a flywheel leg curl. Sixteen amateur university sports athletes participated in the study. The study evaluated the influence of different inertial loads (0.083, 0.132, 0.182, 0.266, and 0.350 kg·m^2) on concentric power, eccentric power, and eccentric overload ratio (eccentric peak power/concentric peak power ∗ 100) using a parallel-group design. Both males and females (n = 16) took part in the intervention. The study discussed that manipulating the inertial load can modify the power generated, with lower inertial loads (0.083 and 0.132 kg·m^2) leading to higher concentric and eccentric power output; however, there was no significant change (<0.05) in eccentric overload between inertial loads. Males did, however, show a more significant difference from the lowest to the higher inertia (ES = 1.37 to 1.51) than females (ES = 1.41 to 1.47), with the authors attributing this to lower strength levels of the female participants, meaning they used a higher relative load and may have a greater resistance to fatigue at higher relative intensities [35]. These findings are not in line with previous research [36,37] which showed that larger inertial loads achieved higher eccentric overloads; however, the studies mentioned both investigated the flywheel squat exercise, so the research may not be comparable.

A more recent hamstring specific intervention [38] investigated the effects of inertial load in a flywheel Romanian deadlift. Fourteen recreationally trained males partook in the study, and they all had a minimum of two years of resistance training experience. No participant had FIT experience; however adequate familiarisation was provided. The inertial loads ranged from low to high (0.025, 0.050, 0.075, and 0.100 kg·m^2). Results again highlighted that a lower inertial load led to higher peak concentric and eccentric peak power outputs, with the lowest load achieving the highest value (0.025 kg·m^2). Contrary to Piqueras-Sanchiz et al. [35], medium to higher loads (0.050, 0.075, and 0.100 kg·m^2) led to higher eccentric overload output, with 0.100 kg·m^2 displaying the largest value. The authors proposed that single-joint exercises such as a flywheel leg curl might recruit fewer muscles than multiple joint exercises such as a flywheel RDL; participants may not have been able to fully resist and break inertial forces, which could lead to a notable decline in eccentric overload. The most effective technique to maximise eccentric overload is to gently resist the force during the first third of the eccentric phase, then maximally decelerate the rotating flywheel and stop at the end range of motion [39]. Research may differ on the effects of higher inertial loads; but it is irrefutable that lower loads result in higher peak power outputs. This has significant practical implications because it may lead to greater performance adaptations. As previously discussed, the load that maximised power output may be the most efficient to develop athletic performance [40–42].

2.3. Volume Prescription

Intensity (inertial load) plays a vital role in training prescription during FIT, but it is matched in importance by volume (number of repetitions used), as it is vital to any training intervention aiming to achieve a specific adaption. To the author's knowledge, no study to date has analysed volume prescription for any hamstring specific FIT exercise. One study [36], however, using a flywheel bilateral squat, investigated the number of repetitions using different inertial loads at which concentric and eccentric peak power maintenance is observed. Participants performed one set of 15 repetitions at varying inertial loads (0.025, 0.050, 0.075 and 0.100 kg·m^2), with each load being tested on a different day over four testing days. The highest value was not found in the first two repetitions for both concentric and eccentric peak power; as in FIT, the first two repetitions are generally used to build momentum of the flywheel. Therefore, the authors recommended that three to four repetitions are warranted to accelerate the flywheel at the start of the set, allowing a build-up to maximal effort. Depending on the inertial load, a range of five to 12 repetitions was advised to maintain power output, but the authors concluded that it is essential to individualise the training volume prescription due to inter-subject variability in power decrements. Previous hamstring specific studies have used a fixed number of repetitions [43–45], or have increased the number of repetitions throughout the intervention [46,47], but no clear criteria for volume prescription has been advised. This highlights that more research into volume prescription is warranted, specifically with regard to the number of repetitions used in hamstring specific FIT, as it is commonly regarded that different training volumes lead to specific neuromuscular adaptations [36] and need to be targeted primarily.

2.4. FIT Technique and Exercise Used

According to previous research [39], the most effective method for providing an eccentric stimulus includes gently resisting the inertial force during the first third of the eccentric action and then exerting total effort to decelerate the revolving flywheel and bring it to a halt after the eccentric phase of the exercise. The subsequent concentric phase is then completed with maximal effort, and the discussed eccentric phase-specific technique is repeated. It should be highlighted that participants may not always adopt this technique [34]. Individuals may attempt to control the velocity of the eccentric phase in anticipation of the more challenging braking stimulus at the end of the movement, which would lead to a decreased eccentric overload [39], perhaps due to an involuntary self-protection mechanism in an effort to avoid peak forces [32]. Avoiding peak forces at the end range may be counterproductive, especially in hamstring training, as this end range is where the hamstrings are most susceptible to injury. In the flywheel leg curl, Tous-Fajardo et al. [32] discovered the opposite, that the window where the more substantial eccentric force was generated occurred in the later stages of a range of motion. Given the complexity of FIT, auto-regulated feedback could aid in readapting performance and getting the athlete comfortable with the correct technique.

Exercise selection may alter the stimulus provided during FIT; it has been previously hypothesised [36] that the number of muscles recruited during FIT may affect performance. Large multi-joint movements such as an RDL may allow participants to halt higher inertial loads more effectively and therefore achieve more significant eccentric overload at these high loads compared to single multi-joint movements such as a leg curl which recruits less musculature. Exercise selection may also affect the musculature that is predominantly used. For example, the flywheel leg curl has been previously shown to recruit the medial hamstring muscles [48] preferentially; hip-dominant exercises (such as an RDL) preferentially recruit the BFlh [49], which is the most frequently injured muscle of the hamstring complex [50]. It may be beneficial to target the most frequently injured muscle, but to the authors' knowledge, no study has directly compared a flywheel hip dominant exercise to a knee dominant exercise regarding hamstring adaptations. Further research is needed before any conclusions can be made.

2.5. Efficacy of FIT on Hamstring Strength Development

In sports with high intensity running demands, such as soccer [50] and Australian rules football [24], eccentric hamstring weakness has been identified as a risk factor for future HSI. Timmins et al. [50] assessed the eccentric hamstring strength of 152 professional soccer players in a sizeable prospective study. Players with eccentric hamstring strength below 4.35 $N \cdot kg^{-1}$ were 4.4 times more likely (RR; 95% CI 1.1 to 17.5) to sustain an HSI in the following season than stronger players. In agreement with this, Opar et al. [24] found that eccentric strength below 256 Newtons (N) at the start of pre-season and 279 (N) at the end of pre-season were said to associate with an increased risk of HSI (2.7- and 4.3-fold, respectively) in a population of 210 elite Australian footballers. Such research highlights the importance of eccentric hamstring strength in HSI and may explain why it is so prevalent in injury prevention programs and in the research. Flywheel inertia training has grown in popularity in recent years, and if performed appropriately, can result in moments of eccentric overload. This eccentric overload may have the potential to improve hamstring strength and warrants investigation.

Askling and colleagues [43] assessed isokinetic concentric and eccentric strength in elite-level male soccer players ($n = 30$). Players were randomly assigned to an intervention ($n = 15$) and to a control group ($n = 15$). Both groups followed the same training regime, with the intervention group performing extra-specific hamstring-specific FIT. The intervention group performed 16 sessions which included four sets of eight repetitions of a flywheel leg curl; the first set was submaximal and used as a warm-up. The training session fell on every fifth day for the first four weeks, then every fourth day for the last six weeks, lasting ten weeks in total. The intervention group showed a significant (<0.05) increase in both concentric and eccentric knee flexor peak torque, with the control group exhibiting no meaningful change. The increase in both concentric and eccentric strength was similar, 15 and 19%, respectively. The study suggested that eccentric overload training may cause greater hamstring strength adaptions than concentric training on its own. It should be noted that there were no familiarisation sessions held during the training period. This is significant because, as previously stated, 2–4 familiarisation sessions are recommended before adequate technique is achieved. The first sessions of the study may have only gained participant experience with the device and not actually achieved the desired strength training stimulus.

Timmins et al. [47] compared the effects of a flywheel RDL to an NHE on eccentric hamstring strength in elite-level Australian footballers. Twenty-seven male athletes were randomised into two groups, and the intervention took place over a 39-week period which included both pre-season and in-season. Eccentric hamstring strength was assessed at three-time points: baseline, end of pre-season and the end of the intervention. To create an appropriate stimulus, both training programs were periodised by gradually increasing the volume and additional weight employed throughout training, with both groups being matched for volume and intensity. The NHE group were required to use a weight (5–10 kg) to achieve desired intensity, whereas the RDL group used 0.05 or 0.075 kg·m^2. The intervention showed favorable adaptions for both groups, with both groups increasing their eccentric strength (RDL: mean change 82N, 95%CI 12 to 152N, d = 1.34, $p = 0.026$; NHE: mean change 97N, 95%CI 47 to 146N, d = 1.77, $p = 0.001$). As previously discussed, having eccentric strength values of 279N at the close of pre-season elevated the chance of a future HSI four times in professional Australian Football [24]. Timmins et al. [47] concluded that the eccentric hamstring strength adaptions gained in both groups from the intervention might positively impact HSI risk reduction. This was the first study to investigate a flywheel hip dominant exercise on modifiable risk factors of HSI, and although the results were favorable, more research is necessary.

Contrary to the previously mentioned studies [43,47], Presland et al. [51] found no significant increase in eccentric hamstring strength following a six-week flywheel leg curl protocol. Participants were randomised into two groups, a control ($n = 10$) and an eccentric biased group ($n = 10$). The control group engaged in unilateral training on the flywheel device with their opposite leg acting as a non-exercising control leg. The eccentric biased group also engaged in flywheel training. They performed the concentric phase with both legs but only one leg when undertaking the eccentric phase. The same leg was used throughout the training intervention. Eccentric strength was assessed pre and post-intervention using a previously validated NHE field testing device (NordBord, Vald Performance, Queensland, Australia) [52]. Although no statistical significance increases were reported, some increases were found among both groups, ranging from 33N to 46N. The study discussed possible neural adaptations in the control limbs contributing to these changes in eccentric strength, but this is unclear and needs further investigation. It should be noted that different flywheel exercises and protocols were used compared to previous research [47], so the direct comparison may be unfair. Research indicates that hamstring specific FIT can be utilised for strength development [43,47], but due to the low number of studies and contrary evidence [51], more research is needed before a definite conclusion to be made.

2.6. Fascicle Length

Hamstring strains comprise 37% of all muscle strain injuries, and of those, the majority occur in the biceps femoris long head (BFlh) [50]. Despite a paucity of evidence, it has been suggested that the length of the hamstring muscle fascicle may influence the likelihood of a future HSI [53,54], specifically the BFlh. One previous study [55] found BFlh fascicles to be shorter in previously injured muscles when compared to contralateral uninjured muscles. Adding weight to this concept, Timmins and colleagues [50] revealed that athletes who suffered a previous HSI possessed shorter BFlh fascicles than their uninjured counterparts while adding that non-modifiable risk factors such as age and previous hamstring injury were negatively influenced by shorter BFlh fascicles concerning HSI risk. They concluded that short BFlh fascicles were associated with an increased risk of future HSI in elite soccer players. Shorter fascicles, with fewer in-series sarcomeres, were previously thought to be more prone to being overstretched and damaged by severe eccentric contractions, such as those experienced during the terminal swing phase of high-speed running [53] but this is still debated in the research. Even if the exact mechanism of why shorter fascicle length increases the risk of HSI is unclear, it is commonly regarded that increasing fascicle length, specifically in the BFlh, is a worthwhile venture. Previously, the Nordic hamstring exercise was the most used tool for this [56], but recently FIT has gained interest as a modality to enhance BFlh fascicle length.

Two previously discussed studies [47,51] were in disagreement on FIT for enhancing eccentric hamstring strength, but agreed on its ability to increase BFlh fascicle length. Timmins et al. [47] found increases in both a flywheel RDL group and an NHE group (RDL: d = 1.99, $p < 0.001$; NHE: d = 1.73, $p < 0.001$) when compared to pre-intervention assessments, with the RDL grouping showing a slightly larger effect. This may be partly because hip dominant movements, such as an RDL, preferentially recruit the BFlh compared to knee dominant ones [57]. Presland et al. noted that after six-weeks of training, subjects who did the eccentrically biased flywheel intervention had a significant 14 $\pm$ 5% ($p < 0.001$, d = 1.98) increase BFlh fascicle length. It should be noted that these adaptions declined after a four-week detraining period, which highlighted that continuing eccentric loading is essential for the maintenance of architectural adaptations following flywheel leg-curl training. The lengthening of BFlh fascicles has been hypothesised [58] as playing a role in the effectiveness of eccentric training programs in reducing the risk of future HSI. If correct, then the discussed studies reflect positively on specific hamstring FIT as a modality to decrease HSI risk by positively affecting fascicle length, specifically BFlh.

2.7. HSI Injury Risk Reduction Using FIT

Strength training has been advocated for preventing HSI for many years [59] and forms the basis of most injury prevention programs used today. Askling and colleagues [43] were the first to investigate whether a flywheel strength training program, which emphasised eccentric loading using the leg curl exercise, might affect the occurrence and severity of HSI. Over the ten-week intervention, players from two elite level Swedish soccer teams were randomised into two groups. One group performed specific hamstring FIT (flywheel leg curl), while the control group did not. Six in the intervention group and four in the control group reported an HSI in the previous season. All injuries were recorded during the intervention period, and an injury was included if it occurred during a match or training session and if a player missed a minimum of one match or training session. A significantly (<0.05) lower number of injuries was reported in the intervention group. The authors noted that the positive effects shown in the study advocated the use of hamstring specific FIT in elite soccer but did not fully attribute the preventive effect on HSI to the eccentric overload derived from FIT and suggested that further longitudinal studies are needed for more definite recommendations.

De Hoyo et al. [44] again investigated the effects of FIT eccentric overload training in elite soccer players, but this time in junior players. Thirty-six elite-level junior (U17-19) players were divided into two groups (intervention v control). The intervention group performed 1-2 FIT sessions weekly, including a flywheel half squat and leg curl exercise. Both groups performed the same volume of match play and training sessions over the ten-week intervention. The control group performed no strength training during the entire season. The number of training sessions and matches and the number of muscle injuries per 1000 h of exposure (match and training) were recorded. The study reported an overall reduction of injury following training. The study monitored all injuries, not just hamstring ones. This fact should be considered when disseminating the findings. Considering that the protocol included a flywheel squat exercise, it cannot be determined how much of a role the hamstring specific exercise (leg curl) played in the positive findings. Nevertheless, it can be agreed that FIT in general, regardless of exercise, may be incorporated into injury reduction strategies.

3. Practical Guidelines and Discussion

- To maximise the benefits of FIT, it is advised to first perform several familiarisation sessions. Two to four familiarisation sessions are recommended to obtain a stable measure. Less complex exercises such as a flywheel leg curl may require shorter familiarisation periods when compared to more complex exercises, and more experienced athletes may require shorter familiarisation periods than novice athletes.
- Lower inertial loads (0.025 $kg{\cdot}m^2$) lead to higher peak concentric and eccentric peak power outputs, while medium to higher loads (0.050, 0.075, and 0.100 $kg{\cdot}m^2$) lead to higher eccentric overload output.
- Although there is no hamstring-specific research available, flywheel squat research suggests that a range of five to 12 repetitions was advised to maintain power output depending on the inertial load. Three to four repetitions are warranted to accelerate the flywheel at the start of the set, to build to maximal effort, and should be viewed as waste repetitions and not counted as working repetitions. Due to large subject variability in FIT, it is essential to individualise the training volume prescription.
- Both the flywheel leg curl and RDL have been shown to increase eccentric hamstring strength in elite level athletes, but more research is needed before a definite conclusion can be made.
- Hamstring specific FIT has been shown to increase fascicle length, specifically in the BFlh. Thus, this architectural adaption could help with a risk reduction of future HSI.
- The most effective technique to achieve an eccentric overload comprises gently resisting the inertial force during the first third of the eccentric action and then exerting a full effort braking action to decelerate the revolving flywheel and bring it to a halt at the end range of motion.
- The flywheel leg curl has been shown to affect the reduction of injuries in elite soccer players positively, but more longitudinal studies are warranted. It may also be interesting to investigate other FIT hamstring specific exercises such as the RDL as injury prevention techniques.

Overall, hamstring specific FIT has positive adaptions on hamstring strength and other modifiable HSI risk factors, such as fascicle length. It should be highlighted that there was a dearth of research on these themes accessible for review, and this fact should be considered when processing these conclusions. To gain the desired stimulus from FIT, its determinants should also be considered. This review highlights that athlete familiarisation and inertia load may both influence the adaptations of FIT and should be monitored carefully. Although there is positive research available regarding volume prescription and varying exercises, more is warranted to investigate different hamstring specific exercises over extended intervention periods. This narrative review is not intended to be a definitive guide for FIT, but it may help practitioners implement best practices while using this method to enhance hamstring strength.

Author Contributions: Conceptualization, J.O.B., D.B., C.L. and D.E.; resources, J.O.B.; writing—original draft preparation, J.O.B.; writing—review and editing, J.O.B., D.B., C.L. and D.E.; project administration, J.O.B. All authors have read and agreed to the published version of the manuscript.

Funding: This project was funded by the Institute of Technology Carlow, Presidents Fellowship Fund. Funding Number:232530.

Institutional Review Board Statement: Institute of Technology Carlow Editorial Board, C00232530.

Informed Consent Statement: Informed consent was obtained from all subjects involved in the study.

Data Availability Statement: Data will be made available upon reasonable request to the corresponding author.

Conflicts of Interest: The authors declare that they have no conflicts of interest.

References

1. Ekstrand, J.; Healy, J.C.; Waldén, M.; Lee, J.C.; English, B.; Hägglund, M. Hamstring muscle injuries in professional football: The correlation of MRI findings with return to play. *Br. J. Sports Med.* **2012**, *46*, 112–117. [CrossRef] [PubMed]
2. Orchard, J.; Seward, H. Epidemiology of injuries in the Australian Football League, seasons 1997–2000. *Br. J. Sports Med.* **2002**, *36*, 39–44. [CrossRef] [PubMed]
3. Ekstrand, J.; Hägglund, M.; Waldén, M. Epidemiology of muscle injuries in professional football (soccer). *Br. J. Sports Med.* **2011**, *45*, 1226–1232. [CrossRef] [PubMed]
4. Malliaropoulos, N.G. Non contact Hamstring injuries in sports. *Muscles Ligaments Tendons J.* **2013**, *2*, 309–311. [PubMed]
5. Brooks, J.H.M.; Fuller, C.W.; Kemp, S.P.T.; Reddin, D.B. Incidence, Risk, and Prevention of Hamstring Muscle Injuries in Professional Rugby Union. *Am. J. Sports Med.* **2006**, *34*, 1297–1306. [CrossRef]
6. Malliaropoulos, N.; Papacostas, E.; Kiritsi, O.; Rad, P.-M.; Papalada, A.; Gougoulias, N.; Maffulli, N. Posterior Thigh Muscle Injuries in Elite Track and Field Athletes. *Am. J. Sports Med.* **2010**, *38*, 1813–1819. [CrossRef]
7. Sherry, M.A. A Comparison of 2 Rehabilitation Programs in the Treatment of Acute Hamstring Strains. *J. Orthop. Sports Phys. Ther.* **2004**, *34*, 116–125. [CrossRef]
8. Ramos, G.A.; Arliani, G.G.; Astur, D.C.; de Pochini, A.C.; Ejnisman, B.; Cohen, M. Rehabilitation of hamstring muscle injuries: A literature review. *Rev. Bras. Ortop. Engl. Ed.* **2017**, *52*, 11–16. [CrossRef]
9. Slavotinek, J.P.; Verrall, G.M.; Fon, G.T. Hamstring injury in athletes: Using MR imaging measurements to compare extent of muscle injury with amount of time lost from competition. *Am. J. Roentgenol.* **2002**, *179*, 1621–1628. [CrossRef]
10. Askling, C.M.; Malliaropoulos, N.; Karlsson, J. High-speed running type or stretching-type of hamstring injuries makes a difference to treatment and prognosis. *Br. J. Sports Med.* **2012**, *46*, 86–87. [CrossRef]
11. Chumanov, E.S.; Schache, A.G.; Heiderscheit, B.C.; Thelen, D.G. Hamstrings are most susceptible to injury during the late swing phase of sprinting. *Br. J. Sports Med.* **2012**, *46*, 90. [CrossRef]
12. Orchard, J.W. Hamstrings are most susceptible to injury during the early stance phase of sprinting: Table 1. *Br. J. Sports Med.* **2012**, *46*, 88–89. [CrossRef]
13. Danielsson, A.; Horvath, A.; Senorski, C.; Alentorn-Geli, E.; Garrett, W.E.; Cugat, R.; Samuelsson, K.; Hamrin Senorski, E.H. The mechanism of hamstring injuries—A systematic review. *BMC Musculoskelet. Disord.* **2020**, *21*, 641. [CrossRef]
14. Liu, H.; Garrett, W.E.; Moorman, C.T.; Yu, B. Injury rate, mechanism, and risk factors of hamstring strain injuries in sports: A review of the literature. *J. Sport Health Sci.* **2012**, *1*, 92–101. [CrossRef]
15. Picerno, P. The Hamstrings-Injury-Mechanism Debate: Are We Close to an Agreement? *J. Sport Rehabil.* **2017**, *26*, 120–121. [CrossRef]
16. Green, B.; Bourne, M.N.; Van Dyk, N.; Pizzari, T. Recalibrating the risk of hamstring strain injury (HSI): A 2020 systematic review and meta-Analysis of risk factors for index and recurrent hamstring strain injury in sport. *Br. J. Sports Med.* **2020**, *54*, 1081–1088. [CrossRef]
17. Freckleton, G.; Pizzari, T. Risk factors for hamstring muscle strain injury in sport: A systematic review and meta-analysis. *Br. J. Sports Med.* **2012**, *47*, 351–358. [CrossRef] [PubMed]
18. De Visser, H.M.; Reijman, M.; Heijboer, M.P.; Bos, P.K. Risk factors of recurrent hamstring injuries: A systematic review. *Br. J. Sports Med.* **2011**, *46*, 124–130. [CrossRef] [PubMed]
19. Opar, D.A.; Williams, M.D.; Shield, A.J. Hamstring strain injuries: Factors that Lead to injury and re-Injury. *Sports Med.* **2012**, *42*, 209–226. [CrossRef] [PubMed]
20. Wing, C.; Bishop, C. Hamstring strain injuries: Incidence, mechanisms, risk factors, and training recommendations. *Strength Cond. J.* **2020**, *42*, 40–57. [CrossRef]
21. Schuermans, J.; Van Tiggelen, D.; Danneels, L.; Witvrouw, E. Susceptibility to Hamstring Injuries in Soccer: A Prospective Study Using Muscle Functional Magnetic Resonance Imaging. *Am. J. Sports Med.* **2016**, *44*, 1276–1285. [CrossRef]
22. Goossens, L.; Witvrouw, E.; Vanden Bossche, L.; De Clercq, D. Lower eccentric hamstring strength and single leg hop for distance predict hamstring injury in PETE students. *Eur. J. Sport Sci.* **2014**, *15*, 436–442. [CrossRef]
23. Freckleton, G.; Cook, J.; Pizzari, T. The predictive validity of a single leg bridge test for hamstring injuries in Australian rules football players. *Br. J. Sports Med.* **2013**, *48*, 713–717. [CrossRef]
24. Opar, D.A.; Williams, M.D.; Timmins, R.G.; Hickey, J.; Duhig, S.J.; Shield, A.J. Eccentric hamstring strength and hamstring injury risk in Australian footballers. *Med. Sci. Sports Exerc.* **2015**, *47*, 857–865. [CrossRef]
25. Vicens-Bordas, J.; Esteve, E.; Fort-Vanmeerhaeghe, A.; Clausen, M.B.; Bandholm, T.; Opar, D.; Shield, A.; Thorborg, K. Eccentric Hamstring Strength is Associated with Age and Duration of Previous Season Hamstring Injury in Male Soccer Players. *Int. J. Sports Phys. Ther.* **2020**, *15*, 246–253. [CrossRef]
26. Berg, H.E.; Tesch, P.A.; Allen, W.J.C.; Allen, W.J.C.; De Keijzer, K.L.; Raya-gonzález, J.; Castillo, D.; Coratella, G.; Beato, M.; Allen, W.J.C.; et al. A gravity-independent ergometer to be used for resistance training in space. *Aviat. Space Environ. Med.* **1994**, *21*, 1031–1041. [CrossRef]
27. Dudley, G.A.; Tesch, P.A.; Miller, B.J.; Buchanan, P. Importance of eccentric actions in performance adaptations to resistance training. *Aviat. Space Environ. Med.* **1991**, *62*, 543–550.

28. Beato, M.; Dello Iacono, A. Implementing Flywheel (Isoinertial) Exercise in Strength Training: Current Evidence, Practical Recommendations, and Future Directions. *Front. Physiol.* **2020**, *11*, 569. [CrossRef] [PubMed]
29. Tesch, P.A.; Fernandez-Gonzalo, R.; Lundberg, T.R. Clinical Applications of Iso-Inertial, Eccentric-Overload (YoYo^{TM}) Resistance Exercise. *Front. Physiol.* **2017**, *8*, 241. [CrossRef]
30. Petré, H.; Wernstål, F.; Mattsson, C.M.; Maroto-Izquierdo, S.; García-lópez, D.; Fernandez-Gonzalo, R.; Moreira, O.C.; González-Gallego, J.; de Paz, J.A.; Vicens-Bordas, J.; et al. Skeletal muscle functional and structural adaptations after eccentric overload flywheel resistance training: A systematic review and meta-analysis. *J. Sci. Med. Sport* **2017**, *20*, 943–951. [CrossRef]
31. Maroto-Izquierdo, S.; García-López, D.; De Paz, J.A. Functional and Muscle-Size Effects of Flywheel Resistance Training with Eccentric-Overload in Professional Handball Players. *J. Hum. Kinet.* **2017**, *60*, 133–143. [CrossRef]
32. Tous-Fajardo, J.; Maldonado, R.A.; Quintana, J.M.; Pozzo, M.; Tesch, P.A. The flywheel leg-curl machine: Offering eccentric overload for hamstring development. *Int. J. Sports Physiol. Perform.* **2006**, *1*, 293–298. [CrossRef] [PubMed]
33. Galiano, C.; Floria, P.; Muñoz-López, A.; Nuñez, F.J. Lack of experience in the use the rotational inertia device is a limitation to mechanical squat performance. *Retos* **2021**, *42*, 12–17. [CrossRef]
34. Suchomel, T.J.; Wagle, J.P.; Douglas, J.; Taber, C.B.; Harden, M.; Haff, G.G.; Stone, M.H. Implementing Eccentric Resistance Training—Part 1: A Brief Review of Existing Methods. *J. Funct. Morphol. Kinesiol.* **2019**, *4*, 38. [CrossRef] [PubMed]
35. Piqueras-Sanchiz, F.; Sabido, R.; Raya-González, J.; Madruga-Parera, M.; Romero-Rodríguez, D.; Beato, M.; de Hoyo, M.; Nakamura, F.Y.; Hernández-Davó, J.L. Effects of Different Inertial Load Settings on Power Output Using a Flywheel Leg Curl Exercise and its Inter-Session Reliability. *J. Hum. Kinet.* **2020**, *74*, 215–226. [CrossRef]
36. Sabido, R.; Hernández-Davó, J.L.; Pereyra-Gerber, G.T. Influence of different inertial loads on basic training variables during the flywheel squat exercise. *Int. J. Sports Physiol. Perform.* **2018**, *13*, 482–489. [CrossRef] [PubMed]
37. Martinez-Aranda, L.M.; Fernandez-Gonzalo, R. Effects of inertial setting on power, force, work, and eccentric overload during flywheel resistance exercise in women and men. *J. Strength Cond. Res.* **2017**, *31*, 1653–1661. [CrossRef]
38. Brien, A.J.O.; Browne, D.; Earls, D.; Lodge, C. The effects of varying inertial loadings on power variables in the flywheel romanian deadlift exercise. *Biol. Sport* **2022**, *39*, 499–503. [CrossRef]
39. Berg, H.E.; Tesch, P.A. Force and power characteristics of a resistive exercise device for use in space. *Acta Astronaut.* **1998**, *42*, 219–230. [CrossRef]
40. Cormie, P.; McGuigan, M.R.; Newton, R.U. Developing maximal neuromuscular power: Part 1—Biological basis of maximal power production. *Sports Med.* **2011**, *41*, 17–38. [CrossRef]
41. Suchomel, T.J.; Wagle, J.P.; Douglas, J.; Taber, C.B.; Harden, M.; Haff, G.G.; Stone, M.H. Implementing Eccentric Resistance Training—Part 2: Practical Recommendations. *J. Funct. Morphol. Kinesiol.* **2019**, *4*, 55. [CrossRef]
42. Douglas, J.; Pearson, S.; Ross, A.; McGuigan, M. Eccentric Exercise: Physiological Characteristics and Acute Responses. *Sports Med.* **2017**, *47*, 663–675. [CrossRef]
43. Askling, C.M.; Karlsson, J.; Thorstensson, A. Hamstring injury occurrence in elite soccer players after preseason strength training with eccentric overload. *Scand. J. Med. Sci. Sports* **2003**, *13*, 244–250. [CrossRef]
44. De Hoyo, M.; Pozzo, M.; Sañudo, B.; Carrasco, L.; Gonzalo-Skok, O.; Domínguez-Cobo, S.; Morán-Camacho, E. Effects of a 10-week in-season eccentric-overload training program on muscle-injury prevention and performance in junior elite soccer players. *Int. J. Sports Physiol. Perform.* **2015**, *10*, 46–52. [CrossRef] [PubMed]
45. Stojanović, M.D.M.; Mikić, M.; Drid, P.; Calleja-González, J.; Maksimović, N.; Belegišanin, B.; Sekulović, V. Greater power but not strength gains using flywheel versus equivolumed traditional strength training in junior basketball players. *Int. J. Environ. Res. Public Health* **2021**, *18*, 1181. [CrossRef] [PubMed]
46. Presland, J.D.; Opar, D.A.; Williams, M.D.; Hickey, J.T.; Maniar, N.; Lee Dow, C.; Bourne, M.N.; Timmins, R.G. Hamstring strength and architectural adaptations following inertial flywheel resistance training. *J. Sci. Med. Sport* **2020**, *23*, 1093–1099. [CrossRef] [PubMed]
47. Timmins, R.G.; Filopoulos, D.; Nguyen, V.; Giannakis, J.; Ruddy, J.D.; Hickey, J.T.; Maniar, N.; Opar, D.A. Sprinting, Strength, and Architectural Adaptations Following Hamstring Training in Australian Footballers. *Scand. J. Med. Sci. Sports* **2021**, *31*, 1276–1289. [CrossRef]
48. Fernandez-Gonzalo, R.; Tesch, P.A.; Linnehan, R.M.; Kreider, R.B.; Di Salvo, V.; Suarez-Arrones, L.; Alomar, X.; Men-dez-Villanueva, A.; Rodas, G.; Vlachopoulos, D.; et al. Individual Muscle use in Hamstring Exercises by Soccer Players Assessed using Functional MRI The effect of a high-impact jumping intervention on bone mass, bone stiffness and fitness parameters in adolescent athletes. *Int. J. Sports Med.* **2016**, *37*, 559–564.
49. Bourne, M.N.; Williams, M.D.; Opar, D.A.; Al Najjar, A.; Kerr, G.K.; Shield, A.J. Impact of exercise selection on hamstring muscle activation. *Br. J. Sports Med.* **2017**, *51*, 1021–1028. [CrossRef]
50. Timmins, R.G.; Bourne, M.N.; Shield, A.J.; Williams, M.D.; Lorenzen, C.; Opar, D.A. Short biceps femoris fascicles and eccentric knee flexor weakness increase the risk of hamstring injury in elite football (soccer): A prospective cohort study. *Br. J. Sports Med.* **2016**, *50*, 1524–1535. [CrossRef]
51. Presland, J.D.; Timmins, R.G.; Bourne, M.N.; Williams, M.D.; Opar, D.A. The effect of Nordic hamstring exercise training volume on biceps femoris long head architectural adaptation. *Scand. J. Med. Sci. Sports* **2018**, *28*, 1775–1783. [CrossRef]
52. Opar, D.A.; Piatkowski, T.; Williams, M.D.; Shield, A.J. A Novel Device Using the Nordic Hamstring Exercise to Assess Eccentric Knee Flexor Strength: A Reliability and Retrospective Injury Study. *J. Orthop. Sports Phys. Ther.* **2013**, *43*, 636–640. [CrossRef]

53. Brockett, C.L.; Morgan, D.L.; Proske, U. Predicting Hamstring Strain Injury in Elite Athletes. *Med. Sci. Sports Exerc.* **2004**, *36*, 379–387. [CrossRef] [PubMed]
54. Brockett, C.L.; Morgan, D.L.; Proske, U. Human hamstring muscles adapt to eccentric exercise by changing optimum length. *Med. Sci. Sports Exerc.* **2001**, *33*, 783–790. [CrossRef]
55. Timmins, R.G.; Shield, A.J.; Williams, M.D.; Lorenzen, C.; Opar, D.A. Biceps femoris long head architecture: A reliability and retrospective injury study. *Med. Sci. Sports Exerc.* **2015**, *47*, 905–913. [CrossRef] [PubMed]
56. Cuthbert, M.; Ripley, N.; McMahon, J.J.; Evans, M.; Haff, G.G.; Comfort, P. The Effect of Nordic Hamstring Exercise Intervention Volume on Eccentric Strength and Muscle Architecture Adaptations: A Systematic Review and Meta-analyses. *Sports Med.* **2020**, *50*, 83–99. [CrossRef] [PubMed]
57. Bourne, M.N.; Timmins, R.G.; Opar, D.A.; Pizzari, T.; Ruddy, J.D.; Sims, C.; Williams, M.D.; Shield, A.J. An Evidence-Based Framework for Strengthening Exercises to Prevent Hamstring Injury. *Sports Med.* **2017**, *48*, 251–267. [CrossRef]
58. Bourne, M.N.; Duhig, S.J.; Timmins, R.G.; Williams, M.D.; Opar, D.A.; Al Najjar, A.; Kerr, G.K.; Shield, A.J. Impact of the Nordic hamstring and hip extension exercises on hamstring architecture and morphology: Implications for injury prevention. *Br. J. Sports Med.* **2017**, *51*, 469–477. [CrossRef]
59. Jönhagen, S.; Nemeth, G.; Eriksson, E. Hamstring Injuries in Sprinters: The Role of Concentric and Eccentric Hamstring Muscle Strength and Flexibility. *Am. J. Sports Med.* **1994**, *22*, 262–266. [CrossRef]

Journal of
Functional Morphology and Kinesiology

MDPI

Article

Hippocampal Adaptations to Continuous Aerobic Training: A Functional and Ultrastructural Evaluation in a Young Murine Model

Ida Cariati [1,*,†], Roberto Bonanni [1,†], Gabriele Pallone [2], Manuel Scimeca [3], Claudio Frank [4], Virginia Tancredi [2,5] and Giovanna D'Arcangelo [2,5]

1 Department of Clinical Sciences and Translational Medicine, "Tor Vergata" University of Rome, Via Montpellier 1, 00133 Rome, Italy; roberto.bonanni1288@gmail.com
2 Department of System Medicine, "Tor Vergata" University of Rome, Via Montpellier 1, 00133 Rome, Italy; gabriele.pallone@gmail.com (G.P.); tancredi@uniroma2.it (V.T.); giovanna.darcangelo@uniroma2.it (G.D.)
3 Department of Biomedicine and Prevention, "Tor Vergata" University of Rome, Via Montpellier 1, 00133 Rome, Italy; manuel.scimeca@uniroma2.it
4 UniCamillus-Saint Camillus International University of Health Sciences, Via di Sant'Alessandro 8, 00131 Rome, Italy; claudio.frank@unicamillus.org
5 Centre of Space Bio-Medicine, "Tor Vergata" University of Rome, Via Montpellier 1, 00133 Rome, Italy
* Correspondence: ida.cariati@uniroma2.it
† These authors contributed equally to this work.

Abstract: Aerobic training is known to influence cognitive processes, such as memory and learning, both in animal models and in humans. Particularly, in vitro and in vivo studies have shown that aerobic exercise can increase neurogenesis in the dentate gyrus, improve hippocampal long-term potentiation (LTP), and reduce age-related decline in mnemonic function. However, the underlying mechanisms are not yet fully understood. Based on this evidence, the aim of our study was to verify whether the application of two aerobic training protocols, different in terms of speed and speed variation, could modulate synaptic plasticity in a young murine model. Therefore, we assessed the presence of any functional changes by extracellular recordings in vitro in mouse hippocampal slices and structural alterations by transmission electron microscopy (TEM). Our results showed that an aerobic training protocol, well designed in terms of speed and speed variation, significantly contributes to improving synaptic plasticity and hippocampal ultrastructure, optimizing its benefits in the brain. Future studies will aim to clarify the underlying biological mechanisms involved in the modulation of synaptic plasticity induced by aerobic training.

Keywords: aerobic exercise; synaptic plasticity; hippocampus; training protocols; cognitive decline

Citation: Cariati, I.; Bonanni, R.; Pallone, G.; Scimeca, M.; Frank, C.; Tancredi, V.; D'Arcangelo, G. Hippocampal Adaptations to Continuous Aerobic Training: A Functional and Ultrastructural Evaluation in a Young Murine Model. *J. Funct. Morphol. Kinesiol.* **2021**, *6*, 101. https://doi.org/10.3390/jfmk6040101

Academic Editor: Peter Hofmann

Received: 28 October 2021
Accepted: 6 December 2021
Published: 8 December 2021

1. Introduction

Physical exercise has positive effects on general health and reduces the incidence of pathological conditions such as diabetes, osteoporosis, cardiovascular diseases, obesity, and other chronic disorders [1–3]. The positive effects of exercise on brain activity have long been discussed, although only recently scientific evidence based on neuroimaging approaches demonstrated the effectiveness of physical activity in improving cognitive health across the human lifespan [4,5].

The beneficial effects of exercise, particularly aerobic exercise, on the brain and behavior were initially studied in animal models and focused largely on the impact of exercise on hippocampal structure, which plays a key role in learning and memory formation [6–8]. Evidence has suggested that wheel running and treadmill training improve spatial learning in rodents and promote increased neuron density in the hippocampal areas CA1 and CA3 [9–11]. Furthermore, aerobic exercise is known to increase cell proliferation and neurogenesis in the dentate gyrus, as well as improve synaptic plasticity and spatial learning in both rats and mice [12–14]. Interestingly, exercise-induced changes in the hippocampus were associated with improved performance in spatial memory tasks [15].

Similar results have been found in human studies, showing that aerobic exercise increases hippocampal volume and reduces age-related decline in memory function [16–18]. In addition, several intervention studies have exhibited improved cognitive performance in elderly subjects undergoing a physical activity program that produces significant increases in cardiorespiratory fitness, strongly supporting the impact of training on cognitive processes [19].

Over the decades, our knowledge of the neuronal and molecular processes of memory has greatly improved, providing a basis for the identification of therapeutic strategies to slow and/or prevent age-related cognitive decline in humans [20,21]. Among these, exercise has been suggested as an effective non-pharmacological approach to preserve cognitive function and treat neurodegenerative and/or psychiatric conditions [22]. In this regard, Do et al. recently studied the effects of voluntary exercise on hypothalamic neurodegeneration in a mouse model of Alzheimer's disease, in which metabolic abnormalities, such as increased energy expenditure through enhanced oxygen consumption and increased caloric intake, were observed prior to the accumulation of amyloid plaques [23]. Interestingly, the authors observed a significant reduction in the expression of inflammatory and apoptotic markers in the hypothalamus of mice subjected to 4 weeks of voluntary wheeled exercise, suggesting a hypothalamus-mediated mechanism whereby exercise could counteract Alzheimer's disease-related neurodegeneration [23].

In recent years, several mechanisms have been proposed to explain the positive impacts of aerobic exercise, including increased cerebral blood flow, changes in neurotransmitter release, structural changes in the central nervous system (CNS), and altered arousal levels [24]. A more recent proposal points to neurotrophic factors as possible agonists in facilitating improved motor performance [25]. Among these, brain-derived neurotrophic factor (BDNF) could play a key role, as observed in previous studies showing that motor performance in rat models with middle cerebral artery occlusion was impaired following pharmacological interruption of BDNF production or, conversely, improved when BDNF production was enhanced [26,27].

Importantly, regular physical activity is now generally accepted to promote the release of myokines and metabolites into the circulation, which can cross the blood–brain barrier at the level of brain capillaries and influence the functions of neurons and glial cells, thus modifying neurotransmission in different regions of the brain [28]. In this regard, an important role has recently been attributed to irisin, a myokine produced by cleavage of the precursor fibronectin type III domain-containing 5 (FNDC5) during exercise [29]. Particularly, Lourenco et al. observed a reduced expression of FNDC5/irisin in the hippocampus and cerebrospinal fluid of animal models of Alzheimer's disease, correlated with a significant impairment of long-term potentiation (LTP), a phenomenon of synaptic plasticity, and object recognition memory [30]. Surprisingly, increased FNDC5/irisin levels promoted improved synaptic plasticity and counteracted memory impairment, highlighting the protective role of exercise in neurodegeneration [30].

Despite the latest evidence, prescribing specific exercises to maximize their positive effects on cognitive processes is not yet possible, because the levels of molecules released during muscle contraction change during and after exercise. In addition, it is not yet clear how brain functioning can vary with the type, intensity, and timing of exercise [22].

Based on this evidence, the aim of our work was to verify whether aerobic training can modulate synaptic plasticity in a young murine model, evaluating the presence of any functional changes by extracellular in vitro recordings in mouse hippocampal slices and structural alterations by transmission electron microscopy (TEM). Therefore, we applied two different continuous aerobic training protocols to assess whether any effects observed at the hippocampal level could depend on the use of different training protocols in terms of speed and speed variation.

2. Materials and Methods

2.1. Animals

Eighteen 1-month-old male mice, belonging to the wild-type BALB/c strain, were used, following the procedures established by the European Union Council Directive 2010/63/EU for animal experiments [31]. All experimental protocols were approved by the Italian Ministry of Public Health (authorization no. 86/2018-PR).

Animals were divided into two groups (five mice per group), each subjected to a different aerobic training protocol, and a third control group (eight mice), which did not perform any type of training. The health status of animals was monitored daily by resident veterinarians and experimenters, considering weight, coat and skin condition, and body functions. All experimental animals were kept under the same housing conditions and diet.

2.2. Training Protocols

The two experimental groups underwent aerobic training using a RotaRod (Cat N 47600, Ugo Basile srl, Milan, Italy). It features 5 cylinders with a diameter of 3 cm and a circumference of 9.42 cm covered by rubber to ensure an optimal grip for the rodents. A total of 6 panels with a diameter of 25 cm divided the 5 lanes, each with a 57 mm width, allowing 5 animals to run simultaneously. An attached display showed the types and speeds of rotation, the time elapsed since the start of the training session, and the time since the last fall. Finally, a control panel allowed the angular speed to be varied within a range (2–80 laps for minute, RPM) and the time intervals for the increasing speed modes from 6 sec to 10 min.

We administered two aerobic training protocols, progressive continuous (PC) and uniform continuous (UC), differing in terms of speed and speed variations, as described in our previous work [32] and summarized in Table 1. The PC protocol consisted of 18 min of training at a gradual speed of rotation, increasing from low to high intensity (10–32 RPM). During the UC training protocol, a rate of 13 RPM was set for 26 min. Training sessions were conducted three times a week for 12 weeks, for a total of 36 days of activity. The animals were raised on a light/dark cycle of 12/12 h, and training was carried out in the morning, between 10:00 and 11:00 a.m.

Table 1. A schematic description of the two different aerobic exercise protocols used to train mice.

	PC Protocol	UC Protocol
Main features	Incremental speed changes with gradually increasing exercise intensity. Intensity increases in 2 RPM intervals from 10 to 32 RPM, with 12 speed changes	Single session training at 9 RPM, without speed changes
Training session duration	18 min	26 min
Weekly frequency	3 times a week	3 times a week
Training period	12 weeks	12 weeks

PC: progressive continuous; UC: uniform continuous; RPM: laps per minute.

2.3. Extracellular Recordings in Mouse Hippocampal Slices

The animals belonging to the different experimental groups were sacrificed after 12 weeks of training, as were the sedentary animals. All efforts were made to minimize the number of animals used and their suffering. Under anesthesia with halothane (2-Brom-2-chlor-1,1,1-trifuor-ethan), mice were sacrificed, and their brains were quickly removed and placed in cold, oxygenated artificial cerebral spinal fluid (ACSF) containing the following (in mM): NaCl, 124; KCl, 2; KH_2PO_4, 1.25; $MgSO_4$, 2; $CaCl_2$, 2; $NaHCO_3$, 26; and glucose, 10. The hippocampus was rapidly dissected and cut transversely into 450 μm thick slices using a McIlwain tissue chopper (Mickle Laboratory Engineering Co., Gomshall, UK). Then, hippocampal slices were transferred to a tissue chamber, where they were laid in an interface between oxygenated ACSF and humidified gas (95% O_2, 5% CO_2) at 32–34 °C (pH = 7.4), constantly superfused at flow rate of 1.2 mL/min.

Extracellular recordings of the population spikes (PSs) were made in the stratum pyramidale of the CA1 subfield, with glass microelectrodes filled with 2 M NaCl (resistance 5–10 MΩ). Orthodromic stimuli (10–500 mA, 20–90 ms, 0.1 Hz) were delivered through a platinum electrode placed in the stratum radiatum (Schaffer collaterals). The test stimulus intensity of 50 ms square pulses was adjusted to give a PS of 2–4 mV at 0.03 Hz. The PS amplitude was calculated every minute as the average of six recordings performed every 10 s. A high-frequency stimulation (HFS, 100 Hz, 1 s), after the recording of stable signals (15–20 min), was given to assess changes in PS amplitude, which was expressed as a percentage of the basal PS amplitude. Signals were fed to an Axoclamp-2A amplifier (Foster City, CA, USA), acquired through a digital/analogic system (Digidata 1440A, Axon Instruments, Foster City, CA, USA) and analyzed with pCLAMP10 software (Axon Instruments, Foster City, CA, USA).

2.4. TEM Evaluation

For TEM evaluation, 1 mm^3 of hippocampal tissue from cerebral biopsies was fixed in 4% paraformaldehyde and post-fixed in 2% osmium tetroxide [33]. After washing with 0.1 M phosphate buffer, the sample was dehydrated by a series of incubations in 30%, 50%, and 70% ethanol. Dehydration was continued by incubation steps in 95% ethanol, absolute ethanol, and propylene oxide, after which samples were embedded in Epon (Agar Scientific Ltd., Parsonage Lane, Stansted, Essex CM24 8GF, UK) [34]. Ultra-thin sections, 80 nm thick, were mounted on copper grids and examined with a transmission electron microscope (Model JEM-1400 series 120 kV, JEOL USA, Inc. 11 Dearborn Road Peabody, MA, USA).

2.5. Statistical Analysis

Statistical analysis was performed using GraphPad Prism 8 Software (Prism 8.0.1, La Jolla, CA, USA). For electrophysiological experiments, data were expressed as the mean ± SEM, with n representing the number of slices analyzed. Data were compared with two-way ANOVA and Tukey's multiple comparison tests and were considered significantly different if $p < 0.05$.

3. Results

3.1. Synaptic Plasticity Following Continuous Aerobic Training

The effects of two continuous aerobic training protocols, differing in terms of speed and speed variation, on the synaptic plasticity expression were analyzed in the CA1 region of hippocampal slices from trained mice compared to sedentary mice which did not perform any type of training.

Figure 1a shows how the influence of training on synaptic plasticity varied depending on the protocol administered. Particularly, we obtained optimal results for the PC training protocol, which seemed to positively modulate synaptic plasticity throughout the electrophysiological recording, with significantly higher PS amplitude values than those of the other experimental groups. In contrast, no improvement in synaptic plasticity was observed in the hippocampal slices of mice trained with the UC protocol, which was inhibited in the first twenty minutes after HFS, whereas PS amplitude values remained stable until the end of the electrophysiological recording with values similar to those of the CTRL group.

Figure 1b shows the following PS amplitude values at four different experimental times and their significance: basal synaptic transmission (BST), before HFS (CTRL: 100.7 ± 0.4, PC-trained: 101.0 ± 0.2, UC-trained: 102.1 ± 0.3); at min 15, immediately after HFS (CTRL: 321.0 ± 14.3, PC-trained: 374.2 ± 13.0, UC-trained: 257.0 ± 18.7; CTRL vs. PC-trained, ** $p < 0.01$; CTRL vs. UC-trained, *** $p < 0.001$; PC-trained vs. UC-trained, **** $p < 0.0001$); at min 45 (CTRL: 224.0 ± 9.4, PC-trained: 270.0 ± 13.0, UC-trained:

209.7 ± 15.7; CTRL vs. PC-trained, * $p < 0.05$; PC-trained vs. UC-trained, ** $p < 0.01$); and at min 65 (CTRL: 218.1 ± 10.0, PC-trained: 248.6 ± 11.5, UC-trained: 188.7 ± 14.7; PC-trained vs. UC-trained, ** $p < 0.01$).

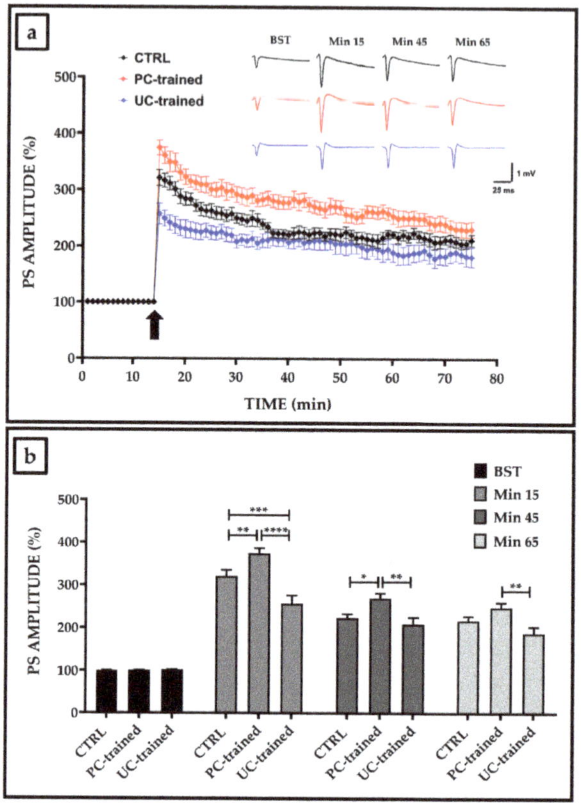

Figure 1. Synaptic plasticity in the CA1 hippocampal subfield of trained and sedentary mice. (**a**) Percentage population spike (PS) amplitude as a function of time after high-frequency stimulation (HFS), applied at time t = 15 (arrow), is shown in CTRL (black line, $n = 15$), in PC-trained (red line, $n = 9$), and in UC-trained (blue line, $n = 8$) mice slices. The insert shows representative recordings obtained from slices of each experimental group. The first curve of each group refers to the basal synaptic transmission (BST) and it was recorded before the HFS application, whereas the other curves refer to PS at times 15, 45 and 65 min after the HFS. (**b**) The PS amplitude values of BST, at min 15 (immediately after HFS), at min 45 and at min 65 from the HFS, are shown for each experimental group. Bars in the plot are means ± SEM of values obtained from different slices. Note that a significant statistical difference was reported between trained and control groups at min 15 (CTRL vs. PC-trained, ** $p < 0.01$; CTRL vs. UC-trained, *** $p < 0.001$; PC-trained vs. UC-trained, **** $p < 0.0001$), at min 45 (CTRL vs. PC-trained, * $p < 0.05$; PC-trained vs. UC-trained, ** $p < 0.01$) and at min 65 (PC-trained vs. UC-trained, ** $p < 0.01$).

3.2. Ultrastructural Hippocampal Evaluation of the Sedentary and Trained Mice

TEM evaluation was performed to assess the presence of any relevant differences in hippocampal slices taken from sedentary and trained mice.

Ultrastructural analysis of the hippocampus of the CTRL group (Figure 2a–c) showed normal tissue organization with well-preserved nerve and glial cells. Nerve extensions were well represented, rich in neurotubules and neurofilaments, with slight vacuolization at the axonal level. In addition, synapses were well represented and with well-preserved morphology.

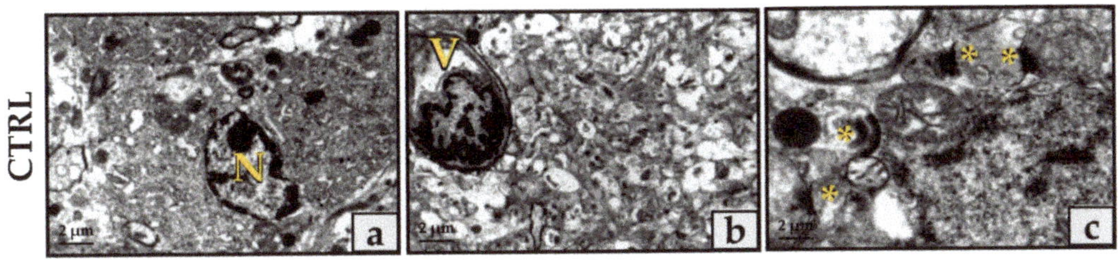

Figure 2. Ultrastructural evaluation by transmission electron microscopy (TEM) of the hippocampus of sedentary mice. (**a**) TEM evaluation of the hippocampus of CTRL mice not subjected to continuous aerobic training showed normal tissue organization with well-preserved nerve and glial cells. (**b**) Nerve processes were well represented, rich in neurotubules and neurofilaments, with slight vacuolization at the axonal level. (**c**) Synapses were well represented with well-preserved morphology. Scale bar represents 2 μm (N: nucleus; V: vessel; *: synapse).

Continuous aerobic training influenced hippocampal structure differently depending on the type of protocol performed by the animals.

Particularly, ultrastructural analysis of the hippocampus of PC-trained mice (Figure 3a–d) showed features very similar to those of the CTRL group, with well-organized neuronal and glial cells and nerve processes rich in neurotubules and neurofilaments. In addition, brain tissue showed numerous highly preserved myelin bundles, and mitochondria were free of morphological changes. In contrast, some morphological changes were found in the brain tissue of UC-trained mice (Figure 3e–h), including a slight vacuolization caused by axonal swelling and a reduction in the number of neurotubules and neurofilaments. Finally, TEM evaluation showed a reduced number of myelin bundles and frequent mitochondrial alterations.

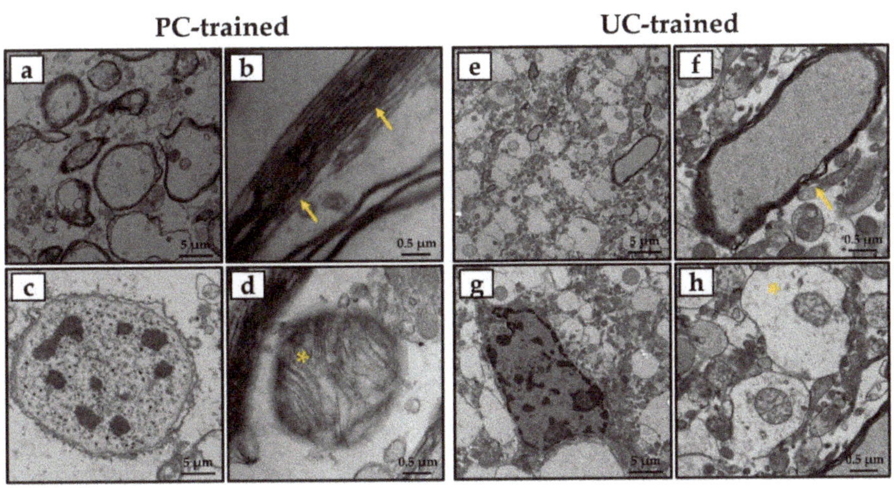

Figure 3. Transmission electron microscopy (TEM) evaluation of the hippocampus of trained mice. (**a**,**c**) Ultrastructural analysis of the hippocampus of PC-trained mice showed well-preserved tissue organization, with well-organized neuronal and glial cells and nerve processes rich in neurotubules and neurofilaments. (**b**,**d**) Numerous highly preserved myelin bundles (arrows) and mitochondria without morphological changes (asterisk) were observed. (**e**,**g**) Ultrastructural analysis of the hippocampus of UC-trained mice showed some morphological changes, such as slight vacuolization at the axonal level and few neurotubules and neurofilaments. (**f**,**h**) A reduced number of myelin bundles (arrow) and frequent mitochondrial alteration (asterisk) were detected. Scale bars represent 5 or 0.5 μm.

4. Discussion

Regular exercise induces profound health benefits for the body through mechanisms involving various physiological adaptations, including neural, immunological, vascular, and metabolic systems [35,36]. Interestingly, emerging data from studies in animal models and humans indicate that aerobic exercise benefits brain function and may prevent or delay the onset of neurodegenerative conditions by inducing structural and functional changes in the hippocampus, an area of the brain important for learning and memory [37,38]. Indeed, synaptic changes, which underlie cognitive processes, are known to depend on physiological mechanisms such as LTP, which is particularly present in the hippocampus [39]. Furthermore, it has been reported that the improvement in synaptic plasticity depends on the type of training provided [34,40]. Therefore, to better understand the mechanisms underlying the effects of aerobic exercise on the hippocampus and more generally on synaptic plasticity, in the present study we subjected young mice to two training protocols, PC and UC, differing in speed and speed variation.

First, we performed a functional evaluation by analyzing the effects of aerobic training on the synaptic plasticity expression by means of in vitro extracellular recordings in the CA1 region of mouse hippocampal slices. Our results showed that only the PC training protocol appeared to exert positive effects on synaptic plasticity throughout the electrophysiological recording, because we observed a significant increase in PS amplitude values after HFS compared to the other experimental groups. These data are in agreement with the results of our previous study, in which the administration of a PC training protocol has been shown to positively modulate hippocampal plasticity not only in young mice, but also reverses the blockage of the LTP induction phase typical of aged mice [41]. Additional scientific evidence confirms the beneficial effects of aerobic training on hippocampal synaptic plasticity. For example, Li et al. recently evaluated the effectiveness of a four-week aerobic training protocol on memory and the expression of proteins involved in synaptic plasticity in diabetic mice [42]. In addition to observing a significant reduction in fasting blood glucose and an improvement in insulin resistance, the authors found an increase in proteins associated with synaptic plasticity, pointing to aerobic exercise as a valid strategy to counteract the cognitive decline that characterizes diabetic mice [42].

In contrast, the UC training protocol did not induce any improvement in synaptic plasticity compared to sedentary mice of the same age. Particularly, PS amplitude values were significantly reduced in the first twenty minutes after HFS, whereas they reached values comparable to those of the CTRL group in the remaining time of electrophysiological recording. Notably, although the two trained groups did not differ significantly from the sedentary group after 65 min, a significant difference between them was found at the end of the electrophysiological recordings. This result suggests the importance of designing an appropriate training protocol to optimize the beneficial effects at the hippocampal level.

Electrophysiological data were confirmed by TEM evaluation, which showed that synaptic plasticity was affected differently depending on the type of protocol performed by the animals. Specifically, ultrastructural analysis of the hippocampus of PC-trained mice showed features very similar to those of the CTRL group, highlighting the presence of well-organized neuronal and glial cells and nerve processes rich in neurotubules and neurofilaments. Synapses were also well represented and with well-preserved morphology, in addition to the presence of numerous highly preserved myelin bundles and mitochondria without morphological changes. In contrast, the hippocampal tissue of the UC-trained mice exhibited some morphological changes, such as slight axonal vacuolization and a reduction in the number of neurotubules and neurofilaments, as well as a reduced number of myelin bundles and frequent mitochondrial changes.

In agreement with other experimental evidence, our results show that the benefits of exercise on cognitive function and neuroplasticity depend on the type of training protocol used. The underlying molecular and cellular mechanisms are not yet known. However, most scientific evidence agrees that the benefits of aerobic exercise may depend on an increase in growth factors and the increased expression of markers of synaptic plasticity, such as synaptophysin and postsynaptic density protein 95 (PSD-95) in the hippocampus [43]. In this context, the PI3K/AKT/mTOR pathway could play a crucial role, because exercise-induced activation of this pathway has been reported to promote the expression of PSD-95, improving memory performance [44,45]. Studies in rodents have also shown that early exercise increases axonal and neuronal density and improves the expression of BDNF and its tropomyosin-related receptor kinase B (TrkB) in hippocampal formation [11,46]. In agreement, Redila et al. observed that young, physically active rats show increased neurogenesis and dendritic arborization in the dentate gyrus compared to sedentary rats [47]. Interestingly, Serra and colleagues have suggested that exercise increases the expression of neurotrophic factors and stimulates neuronal growth, resulting in a neural reserve to be used in later life [48]. This hypothesis is supported by previous research in humans, which has shown a correlation between physical activity at an early age and long-term benefits on brain function [49].

5. Conclusions

Our data show that the use of an aerobic training protocol, such as the PC protocol, properly designed in terms of speed and speed variation, helps to maintain brain health and cognition. Interestingly, an adequate aerobic training protocol can induce important structural and functional changes in the hippocampus, the brain area responsible for learning and memory. This underlines the importance of physical exercise in counteracting age-related cognitive decline and suggests its key role in preventing the onset of cognitive impairment. Further studies will be required to understand the underlying cellular and molecular mechanisms, as well as the role of key biochemical mediators involved in the modulation of synaptic plasticity induced by aerobic training.

Author Contributions: Conceptualization, I.C., R.B., V.T. and G.D.; investigation, I.C., R.B. and G.P.; data curation, I.C., R.B. and M.S.; writing—original draft preparation, I.C. and R.B.; writing—review and editing, C.F., V.T. and G.D.; supervision, V.T. and G.D. All authors have read and agreed to the published version of the manuscript.

Funding: This research received no external funding.

Institutional Review Board Statement: Not applicable.

Informed Consent Statement: Not applicable.

Data Availability Statement: The data presented in this study are available on request from the corresponding author.

Acknowledgments: The authors acknowledge the Centre of Space Bio-medicine, "Tor Vergata" University of Rome for supporting this study.

Conflicts of Interest: The authors declare no conflict of interest.

References

1. Warburton, D.E.R.; Bredin, S.S.D. Health benefits of physical activity: A systematic review of current systematic reviews. *Curr. Opin. Cardiol.* **2017**, *32*, 541–556. [CrossRef]
2. Cariati, I.; Bonanni, R.; Onorato, F.; Mastrogregori, A.; Rossi, D.; Iundusi, R.; Gasbarra, E.; Tancredi, V.; Tarantino, U. Role of physical activity in bone-muscle crosstalk: Biological aspects and clinical implications. *J. Funct. Morphol. Kinesiol.* **2021**, *6*, 55. [CrossRef] [PubMed]
3. Bang, D.-H.; Son, Y.-L. Effect of intensive aerobic exercise on respiratory capacity and walking ability with chronic stroke patients: A randomized controlled pilot trial. *J. Phys. Ther. Sci.* **2016**, *28*, 2381–2384. [CrossRef] [PubMed]
4. Cassilhas, R.C.; Tufik, S.; de Mello, M.T. Physical exercise, neuroplasticity, spatial learning and memory. *Cell. Mol. Life Sci.* **2016**, *73*, 975–983. [CrossRef]
5. Gomez-Pinilla, F.; Hillman, C. The influence of exercise on cognitive abilities. *Compr. Physiol.* **2013**, *3*, 403–428. [CrossRef] [PubMed]
6. van Praag, H.; Shubert, T.; Zhao, C.; Gage, F.H. Exercise enhances learning and hippocampal neurogenesis in aged mice. *J. Neurosci.* **2005**, *25*, 8680–8685. [CrossRef]

7. Clark, P.J.; Brzezinska, W.J.; Thomas, M.W.; Ryzhenko, N.A.; Toshkov, S.A.; Rhodes, J.S. Intact neurogenesis is required for benefits of exercise on spatial memory but not motor performance or contextual fear conditioning in C57BL/6J mice. *Neuroscience* **2008**, *155*, 1048–1058. [CrossRef]
8. Van der Borght, K.; Havekes, R.; Bos, T.; Eggen, B.J.L.; Van der Zee, E.A. Exercise improves memory acquisition and retrieval in the Y-maze task: Relationship with hippocampal neurogenesis. *Behav. Neurosci.* **2007**, *121*, 324–334. [CrossRef]
9. Anderson, B.J.; Rapp, D.N.; Baek, D.H.; McCloskey, D.P.; Coburn-Litvak, P.S.; Robinson, J.K. Exercise influences spatial learning in the radial arm maze. *Physiol. Behav.* **2000**, *70*, 425–429. [CrossRef]
10. Vaynman, S.; Ying, Z.; Gomez-Pinilla, F. Hippocampal BDNF mediates the efficacy of exercise on synaptic plasticity and cognition. *Eur. J. Neurosci.* **2004**, *20*, 2580–2590. [CrossRef] [PubMed]
11. Uysal, N.; Tugyan, K.; Kayatekin, B.M.; Acikgoz, O.; Bagriyanik, H.A.; Gonenc, S.; Ozdemir, D.; Aksu, I.; Topcu, A.; Semin, I. The effects of regular aerobic exercise in adolescent period on hippocampal neuron density, apoptosis and spatial memory. *Neurosci. Lett.* **2005**, *383*, 241–245. [CrossRef]
12. Farmer, J.; Zhao, X.; van Praag, H.; Wodtke, K.; Gage, F.H.; Christie, B.R. Effects of voluntary exercise on synaptic plasticity and gene expression in the dentate gyrus of adult male Sprague-Dawley rats in vivo. *Neuroscience* **2004**, *124*, 71–79. [CrossRef]
13. Kronenberg, G.; Bick-Sander, A.; Bunk, E.; Wolf, C.; Ehninger, D.; Kempermann, G. Physical exercise prevents age-related decline in precursor cell activity in the mouse dentate gyrus. *Neurobiol. Aging* **2006**, *27*, 1505–1513. [CrossRef]
14. Liu, H.; Zhao, G.; Cai, K.; Zhao, H.; Shi, L. Treadmill exercise prevents decline in spatial learning and memory in APP/PS1 transgenic mice through improvement of hippocampal long-term potentiation. *Behav. Brain Res.* **2011**, *218*, 308–314. [CrossRef]
15. Herting, M.M.; Nagel, B.J. Aerobic fitness relates to learning on a virtual Morris Water Task and hippocampal volume in adolescents. *Behav. Brain Res.* **2012**, *233*, 517–525. [CrossRef]
16. Chaddock, L.; Hillman, C.H.; Buck, S.M.; Cohen, N.J. Aerobic fitness and executive control of relational memory in preadolescent children. *Med. Sci. Sports Exerc.* **2011**, *43*, 344–349. [CrossRef] [PubMed]
17. Erickson, K.I.; Voss, M.W.; Prakash, R.S.; Basak, C.; Szabo, A.; Chaddock, L.; Kim, J.S.; Heo, S.; Alves, H.; White, S.M.; et al. Exercise training increases size of hippocampus and improves memory. *Proc. Natl. Acad. Sci. USA* **2011**, *108*, 3017–3022. [CrossRef]
18. Etnier, J.L.; Nowell, P.M.; Landers, D.M.; Sibley, B.A. A meta-regression to examine the relationship between aerobic fitness and cognitive performance. *Brain Res. Rev.* **2006**, *52*, 119–130. [CrossRef]
19. Bherer, L.; Erickson, K.I.; Liu-Ambrose, T. A review of the effects of physical activity and exercise on cognitive and brain functions in older adults. *J. Aging Res.* **2013**, *2013*, 657508. [CrossRef]
20. Erickson, K.I.; Hillman, C.; Stillman, C.M.; Ballard, R.M.; Bloodgood, B.; Conroy, D.E.; Macko, R.; Marquez, D.X.; Petruzzello, S.J.; Powell, K.E. Physical activity, cognition, and brain outcomes: A review of the 2018 physical activity guidelines. *Med. Sci. Sports Exerc.* **2019**, *51*, 1242–1251. [CrossRef] [PubMed]
21. Liu, Y.; Yan, T.; Chu, J.M.-T.; Chen, Y.; Dunnett, S.; Ho, Y.-S.; Wong, G.T.-C.; Chang, R.C.-C. The beneficial effects of physical exercise in the brain and related pathophysiological mechanisms in neurodegenerative diseases. *Lab. Investig.* **2019**, *99*, 943–957. [CrossRef] [PubMed]
22. Cariati, I.; Masuelli, L.; Bei, R.; Tancredi, V.; Frank, C.; D'Arcangelo, G. Neurodegeneration in niemann-pick type c disease: An updated review on pharmacological and non-pharmacological approaches to counteract brain and cognitive impairment. *Int. J. Mol. Sci.* **2021**, *22*, 6600. [CrossRef]
23. Do, K.; Laing, B.T.; Landry, T.; Bunner, W.; Mersaud, N.; Matsubara, T.; Li, P.; Yuan, Y.; Lu, Q.; Huang, H. The effects of exercise on hypothalamic neurodegeneration of Alzheimer's disease mouse model. *PLoS ONE* **2018**, *13*, e0190205. [CrossRef]
24. Gligoroska, J.P.; Manchevska, S. The effect of physical activity on cognition—Physiological mechanisms. *Mater. Sociomed.* **2012**, *24*, 198–202. [CrossRef]
25. Mackay, C.P.; Kuys, S.S.; Brauer, S.G. The effect of aerobic exercise on brain-derived neurotrophic factor in people with neurological disorders: A systematic review and meta-analysis. *Neural Plast.* **2017**, *2017*, 4716197. [CrossRef] [PubMed]
26. Ploughman, M.; Windle, V.; MacLellan, C.L.; White, N.; Doré, J.J.; Corbett, D. Brain-derived neurotrophic factor contributes to recovery of skilled reaching after focal ischemia in rats. *Stroke* **2009**, *40*, 1490–1495. [CrossRef] [PubMed]
27. Schäbitz, W.-R.; Berger, C.; Kollmar, R.; Seitz, M.; Tanay, E.; Kiessling, M.; Schwab, S.; Sommer, C. Effect of brain-derived neurotrophic factor treatment and forced arm use on functional motor recovery after small cortical ischemia. *Stroke* **2004**, *35*, 992–997. [CrossRef]
28. Pedersen, B.K. Physical activity and muscle-brain crosstalk. *Nat. Rev. Endocrinol.* **2019**, *15*, 383–392. [CrossRef]
29. Boström, P.; Wu, J.; Jedrychowski, M.P.; Korde, A.; Ye, L.; Lo, J.C.; Rasbach, K.A.; Boström, E.A.; Choi, J.H.; Long, J.Z.; et al. A PGC1-α-dependent myokine that drives brown-fat-like development of white fat and thermogenesis. *Nature* **2012**, *481*, 463–468. [CrossRef]
30. Lourenco, M.V.; Frozza, R.L.; de Freitas, G.B.; Zhang, H.; Kincheski, G.C.; Ribeiro, F.C.; Gonçalves, R.A.; Clarke, J.R.; Beckman, D.; Staniszewski, A.; et al. Exercise-linked FNDC5/irisin rescues synaptic plasticity and memory defects in Alzheimer's models. *Nat. Med.* **2019**, *25*, 165–175. [CrossRef]
31. Louhimies, S. Directive 86/609/EEC on the protection of animals used for experimental and other scientific purposes. *Altern. Lab. Anim.* **2002**, *30*, 217–219. [CrossRef] [PubMed]
32. Pallone, G.; Palmieri, M.; Cariati, I.; Bei, R.; Masuelli, L.; D'Arcangelo, G.; Tancredi, V. Different continuous training modalities result in distinctive effects on muscle structure, plasticity and function. *Biomed. Rep.* **2020**, *12*, 267–275. [CrossRef]

33. Palmieri, M.; Cariati, I.; Scimeca, M.; Pallone, G.; Bonanno, E.; Tancredi, V.; D'Arcangelo, G.; Frank, C. Effects of short-term aerobic exercise in a mouse model of Niemann-Pick type C disease on synaptic and muscle plasticity. *Ann. Ist. Super. Sanita* **2019**, *55*, 330–337. [CrossRef] [PubMed]
34. Cariati, I.; Scimeca, M.; Tancredi, V.; D'Amico, A.G.; Pallone, G.; Palmieri, M.; Frank, C.; D'Arcangelo, G. Effects of different continuous aerobic training protocols in a heterozygous mouse model of Niemann-Pick type C disease. *J. Funct. Morphol. Kinesiol.* **2020**, *5*, 53. [CrossRef]
35. Guiney, H.; Machado, L. Benefits of regular aerobic exercise for executive functioning in healthy populations. *Psychon. Bull. Rev.* **2013**, *20*, 73–86. [CrossRef]
36. Ruegsegger, G.N.; Booth, F.W. Health benefits of exercise. *Cold Spring Harb. Perspect. Med.* **2018**, *8*, a029694. [CrossRef]
37. Zhang, D.; Lu, Y.; Zhao, X.; Zhang, Q.; Li, L. Aerobic exercise attenuates neurodegeneration and promotes functional recovery —Why it matters for neurorehabilitation & neural repair. *Neurochem. Int.* **2020**, *141*, 104862. [CrossRef]
38. Firth, J.; Stubbs, B.; Vancampfort, D.; Schuch, F.; Lagopoulos, J.; Rosenbaum, S.; Ward, P.B. Effect of aerobic exercise on hippocampal volume in humans: A systematic review and meta-analysis. *Neuroimage* **2018**, *166*, 230–238. [CrossRef]
39. Bettio, L.; Thacker, J.S.; Hutton, C.; Christie, B.R. Modulation of synaptic plasticity by exercise. *Int. Rev. Neurobiol.* **2019**, *147*, 295–322. [CrossRef]
40. Cariati, I.; Bonanni, R.; Pallone, G.; Annino, G.; Tancredi, V.; D'Arcangelo, G. Modulation of synaptic plasticity by vibratory training in young and old mice. *Brain Sci.* **2021**, *11*, 82. [CrossRef]
41. De Domenico, E.; D'Arcangelo, G.; Faraoni, I.; Palmieri, M.; Tancredi, V.; Graziani, G.; Grimaldi, P.; Tentori, L. Modulation of GDF11 expression and synaptic plasticity by age and training. *Oncotarget* **2017**, *8*, 57991–58002. [CrossRef] [PubMed]
42. Li, J.; Liu, Y.; Liu, B.; Li, F.; Hu, J.; Wang, Q.; Li, M.; Lou, S. Mechanisms of aerobic exercise upregulating the expression of hippocampal synaptic plasticity-associated proteins in diabetic rats. *Neural Plast.* **2019**, *2019*, 7920540. [CrossRef] [PubMed]
43. Cassilhas, R.C.; Lee, K.S.; Fernandes, J.; Oliveira, M.G.M.; Tufik, S.; Meeusen, R.; de Mello, M.T. Spatial memory is improved by aerobic and resistance exercise through divergent molecular mechanisms. *Neuroscience* **2012**, *202*, 309–317. [CrossRef]
44. Kang, E.B.; Cho, J.Y. Effects of treadmill exercise on brain insulin signaling and β-amyloid in intracerebroventricular streptozotocin induced-memory impairment in rats. *J. Exerc. Nutr. Biochem.* **2014**, *18*, 89–96. [CrossRef]
45. Pazini, F.L.; Cunha, M.P.; Rosa, J.M.; Colla, A.R.S.; Lieberknecht, V.; Oliveira, Á.; Rodrigues, A.L.S. Creatine, similar to ketamine, counteracts depressive-like behavior induced by corticosterone via PI3K/Akt/mTOR pathway. *Mol. Neurobiol.* **2016**, *53*, 6818–6834. [CrossRef]
46. Gomes da Silva, S.; Unsain, N.; Mascó, D.H.; Toscano-Silva, M.; de Amorim, H.A.; Silva Araújo, B.H.; Simões, P.S.R.; da Naffah-Mazzacoratti, M.G.; Mortara, R.A.; Scorza, F.A.; et al. Early exercise promotes positive hippocampal plasticity and improves spatial memory in the adult life of rats. *Hippocampus* **2012**, *22*, 347–358. [CrossRef]
47. Redila, V.A.; Christie, B.R. Exercise-induced changes in dendritic structure and complexity in the adult hippocampal dentate gyrus. *Neuroscience* **2006**, *137*, 1299–1307. [CrossRef] [PubMed]
48. Serra, F.T.; Carvalho, A.D.; Araujo, B.H.S.; Torres, L.B.; Cardoso, F.D.S.; Henrique, J.S.; Placencia, E.V.D.; Lent, R.; Gomez-Pinilla, F.; Arida, R.M.; et al. Early exercise induces long-lasting morphological changes in cortical and hippocampal neurons throughout of a sedentary period of rats. *Sci. Rep.* **2019**, *9*, 13684. [CrossRef] [PubMed]
49. Dik, M.; Deeg, D.J.H.; Visser, M.; Jonker, C. Early life physical activity and cognition at old age. *J. Clin. Exp. Neuropsychol.* **2003**, *25*, 643–653. [CrossRef]

MDPI AG
Grosspeteranlage 5
4052 Basel
Switzerland
Tel.: +41 61 683 77 34

Journal of Functional Morphology and Kinesiology Editorial Office
E-mail: jfmk@mdpi.com
www.mdpi.com/journal/jfmk

www.ingramcontent.com/pod-product-compliance
Lightning Source LLC
Chambersburg PA
CBHW061553220926
51850CB00005B/323
* 9 7 8 3 7 2 5 8 1 9 1 9 5 *